2013 International Conference on Science & Engineering in Mathematics, Chemistry and Physics

(ScieTech 2013)

Journal of Physics: Conference Series Volume 423

Jakarta, Indonesia
24-25 January 2013

ISBN: 978-1-62748-300-1
ISSN: 1742-6588

Printed from e-media with permission by:

Curran Associates, Inc.
57 Morehouse Lane
Red Hook, NY 12571

Some format issues inherent in the e-media version may also appear in this print version.

Printed by Curran Associates, Inc. (2013)

For permission requests, please contact the Institute of Physics
at the address below.

Institute of Physics
Dirac House, Temple Back
Bristol BS1 6BE UK

Phone: 44 1 17 929 7481
Fax: 44 1 17 920 0979

techtracking@iop.org

Additional copies of this publication are available from:

Curran Associates, Inc.
57 Morehouse Lane
Red Hook, NY 12571 USA
Phone: 845-758-0400
Fax: 845-758-2634
Email: curran@proceedings.com
Web: www.proceedings.com

Papers

Invited and Plenary Speakers

Ttile : Fabrication methodology of MEMS micro-sensors
Speaker : Prof. Jamil Akhtar, Ph.D
Affiliation : Sensors and nanotechnology group, CSIR-CEERI, Pilani-333 031, INDIA

Abstract :

A MEMS sensor consists of major two types of components; mechanical and electronic, with a mutual compatibility through basic physical mechanism in order to transform mechanical energy into electronic or/and vice versa. The mechanical component of the sensor definitely requires a 3-D shape which is carried out by employing controlled micromachining of base material. The electronic component can be realised using microelectronics and therefore requires relatively less efforts. However a compatibility issue of involved processes for realising mechanical and electronic components is always there for improved repeatability and yield.

Anisotropic etching of crystalline silicon has paved way for atomic resolution (ultimate) bulk micromachining to realise 3-D tiny structuresbut at the cost of surface planarity. Influenced with microelectronics, surface micromachining led to provide IC compatible fabrication of microstructures while maintaining surface planarity but limited tospecialized structures. Combination of bulk and surface micromachining is smartly used taking advantage of both in order to realise micro/nano 3-D structures with less significant loss of surface planarity and equally applicablefor wider usage. This has been termed as front-side later etching technique. Deep reactive ion etching (DRIE) is used as alternate of wet anisotropic etching and provides flexibility in terms of etched geometry with enhanced process complexity. Circular geometries are possible to realise by using DRIE while wet anisotropic etching ceases to such shapes. Ultrasonic milling can be explored as low cost alternative of DRIE process for circular shapes.

Title` : The use of biotechnology in apricot
Speaker : Prof. Sezai Ercisli, Ph.D
Affiliation : Department of Horticulture, Faculty of Agriculture, Ataturk University 25240 Erzurum-Turkey

Abstract :
Plant biotechnology as an interdisciplinary science is able to provide impulses and solutions to agricultural challenges, particularly in the case of woody fruit crops, e.g. by rapid propagation of selected cultivars, conservation of valuable germplasm, phytosanitary and genetic improvement and safeguarding human health, not only through nutritional, but also through ecological aspects. Since many of these aspects, particularly in the case of perennial crop plants are long term efforts, the application of biotechnological methods can make considerable contributions on apricot improvements, including:

• The application of molecular markers for the identiï¬• cation and conservation of valuable apricot genetic resources
• Pathogen detection and elimination byin vitromethods
• New breeding goals in apricot trees, e.g. resistance breeding against PPV
• Molecular determination of S allele compositions for determining cross incompatibility groups in apricots
• Molecular determination of aroma developing in apricot (gene expression)

More recently most of biotechnological approach applied in disease resistance and characterization of genetic resources. The breeding and cultivation of PPV virus resistant apricots is a major contribution to the control of viral diseases, since there do not exist chemical control strategies.

Ttile: : Warped product CR-submanifolds of Lorentzian −Kenmotsu manifolds
Speaker : Prof. Siraj Uddin, Ph.D
Affilaition : Institute of Mathematical Sciences, Faculty of Science, University of Malaya, Kuala Lumpur, Malaysia.

Abstract :

Warped product manifolds provide excellent setting to model spacetime near black holes or bodies with large gravitational force (J.K. Beem, P.E. Ehrlich, K. Easley, Marcel Dekker, Global Lorentzian geometry New York, 1996., Bishop, R.L.; O'Neill, B.: Manifolds of Negative curvature, Trans. Amer. Math. Soc. 145 (1969) 1-49; Hong, S.T.: Warped products and black holes, Nuovo Cim. J.B 120 (2005) 1227-1234). Recently, Results are published exploring the existence (or non-existence) of warped product submanifolds in Kaehlerian and contact settings (cf. B.Y. Chen, Geometry of warped product CR-submanifolds in Kaehler Manifolds, Monatsh. Math. 133 (2001) 177-195, M.I. Munteanu, A note on doubly warped contact CR-submanifolds in trans-Sasakian manifolds, Acta Math. Hungar. 16 (2007), 121-126; B. Sahin, Non existence of warped product semi-slant submanifolds of Kaehler manifolds, Geometriae Dedicata. 117 (2006) 195-202 etc.). To continue the sequel, we have considered warped product submanifolds of Lorentzian manifolds especially Lorentzian -Kenmotsu manifoolds in the sense of B.Y.Chen, Differential Geometry of Real Submanifolds in a Kaehler Manifold, Monatsh. Math. 91 (1981) 257-275 and provide a generalization of CR-submanifolds. It is shown that Lorentzian manifolds do not admit non trivial warped product CR-submanifolds of the type N ×fNT , where NT and N are invariant and anti-invariant submanifolds of Lorentzian manifolds (e.g. Lorentzian -Kenmotsu). However, we discuss non trivial CR-warped products (obtained by reversing the two factors of warped product CR-submanifolds) exist. Some interesting results on the geometry of these submanifolds are duscussed in the talk.

Ttile : Antibacterial activity of methanol extracts of Salvia species from Turkey

Speaker :Bulent Cetin, Ph.D

Affilaition : Department of Food Engineering, Faculty of Agriculture Ataturk University 25240 Erzurum-Turkey

Abstract :

The crude methanol extracts of aerial parts of six Salvia species (Salvia pachystachys, Salvia virgata, Salvia verticillata, Salvia limbata, Salvia nemorosaand Salvia microstegia) have been investigated for their antibacterial activity by using disk diffusion assays against 18 bacteria strains belonging to 13 bacteria species. The minimum inhibitory concentrations (MIC) of single compounds were determined by the microbroth dilution method. The results showed that methanol extracts of six Salvia species tested exhibited significant antibacterial activity. The results provide evidence that Salvia species used except Salvia nemorosa might indeed be potential sources of new antibacterial agents. The highest antibacterial activity was obtained from Salvia virgata. This specie was active against Proteus vulgaris, Yersinia enterocolitica, Pseudomonas putida, Escherichia coli, Streptococcus pyogenes, Staphylococcus aureusand Pseudomonas florescens.

Title : Identification of presence Salmonella spp with conventional culture technique in chicken carcasses and meat samples in Erzurum

Speaker : Neslihan Dikba, Ph.D
Affiliation : Atatürk University, Faculty of Agriculture, Department of Agricultural Biotechnology, 25240, Erzurum-Turkey

Abstract :

In this study, the incidence of Salmonella spp. on chicken samples was investigated in Erzurum province of Turkey. A total of 120 samples (40 samples of the chicken wing, 40 samples of chickenbreast meat, 40 samples of the chicken leg) obtained from retail market in the region. The presence of Salmonella spp.in the chicken various cutswas determined by conventional technique and the Salmonella spp. wasisolated and identificated by ISO 6579 references method.In this study,37(31%) Salmonella strains were isolated from 120 chicken cuts. The number of Salmonella serotypes, whichisolated from chickensamples were 13 (33%), 15 (38%) and 9 (23%) in chicken wings, legs and breast meats,respectively.This contamination level can be explained by poor hygiene of personal,equipment, and operating place.

It was concludedthat the useof bothmicrobiological and serological methods could be more useful for the detection of Salmonella infections. As result of study, it's suggested thatdevelopÄ±ng an HACCP program for the products is likely to be very useful ineliminating the Salmonella spp from chicken cuts.

Title : The quantum Hall effect: The symmetry and the velocity

Speaker : Prof. Keshav N. Shrivastava, Ph.D

Affiliation : Department of Physics, University of Malaya, Kuala Lumpur 50603 Malaysia and School of Physics, University of Hyderabad, Hyderabad 500046, India

Abstract :

The complete and the correct interpretation of all of the fractions found in the quantum Hall effect of AlGaAs/GaAs heterostructures is presented by use of a specific treatment of spin. Hence the flux quantization depends on spin and the Lande's g value expression is replaced by a linear formula. The fractions occur in pairs of left and right helicities and the velocity of the particles is directed such that if one particle moves upstream then its conjugate moves downstream. The change in the sign of the spin is equivalent to the change in the sign of the velocity.

2013 International Conference on Science & Engineering in Mathematics, Chemistry and Physics (SCITECH 2013)

MEMS technology

Submanifolds

Plant material

The Aston Rasuna Hotel – The Bridge Meeting Room
Jakarta, 24 January 2013

Peer review statement

All papers published in this volume of *Journal of Physics: Conference Series* have been peer reviewed through processes administered by the proceedings Editors. Reviews were conducted by expert referees to the professional and scientific standards expected of a proceedings journal published by IOP Publishing.

Journal of Physics: Conference Series **423** (2013) 012001 doi:10.1088/1742-6596/423/1/012001

Application of artificial bee colony (ABC) algorithm in search of optimal release of Aswan High Dam

Md. S. Hossain[1] and A. El-shafie[2]
[1]PhD student. Dept. of civil and structural engineering, UKM, Malaysia
[2]Assoc. Professor. Dept. of civil and structural engineering, UK,Malaysia

E-mail: shabbier@eng.ukm.my; elshafie@eng.ukm.my

Abstract. The paper presents a study on developing an optimum reservoir release policy by using ABC algorithm. The decision maker of a reservoir system always needs a guideline to operate the reservoir in an optimal way. Release curves have developed for high, medium and low inflow category that can answer how much water need to be release for a month by observing the reservoir level (storage condition). The Aswan high dam of Egypt has considered as the case study. 18 years of historical inflow data has used for simulation purpose and the general system performance measuring indices has measured. The application procedure and problem formulation of ABC is very simple and can be used in optimizing reservoir system. After using the actual historical inflow, the release policy succeeded in meeting demand for about 98% of total time period.

1. Introduction

A guideline is essential for a reservoir system to operate it to get the maximum benefits. Here the 'benefits' terms may be defined as fulfilling different wants such as – irrigation, hydropower generation, water supply for domestic and industrial uses etc. Many researches proposed different methodology with different mathematical formulation regarding reservoir optimization model. Evolutionary methods and swarm intelligences becomes very much popular these days in this research field. The most popular and vastly used optimization tool in reservoir system operation modelling is genetic algorithm (GA). Both real valued and binary coded GA has been successfully applied in search of reservoir release policy ([1], [2], [3] and [4]). Selection, cross-over and mutation are the basic operator of a GA procedure. Particle swarm optimization (PSO) got popularity as it free from the complexities of using those operators of GA. According to the study of Kumar & Reddy [5], a modified version of PSO (elitist-mutated PSO) performed better than GA and standard PSO in reservoir release policy.

The application of artificial intelligences getting extensive day by day in every section (like rainfall forecasting engineering [6-7], Global positioning system [8], nonlinear blind signal separation [9], designing spread footing [10] etc.) including reservoir release optimization. In this study we proposed artificial bee colony (ABC) algorithm to optimize the reservoir release policy. ABC is a population based swarm intelligence that mimics the foraging behaviour of natural honey bees. Karaboga [11] firstly proposed the ABC algorithm in optimizing numerical functions and after that several studies proves that ABC shows better performances than GA and PSO algorithm ([12], [13], and [14]). Here we developed an optimum release curve for each month that can provide the amount of water to be release for a known storage condition. Three different inflow categories (high, medium and low) have considered in search of more accurate release volume. Total 36 curves (12 moths x 3 inflow category) have generated in this purposes using ABC optimization algorithm. Each curve shows the amount of release for a definite time period. To verify the model, we used the periodical reliability approach [15], where the number of satisfied period is counted to measure the system performances. Vulnerability and resiliency of the simulation model also measured in this aspect.

2. Artificial bee colony (ABC)

2.1 Foraging principles of natural honey bees

Tereshko [16] identified two major components regarding the process of collecting nectar of honey bees as – 'recruitment' and 'abandonment'. Recruitment refers the involvement of the bees in the process and abandonment refers leaving a food source after using it. The honey bees are identical in shape and size but we can categorize these bees in terms of their working pattern or responsibility during the process they carry. The bees those are currently exploiting a food source may called as 'employer bees'. The employer bees also carry the information about the sources and after returning to the hive they share the information with other bees waiting in the hive. The bees those are waiting for the information of employer is known as 'onlooker bees'. To share the information an employer uses a unique technique known as 'waggle dance'. The waggle dance is a physical movement of the bees and is essential for the foragers as it exhibit some important information about the food sources like – direction, distance and quality/quantity of nectar. After a certain collecting period some food sources becomes useless and here comes the abandonment process. The bees those leave the hive to discover new food sources called as 'scout bees'.

Figure 1 would help here to understand the basic of foraging process. In figure – 1 let consider an employer bee currently exploiting a food source. After his returning to the hive through E he might have three options to choose. Path E1 that lead him to keep continue to collecting, path E2 if he decided to share the information by dancing and E3 if abandonment of the food source has done. On the other hand an onlooker waits in the hive and observes (position – O) the waggle dance to get recruited. The path S denotes the random search of undiscovered food source and explored by scout bees.

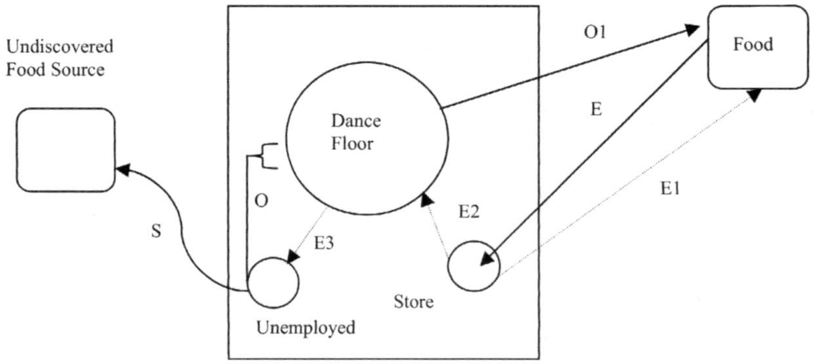

Figure 1. Foraging behaviour of honey bees

2.2 ABC algorithm in reservoir release policy

The basic ABC algorithm can be divided into three phase – Employed bee phase, Onlooker bee phase and Scout bee phase. At first the food sources need to be selected. A set of possible solution of a problem can be taken as the food sources. So if we have a D dimensional problem (where $i=1,2,...,D$) then the decision variables (x_i) act as the members of a single food sources. The value of these decision variables are randomly generated within the variable bounds if they have any. For a typical monthly water supply problem of a reservoir, a set of release options (consisting 12 values for each month) can be taken as food sources. That means a single source contain 12 consecutive monthly release options. The employed bee tested the fitness values of pre-defined no. of food sources and records the information about them. The quality of each source is the fitness value of the objective function considered to minimize or maximize according to service criteria of the reservoir. Employed bees select a food source and randomly choose a candidate solution within the source to update it providing new values by using equation 1.

$$v_{ij} = x_{ij} + \phi_{ij}(x_{ij} - x_{ik}) \tag{1}$$

Here, x_{ij} represents the current candidate solution of a source and x_{ik} represent other randomly chosen solution but must be taken from different neighbour source. $\emptyset_{ij}$ is taken as a random number between [-1 to 1]. In case of any violation of variable limits (after using equation 1), the updated food sources are needed to crop to it allowable ranges. The qualities (fitness values) of the new food sources are checked again with the previously stored optimum solution and the decision for survival of a particular source solely depends on the fitness of the solutions.

The onlookers begin to work with the information provided by the employed bees about the food sources. Onlookers select the sources according to their probability of being selected. The probability of each source depends on their fitness value and for total N no. of sources it can be expressed as equation 2.

$$P_i = \frac{f(x_i)}{\sum_{n=1}^{N} f(x_n)} \tag{2}$$

In equation 2, the probability (P_i) of any i'th source is the ratio of its individual fitness to the sum of the fitness of all sources. After selection, onlookers use equation 1 to maintain the diversion of the sources. The scout bees create new sources randomly to replace the abandoned food sources (the source posing weak fitness value).

3. Case study: Aswan High Dam (AHD)

Aswan high dam is located across the river Nile in Aswan, Egypt. The AHD is the most important achievement for the people of Egypt as it serves them by providing hydropower, irrigation water, annual flood control and water supply for domestic/industrial uses. The river Nile annually supplies an average of $84\text{x}10^9$ m^3 of water to the reservoir. All important data used in this study (like – inflow, monthly demand, evaporation and other losses) are adopted from the study of El-Shafie and El-Manadely [17].The operational zone and water level/storage relations of AHD are given in figure 2. The relationship of storage-(S)/elevation-(H) and storage-(S)/surface area-(A) are given as equation 3 and 4 [17].

$$H = 79.97 + 0.0369S + 18.87\ln(S) \tag{3}$$

$$A = -3164.28 + 25.49S + 1092.92\ln(S) \tag{4}$$

In this study equation 4 has used to calculate evaporation volume from monthly evaporation rate. Also the seepage (approx. 0.08 BCM/month) and rock absorption losses (approx. 0.125 BCM/month) has used to calculate total water loss. The inflow pattern and targeted demand has shown in table 1.

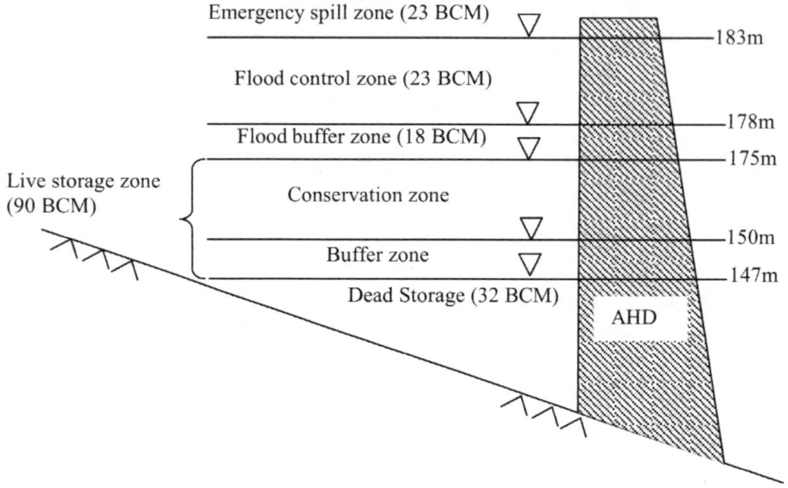

Figure 2. Water level and operational zones of AHD.

For the safe maintenance of the dam, water level for any time should maintain the range of 147 – 183 m. So in terms of storage, the bound for storage capacity is considered as 32 – 162 BCM. According to the dam authority there is another criteria should be followed in the model that, at the end of the month of July the water level should be below 175 m (approx. 122 BCM). For release bounds, 7.5 BCM is taken as the maximum allowable release in a month.

Table 1. Monthly inflow and water demand (in BCM) for AHD.

Months	High	Medium	Low	Demand
Jan	4.8	3.15	1.9	3.5
Feb	3.7	1.95	0.8	3.8
Mar	3.5	1.7	0.55	4.4
Apr	2.7	1.15	0.3	4.1
May	2.5	1.35	0.65	5.1
Jun	2.8	1.65	0.9	6.3
Jul	7.7	4.75	2.8	6.8
Aug	27.5	20.4	15.05	5.9
Sep	31	24.05	18.55	4.5
Oct	21.2	15.6	11.3	3.9
Nov	10.9	7.3	4.75	3.8
Dec	6.5	4.3	2.7	3.7

4. Model formulation

Using the storage bounds, ten discrete values have taken as initial storage. So the model runs each time with different initial storage and inflow conditions. Minimizing the water deficit has considered as the objective function of the study and given in equation 5.

$$\min f(x) = \sum_{t=1}^{12} \left(D_t - x_t\right)^2 \tag{5}$$

Here D_t is the targeted demand and x_t is the release for a time period t. For calculating the final storage for a month or initial storage for a consecutive month water mass balance relation has used (equation 6).

$$S_{(t+1)} = S_t + \text{Inflow}_t - \text{Release}_t - \text{Losses}_t \qquad (6)$$

Penalty function method used to handle the reservoir bounds and other constraints. The penalty terms for release and storage bounds are given as follows-

$$penalty1 = \begin{cases} 0 & \text{if } S_t > S_{min} \\ C_1(S_{min} - S_t)^2 & \text{if } S_t < S_{min} \end{cases} \; ; \; penalty2 = \begin{cases} 0 & \text{if } S_t < S_{max} \\ C_2(S_t - S_{max})^2 & \text{if } S_t > S_{max} \end{cases}$$

$$penalty_{AHD} = \begin{cases} 0 & \text{if } S_{July} < 122 \text{ BCM} \\ C(S_{July} - 122)^2 & \text{if } S_{July} > 122 \text{ BCM} \end{cases}$$

5. Results and discussion

Total 36 release curve has developed for every month of a year with three different inflow categories. Such curve for the month of January has given in figure 3. From the figure we can see that a release in high inflow period can meet the demand even in low storage capacity but as the storage is getting higher value, oversupply is required to keep the storage in safe range.

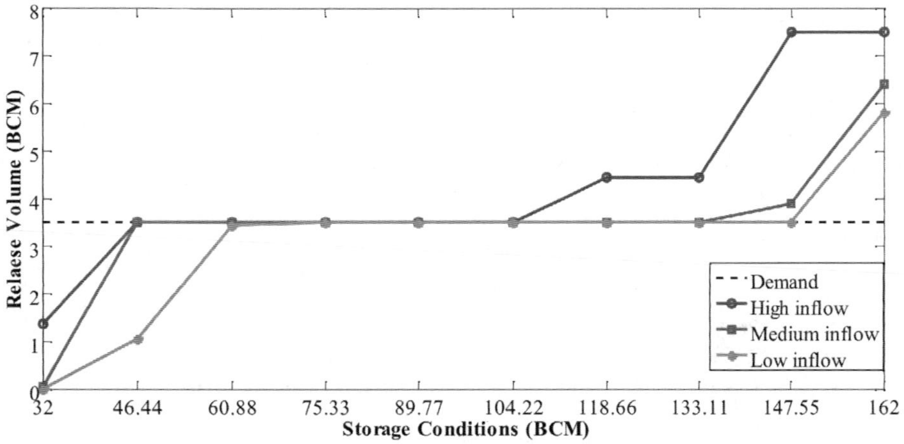

Figure 3. Release options for the month of January with different inflow conditions.

The fitness value through all iteration has given in figure 4. According to the figure 4, ABC optimization reached to the optimal solution very fast. For verification of the model, 18 years (from Aug 1980 to Jul 1998) of historical inflow data has used as an input of the release curves. The simulation results for ABC release has given in figure 5. In figure 5 we can see that how the release options given by ABC model minimize the deficit and close to demand for every month. From the simulation results, performance measuring indices has measured [18]. These indices can be given as – reliability (number of time period able to meet the demand over total time period), resiliency (maximum consecutive shortage period) and vulnerability (worst shortage ever) has calculated and given in table 2.

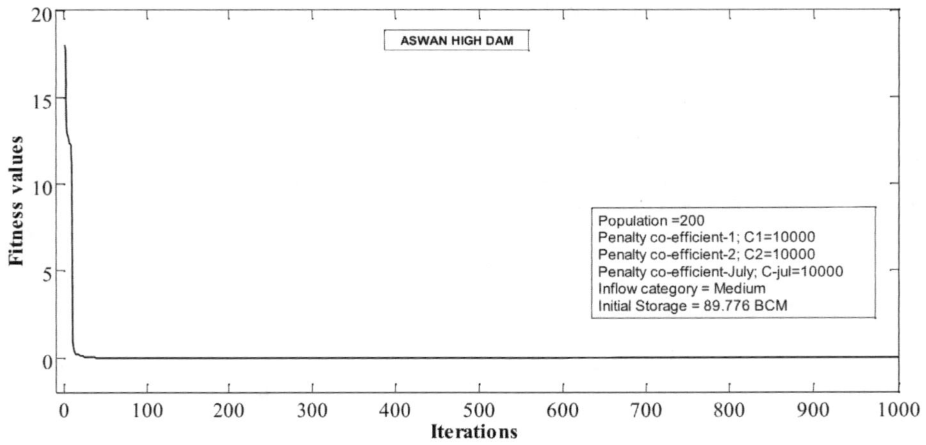

Figure 4. Objective function values through 1000 iterations

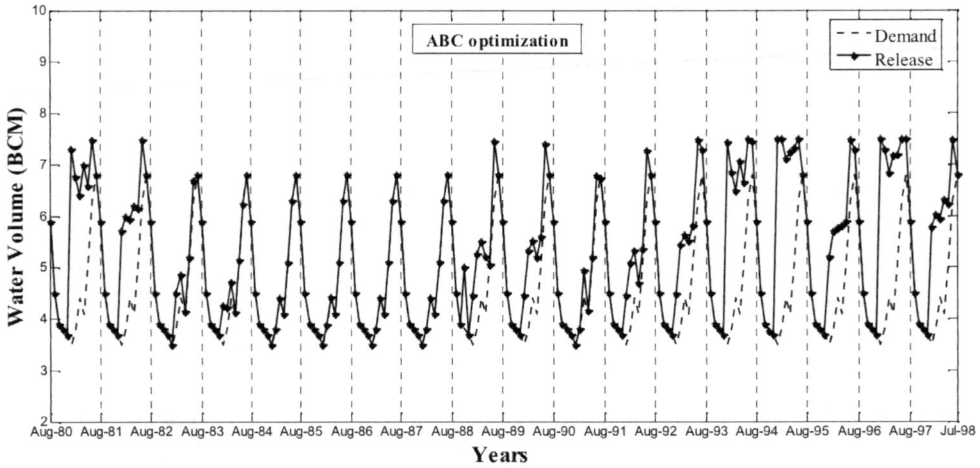

Figure 5. Simulation results of ABC release policy using historical inflow

Table 2. System performance indices

Performance measuring indices	ABC release policy
Reliability	98.14 %
Resiliency (months)	1
Vulnerability (% of demand)	1.11

In table 2 we can see that the release policy developed from ABC model satisfied the demand more than 98% of the total historical time period. Also the ability of getting back from a shortage period is very good as the resiliency is 1. Only 1.11% of demand is the worst shortage scenario that the model has experienced.

6. Conclusion

ABC is very fast and the results are very much reliable in finding optimal solutions, as we can see in figure 4 and 5. The simplicity of the ABC optimization technique is the main attraction over GA and PSO algorithm in finding release policy of a reservoir. Less parameter handling, simple analyzing of the problem and also very important - the practical application in a reservoir system suggest to consider ABC algorithm in this research field.

References

[1] Chang F-J and Chen L 1998 Real-coded genetic algorithm for rule-based flood control reservoir management. *Water Res. Man.* **12** 185-198

[2] Wardlaw R and Sharif M 1999 Evaluation of genetic algorithms for optimal reservoir system operation. *J. Water Res. Plan. and Man.* **125** 25-33

[3] Ahmed J A and Sarma A K 2005 Genetic algorithm for optimal operating policy of a multipurpose reservoir. *Water Res. Man.* **19** 145-161

[4] Jothiprakash V and Shanthi G 2006 Single reservoir operation policies using genetic algorithm. *Water Res. Man.* **20** 917-929

[5] Kumar D N and Reddy M J 2007 Multipurpose reservoir operation using particle swarm optimization. *J. Water Res. Plan. and Man.* **133** 192-201

[6] El-Shafie Amr H, El-Shafie A, El Mazoghi H G, Shehata A and Taha M R 2011 Artificial neural network technique for rainfall forecasting applied to Alexandria, Egypt. *Int. J. Phy. Sci.* **6** 1306-1316

[7] El-Shafie A, Jaafer O and Seyed A. 2011 Adaptive neuro-fuzzy inference system based model for rainfall forecasting in Klang River Malaysia. *International Journal of Physical Sciences.* **6** 2875-2888

[8] Noureldin A, El-Shafie A and Bayoumi M 2011 GPS/INS integration utilizing dynamic neural networks for vehicular navigation. *Inf. Fusion.* **12** 48-57

[9] Khor LC, Woo WL and Dlay SS 2005 Nonlinear Blind Signal Separation with Intelligent Controlled Learning. *IEE Proc. on Vision, Image and Signal Processing.* **152** 297-306

[10] Khajehzadeh M, Taha MR, El-Shafie A and Eslami M 2011 Modified particle swarm optimization for optimum design of spread footing and retaining wall. *Journal of Zhejiang University: Science A* **12** 415-427

[11] Karaboga D 2005 An idea based on honey bee swarm for numerical optimization. *Technical report –TR06.* (Kayseri, Turkey)

[12] Karaboga D and Basturk B 2007 A powerful and efficient algorithm for numerical function optimization: artificial bee colony (ABC) algorithm. *J. Glob. Opt.* **39** 459-471

[13] Karaboga D and Basturk B 2008 On the performance of artificial bee colony (ABC) algorithm. *App. Soft Com.* **8** 687-697

[14] Karaboga D and Akay B 2009 A comparative study of artificial bee colony algorithm. *App. Mat. and Com.* **214**108-132

[15] Wurbs R A 1996 *Modelling and Analysis of Reservoir System Operations* (Prentice-Hall, Inc. NJ)

[16] Tereshko V 2000 Reaction-diffusion model of a honeybee colony's foraging behavior. *In: Proc. PPSN VI: Parallel problem solving from nature, Lecture notes in computer science, Springer*. 807-816.

[17] El-Shafie A H & El-Manadely M S 2011 An integrated neural network stochastic dynamic programming model for optimizing the operation policy of Aswan High Dam. *Hydrology Research.* **42**, 50-67.

[18] Hashimoto T, Stedinger J R, & Loucks D P 1982 Reliability, resiliency, and vulnerability criteria for water resource system performance evaluation. *Water Resources Research.* **18**, 14-20.

Journal of Physics: Conference Series **423** (2013) 012002 doi:10.1088/1742-6596/423/1/012002

Study on the Rationality and Validity of Probit Models of Domino Effect to Chemical Process Equipment caused by Overpressure

Dongliang Sun[1], Guangtuan Huang[1,*] , Juncheng Jiang[2], Mingguang Zhang[2], Zhirong Wang[2]

[1]State Environmental Protection Key Laboratory of Environmental Risk Assessment and Control on Chemical Process, School of Resources and Environmental Engineering, East China University of Science and Technology, No. 130 Meilong Road, Xuhui District, Shanghai 200237, China

[2]Jiangsu Key Laboratory of Urban and Industrial Safety, Institute of Safety Engineering, School of Urban Construction and Safety Engineering, Nanjing University of Technology, No. 200 North Zhongshan Road, Nanjing 210009, Jiangsu, China

E-mail: dongliangsun@ecust.edu.cn

Abstract. Overpressure is one important cause of domino effect in accidents of chemical process equipments. Some models considering propagation probability and threshold values of the domino effect caused by overpressure have been proposed in previous study. In order to prove the rationality and validity of the models reported in the reference, two boundary values of three damage degrees reported were considered as random variables respectively in the interval [0, 100%]. Based on the overpressure data for damage to the equipment and the damage state, and the calculation method reported in the references, the mean square errors of the four categories of damage probability models of overpressure were calculated with random boundary values, and then a relationship of mean square error vs. the two boundary value was obtained, the minimum of mean square error was obtained, compared with the result of the present work, mean square error decreases by about 3%. Therefore, the error was in the acceptable range of engineering applications, the models reported can be considered reasonable and valid.

1. Introduction

Many types of equipment containing dangerous materials may exist in industrial installations: tanks. If the equipment was damaged due to critical conditions of pressure, mechanical impacts or structural cracking, a catastrophic sequence may rise, a sudden explosion, for instance. Then overpressure, heat radiation, and many fragments can be generated and threaten other equipment erected in their vicinity,

[*] To whom any correspondence should be addressed.

when the objects are destroyed, new accidents may take place, therefore, domino effect is resulted in and make accident consequence more severe [1-11].

In recent years, a series of research work has identified that overpressure in explosion accident is one important cause of domino effect to chemical process equipments. Some models considering propagation probability and threshold values of the domino effect caused by overpressure have been proposed in previous study [1-19]. With scarce data and some oversimplified assumption, these models almost only related damage to peak static overpressure [1-8]. In the present study, in order to prove the rationality and validity of the models developed by Zhang, two boundary values of three damage degrees were considered as random variables respectively in the interval [0, 100%]. Based on the overpressure data for damage to the equipment and the damage state, and the calculation method reported in the references, the mean square errors of the four categories of damage probability models of overpressure were calculated with random boundary value, and then a relationship of mean square error vs. the two boundary value was obtained, in order to find the minimum of mean square error, and then it was compared with the previous result, to prove whether the previous result was in the range of error allowed.

2. Overview of previous study

In previous work, based on damage phenomenon and relevant threshold data of peak static overpressure from past accidents, many results were obtained, see Table 1 [1-8]: damage probability models of overpressure generally went through three stages: Probit model—Escalation thresholds—Probit model. So far, the models developed by Cozzani and Zhang were more advanced than others, especially the latter, which was obtained with more sufficient data of damage caused by overpressure, and rational method of probability analysis [5-8]. Based on the results of Cozzani, more data about damage to chemical process equipment were gathered and analyzed, the damage degrees of the four categories of vessels (atmospheric, pressurized, elongated, small) were divided into three classes respectively: *DS1LI1*, *DS2LI2*, *DS2LI3*, and then entire probability range of 0–100% was also divided into three ranges: the range of 0–30% was assumed to data in *DS1LI1* damage state; the range of 30–70%was assumed to data in *DS2LI2* damage state; the range of 70–100% was assumed to data in *DS2LI3* damage state. Finally, the damage probabilities were calculated, quantitative relationship between damage probabilities and damage degrees of equipment were built respectively for the four categories of vessels, and the mean square errors of the four models decrease from 39.1% to 12.5%, the improvements of present models were evidenced through comparison with literatures [5-8]. However, in the method of probability analysis above, the boundary values of the three damage degrees 30% and 70% were set arbitrarily, if the boundary values were set to other values, whether the mean square errors of the four models would be less than 12.5%, and whether more reasonable and realistic models could be obtained?

Table 1. Probit models reported in the literatures.

Author	Categorization of process plants	Damage Probability model	Threshold values of peak static overpressure
Eisenberg et al. [1]	No categorization	$Y = -23.8 + 2.92\ln(\Delta P)$	
Bagster and Pitblado [2]	No categorization	$F_d = (1 - r/r_{th})^2$, r: distance from explosion center; r_{th}: distance from explosion center at which static overpressure equals ΔP_{th}	$\Delta P_{th} = 36$kPa
Khan and Abbasi [3]	No categorization	if $\Delta P < 70$kPa, $F_d = 0$; if $\Delta P > 70$kPa, $Y = -23.8 + 2.92\ln(\Delta P)$	

Gledhill and Lines [4]	Atmospheric	if $\Delta P < \Delta P_{th}$, $F_d = 0$	$\Delta P_{th} = 7$kPa
	Pressurized	if $\Delta P \geq \Delta P_{th}$, $F_d = 1$	$\Delta P_{th} = 38$kPa
Cozzani and Salzano [5-7]	Atmospheric	$Y = -18.96 + 2.44\ln(\Delta P)$	
	Pressurized	$Y = -42.44 + 4.33\ln(\Delta P)$	
	Elongated	$Y = -28.07 + 3.16\ln(\Delta P)$	
	Small	$Y = -17.79 + 2.18\ln(\Delta P)$	
Zhang etal. [8]	Atmospheric	$Y = -9.36 + 1.43\ln(\Delta P)$	
	Pressurized	$Y = -14.44 + 1.82\ln(\Delta P)$	
	Elongated	$Y = -12.22 + 1.65\ln(\Delta P)$	
	Small	$Y = -12.42 + 1.64\ln(\Delta P)$	

F_d: failure probability; Y: probit values corresponding to failure probability; ΔP: peak static overpressure (Pa).

3. Methods

3.1. Random variables

Based on the statement above, in order to prove the rationality of the models developed by Zhang, the boundary values above were considered as random variables x, y, and supposed to followed uniform distributions in the interval [0, 100%] respectively, and $x < y$:

$$\begin{aligned} x &= u_i \\ y &= v_i \end{aligned} \tag{1}$$

where u_i, v_i are uniform random number respectively in [0, 1].

3.2. Damage probabilities

In each damage degree, the damage probabilities of the four categories of vessels caused by overpressure can be obtained by the Eq. (2), whose initial form was built by Zhang in Ref. [8]:

$$P = \begin{cases} x \times \dfrac{\Delta P}{\Delta P_{\max 1}}, DS1LI1 \\ x + (y - x) \times \dfrac{\Delta P}{\Delta P_{\max 2}}, DS2LI2 \\ y + (1 - y) \times \dfrac{\Delta P}{\Delta P_{\max 3}}, DS2LI3 \end{cases} \tag{2}$$

where P is probability value, ΔP is overpressure value and $\Delta P_{\max i}$ (i = 1, 2, 3) is a threshold value of overpressure in each damage state. The coefficient value x and y were decided based on some assumptions stated below:

–In *DS1LI1* damage state, damage probability value x was assumed to be correspondent to the highest overpressure value ($\Delta P_{\max 1}$) of *DS1LI1* damage state.

–In *DS2LI2* damage state, damage probability value y was assumed to be correspondent to the highest overpressure value ($\Delta P_{\max 2}$) of *DS2LI2* damage state.

–In *DS2LI3* damage state, damage probability value 100% was assumed to be correspondent to the highest overpressure value ($\Delta P_{\max 3}$) of *DS2LI3* damage state.

In Eq. (2), we can see that, $\Delta P/\Delta P_{\max 1} < 1$, $\Delta P/\Delta P_{\max 2} < 1$ and $\Delta P/\Delta P_{\max 3} < 1$, P follows uniform distributions respectively in the interval [0, x], [x, y] and [y, 1], with respect to *DS1LI1*, *DS2LI2* and *DS2LI3*.

3.3. Mean square errors

Based on ΔP values, damage degree for each kind of vessel reported in Table 2-5, the damage probabilities can be calculated by Eq. (2) for each group of x and y, and then probit values can be calculated from probability data. Therefore, probit models for different categories were obtained by least square regression, as well as mean square error of the models. In a word, the mean square error of the four categories of vessels could be calculated with x and y. If the mean square error less than 12.5% could be obtained by building the relationship of mean square error vs. the two boundary value, more reasonable models could be obtained.

Table 2. ΔP values, damage phenomenon and degree for atmospheric vessels.

ΔP (kPa)	Damage phenomenon	Damage degree
5.17	Minor damage, cone roof tank (100% filled)	DS1LI1
5.17	Minor damage, cone roof tank (50% filled)	DS1LI1
6.10	1% structural damage of equipment	DS1LI1
10.00	Failure of atmospheric equipment	DS2LI2
14.00	Minor damage of atmospheric tank	DS1LI1
18.70	Minor damage, floating roof tank (50% filled)	DS1LI1
18.70	Catastrophic failure, cone roof tank (50% filled)	DS2LI3
20.00	Deformation of atmospheric tank	DS1LI1
20.40	50% structural damage of equipment	DS2LI2
24.00	20% structural damage of steel floating roof tank	DS2LI2
25.00	Atmospheric tank destruction	DS2LI3
27.00	Failure of steel vessel	DS1LI1
34.00	99% structural damage of equipment	DS2LI3
42.51	Catastrophic failure, cone roof tank (100% filled)	DS2LI3
136.00	Structural damage, low pressure vessel	DS2LI3
136.05	Catastrophic failure, floating roof tank (50%filled)	DS2LI3
136.05	Catastrophic failure, floating roof tank (100% filled)	DS2LI3
136.10	99% structural damage of floating roof tank	DS2LI3
137.00	99% damage (destruction) of floating roof petroleum tank	DS2LI3
7.00	Collapse of atmospheric tank roof	DS1LI1
7.00	Partial damage to atmospheric tank	DS1LI1
10.00	Fixed roof tank damage	DS1LI1
10.00	50% damage to atmospheric tank	DS2LI2
20.00	Displacement of steel supports	DS1LI1
20.00	100% damage, atmospheric tank	DS2LI3
35.00	80% damage of process plant	DS2LI3
21.00	Destruction of fixed roof atmospheric tank	DS2LI3
42.51	Minor damage, floating roof tank (100% filled)	DS1LI1
45.00	Catastrophic failure, floating roof tank	DS2LI3
7.00	Failure of connection	DS1LI1

Journal of Physics: Conference Series **423** (2013) 012002 doi:10.1088/1742-6596/423/1/012002

20.00	Tubes deformation	DS1LI1
22.10	Minor damage, pipe supports	DS1LI1
37.42	Catastrophic failure, pipe supports	DS2LI2
42.00	Tubes failure	DS2LI2

Table 3. ΔP values, damage phenomenon and degree for pressurized vessels.

ΔP (kPa)	Damage phenomenon	Damage degree
30.00	Failure of pressure vessel	DS1LI1
39.00	Structural damage to pressure vessel	DS2LI2
39.12	Minor damage, pressure vessel horizontal	DS1LI1
42.00	Pressure vessel deformation	DS1LI1
52.72	Minor damage, tank sphere	DS1LI1
53.00	Pressure vessel failure	DS2LI2
53.00	Failure of spherical pressure vessel	DS2LI2
55.00	20% structural damage of spherical steel petroleum tank	DS2LI2
61.22	Catastrophic failure, pressure vessel horizontal	DS2LI3
81.63	Minor damage, pressure vessel vertical	DS1LI1
83.00	20% damage of vertical cylindrical steel pressure	DS2LI2
88.44	Catastrophic failure, pressure vessel vertical	DS2LI3
95.30	99% structural damage of vertical, steel pressure vessel	DS2LI3
97.00	99% damage of vertical cylindrical steel pressure vessel	DS2LI3
108.84	Catastrophic failure, tank sphere	DS2LI3
108.90	99% structural damage of spherical, pressure steel vessel	DS2LI3
110.00	99% damage of spherical steel petroleum tank	DS2LI3
38.00	Partial damage of pressure vessel	DS2LI2
70.00	Failure of pressurized storage sphere	DS2LI2
7.00	Failure of connection	DS1LI1
20.00	Displacement of steel supports	DS1LI1
20.00	Tubes deformation	DS1LI1
22.10	Minor damage, pipe supports	DS1LI1
37.42	Catastrophic failure, pipe supports	DS2LI2
42.00	Tubes failure	DS2LI2

Table 4. ΔP values, damage phenomenon and degree for elongated vessels.

ΔP (kPa)	Damage phenomenon	Damage degree
17.00	Minor damage, distillation tower and cylindrical steel vertical structure. Failure of part of equipment	DS1LI1
29.00	Distillation tower and cylinder steel vertical structure	DS1LI1
35.71	Minor damage, fractionation column	DS1LI1
38.00	Deformation of non-pressure equipment	DS1LI1
42.52	Minor damage, extraction column	DS1LI1
45.92	Catastrophic failure, fraction column	DS2LI3
47.00	Failure of non-pressure	DS2LI2
69.73	Catastrophic failure, extraction column	DS2LI3
35.00	Damage to fractionating column	DS1LI1

14.00	Minor damage of cooling tower	DS1LI1
7.00	Failure of connection	DS1LI1
20.00	Displacement of steel supports	DS1LI1
20.00	Tubes deformation	DS1LI1
22.10	Minor damage, pipe supports	DS1LI1
37.42	Catastrophic failure, pipe supports	DS2LI2
42.00	Tubes failure	DS2LI2

Table 5. ΔP values, damage phenomenon and degree for small equipments.

ΔP (kPa)	Damage phenomenon	Damage degree
25.30	Minor damage, reactor chemical	DS1LI1
49.32	Minor damage, heat exchanger	DS1LI1
59.52	Catastrophic failure, reactor chemical	DS2LI3
59.52	Catastrophic failure, heat exchanger	DS2LI3
76.53	Catastrophic failure, reactor cracking	DS2LI3
81.63	Minor damage, pump	DS1LI1
108.84	Catastrophic failure, pump	DS2LI3
18.70	Minor damage, reactor cracking	DS1LI1
7.00	Failure of connection	DS1LI1
20.00	Displacement of steel supports	DS1LI1
20.00	Tubes deformation	DS1LI1
22.10	Minor damage, pipe supports	DS1LI1
37.42	Catastrophic failure, pipe supports	DS2LI2
42.00	Tubes failure	DS2LI2

4. Analysis of the rationality and accuracy of the models

By the calculation method discussed above, the mean square errors were obtained through a certain number of simulations, and then a relationship of mean square error vs. x and y could be obtained by data fitting. However, the accuracy of the regression depends on the number of the data, which is referred to the number of the simulations. Finally, the relationship is expressed as follows and $R = 0.9943$:

$$M = -0.05237\ln x + 0.052176\tan y + 0.027957\cos(xy) + \frac{0.000493}{\tan x} + \frac{0.00211}{\tan y} - \frac{0.00047}{\tan x \cos y} \quad (3)$$

where M is mean square error of the four obtained models. In the process of simulation, the coefficients were set to a, b, c, d, e, and f (in Eq. (3), a = -0.05237, b = 0.052176, c = 0.027957, d = 0.000493, e = 0.00211, f = -0.00047), they changed with the number of simulations, see Fig .1: a number of 1200 simulations could provide a good convergence for the value of the coefficients.

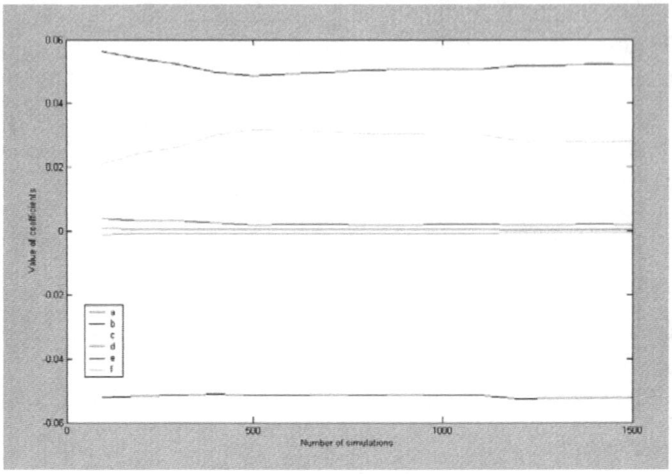

Figure 1. The changes of the coefficients of the relationship with simulations.

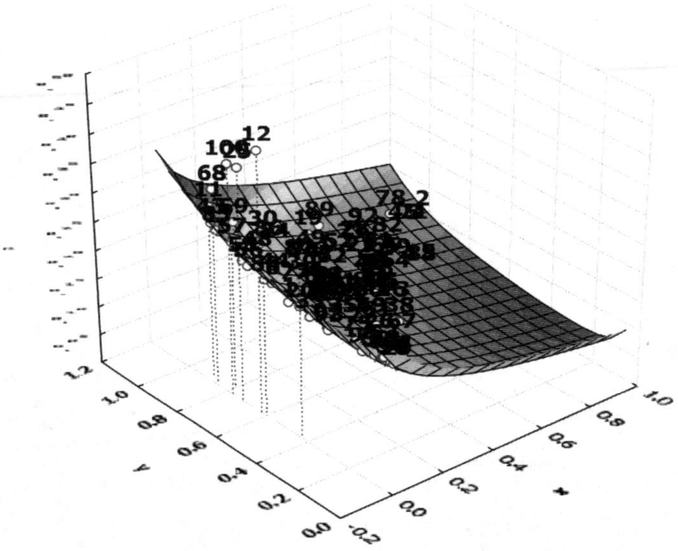

Figure 2. The changes of the coefficients of the relationship with simulations.

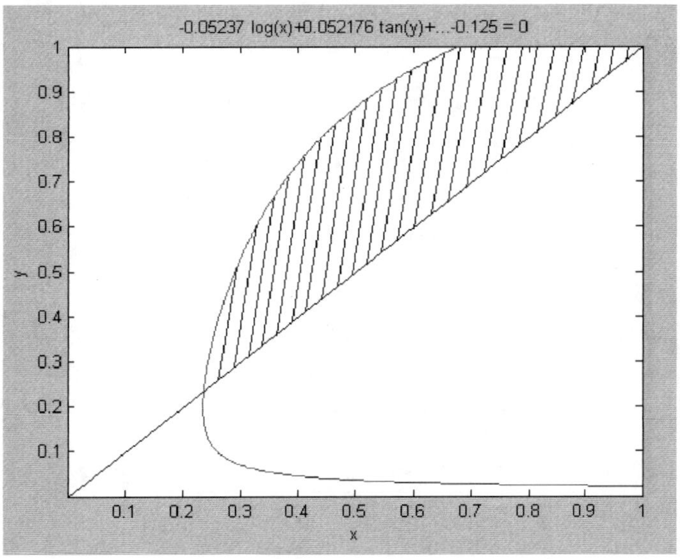

Figure 3. The changes of the coefficients of the relationship with simulations.

Eq. (3) was a binary function, whose curve was shown in Fig. 2. Based on the trends of the curve, the collection of x and y, in which the value of M was less than 12.5% was shown in the shadow of Fig. 3: in this part, $x < y$, the blue curve was the point collection of x and y in which M=12.5%. In the shadow part of Fig. 3, the minimum of M was 9.06%, compared with 12.5% obtained by Zhang, mean square error decreased by about 3%. Therefore, the error was in the acceptable range of engineering applications (generally 5%), the models developed by Zhang could be considered reasonable and valid.

5. Conclusion

The present work was to prove the rationality and validity of the probit models for damage to chemical process equipment caused by overpressure:

–Damage probability models of overpressure generally went through three stages: Probit model—Escalation thresholds—Probit model, and so far, the models developed by Cozzani and Zhang were more advanced than others, especially the latter, which was obtained with more sufficient data of damage caused by overpressure, and rational method of probability analysis.

–Damage probabilities of vessels, P followed uniform distributions respectively in the interval [0, x], [x, y] and [y, 1], with respect to DS1LI1, DS2LI2 and DS2LI3.

–The relationship of mean square error vs. the two boundary value was built, and regression coefficient R was more than 0.99.

–For building the relationship by data fitting, a number of 1200 simulations (1200 groups of x and y) could provide a good convergence for the value of the coefficients of the equation.

–Based on the trends of the curve of the relationship erected by data fitting, the collection of x and y, in which the value of M was less than 12.5% was obtained, the minimum of M was 9.06%, compared with the result obtained by Zhang, mean square error decreased by about 3%. Therefore, the error was in the acceptable range of engineering applications (generally 5%), the models developed by Zhang could be considered reasonable and valid.

Acknowledgements

The financial support of China Postdoctoral Science Foundation-funded project (2012M520851), the Shanghai Leading Academic Discipline Project (B506), National Natural Science Foundation of China

(No. 71001051, 50904037), and Research and Innovation Plan for Graduates of Colleges and Universities in Jiangsu Province (CX09B_142Z) are gratefully acknowledged.

References

[1] Eisenberg N A, Lynch C J and Breeding R J 1975 *Vulnerability model: a simulation system for assessing damage resulting from marine spills*, Rockville, MD, Enviro Control Inc., Report CG-D-136-75
[2] Bagster D F and Pitblado R M 1991 *Process Saf. Environ. Prot.* **69** 195–9
[3] Khan F I and Abbasi S A 1998 *Process Saf. Prog.* **17** 107–23
[4] Gledhill J and Lines I 1998 *Development of methods to assess the significance of domino effects from major hazard sites*, Health and Safety Executive, CR Report 183
[5] Cozzani V and Salzano E 2004 *J. Hazard. Mater.* **A107** 67–80
[6] Cozzani V and Salzano E 2004 *J. Hazard. Mater.* **A107** 81–94
[7] Cozzani V and Salzano E 2004 *J. Loss Prevent. Process Ind.* **17** 437–47
[8] Zhang M G and Jiang J C 2008 *J. Hazard. Mater.* **158** 280–6
[9] Antonioni G, Spadoni G and Cozzani V 2009 *J. Loss Prevent. Process Ind.* **22** 439–49
[10] Cozzani V, Antonioni G and Spadoni G 2006 *J. Loss Prevent. Process Ind.* **19** 463–77
[11] Cozzani V, Gubinelli G, Antonioni G, Spadoni G and Zanelli S 2005 *J. Hazard. Mater.* **A127** 14–30
[12] Darbra R M, Palacios A and Casal J 2010 *J. Hazard. Mater.* **183** 565–73
[13] Kozanoglu B, Zάrate L, Gómez-Mares M and Casal J 2011 *J. Hazard. Mater.* **197** 104–8
[14] Palacios A and Casal J 2011 *Fuel.* **90** 824–33
[15] Pavlova Y and Reniers G 2011 *J. Hazard. Mater.* **186** 401–6
[16] Reniers G and Amyotte P 2012 *J. Loss Prevent. Process Ind.* **25** 227–31
[17] Reniers G, Cuypers S and Pavlova Y 2012 *J. Hazard. Mater.* **209-210** 164–76
[18] Reniers G L L, Audenaert A, Pauwels N and Soudan K 2011 *J. Hazard. Mater.* **186** 779–87
[19] Reniers G L L, So"rensen K and Dullaert W 2012 *Reliab. Eng. Syst. Safe.* **98** 35–42

ScieTech 2013 IOP Publishing
Journal of Physics: Conference Series **423** (2013) 012003 doi:10.1088/1742-6596/423/1/012003

The quantum Hall effect: The symmetry and the velocity

Keshav N. Shrivastava

Department of Physics, University of Malaya, Kuala Lumpur 50603 Malaysia and

School of Physics, University of Hyderabad, Hyderabad 500046, India

keshav1001@yahoo.com

Abstract. The complete and the correct interpretation of all of the fractions found in the quantum Hall effect of AlGaAs/GaAs heterostructures is presented by use of a specific treatment of spin. Hence the flux quantization depends on spin and the Lande's g value expression is replaced by a linear formula. The fractions occur in pairs of left and right helicities and the velocity of the particles is directed such that if one particle moves upstream then its conjugate moves downstream. The change in the sign of the spin is equivalent to the change in the sign of the velocity.

1. Introduction

The Hall resistivity is a linear function of magnetic field. In real systems it is found to have fractional values which require understanding. We explain the quantum Hall effect in GaAs heterostructures and in graphene. We find that the size of the devices is often only a few nm and temperatures of measurements are quite low such as mK. There has been considerable effort to understand the underlying quantum mechanics in terms of wave functions [1,2].

In this paper, we explain the general theory of the quantum Hall effect on the basis of angular momentum theory of quantum mechanics. We explain that electrons of positive as well as negative velocities occur and helicity is used to connect the spin with the velocity. The electrons occur as single particles which explains the "principal fractions". There are resonances as well as "two-particle" states. A small part of the data requires that "electron clusters" occur which are subject to spin waves and rotations.

2. Theory of quantum Hall effect

We define the cyclotron frequency as,

$$g\mu_B B = \hbar\omega \qquad (1)$$

The Bohr magneton is,

$$\mu_B = \frac{e\hbar}{2mc} \tag{2}$$

Hence,

$$\frac{g}{2}\frac{eB}{mc} = \omega_c \,. \tag{3}$$

Since for L=0, g=2, it is a common practice to define the cyclotron frequency as,

$$\omega_c = \frac{eB}{mc} \,. \tag{4}$$

The electrons in a magnetic field behave like harmonic oscillators. Hence, the energy of a state is given by,

$$E_n = \hbar\omega_c(n+\frac{1}{2}) = \hbar\frac{eB}{mc}(n+\frac{1}{2}) \tag{5}$$

It is proper to include the factor of g/2 so that the above energy is modified to,

$$E_n = \frac{1}{2}g\hbar\frac{eB}{mc}(n+\frac{1}{2}). \tag{6}$$

The same effect can be obtained by replacing e by $e^* = (1/2)ge$, so that the energy can be written as,

$$E_n = \hbar\frac{e^*B}{mc}(n+\frac{1}{2}). \tag{7}$$

The transition can occur from E_n to $E_{n'}$ The Hall effect resistivity is,

$$\rho_{xy} = \frac{B}{nec} = \frac{B}{ne^*c} \,. \tag{8}$$

Hence the effective charge can be measured. The flux quantization is given by,

$$B.A = n'\frac{hc}{e}. \tag{9}$$

Substituting this field in the Hall effect formula gives,

$$\rho_{xy} = \frac{n'hc}{ne^*ceA} = \frac{n'}{nA}\frac{h}{ee^*} = n''\frac{h}{\frac{1}{2}ge^2} \tag{10}$$

which gives the quantum Hall effect. We do not use the Lande's formula but suggest a formula linear in the angular momentum. This formula gives,

$$g = \frac{2j+1}{2l+1} \tag{11}$$

Here, $j = l \pm s$ is the total angular momentum quantum number. The effective charge is thus,

$$\nu_\pm = \frac{1}{2}g_\pm e = \frac{l+\frac{1}{2}\pm s}{2l+1}. \tag{12}$$

For positive sign and s=1/2,

$$\nu_{+} = \frac{l+1}{2l+1} \tag{13}$$

and for the negative sign and s=1/2,

$$\nu_{-} = \frac{l}{2l+1}. \tag{14}$$

Here ,

$$\nu_{\pm} = (1/2)g \tag{15}$$

gives the correct fractional charge as, for L=1, $\nu_{-} = 1/3$ and $\nu_{+} = 2/3$, etc. These values agree with experimental data. We call these as "principal fractions". Instead of the principal values, we can also generate the resonances, $\nu_1 - \nu_2$. This process produces some more fractions not already present in the "principal fractions", except in the case when we consider the energy level difference $E_1 - E_2$ with $E_2 = 0$. Indeed, there is a zero energy state for l=0, with negative sign, s=1/2, $\nu_{-} = 0$. Many resonances are in fact present in the experimental data and indeed predicted from the linear theory. At low temperatures, the excitation populations are small so that interactions are minimized. Hence we predict the "resonances". We are thus able to produce a large number of principal fractions and resonances so that "two-particle states" occur. For these particles we have $\nu_1 + \nu_2$ so that these processes produce more fractions than are found in resonances. Hence we have (i) principal fractions, (ii) resonances and the (iii) two-particle states. The real material is often having electron clusters so that the spin need not be ½. The spin
becomes

$$\text{NS}=(N_{\uparrow} - N_{\downarrow})S. \tag{16}$$

Hence the spin may be 3/2, 5/2, etc. For example for

$$l = 0 \tag{17}$$

the fractional charge can occur at,

$$e^{*} = \frac{1}{2} \pm s \tag{18}$$

For finite L,

$$e^{*}/e = \frac{l + \frac{1}{2} \pm s}{2l+1} \tag{19}$$

For S=0 which can arise in pairs

$$N_{\uparrow} - N_{\downarrow} = 0 \tag{20}$$

so that the effective charge becomes ,

$$e^{*}/e = \frac{1}{2}\frac{2l+1}{2l+1} = \frac{1}{2}. \tag{21}$$

which has even denominator. For S=1,

$$e^{*}/e = \frac{l+\frac{1}{2}\pm 1}{2l+1} = \frac{l+\frac{3}{2}}{2l+1}, \frac{l-\frac{1}{2}}{2l+1} \tag{22}$$

which has even denominators and occurs in pairs.

3. Spin deviation

For a large cluster, there is a small spin deviation so that the spin ceases to be an integer. For example,

$$S - \delta S = 2 - 0.02 = 1.98 \tag{23}$$

which has 1 % spin deviation. All of the data is correctly explained on the basis of this theory and we have written down the interpretation of 101 fractions satisfactorily. The fractions of charge which are measured from the plateaus in the Hall resistivity ρ_{xy} are usually the same as those measured from the minima in ρ_{xx}. In the case of electron clusters, the fractions derived from xx direction are slightly different from those found from xy values. The effective charge of the electron becomes anisotropic due to the spin wave propagation in the micro-cluster of electrons. The wave vector of the spin waves appears in the effective value of the spin. There is an explicit dependence of the charge on the spin which modifies the condition of the flux quantization and leads to an anisotropic charge. In the Hall effect, the resistivity is a linear function of magnetic field. When field is quantized there occur plateaus in the ρ_{xy} and minima in ρ_{xx}. The plateaus at the integer multiples of hc/e were first detected by von Klitzing et al [3]. They found that the resistivity is quantized at $\rho_{xy} = h / ie^2$ where i is an integer. Later on Tsui et al [4] working at still lower temperatures and higher magnetic fields discovered that i need not be an integer as it becomes a fraction such as 1/3 or 2/3. At this stage, Laughlin wrote a wave function [1] by using the first principles and suggested that the charge may become fractional such as e/3 or e/5 due to electron correlations. Many efforts have been made to suggest composite fermions, parafermions and non-Abelian fermion pairs to explain the fractionally charged quasiparticles. Instead of starting from a wave function, we looked at the factors which multiply the Bohr magneton which change the effective charge of the electron. We found that there is a possibility of a spin-charge coupling which changes the charge of a particle [2]. The factor of $g = (2j+1)/(2l+1)$ multiplies the Bohr magneton and hence changes the charge of the electron to $e^* = (1/2)ge$ which is e for g=2. In actual samples which have a layer of GaAs over Al doped GaAs, electron clusters occur. In the clusters the spin need not be ½. For example, it can be 3/2 if three electrons are aligned parallel to each other as in a ferromagnet. Hence the spin becomes large, such as NS. It is also possible that the electrons are aligned in such a way that there are $N_\uparrow$ electrons with spin up and there are $N_\downarrow$ electrons with spin down so that the spin of the electrons in the cluster becomes NS with $N = N_\uparrow - N_\downarrow$. Hence the idea of "clusters" of electrons explains the data. In this way, all of the 101 fractions can be explained. Instead of detecting one fraction at a time, some times a full series of fractions can be written down. Such series can be generated by using higher than s=1/2 and by varying the orbital angular momentum quantum number , l. These series of fractions are observed in graphene and correctly predicted [2]. Recently, Kumar et al [6] have found that a fraction which shows one value when measured from the plateau in the ρ_{xy} need not show the same value when measured from the minimum in ρ_{xx}. We show that there are spin waves in small clusters of electrons which create a finite "spin deviation" giving rise to anisotropy in the fractional charge. The spin deviation depends on the spin wave vector and makes the spin value anisotropic which due to spin-charge coupling makes the charge anisotropic and hence the effective charge along xy plane is not equal to that in the xx direction. We find that a small difference occurs in the value of the fraction in going from xx to xy value.The Hamiltonian of the ferromagnetic spin waves with exchange interaction is given by,

$$\mathrm{H} = -J\Sigma_{j\delta} S_j . S_{j+\delta} - g\mu_B H \Sigma_j S_{jz} \tag{24}$$

where J is the exchange interaction and H is the magnetic field. S_j are the spin operators at the jth site and the summation can be carried out to nearest neighbors, $S_{j+\delta}$. The number of nearest neighbors is z so that the Fourier transforms of spin operators requires,

$$\gamma_k = \frac{1}{z}\Sigma_\delta \exp(ik.\delta) \tag{25}$$

Leaving out the magnetic field dependent term, the unperturbed frequency of a magnon is,

$$\hbar\omega_k = 2JzS(1-\gamma_k) \tag{26}$$

Hence, the spin has been changed from S to S(1-γ_k) in going from the site variables to spin wave variables. This is called the spin deviation and it amounts to a few per cent in real materials. For $k.\delta$ <<1, exp(ik. δ) has only the sine term and the cos term is zero so that for small wave vectors,

$$z(1-\gamma_k) \approx \frac{1}{2}\Sigma_k (k.\delta)^2 \tag{27}$$

so that

$$\hbar\omega_k \approx JS\Sigma_k (k.\delta)^2 \tag{28}$$

which for cubic lattices is 2JS(k.a)^2 where a is the lattice constant. Hence, the spin is changed to S(k.a)^2. In the case of two-sublattice antiferromagnets, $\hbar\omega_k \approx 4\sqrt{3}JSka.$ In a cluster, in the a-b plane, the frequency of a magnon is given by the expression 2JS[(k.a)^2 +(k.b)^2] whereas along the x direction it appears as 2JS(k.a)^2. Hence, the spin value becomes anisotropic.
The Hall effect resistivity is,

$$\rho_{xy} = \frac{h}{\frac{1}{2}ge^2} = \frac{h}{\frac{l+\frac{1}{2}\pm s}{2l+1}e^2} \tag{29}$$

In the case of an electron cluster, the value of S along xx direction is NS which due to spin waves becomes $S(k.a)^2 N$ whereas along the xy direction it will be $\frac{1}{2}S[(k.a)^2 + (k.b)^2]N.$ Hence the value of (1/2)g depends of the direction. The value of the filling factor $\nu = (1/2)g$ also depends on the direction. Hence ν_{xx} need not be equal to ν_{xy} where ν_{xx} is deduced from ρ_{xx} and ν_{xy} from ρ_{xy}. Kumar et al [6] have found that in the xx direction $\nu = 2.463 \pm 0.002$ whereas in the xy direction $\nu = 2.461$. Hence, the value deduced from the xy is slightly different from that deduced from the xx which we assign to the value of the spin due to spin waves. For L=0, the filling fraction is $\frac{1}{2} \pm s = \frac{2463}{1000}$.
Hence $\pm s = 1.963$ which shows 1.8 % spin deviation compared with S=2. The Hall effect experiment is performed on a polycrystalline heterostructure film so that the crystallographic directions are not the same as those used in the Hall effect.

4. Helicity and velocity

In single particles, the helicity is defined as the projection of its spin along the direction of motion. It is impossible to change the helicity of a particle because it would then become its antiparticle. In the solid state it is possible to consider a cluster of particles which permits the observation of a

phenomenon not found in single particles. Similarly, in molecules it is possible to see some phenomenon which is not seen in single particles. The wave function of a particle may be made of a linear combination of two particles which differ in helicity. The helicity of one particle may be positive while that of the other may be negative and we can make a third particle which is a combination of the first and the second particle. A particle of mixed helicity may mean that it is made of particles of different helicities. It does not mean that a particle of positive helicity can change itself into that of a negative helicity. This means that "reversed helicity" does not occur. Under the rotation it is possible to change the direction of the velocity and hence the helicity of a particle in a molecule or in a solid. In the quantum Hall effect, the resistivity is quantized at the plateau. The usual Hall effect occurs with the resistivity linearly proportional to the magnetic field, $\rho_{xy} = B/nec$ where B is the magnetic induction, e is the electron charge, c is the velocity of light and n is the electron concentration, the number of electrons per unit area for a slab of a metal. The flux quantization leads to plateaus at which $\rho_{xy} = h/ie^2$ where h is the Planck's constant and i is an integer. When angular momentum is considered, i becomes a fraction. Due to particle-antiparticle symmetry the fractions occur in pairs, one for the positive spin and the other for the negative spin. For single particles, the resistivity at the plateau is given
by,

$$\rho_{xy} = \frac{h}{\frac{1}{2} g_{\pm} e^2}$$

where g=(2j+1)/(2*l*+1) and j= *l* ± s. The ± sign in the expression for j introduces the particle-antiparticle symmetry. Since the signs occur in the spin, the values of the resistivity correspond to two helicities. The values of (1/2)$g_\pm$ for the two helicities give two different plateaus in the resistivity. The spin as well as the helicities thus enter in the problem of Hall resistivity. The plateaus which occur for a single value of *l* and s are arising from the "principal fractions". The two-particle states occur for $\nu_1+\nu_2$ and resonances occur at $\nu_1-\nu_2$ where $\nu = (1/2)g_\pm$ and the effective charge of a particle becomes e*= (1/2)ge. In $Al_xGa_{1-x}As/GaAs$ heterostructures, clusters of electrons are often formed in which case the spin is different from 1/2. The large values of spin are required to understand the plateaus which depend on the sample and show spin wave reduction in the value of the spin. The spin larger than 1/2 occurs in clusters which are subject to rotation. Hence , "roton" type effect is seen in the quantum Hall effect of clusters of electrons. The single-particle states with spin 1/2 occur at (1/2)$g_\pm$. The positive sign corresponds to + helicity (right handed) and the negative sign corresponds to - helicity (left handed). For *l*=0 the above expression reduces to (1/2)±s so that the effective charge for s=1/2 becomes 1 and corresponds to + helicity and negative sign gives zero charge and corresponds to - helicity. For *l* =1, the denominator is 2 *l*+1 =3 and the numerator gives (3/2)±s which is 2 for the + sign and 1 for the negative sign. Hence 1/3 belongs to negative helicity and 2/3 belongs to positive helicity. These are single-particle states which can not rotate. There is no doubt that all of the predicted series of fractions *l*/(2l+1) as well as (*l*+1)/(2l+1) are the same as in the experimental data. The two particle state of charge 1/3 occurs at (1/3)+(1/3)=2/3 with negative helicity. The one-particle state at 2/3 has positive helicity. Hence the two particle state at 2/3 and the one-particle state also at 2/3 are degenerate. Hence a degenerate state of mixed helicities is predicted at 2/3. The two-particle state at 2/3 can rotate so that the direction of the velocity changes. The state at 1/3 has negative helicity. The two-particle state at (1/3)+1/3)=2/3 also has negative helicity. Hence 2/3 corresponds to mixed helicity. . The sign of the charge is that of the electron charge in both cases. When a particle has energy proportional to 2/3, then two-particle state occurs at (2/3)+(2/3)=4/3. Now there is a particle at 4/3 with positive helicity and it is made of two-particles of charge 2/3 each. The particle of charge 1/3 is of negative helicity. The three-particle state occurs at (4/3)+(1/3)=5/3. It is made of particles of positive as well as negative helicities. By rotation, the direction of velocity changes. One 2/3 becomes

1/3 and one 1/3 becomes 2/3. Hence, (2/3)+(2/3)+(1/3)= (2/3)+(1/3)+(2/3)=5/3 so that charge is conserved and the helicities of some of the charges are reversed except that one of the values should not rotate. We started with the odd number of particles and reversed the helicities of only two values. It is possible to start with
even number of values in which we reverse the helicities of even number of charges so that we reverse the helicities without loosing the charge conservation. The state 7/3 is made of (2/3)+(2/3)+(1/3)+(2/3)=7/3. This is a cluster of 4 electrons. Upon rotation 1/3 changes into 2/3 and 2/3 changes to 1/3 so that 7/3 is an example of mixed helicities as well as "reversed helicities".

5. Noise power and helicity

The concept of holes is obtained from the absence of electrons from the valence band. Hence, the holes occur in the band theory of solids. Since the electrons are negatively charged, the holes are positively charged and there is a one to one correspondence due to the conservation of energy. When light shines on the solid which has a periodic potential, a hole with positive charge is created in the valence band along with an electron in the conduction band. In the periodic potential, the 1/3 charge is replaced by ± 1/3 and 2/3 is replaced by ± 2/3 and so on and so forth. For electronic states of fractional charges such as 2/3, 3/5, 8/3, 5/2 etc. there are hole states of opposite charge. Unfortunately, the experimentalist neither identify the spin nor the symmetries of the fractions. So they are unable to identify the conjugate pairs. The noise measurements [7] indicate that there are neutral modes which accompany the charged modes. The noise power at a point is of the form,

$$S = 2g_Q \int_o^\infty f(1-f)dE$$

where f(1-f) is a fermion factor which arises from the creation of an electron at one site accompanied with the destruction of an electron at another site and g_Q is the conductance of the material at the quantum contact point. The thermal excess noise power is of the form,
$S_{excess}=2\Delta\varepsilon_F g_Q t(1\text{-}t)[\coth(\Delta\varepsilon_F/2k_BT) +2k_BT/\Delta\varepsilon_F]$
where $\Delta\varepsilon_F$ is the change in the Fermi energy, t(1-t) is the transmission probability and T is the temperature. The conductance at a point at 1/3 is , for example,

g_Q= (1/3) e^2/h.

For ℓ =0, s=+1/2, (1/2)g_= (1/2)-s=0 which is electrically neutral or a neutral particle. Two particles of charge 1/3 each and one hole of charge -2/3 create a zero charge. Similarly, two holes of charge -1/3 each and one particle of charge 2/3 can produce charge neutrality. In the experimental data [8] of AlGaAs/GaAs also there is a zero-energy state with g=0. The excess noise power has the factor g_Q=(1/3)e^2/h accompanied with (2/3)e^2/h which differ in helicity. Hence, the noise power changes upon change in the helicity. Some times the experimentalists [9,10] use the idea of chirality which is similar to that of the helicity defined in the present work as the sign of p.s. The sign of spin s becomes the sign of the velocity. So that the particles of + helicity do not move in the same direction as those of - helicity. By applying a heater on the sample we can see the electrons move with fractional charges and the electrons of opposite helicities move in opposite directions. The particles which are clusters of electrons with mixed helicity rotate with velocity in two directions.

6. Conclusions

We find that every fraction at which a plateau occurs in the Hall resistivity is subject to helicity. There are both right as well as left handed fractions in the quantum Hall effect. We find that mixed helicity as well as "reversed helicity" occurs in various fractions in the quantum Hall effect. The activation

Journal of Physics: Conference Series **423** (2013) 012003 doi:10.1088/1742-6596/423/1/012003

energy is found to become double valued. The doubling occurs in the conductivity so that the noise power changes upon change in the helicity. All of the experimental data[11,12] is consistent with the angular momentum theory [13,14].

References

[1] Laughlin R B,1983, *Phys. Rev. Lett.* **50** 1395.
[2] Shrivastava K N, 2011 *Internatl J. Mod. Phys. B* **25** 1301-1357.
[3] Klitzing K v, Dorda G and Pepper M, 1980, *Phys. Rev. Lett.* **45** 494-497.
[4] Tsui, D C, Stormer H L and Gossard A C, 1982, *Phys. Rev. Lett.* **48** 1559.
[5] Shrivastava K N, 2009, *AIP Conf. Proc.* **1150** 59
[6] Kumar A , Csáthy G S, Manfra M J, Pfeiffer L N and West K W, 2010 *Phys. Rev. Lett.* **105** 246808.
[7] Gross Y, Dolev M, Heiblum M, Umansky V and Mahalu D, 2012 *Phys. Rev. Lett.* **108** 226801.
[8] Shrivastava K N, 2011, *Internatl. J. Nanoscience* **10** 507-514.
[9] Huynh P-A, Portier F, Sueur H le, Faini G, Gennser U, Mailly D, F. Pierre F, Wegscheider W and Roche P,2012, *Phys. Rev. Lett.***108** 256802.
[10] Granger G, Eisenstein J P and Reno J L, 2009, *Phys. Rev. Lett.* **102** 086803.
[11] Pan W, Baldwin K W, West K W, Pfeiffer L N and Tsui D C, 2012, *Phys. Rev. Lett.* **108** 216804.
[12] Stormer H L,1999, *Rev. Mod. Phys.***71** 875.
[13] Shrivastava K N, 1986, *Phys. Lett. A* **113** 435.
[14] Shrivastava K N, 2004, *Phys. Lett. A* **326** 469.

Effect of precursor medium on morphological and optical characterization of spray deposited EuO thin films

M M Betkar[1], **G D Bagde**[2]
[1,2]Physics Research Centre, Mahatma Gandhi College, Ahmedpur, Dist. Latur, MS, India. Pin-431515
E. mail: betkarmm@rediffmail.com

Abstract
Europium chalcogenide (EuO) thin films have been deposited by spray pyrolysis technique on non-conductive bare soda lime glass substrates in aqueous and non-aqueous precursor mediums. The films synthesized are characterised by XRD, SEM, EDAX and UV-Visible spectrometry. The effect of precursor medium on morphological and optical properties has been investigated. The XRD studies reveal that, the films are polycrystalline in nature and the deposited material was europium oxide. It was found that, morphological and optical properties are strongly related to the precursor medium. The SEM analysis conformed the formation of small spherical and square shaped grains on overall substrate.

Keywords
EuO; precursor medium; band gap; EDAX; XRD; spray pyrolysis

1. Introduction

One of the europium chalcognides in perticular europium monoxide (EuO) is a facinating compound having ferromagnetic semiconducting property. It is a divalent material but its compounds can occurs in divalent and also in trivalent form of configurations. The europium chalcogenide materials initiate a simple crystal and magnetic structure and those are fit for their research studies. The ferromagnetic property has been found in some divalent europium compounds [1,2]. The europium chalcogenide series exhibits a simple rocksalt structure. Out of those the oxides, sulphides, and selenides are of ferromagnetic nature. The magnetic and electronic properties of europium chalcogenides Eus, EuO and EuSe have been studied with more attention. Moreover the EuO is a model compound for a whole class of compounds with novel properties [3-5]. To the best of our knowledge, there is no report available on europium chalcogenides thin films by spray pyrolysis deposition technique. The morphological and optical properties of europium monoxide (EuO) by spray deposition have not reported so far. A number of methods of thin film deposition such as electro-deposition, vacuum evaporation, chemical bath deposition, chemical vapour deposition, spray pyrolysis etc are employed [6].

These techniques have major influence on morphological, structural and optical properties of synthesized thin films [7]. The surface morphology of the films varies with the precursor concentrations, volume ratio also with film deposition time [8-9].

In the present work, europium monoxide (EuO) thin films are deposited by spray pyrolysis method in aqueous and non-aqueous precursor mediums. Spray pyrolysis is a non vacuum, simple and low cost technique, having large area coating capacity and can be optimised for a wide control on morphological and optical properties of thin films. The synthesized films are characterised by X-ray diffraction (XRD), UV-Visible spectrometry, SEM and EDAX analysis. The results of morphological and optical characterisation are outlined.

---

[1]Corresponding author.

2. Experimental

The europium chalcogenide (EuO) thin films are deposited onto non conductive bare soda lime glass substrates. The europium monoxide Eu_2O_3 was dissolved in 1 M hydrochloric acid. For aqueous precursor medium deionised water was employed and for non-aqueous medium methanol was employed. Aqueous and non-aqueous solutions of Eu_2O_3 at 0.05 M was used as a source of EuO. Magnetic stirrer was used to stir the precursors at the rate 550 rpm for 40 min. The experimental substrates were cleaned with 0.5 M hydrochloric acid and also with standard laboratory detergent Labonate. Substrates were ultrasonically cleaned with double distilled hot water for 10 min and dried well with hot air hand drier before deposition process. The film deposition for aqueous precursor medium was carried out at temperatures 523, 548 , 573, 598 and 623 K and that of for non-aqueous medium was carried out at 423, 443, 463 and 483 K. The deposition was carried out on spray pyrolysis equipment (Model No. Holmark HO-TH-04). All other deposition parameters; pressure on carrier air (30 psi), substrate to spray nozzle distance (17 cm), spray duration (2 min) and precursor quantity (2 ml/min) were kept constant throughout the experiment. The carrier gas was air. The chemicals used were of analytical reagent grade (of 99% purity).

3. Results and Discussion

3.1. Structural and morphological characteristics

The structural characterization of EuO thin films was carried out on X-ray diffractometer model MiniFlex2, with Cu/ 30 kV /15 mA and k_α radiation (wavelength $\lambda = 0.1542$ nm). The XRD patterns for the spray deposited EuO films on bare soda lime non conductive glass substrates for aqueous and non-aqueous precursor mediums are envisaged in following figures 1 and 2, respectively.

In figure 1, the XRD patterns are shown at temperatures (a) 523, (b) 548 and (c)573 K also that of in figure 2 are that of at (a) 423, (b) 443 and (c) 463 K.

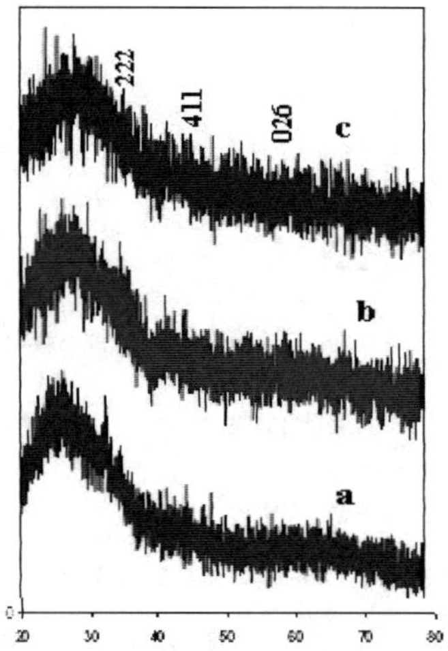

Figure 1 XRD Pattern for aqueous medium

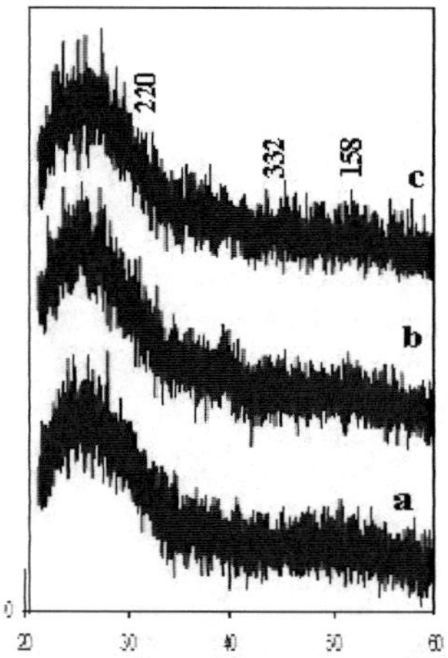

Figure 2 XRD Pattern for non-aqueous medium

The XRD different peaks in the diffractogram were listed and the corresponding values of interplanar spacing 'd' were calculated. The 'd' values are compared with the standard one.[10] The diffraction peaks of EuO films for aqueous precursor medium are originated at 2θ values of 27.980,

35.280 and 53.160 corresponding to the hkl planes (222), (411) and (026) respectively. The diffraction peaks of EuO films for non-aqueous precursor medium are originated at 2 θ values of 22.340, 37.960 and 62.000 corresponding to the hkl planes (220), (332) and (158) respectively. The XRD peaks in figures 1 and 2 indicate well crystallized EuO thin films. It conforms the polycrystalline nature and are of simple cubic structure. The optimized temperature for deposition of good quality EuO thin films for aqueous precursor medium is found as 550 K and that of for non-aqueous medium is 423 K. The height of peak in X-ray diffraction pattern at (222) plane deposited in aqueous precursor medium is found more sharper and FWHM data conforms the improvement of crystallite at temperature 523 K and that of for non-aqueous medium at (220) plane is found at temperature 445 K.

X-ray diffraction patterns of EuO thin films for aqueous and non-aqueous mediums are evaluated using FWHM data and Debye-Scherrer formula for the calculation of the crystallite size.
Figure 3 displays the variation of film thickness with deposition temperature for aqueous precursor medium and figure 4 displays the variation that of for non-aqueous precursor medium. The thickness of thin films (t) was calculated by weight difference method, by using formula,

$$t = \frac{m}{A\rho}$$

where m is the mass, A is the area and ρ is the density of the material in bulk form.

It is observed from figure 2 and 3 that, for aqueous precursor medium the film thickness increases with substrate temperature upto 550 K and that of for non-aqueous precursor medium increases upto 450 K. After the substrate temperatures 550 K and 450 K the film thickness starts to decrease in aqueous and non-aqueous precursor mediums respectively.

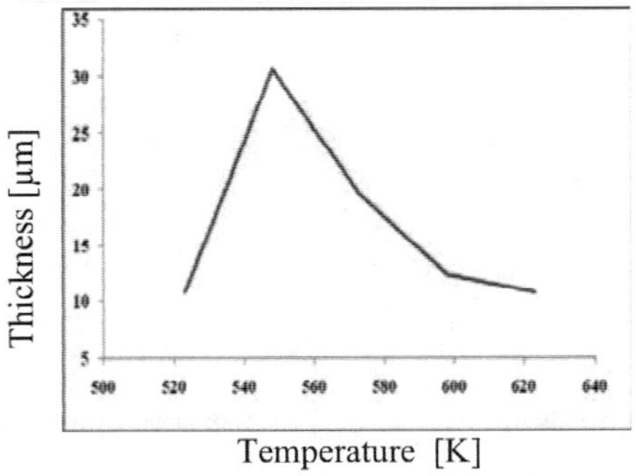

Figure 3 Variation of crystalline size with temperature for aqueous precursor medium

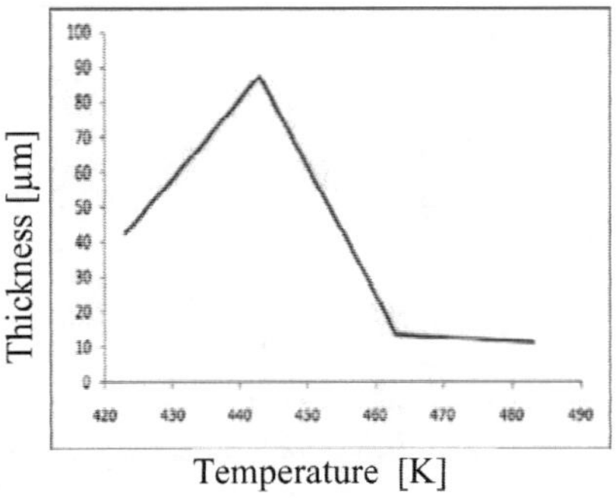

Figure 4 Variation of crystalline size with temperature for non-aqueous precursor medium

Figures 5 and 6 shows SEM images of EuO thin films deposited for aqueous and non-aqueous precursor mediums respectively.

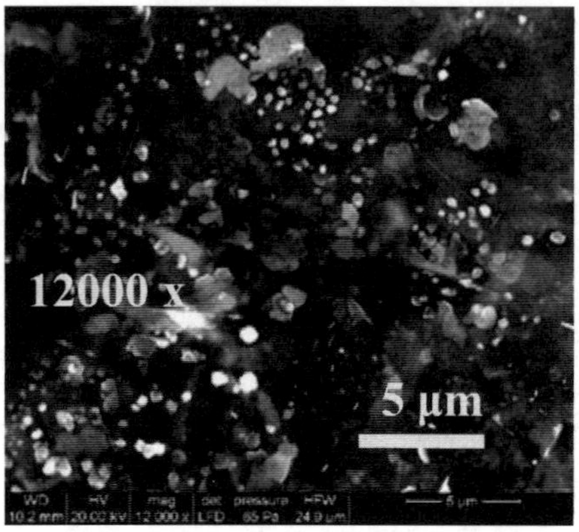

Figure 5 SEM photograph for aqueous medium

Figure 6 SEM photograph for non-aqueous medium

Thin films are analysed on Electron Microscope SEM Model: Quanta 200 ESEM System, manufactured by Icon Analytical Equipment Pvt. Ltd, Mumbai. Figure 5 is the 12000**X** magnified micrograph of thin films for aqueous precursor medium, indicates the uniform film deposition and continuous coverage of the substrate surface with small spherical nanoparticles. Figure 6 is the 10000**X** magnified micrograph of thin films for non-aqueous precursor medium, indicates the uniform film deposition and continuous coverage of the substrate surface but with squared shaped nanoparticles. The microgram study reveals the effect of precursor medium on morphology of thin films. As a result the change in precursor medium from aqueous to non-aqueous changes the shapes of nanoparticles from spherical to square, films grow by spray pyrolysis technique of experimental precursor leads to a perfect flat surface of the films.[12-13]

The figure 7 and 8 displays the EDAX pattern of synthesized EuO thin films deposited for aqueous and non-aqueous precursor mediums respectively.

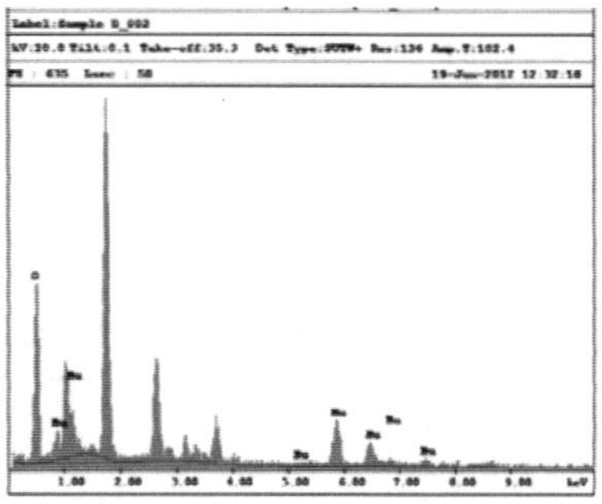

Figure 7 EDAX pattern for aqueous precursor medium

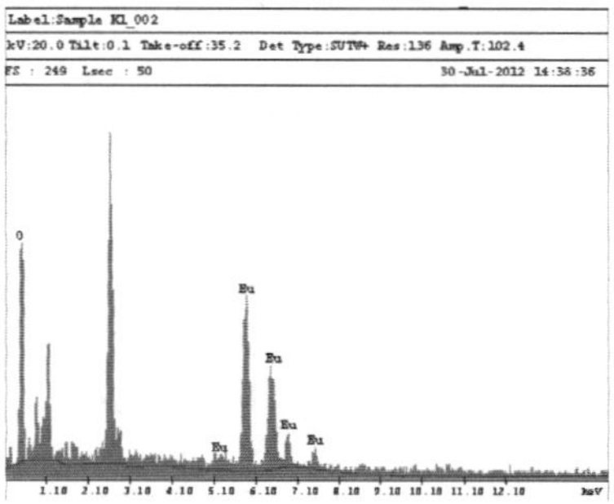

Figure 8 EDAX pattern for non-aqueous precursor medium

The pattern covers the peaks of Eu and O with varying intensity. The undefined peaks are assigned to the glass substrate. This conforms the deposition of Eu and O. The study revels that the doping percentage of europium increases in non-aqueous precursor medium as compared with aqueous precursor medium.

The elemental composition for aqueous precursor medium was analysed for Eu with weight 66.24% at 17.12 % and that of for O was wt. 25.62 % at 82.88%. The net intensities for Eu and O was 20.54 and 25.62 respectively, as shown in following table 1.

Table 1. The elemental compositions of Eu and O for aqueous precursor medium.

ELEMENT	Wt %	At %	K ratio	Net Int.
Eu	66.24	17.12	0.5868	20.54
O	25.62	82.88	0.1334	25.62

The elemental composition for non-aqueous precursor medium was analysed for Eu with weight 76.15% at 25.16% and that of for O was wt. 23.85% at 74.84%. The net intensities for Eu and O was 24.54 and 16.96 respectively, as shown in following table 2.

Table 2. The elemental compositions of Eu and O for non-aqueous precursor medium.

ELEMENT	Wt %	At %	K ratio	Net Int.
Eu	76.15	25.16	69.59	24.54
O	23.85	74.84	0.0878	16.96

3. 2. Optical characterisation

The position of the absorption coefficient edge is accessed for the band gap energy calculation from the optical transmittance measurements of the EuO thin films. The absorption coefficient α was calculated by using the following equation,

$$\alpha = \frac{A}{h\nu}(h\nu - Eg)^{1/2}$$

Where, A is a slope of curve,
Eg is band gap energy.

Figures 9 and 10 shows, Tauc's plot between, the calculated values of the absorption coefficients $(\alpha h\nu)^2$ and $h\nu$, where $h\nu$ is the photon energy and α is the optical absorption coefficient of the material.

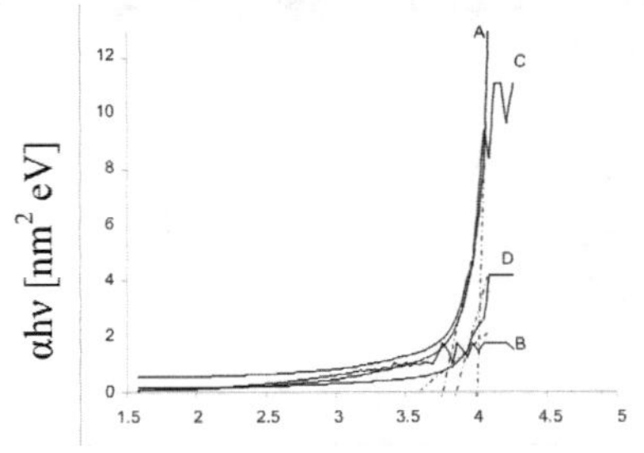

Figure 9 Plot of $(\alpha h\nu)^2$ versus energy $h\nu$ for aqueous precursor medium

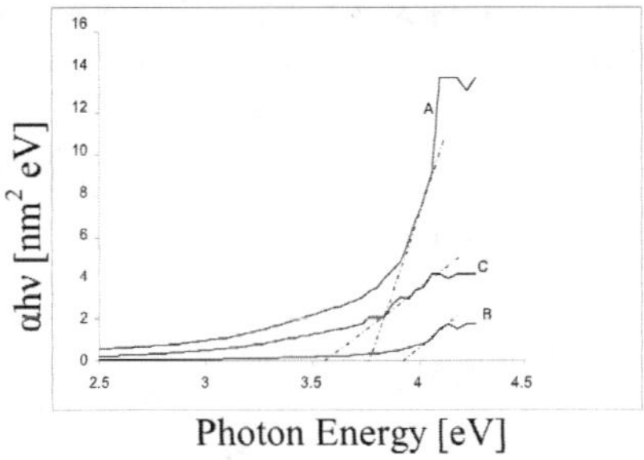

Figure 10 Plot of $(\alpha h\nu)^2$ versus energy $h\nu$ for non-aqueous precursor medium

The band gap energy value of the EuO films deposited for aqueous precursor medium, is calculated by extrapolating of the plots to the X-axis.

Figure 9 shows a plot for the optical band gap energy for aqueous precursor medium, deposited at (A) 523 K, (B) 548 K , (C) 573 K and (D) 598 K , the energy band gap are found as 3.6, 3.7, 3.75 and 3.8 eV respectively and figure 10 shows that of for non-aqueous precursor medium, deposited at (A) 423 K, (B) 443 K and (C) 483 K, and the band gap energies are found as 3.6, 3.75, and 3.8 eV respectively. It is found that, the precursor medium have no major influence on band gap energy of thin films. The band gap value are in concord with the value reported. [11]

4. Conclusion

Spray pyrolysis technique was successfully employed to deposit europium monoxide (EuO) thin films on non-counducting bare soda lime glass substrates for aqueous and non-aqueous precursor mediums. The XRD study of the films conforms the cubic structure. Structural parameters like crystallite size, thickness are calculated and are found increased in non-aqueous precursor medium with conforming the formation of well-crystallized EuO films. The crystallinity of the thin films found increased in non-aqueous medium. Optical transmittance measurements revels that the change in precursor medium have no major influence on optical band gap energy. The optimized EuO thin film has highly crystallized.

Acknowledgements

One of the authors, MMB is thankful to University Grants Commission, New Delhi, India, for the award of Teacher Fellowship under 'Faculty Improvement Programme', and also thankful to S.R.T. M. University recognised, Physics Research Centre, Mahatma Gandhi College, Ahmedpur, Dist. Latur, MS, India, for providing research facility for spray deposition and characterizations.

References

[1] Paul L, Walter R, Lambrecht L 2006 *J. Phys.Condens. Matter* **18** 11333-11345
[2] Mc Guire T R, Shafer M W 1964 *J. of Appl. Physics* **35** 3
[3] Kathalingam A, Ambika N, Kim M R, Elanchezhiyan, Chae J Y S, Rhee K 2010 *J. Materials Science Poland* **28** 2
[4] Aitken J A, Cowen J A, Kanatzidis M G 1998 *Chem. Mater* **10**
[5] Savne A, Santi G, Szotek Z , Temmerman W M , Strange P, Horne M, Vaitheeswaran G, KanchanaV, Petit L, Winter H 2004 *Phys. Stat. Sol.* **B 14** 241
[6] Holroyda J, Idzerda Y U 2004 *J. of Appl. Physics* **95** 11
[7] Zhengwen L, Roy G G 2006 *Chem. Vap. Deposition* **12** 435–441
[8] Bakry A M 2008 *Egypt. J. Solids* **1** 31
[9] Jennifer A A, Jerry A C, Mercouri G K 1998 *Chem. Mater* **10** 3928-3935
[10] Leea C H, Junga K Y, Choib, J G, Kangc Y C 2005 *Materials Science and Engineering* **B 11** 6
[11] Ashour A 2003 *Turkish j. Physics* **27**
[12] Wagner C Z , Electrichem Z, Bunsenges B 1961 *Phys. Chem.* **65** 581
[13] JCPDF-WIN Diffraction Data Card No. **76-0154**
[14] Eastman D E, Hotzberg F, Methfessel 1969 *Physical Review letters* **23** 5

Journal of Physics: Conference Series **423** (2013) 012005 doi:10.1088/1742-6596/423/1/012005

Synthesis and Characterization of Filtered-cathodic-vacuum-arc-deposited TiO_2 Films for Photovoltaic Applications

C Aramwit[1], S Intarasiri[2,3], D Bootkul[3,4], U Tippawan[1,3], B Supsermpol[4,5], N Seanphinit[4,5], W Ruangkul[3] and LD Yu[1,3,*]
[1] Plasma and Beam Physics Research Facility, Department of Physics and Materials Science, Faculty of Science, Chiang Mai University, Chiang Mai 50200, Thailand
[2] Science and Technology Research Institute, Chiang Mai University, Chiang Mai 50200, Thailand
[3] Thailand Center of Excellence in Physics, Commission on Higher Education, 328 Si Ayutthaya Road, Bangkok 10400, Thailand
[4] Departments of General Science, Faculty of Science, Srinakharinwirot University, Bangkok 10110, Thailand
[5] Western Digital Thailand Co. Ltd., Ayutthaya 13160, Thailand

* Email: yuld@fnrf.science.cmu.ac.th

Abstract. Titanium dioxide (TiO_2) is well-known as a photovoltaic and photocatalytic material. For improvement in the dye-sensitized solar cell (DSSC) performance efficiency, the photocatalyst TiO_2 layer would be desired in nanoporous anatase. In this research, TiO_2 films were synthesized on glass or p-type silicon substrate using our in-house Filtered Cathodic Vacuum Arc Deposition (FCVAD) system. The deposition was operated at varied oxygen (O_2) partial pressures of 10^{-4}, 10^{-3}, 10^{-2} to 10^{-1} torr with fixed 0 or 250-V bias and 600-V arc for 10 or 20 minutes. The film transparency increased with increasing of the O_2 pressure, indicating increase in the structure required for applications in dye-sensitized solar cells. The films were characterized using the Energy-Dispersive X-ray spectroscopy (EDS) and Raman spectroscopy techniques. The EDS confirmed that the transparent deposited films contained stoichiometric titanium and oxygen under the medium O_2 pressure. Raman spectra confirmed that the films were TiO_2 containing some rutile but no anatase which needed annealing to form. Scanning electron microscopy (SEM) and atomic force microscopy (AFM) were used for evaluation of the film's surface morphology and thickness. The result showed that increasing of the O_2 pressure decreased the thickness to a nanoscale but increased the amount of TiO_2.

1. Introduction

In the world trend of developing green energy, solar cells are the first choice. Among types of the solar cells, dye-sensitized solar cells (DSSC) [1,2,3] developed in recent two decades are considered to be the third generation solar cells because of their lower cost and higher efficiency compared with other traditional solid-state-semiconductor solar cells [4]. A DSSC basically consists of six components, i.e. normal glass, conducting glass, wide-band semiconductor, dye, electrolyte and a metal counter electrode. The semiconductor material popularly used is titanium dioxide (TiO_2) in anatase [2]. Critical factors to increase the performance efficiency of DSSC include the porosity and the electron transportation property of the TiO_2. Highly nanostructured TiO_2 can greatly increase the material porosity so that the material surface area can be increased to let more electrons be injected. Metal or nonmetal-doped TiO_2 can further improve the material electron transportation property [5,6,7]. DSSCs are currently very attractive to the market owing to the high ratio between efficiency and cost, good workability under weak sunlight conditions and higher temperatures, and strong mechanical structures.

Filtered cathodic vacuum arc deposition (FCVAD) is a simple and effective plasma technique to make metallic as well as carbon thin films [8,9,10]. Owing to many advantages [11], such as a very high deposition rate, a good control of the deposition parameters, and an excellent quality and good adhesion of the coating, FCVAD has been utilized to TiO_2-film coating for many applications, including optics, photocatalysis, photo energy conversion and biocompatibility. Titanium oxide has three major crystalline phases: rutile, anatase and brookite [12], but only the anatase is good for photo energy conversion [3]. Moreover, nanostructure of TiO_2 greatly affects the photo energy conversion [2]. Formations of TiO_2 (anatase) and its nanostructure by using FCVAD depend on the deposition conditions, including oxygen (O_2) partial working pressure, substrate temperature, substrate bias, deposition time, arc current, and pulse parameters (e.g. the pulse width and frequency), etc. [11,12]. The microstructure of FCVAD-TiO_2 is generally amorphous at the substrate temperature below 330°C under the O_2-pressure lower than about 3×10^{-4} torr without bias [13,14]. At a higher substrate temperature and using bias, anatase can possibly be formed. However, there are many controversial experimental results to show different phase formation under similar conditions [15-19]. Therefore, using FCVAD to form TiO_2 (anatase) is still a complex issue because of so many conditions or parameters involved. In physics principle, elevated substrate temperature and bias increase the ion energy so as to increase activation of the phase transformation. O_2-pressure controls the TiO_2 phase stability [20] and it cooperates with the arc current and the pulse parameters [21] to control the amounts of two elements deposited and thus the phase formation. Moreover, the microstructure of the deposited film is governed by the famous structural zone model for coating growth [22-24]. According to the model, lower temperature and higher working gas pressure could result in porous structure. This really leads to a challenge that at what appropriate working temperature and gas pressure the deposited TiO_2 film has both porosity and anatase. The present research was aimed a systematic investigation on the effect on microstructure of the TiO_2 films from FCVAD. But, this report presents preliminary results of the study in the effect of the O_2-pressure. The laboratory of the Plasma and Beam Physics Research Facility at Chiang Mai University which has been the unique comprehensive ion beam and plasma research laboratory in the ASEAN region has had a long history of working on ion beam and plasma technology. We have a number of ion beam accelerators with the beam energy ranging from 30 keV to 2 MeV and plasma facilities with the working pressure ranging from low to atmospheric pressures. Our FCVAD system has been in-house developed in the cooperation with the Lawrence Berkeley National Laboratory and applied for various applications [25-27]. To contribute to green energy development we are utilizing the novel FCVAD technique for advanced TiO_2 film coating.

2. Methodology

Before deposition of the films, substrates of glass and p-type silicon in sizes of 1×1 cm^2 (for analysis) or 2×8 cm^2 (for measurements of macro-properties) were cleaned by boiling in acetone with applying an ultrasonic frequency of 35 kHz at 60°C for 6 minutes. Then, they were dried by blowing pure nitrogen gas flow. The in-house developed FCVAD system (Figure. 1) was used for film deposition. Ti was the cathodic material which was made into a small rod (about 3 mm in diameter and a few cm in length) to put in the cathode house. Pulsed high-temperature arc that was generated by a-few-kV voltage between the cathode and the anode melted and vaporized the cathodic metal to form Ti-vapor plasma. Ti ions in the plasma were guided through a 90°-solenoid filter to the substrate surface for deposition. The prepared and cleaned substrate was mounted on the target holder which was connected to a bias power supply for deposition. The deposition chamber atmosphere and pressure were controllable during the deposition by controlling the gas inlet. In our experiment, O_2 gas was used for forming metal oxide. The working pressure with O_2 inlet was varied from the base pressure of 5×10^{-5} torr to 1×10^{-4}, 1×10^{-3}, 1×10^{-2} and 1×10^{-1} torr, respectively. The arc voltage was 600 V, the bias was 0 or -250 V, and the deposition time was 10 and 20 minutes, respectively.

The FCVAD-synthesized films on glass were characterized for their morphology by scanning electron microscopy (SEM) and the composition by energy dispersive x-ray spectroscopy (EDS) which had been widely used as a characterizing method to identify elements in deposited films. The titanium on the substrate would emit X-rays at energy around 4.5 keV and 0.45 keV, while oxygen would emit X-ray at 0.52 keV [28]. However, the deposited film on the glass substrate was not suitable to using Raman spectroscopy due to the imperfection of glass which could produce a large amount of noises in the spectrum. Therefore, films deposited on Si in the same deposition condition were analyzed using Raman spectroscopy for phase formations. Raman spectroscopy has often been used to investigate thin film's structure. The vibration of film's structure has its own frequency. The TiO_2 in the rutile structure responds to the wave numbers 143 cm^{-1}, 447 cm^{-1}, 612 cm^{-1} and 826 cm^{-1}, different from the anatase structure which responds to 144 cm^{-1}, 197 cm^{-1}, 399 cm^{-1}, 513 cm^{-1}, 519 cm^{-1} and 639 cm^{-1} [29]. The film surface morphology and thickness were measured using atomic force microscopy (AFM). To measure the film thickness, before deposition a tape was used to cover a part of the substrate and after deposition the tape was removed to form a step between the deposited and undeposited areas for AFM to scan.

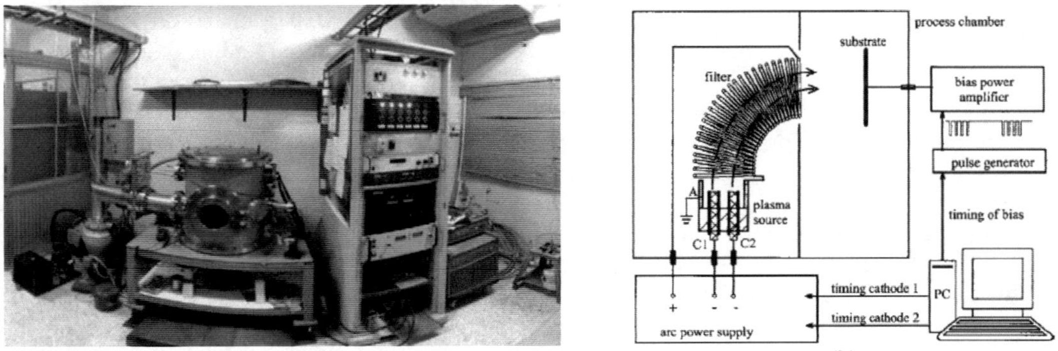

(a) (b)

Figure 1. The in-house developed FCVAD system at Chiang Mai University. (a) Photograph of the entire FCVAD system. (b) Schematic diagram of the system inside the chamber. Here it shows only one of three technical configurations, dual-cathode source unit.

Table 1. A summary of the deposition and characterization of the 10-min deposited TiO_2 film samples. Y: yes. N: no. Pressure unit is in torr.

Sample No.	Substrate	Base pressure	O_2 pressure	SEM	EDS	Raman analysis
1	Glass	$5x10^{-5}$	0	Y	Y	N
2	Glass	$5x10^{-5}$	$1x10^{-4}$	Y	Y	N
3	Glass	$5x10^{-5}$	$1x10^{-3}$	Y	Y	N
4	Glass	$5x10^{-5}$	$1x10^{-2}$	Y	Y	N
5	Glass	$5x10^{-5}$	$1x10^{-1}$	Y	Y	N
6	Si	$5x10^{-5}$	0	N	N	Y
7	Si	$5x10^{-5}$	$1x10^{-4}$	N	N	Y
8	Si	$5x10^{-5}$	$1x10^{-3}$	N	N	Y
9	Si	$5x10^{-5}$	$1x10^{-2}$	N	N	Y
10	Si	$5x10^{-5}$	$1x10^{-1}$	N	N	Y

3. Results and Discussion

The deposition and characterization of the TiO_2 films are summarized in Table 1 under the conditions of 10-minutes deposition, 600-V arc voltage and -250-V bias. Figure 2 shows the SEM images of the film surface morphology and the film cross section. Compared with the glass control surface (without coating) (Figures 2a and 2b), the pure Ti film (without O_2 inlet) (Figures 2c and 2d) is seen obviously to be present with a film thickness of about 3 μm. The Ti film seems poor in adhesion to the substrate and integration to a large-area solid. However, after O_2 was introduced, even a small amount (at an O_2-partial pressure of 10^{-4} torr), the film (Figures 2e, 2f, and 2g) looks more condense and solid compared with the pure Ti film. Because of the densification, the film thickness is less than 3 μm, around 2-3 μm. With increase in the working pressure or O_2 inlet, the film surface becomes smoother (Figure 2h). On the other hand, the film transparency increases as increasing of the O_2 pressure, as shown in Figure 3. High film transparency is one of the requirements for the DSSC to receive more sunlight.

The film surface morphology observed by AFM scanning in comparison with the substrate surface is shown in Figure 4. The result is in consistence with the SEM-observed result. The film thickness measured by AFM scanning across the boundary between the film and the substrate surfaces is shown in Figure 5. The film thickness decreases as increasing of the pressure and reaches a nanoscale at 10^{-2} torr. The film thickness decrease is cooperated with the film transparency increase as increasing of the pressure. The results on the morphology and transparency indicated that a transformation in the film material characteristics occurred when the O_2 inlet increased in the deposition.

The oxygen content in the film was measured using EDS (Figure 6) and the result is shown in Figure 7. The O_2 content in the Ti-O film due to the O_2 inlet was calculated from the measured O_2 content in various pressures minus the original O_2 content in the glass (mainly silicon oxide) measured from the condition of using the base pressure. It is seen that the O_2 content in the film increases as increasing of the partial O_2 pressure and the optimized stoichiometric ratio of Ti to O which should be 1:2 for the TiO_2 formation is nearly achieved at the working pressure of 1×10^{-3} torr.

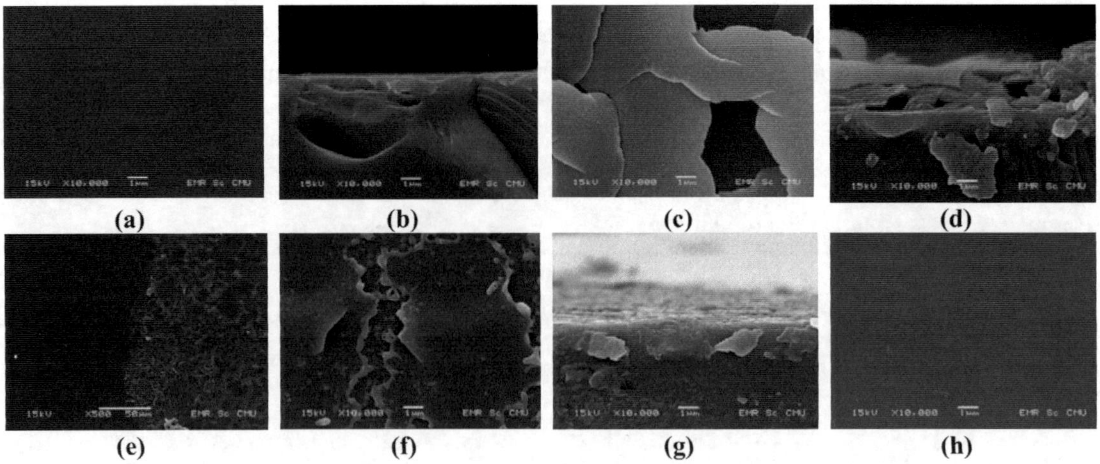

Figure 2. SEM images of the films. (a) (top view) and (b) (cross section): glass without film. (c) (top view) and (d) (cross section): Ti film deposited at the base pressure. (e), (f) (top views) and (g) (cross section): pressure 1×10^{-4} torr. (e) shows a border section between the coated (right) and uncoated (left) areas. (h) (top view): pressure 1×10^{-1} torr.

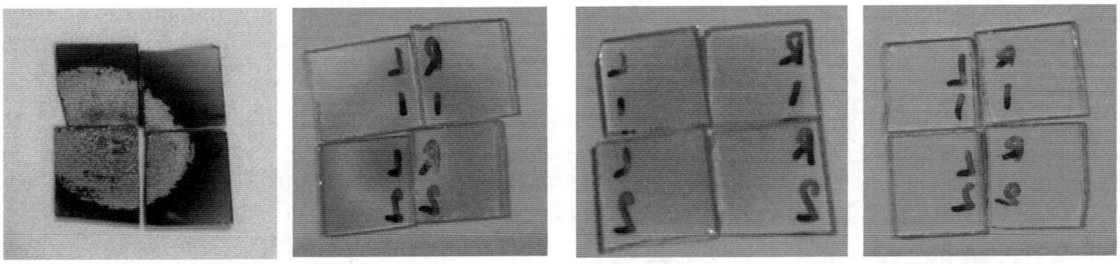

(a) (b) (c) (d)

Figure 3. The film transparency at pressure (a) 1×10^{-4}, (b) 1×10^{-3}, (c) 1×10^{-2}, (d) 1×10^{-1} torr.

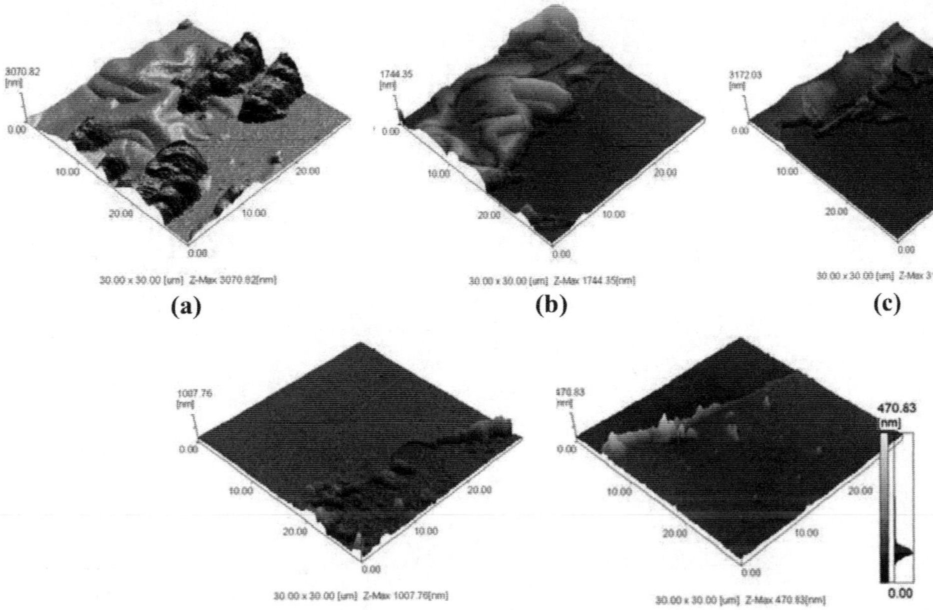

(a) (b) (c)

(d) (e)

Figure 4. AFM-measured film surface morphology across the boundary between the film and the substrate as a function of the working pressure. (a) Base pressure 5×10^{-5} torr. (b) 1×10^{-4} torr. (c) 1×10^{-3} torr. (d) 1×10^{-2} torr. (e) 1×10^{-1} torr.

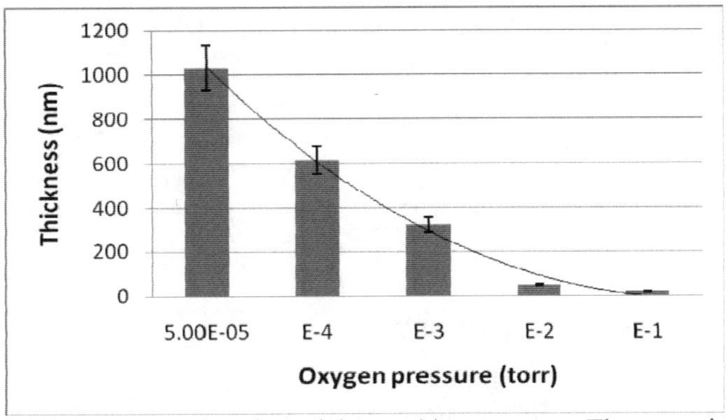

Figure 5. The film thickness as a function of the working pressure. The curve is only a guideline.

Journal of Physics: Conference Series **423** (2013) 012005 doi:10.1088/1742-6596/423/1/012005

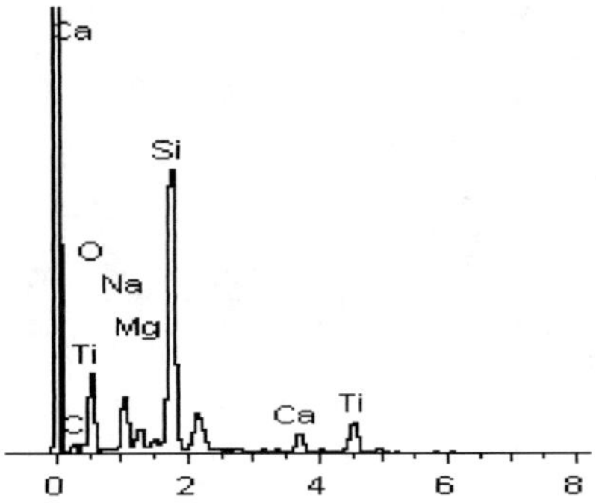

Figure 6. An example of the EDS spectrum from the film deposited at 1×10^{-4} torr. The vertical axis is the X-ray intensity and the horizontal axis is the X-ray energy in keV. The dispersed energy peak intensities can be converted to the percentages of elements in samples such as those of Ti and O as shown in Figure 7.

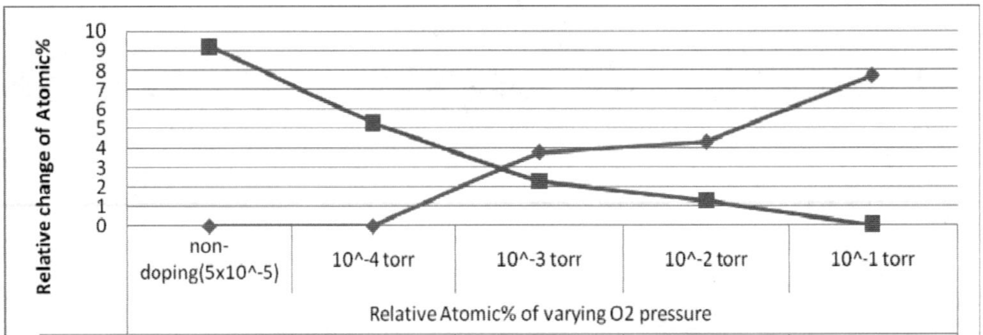

	non-doping(5x10^-5)	10^-4 torr	10^-3 torr	10^-2 torr	10^-1 torr
O K	0	0	3.77	4.28	7.71
Ti K	9.22	5.29	2.28	1.28	0.02

Figure 7. The Ti and O contents in the Ti-O film calculated from the EDS spectra as a function of the working pressure.

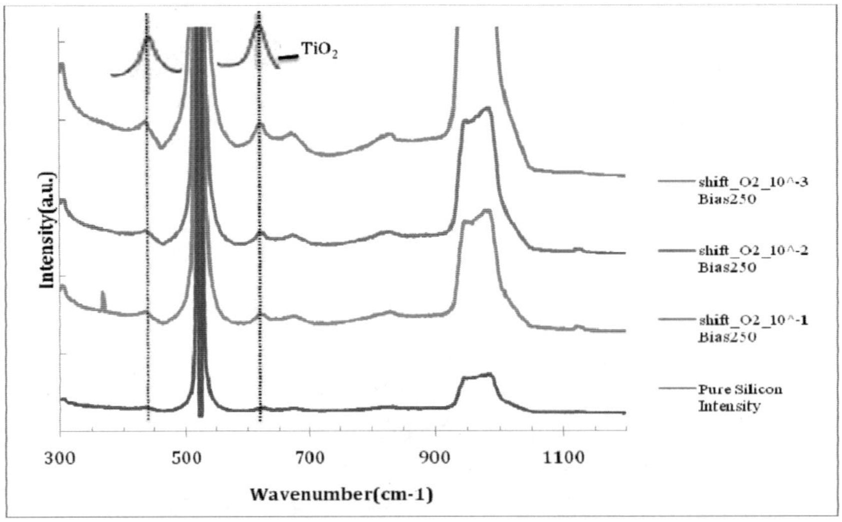

Figure 8. Preliminary Raman spectra of the deposit TiO_2 films. Note: Raman spectroscopy analysis is still ongoing to confirm the results since some confusions exist in the spectra.

The FCVAD-formed film microstructure was analyzed using the Raman spectroscopy as shown in Figure 8. Three rutile characteristic peaks at 447, 612 and 826 cm^{-1} can be identified from the spectra, but no anatase characteristic peaks (e.g. 399 and 639 cm^{-1}) are seen. These rutile peaks increase their intensities as the O_2 pressure decreases to 1×10^{-3} torr. This result tells that with the deposition conditions used, the film microstructure contains the rutile phase of TiO_2 but no anatase phase. This is not completely in agreement with previously reported results, which showed that in similar conditions the Ti-O film was amorphous and definitely no anatase which must be formed with annealing at a temperature higher than 330°C [13,14]. Our result showed that although there was no anatase, rutile was present besides amorphous structure.

4. Conclusion

The report presents preliminary results from our study on using the FCVAD technique to deposit TiO_2 films for green energy applications. Among a number of technical parameters of FCVAD, only a limited number of conditions were tested, such as varied O_2 pressure and deposition time with and without bias at room temperature. From the investigation we mainly concluded that at the O_2 pressure of about 1×10^{-3} torr hundred-nanometer-thick TiO_2 films with the optimized stoichiometric ratio containing the rutile structure could be synthesized. The film was transparent and solid enough. Since the film did not contain anatase which was the desired phase, elevated temperature or annealing was needed. This still remained challenged as higher temperature treatment might be harmful to create porous structure which was also desired. Therefore, further investigations on the film characteristics and structure with more varied parameters such as varied bias and substrate temperature and using more advanced analysis techniques such as Rutherford backscattering spectrometry (RBS) and positron annihilation spectroscopy (PAS) of ion beam analysis techniques are being carried out.

5. Acknowledgements

The study was supported by the Development and Promotion of Science and Technology Talented Project (DPST) scholarship, Thailand, the Thailand Center of Excellence in Physics (ThEP), and the International Atomic Energy Agency (IAEA).

6. References

[1] O'Regan B, Grätzel M 1991 A low-cost, high-efficiency solar cell based on dye-sensitized colloidal TiO_2 films *Nature* **353** 737–740
[2] Grätzel M 2003 *J Photochem and Photobio C: Photochemistry Reviews* **4** 145–153
[3] Grätzel M 2005 *Inorg Chem* **44** 6841-6851
[4] Lewis NS and Crabtree G 2005 *Basic Research Needs for Solar Energy Utilization: Report of the Basic Energy Sciences Workshop on Solar Energy Utilization*, April 18-21, 2005, (Office of Basic Energy Science, US Department of Energy, Washington DC)
[5] Cheng X, Zeng P, Hu S, Kuang T, Xie G, Gao F 2006 *Rare Metals* **25** 190–194
[6] Guo W, Wu L, Chen Z, Boschloo G, Hagfeldt A, Ma T 2011 *J Photochem and Photobio A: Chemistry* **219** 180-187
[7] Chou TP, Zhang Q, Russo B, Cao G 2008 *J of Nanophot* **2** 023511
[8] Wright AW 1877 *Am. J. Sci. Arts* **13** 49
[9] Edison TA *US Patents* 526,147 (1884) and 484,582 (1892)
[10] Martin PJ, Bendavid A 2001 *Thin Solid Films* **394** 1–14
[11] Zhirkov IS, Paternoster C, Delplancke-Ogletree MP 2011 Titanium oxide thin film deposition by pulsed arc vacuum plasma *J of Phys: Conf. Series* **275** 012019
[12] Tay BK, Zhao ZW, Chua DHC 2006 Review of metal oxide films deposited by filtered cathodic vacuum arc technique *Materials Science and Engineering R* **52** 1–48
[13] Takikawa H, Matsui T, Sakakibara T, Bendavid A, Martin PJ 1999 *Thin Solid Films* **348** 145
[14] Zhao ZW, Tay BK, Yu GQ 2004 *Appl Opt* **43** 1281
[15] Zhang F, Liu X 1998 *Thin Solid Films* **326** 171

[16] Zhang F, Wang X, Li C, Wang H, Chen L, Liu X 1998 *Surf Coat Technol* **110** 136
[17] Bendavid A, Martin PJ, Jamting A, Takikawa H 1999 *Thin Solid Films* **355** 6
[18] Bendavid A, Martin PJ, Takikawa H 2000 *Thin Solid Films* **360** 241
[19] Zhao ZW, Tay BK, Lau SP, Yu GQ 2004 *J Cryst Growth* **268** 543
[20] Shen YM, Yu H, Yao JK, Shao SY, Fan ZX, He HB, Shao JD 2008 Investigation on properties of TiO_2 thin films deposited at different oxygen pressures *Optics & Laser Technology* **40** 550–554
[21] Bendavid A, Martin PJ, Preston EW 2008 The effect of pulsed direct current substrate bias on the properties of titanium dioxide thin films deposited by filtered cathodic vacuum arc deposition *Thin Solid Films* **517** 494-499
[22] Movchan BA, Demchishin AV 1969 *Phys Met Metallogr* **28** 83-90
[23] Thornton JA 1974 *J Vac Sci Techno* **11** 666-70
[24] Thornton JA 1977 *Ann Rev Mater Sci* **7** 239-60
[25] Anders A, Pasaja N, Sansongsiri S 2007 Filtered cathodic arc deposition with ion-species-selective bias *Review of Scientific Instruments* **78** 063901-5
[26] Pasaja N, Sansongsiri S, Intarasiri S, Vilaithong T, Anders A 2007 Mo-containing tetrahedral amorphous carbon deposited by dual filtered cathodic vacuum arc with selective pulsed bias voltage *Nucl Instru Meth B* **259** 867-870
[27] Medhisuwakul M, Pasaja N, Sansongsiri S, Kuhakan J, Intarasiri S, Yu LD 2012 *Surf Coat Technol* in press
[28] X-Ray Data Booklet: Photon energies, in electron volts, of principal K-, L-, and M-shell emission lines, http://xdb.lbl.gov/Section1/Table_1-2.pdf, accessed 1 June, 2012.
[29] Kalantar-zadeh K et al 2009 Nanoporous titanium oxide synthesized from anodized filtered cathodic vacuum arc Ti thin films *Thin Solid Films* **518** 1180-1184

Synthesis and Antimicrobial activity of some new Schiff bases

Karamunge K G[1], Vibhute Y B[2]

[1]Department of Chemistry, Baliram Patil College, Kinwat-431804 (MS) India
[2]P. G. Department of Chemistry, Yeshwant College, Nanded-431 605 (MS) India.

E-mail: khushalgk@rediffmail.com

Abstract New Schiff bases derived from 4iodoaniline, 4-chloroniline, p- toluidine and halogenohydroxy substituted acetophenone have been synthesized. All the compounds have been characterized by IR, H[1]NMR and mass spectral, halogen, nitrogen analysis. The synthesized compounds have been screened for antimicrobial activity.

Keywords: Synthesis, halogenohydroxy acetophenone, Schiff bases, antibacterial, antifungal activity

1. Introduction

Compound containing an azomethine group (-CH = N) are known as Schiff bases. Schiff bases[1] are the important compound owing to their wide range of biological activities and industrial application. They have been found to posses the pharmacological activities such as antibacterial,[2-4] antifungal,[2,4-6] anticancer,[2,3] anti HIV,[7] antitubercular[8], antiviral,[2,9] antiinflammetry,[10] anticonvulsant[11] and antimalarial.[12] Schiff bases are used as starting material for the synthesis of various heterocyclic compounds.

The presence of various substituents in the phenyl rings of aromatic Schiff bases[13] are responsible for antifungal activity, which can be changed depending upon the type of substituent present on the aromatic rings.

In view of these above biological importance of Schiff bases, we have synthesized some new Schiff bases evaluated for their bioactivity.

New Schiff bases were synthesized by refluxing the reaction mixture of 4-iodoaniline, 4-chloroaniline, p-toluidine with halogenohydroxy substituted acetophenones in methanol [14] in the presence of glacial acetic acid .

All the synthesized compound were characterized on the basis of their IR, H[1]NMR, mass spectral and halogen, nitrogen analysis. The antibacterial and antifungal activity of compounds was evaluated by agar diffusion method and poison plate method respectively.

2. Materials and Method

2.1. Experimental

All the chemicals used were of analytical grade (AR). Melting points of the synthesized compounds were determined by open capillary and are uncorrected. The purity of the compounds was checked by TLC using silica gel plates and spots were developed in iodine chamber, IR spectra were recorded with KBr pellets on Shimadzu FTIR model 8400 spectrophotometer, H[1]NMR spectra in $CdCl_3$ on a Varian NMR mercury-300 instrument, Mass spectra were recorded on VG 7070H mass spectrometer.

2.1.1. General procedure for synthesis of Schiff bases

A mixture of 4-iodoaniline, 4-chloroniline / p-toluidine (0.01 mol) and halogenohydroxy substituted acetophenone (0.01 mol) in 15 ml of methanol and (0.5ml) glacial acetic acid was refluxed for 4 hrs, then the reaction mixture was left stand for overnight. The solid separates out. Separated solid was filtered, dried and recrystallised by ethanol.

Scheme

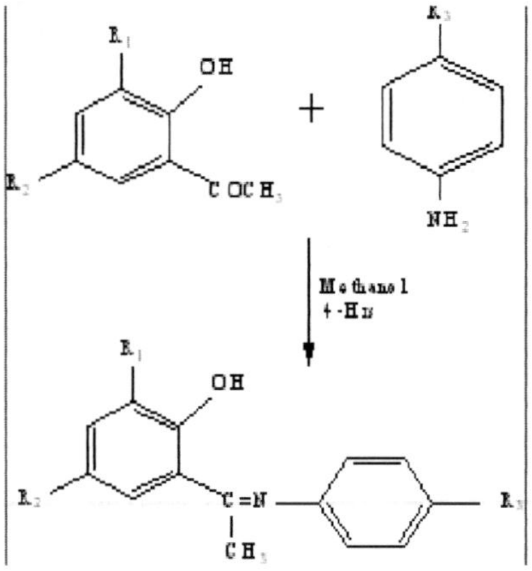

Comp.	R_1	R_2	R_3
Ia	I	Cl	I
Ib	Br	Cl	I
Ic	I	CH_3	I
Id	Br	CH_3	I
IIa	I	Cl	CH_3
IIb	Br	Cl	CH_3
IIc	I	CH_3	CH_3
IId	Br	CH_3	CH_3
IIIa	I	Cl	Cl
IIIb	Br	Cl	Cl
IIIc	I	CH_3	Cl
IIId	Br	CH_3	Cl

Synthesis of Schiff bases (Ia-IIId)

1-(5'-chloro-2'-hydroxy-3'-iodophenyl) ethylidene-4-iodoaniline (Ia) ;-
Yield 77 %, Greenish yellow, m.p. 188 ^{0}C; IR (KBr cm-1): 1612 (-C=N), 1550(C=C), 3213-3065 (ph-OH). ^{1}HNMR (δ, ppm): 2.35(s, 3H, CH_3), 6.67-7.86 (m, 6H, Ar-H).12.95 (s,1H,OH);MS (m/z): 497. Calcd.for $C_{14}H_{10}ONClI_2$ **:** I and Cl, 58.19; N, 2.81 Found I and Cl, 58.26, N, 2.89.

1-(3'-bromo-5'-chloro-2'-hydroxyphenyl) ethylidene-4-iodoaniline (Ib) ;-
Yield **:** 80 %, Yellow, m.p. 180^{0}C; IR (KBr cm-1): 1612 (-C=N), 1550(C=C), 3213-3065 (ph-OH). ^{1}HNMR (δ, ppm): 2.35(s, 3H, CH_3), 6.67-7.86 (m, 6H, Ar-H), 11.55 (s,1H, OH); MS (m/z): 450 Calcd.for $C_{14}H_{10}ONClBrI$: Cl, Br and I, 53.83, N, 3.11 Found Cl, Br and I, 53.76, N, 3.17.

1-(2'-hydroxy-3'-iodo-5'-methylphenyl) ethylidene-4-iodoaniline (Ic) ;-
Yield 75 %, Yellowish white, m.p. 150 ^{0}C; IR(KBr cm-1): 1628 (-C=N), 1550(C=C), 3243-3015 (ph-OH); ^{1}H NMR (δ, ppm): 2.45(s, 3H, CH_3), 2.36(s, 3H, CH_3),6.60-7.81 (m,6H,Ar-H); 12.25 (s, 1H, OH); MS (m/z): 477 Calcd.for $C_{15}H_{13}ONI_2$ **:** I, 53.25, N, 2.93. Found I, 53.32, N, 2.85

1-(5'-chloro-2'-hydroxy-3'-iodophenyl) ethylidene-4-methylaniline (IIa) ;-
Yield 70 %, Yellow, m.p. 137 ^{0}C IR (KBr cm-1): 1612 (-C=N), 1506(C=C), 3450-3030 (ph-OH); ^{1}H NMR (δ, ppm): 2.38(s, 3H, CH_3), 2.37(s, 3H, CH_3), 6.83-7.84 (m, 6H, Ar-H). 12.40 (s,1H,OH); MS (m/z): 385 Calcd.for $C_{15}H_{13}OClNI$: Cl and I, 42.21, N, 3.64 Found: Cl and I, 42.28. N, 3.67

1-(3'-bromo-5'-chloro-2'-hydroxyphenyl) ethylidene-4-methylaniline (IIb);-
Yield 78 %, Yellow, m.p. 105 ^{0}C IR (KBr cm-1): 1645 (-C=N), 1435(C=C), 3080 (ph-OH).^{1}H NMR (,δ ppm): 2.37(s, 3H, CH_3), 2.31(s, 3H, CH_3),6.683-7.75 (m, 6H, Ar-H), 12.86 (s, 1H, OH).MS (m/z): 338 Calcd.for $C_{15}H_{13}ONClBr$: Cl and Br, 34.17, N, 4.14. Found : Cl and Br, 34.22, N, 4.08

1-(2'-hydroxy-3'-iodo-5'-methylphenyl) ethylidene-4-methylaniline (IIc);-
Yield 75 %, Dark yellow m.p. 110 ^{0}C IR (KBr cm-1): 1597. (-C=N), 1504(C=C), 2914 (ph-OH). ^{1}H NMR (δ , ppm): 2.66 (s, 3H, CH_3), 2.38 (s, 3H, CH_3), 2.37 (s, 3H, CH_3), 6.84-7.71 (m, 6H, Ar-H). 12.65 (s,1H,OH), MS (m/z): 365 Calcd.for $C_{16}H_{16}ONI$: I, 34.79 N, 3.84. Found : I, 34.85, N, 391.

1-(5'-chloro-2'-hydroxy-3'-iodophenyl) ethylidene-4-chloroaniline (IIIa) ;-
Yield 80 %, pale yellow, m.p. 112 ^{0}C; IR (KBr cm-1): 1614 (-C=N), 1552(C=C), 3220-3065 (ph-OH). ^{1}HNMR (δ, ppm): 2.36(s, 3H, CH_3), 6.70-7.85 (m, 6H, Ar-H).12.85 (s,1H,OH); MS (m/z): 406. Calcd.for $C_{14}H_{10}ONCl_2I$ **:** I and Cl, 48.77; N, 2.81 Found I and Cl, 48.15, N 2.35.

1-(3'-bromo-5'-chloro-2'-hydroxyphenyl) ethylidene-4-chloroaniline (IIIb) ;-
Yield **:** 75 %,pale Yellow, m.p. 141^{0}C; IR (KBr cm-1): 1615 (-C=N), 1548(C=C), 3218-3060 (ph-OH). ^{1}HNMR (δ, ppm): 2.32(s, 3H, CH_3), 6.72-7.89 (m, 6H, Ar-H), 12.35 (s,1H, OH); MS (m/z): 359 Calcd.for $C_{14}H_{10}ONCl_2Br$: Cl, and Br,42.06, N, 3.11 Found : Cl, and Br,41.88, N 2.82.

1-(2'-hydroxy-3'-iodo-5'-methylphenyl) ethylidene-4-chloroaniline (IIIc) ;-Yield 70 %, Yellowish white, m.p. 98 ^{0}C; IR(KBr cm-1): 1625 (-C=N), 1545=C), 3243-3015 (ph-OH); ^{1}H NMR (δ, ppm): 2.45(s, 3H, CH_3), 6.60-7.81 (m,6H,Ar-H); 12.25 (s, 1H, OH); MS (m/z): 385 Calcd.for $C_{15}H_{13}ONICI$: I and Cl, 42.21,N,3.64
Found: I and Cl, 41.98, N,3.44.

2.1.2. Biological activity
Antimicrobial (*Escherishia coli, Salmonella typhi, Staphylococcus aureus and Bacillus subtilis*) activity was evaluated by Agar cup method[15,16] and antifungal (*Aspergillus niger, Penicillium chrysogenum, Fusarium moneliforme and Aspergillus flavus*) activity was evaluated by Poison plate method[17] Nutrient ager was used as culture for antibacterial activity and potato dextrose ager was used as culture for antifungal activity and DMSO was used to dissolved compounds. The results of the antibacterial and antifungal activity is shown in **table 1**. Result shows that compounds Ib, Id, are having significant activity for Bacillus subtilis. Compounds Ib, Id and IIa showed significant activity for Escherishia coli.

The compounds Ia, Ib, Id, IIa and IIb showed antifungal activity. In general compounds having chloro, bromo or iodo substitutent in their structure are having significant antibacterial and antifungal activity.

Table 1. Antibacterial and antifungal activity of the Schiff bases

Sr.	Antibacterial activity Zone of inhibition in mm				Antifungal activity			
	E. coli	**S. typhi**	**S. aureus**	**B. subtilis**	**A. niger**	**P. chryso genum**	**F. moneli forme**	**A. flavus forme**
Ia	-ve	-ve	-ve	-ve	-ve	-ve	-ve	RG
Ib	16	-ve	-ve	17	RG	-ve	-ve	RG
Id	12	-ve	-ve	15	RG	-ve	-ve	RG
IIa	11	-ve	-ve	-ve	-ve	-ve	-ve	--
IIb	-ve	-ve	-ve	-ve	RG			+ve
Penicillin	13	18	36	16	--	--	--	--
Grysofulvin	--	--	--	--	-ve	-ve	-ve	-ve

-ve → No Antibacterial activity.

-ve → No growth, Antifungal observed.

RG → Reduced growth (less active).

+ve → Fungal growth, No antifungal activity.

3. Result and discussion

New Schiff bases were obtained by condensing halogenohydroxy substituted acetophenone with 4-iodoaniline, 4-chloroniline, p-toluidine. The IR spectra of the synthesized compounds showed the presence of C=N stretching bands at 1645-1597 cm^{-1}. ^1H NMR spectra of the synthesized compounds, the protons of azomethine compounds have given δ, ppm 2.30-2.37 due to $-N=C-CH_3$ for all compounds. The element analyses of the compounds are in agreement with the composition suggested for the compounds. Mass spectroscopy was performed of all the Schiff bases to determine its molecular weight. Molecular ion peak of Schiff bases was confirming its formula weight. Which are same as the calculated molecular ion value.

In accordance with the data obtained from antimicrobial activity, all the synthesized Schiff bases of halogenohydroxy acetophenone have shown good activity against the tested microbes. Among these Schiff bases of halogenohydroxy acetophenone Ib, Id, and IIa has shown good activity against all the tested bacteria and fungi.

4. Conclusions

The research study reports the successful synthesis of new Schiff bases. Schiff bases were physically and chemically characterized through elemental analysis, IR, ^1H NMR, and Mass spectral data.

The antimicrobial activity was tested of the Schiff bases on the strains *E. coli, S. typhi, S. aureus and B. subtilis, A. niger, P. chrysogenum, F. moneliforme and A. flavus,* and it was compared to the standard Penicillin and Grysofulvin. The tested compound Ib, Id and Ie are very active against both gram positive and gram negative bacteria. The compounds Ia, Ib, Id, IIa and IIb showed antifungal activity.

Acknowledgements

The authors are also thankful to University Grants Commission, New Delhi , India and WRO, Pune, for sanctioning Research Project to K.K.G., Director of IICT, Hyderabad, Department of

Journal of Physics: Conference Series **423** (2013) 012006 doi:10.1088/1742-6596/423/1/012006

Chemistry Pune, Pune University Pune, for providing spectra and Mrs Shete, Head Department of Microbiology, N.S.B. College Nanded, for providing antimicrobial activity.

References:

[1] Wang L, Feng Y, Xue J and Li Y *2008 J. Serb. Chem Soc.* **73** 1-6
[2] Cerhiaro G, and Ferreira A M D 2006 *Journal of the Brrazilian Chemical Society* **17(8)** 1473-1485
[3] Pandeya S N, Smitha S, Jyoti M and Sridhar S K 2005 *Acta Pharmaceutica* **55** 27-46
[4] Sharma V K, Srivastava S and Srivastava A 2006 *Polish Journal of Chemistry* **80** 387-396
[5] Sharma V K, Srivastava A and Srivastava S 2006 *Journal of the Serbian Chemical Society* **71 (8-9)** 917-928
[6] Pandey S N, Lakshmi V S and Pandey A 2003 *Indian J. Pharm. Sci.* **65** 213-222
[7] Bal T R, Anand B, Yogeeswari P and Sriram D 2005 *Bioorganic and Medicinal Chemistry Letters* **15(20)** 4451-4455
[8] Bhat M A, Imran M, Khan S A and Siddiqui N 2005 *J. Pharm Sci* **67** 151-159
[9] Karthikeyan M S, Dasappa J P, Boja P, Subrahmanya, Bhat K, Bantwal S H 2006 *Bioorg and Med Chem* **14** 7482-7489
[10] Wadher S J, Puranik M P, Karande N A and. Yeole P G 2009 *International Journal of Pharm Tech Research* **1** 22-33
[11] Chinnasamy R P, Sundararajan R and Govindaraj S 2010 *International Journal of Pharmacy and Pharmaceutical* **2 (4)** 177-181
[12] Li Y, Yang Z S, Zhang H, Cao B J and Wang F D 2003 *Bio org and Med Chem* **11** 4363-4368
[13] Manrao M R, Gill K K, Sharma J R, Kalsi P S 2000 *Pestic. Res. J.* **12** 239-241
[14] Sridhar S K, Sarvanan M and Ramesh A 2001 *Eur. J. Med. Chem.* **36** 615
[15] Collins C H 1967 *Microbiological Methods* (Butterworth, London) p 364
[16] Godkar P B *Text Book of Medical Laboratory Technology* 1996 (Bhalavi Publication House, Mumbai, India) p 326
[17] Robert Cruickshant *A Practice of Medicinal Microbiology Vol. II*

Journal of Physics: Conference Series **423** (2013) 012007 doi:10.1088/1742-6596/423/1/012007

Synthesis and biological activity of pyrimido [1, 2-*a*] quinoline moiety and its 2-substituted derivatives

A.G. Jadhav and N. K. Halikar
Department of chemistry,
Mahatma Gandhi Mahavidyalaya, Ahmedpur (MS) India.

Email: aanil_kumar2006@rediffmail.com

Abstract. 2-Amino-3-cyano quinoline (**1**) and bis (methylthio) methylene malononitrile (**2**) were refluxed in N,N-dimethyl formamide (DMF) in presence of catalytic amount of anhydrous potassium carbonate to afforded 3, 11-dicyano-4-imino-2-methylthio -*4H*-pyrimido [1, 2-*a*] quinoline (**3**). The latter were further reacted with different substituted aniline, phenol, hetryl amine and compound containing active methyl group. Afforded to 3, 11-dicyano-4-imino -*4H*-pyrimido [1, 2-*a*] quinoline and their 2-substuited derivatives **(4a-7c)**. All these newly synthesized compounds were characterized by elemental analysis and spectral data, and screened for their antimicrobial activities.

1. Introduction

The quinoline skeleton is often used for the design of many synthetic compounds with diverse pharmacological properties. In 1820, quinine was isolated as the active ingredient from the bark of cinchona tree and successively replaced the crude bark for the treatment of malaria. Despite its relatively low efficacy and tolerability, quinine still plays an important role in the treatment of multi resistant malaria[1]. Quinoline and its derivatives are receiving important due to their due to wide range of biological activities as a drug analgesics[2],antiamoebic [3-6], tryphocidal[7], antiseptic[8] and antiserotonin[9].In addition to these, derivatives also exhibit good antimalarial [10,11], antitubercular[12], antibacterial[13], antihistaminic [14], antineurodegerative[15], anticonvulsant [16], antitumor [17], anticancers [18-20] and antiallergics [21] activities.

In view of these observations, such as quinoline fused with pyrimidine ring would exhibit some interesting pharmacological activities, further, the ring anellation to amino groups containing nitrogen heterocycles with ketene dithioacetals as reagent has reported[22-23]. Recently we report one pot synthesis of 3, 11-Dicyano-4-imino-2-methylthio-*4H*-pyrimido [1, 2-*a*] quinoline and its reaction with selected nucleophiles [24]. In continuation of this further development of milder condition with better yields. The compound (**3**) was prepared by the reaction of 2-anino-3-cyano-quinoline (**1**) reaction with bis (methylyhio) methylene malononitrile (**2**) in presence of N,N'-dimethyl formamide and catalytic amount of Potassium carbonate **scheme-1**. Compound (**3**) possesses an active methylthio group at the 2-positation that is activated by the ring 1-nitrogen atom and the electron withdrawing 3-cyano group. Compound **(3)** was reacted with selected N-, O-, and C-nucleophiles like aryl amines, substituted phenols, heteryl amines and compound containing active methylene group. Hence compound (**3**) independently reaction with different substituted anilines, substituted phenols, active methylene compounds and hetryl amines in presence of N,N'-dimethyl formamide and catalytic amount of Potassium carbonate afforded new compounds 4a-c,5a-c,6a-c,7a-c.

2. Experimental Section

Melting points were determined by in an open capillary method and are uncorrected. The chemicals and solvents used for laboratory grade and were purified. IR spectra were recorded (in KBr pallets) on Shimadzu spectrophotometer. ^{1}H NMR spectra were recorded (in DMSO-d6) on Avance-300 MHz spectrometer using TMS as an internal standard. The mass were recorded on EI-Shimadzu GC-MS spectrometer. Elemental analyses were performed on a Heraeus CHN-O rapid analyzer.

3, 11-Dicyano-4-imino-2-methylthio-*4H*-pyrimido [1, 2-*a*] quinoline (3)

A mixture of 2-anino-3-cyano-quinoline **(1)** (2.91g, 0.01 m mol) and bis (methylthio) methylene malononitrile **(2)** (1.70 g, 0.01 m mol) was refluxed in the presence of N, N'-dimethyl formamide catalytic amount of anhydrous Potassium carbonate for 4 hrs. The reaction mixture was cooled to room temperature and poured in to ice cold water. The separated solid was filtered, washed with water and recrystallized from N, N ' –dimethyl formamide-ethanol mixture to afford compound **(3)**.
Brown powder, (88% yield), Mp: above 300 °C; EI-MS (m/z-RA%) : 291 (M^+) ,IR (cm^{-1}, KBr) : 3236 (=NH), 2216-2218 (CN), ^{1}H NMR (DMSO-d_6, ppm) : 2.4 (s, 3H, SCH_3), 6.3-7.8 (m, 5H, Ar-H),8.8(s,1H,=NH), ^{13}C NMR (300 MHz, DMSO-d_6, ppm) 15,80,96, 110,115.9, 115.9, 116.2, 118.7,124.7,124.9,127.2,128.7,139.2,156.7,164.9; Anal. Calcd. For: $C_{15}H_9N_5S$; C, 61.84; H, 3.11; N, 24.04; Found: C, 61.38; H, 2.83; N, 23.62.

3, 11-Dicyano-4-imino-2-(4'-chloro aniline/4'methoxy aniline/4'methyl aniline)-*4H*-pyrimido [1, 2-*a*] quinoline(4a-c)

A mixture of compound **(3)** (2.91 g, 0.01 mol), N, N' –dimethyl formamide and catalytic amount anhydrous Potassium carbonate and appropriate substituted anilines (0.01m mol) was refluxed for 4 hrs. the reaction mixture was cooled to room temperature and poured into ice cold water. The products 4a-c thus obtained were recrystallized from N, N' –dimethyl formamide-ethanol solvent.
3, 11-Dicyano-4-imino-2-(4'-chloro aniline)-*4H*-pyrimido [1, 2-*a*] quinoline **(4a)**

Brown powder, (72% yield), Mp: 268-69 °C; EI-MS (m/z-RA%) : 370 (M^+) ,IR (cm^{-1}, KBr) : 3284 (=NH), 2210-2212 (CN), ^{1}H NMR (DMSO-d_6, ppm) : 4.2 (-NH), 6.4-7.5 (m, 9H, Ar-H),9.3(s,1H,=NH), Anal. Calcd. For: $C_{20}H_{13}ClN_6$; C, 64.43; H, 3.51; N, 22.54; Found: C, 63.87; H, 2.98; N, 22.02.
3, 11-Dicyano-4-imino-2-(4'methoxy aniline)-*4H*-pyrimido [1, 2-*a*] quinoline **(4b)**

Brown powder, (60% yield), Mp: 284 °C; EI-MS (m/z-RA%) : 366 (M^+) ,IR (cm^{-1}, KBr) : 3262 (=NH), 2195-2198 (CN), ^{1}H NMR (DMSO-d_6, ppm) : 3.7 (s,3H,OCH_3), 4.3 (-NH), 6.3-7.8 (m, 9H, Ar-H),8.9(s,1H,=NH), Anal. Calcd. For: $C_{21}H_{14}N_6O$; C, 68.84; H, 3.85; N, 22.94;Found: C, 68.14; H, 3.25; N, 22.36.
3, 11-Dicyano-4-imino-2-(4'methyl aniline)-*4H*-pyrimido [1, 2-*a*] quinoline **(4c)**

Brown powder, (75% yield), Mp: 273-75 °C; EI-MS (m/z-RA%) : 351 (M^+ +1) ,IR (cm^{-1}, KBr) : 3236 (=NH), 2210-2212 (CN), ^{1}H NMR (DMSO-d_6, ppm) : 2.4 (s,3H,CH_3), 4.4 (-NH), 6.5-7.8 (m, 9H, Ar-H),9.4(s,1H,=NH), Anal. Calcd. For: $C_{21}H_{14}N_6$; C, 71.99; H, 4.03; N, 23.99;Found: C, 71.24; H, 3.42; N, 23.22.

3, 11-Dicyano-4-imino-2-(4'-chloro phenol/4'methoxy phenol /2'-chloro phenol)-*4H*-pyrimido [1, 2-*a*] quinoline(5a-c)

A mixture of compound **(3)** (2.91 g, 0.01 mol), N, N' –dimethyl formamide and catalytic amount anhydrous Potassium carbonate and appropriate substituted phenol (0.01m mol) was refluxed for 4 hrs. The reaction mixture was cooled to room temperature and poured into ice cold water. The products 5a-c thus obtained were recrystallized from N, N' –dimethyl formamide-ethanol solvent.
3, 11-Dicyano-4-imino-2-(4'-chloro phenol)-*4H*-pyrimido [1, 2-*a*] quinoline (**5a**)

Brown powder, (60% yield), Mp: 259-61 °C; EI-MS (m/z-RA%) : 371 (M^+) ,IR (cm^{-1}, KBr) : 3298 (=NH), 2222-2224 (CN), ^1H NMR (DMSO-d_6, ppm) : 6.3-7.8 (m, 9H, Ar-H), 9.1(s,1H, =NH), Anal. Calcd. For: $C_{20}H_{10}ClN_5O$; C, 64.61; H, 2.71; N, 18.84; Found: C, 64.02; H, 2.11; N, 18.31.

3, 11-Dicyano-4-imino-2-(4'-methoxy phenol)-*4H*-pyrimido [1, 2-*a*] quinoline (**5b**)

Brown powder, (72% yield), Mp: 259-61 °C; EI-MS (m/z-RA%) : 367 (M^+) ,IR (cm^{-1}, KBr) : 3265 (=NH), 2214-2218 (CN), ^1H NMR (DMSO-d_6, ppm) : 3.6 (s,3H,OCH_3), 6.4-7.9 (m, 9H, Ar-H), 8.8(s,1H, =NH), Anal. Calcd. For: $C_{21}H_{13}N_5O_2$; C, 68.66; H, 3.57; N, 19.06;Found: C, 68.26; H, 3.02; N, 18.48.

3, 11-Dicyano-4-imino-2-(2'- chloro phenol)-*4H*-pyrimido [1, 2-*a*] quinoline (**5c**)

Brown powder, (68% yield), Mp: 276 °C; EI-MS (m/z-RA%) : 372 (M^++1) ,IR (cm^{-1}, KBr) : 3284 (=NH), 2210-2212 (CN), ^1H NMR (DMSO-d_6, ppm) : 6.2-7.5 (m, 9H, Ar-H), 8.9(s,1H, =NH), Anal. Calcd. For: $C_{20}H_{10}ClN_5O$; C, 64.61; H, 2.71; N, 18.84; Found: C, 64.02; H, 2.11; N, 18.31.

3,11-Dicyano-4-imino-2-(pyrrolidino/piperidino/morpholino)-*4H*-pyrimido[1,2-*a*]quinoline (**6a-c**)

A mixture of compound **(3)** (2.91 g, 0.01 mol), N, N' –dimethyl formamide and catalytic amount anhydrous Potassium carbonate and appropriate substituted hetryl amine (0.01m mol) was refluxed for 4 hrs. The reaction mixture was cooled to room temperature and poured into ice cold water. The products 6a-c thus obtained were recrystallized from N, N' –dimethyl formamide-ethanol solvent.
3, 11-Dicyano-4-imino-2-(pyrrolidino)-*4H*-pyrimido [1, 2-*a*] quinoline (**6a**)

Brown powder, (58% yield), Mp: 288 °C; EI-MS (m/z-RA%) : 314 (M^+) ,IR (cm^{-1}, KBr) : 3264 (=NH), 2202-2204 (CN), ^1H NMR (DMSO-d_6, ppm) : 1.6 (t,4H, CH_2), 2.7 (t,4H, -NCH_2-), 6.3-7.8 (m, 5H, Ar-H), 9.1(s,1H, =NH), Anal. Calcd. For: $C_{18}H_{14}N_6$; C, 68.78; H, 4.49; N, 26.74; Found: C, 68.21; H, 4.01; N, 26.13.

3, 11-Dicyano-4-imino-2-(piperidino)-*4H*-pyrimido [1, 2-*a*] quinoline (**6b**)

Brown powder, (68% yield), Mp: 298 °C; EI-MS (m/z-RA%) : 328 (M^+) ,IR (cm^{-1}, KBr) : 3248 (=NH), 2211-2213 (CN), ^1H NMR (DMSO-d_6, ppm) : 1.5 (t,6H, three -CH_2-), 2.8 (t,4H, two-NCH_2-), 6.3-7.6 (m, 5H, Ar-H), 8.9(s,1H, =NH), Anal. Calcd. For: $C_{19}H_{16}N_6$; C, 69.50; H, 4.91; N, 25.59;Found: C, 69.01; H, 4.18; N, 25.02.

3, 11-Dicyano-4-imino-2-(morpholino)-*4H*-pyrimido [1, 2-*a*] quinoline (**6c**)

Brown powder, (64% yield), Mp: 292-94 °C; EI-MS (m/z-RA%) : 330 (M^+) ,IR (cm^{-1}, KBr) : 3284 (=NH), 2196-2198 (CN), ^1H NMR (DMSO-d_6, ppm) : 2.7 (t,4H, -NCH_2),3.7(t,4H, OCH_2), 6.2-7.5 (m, 5H, Ar-H), 8.9(s,1H, =NH), Anal. Calcd. For: $C_{18}H_{14}N_6O$; C, 65.44; H, 4.27; N, 25.44; Found: C, 65.01; H, 3.89; N, 24.94.

3, 11-Dicyano-4-imino-2-(α-ethylacetoacetyl /α-ethylcyanoacetyl /α-malononitriyl)-*4H*-pyrimido [1, 2-*a*] quinoline (7a-c)

A mixture of compound **(3)** (2.91 g, 0.01 mol) N, N' –dimethyl formamide and catalytic amount anhydrous Potassium carbonate and appropriate substituted active methylene group (0.01m mol) was refluxed for 4 hrs. The reaction mixture was cooled to room temperature and poured into ice cold water. The products 7a-c thus obtained were recrystallized from N, N' –dimethyl formamide-ethanol solvent.
3, 11-Dicyano-4-imino-2-(α-ethylacetoacetyl)-*4H*-pyrimido [1, 2-*a*] quinoline (**7a**)

Brown powder, (65% yield), Mp: 288 °C; EI-MS (m/z-RA%) : 373 (M^+) ,IR (cm^{-1}, KBr) : 3288 (=NH), 2196-2197 (CN), ^1H NMR (DMSO-d_6, ppm) : 1.4 (s,3H,-CH_3), 2.2 (s,3H,$COCH_3$), 3.9 (s,1H,-CH),4.1 (s,2H,-CH_2), 6.4-7.7 (m, 5H, Ar-H), 8.9(s,1H, =NH), Anal. Calcd. For: $C_{20}H_{15}N_5O_3$; C, 64.34; H, 4.05; N, 18.76; Found: C, 63.76; H, 3.64; N, 18.26.

3, 11-Dicyano-4-imino-2-(α-ethylcyanoacetyl)-*4H*-pyrimido [1, 2-*a*] quinoline (**7b**)

Brown powder, (63% yield), Mp: 298 °C; EI-MS (m/z-RA%) : 356 (M^+) ,IR (cm^{-1}, KBr) : 3273 (=NH), 2216-2218 (CN), ^1H NMR (DMSO-d_6, ppm) : 1.4 (s,3H,-CH_3), 3.9 (s,1H,-CH), 4.2(s, 2H, -OCH_2-), 6.4-7.8 (m, 5H, Ar-H), 9.3(s,1H, =NH), Anal. Calcd. For: $C_{19}H_{12}N_6O_2$; C, 64.04; H, 3.39; N, 23.58; Found: C, 63.62; H, 2.46; N, 23.01.

3, 11-Dicyano-4-imino-2-(α-malononitriyl)-*4H*-pyrimido [1, 2-*a*] quinoline (**7c**)

Brown powder, (57% yield), Mp: 290°C; EI-MS (m/z-RA%) : 309 (M^+) ,IR (cm^{-1}, KBr) : 3284 (=NH), 2214-2218 (CN), ^1H NMR (DMSO-d_6, ppm) 4.3 (s,1H,-CH), 6.3-7.7 (m, 5H, Ar-H), 8.9(s,1H, =NH), Anal. Calcd. For: $C_{17}H_7N_7$; C, 66.02; H, 2.28; N, 31.70 Found: C, 65.54; H, 1.58; N, 31.12.

3. Result and Discussion

In the present investigation, we have developed new methodology towards the synthesis of 3, 11-Dicyano-4-imino-2-methylthio-*4H*-pyrimido [1, 2-*a*] quinoline and their substituted derivatives. **(3)** Our method gives single product with high yield. The reaction started with 2-amino-3-cyano quinoline **(1)** and bis (methylthio) methylene malononitrile **(2)** were refluxed in N, N' –dimethyl formamide and catalytic amount anhydrous Potassium carbonate to afford **(3) scheme-1.**

Compound **3** posses a replaceable active methylthio group at 2- position which is activated by ring 1-nitrogen atom and electron withdrawing group 3-cyano group. Compound **(3)** reacted with selected N-, O-, C- nucleophiles like aryl amines hetryl amines, substituted phenols and compounds containing an active methylene group. The compound (3) on independently reaction with p-chloro aniline, p-methoxy aniline, p-methyl aniline in N, N' –dimethyl formamide and catalytic amount anhydrous Potassium carbonate, afforded 3, 11-Dicyano-4-imino-2-(4'-chloro aniline/4'methoxy aniline/4'methyl aniline)-*4H*-pyrimido [1, 2-*a*] quinoline(4a-c) respectively. **Scheme-2**

3, 11-Dicyano-4-imino-2-(4'-chloro phenol/4'methoxy phenol /2'-chloro phenol)-*4H*-pyrimido [1, 2-*a*] quinoline(5a-c) respectively **scheme-3** were obtained by condensation of **(3)** with p-chloro phenol, p-methoxy phenol, o-chloro phenol in N, N' –dimethyl formamide and catalytic amount anhydrous Potassium carbonate .

Under similar experimental condition compound **(3)** reacted independently with hetryl amines like pyrolidine, piperidine, morpholine to yield 3,11-Dicyano-4-imino-2-(pyrrolidino/piperidino/morpholino)-*4H*-pyrimido[1,2-*a*]quinoline (6a-c) respectively **scheme-4**

Compounds on reaction with ethyl acetoacetate, ethyl cyano acetate, malanonitrile in presence of N, N' –dimethyl formamide and catalytic amount anhydrous Potassium carbonate yielded compounds 3, 11-Dicyano-4-imino-2-(α-ethylacetoacetyl /α-ethylcyanoacetyl /α-malononitriyl)-*4H*-pyrimido [1, 2-*a*] quinoline (7a-c) respectively **scheme-5**.

Compounds 4a-c, 5a-c, 6a-c, 7a-c show absorption bands in their IR spectra in the range of 3200 cm^{-1} to 3410 cm^{-1} and 2190 cm^{-1} to 2230 cm^{-1} due to =NH and CN stretching respectively, ^{1}H – NMR and Mass spectral data are also in agreement with structures of newly synthesized compounds 4a-c, 5a-c, 6a-c, 7a-c.

The structures of these newly synthesized compounds were established on the basis of elemental analysis, IR, PMR and MASS spectral data, spectral studies of all compounds shows that compounds are stable & do not exhibit any tautomerism.

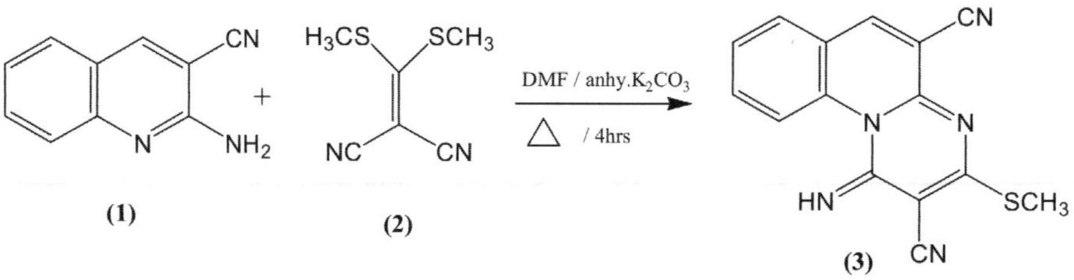

Scheme-1

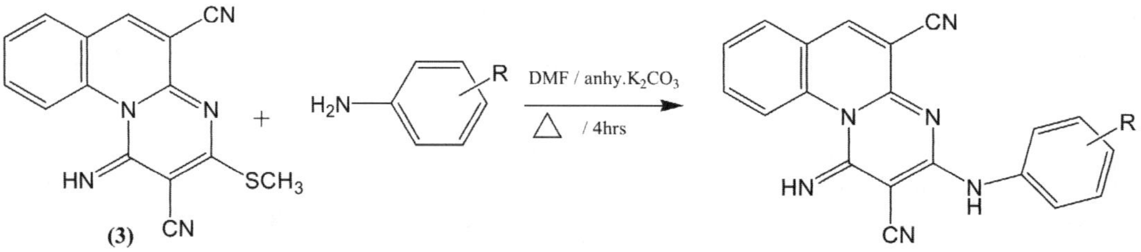

Scheme-2

(4a-c)

R=4'-Cl ,4'-CH_3 ,4'-OCH_3

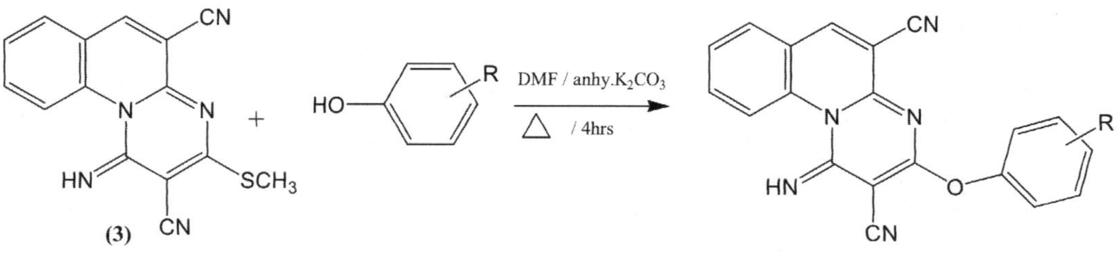

Scheme-3

(5a-c)

R=4'-Cl ,4'-OCH_3 ,2'-Cl

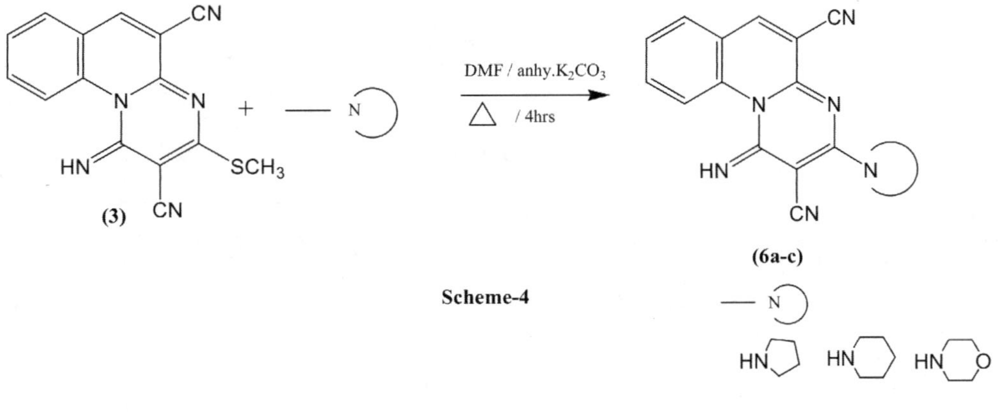

Scheme-4

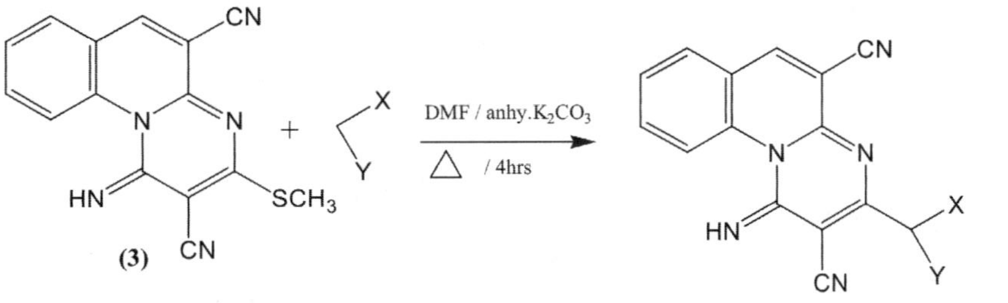

(7a-c)

Scheme-5

X	Y
$COOC_2H_5$	$COCH_3$
$COOC_2H_5$	CN
CN	CN

4. Antimicrobial activity

All the compounds were for their antifungal and antibacterial activity against species *Aspergillus flavus, Aspergillus Niger* and *E. coli* and B. *Subtilis by* paper disc diffusion method[25]. Control for fungal and bacterial species using Disc diffusion method, solvent DMSO was used, Fungi- Czapek's dox agar, Bacteria- Nutrient agar. Incubation period for fungi 4 days (24+/- 2 ^{0}C) and for bacteria 24 hrs (37 ^{0}C). The synthesized compounds exhibited zone of inhibition of 10-31 mm in diameter, where as standard Fluconazole exhibited zone of inhibition of 26 and 25 mm . Streptomycin exhibited zone of inhibition of 32 and 30 mm in diameter against *E. coli* and *B. Subtilis* respectively.

Table 1: Evaluation of antimicrobial activity of newly synthesized compound

Sr. No.	code	Zone of inhibition in mm			
		Fungal species		Bacterial species	
		Af	*An*	*Ec*	*Bs*
1	3	12	15	11	22
2	4a	13	--	14	17
3	4b	10	12	10	16
4	4c	12	14	12	17
5	5a	13	12	12	15
6	5b	14	12	14	12
7	5c	--	14	16	20
8	6a	14	11	13	12
9	6b	16	18	08	13
10	6c	20	16	15	14
11	7a	16	13	13	06
12	7b	17	15	15	17
13	7c	16	15	14	18
Positive control		26		25	
		Fluconazole		**Streptomycin**	

5. Conclusion

In this communication all the newly synthesized compounds reported first time and describe the simple route of their synthesis in mild condition with good yield. The present study showed that all the newly synthesized compounds were exhibiting significant antibacterial and antifungal activities. However, further studies are required to establish the mechanism of action of the title compounds. From the screening data it was found some derivative have encouraging antifungal activity, which need to be further investigation to get better antifungal and antibacterial agents in feature.

Acknowledgements

The authors are grateful to Dr. G. D. Bagde, Principal, Mahatma Gandhi Mahavidyalaya, Ahmedpur, for providing laboratory facilities and Director, Indian Institute of Chemical Technology, Hyderabad, for providing spectra and also thanks Principal Dr. Kalamse, Science College Nanded, providing biological activity.

References

1. Wiesner, J.; Ortmann, R.; Jomaa, H.; Schlitzer, M. *Angew. Chem .Int.Ed.,* 2003, **43,** 5274.
2. Sevim R and Ş. Güniz Küçükgüzel , *Molecules* 2007**,** **12***,* 1910-1939.
3. Haythem A. S, Ibrahim M. M and Mohammad S.M, *Molecules* 2009*,* **14***,* 1483-1494;
4. Dhawan K.N, Jaju B. P. and Gupta G. P. , *Ind. J. Microbiology* ,Psychopharmacology, 1970, Volume 18, Number 1, Pages 94-98

5. Yadava, J.N.S.; Datta, G. P. *Ind. J. Med. Res.*, 1973, **61**, 971.
6. Chakravarti, S. S. ; Sen, P. K. ; Choudhari, S. and Das, M. *Ind. J. Chem.,* 1985, **248,** 737.
7. Browning, C. H. *Proc. R. Soc. London,* 1932, **372**.
8. Browning, C. H, *J. Path. Bact.,* 1924**, 27**, 121.
9. Barbara.M. K, Gordan B. B, Michel D. E, Karl H. R. *Antimicrobial Agents and Chemotharpy*, June 1997, p. 1369–1374
10. Freek, L. A.; Martin, J.L.; Rowlett, Willson. *J. Am. Chem. Soc.,* 1946, **68,** 1285.
11. R.J. Rowlett Jr., Robert E. Lutz**.** *J. Am. Chem. Soc.,* 1946, **68**, 1288-191.
12. I. V. Ukrainets, Liu Yangyang, A. A. Tkach and A. V. Turov, *Chemistry of Heterocyclic Compounds* 2009, **45**, No.7, 802-808,
13. Maher F. El-Z, Abd El-H. N. Ahmed, Farghaly A. O, Mohamed A. Abd-Alla, *Journal of Chemical Technology and Biotechnology,* 1992**, 53**, Issue 4, 329–336,**.**
14. Chi-Hoon Lee and Hoi-Seon Lee. , *J. Korean Soc. Appl. Biol.Chem.* 2011**, 54(1),** 118-123
15. Hailin Z., a Lev M. Weiner, a Orit Bar-Am, b Silvina E, c Z. Ioav Cabantchik, c A. Warshawsky, a . B. H. Youdimb and Mati Fridkina, *Bio. & Med Chemistry* 2005**, 13** 773– 783.
16. Michael, R.; Ian, S.; Paul, L.; Graeme, S.; Angela, M.; Lesley R.; Raymond, B.; John, A.*; J. Med. Chem.* 1993**, 36 (22),** 3386-96**;** *Chem. Abstr.*, 1993, **119**, 249817y.
17. Wentland, M.P. ; Aldous, S. C.; Gruett, M. D.; Perni, R. B.; Powles, R.G.; Danz, D.W.; Kingbeil, K.M., Peverly, A. D.; Robinson, R.G.; Corbett, J.H.; Rake, J.B.; Coughlin, S.A. *Bioorganic and Medicinal Chem. Letters*, 1995, **5 (4)**, 405-410.
18. Bol, S.A.; Horibeck, J.; Markovic, J.; Boer, de J.G.; Turesky, R.J.; Constable. *Carcinogenesis,* 2000, **21***,* 1-6.
19. El-Sabbagh, H.I. ; Abadi, A.H. ; Al-Khawad, I.E. ; Al-Rashood, K.A. *Arch. Pharm. Pharm. Med. Chem*., 1990, **333,** 19-24
20. Althuis, T.H.; Moore, P.F.; Hess, H.J. *J. Med. Chem*., 1979, **22 (1),** 44-48.
21. Pingle, M. S;. Vartale, S. P.; Bhosale, V. N.; Kuberkar, S.V. *Arkivoc*, 2006, **(x),** 190-198.
22. Okram, Mukherjee, singh; Mohd., Farooque, Ahmed., *Indian J. of Chemistry*, 2004, **43B**,1561.
23. Chowdhury, A.Z.M.; Shibata, Y. *Chem. Pharm. Bull.*, 2001,**49 (4),** 391.
24. Sambhaji P.Vartale ,Vijay N. Bhosale , Sandeep V.Khansole Ramdas N .Katapalle *Letters in Organic Chemistry,* **2009**, **6**, 244-248.
25. Ananthanarayan R& Panikar J C K, *Textbook of microbiology* (orient Longman) , 1999, 578.

Journal of Physics: Conference Series **423** (2013) 012008 doi:10.1088/1742-6596/423/1/012008

Power functions of the Shewhart control chart

M B C Khoo

School of Mathematical Sciences, Universiti Sains Malaysia, 11800 Minden, Penang, Malaysia

Email: mkbc@usm.my

Abstract. The Shewhart control chart is used to check for a lack of control (a shift in the process). However it is insensitive to small shifts. To increase the sensitivity of the Shewhart control chart, some actions can be taken. These include reducing the width of the control limits, increasing the subgroup size to reduce the variance of the sample mean and making use of detection rules to increase the sensitivity of the chart. All of these actions will influence the power functions of the Shewhart control chart. A probability table showing the probabilities of detecting sustained shifts in the process mean, calculated using the formulae have been recommended. However, the calculations of the probabilities using the formulae are complicated and time consuming. In this paper, a Monte Carlo simulation using the Statistical Analysis System (SAS) is conducted to compute the necessary probabilities. These probabilities are closed to that obtained by using the formulae. The Monte Carlo simulation method is recommended as a better way to calculate the probabilities, as it provides savings, in terms of time and cost. Besides, the Monte Carlo simulation method also provides a higher flexibility in calculating the probabilities, for different combinations of the detection rules. The results obtained will assist practitioners in designing and implementing the Shewhart chart effectively.

1. Introduction

The control charting method for controlling the quality of a process in manufacturing was introduced by Dr. Walter A. Shewhart in the 1920s. According to [1], Shewhart defined the product attributes, types of product variations and suggested methods on how to collect, plot and analyze data. In fact, control charts used by quality practitioners from all around the world are based on the process control methods provided by Shewhart.

The Shewhart control chart plays an important role in Statistical Process Control (SPC). It is able to detect the occurrence of assignable causes quickly in a process control so that investigations of the process and corrective actions can be made promptly before many defective products are produced [2]. The main objective of applying the Shewhart chart is not to detect defective products but to prevent failure so that production of low quality products will not happen. This move can help in cost savings, as well as helping to produce high quality products.

The Shewhart chart is powerful in the detection of large shifts in the process mean. Nonetheless, it exhibits a lack of sensitivity to detect small shifts. Many researches have been made by quality experts to improve the chart's sensitivity. Decreasing the variance of the sample mean by increasing the subgroup size, employing a narrower width of the control limits and applying sensitizing rules, such as those by [3], [4], [5], and [6], to the Shewhart chart are some of the popular approaches made to enhance the performance of the chart. All these approaches can influence the power functions of the Shewhart chart. A probability table showing the probabilities of detecting sustained shifts in the process mean, computed by means of formulae was presented by [7] to serve as guidelines to quality practitioners in constructing the Shewhart chart. In this paper, we present a Monte Carlo simulation method to obtain similar results. This research is motivated by the fact that the Shewhart chart is the most well known control charting tool among practitioners and that an in-depth understanding of the power function of this chart will enable a more efficient use of the chart in process monitoring.

This paper is organized as follows: The Shewhart control chart is presented in Section 2. In Section 3, the power functions of the Shewhart chart are discussed. The probability table of the power functions of the Shewhart $\bar{X}$ chart is explained in Section 4. In Section 5, performance comparisons are made between the probabilities obtained via formulae and that obtained by means of Monte Carlo simulation. Finally, conclusions are drawn in Section 6.

2. The Shewhart control chart

The Shewhart control chart is a graphical plot that helps in the study on how a process changes over time. The Shewhart chart consists of three lines that are the lower control limit (LCL), upper control limit (UCL) and center line (CL). The construction of the Shewhart chart is based on statistical principles. Typically, the UCL is plotted $+3\sigma$ above the CL whereas the LCL is plotted -3σ below the CL. Here, CL represents the mean of the sample data. We intend to detect a process that is out-of-control when there are sample points plotting beyond the control limits, but when the process is in-control, we wish that the probability of detecting a false alarm is as small as possible [8].

Assume that a quality characteristic follows a normal distribution with mean μ and standard deviation σ, where both μ and σ are known. If $X_1, X_2, ..., X_n$ is a sample of size n, then the average of this sample is [9]

$$\bar{X} = \frac{X_1 + X_2 + ... + X_n}{n} \tag{1}$$

and we know that $\bar{X}$ is normally distributed with mean μ and standard deviation $\sigma_{\bar{X}} = \frac{\sigma}{\sqrt{n}}$. Besides, the probability is $1-\alpha$ that any sample mean will fall between

$$\mu + Z_{\alpha/2} \cdot \frac{\sigma}{\sqrt{n}} \tag{2a}$$

and

$$\mu - Z_{\alpha/2} \cdot \frac{\sigma}{\sqrt{n}}. \tag{2b}$$

Hence, if μ and σ are known, equations (2a) and (2b) can be used as upper and lower control limits of the Shewhart $\bar{X}$ chart. As mentioned earlier, it is customary to replace $Z_{\alpha/2}$ by 3 so that the $\pm 3\sigma$ limits are used. The process mean is out-of-control if the sample mean falls outside these limits. In practice, μ and σ are usually unknown. Consequently, these parameters are estimated from a set of in-control Phase-I data. These estimates should usually be based on at least 20 to 25 samples.

3. Power functions of the Shewhart control chart

In statistical quality control, an inference on the mean of a population can be summarized as the following hypotheses testing:

$$\begin{aligned} H_0 &: \mu = \mu_0, \\ H_1 &: \mu \neq \mu_0. \end{aligned} \tag{3}$$

The definition of power is given by [9]

$$\begin{aligned}\text{Power} &= 1-\beta \\ &= P\{\text{reject } H_0 \mid H_0 \text{ is false}\},\end{aligned} \tag{4}$$

where β is the probability of a Type-II error. In industries, the main goal of using a control chart is to minimize the number of defective products produced prior to the detection of a shift in the process mean [10]. Hence, it is vital to identify and understand the methods that help to increase the power of a control chart so that a quicker detection of process shifts can be attained.

It is found that the subgroup size, n is used in the calculation of the control limits and sample standard deviation and for this reason, n is one of the factors that has a significant impact on the power of a control chart. As n increases, the power of a control chart increases. The application of detection rules can help to enhance the sensitivity of a control chart in detecting process mean shifts. Some of the recent works on detection rules, also called runs rules were made by [11], [12], [13], [14] and [15].

[9] and [16] provided an overall Type-I error probability when more detection rules are used. The overall Type-I error probability is

$$\alpha = 1-\prod_{i=1}^{r}(1-\alpha_i), \tag{5}$$

where
α = Type-I error probability,
r = number of detection rules used,
α_i = Type-I error probability of the ith rule, for $i = 1,2,...,r$.
Note that equation (5) assumes that the r detection rules are independent of one another.

[7] had estimated the probability of the occurrence of a Type-I error in the first k subgroups by using the following four detection rules given in table 1. The four detection rules considered by [7] are as follows:

Rule 1: Process is out-of-control when one or more points fall above the $+3\sigma$ control limit.
Rule 2: Process is out-of-control when 2 of 3 consecutive points fall in the same zone and these points are between the $+2\sigma$ and $+3\sigma$ control limits.
Rule 3: Process is out-of-control when 4 of 5 consecutive points fall in the same zone and these points are beyond the $+1\sigma$ control limits.
Rule 4: Process is out-of-control when 8 consecutive points fall in the same zone, either above or below the CL.

Using more detection rules will increase the sensitivity of a control chart but it will also inflate the overall Type-I error probability.

Table 1. The probabilities of a Type-I error for several combinations of detection rules 1 to 4 [7].

Detection rules	Number of subgroups									
	k = 1	k = 2	k = 3	k = 4	k = 5	k = 6	k = 7	k = 8	k = 9	k = 10
1	0.003	0.005	0.008	0.011	0.013	0.016	0.019	0.021	0.024	0.027
1 and 2	0.003	0.006	0.011	0.015	0.020	0.024	0.028	0.034	0.039	0.043
1, 2 and 3	0.003	0.006	0.011	0.016	0.025	0.032	0.040	0.050	0.060	0.060
1, 2, 3 and 4	0.003	0.006	0.011	0.016	0.025	0.032	0.040	0.060	0.070	0.080

4. Probability table for the power functions of a Shewhart $\overline{X}$ chart

In this section, the method by means of formulae to compute the probabilities of several combinations of detection rules applied on the Shewhart $\overline{X}$ chart will be discussed. The probability table given by [7] will also be discussed.

4.1. Formulae and probability computation for the power functions

The symbols that will be used in the following discussion are defined as follows:

$k =$ number of subgroups.
$PDS(k) =$ probability of detecting an out-of-control signal within k subgroups.
$p_k =$ probability of detecting an out-of-control signal at the kth subgroup.
The four detection rules employed by [7] as stated in Section 3 are still considered in the following discussion:

4.1.1. The use of detection rule 1. Assume that "a" is the probability that an $\overline{X}$ sample falls above the $+3\sigma$ limit.

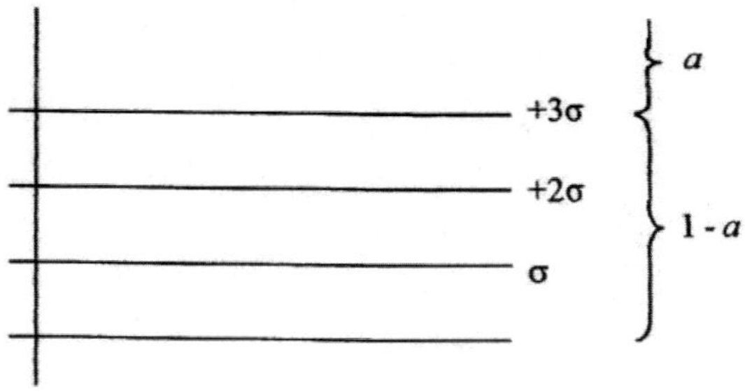

Figure 1. Probabilities of an $\overline{X}$ sample on the Shewhart chart, based on detection rule 1.

A general formula for detecting an out-of-control signal within k subgroups is given by [7]

$$\begin{aligned} PDS(k) &= \sum_{i=1}^{k} p_i \\ &= a + a(1-a) + a(1-a)^2 + a(1-a)^3 + \ldots + a(1-a)^{k-1} \\ &= \sum_{i=1}^{k} a(1-a)^{i-1} \\ &= 1-(1-a)^k . \end{aligned} \tag{6}$$

Note that the case discussed here is only valid for the one sided control chart (upper sided).

4.1.2. The use of detection rules 1 and 2. Assume that
$a =$ the probability that an $\overline{X}$ sample falls above the $+3\sigma$ limit,
$b =$ the probability that an $\overline{X}$ sample falls between the $+2\sigma$ and the $+3\sigma$ limits,
$c =$ the probability that an $\overline{X}$ sample falls between the CL and the $+2\sigma$ limit.

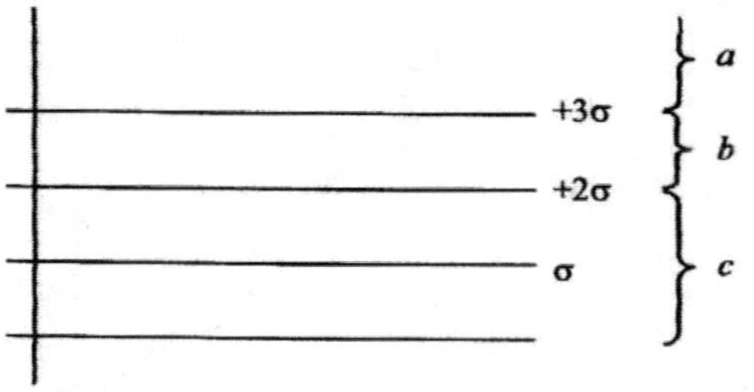

Figure 2. Probabilities of an $\overline{X}$ sample on the Shewhart chart, based on detection rules 1 and 2.

When detection rule 2 is applied, at least 2 subgroups are required in the analysis. The probability obtained will be similar to that in equation (6), except for cases where $k=2$ or more.

For case $k=2$, p_2 and $PDS(2)$ can be computed using [7]

$$p_2 = ca + ba + b^2 \tag{7}$$

and

$$\begin{aligned} PDS(2) &= p_1 + p_2 \\ &= a + ca + ba + b^2 \\ &= a + b^2 + (b+c)a. \end{aligned} \tag{8}$$

For case $k=3$, p_3 and $PDS(3)$ are given by [7]

$$p_3 = 2abc + 2b^2c + ac^2 \tag{9}$$

and

$$\begin{aligned} PDS(3) &= p_1 + p_2 + p_3 \\ &= a + b^2 + (b+c)a + 2abc + 2b^2c + ac^2 \\ &= a + b^2 + ac + ab(1+2c) + 2b^2c + ac^2. \end{aligned} \tag{10}$$

From a similar manner, p_k, for $k = 4, 5, \ldots, 10$ can be obtained using the following equations [7]:

$$p_4 = 3abc^2 + 2b^2c^2 + ac^3. \tag{11}$$

$$p_5 = ab^2c^2 + b^3c^2 + 4abc^3 + 2b^2c^3 + ac^4. \tag{12}$$

$$p_6 = 3ab^2c^3 + 3b^3c^3 + 5abc^4 + 2b^2c^4 + ac^5. \tag{13}$$

$$p_7 = 6ab^2c^4 + 5b^3c^4 + 6abc^5 + 2b^2c^5 + ac^6. \tag{14}$$

$$p_8 = ab^3c^4 + b^4c^4 + 10ab^2c^5 + 5b^3c^5 + 7abc^6 + 4b^2c^6 + ac^7. \tag{15}$$

$$p_9 = 4ab^3c^5 + 4b^4c^5 + 15ab^2c^6 + 9b^3c^6 + 8abc^7 + 2b^2c^7 + ac^8. \tag{16}$$

$$p_{10} = 11ab^3c^6 + 10b^4c^6 + 21ab^2c^7 + 10b^3c^7 + 8abc^8 + 2b^2c^8 + ac^9. \tag{17}$$

4.1.3. The use of detection rules 1, 2 and 3. Assume that
$a =$ the probability that an $\bar{X}$ sample above the $+3\sigma$ limit,
$b =$ the probability that an $\bar{X}$ sample falls between the $+2\sigma$ and the $+3\sigma$ limits,
$c =$ the probability that an $\bar{X}$ sample falls between the $+\sigma$ and the $+2\sigma$ limits,
$d =$ the probability that an $\bar{X}$ sample falls between the CL and the $+\sigma$ limit.

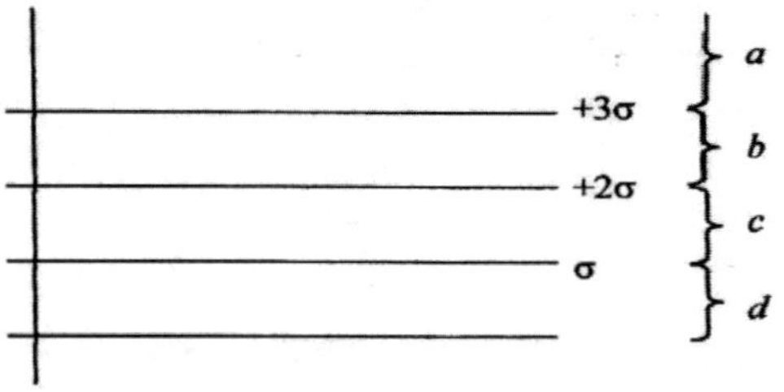

Figure 3. Probabilities of an $\bar{X}$ sample on the Shewhart chart, based on detection rules 1, 2 and 3.

When detection rule 3 is used, at least 4 subgroups are required in the analysis. Thus, the probabilities obtained are similar to that in the equations given in Section 4.1.2, except for cases where $k=4$ or more. When $k \geq 4$, p_k can be computed using the equations as follows [7]:

$$p_4 = 3abc^2 + 6abcd + 3abd^2 + 3ac^2d + 3acd^2 + ac^3 + ad^3 + 3b^2c^2 + 4b^2cd + 2b^2d^2 + c^4 + 4bc^3. \quad (18)$$

$$p_5 = 2ab^2cd + 2b^3cd + 12abc^2d + 14b^2c^2d + 16bc^3d + 4ac^3d + 4c^4d + ab^2d^2 + b^3d^2 + 12abcd^2 + 6b^2cd^2 + 6ac^2d^2 + 4abd^3 + 2b^2d^3 + 4acd^3 + ad^4. \quad (19)$$

$$p_6 = 9ab^2cd^2 + 9b^3cd^2 + 30abc^2d^2 + 20b^2c^2d^2 + 10ac^3d^2 + 16bc^3d^2 + 4c^4d^2 + 3ab^2d^3 + 3b^3d^3 + 20abcd^3 + 8b^2cd^3 + 10ac^2d^3 + 5abd^4 + 2b^2d^4 + 5acd^4 + ad^5. \quad (20)$$

$$P_7 = 19ab^2c^2d^2 + 18b^3c^2d^2 + 24abc^3d^2 + 25b^2c^3d^2 + 15bc^4d^2 + 6ac^4d^2 + 3c^5d^2 + 24ab^2cd^3 + 20b^3cd^3 + 60abc^2d^3 + 28b^2c^2d^3 + 20ac^3d^3 + 16bc^3d^3 + 4c^4d^3 + 6ab^2d^4 + 5b^3d^4 + 30abcd^4 + 10b^2cd^4 + 15ac^2d^4 + 6abd^5 + 2b^2d^5 + 6acd^5 + ad^6. \quad (21)$$

The probabilities p_8, p_9 and p_{10} are computed using Monte Carlo simulation.

4.1.4. The use of detection rules 1, 2, 3 and 4. When detection rule 4 is employed, at least 8 subgroups are needed in the analysis. Thus, the probabilities obtained are similar to that in the equations given in Section 4.1.3, except for cases where $k = 8$ or more. Due to the complexity of the equations, [7] confirmed that the computation needs more than 12,000 combinations to detect a shift and thus, instead of using formulae, the Monte Carlo simulation method will be used to calculate the probabilities for $k = 8$, 9 and 10.

4.2. Probability tables of the power functions

A probability table to show the probability of detecting a sustained shift in the process mean, calculated using the formulae has been recommended by [7]. The characteristics of the probability table include:

- Size of the shift:
  11 sizes of shifts are considered in the probability table. They are 0.42σ, 0.67σ, 0.95σ, 1.25σ, 1.52σ, 1.72σ, 1.96σ, 2.16σ, 2.48σ, 2.75σ and 3.00σ.
- Number of subgroups, k:
  The number of subgroups displayed in the probability table ranges from $k = 1$ to $k = 10$.
- Subgroup size, n:
  The subgroup sizes shown by [7] in the probability table are $n = 1$ to $n = 10$, $n = 12$, $n = 15$ and $n = 20$.
- Detection rules:
  Several combinations of the detection rules have been used to obtain the necessary probabilities.

5. Performance comparison

From the framework of the probability table and the formulae (see also Sections 4.1.1, 4.1.2, 4.1.3 and 4.1.4) given in [7], computer programs are written in the Statistical Analysis System (SAS) software to compute the probabilities of detecting a sustained shift in the mean within certain number of subgroups following the shift. Tables 2 to 9 show the results obtained.

From tables 2 to 9, we notice that

- as the shift size increases, the probability of detecting a process mean shift increases, regardless of the values of n and k, and the combination of the detection rules being used.
- the probability of detecting a process mean shift within k subgroups increases as k increases, for all n, shift sizes and combinations of the detection rules.
- the control chart will become more sensitive if we use more detection rules, for any k, n and sizes of shifts.

Furthermore, if we compare table 2 with table 6, we see that the probabilities in table 6 are generally greater than the corresponding probabilities in table 2. Similar trends are observed if we compare the probabilities in table 3 with table 7, table 4 with table 8 and table 5 with table 9. This means that the power of a control chart increases with an increase in the subgroup size, irrespective of the value of k, size of the shift and detection rules being used.

Table 2. Probabilities of detecting a sustained shift in the process mean within k subgroups, following a shift, for $n = 2$, using detection rule 1 only.

Shift size	$k = 1$	$k = 2$	$k = 3$	$k = 4$	$k = 5$	$k = 6$	$k = 7$	$k = 8$	$k = 9$	$k = 10$
0.42σ	0.010	0.017	0.024	0.033	0.041	0.048	0.056	0.064	0.070	0.078
0.67σ	0.022	0.041	0.058	0.078	0.096	0.116	0.132	0.150	0.165	0.186
1.52σ	0.197	0.351	0.477	0.580	0.665	0.732	0.789	0.827	0.863	0.893
1.72σ	0.289	0.488	0.629	0.735	0.812	0.867	0.902	0.928	0.948	0.963
3.00σ	0.893	0.988	0.999	1.000	1.000	1.000	1.000	1.000	1.000	1.000

Table 3. Probabilities of detecting a sustained shift in the process mean within k subgroups, following a shift, for $n = 2$, using detection rules 1 and 2.

Shift size	$k = 1$	$k = 2$	$k = 3$	$k = 4$	$k = 5$	$k = 6$	$k = 7$	$k = 8$	$k = 9$	$k = 10$
0.42σ	0.010	0.021	0.037	0.054	0.071	0.087	0.104	0.118	0.133	0.147
0.67σ	0.022	0.054	0.097	0.142	0.184	0.219	0.251	0.289	0.320	0.354
1.52σ	0.197	0.481	0.696	0.811	0.877	0.923	0.956	0.973	0.982	0.988
1.72σ	0.289	0.626	0.831	0.910	0.954	0.978	0.989	0.994	0.998	0.999
3.00σ	0.893	0.998	1.000	1.000	1.000	1.000	1.000	1.000	1.000	1.000

Table 4. Probabilities of detecting a sustained shift in the process mean within k subgroups, following a shift, for $n = 2$, using detection rules 1, 2 and 3.

Shift size	$k = 1$	$k = 2$	$k = 3$	$k = 4$	$k = 5$	$k = 6$	$k = 7$	$k = 8$	$k = 9$	$k = 10$
0.42σ	0.010	0.021	0.037	0.064	0.112	0.143	0.171	0.196	0.227	0.260
0.67σ	0.022	0.054	0.097	0.172	0.285	0.348	0.403	0.462	0.507	0.558
1.52σ	0.197	0.481	0.696	0.882	0.960	0.980	0.991	0.997	0.998	0.999
1.72σ	0.289	0.626	0.831	0.951	0.991	0.997	0.999	1.000	1.000	1.000
3.00σ	0.893	0.998	1.000	1.000	1.000	1.000	1.000	1.000	1.000	1.000

Table 5. Probabilities of detecting a sustained shift in the process mean within k subgroups, following a shift, for $n = 2$, using detection rules 1, 2, 3 and 4.

Shift size	$k = 1$	$k = 2$	$k = 3$	$k = 4$	$k = 5$	$k = 6$	$k = 7$	$k = 8$	$k = 9$	$k = 10$
0.42σ	0.010	0.021	0.037	0.064	0.112	0.143	0.171	0.233	0.275	0.320
0.67σ	0.022	0.054	0.097	0.172	0.285	0.348	0.403	0.541	0.593	0.643
1.52σ	0.197	0.481	0.696	0.882	0.960	0.980	0.991	0.999	1.000	1.000
1.72σ	0.289	0.626	0.831	0.951	0.991	0.997	0.999	1.000	1.000	1.000
3.00σ	0.893	0.998	1.000	1.000	1.000	1.000	1.000	1.000	1.000	1.000

Table 6. Probabilities of detecting a sustained shift in the process mean within k subgroups, following a shift, for $n = 10$, using detection rule 1 only.

Shift size	$k=1$	$k=2$	$k=3$	$k=4$	$k=5$	$k=6$	$k=7$	$k=8$	$k=9$	$k=10$
0.42σ	0.047	0.090	0.132	0.173	0.210	0.253	0.287	0.318	0.350	0.383
0.67σ	0.189	0.340	0.466	0.568	0.652	0.715	0.769	0.810	0.843	0.872
1.52σ	0.967	0.999	1.000	1.000	1.000	1.000	1.000	1.000	1.000	1.000
1.72σ	0.993	1.000	1.000	1.000	1.000	1.000	1.000	1.000	1.000	1.000
3.00σ	1.000	1.000	1.000	1.000	1.000	1.000	1.000	1.000	1.000	1.000

Table 7. Probabilities of detecting a sustained shift in the process mean within k subgroups, following a shift, for $n = 10$, using detection rules 1 and 2.

Shift size	$k=1$	$k=2$	$k=3$	$k=4$	$k=5$	$k=6$	$k=7$	$k=8$	$k=9$	$k=10$
0.42σ	0.047	0.132	0.234	0.317	0.388	0.452	0.519	0.572	0.619	0.656
0.67σ	0.189	0.466	0.682	0.797	0.876	0.921	0.950	0.966	0.979	0.988
1.52σ	0.967	1.000	1.000	1.000	1.000	1.000	1.000	1.000	1.000	1.000
1.72σ	0.993	1.000	1.000	1.000	1.000	1.000	1.000	1.000	1.000	1.000
3.00σ	1.000	1.000	1.000	1.000	1.000	1.000	1.000	1.000	1.000	1.000

Table 8. Probabilities of detecting a sustained shift in the process mean within k subgroups, following a shift, for $n = 10$, using detection rules 1, 2 and 3.

Shift size	$k=1$	$k=2$	$k=3$	$k=4$	$k=5$	$k=6$	$k=7$	$k=8$	$k=9$	$k=10$
0.42σ	0.047	0.132	0.234	0.389	0.566	0.645	0.715	0.771	0.817	0.855
0.67σ	0.189	0.466	0.682	0.868	0.957	0.979	0.990	0.995	0.998	0.999
1.52σ	0.967	1.000	1.000	1.000	1.000	1.000	1.000	1.000	1.000	1.000
1.72σ	0.993	1.000	1.000	1.000	1.000	1.000	1.000	1.000	1.000	1.000
3.00σ	1.000	1.000	1.000	1.000	1.000	1.000	1.000	1.000	1.000	1.000

Table 9. Probabilities of detecting a sustained shift in the process average within k subgroups, following a shift, for $n = 10$, using detection rules 1, 2, 3 and 4.

Shift size	$k=1$	$k=2$	$k=3$	$k=4$	$k=5$	$k=6$	$k=7$	$k=8$	$k=9$	$k=10$
0.42σ	0.047	0.132	0.234	0.389	0.566	0.645	0.715	0.846	0.883	0.914
0.67σ	0.189	0.466	0.682	0.868	0.957	0.979	0.990	0.999	0.999	0.999
1.52σ	0.967	1.000	1.000	1.000	1.000	1.000	1.000	1.000	1.000	1.000
1.72σ	0.993	1.000	1.000	1.000	1.000	1.000	1.000	1.000	1.000	1.000
3.00σ	1.000	1.000	1.000	1.000	1.000	1.000	1.000	1.000	1.000	1.000

To compare between Wheeler's formulae method [7] and the Monte Carlo simulation method, the percentage of difference is computed. Comparisons in terms of percentages are made for each probability table using the following formula:

$$\text{Percentage of difference} = \left|\frac{s-t}{t}\right| \times 100\%, \tag{22}$$

where
s = probability obtained from Monte Carlo simulation,
t = probability calculated using formula.
The results obtained are shown in tables 10 to 17. From tables 10 to 17, it is obvious that the probabilities obtained by Monte Carlo simulation are closed to that obtained by means of formulae.

The boldfaced entries in tables 10 to 17 denote cases, where differences in the probabilities between the two methods exist. Note that the differences in the probabilities between the two methods are small.

Table 10. Percentages of difference in probabilities, for subgroup size, $n = 2$ using detection rule 1 only.

Shift size	$k = 1$	$k = 2$	$k = 3$	$k = 4$	$k = 5$	$k = 6$	$k = 7$	$k = 8$	$k = 9$	$k = 10$
0.42σ	**20%**	**5%**	**2%**	**3%**	**1%**	**3%**	**3%**	**1%**	0%	0%
0.67σ	**9%**	**1%**	**2%**	**1%**	0%	**1%**	0%	0%	**1%**	**2%**
1.52σ	**1%**	**1%**	**1%**	**1%**	0%	0%	0%	0%	0%	0%
1.72σ	**1%**	0%	**1%**	**1%**	0%	0%	0%	0%	0%	0%
3.00σ	0%	0%	0%	0%	0%	0%	0%	0%	0%	0%

Table 11. Percentages of difference in probabilities, for subgroup size, $n = 2$ using detection rules 1 and 2.

Shift size	$k = 1$	$k = 2$	$k = 3$	$k = 4$	$k = 5$	$k = 6$	$k = 7$	$k = 8$	$k = 9$	$k = 10$
0.42σ	**20%**	**2%**	**5%**	**1%**	**1%**	0%	**2%**	**4%**	**4%**	**3%**
0.67σ	**9%**	**3%**	**5%**	**1%**	**1%**	0%	**1%**	**3%**	**3%**	**2%**
1.52σ	**1%**	**1%**	**1%**	0%	**1%**	0%	0%	0%	0%	0%
1.72σ	**1%**	**1%**	**1%**	**1%**	0%	0%	0%	0%	0%	0%
3.00σ	0%	0%	0%	0%	0%	0%	0%	0%	0%	0%

Table 12. Percentages of difference in probabilities, for subgroup size, $n = 2$ using detection rules 1, 2 and 3.

Shift size	$k = 1$	$k = 2$	$k = 3$	$k = 4$	$k = 5$	$k = 6$	$k = 7$	$k = 8$	$k = 9$	$k = 10$
0.42σ	**20%**	**2%**	**5%**	**2%**	**3%**	**1%**	**1%**	**2%**	**1%**	0%
0.67σ	**9%**	**3%**	**5%**	**2%**	0%	**1%**	**2%**	**2%**	**3%**	**2%**
1.52σ	**1%**	**1%**	**1%**	0%	0%	0%	0%	0%	0%	0%
1.72σ	**1%**	**1%**	**1%**	0%	0%	0%	0%	0%	0%	0%
3.00σ	0%	0%	0%	0%	0%	0%	0%	0%	0%	0%

Table 13. Percentages of difference in probabilities, for subgroup size, $n = 2$ using detection rules 1, 2, 3 and 4.

Shift size	$k = 1$	$k = 2$	$k = 3$	$k = 4$	$k = 5$	$k = 6$	$k = 7$	$k = 8$	$k = 9$	$k = 10$
0.42σ	**20%**	**2%**	**5%**	**2%**	**3%**	**1%**	**1%**	**7%**	**5%**	0%
0.67σ	**9%**	**3%**	**5%**	**2%**	0%	**1%**	**2%**	**2%**	**1%**	**3%**
1.52σ	**1%**	**1%**	**1%**	0%	0%	0%	0%	0%	0%	0%
1.72σ	**1%**	**1%**	**1%**	0%	0%	0%	0%	0%	0%	0%
1.96σ	0%	**1%**	0%	0%	0%	0%	0%	0%	0%	0%
3.00σ	0%	0%	0%	0%	0%	0%	0%	0%	0%	0%

Table 14. Percentages of difference in probabilities, for subgroup size, $n = 10$ using detection rule 1 only.

Shift size	$k = 1$	$k = 2$	$k = 3$	$k = 4$	$k = 5$	$k = 6$	$k = 7$	$k = 8$	$k = 9$	$k = 10$
0.42σ	0%	**3%**	**2%**	**2%**	**2%**	0%	0%	**1%**	**1%**	0%
0.67σ	0%	**1%**	0%	0%	0%	0%	0%	0%	**1%**	**1%**
1.52σ	0%	0%	0%	0%	0%	0%	0%	0%	0%	0%
1.72σ	0%	0%	0%	0%	0%	0%	0%	0%	0%	0%
1.96σ	0%	0%	0%	0%	0%	0%	0%	0%	0%	0%

Table 15. Percentages of difference in probabilities, for subgroup size, $n = 10$ using detection rules 1 and 2.

Shift size	$k = 1$	$k = 2$	$k = 3$	$k = 4$	$k = 5$	$k = 6$	$k = 7$	$k = 8$	$k = 9$	$k = 10$
0.42σ	0%	**1%**	**1%**	**1%**	**1%**	**1%**	**1%**	**1%**	**1%**	**1%**
0.67σ	0%	**1%**	**1%**	0%	**1%**	0%	0%	0%	0%	0%
1.52σ	0%	0%	0%	0%	0%	0%	0%	0%	0%	0%
1.72σ	0%	0%	0%	0%	0%	0%	0%	0%	0%	0%
1.96σ	0%	0%	0%	0%	0%	0%	0%	0%	0%	0%

Table 16. Percentages of difference in probabilities, for subgroup size, $n = 10$ using detection rules 1, 2 and 3.

Shift size	$k = 1$	$k = 2$	$k = 3$	$k = 4$	$k = 5$	$k = 6$	$k = 7$	$k = 8$	$k = 9$	$k = 10$
0.42σ	0%	**1%**	**1%**	0%	0%	**1%**	**1%**	0%	0%	**1%**
0.67σ	0%	**1%**	**1%**	0%	0%	0%	0%	**1%**	0%	0%
1.52σ	0%	0%	0%	0%	0%	0%	0%	0%	0%	0%
1.72σ	0%	0%	0%	0%	0%	0%	0%	0%	0%	0%
1.96σ	0%	0%	0%	0%	0%	0%	0%	0%	0%	0%

Table 17. Percentages of difference in probabilities, for subgroup size, $n = 10$ using detection rules 1, 2, 3 and 4.

Shift size	$k = 1$	$k = 2$	$k = 3$	$k = 4$	$k = 5$	$k = 6$	$k = 7$	$k = 8$	$k = 9$	$k = 10$
0.42σ	0%	**1%**	**1%**	0%	0%	**1%**	**1%**	**1%**	0%	0%
0.67σ	0%	**1%**	**1%**	0%	0%	0%	0%	0%	0%	0%
1.52σ	0%	0%	0%	0%	0%	0%	0%	0%	0%	0%
1.72σ	0%	0%	0%	0%	0%	0%	0%	0%	0%	0%
1.96σ	0%	0%	0%	0%	0%	0%	0%	0%	0%	0%

6. Conclusions

The Shewhart chart is the simplest and most popular process monitoring tool in SPC. In view of this, tables 2 – 17 are provided to help practitioners in a quick and effective design and implementation of the Shewhart chart, for a more efficient process monitoring system. The approach by means of formulae is extremely complicated, where it becomes impossible when detection rule 4 is employed (see Section 4.1.4). The results of this study given in tables 2 – 17 show that the Monte Carlo simulation method is an easier and quicker way to calculate the probabilities of detecting a sustained shift in the process mean within k subgroups, for the various detection rules, as compared to the calculation of these probabilities using the formulae proposed by [7]. On the contrary, the Monte Carlo simulation method is not only simple, but it is also accurate and less time consuming. Furthermore, a higher flexibility in obtaining the probabilities for different combinations of the detection rules can be achieved by means of the Monte Carlo simulation method.

Acknowledgement
This research is supported by the Universiti Sains Malaysia, Fundamental Research Grant Scheme (FRGS), no. 203/PMATHS/6711232.

References

[1] Berger R W and Hart T 1986 *Statistical Process Control, A guide for Implementation* (Marcel Dekker, Inc)
[2] Grant E L and Leavenworth R S 1980 *Statistical Quality Control* 5th ed (New York: McGraw-Hill Book Company)
[3] Lim T J and Cho M 2009 Design of control charts with m-of-m m runs rules *Qual. Reliab. Eng. Int.* **25(8)** 1085-1101
[4] Kim Y B, Hong J S and Lie C H 2009 Economic-statistical design of 2-of-2 and 2-of-3 runs rule scheme *Qual. Reliab. Eng. Int.* **25(2)** 215-228
[5] Zarandi M H F, Alaeddini A and Turksen I B 2008 A hybrid fuzzy adaptive sampling - run rules for Shewhart control charts *Inf. Scien.* **178(4)** 1152-70
[6] Antzoulakos D L and Rakitzis C A 2008 The modified r out of m control chart *Commun. Stat. - Simul. Comput.* **37(2)** 396-408
[7] Wheeler D J 1983 Detecting a shift in process average - tables of the power function for X charts *J. Qual. Technol.* **15(4)** 155-170
[8] Ryan T P 1989 *Statistical Methods for Quality Improvement.* (New York: John Wiley & Sons)
[9] Montgomery D C 2001 *Introduction to Statistical Quality Control* 4th ed (New York: John Wiley & Sons)
[10] Guldner F J 1986 *Statistical Quality Assurance.* (Delmar Publishers Inc)
[11] Low C K, Khoo M B C, Teoh W L and Wu Z 2012 The revised m-of-k runs rule based on median run length *Commun. Stat. - Simul. Comput.* **41(8)** 1463-77
[12] Antzoulakos D L and Rakitzis C A 2008 The revised m-of-k runs rule *Qual. Eng.* **20(1)** 175-181
[13] Khoo M B C and Ariffin K N 2006 Two improved runs rules for the Shewhart $\overline{X}$ control chart *Qual. Eng.* **18(2)** 173-178
[14] Khoo M B C 2003 Design of runs rules schemes *Qual. Eng.* **16(1)** 27-43
[15] Klein M 2000 Two alternatives to the Shewhart $\overline{X}$ control chart *J. Qual. Technol.* **32(4)** 427-431
[16] Duncan A J 1986 *Quality Control and Industrial Statistics* 5th ed (Homewood, IL: Richard D Irwin)

Analysis of Fluconazole in Human Urine Sample by High Performance Liquid Chromatography Method

D Hermawan [1,2], N A Md Ali [1], W A Wan Ibrahim[1], M M Sanagi[1]

[1] Separation Science and Technology Group, Ibnu Sina Institute for Fundamental Sciences Studies, Faculty of Science, Universiti Teknologi Malaysia, 81310 Johor Bahru, Malaysia
[2] Chemistry Study Program, Faculty of Science and Engineering, Jenderal Soedirman University, Purwokerto, Indonesia

[*]Email: dadan@ibnusina.utm.my ; dadanphd@gmail.com

Abstract. A method for determination of fluconazole, antifungal drug in human urine by using reversed-phased high performance liquid chromatography (RP-HPLC) with ultraviolet (UV) detector was developed. Optimization HPLC conditions were carried out by changing the flow rate and composition of mobile phase. The optimum separation conditions at a flow rate 0.85 mL/min with a composition of mobile phase containing methanol:water (70:30, v/v) with UV detection at a wavelength 254 nm was able to analyze fluconazole within 3 min. The excellent linearity was obtained in the range of concentration 1 to 10 µg/mL with $r^2 = 0.998$. The limit of detection (LOD) and limit of quantitation (LOQ) were 0.39 µg/mL and 1.28 µg/mL, respectively. Solid phase extraction (SPE) method using octadecylsilane (C18) as a sorbent was used to clean-up and pre-concentrated of the urine sample prior to HPLC analysis. The average recoveries of fluconazole in spiked urine sample was 72.4% with RSD of 3.21% (n=3).

Keywords: Antifungal drug, Fluconazole, Human urine, RP-HPLC, SPE

---

[1] Separation Science and Technology Group, Ibnu Sina Institute for Fundamental Sciences Studies, Faculty of Science, Universiti Teknologi Malaysia, 81310 Johor Bahru, Malaysia
[2] Chemistry Study Program, Faculty of Science and Engineering, Jenderal Soedirman University, Purwokerto, Indonesia

1. Introduction

An antifungal drug had widely used as a medication to treat fungal infection such as a ringworm, candiasis, foot fungus and many others. Until 1970, antifungal agents had many problems included toxicity, narrow activity spectrum or low pharmacological activity. But, the patients who acquire immunodeficiency syndrome (AIDS), undergoing anti-cancer chemotherapy or having suffered an organ transplant are necessary to take effective drug for the bacterial and fungal disease treatment. It is because the immune system of the cancer patients who is receiving chemotherapy usually at high risk of developing invasive fungal infections. Hence, this has resulting in increasing clinical demand for newer and safer antifungal drugs which arrived infection focus, efficient and lower toxicity [1].

Fluconazole is a former synthetic triazole antifungal agents which available for over two decades that safe and effective in cancer patients. According to the clinical pharmacokinetics of fluconazole, it has been indicated effective at wide range of body sites because of its large distribution volume, long half life and also rapid absorption after oral administration. In the pharmaceutical analysis, high performance liquid chromatography (HPLC) is the most extensively technique used [2-5]. The application of modern HPLC toward quantification of various compounds is very applicable because it can perform a separation, identification and also purification. The analysis using HPLC also perform in a shorter time and the most important thing that HPLC is characterized by high resolution, selectivity, sensitivity, precision and accuracy. In this report, solid phase extraction (SPE) coupled with HPLC was successfully applied to the determination of fluconazole in human urine sample.

2. Material and Method

2.1. Chemical and Reagents

Fluconazole (99.1% purity) was purchased from Sigma (St Louis, MO, USA). Methanol (HPLC) grade was purchased from J. T. Baker, USA. Purified water was collected from a Millipore Simplicity (Simpak®2). All solvents are filtered through 0.45 μm membrane prior to usage. Human urine samples were collected from healthy drug-free volunteers.

2.2. Chromatographic Instrumentation

The HPLC system was an Agilent 1100 Series (Germany) and consisted of a Model Agilent 1100 pump, an on-line solvent vacuum degasser, an auto sampler with 20 μL injection loop, and an Agilent 1100 Series UV detector. The analytical column LichroCART® C_{18} (150 mm x 4.6 mm x 5 μm particles) was obtained from Germany. The Agilent Chem Station software was used to record chromatograms and calculate peak area. The mobile phase consisted of a mixture of methanol and water (70:30 v/v). The HPLC system was operated isocratically at a flow rate of 0.85 mL/min and the effluent of the column was monitored at a wavelength of 254 nm. An Agilent 1100 Series UV Spectrophotometer (Germany) was used to record the absorption maxima of analyte.

2.3. Preparation of Standards

Stock solution (2000 μg/mL) of fluconazole was prepared by dissolving ten mg accurately weighed fluconazole in methanol as a solvent. For the quantitation of fluconazole, an appropriate standard working solution 100, 50, 20 and 10 μg/mL were prepared by diluting a stock solution in methanol. Serial dilutions from 10 μg/mL of standard working solution were made to obtain four different concentrations in the range of 0.5 to 10 μg/mL using methanol. Three replicates for each concentration were used for calibration standards to determine the limit of detection, linearity range and correlation coefficients. The stock and standard working solutions were sealed and stored in the refrigerator at temperature 5°C prior to use.

2.4. Sample Preparation

The sample preparation involved a spiking the human urine sample with standard fluconazole solutions. Sample was spiked with 5 ppm of standard fluconazole after the blank sample was confirmed that no fluconazole was detected in the sample using HPLC-UV. For the sample pre-treatment, the human urine was filtered through a Buchner funnel prior extraction with solid phase extraction and the spiked sample was added with 1.0 M sodium hydroxide to maintain the basic condition of drug. After that, the spiked sample human urine was introduced to the C_{18} extraction cartridge. The eluent was reconstituted with 500 μL of methanol and 20 μL was injected into HPLC-UV for analysis.

2.5. Extraction Procedure

The isolation of fluconazole from human urine was conducted by using solid phase extraction (SPE). The SPE C_{18} cartridge with a capacity of 3 mL was placed in a 12-port of SPE vacuum manifold system. The vacuum pressure was adjusted to -0.5 Hg/bar and each column was activated by preconditioning with 5 mL of methanol followed by 5 mL of deionized water. The approximately 1.0 mL of spiked sample was introduced into the cartridge under a vacuum at a flow rate 0.5 mL/min. The solid phase in the cartridge was not allowed to become dry at any moment. Afterwards, analyte was eluted with 3 mL methanol. The eluent was concentrated under a stream of nitrogen to dryness until all the solvent evaporated. Finally, the analyte residue was dissolved in 500 μL methanol and aliquot (200 μL) was injected into HPLC-UV. Analysis of a blank sample and spiked sample were performed for a comparison purpose.

3. Results and Discussions

3.1 Effect of Mobile Phase

Addition of organic modifier increase the strength of eluting, varying it composition will affect the elution time for analyte. The results shows that retention time will be longer when the percentage of water higher than organic modifier or in other word, retention time of analyte will be shorter with increasing of organic modifier. Fluconazole was a polar compound will have a strong interaction with the high percentage of organic solvent which was methanol. Methanol lowering the polarity of mobile phase cause the analyte less retained in mobile phase. Analyte will be eluted faster and resulting in the shorter retention time.

A column with a high number of theoretical plates will have a narrower peak at a given retention time than a column with a lower N number. The effect of organic modifier (methanol) on some basic compound was gave a significantly better peak shape [6]. This result may become significance in this study because the properties of fluconazole which was also a basic compound. The higher percentage of organic modifier in mobile phase may results a narrow peak. The sharp peak, which present in the density profiles of water, result from solvent molecules directly hydrogen bonding with surface silanols. The result obtained, the peak shape of analyte will become narrower due to high value of N when the percentage of water was increased. This result was different from the literature stated above and it was due to a different protonation of analyte in mobile phase which gave a possible influential factor of different in a peak shape.

3.2 Effect of Flow Rate

Flow rates of mobile phase have important effect on the analyte retention time. Higher flow rates lead to a shorter retention time and vice versa. The result shows that the retention time was decrease as the mobile phase flow rate was increased. A further improvement in separation may be possible by varying column condition such as flow rate to improve the column plate number, N. An increase in N leads to an increase in resolution and usually longer run time. Conversely, a decrease in N can provide a shorter run time. The result shows that the high flow rate was decreased the value of N which means that the resolution was not good and gave a slightly broad peak.

3.3 Linearity and Recovery
The HPLC method assay for determination of fluconazole was linear from 1 to 10 µg/mL and show a good correlation coefficients, r^2 0.998. The limit of detection and limit of quantification of fluconazole was 0.39 µg/mL and 1.28 µg/mL respectively. The low limit of detection showed the method developed is sensitive and sufficient to determine fluconazole in urine samples for routine analysis.

The absolute recovery of fluconazole from human urine sample was estimated by comparing the peak height obtained from chromatogram of standards with those obtained by the spiked urine sample with known concentration of fluconazole (figure 1). The recovery of fluconazole obtained in urine sample was 72.4% with RSD 3.27 % (n=3).

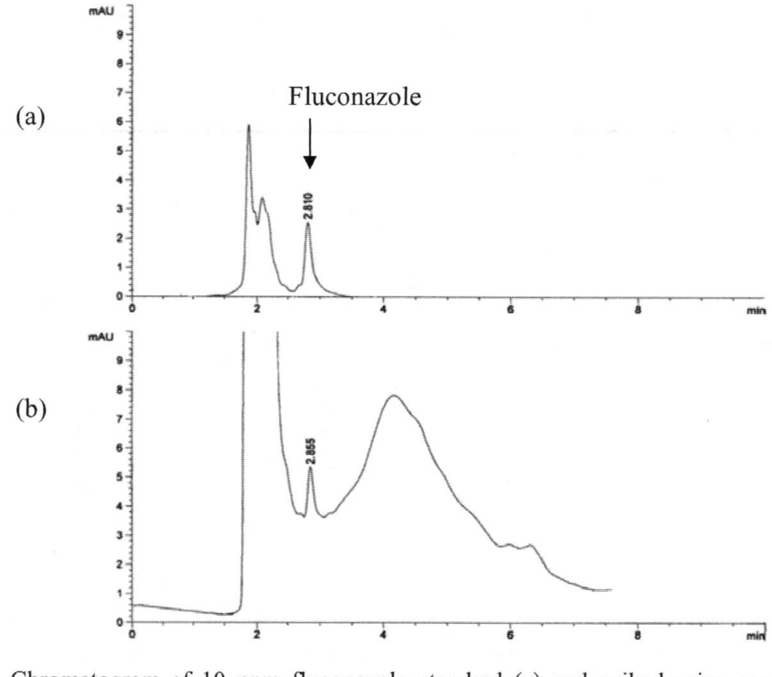

Figure 1. Chromatogram of 10 ppm fluconazole standard (a) and spiked urine sample after SPE procedure (b). HPLC conditions: methanol:water (70:30,v/v) as mobile phase, flow rate of 0.85 mL/min and UV detection at 254 nm. Column: LichroCART® C_{18} (150 mm x 4.6 mm x 5 µm particles).

4. Conclusions
In this study, high performance liquid chromatography (HPLC) method for analysis of fluconazole in human urine has been developed. The HPLC system was performed with C_{18} (150 mm x 4.6 mm x 5 µm particles) column using methanol:water (70:30,v/v) as a optimum mobile phase. The calibration graph was obtained in the range of 1 to 10 ppm. The correlation coefficient from the graph is 0.998. The limit of detection (LOD) for fluconazole is 0.39 ppm while limit of quantifying (LOQ) is 1.28 ppm. Solid phase extraction (SPE) using octadecylsilane (C_{18}) was used for sample clean-up and pre-concentration method. The SPE coupled with HPLC-UV method has been successfully applied to determination of fluconazole in urine sample with percentage of recovery = 72.4% (RSD = 3.27%) with relatively shorter analysis time (3 min) compared to the previous method (10 min) [7].

References

[1] Crego A L *et al* 2001 *J. Chromatogr. A* **917** 337
[2] Koks C H W *et al* 1995 *J. Chromatogr. B* **663** 345
[3] Cociglio M *et al* 1996 *J. Chromatogr. B* **686** 11
[4] Compas D *et al* 1996 *J. Chromatogr. B* **687** 453
[5] Gubbins P O *et al* 1998 *J. Pharm. Biomed. Anal* **16** 1005
[6] McCalley D V 1995 *J. Chromatogr. A* **708** 185
[7] Baranowska I *et al* 2006 *Anal. Chim. Act* **570** 46

ScieTech 2013 IOP Publishing
Journal of Physics: Conference Series **423** (2013) 012010 doi:10.1088/1742-6596/423/1/012010

The Adequateness of Wavelet Based Model for Time Series

Rukun S[1], Subanar[2], Dedi Rosadi[2] and Suhartono[3]

[1] Statistics Department of Undip Semarang
[2] Mathematics Department of UGM Yogyakarta
[3] Statistics Department of ITS Surabaya

E-mail: rukuns@yahoo.com, subanar@yahoo.com, dedirosadi@ugm.ac.id, gmsuhartono@gmail.com

Abstract. In general, time series is modeled as summation of known information i.e. historical information components, and unknown information i.e. random component. In wavelet based model, time series is represented as linear model of wavelet coefficients. Wavelet based model captures the time series feature perfectly when the historical information components dominate the process. In other hand, it has low enforcement when the random component dominates the process. This paper proposes an effort to develop the adequateness of wavelet based model, when the random component dominated the process. By weighted summation, the data is carried to the new form which has higher dependencies. Consequently, wavelet based model will work better. Finally, it is hoped that the better prediction of wavelet based model will be carried to the original prediction in reverting process.

1. Introduction

George E. P. Box and Gwilym M. Jenkins have become the pioneers of time series analyses with their famous book *Time Series Analysis: Forecasting and Control* (1976). In general the model is called an Autoregressive Integrated Moving Average model (ARIMA). Box-Jenkins method gives a satisfied approximation and forecasting models when the process is in stationary condition. In the opposite, this model can't capture non stationary behaviors well. Engel [6] has proposed a parametric model for non stationary time series which is called Autoregressive Conditional Heteroskedasticity model (ARCH). Bollerslev [2] develops ARCH model become to Generalized Autoregressive Conditional Heteroscedasticity model (GARCH). Almost all economics data as well as stock trading, money exchange and commodity trading are non stationer and non linear. So, it is interesting to explore some solution on non stationer and non linear phenomenon.

Beside of parametric models, there are also available some models in non parametric sense. This method not need special assumption about population distribution, so that easier to be practiced. In general non parametric methods are chosen by the reason of practicability. In 1999, Haykin [8] wrote a useful book about neural network (NN) methods. Some applications of NN had be pointed by Samarasinghe [11]. Wavelet based method becomes the recently non parametric model for time series (Murtagh et al. [9]). Ciancio [4] pointed the use of discrete wavelet transform for time series modeling. Rukun et al. [12] give a consideration of wavelet method for heteroscedastic time series modeling. In the fact, wavelet model captures time series feature well when the historical information components dominate the process. Pre-processing and post-processing are needed for adequateness of wavelet based model when the random component dominates the process. This solution will be discussed in this paper. Furthermore, all of these non parametric models must be done along with intensives computation. R system is used as the main computation tool in this paper, and supported with wavelet package by Aldrich [1].

Let $\{X_t\}_{t \in Z}$ is representing a time series data. The value at time $t+1$, that is X_{t+1}, depends to historical information component and a random component as described in equation (1).

$$X_t = f(X_{t-i}, \epsilon_{t-j}) + \epsilon_t, \quad i = 1, 2, .., p \quad j = 1, 2, .., q \tag{1}$$

For instance, the model of AR(1) indicate that the value at time t depends to a historical data X_{t-1} and a random component ϵ_t as mentioned in equation (2).

$$X_t = \phi X_{t-1} + \epsilon_t, \quad -1 < \phi < 1, \quad \epsilon_t \sim N(0, \sigma^2) \tag{2}$$

It has been understood that, the random component will dominate the process when $|\phi|$ decreasing tends to 0 and then the process tends to stationer or white noise process (see figure 1). In other hand, the historical information component will dominate the process when $|\phi|$ tends to 1 , so that the process tends to non stationer (see figure 2). Capability of wavelet based model for capturing time series feature will be discussed in section 3. It will be seen that this feature will be captured by wavelet perfectly when historical information component dominates the process. The capability will decrease when the random component dominates the process. Furthermore, the effort for wavelet based model capability development will be discussed in section 4.

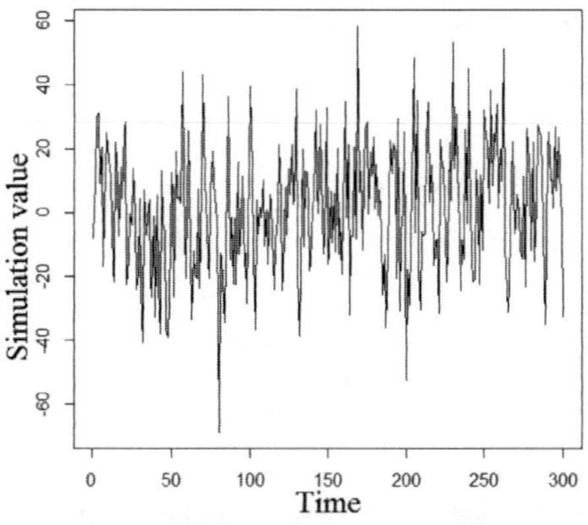

Figure 1. Simulation Data of AR(1) for $\phi = 0.2$ and $\sigma^2 = 400$.

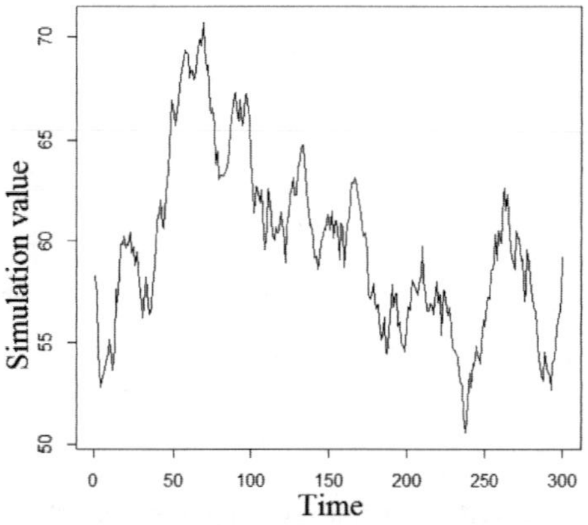

Figure 2. Simulation Data of AR(1) for $\phi = 0.9999$ and $\sigma^2 = 1$

There are some measurements which can be used for selecting the adequate model of time series estimation. Wei [14] pointed that model selection can be done based on forecast residuals. For instances, Mean Absolute Error (MAE) and Mean Absolute Percentage Error (MAPE) as represented in equation (3) and equation (4).

$$MAE = \frac{1}{n} \sum_{t=1}^{n} |\epsilon_t| \tag{3}$$

$$MAPE = \left(\frac{1}{n}\sum_{t=1}^{n}\left|\frac{\epsilon_t}{X_t}\right|\right)100\% \tag{4}$$

It can be seen in section 3 that the domination of random component will increase together with MAPE values. It is hoped that MAPE value can be decreased by a method which is proposed in section 4.

2. Wavelet and Filtering

Wavelet is a small wave function which can build a basis for $L_2(R)$, so that every function $f \in L_2(R)$ can be expressed by a linear combination of wavelets as described in equation (5).

$$\begin{aligned} f(t) &= \sum_{k\in Z} c_{J,k}\phi_{J,k}(t) + \sum_{j\leq J}\sum_{k\in Z} w_{j,k}\psi_{j,k}(t) \\ &= S_J + D_J + D_{J-1} + \ldots + D_1 \end{aligned} \tag{5}$$

where ϕ and ψ is a father and mother wavelet respectively with dilation and translation indexes

$$\phi_{j,k}(t) = 2^{-\frac{j}{2}}\phi(2^{-j}t - k) \tag{6}$$

$$\psi_{j,k}(t) = 2^{-\frac{j}{2}}\psi(2^{-j}t - k) \tag{7}$$

In the discrete version, wavelet can construct filters so that every discrete realizations of $f \in L_2(R)$ can be transformed into decomposition form of scaling component or smooth component (S) and detail components (Daubechies [5]). This process is called discrete wavelet transform (DWT). There are two kinds of discrete wavelet transforms, i.e. decimated discrete wavelet transforms (DWT) and undecimated wavelet transforms (UDWT). Percival and Walden [10] pointed that UDWT has some superiorities in time series analyses compared to DWT. UDWT has shift invariant properties, so that the choice of starting point in time series is not influencing the result of transformation. In every transformation level, it produces wavelet coefficients length which is equal to the length of time series. Finally, it is able to be applied to arbitrary length of data series.

Let $\mathbf{u} = [u_0, u_1, \ldots, u_{L-1}]$ is a DWT wavelet filter. The DWT scaling filter $\mathbf{v}$ can be derived by formulation appeared in equation (8)

$$v_i = (-1)^{i+1}u_{L-1-i} \tag{8}$$

For example the Haar filter and its scaling filter can be described in equation (9)

$$\mathbf{u} = \left[-\frac{1}{\sqrt{2}}, \frac{1}{\sqrt{2}}\right], \quad \mathbf{v} = \left[\frac{1}{\sqrt{2}}, \frac{1}{\sqrt{2}}\right] \tag{9}$$

Furthermore, wavelet filters for UDWT are symbolized by $\mathbf{h}$ and $\mathbf{g}$. The relationship between DWT filters and UDWT filters can be formulated as $\mathbf{h} = \mathbf{u}/\sqrt{2}$ and $\mathbf{g} = \mathbf{v}/\sqrt{2}$. It means the Haar filters for UDWT are $\mathbf{h} = [-\frac{1}{2}, \frac{1}{2}]$, and $\mathbf{g} = [\frac{1}{2}, \frac{1}{2}]$.

Suppose $\mathbf{X} = \{X_t\}$ is a time series which is constructed from discrete time realizations of $f \in L_2(R)$. The scenario of UDWT at $j = 1$ can be shown in Figure 3 as found in Fugal [7]. The wavelet coefficient $\mathbf{w}_{1,k}$ is resulted by convolution of time series $\mathbf{X}$ and $\mathbf{h}$. The first detail component $\mathbf{D_1}$ is resulted by convolution of $\mathbf{w}_{1,k}$ and $\mathbf{h}'$ where $\mathbf{h}'$ is time reverse version of $\mathbf{h}$. The scaling coefficient $\mathbf{c}_{1,k}$ is resulted by convolution of $\mathbf{X}$ and $\mathbf{g}$. The first smooth component $\mathbf{S_1}$ is resulted by convolution of $\mathbf{c}_{1,k}$ and $\mathbf{g}'$ where $\mathbf{g}'$ is time reverse version of $\mathbf{g}$. Furthermore $\hat{\mathbf{X}} = \mathbf{S_1} + \mathbf{D_1}$ will equal to $\mathbf{X}$ regard to wavelet filtering. A higher level of UDWT can be constructed by splitting the scaling coefficient $\mathbf{c}_{j,k}$ into $\mathbf{c}_{j+1,k}$ and $\mathbf{w}_{j+1,k}$. The UDWT algorithm at j=2 can be shown in Figure 4. Furthermore $\hat{\mathbf{X}} = \mathbf{S_2} + \mathbf{D_2} + \mathbf{D_1}$ will equal to $\mathbf{X}$ regard to wavelet filtering.

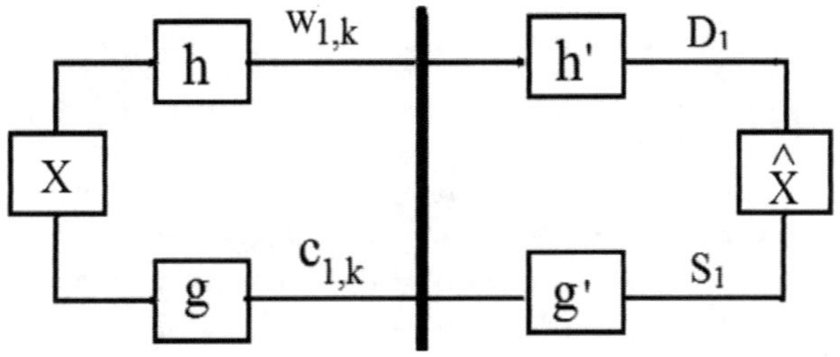

Figure 3. UDWT algorithm for producing wavelet coefficients at $j = 1$

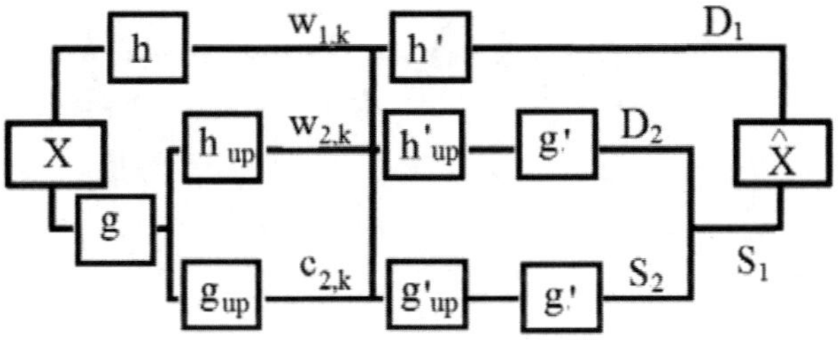

Figure 4. UDWT algorithm for producing wavelet coefficients at $j = 2$

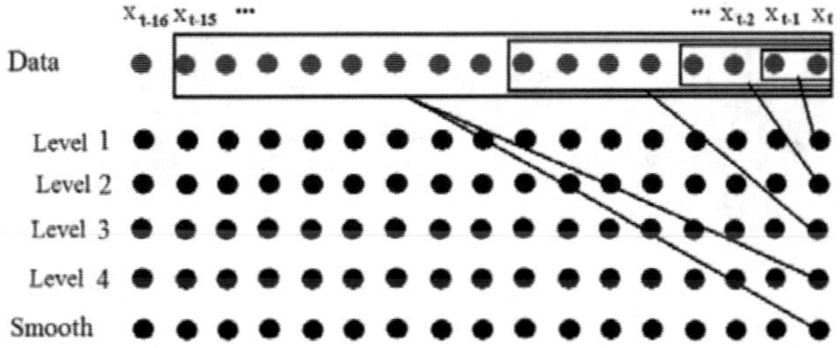

Figure 5. Computation of last coefficients at $j = 4$

3. Wavelet Based Prediction

Prediction of X at time $t+1$ will refer to realization of X in the past which represented by wavelet coefficients resulted from decomposition. When the Haar wavelet is used on this necessity, the computational process of last coefficients can be shown in Figure 5. Murtagh et al. [9] and Starck et al. [13] proposed that the wavelet coefficients at each level j which will be used for forecasting at time $t+1$ have the form $w_{j,N-2^j(k-1)}$ and $c_{J,N-2^J(k-1)}$ (see Figure 6). The forecasting model is expressed in equation (10)

$$\hat{X}_{N+1} = \sum_{j=1}^{J} \sum_{k=1}^{A_J} \hat{a}_{j,k} w_{j,N-2^j(k-1)} + \sum_{k=1}^{A_{J+1}} \hat{a}_{J+1,k} c_{J,N-2^J(k-1)} \tag{10}$$

The highest level of decomposition is indicated by J, and A_j is indicate the number of coefficients which chosen at level j. For example, if $J = 4$ and $A_j = 2$ for $j = 1, 2, 3, 4$ then (10) can be expressed as (11)

$$\hat{X}_{N+1} = \hat{a}_{1,1} w_{1,N} + \hat{a}_{1,2} w_{1,N-2} + \hat{a}_{2,1} w_{2,N} + \hat{a}_{2,2} w_{2,N-4} +$$

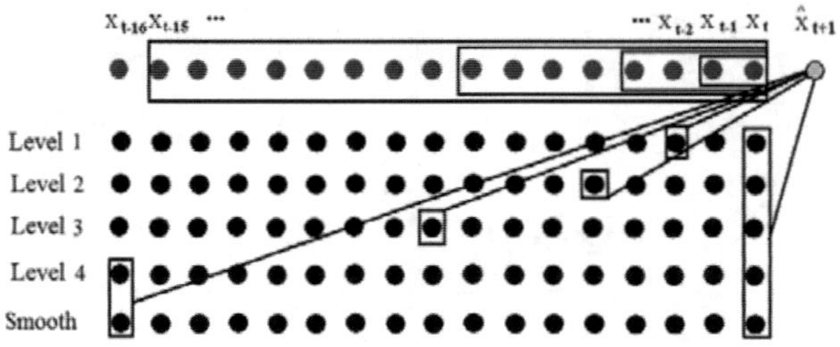

Figure 6. Wavelet coefficients which are used on prediction at $j = 4$

$$\hat{a}_{3,1}w_{3,N} + \hat{a}_{3,2}w_{3,N-8} + \hat{a}_{4,1}w_{4,N} + \hat{a}_{4,2}w_{4,N-16} + \hat{a}_{5,1}c_{4,N} + \hat{a}_{5,2}c_{4,N-16} \quad (11)$$

The model which is established in equation (11), will be implemented to the AR(1) simulation data sets in Figure 1. Plots of original data and predictions can be seen in Figure 7 and Figure 8. It is seen that the model with low value of $|\phi|$ more difficult to be captured by wavelet based model (Figure 7).

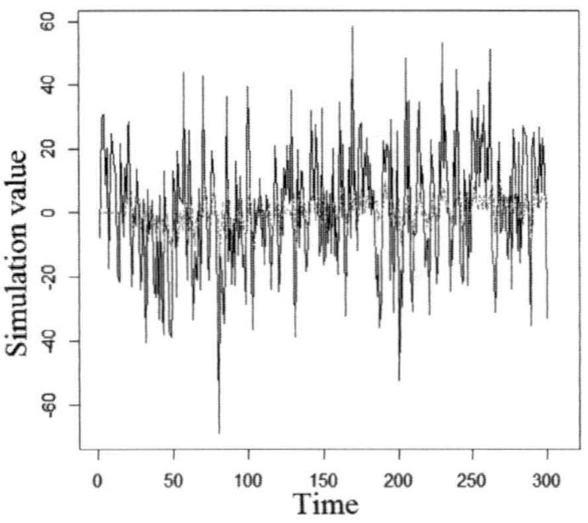

Figure 7. AR(1) plot of original data (black line) and its wavelet based predictions (red line) at $j = 4$, $\phi = 0.2$.

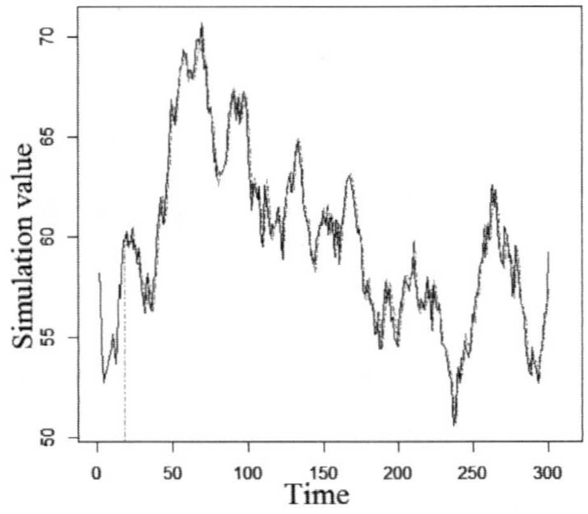

Figure 8. AR(1) plot of original data (black line) and its wavelet based predictions (red line) at $j = 4$, $\phi = 0.9999$.

4. Adequateness Development of Wavelet Based Model

Wavelet based model makes time series predictions based on dependencies of series to the historical information. The data pattern will be captured well when the series have high dependencies to the historical information. It is indicated by the high value of $|\phi|$. Wavelet based model will tracks the data pattern inaccurately when the series have high randomness. It means adequateness of wavelet based model will increase along with increasing of time series dependencies.

Suppose $\{X_t\}$, $t = 1, 2, \ldots n$ is a time series. Three transformations form will be proposed so that these results have high enough dependencies to the historical informations. Furthermore,

Journal of Physics: Conference Series **423** (2013) 012010 doi:10.1088/1742-6596/423/1/012010

the transformation results become new inputs for wavelet based modeling. The prediction results of wavelet based model will be reverted to original data predictions. It is hoped that the good prediction of wavelet based model will influence to the original data prediction.

4.1. 2-step Weighted Summation

The first, it will be constructed a new series which expresses the weighted summation of series $\{Y_t\} = p_1X_t + p_2X_{t-1}$. The dependencies to the historical information will increase in the new series.

Suppose $\{X_t\},\ t = 1, 2, \ldots n$ is a random process so that $E(X_t) = 0$, $E(X_tX_s) = 0$ for $t \neq s$, and $E(X_t^2) = \sigma^2 < \infty$. If $Y_t = p_1X_t + p_2X_{t-1},\ p_1, p_2 > 0, p_1 + p_2 = 1$ then $E(Y_tY_s) > 0$ for $|t - s| = 1$. Suppose $t < s$.

$$\begin{aligned} E(Y_tY_s) &= E(p_1X_t + p_2X_{t-1})(p_1X_{t+1} + p_2X_t) \\ &\geq p_1p_2E(X_t^2) \\ &> 0 \quad \square \end{aligned}$$

4.2. k-step Weighted Summation

If the dependencies to the historical informations has not sufficient for wavelet based model yet, then it can be constructed the second opinion form $Y_t = \sum_{i=0}^{k} X_{t-i}$. The dependencies to the historical information will increase in the new series. Suppose $\{X_t\},\ t = 1, 2, \ldots n$ is a random process so that $E(X_t) = 0$, $E(X_tX_s) = 0$ for $t \neq s$, and $E(X_t^2) = \sigma^2 < \infty$. If $Y_t = \sum_{i=0}^{k} p_iX_{t-i},\ p_i > 0, \sum_{i=0}^{k} p_i = 1$ then $E(Y_tY_s) > 0$ for $|t - s| \leq k$.

Suppose $t < s$ and $p = \min_{0 \leq i \leq k}[p_i]$.

$$\begin{aligned} E(Y_tY_s) &= E(p_1X_t + \ldots + p_kX_{t-k})(p_1X_s + \ldots + p_kX_{s-k}) \\ &\geq p(k - |t - s| + 1)E(X_t^2) \\ &> 0 \quad \square \end{aligned}$$

4.3. Partial Sum Weighted Summation

The third alternative to increase the dependencies to the historical informations can be done by calculating the weighted partial sum of series $\{X_t\}$ as described in equation (12)

$$Y_t = \frac{\sum_{i=1}^{t} p_iX_i}{\sum_{i=1}^{t} p_i} \tag{12}$$

The dependencies to the historical information will increase in the new series.

Suppose $\{X_t\},\ t = 1, 2, \ldots n$ is a random process so that $E(X_t) = 0$, $E(X_tX_s) = 0$ for $t \neq s$, and $E(X_t^2) > 0$. If $\{Y_t\}$ represents the weighted partial sum of $\{X_t\}$ then $E(Y_tY_s) > 0$. Suppose $t < s$.

$$\begin{aligned} E(Y_tY_s) &= E\left(\frac{\sum_{k=1}^{t} p_kX_k \sum_{k=1}^{s} p_kX_k}{\sum_{k=1}^{t} p_k \sum_{k=1}^{s} p_k}\right) \\ &\geq E\left(\frac{\sum_{k=1}^{t} p_kX_k}{\sum_{k=1}^{s} p_k}\right)^2 \\ &\geq \frac{\sum_{k=1}^{t} p_k^2E(X_k^2)}{(\sum_{k=1}^{s} p_k)^2} \\ &> 0 \quad \square \end{aligned}$$

Table 1. MAPE Simulation Result.

Mean Absolute Percentage Error			
ϕ	AR Model	Wavelet Model	Weighted Wavelet Model
0.999	0.3002462	0.2273194	0.2231491
0.7	2.239669	2.001092	2.121224
0.6	1.911261	1.661402	1.594556
0.4	1.558648	1.403207	1.430252
0.2	2.565376	2.357589	2.196195
0.1	1.027824	1.242509	1.228172
0.0001	1.078761	1.544106	1.528038

4.4. Simulation Result

For example of the proposed methods, it is made a 4-step weighted summation of AR(1) simulation data for various value of ϕ and standard deviation equal to 7. The result can be shown in table 1. Actually, various value of ϕ and σ had been tried. In general, wavelet model and weighted summation model have lower MAPE for ϕ tend to 1.

5. Concluding Remarks

The appearing of wavelet based model becomes an interesting tool in time series prediction. Beside of easy to use, as a part of non parametric methods, it does not need special assumption about the population distribution. The wavelet based model has superiority to capture non stationer and non linear pattern, thus it becomes a recently alternative solution in time series modeling. In other hand, the Box-Jenkins model has superiority to capture stationer pattern. In general the MAPE of wavelet model is lower than the MAPE of AR model for ϕ high enough. The example can be shown in table 1. It is not easy to choose the summation weight which will raise ϕ value as well as capable to track the data pattern. The effort aiming to increase the adequateness of wavelet based model has put the little result, but still needs some exploration to gain satisfied achievement.

References

[1] Aldrich E 2009 *A package of Functions for Computing Wavelet Filters, Wavelet Transforms and Multiresolution Analyses* http://www.ealdrich.com/wavelets/.
[2] Bollerslev T 1986 generalized autoregressive conditional heteroskedasticity *J. of Econometrics* **31** 307-27
[3] Box G E P and Jenkins G M 1976 *Time Series Analysis: Forecasting and Control* (San Francisco: Holden-Day)
[4] Ciancio A 2007 analysis of time series with wavelets *Int. J. of Wavelets, Multiresolution and Information Processing* **5**(2) 241-56
[5] Daubhechies I 1992 *Ten Lecture on Wavelets* (Philadelphia: SIAM)
[6] Engel R F 1982 autoregressive conditional heteroscedasticity with estimates of the variance of united kingdom inflation, *J. Econometrica* **50** 987-1008
[7] Fugal D L 2009 *Conceptual Wavelets in Digital and Signal Processing* (Space and Signals Technologies LLC)
[8] Haykin S 1999 *Neural Networks: A Comprehensive Foundation* Prentice Hall
[9] Murtagh F, Starck J L and Renaud O 2004 on neuro wavelet modeling *Decision Support System* **37** 475-90
[10] Percival D B and Walden A T 2000 *Wavelet methods for time series analysis* (Cambridge: CU Press)
[11] Samarasinghe S 2006 *Neural Network for Applied Science and Engineering* (New York: Auerbach Pub)
[12] Rukun S, Subanar, Rosadi D, and Suhartono 2011 heteroscedastic time series model by wavelet transform *Proc. of "The 6th SEAMS-UGM Conference 2011"*

[13] Starck J L, Jalal F and Murtagh F 2007 the undicimated wavelet decomposition and its reconstruction, *IEEE on Image Processing* **16**(2) 297-309

[14] Wei William W S 1994 *Time Series Analysis: Univariate and Multivariate methods* (Canada: Addison-Wesley)

Journal of Physics: Conference Series **423** (2013) 012011 doi:10.1088/1742-6596/423/1/012011

Regression Test on the Rotational Speed between Two Loads as the Preparation for Braking System

B S R. Purwanti [1], **F Yusivar** [2], **I Garniwa M K** [3]
[1] Student of Doctoral Program, University of Indonesia, Electrical Engineering Department DTE-FTUI, INA
[2,3] University of Indonesia, Electrical Engineering Department, DTE-FTUI, INA

E-mail: bsrp_81@yahoo.co.id

Abstract. This paper is preparing the mathematic model of braking control, continuously of determination the error (e), delta error (de) of speed reduction [9]. Load-1 and Load-2 are driven by an electric motor located on the same shaft. Both loads are driven clock wise (CW), counter clock wise (CCW) by an asynchronous three-phase motor (M3). The mass of each load is also differentiated to simulate slip phenomena. Rotational speed of M3 is equal to Load-1, detected by Sensor-1, while speed rotation of Load-2 is detected by Sensor-2. The rotation for Load-1 and Load-2 can be adjusted on several position H_j (j = 1, 2, 3). Once Load-1 and Load-2 reach a constant speed, current source will be disconnected. Speed reduction from ($\omega \pm 1475$ rpm) to stagnant ($\omega = 0$ rpm) on Load-1 and Load-2 is considered time function. Data collected from both load (ω (t)) known as e, de; on each position of H_j. It uses covariance analysis to make sure that both loads are concurrent with each other against time difference. The objective of this research is to determine slip phenomena of speed reduction of each load. The expectations are to generate smoother braking and minimize the time needed when implemented with ANFIS.

1. Introduction

Braking is the process of reducing speed, usually performed in an attempt to avoid collision/crash. Reliability is critical in braking to avoid deadly collision and reducing incidents between vehicles. This research is aimed to develop a braking system that is appropriate to the customer's demand and need. The implementations of braking system or speed reduction can be found everywhere, for example in industry, in the street, in buildings, even at home. Belt Conveyor (industries), elevator, escalator (buildings), blender, mixer, vacuum cleaner (household appliances), and many others. Reducing speed from high to naught (stagnant) by counteracting the momentum/force resulted from translation. Implementation of braking system has been studied by various methods. Simulation and testing of hydraulic antiskid braking system based Adhesion Creep Behaviour [1] has been implemented on a modern railway. Zuo Jianyong used the pneumatic method, sub model showed that the output data antiskid braking is graded; it is recommended to use Fuzzy Control Method.

Land vehicles demand a comfortable ride with reliable braking system, assuring safety from any accidents. The braking system has to adapt with various load/weight changes. The response time, velocity, and acceleration, which are generated by external loads cause waste in energy usage. In addition to external disturbances on the system, the characteristics, and behaviour of dynamic braking,

[1] *Lecturer on the Electronic Department, Polytechnique State of Jakarta, Depok, INA*

the energy savings can also be traced. This research will study the slip between two loads due to several load masses. The system is designed to get the slip value between two Loads due to differences in rotational speed. The results are statistically tested for reliability.
Antiskid braking concept known as ABS (Antilock Breaking System) is very interesting and unique. ABS system requires precise control and effectiveness in order to obtain braking perfection. Fuzzy intelligent control by Takagi-Sugeno in ABS system, minimize slip by regulating current-torque. FLC is claimed to optimize linguistic switching, thus generating better control. A good control system will include various aspects [2], such as energy conversion and energy balance. Yimin G, made a statement about energy conversion from electrical into kinetic energy to drive the railway.
Reliable braking system ensures safety, comfort, and ultimately has smoother operation [3]. A smooth braking action designed to enhance passenger's comfort can be achieved by minimizing slip due to dynamic force and friction between surfaces [4]. With bigger loads, the higher the speed, mean higher moment of inertia.
Driving force produced by Three Phases Induction Motor Asynhronous. Translational velocity is affected by the load (the car and its contents) and the coefficient of friction between the wheel-rail. Movers "railway" is a Induction Motor or Three Phase. Formed a separate cage induction motor with a local vector field (d, q) is [5, 6, 7].
Given the speed is influenced by driving force (electric motors), speed reduction can be done with disconnecting the current. This study will simulate slip between two rotating objects, Load-1 and Load-2. Both Loads are connected by a transmission and rotated by an Asynchronous Three-Phase Motor. The short-term objective of this research is to determine the differences in rotational speed between two Loads due power disruption. Slip value is very small but very important, ultimately to ensure the safety from any accidents. Statistical test is done to understand the ratio and the effect of rotational speed on Load-1 and Load-2. The difference in rotational speed between Load-1 and Load-2 caused slip in three different transition positions. This research will serve as the preparation for ANFIS based braking system for three-phase motor. The system will test the value of rotational speed covariant with two Loads. The long-term goal is to reduce the slip value due to weight (loading) to minimum.

2. Speed Reduction between Two Load for the Phenomenon of Slip

Speed reduction was smooth in the railway stabilize the braking system; the realization did not occur buffeting. Pounding in the braking process perceived as an inconvenience passengers. Significant speed reduction due to braking in the form of data on the curve visualization rail with wheel railway carriages loaded and its contents. Researcher [7] has said that the wheel speed will tend to be lower than the vehicle speed. The other opinion in [8] purpose a gain scheduling control structure base on slip and maximum frictions are not directly measurable but can estimated.

2.1. The Slip from difference of speed

The road adhesion coefficient (or coefficient of friction) [7] is the proportion of the road friction force to the normal load of the vehicle, and it is a nonlinear function of some physical variables; including wheel slip (1).

$$v_s = \frac{v - \omega r}{v} \tag{1}$$

Where v_s is the slip value, r is the radius of wheel, ω is the rotational speed of the wheel, and v is the longitudinal speed of the vehicle. Form the equation (1), v_{slide} is the speed slip and called the negative slip (2).

$$v_{slide} = \omega r - v \tag{2}$$

In general, equation (2) we can be stated as equation (3):

$$\Delta v_{slide} = \Delta(\omega r - v) \tag{3}$$

Researcher [7] has said that the wheel speed will tend to be lower than the vehicle speed. The other opinion in [8] purpose a gain scheduling control structure base on slip and maximum frictions are not directly measurable but can estimated.

2.2. Illustration of Railway

Reducing slip promoted smoother braking and improves railway safety and passenger's comfort. The research will be conducted in four stages [9], currently this paper only covered the first phase (figure 1) shown the phenomeon of slip.

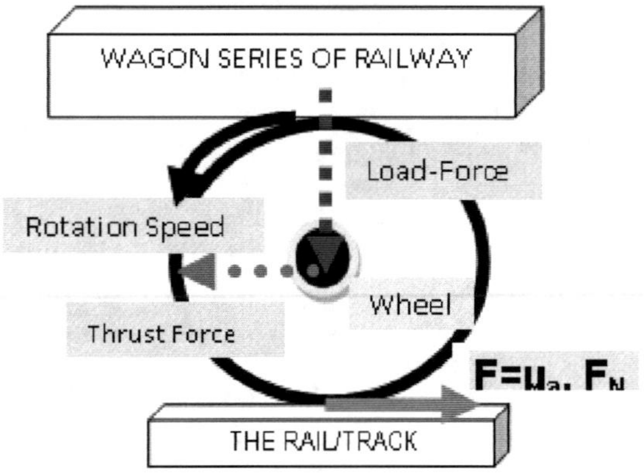

Figure 1 Illustration of Railway [9]

Translational speed is affected by load (weight) and friction-force, where (μ_a) coefficient of friction between the wheel and the track (rail).

The components of Braking Motion are:

- The wheels spin with particular thrust affects translational speed.
- Thrust is generated by Asynchronous Three-phase Induction Motor.

Real time observation for railway's braking system is difficult to be done. Braking system simulator is required in order to obtain the phenomenon (figure 2). All wheels are driven by M3 with with certain torque (force) to achieve certain speed.

2.3. The Railway Illustration

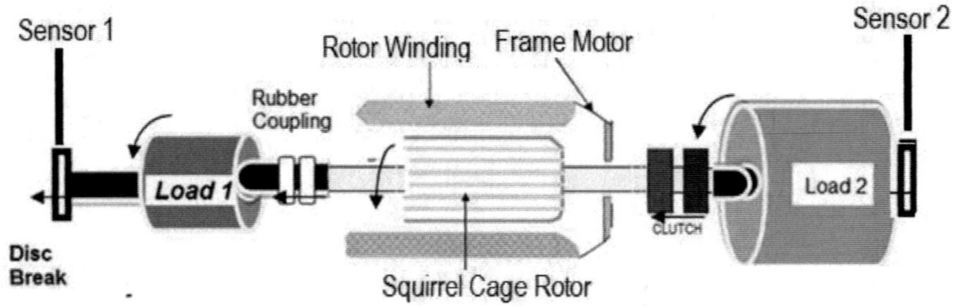

Figure 2 Block Diagram System [9]

Actual design was activated so that rotational speed of both loads could be measured using optocoupler sensors. Optocoupler sensors determined rotational speed of Load-1 and Load-2. Once it reached a steady speed, current source was disconnected. The disconnection reduced the speed of the three-phase induction motor, which also reduced rotational speed of Load-1 and Load-2. The rotational speed of Load-1 and Load-2 were conditioned of in third positions different of clutch (H_j, with j= 1, 2, 3). The clutch positions (H_j) were conducted in clock wise (CW) and counter clock wise (CCW). Position of H_j conducted from G_4, it is part of the fourth position of G_i (i= 1, 2, 3, 4). The e values and de values in all positions (G_i, i=1, 2, 3, 4) are minimum 0.00002 $\Leftrightarrow$ 0, maximum 0.987 $\Leftrightarrow$ 1 and applicable in CW and CCW [9].

2.4. Identity and Design of Simulation

Rotational speed of both loads can be determined (see figure 1), Load-1 and Load-2 are driven by three phase induction motor. In the event of braking, rotational Load-1 affects rotational speed Load-2 and vice versa. Illustration model for study any slips between two Loads (see figure 1) and Block Diagram in (see figure 2) is used to realize braking system with ANFIS based ICS. ANFIS requires e and de in interval [0, 1], thus this research is aimed to determine the value of e and de, similar to input-output on Fuzzy Control System.

Two optocoupler speed sensors (see figure 3a, figure 3b) are needed to detect the rotational speed of Load-1, Three-phase induction motor, and Load 2. Each sensor has acrylic-disc with 36 holes as trigger (see figure 3b). The value of e and de are very influential to the braking process with ICS. ANFIS based ICS is expected to prevent slip and buffeting occurred during braking process. Slip phenomenon generated by comparing the rotational speed of the two Loads as the result of current termination. Load-1, consists of solid cylinders. The transmission acts as a transition block, placed between Three-phase induction motor and Load-2, to allow the researcher change the clutch position. Three-phase induction motor is positioned on the same axle as Load-1. The connecting coupler for Load 2 is located on the different axle. Mass of Load 2 is designed to be adjustable, allowing the addition of weight to Load-2. Load 2 acts as an absorber for mechanical/electrical load.

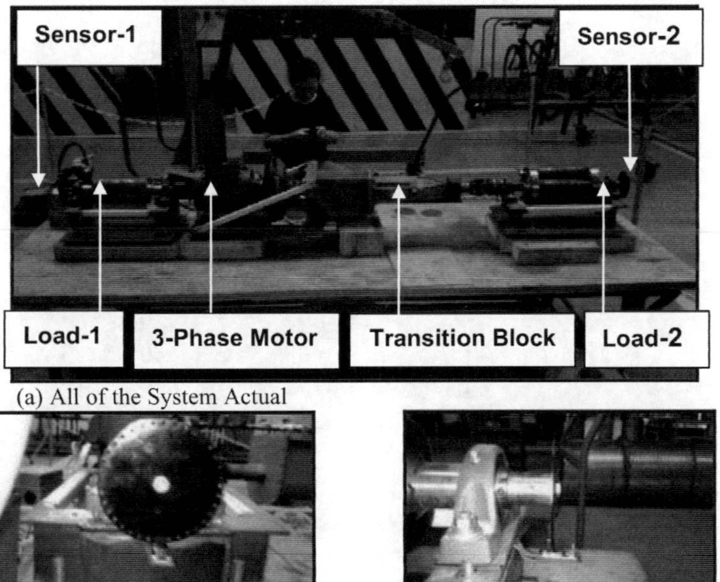

(a) All of the System Actual

(b) Optocoupler Sensor (Front View) **(c)** Optocoupler Sensor (Side View)

Figure 3 Simulator of Breaking System

3. Braking System Based on Intelligent Control

This research studies the mathematical model of braking system in actual plant (see figure 3). Actual plant is set of tools designed as braking system simulator on *railway*. Asynchronous Three-Phase Induction Motor drives Load-1 and Load 2. The objective of this study is to observe the speed reduction of Load-1 and Load 2 in Clock Wise (CW) and Counter Clock Wise (CCW). Then data between both Loads will be analyzed to estimate the effect of initial velocity against braking.

3.1. Methodologies

Smoother speed reduction in railway will stabilize braking system; the realization would be the absence of buffeting. Pounding in the braking process is considered as disturbance for the passengers. Especially in case of significant speed reduction due to braking in the form of graded data on the curve (mathematically). A model is made to simulate the rail-track and the wheels. Speed difference between the two Loads, the existence of friction force (see figure1), will generate slip.

3.2. Data Collection and Sampling and Result of Experiment

Table 1 Data Sample of CW Rotation Speed

Load:	CW1		CW2		CW3	
Second of:	**1**	**2**	**1**	**2**	**1**	**2**
1	0	0	0	0	0	0
2	0	0	0	0	0	0
3	10	8	191	190	16	16
4	845	845	1236	1236	860	860
5	1451	1451	1475	1475	1461	1461
	Some data have been hidden					
53	216	216	1	1	148	0
54	165	163	0	0	101	0
55	116	115	0	0	50	0
56	71	73	0	0	6	0

Table 2 Data Sample of CCW Rotation Speed

Load:	CW1		CW2		CW 3	
Second of	**1**	**2**	**1**	**2**	**1**	**2**
1	0	0	0	0	0	0
2	0	0	0	0	0	0
3	0	0	113	120	618	618
4	243	243	1143	1143	1431	1433
5	1285	1285	1475	1475	1476	1476
.	**Some data have been hidden**					
58	158	158	6	0	941	81
59	110	111	0	0	878	55
60	66	68	0	0	815	35
61	31	41	0	0	751	21

The experimental results of the motor rotation speed from 0 to ± 1476 rpm (acceleration) and toward zero velocity (deceleration). Some sample data acceleration and load deceleration round two (see table 1, table 2). Braking process occur a reduction of speed, namely the speed of 1475 rpm ± towards zero, identical to deceleration events. Data selected special deceleration events at three positions transitions G_j, with $j = 1, 2, 3$, as the braking process variables.

3.3. Speed Rotation of the CW and CCW

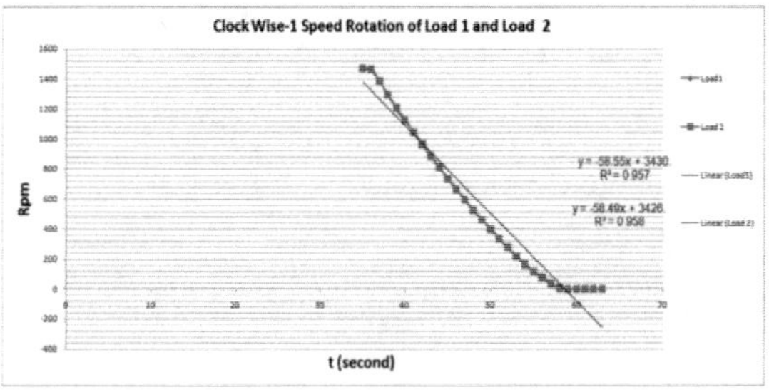

Figure 3 CW1 Speed Rotation of Load 1, 2

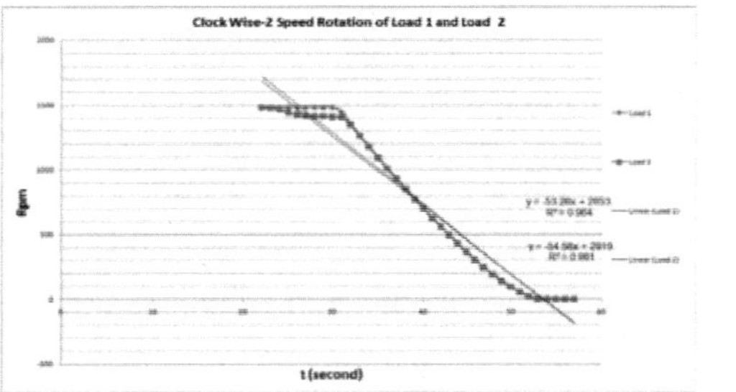

Figure 4 CW2 Speed Rotation of Load 1, 2

The experimental results are to group become CW rotation (see figure 3, figure 4, and figure 5) and CCW rotation (see figure 6, figure 7, and figure 8). Regression equations for entire round of Load 1 and Load 2 Clock Wise (CW) rotation speed.

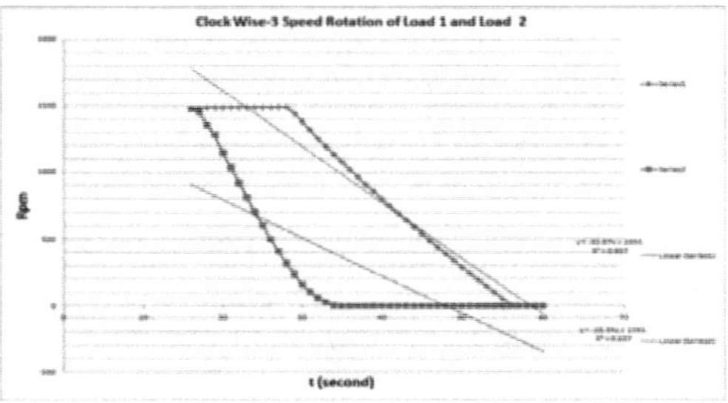

Figure 5 CW3 Speed Rotation of Load 1, 2

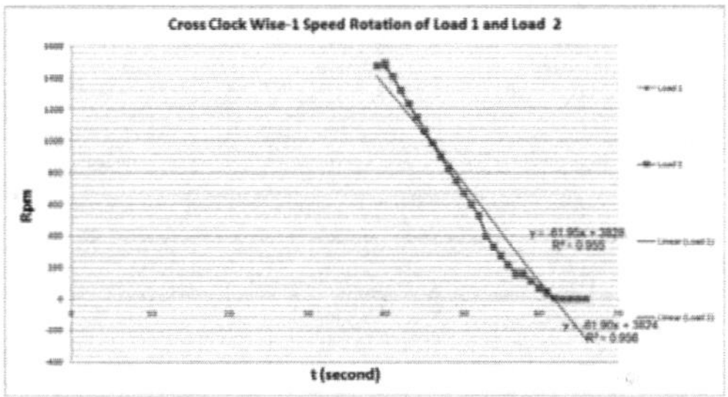

Figure 6 CCW1 Speed Rotation of Load 1, 2

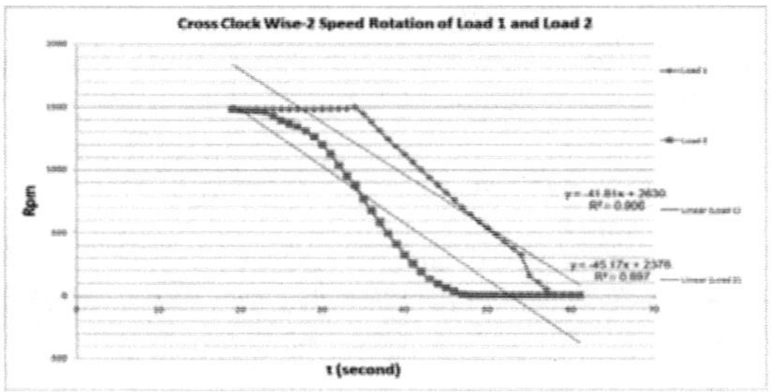

Figure 7 CCW2 Speed Rotation of Load 1, 2

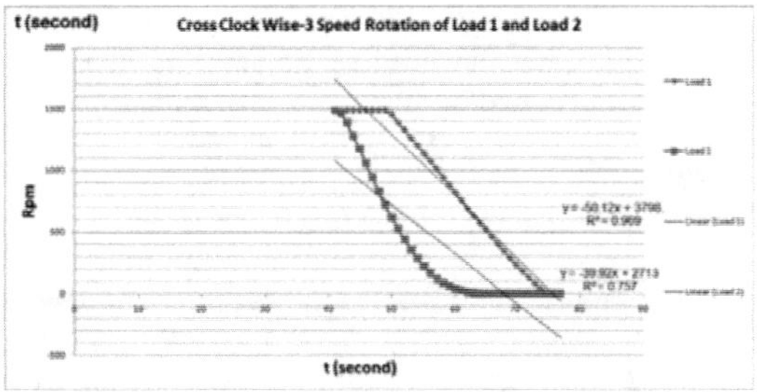

Figure 8 CCW3 Speed Rotation of Load 1, 2

3.4. Regression of CW and CCW Speed Rotation

Regression Equations of Load 1, Load 2 with three conditions are CW-CCW1, CW-CCW2, and CW-CCW3 (see table 3). In accordance with the objectives research, analysis and testing of the data discussed specifically for section 3. 4. Processing data using with regression test to examine the relationship whether or not the correction of the measured variables, namely Load-1 to Load-2 with CW1, CW2, CW3 (see table 3).

Figure 3, figure 6 there was no difference, almost harmonious whole. Rated slip between Load-1 and Load-2, CW and CCW from zero. Figure 4 there was a slight difference, and different overall, rated slip between Load-1 and Load-2 small slip value CW and CCW large enough. Figure 5, figure 8 there was different overall, rated slip Load-1 and Load-2 large CW and CCW, value of regression (R) Load-1, Load-2 for CW and CCW$\geq 90\%$.

Table 3 Regression Equation of CW and CCW Speed Rotation Load 1 and Load 2

Position of Transition	Regression and R	
	Load 1	**Load 2**
Clock Wise-1	$y=-58.55x+3450,\ R=0.957$	$y=-58.49x+3426,\ R=0.958$
Clock Wise 2	$y=-54.64x+2922,\ R=0.960$	$y=-53.26x+2853,\ R=0.964$
Clock Wise 3	$y=-41.97x+2451,\ R=0.957$	$y=-28.444x+1361,\ R=0.637$
Counter Clock Wise-1	$y=-61.90x+3824,\ R=0.956$	$y=-61.95x+3828,\ R=0.955$
Counter Clock Wise-2	$y=-41.81x+2630,\ R=0.906$	$y=-45.17x+2378,\ R=0.897$
Counter Clock Wise-3	$y=-50.12x+3798,\ R=0.969$	$y=-39.92x+2713,\ R=0.757$

4. Conclusion

The slip phenomenon was founded in the CW and CCW rotation of the reduction speed, and then could be continued to experiment with to looking for the slip values. The deceleration system, Load-2 as the variable dependent to the power of Load-1 in CCW model, does not influence to the CW model

5. References

[1]. Yu Zhang; Zhenhua Jiang; Xunwei Yu. Indirect Field-Oriented Control of Induction Machines Based on Synergetic Control Theory 2008 *Proc. of the IEEE Int. Conf. on Power and Energy Society General Meeting-Conversion and Delivery of Electrical Energy in the 21st Century* **20**-4 1-7

[2]. Yimin G. et al 2003 Electric Motor Flywheel/Characterization of Generator of Hybrid Vehicles. *Proc. of the IEEE 58th Vehicular Technology Conference, IEEE Explore Press USA* DOI: 10.1109/ VETECF1286291 3321-25

[3]. Purwanti, B S R, Feri Y, Iwa G. 2011. COG FLC Implementation for Input-Output Value Calculation. *Proc. of the 3th Int. Conf. Math and Natural Sci. (ICMNS 2010) ITB* 1011-18

[4]. Yesim Oniz , Okyay Kaynak 2009 A Dynamic Method to Forecast the Wheel Slip for Antilock Braking System and Its Experimental Evaluation *IEEE Transactions On Systems, Man, And Cybernetics—Part B: Cybernetics* **39**-2 551-60

[5]. Kais Jamoussi, M. Ouali,, Haseen Charradi 2007 A Sliding Mode Speed Control of on Induction Motor. *American J of Appl. Sci.* **12**-4 987-94

[6]. M. Ouali, M. B. A. Kamaoun 1997 Field Oriented Control Induction Machine and Control by Sliding Mode. Simulation Practice and Theory **5** 121-36

[7]. A. Mansouri, M. Chenafa, A. Bouhenna, E. Eftien. 2004. Powerful Non Linier with Field Oriented Control of an Induction Motor Int. *J. Appl. Math. Comp. Sci.***2**-14 229-32

[8]. Solyom, S. and Rantzer,A 2003 *ABS A Design Model and Control Structure* in Rolf Johansson and Anders Rantzers (Ed). *Nonlinear and Hybrid Control System*

[9]. Purwanti, B. S. R. Yusivar, F. Garniwa, I. M. K. 2012. Determination the Error and Delta Error for Braking Control System of Three Phases Motor *Proc. Int. Con. on Math. Stat and Its Applications (ICMSA) 2012,* ITS AM3 29-38.

Journal of Physics: Conference Series **423** (2013) 012012 doi:10.1088/1742-6596/423/1/012012

B-spline Collocation with Domain Decomposition Method

M I P Hidayat[1], B Ariwahjoedi[2] and S Parman[1]
[1] Department of Mechanical Engineering, Universiti Teknologi PETRONAS, Bandar Seri Iskandar, 31750 Tronoh, Perak Darul Ridzuan, Malaysia
[2] Department of Fundamental and Applied Science, Universiti Teknologi PETRONAS, Bandar Seri Iskandar, 31750 Tronoh, Perak Darul Ridzuan, Malaysia

E-mail:
irfan@mat-eng.its.ac.id, bambang_ariwahjoedi@petronas.com.my, setyamartana@petronas.com.my

Abstract. A global B-spline collocation method has been previously developed and successfully implemented by the present authors for solving elliptic partial differential equations in arbitrary complex domains. However, the global B-spline approximation, which is simply reduced to Bezier approximation of any degree p with C^0 continuity, has led to the use of B-spline basis of high order in order to achieve high accuracy. The need for B-spline bases of high order in the global method would be more prominent in domains of large dimension. For the increased collocation points, it may also lead to the ill-conditioning problem. In this study, overlapping domain decomposition of multiplicative Schwarz algorithm is combined with the global method. Our objective is two-fold that improving the accuracy with the combination technique, and also investigating influence of the combination technique to the employed B-spline basis orders with respect to the obtained accuracy. It was shown that the combination method produced higher accuracy with the B-spline basis of much lower order than that needed in implementation of the initial method. Hence, the approximation stability of the B-spline collocation method was also increased.

Introduction

1.1. Overview

The method of collocation is attractive for solving differential equations due to its ease of implementation and efficiency. The method applies directly to the original differential equations that describe the physics of the problem. In the last decades, collocation methods using B-spline functions for the numerical solutions of various types of partial differential equations have been paid attention and becoming an active research area. The spline collocation methods are interesting due to their high accuracies. There are many variants of spline collocation methods, which are mainly related to the locations of the collocation points, such as nodal, orthogonal and modified collocation (see the survey of spline collocation methods by Fairweather and Meade [1], Bialecki and Fairweather [2]). Other variants of spline collocation methods have been investigated as well: Botella [3], Greville [4, 5] and

Moment [6] collocations. The accuracy of the spline collocation methods depends on the order of B-spline basis, regularity or continuity, and location of the collocation points [3-6]. Recently, B-spline collocation methods have been applied successfully in various areas including fluid dynamics [3, 4, 7-9], computational aero-acoustics [10], wave equation [11], non-linear analysis of laminated panels [12], vibration analysis [13, 14], and non-linear two-point boundary value problems [15].

1.2. Motivation
However, there are serious shortcomings in the implementation of the B-spline collocation methods. The fact that basis needed for approximation is a tensor product operation from one-dimensional B-spline basis functions has restricted the method extension to applications in higher dimensions. Moreover, the tensor product structure that is essentially rectangular in nature has also further limited the approximation to rectangular problem domain or other domains in which a rectangular structure can be envisaged. Obviously, this produces a quite severe limitation with respect to the flexibility and suitability of the method for applications in non-rectangular or irregular domains. In [16], it has been stated that the B-spline collocation methods have not yet reached the level of flexibility of the isoparametric formulation of the finite element method for arbitrary geometries.

1.3. Newer Strategy
A global B-spline collocation method has been previously developed and successfully implemented for solving elliptic partial differential equations in arbitrary complex domains. The implementation of global collocation has been inspired by successful application of radial basis functions (RBF) collocation in non-uniform grids [17]. However, the global B-spline approximation, which is simply reduced to Bezier approximation of any degree p with C^0 continuity, has led to the use of B-spline basis of high order in order to achieve high accuracy. The need for B-spline bases of high order in the global method would be more prominent in domains of large dimension. For the increased collocation points, it may also lead to the ill-conditioning problem.

In this study, overlapping domain decomposition of multiplicative Schwarz algorithm [18] is combined with the global method. It is aimed that the obtained B-spline approximation can be more robust, but still preserving great flexibility of the initial method in dealing with arbitrary complex domains. Hence, our objective is two-fold that improving the accuracy with the combination technique, and also investigating influence of the combination technique to the employed B-spline basis orders with respect to the obtained accuracy.

1.4. Related Works
The present method, in the absence of triangulation, may be also related to the work of Hu, Han and Lai [19] in the sense that the constructed approximation rely upon the use of splines of various degrees. In the authors work, which is an extension of Awanou, Lai and Wenston s spline method [20], the method of bivariate splines of various degrees constructed from Bernstein polynomials of degree p over triangles was used for the numerical solution of Poisson and biharmonic equations in an adaptive manner. High order approximation is used to get a better approximation power, much similar also with p-adaptation in hp-finite element methods [21, 22]. In general, the method of bivariate splines falls into high order method.

B-spline Collocation
Consider the following boundary value problem:

$$\begin{aligned} Lu(\mathbf{x}) &= f(\mathbf{x}) \quad \forall \mathbf{x} \in \Omega \\ B^h u(\mathbf{x}) &= h(\mathbf{x}) \quad \forall \mathbf{x} \in \partial\Omega^h \\ B^g u(\mathbf{x}) &= g(\mathbf{x}) \quad \forall \mathbf{x} \in \partial\Omega^g \end{aligned} \tag{1}$$

where Ω is the problem domain, $\partial\Omega^h$ is the Dirichlet boundary, $\partial\Omega^g$ is the Neumann boundary, $\partial\Omega = \partial\Omega^h \cup \partial\Omega^g$ is the problem domain boundary, L is the Laplace operator (∇^2) in Ω, B^h is the differential operator on $\partial\Omega^h$, B^g is the differential operator on $\partial\Omega^g$, $\mathbf{x} \in R^d$.
Let also $\{CP_i = (\xi_i)\}_{i=1}^{NC}$ be NC collocation points in Ω, of which $\{(\xi_i)\}_{i=1}^{N_I}$ are interior points, $\{(\xi_i)\}_{i=1}^{N_h}$ are boundary points on $\partial\Omega^h$ and $\{(\xi_i)\}_{i=1}^{N_g}$ are boundary points on $\partial\Omega^g$, hence $NC = N_I + N_h + N_g$.

The approximate solution $\hat{u}(\mathbf{x})$ for the problem (1) is to be obtained with the collocation using B-splines. B-spline basis construction is briefly described in what follows.

Consider one-dimensional B-spline basis function construction given by the Cox-de Boor recursion formula [23, 24]:

$$N_{i,1}(t) = \begin{cases} 1 & \text{if } \tau_i < \tau < \tau_{i+1} \\ 0 & \text{otherwise} \end{cases} \quad (2)$$

$$N_{i,1}(t) = \frac{t - t_i}{t_{i+k-1} - t_i} N_{i,k-1}(t) + \frac{t_{i+k} - t}{t_{i+k} - t_{i+1}} N_{i+1,k-1}(t) \quad \text{for} \quad 2 \le k \le n+1 \quad (3)$$

In the formula, the convention $0/0 = 0$ is used for the division calculation. Thus, $b = (N_{0,k}, \ldots, N_{n,k})$ are the univariate B-spline bases for the space C of polynomial functions of degree less than or equal to $k-1$ on interval $I = [a_1, b_1] = [\tau_0, \tau_{n+k}]$. It is also clear from (3) that for $k > 1$ the $N_{i,k}(t)$ are a linear combination of two $(k-1)$ order basis functions.

The relationship between the number of knots in the knot vector $(\tau_i)_{i=0}^{n+k}$ (mk), control points $(n+1)$ and B-spline order k is also given as follows:

$$mk = n + k + 1 \quad (4)$$

For higher dimensional problems, the basis functions are constructed by taking the product of the one-dimensional B-spline basis functions. For approximation in 2D, the basis functions are given by:

$$B := (N_{i,k} \otimes M_{j,l})_{i=0,j=0}^{n,m} \quad (5)$$

where B are the tensor product B-spline bases for the space D of tensor product polynomial splines on the tensor product domain $[a_1, b_1] \times [a_2, b_2] = [\tau_0, \tau_{n+k}] \times [\varsigma_0, \varsigma_{m+l}]$. $N_{i,k}$ and $M_{j,l}$ are the B-spline basis functions of orders k and l.

Note that the tensor product B-spline bases have the partition of unity property:

$$\sum_{i=0,j=0}^{n,m} N_{i,k}(t)\, M_{j,l}(s) = 1 \quad (6)$$

In addition, continuity of a B-spline curve depends generally on its order. A B-spline curve of order k has in general C^{k-2} continuity. There is however a continuity constraint with respect to the

Journal of Physics: Conference Series **423** (2013) 012012 doi:10.1088/1742-6596/423/1/012012

multiplicity (mp) of a knot in the knot vector. It is called as knot continuity, where increasing the multiplicity (mp) of a knot reduces the continuity of the curve to C^{k-mp-1} at that knot. It also means only $p-mp$ derivatives exist for a knot of multiplicity mp , where p is the B-spline degree [25].

In this study a special class of B-spline approximation obtained by using an open-uniform knot vector with full multiplicity of its end knots (no interior knots) is used for the collocation. Such a B-spline approximation is chosen because it will produce global approximation in order to deal with arbitrary complex domains with arbitrary distribution of knot points. It is noted here the B-spline approximation is simply reduced into Bezier approximation of any degree p with C^0 continuity [26].

Thus, we have here for the collocation (1) the following two-dimensional B-spline approximation:

$$\hat{u}(x,y)=\sum_{i=0,j=0}^{n,m} N_{i,k}(x)M_{j,l}(y)\alpha_{i,j} \tag{7}$$

The B-spline approximation can be simplified to be much similar with shape functions in mesh-less and finite element methods [27] as:

$$\hat{u}(x,y)=\sum_{a=1}^{N_B} N^a(x,y)\alpha_a \tag{8}$$

where $N^a(x,y)=N_{i,k}(x)M_{j,l}(y)$, α_a are the coefficients of approximation related to $N^a(x,y)$ and N_B is the number of B-spline bases used in approximation.

Overlapping Domain Decomposition Method

By overlapping domain decomposition method, a given computational domain is partitioned into smaller overlapping sub-domains [18, 28, 29]. A typical option to accomplish such a decomposition task is by partitioning the computational domain Ω into K non-overlapping sub-domains and then extending from the shared boundaries of neighbouring sub-domains Ω_i and Ω_j an amount of overlapping portions o_{ij} to a larger domain Ω_i. The artificial interior boundaries are hence introduced here. This schematic is illustrated in Figure 1.

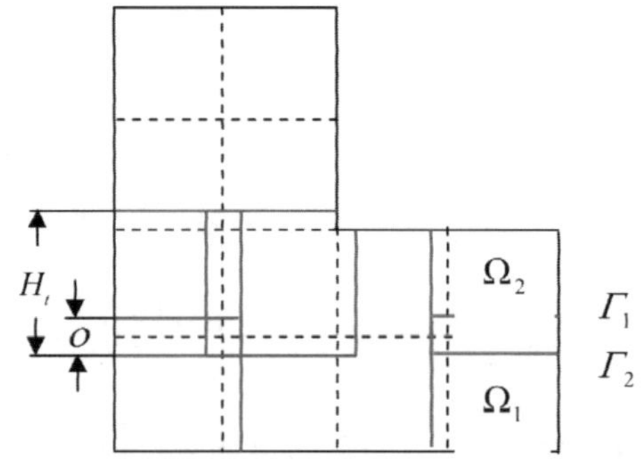

Figure 1. Schematic of domain decomposition of Ω into smaller

overlapping sub-domains Ω_i. The artificial boundaries Γ_i are part of the boundary of Ω_i that is interior of Ω.

Denote Ω_i, $\partial\Omega_i \setminus \Gamma_i$ and Γ_i the extended sub-domain, the natural boundary and the artificial interior boundary overlapped with other neighbouring sub-domains, respectively. Let also $\overline{\Omega}_i = \Omega_i \cup \partial\Omega_i \setminus \Gamma_i \cup \Gamma_i$ denote the closed sub-domain, S denote the operator of artificial interior boundary conditions and Λ_i be the artificial interior boundary value of sub-domain $\overline{\Omega}_i$ extracted from the neighbouring sub-domains. The original global problem is now reformulated as a series of sub-domain problems:

$$\begin{aligned} Lu(\mathbf{x}) &= f_i \quad \text{in } \Omega_i \\ Bu(\mathbf{x}) &= b_i \quad \text{on } \partial\Omega_i \setminus \Gamma_i \\ Su(\mathbf{x}) &= \Lambda_i \quad \text{on } \Gamma_i \end{aligned} \tag{9}$$

B, S specify Dirichlet, Neumann or mixed (Robin) boundary condition operator and $i = 1,2,...,K$.

Communication and transmission of information among the sub-domains is carried out iteratively in the overlapping regions o_{ij} i.e. through the artificial interior boundaries Γ_i, which is described in the following. For each sub-domain the B-spline approximation is obtained as follows:

$$\begin{bmatrix} \mathbf{A}_{\overline{\Omega}_i \setminus \Gamma_i} \\ \mathbf{A}_{\Gamma_i} \end{bmatrix} \boldsymbol{\alpha}_i = \begin{Bmatrix} \mathbf{F}_{\overline{\Omega}_i \setminus \Gamma_i} \\ \mathbf{F}_{\Gamma_i} \end{Bmatrix} \tag{10}$$

where: $\mathbf{A}_{\overline{\Omega}_i \setminus \Gamma_i}$ are matrices representing the operators acting on $\overline{\Omega}_i$ excluding Γ_i, $\mathbf{F}_{\overline{\Omega}_i \setminus \Gamma_i}$ are the corresponding right-hand side vectors, in which the natural boundary conditions are imposed. The sub-matrices $\mathbf{A}_{\Gamma_i}$ and vectors $\mathbf{F}_{\Gamma_i}$ are produced from the artificial interior boundary conditions.

Solving further equation (10), it is obtained:

$$\boldsymbol{\alpha}_i = \mathbf{A}_i^{-1} \mathbf{F}_i \tag{11}$$

where: $\mathbf{A}_i = \left[\mathbf{A}_{\overline{\Omega}_i \setminus \Gamma_i} \; \mathbf{A}_{\Gamma_i}\right]^T$, $\mathbf{F}_i = \left\{\mathbf{F}_{\overline{\Omega}_i \setminus \Gamma_i} \; \mathbf{F}_{\Gamma_i}\right\}^T$ and $\boldsymbol{\alpha}_i$ are the vector of coefficients of approximation of each sub-domain. Note that $\mathbf{F}_i$, containing the artificial interior boundary conditions, will continuously change with the updating results of its neighbouring sub-domains.

The approximated values of the unknown function in sub-domain $\overline{\Omega}_i$ are:

$$\phi_i(\mathbf{x}) = G_i(\mathbf{x})\, \boldsymbol{\alpha}_i \tag{12}$$

where: $G_i(\mathbf{x}) = \left[N^a(\mathbf{x}_1), N^a(\mathbf{x}_2), \ldots, N^a(\mathbf{x}_{Ns_i})\right]^T$ and Ns_i is the total number of collocation points in sub-domain $\overline{\Omega}_i$.

It can be seen that the approximated values of the unknown function in sub-domain $\overline{\Omega}_i$ are involving update in the artificial interior boundary conditions. In addition, depending on how artificial interior boundary conditions $\mathbf{F}_i$ are updated, there are two types of overlapping domain decomposition methods i.e. additive Schwarz and multiplicative Schwarz methods. For clarity, both these overlapping domain decomposition techniques are described here.

In the overlapping additive Schwarz method, the artificial interior boundary values of each sub-domain problem at the $(n+1)$th step are updated from the results of the n th step. This is illustrated as follows: Let Ω be partitioned into two sub-domains Ω_1 and Ω_2, where $\Omega_1 \cap \Omega_2 \neq \varnothing$. The additive Schwarz domain decomposition can be written as:

$$\begin{aligned} L\phi_1^n &= f \quad \text{in } \Omega_1 \\ B\phi_1^n &= b \quad \text{on } \partial\Omega_1 \setminus \Gamma_1 \\ \phi_1^n &= \phi_2^{n-1} \quad \text{on } \Gamma_1 \end{aligned} \tag{13}$$

$$\begin{aligned} L\phi_2^n &= f \quad \text{in } \Omega_2 \\ B\phi_2^n &= b \quad \text{on } \partial\Omega_2 \setminus \Gamma_2 \\ \phi_2^n &= \phi_1^{n-1} \quad \text{on } \Gamma_2 \end{aligned} \tag{14}$$

On the other hand, in the overlapping multiplicative Schwarz method, each sub-domain problem is solved sequentially, where each sub-domain uses the most recent updates from its neighbouring sub-domains. The multiplicative Schwarz domain decomposition is then illustrated as:

$$\begin{aligned} L\phi_1^n &= f \quad \text{in } \Omega_1 \\ B\phi_1^n &= b \quad \text{on } \partial\Omega_1 \setminus \Gamma_1 \\ \phi_1^n &= \phi_2^{n-1} \quad \text{on } \Gamma_1 \end{aligned} \tag{15}$$

$$\begin{aligned} L\phi_2^n &= f \quad \text{in } \Omega_2 \\ B\phi_2^n &= b \quad \text{on } \partial\Omega_2 \setminus \Gamma_2 \\ \phi_2^n &= \phi_1^{n} \quad \text{on } \Gamma_2 \end{aligned} \tag{16}$$

Both the multiplicative and additive Schwarz domain decomposition methods can be expressed in terms of the Dirichlet artificial interior boundary conditions as follows [28]:

Algorithm (*Overlapping Domain Decomposition Method*)
$\phi \leftarrow$ Guess initial solution on the artificial boundaries
For i = 1,2,..., until convergence
    For j = 1,2,..., K Do:
        Solve equation (17), get $\boldsymbol{\alpha}_i$

$\phi_i \leftarrow G_i \alpha_i$

If (Multiplicative method) Update $\mathbf{F}_i$ for each $\overline{\Omega}_i$ part of Ω from neighbouring domains

End Do

If (Additive method) Update all $\overline{\Omega}_i$ from previous steps

End Do

Numerical Results and Discussion

We consider the following elliptic PDE:

$$\nabla^2 u = -\frac{751\pi^2}{144}\sin\left(\frac{\pi x}{6}\right)\sin\left(\frac{7\pi x}{4}\right)\sin\left(\frac{3\pi y}{4}\right)\sin\left(\frac{5\pi y}{4}\right)+\frac{7\pi^2}{12}\cos\left(\frac{\pi x}{6}\right)\cos\left(\frac{7\pi x}{4}\right)\sin\left(\frac{3\pi y}{4}\right)\sin\left(\frac{5\pi y}{4}\right) + \frac{15\pi^2}{8}\sin\left(\frac{\pi x}{6}\right)\sin\left(\frac{7\pi x}{4}\right)\cos\left(\frac{3\pi y}{4}\right)\cos\left(\frac{5\pi y}{4}\right) \quad (17)$$

which is applied on the L-shaped domain as shown in Figure 2 and has the exact solution as follows:

$$u(x, y) = \sin\left(\frac{\pi x}{6}\right)\sin\left(\frac{7\pi x}{4}\right)\sin\left(\frac{3\pi y}{4}\right)\sin\left(\frac{5\pi y}{4}\right) \quad (18)$$

For clarity, the PDE solution on the L-shaped domain is further depicted in Figure 3. As can be seen, the solution shows wave characteristic on the domain with peak values near the corner point (2, 2).

B-spline approximations in both the global domain and domain decomposition application are compared here. In implementation, evaluation of B-spline basis functions can be either carried out symbolically or directly. In the symbolic manner, B-spline basis functions (including their derivatives) are first constructed and expressed symbolically, then followed by substitution of collocation points. Here, to clearly assess the performance of DD technique the symbolic manner is used. All numerical computations were carried out in MATLAB 2006 environment and run on Toshiba Satellite with OS Windows Vista Basic, processor of Intel Pentium Dual-Core and RAM of 1 GB.

For comparison of numerical accuracies, the following $L2$ error is used:

$$E_n = \sqrt{\frac{\sum_{i=1}^{NC}[u(\mathbf{x}_i) - \hat{u}(\mathbf{x}_i)]^2}{\sum_{i=1}^{NC} u(\mathbf{x}_i)^2}} \quad (19)$$

where NC is the number of collocation points, $u(\mathbf{x}_i)$ and $\hat{u}(\mathbf{x}_i)$ represent the exact and numerical solution, respectively. In addition, a fixed error tolerance 1×10^{-7} and zero initial guess on the artificial interior boundaries are used in all numerical simulations. Moreover, equation (11) is solved by using QR factorization technique [30].

Note that the number of sub-domains of 3 and 4 are respectively used in the L-shaped domain for the implementations of domain decomposition method. Several numerical simulations are carried out to examine the influence of the number of sub-domains as well as that of the amount of overlapping.

Table 1 first describes the B-spline approximation on the one large domain. In Table 1, it is shown that while the global collocation effectively solves the PDE problem with arbitrary knot points, it leads to the use of B-spline bases of high orders. This may be explained from the general continuity posed by the global B-spline approximation which is C^0 only.

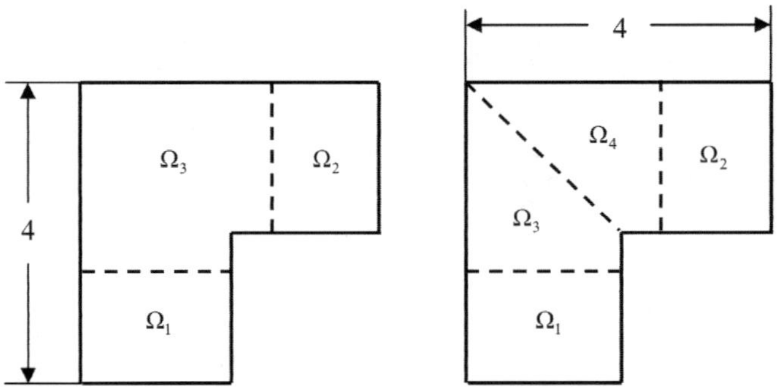

(a) (b)

Figure 2. L-shaped domain with: (a) 3 sub-domains, and (b) 4 sub-domains, used in the overlapping domain decomposition method

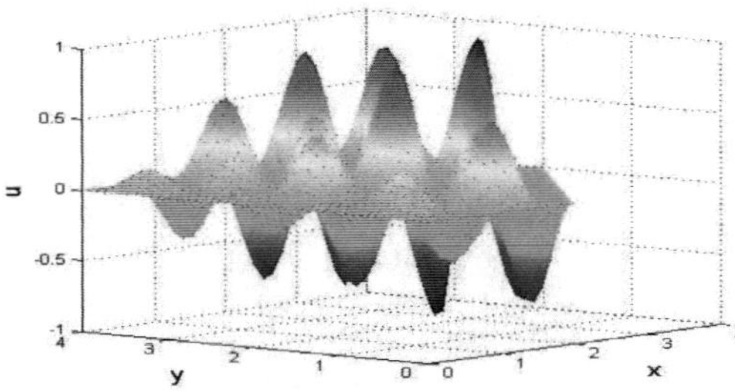

Figure 3. Exact solution of the PDE problem on L-shaped domain.

Table 1. Accuracies of B-spline collocation in global domain for different number of collocation points and B-spline order.

Number of collocation points	Number of domain	B-spline order k	L2 error	Elapsed time (sec)
721	1	23	5.086e 2	84
		25	2.759e 2	116
		27	4.634e 3	154
1155	1	23	3.813e 3	89
		25	5.568e 4	119
		27	3.093e 4	163
1791	1	23	2.250e 4	98
		25	1.971e 5	126
		27	1.757e 6	180

Journal of Physics: Conference Series **423** (2013) 012012 doi:10.1088/1742-6596/423/1/012012

Table 2. Influence of the amount of overlapping in B-spline collocation with domain decomposition.

Number of collocation points	Number of sub-domains	Overlapping ratio o/H_t	B-spline order k	L2 error	Iteration steps	Elapsed time (sec)
1155	3	6.3%	16	6.031e 4	30	48
			18	2.917e 5	27	66
			20	8.797e 6	34	99
		12%	16	6.802e 4	12	46
			18	5.272e 5	11	63
			20	7.674e 6	12	93
1791	3	6.3%	16	9.484e 4	30	57
			18	1.917e 5	27	76
			20	1.943e 6	33	113
		12%	16	7.400e 4	12	51
			18	4.130e 5	12	72
			20	2.330e 6	14	100

Table 3. Influence of the number of sub-domains in B-spline collocation with domain decomposition.

Number of collocation points	Number of sub-domains	Overlapping ratio o/H_t	B-spline order k	L2 error	Iteration steps	Elapsed time (sec)
1791	3	12%	16	7.400e 4	12	51
			18	4.130e 5	12	72
			20	2.330e 6	14	100
	4	12%	16	1.928e 4	23	63
			18	1.740e 5	27	93
			20	5.283e 6	33	132
	4	18%	16	3.450e 4	15	61
			18	1.753e 5	18	86
			20	9.352e 7	19	129

From the results, important features of the combination technique can be described as follows:

(1) As the amount of overlapping increases, the number of iteration steps needed in the multiplicative Schwarz method decreases. This may be understood that more overlapping requires more information exchange, thus allowing faster convergence.

(2) It can be seen that, compared with the global collocation, B-spline order is greatly reduced in the combination technique, thus avoiding ill-conditioning problem and producing increased accuracies. A decrease of B-spline order up to 7 was observed with the combination technique. For example: B-spline collocation with domain decomposition for the number of collocation points 1155, overlapping ratio 6.3% and 3 sub-domains employed the B-spline order of 16, in comparison to the global B-spline collocation of order 23.

(3) The use of more collocation points in the B-spline collocation with domain decomposition can increase the accuracy, as it is also observed in the application of global B-spline collocation. Stable approximation is thus produced with the combined techniques.

Conclusions

In this study, overlapping domain decomposition method of multiplicative Schwarz type has been combined with the global B-spline collocation method. It was shown that the combination method produced higher accuracy with the B-spline basis of much lower order than that needed in implementation of the initial method. In addition, as the approximation stability was also increased, the combined techniques are very promising for advanced problems and applications.

More variations in number of sub-domains/amount of overlapping can be implemented. Other domain decomposition methods are also to be investigated, which are subjects of further publications.

Acknowledgment

The grant provided by Universiti Teknologi PETRONAS is gratefully acknowledged.

References

[1] Fairweather G and Meade D 1989 A survey of spline collocation methods for the numerical solution of differential equations Mathematics for large scale \computing (Lecture notes in pure and applied mathematics vol 120) ed JC Diaz (New York: Marcel Dekker)

[2] Bialecki B and Fairweather G 2001 Orthogonal spline collocation methods for partial differential equations J. Comput. Appl Math.128 pp 55 –82

[3] Botella O 2002 On a collocation B-spline method for the solution of the Navier Stokes equations *Comput. Fluids* **31** pp 397 420

[4] Johnson R W 2005 Higher Order B-Spline Collocation at the Greville Abscissae *Appl. Num. Math.* **52** pp 63 75

[5] Johnson R W 2005 A B-spline collocation method for solving the incompressible Navier Stokes equations using an ad hoc method *Comput. Fluids* **34** pp 121 49

[6] Jator S and Sinkala Z 2007 A high order B-spline collocation method for linear boundary value problems *App. Math. Comp.* **191** pp 100 16

[7] Botella O 2003 A high-order mass-lumping procedure for B-spline collocation method with application to incompressible flow simulations *Int. J. Numer. Meth. Eng.* **41** pp 1295-318

[8] Botella O and Shariff K 2003 B-spline methods in fluid dynamics *Int. J. Comp. Fluid Dyn.* **17** pp 133-49

[9] Morinishi Y, Tamano S and Nakabayashi K A DNS algorithm using B-spline collocation method for compressible turbulent channel flow *Comp. Fluids.* **32** pp 751-76

[10] Widjaja R, Ooi A, Chen L and Manasseh 2005 Computational aeroacoustics using the B-spline collocation method *C. R. Mecanique* **333** 726 31

[11] Saka B and Dag I Quartic 2007 B-spline collocation algorithms for numerical solution of the RLW equation *Numer. Meth. Partial Differential* **23** pp 731 51

[12] Patlashenko I and Weller T 1995 Two-dimensional spline collocation method for nonlinear analysis of laminated panels *Comput. & Struc.* **57** 131-39

[13] Viswanathan K K and Navaneethakrishnan P.V 2003 Free vibration study of layered cylindrical shells by collocation with splines, Journal of Sound and Vibration **260** pp 807 27

[14] Hsu M H 2009 Vibration analysis of pre-twisted beams using the spline collocation method *Journal of Marine Science and Technology* **17**(2) pp 106-15

[15] Rashidinia J and Ghasemi M 2011 B-spline collocation for solution of two-point boundary value problems *J. Comp. App. Math.* 235 pp 2325-42

[16] Botella O and Shariff K 2003 Solving the navier-stokes equations using B-spline numerical methods *Numerical Simulations of Incompressible Flows* ed MM Hafez pp 277-87

[17] Kansa E J 1990 Multiquadric a scattered data approximation scheme with applications to computational fluid dynamics II *Comput. Math. Appl.* **19** pp 147 61

[18] Schwarz H A 1870 Ü ber einen Grenz ü bergang durch alternierendes Verfahren *Ges. Math. Abhandlungen* Bd. 1 pp 133-43

[19] Hu X L, Han D F and Lai M J 2007 Bivariate splines of various degrees for numerical solution of partial differential equations *SIAM J. Sci. Comput.* **29**(3) 13384-54

[20] Awanou G, Lai M J and Wenston P 2006 The multivariate spline method for scattered data fitting and numerical solutions of partial differential equations Wavelets and Splines ed G. Chen and M. J. Lai (Brentwood, TN: Nashboro Press) pp 24 76

[21] Melenk J M 2002 hp-Finite Element Methods for Singular Perturbations *Lecture Notes in Math. 1796* (Berlin: Springer-Verlag)

[22] Solin P, Segeth K and Dolezel I 2003 *High-Order Finite Element Methods* (Boca Raton, FL:Chapman & Hall/CRC Press)

[23] de Boor C 1972 On calculating with B-splines *J. Approx. Theory.* **6**(l) pp 50-62

[24] Cox M 1972 The numerical evaluation of B-spline *J. Inst. Maths. Applics.* **10** pp 134-49

[25] Piegl L and Tiller W 1995 *The NURBS book* (New York: Springer)

[26] Farin G 2002 *Curves and Surfaces for Computer Aided Geometric Design* (San Diego, CA: Academic Press)

[27] Liu Y, Sun L, Xu F, Liu Y and Cen Z 2011 B spline-based method for 2-D large deformation analysis *Eng. Anal. Bound. Elem.* **35** pp 761 67

[28] Zhou X, Hon Y C and Li J 2003 Overlapping domain decomposition method by radial basis functions *Appl. Numer. Math.* **44** pp 241 55

[29] http://www.ddm.org/

[30] MATLAB Help Documentation 2010 The MathWorks, Inc, USA

New Inequalities for Numerical Radius of Hilbert Space Operator And New Bounds For The Zeros Of Polynomials.

Mohammad Al-Hawari [1] and Abdullah Ahmed Aldahash [2]

[1] Associate Professor of Mathematics, College of Applied Medical Sciences, Majmaah University, Kingdom of Saudi Arabia, P.O. Box 1405, Almajmaah Zipcode 11952

[2] Associate Professor of Mathematics Education, Dean of the College community, College of community, Majmaah University

E-mail: m.alhawari@mu.edu.sa [1]

Abstract: It is shown that if A^r does not converge to the zero operator on a complex Hilbert space for some $r \geq 1$, then

$$w(A) \leq \frac{1}{2}\left(\|A^r\|^{\frac{1}{r}} + \|A^2\|^{\frac{1}{2}} \right).$$

And if there exists r such that A^r is a zero operator on a complex Hilbert space for all $r \geq 2$, then

$$w(A) = \frac{1}{2}\|A\|,$$

where $w(.)$ and $\|.\|$ are the numerical radius and the usual operator norm,respectively.Also, the previous inequalities are better than any other classical inequalities.

Key words: Numerical radius, Operator norm, Cauchy- Schwarz inequality.
AMS: 47A12,47A30,47A63.

1. Introduction

Let $B(\mathcal{H})$ denote the C^* - algebra of all bounded linear operators on a complex Hilbert space $\mathcal{H}$ with inner product $\langle \cdot, \cdot \rangle$. For $A \in B(\mathcal{H})$, let $w(A)$ and $\|A\|$ denote the numerical radius and the operator norm of A , respectively. It is well-known that $w(\cdot)$ defines a norm on $B(\mathcal{H})$, and that for every $A \in B(\mathcal{H})$,

$$\frac{1}{2}\|A\| \leq w(A) \leq \frac{1}{2}\left(\|A\| + \|A^2\|^{\frac{1}{2}}\right) \leq \|A\|.$$

For basic properties of the numerical radius, we refer to [4] and [7].
It has been recently shown in [5], that if A^2 does not converge to the zero operator in $M_n(\mathbb{C})$, then

$$w(A) \leq \|A^2\|^{\frac{1}{2}},$$

Also, it has been proved in [6]; that if $A, B \in B(\mathcal{H})$ for $0 < \alpha < 1$, then for $r \geq 1$ we have

$$w(A) \leq 2^{\frac{-1}{r}-1}(\||A|^{2\alpha r}\| + \||A^*|^{2(1-\alpha)r}\|)^{\frac{1}{r}} + \frac{1}{2}\|A^2\|^{\frac{1}{2}}.$$

In the next section of this paper, we establish inequalities that refines the inequalities (1) and (2) using some classical inequalities for nonnegative real numbers and some operator inequalities. To prove our generalized numerical radius inequalities, we need several well-known lemmas. The following lemma follows from spectral theorem for positive operators and Jensen's inequality [1]
.

Lemma 1.1. Let $A \in B(\mathcal{H})$ be positive operator and let $x \in \mathcal{H}$ be any unit vector. Then

$$\langle Ax, x\rangle^r \leq \langle A^r x, x\rangle \text{ for all } r \geq 1,$$

$$\langle A^r x, x\rangle \leq \langle Ax, x\rangle^r \text{ for all } 0 \leq r < 1.$$

Definitions 1.1[3] If $A \in B(\mathcal{H})$, then
(I) The spectral norm (or the operator norm) is defined by

$$\|A\| = \max\{\frac{\|Ax\|}{\|x\|} : \|x\| \neq 0\}, \|A\| = \max\{\|Ax\| : \|x\| = 1\}.$$

(II) The numerical radius of A is defined by

$$w(A) = \max\{| \langle Ax, x\rangle | : x \in H, \|x\| = 1\}$$

In the following Lemma we present a sharper numerical radius inequality for any operator.

Lemma1.2.[6].if $A, B \in B(\mathcal{H})$ for $0 < \alpha < 1$, then for $r \geq 1$ we have

$$w(A) \leq 2^{\frac{-1}{r}-1}(\||A|^{2\alpha r}\| + \||A^*|^{2(1-\alpha)r}\|)^{\frac{1}{r}} + \frac{1}{2}\|A^2\|^{\frac{1}{2}}.$$

2. Numerical radius inequalities

From Lemma1.2 we present several classical and sharp numerical radius inequalities as follows.

Theorem 2.1.if A^r does not converge to the zero operator on a complex Hilbert space for some $r \geq 1$, then

$$w(A) \leq \frac{1}{2}\left(\|A^r\|^{\frac{1}{r}} + \|A^2\|^{\frac{1}{2}}\right).$$

proof: Letting $\alpha = \frac{1}{2}$ and $r \geq 1$ in (8), we obtain the inequality

$$w(A) \leq 2^{\frac{-1}{r}-1}(\||A|^r\| + \||A^*|^r\|)^{\frac{1}{r}} + \frac{1}{2}\|A^2\|^{\frac{1}{2}}.$$

and since,

$$s_1^2(|A|^r) = s_1^2(|A^*|^r),$$

where $s_1(A)$ denotes the largest singular value of A and $s_1(A) = \|A\|$.
Hence,

$$\||A|^r\| = \||A^*|^r\|.$$

So, we get the result.

Corollary 2.1. Letting $r = 1$ in (9), we obtain the inequality

$$w(A) \leq \frac{1}{2}\left(\|A\| + \|A^2\|^{\frac{1}{2}}\right),$$

(that is Kittaneh's inequality see e.g.[1])

Corollary 2.2. Letting $r = 2$ in (9), we obtain the inequality

$$w(A) \leq \|A^2\|^{\frac{1}{2}},$$

A^2 does not converge to the zero operator. (that is Al-Hawari's inequality see e.g.[5]).

If A^2 is a non-zero operator,then the inequality (11) is not true
To see this ,consider

$$A = \begin{bmatrix} t & 1 \\ 0 & 0 \end{bmatrix},$$

then

$$w(A) = \frac{t + \sqrt{1+t^2}}{2} \geq \frac{1}{2},$$

and

$$\| A^2 \| \leq \frac{1}{2}, \text{for taking small } t.$$

The inequality (9) refines the inequality (1), because

$$\| A^r \|^{\frac{1}{r}} \leq \| A \|,$$

for every $A \in B(\mathcal{H})$.

Corollary 2.3. If A^r is the zero operator for all $r \geq 2$, then

$$w(A) \leq \frac{1}{2} \| A^r \|^{\frac{1}{r}} \leq \frac{1}{2} \| A \|,$$

A is a non-zero operator.

Corollary 2.4. Since $w(A) \geq \frac{1}{2} \| A \|$ .So,we get from the inequalities (1) and (13)

$$w(A) = \frac{1}{2} \| A \|.$$

If A^r is the zero operator for all $r \geq 2$. (e.g.see [1]).

It should be mentioned that it is not true if A^r is the zero operator for some $r \geq 3$, then we get the inequality (14) .

To see this, consider

$$A = \begin{bmatrix} 0 & 0 & 0 \\ 1 & 0 & 0 \\ 0 & 1 & 0 \end{bmatrix},$$ then

$$w(A) = \frac{1}{\sqrt{2}} \neq \frac{1}{2}\|A\|,$$

and

$$\|A\| = \|A^2\| = 1,$$

where A^3 is the zero matrix operator.

Open Problems

The first open problem is possible to complement the bound (9) by giving an upper bound estimate for the zeros of

$$p(z) = z^n + a_n z^{n-1} + a_{n-1} z^{n-2} + \ldots + a_2 z + a_1 \text{ with } a_i \in \mathbb{C} \text{ for } 1 \leq i \leq n \text{ and } a_1 \neq 0.$$

The second open problem is possible to complement the upper bound (9) by giving a lower bound estimate for the zeros of p .To see this,observe that the zeros of the polynomial

$$q(z) = \frac{z^n}{a_1} p\left(\frac{1}{z}\right),$$

are the reciprocals of those of p . Thus ,applying the upper bound (9) to the zeros of q yields the desired lower bound estimate for the zeros of p . this enables us to present a new annulus containing the zeros of p .

References

[1] F.Kittaneh (2003), A numerical radius inequality and an estimate for the numerical radius of the Frobenius companion matrix. Studia Math 158, 11 -- 17.

[2] F.Kittaneh (2005), Numerical radius inequalities for Hilbert Space operators. Studia Math 168, 73-80.

[3] Horn , R.A. and Johnson , C.R. (1991), Topics in Matrix Analysis. Cambridge

Journal of Physics: Conference Series **423** (2013) 012013 doi:10.1088/1742-6596/423/1/012013

University press , Cambridge.
[4] K.E.Gustafson , and K.M. Rao, (1997), Numerical Range. Springer--Verlag, New York.
[5] M. Al-Hawari, (2009), New estimate for the numerical radius of a given matrix and new bounds for zeros of polynomials, studia math(Romania)., .
[6] M. Al-Hawari, (2009), Sharper Inequalities for Numerical Radius of Hilbert Space Operators, Int. J. of Math.analysis,21(3), 1021-1025.
[7] M. El-Haddad and F. Kittaneh, (2007), Numerical radius inequalities for Hilbert space operators ii, Studia Math., 182(2),133-140.
[8] R Bhatia, F.Kittaneh, (2008), The matrix arithmetic-geometric mean inequalities. Linear Algebra Appl.428, 2177-2191.
[9] S.S.Dragomir, (2006), Some inequalities for the Eculidean operator radius of two operators in Hilbert spaces, Linear Algebra Appl. 419, 256-264.

SciеTech 2013 IOP Publishing
Journal of Physics: Conference Series **423** (2013) 012014 doi:10.1088/1742-6596/423/1/012014

Aliasing free for evolutionary spectrum of non stationary processes

R Sabre [1], **S Safi** [2], **B Schmitt** [2,]

[1] Laboratory LE2I CNRS 6306, University of Bourgogne, Dijon, France
[2] IMB, University of Bourgogne, Dijon, France

E-mail: r.sabre@agrosupdijon.fr

Abstract. This paper proposes a non parametric discrete estimate of evolutionary spectral density of non-stationary process with continuous time. To solve a problem related to aliasing effects, we assume that the evolutionary spectral density has a compact support and we use a specific uniform sampling depending on the width of the compact support. An asymptotically unbiased and consistent estimate is given.

1. Introduction

Stationarity is an important assumption in the spectral analysis of time series. However, in many applications the signal is modeled as a non-stationary process. Priestley in [9]),[10]),[11]) developed the basic theory of Evolutionary Spectrum ES for non-stationary oscillatory processes. Other studies have been made as in [7], which where based on Wold-Cramér decomposition of the process. The ES is time-dependant and describes the local power-frequency distribution at each instant of time. The application of the ES covers various scientific fields: signal and image processing [4], [1], [20], seismic [18], oceanography, music [2]. The estimation of the evolutionary spectral density is studied in [12], [7], [4], [16], [19] and [8].

In this work, as in [9], [10],[11], we consider a non-stationary oscillatory process $X(t)$ defined by:

$$X\left(t\right)=\int_{-\infty}^{+\infty}e^{it\omega}A_t(\omega)dZ(\omega);\quad t\in R \tag{1}$$

where, $A_t(\omega)$, as function of t, is the generalized Fourier transform whose module has an absolute maximum at origin. Z is a process with orthogonal increments and $E\left[|dZ(\omega)|^2\right]=d\mu(\omega)$, for some positive measure μ. Processes which admit such representation are called oscillatory processes. The evolutionary spectral measure, with respect to the oscillatory family $\mathcal{F}=A_t(w)e^{itw}$, is defined at each instant t by $dH_t(\omega)=|A_t(\omega)|^2\,d\mu\left(\omega\right)$. The choice of the oscillatory process is motivated by its physical interpretation namely the variance of the process interpreted as a measure of the total power of the process at time t: $Var(X(t))=\int_{-\infty}^{+\infty}dH_t(\omega)$ see [9]). The evolutionary spectral density of the process $\{X(t)\}$ is given by $h_t(\omega)$ and defined as follows: $h_t(\omega)=\frac{dH_t(\omega)}{d\omega}$, $\omega\in R$. Priestley in [12] proposed an estimator of $h_t(\omega)$ defined by: $\hat{h}_t(\omega)=\int_{-\infty}^{\infty}w_{T'}(u)|U(t-u,\omega)|^2du$ where $U(t,\omega)=\int_{t-T}^{t}g_uX(t-u)e^{-i\omega(t-u)}du$. The filter g_u and the weight function $w_{T'}(u)$ are satisfying the usual conditions given in [9]. To calculate this estimator, we must observe the process on a time interval $[t-T,t]$ which is impossible in practice.

The first idea is to use the discrete estimator proposed in [12] for the time-discrete non-stationary processes, but this estimator is not appropriate because our process is time-continuous. Indeed, we show in this paper that this estimator converges to $\sum_{k\in Z} h_t(\omega + 2k\pi)$ instead of converging to $h_t(w)$, this phenomenon is called the "aliasing effects" see [6]. Several studies have provided a solution for alaising effects using non-uniform sampling. Except that non-uniform sampling is often perceived as not suitable when we has need to observe the signal at precise instants as it frequently happens in astronomy, medicine or geosciences. It is also not suitable when there are missing observations. These difficulties are illustrated in Lomb [5], Scargle [15] and Tarczynski [17] . Our contribution is to provide a solution for a uniform sampling when the evolutionary spectral density has compact support. Indeed, as in Sabre ([14] and [13]), we assume that the evolutionary spectral density has a compact support i.e. it is vanishing outside an interval $I = [-\Omega, \Omega]$.We construct a periodogram where the process $\{X(t),\ t \in R\}$ is sampled at discrete instants chosen such that the interval between successive observations depends on the width of the interval I. Using usual technique, we give a smoothing estimator of $h_t(w)$ and show that it is asymptotically unbiased and consistent.

The paper is organized as follows: Section 2 shows the existence of the aliasing effects. Section 3 constructs a periodogram and shows that it is an asymptotically unbiased estimator of $h_t(\omega)$ (lemma 2.1, theorem 2.2). Section 4 presents a smoothing estimator in the neighborhood of time-instant t using a weight function and shows that it is an asymptotically unbiased and consistent estimator of the (weighted) average of $h_t(\omega)$ in the neighborhood of the time-instant t (theorems 3.1 and 3.2). Section 5 contains the concluding remarks, the potential applications and open research problems.

2. The aliasing effects

According to [9], a non-stationary oscillatory centered processes with continuous time can be written as (1). where $A_t(\omega) = \int_{-\infty}^{+\infty} e^{iut} dK_\omega(u)$. K_ω is a measure such that: $\int_{-\infty}^{+\infty} |dK_\omega(\theta)| = 1$

To estimate the evolutionary spectral density $h_t(w)$ from a discrete sample, we use the estimator proposed by [9] for process with discrete time defined as follows:

$$I_t(w) = \sum_u g_u X(t-u) e^{-iw(t-u)}$$

where g_u is a filter respecting the usual conditions defined by:

$$g_u = O(e^{-|u|}) \quad (2)$$

$$2\pi \sum_{u=-\infty}^{+\infty} |g_u|^2 = \int |\Gamma(u)|^2 du = 1 \quad (3)$$

Γ is transform Fourier of g: $\Gamma(\omega) = \sum_{u=-\infty}^{+\infty} g_u e^{-iu\omega}$. The width of the filter is described as $\mathcal{B}_g = \sum_{u=-\infty}^{+\infty} |u|\,|g_u| < \infty$

For each family $\mathcal{F} = \{e^{it\omega} A_t(\omega)\}$, the function $\mathcal{B}_{\mathcal{F}}(\omega)$ is defined by: $\mathcal{B}_{\mathcal{F}}(\omega) = \int_{-\infty}^{+\infty} |\theta|\,|dK_\omega(\theta)|$. This function is a measure of the "width" of $|dK_\omega(\theta)|$. Assume that the family $\mathcal{F}$ is semi stationary i.e. the function $\mathcal{B}_{\mathcal{F}}(\omega)$ is bounded for all ω. the function $\mathcal{B}_{\mathcal{F}} = (\sup_\omega \mathcal{B}_{\mathcal{F}}(\omega))^{-1}$ is the characteristic "width" of the family $\mathcal{F}$. The filter g is chosen such that its width $\mathcal{B}_g$ is enough smaller than $\mathcal{B}_{\mathcal{F}}$: $\mathcal{B}_g << \mathcal{B}_{\mathcal{F}}$, and $\frac{\mathcal{B}_g}{\mathcal{B}_{\mathcal{F}}} < \varepsilon$, ε is a small nonnegative number. As in [9],[10], [12], suppose that Γ is pseudo δ-function with respect to h_t in order $\frac{\mathcal{B}_g}{\mathcal{B}_{\mathcal{F}}}$, which means:

$$\left| \int \Gamma(s) h_t(s+k) ds - h_t(k) \int \Gamma(s) ds \right| < \frac{\mathcal{B}_g}{\mathcal{B}_{\mathcal{F}}}$$

.

Let us show that $|I_t(w)|^2$ is an asymptotically unbiased estimator of $\sum_{k\in Z} h_t(2k\pi + w)$ precisely $E|I_t(w)|^2 = \sum_{k\in Z} h_t(2k\pi + w) + O(\varepsilon)$. Indeed, it is easy to show that

$$E\,|I_t(w)|^2 = \int |\Gamma_{t,v+w}(v)|^2\,|A_t(w+v)|^2\,d\mu(w+v)$$

where

$$\Gamma_{t,\lambda,\tau}(\theta) = \sum_{u=-\infty}^{+\infty} g_u e^{-iu\theta}\frac{A_{t-\tau u}(\lambda)}{A_t(\lambda)}. \tag{4}$$

From the theorem 11.2.1 in Priestley [13], we have $\Gamma_{t,v+w}(v) = \Gamma(v) + r(t,w)$. where

$$|r(t,w)| < \frac{\varepsilon}{|A_t(w+v)|}. \tag{5}$$

Therefore $E\,|I_t(w)|^2 = \int |\Gamma(v)|^2\,h_t(w+v)dv + L_1 + L_2 + L_3$, where

$$\begin{aligned} L_1 &= \int \Gamma(v)r^*(t,w)\,|A_t(v+w)|^2\,d\mu(v+w) \\ L_2 &= \int \Gamma^*(v)r(t,w)\,|A_t(v+w)|^2\,d\mu(v+w) \\ L_3 &= \int |r(t,w)|^2\,|A_t(v+w)|^2\,d\mu(v+w) \end{aligned}$$

r^* is the conjugate of r. Since the measure μ is finite and using the inequality (5), we have

$$L_2 \le \varepsilon \int_\Lambda d\mu(v+w) + \varepsilon\int_{-\infty}^{0} |\Gamma(v)|^2 dH_t(v+w)$$

where the set Λ is defined as follows: $\Lambda = \{v; |\Gamma(v)|\,|A_t(v+\omega_0)| \le 1\}$.

$$L_2 \le \varepsilon \int_\Lambda d\mu(v+w) + \varepsilon \sup(h_t)\int_{-\infty}^{+\infty} |\Gamma(v)|^2 dv$$

Thus $L_2 = O(\varepsilon)$. Similarly we show that $L_1 = O(\varepsilon)$ and $L_3 = O(\varepsilon)$. Therefore

$$E\,|I_t(w)|^2 = \sum_{k\in Z}\int_{-\pi}^{\pi} |\Gamma(v+2k\pi)|^2\,h_t(v+2k\pi+w)dv + O(\varepsilon)$$

Since Γ is 2π-periodic, we obtain $E\,|I_t(w)|^2 = \int_{-\pi}^{\pi} |\Gamma(v)|^2 \sum_{k\in Z} h_t(v+2k\pi+w)dv + O(\varepsilon)$ From (3) and (4), we get $E\,|I_t(w)|^2 = \sum_{k\in Z} h_t(2k\pi+w) + O(\varepsilon)$.

3. periodogram and its proprieties

To solve this problem, we assume that the evolutionary spectral density $h_t(\omega)$ is vanishing outside an interval $]-\Omega,\Omega[$, where Ω is a non negative real and we construct the following periodogram:

$$U_t(\omega_0) = \tau^{\frac{1}{2}}\sum_{u=-N}^{+N} g_u X(t-\tau u)\,e^{-i\omega_0(t-\tau u)}; \qquad \omega_0 \in]-\Omega,\Omega[$$

where τ is a real number such as $0 < \tau < \inf\left\{\frac{2\pi}{\Omega+\omega_0}, \frac{4\pi}{\Omega-\omega_0}\right\}$ and g is a filter defined above chosen such that $|\Gamma(v)|^2$ is pseudo δ-function with respect to the function $x \to h_t(\frac{x}{\tau})$ in order ε.

Firstly let us, in the following lemma, give the results about the generalized transform of the filter g_u.

Lemma 3.1 *For all real numbers λ and θ, we have*

$$|A_t(\lambda)|\,|\Gamma_{t,\lambda,\tau}(\theta) - \Gamma(\theta)| \leq \tau\varepsilon.$$

where $\Gamma_{t,\lambda,\tau}(\theta)$ is defined in (12).

The proof of this lemma is similar to that used in Priestley [10].

The following theorem shows that $|U_t(\omega_0)|^2$ is an asymptotically unbiased estimator of $h_t(\omega_0)$.

Theorem 3.2 *Let ω_0 be a real number, we have*

$$\left[E\,|U_t(\omega_0)|^2\right] \simeq h_t(\omega_0) + O((\tau+1)\varepsilon).$$

Proof: We have $U_t(\omega_0) = \tau^{\frac{1}{2}} \int_{-\infty}^{+\infty} \Gamma_{t,\omega,\tau}(\tau\omega - \tau\omega_0) A_t(\omega) e^{it(\omega-\omega_0)} dZ(\omega)$. Let $\omega - \omega_0 = v$, then $U_t(\omega_0) = \tau^{\frac{1}{2}} \int_{-\infty}^{+\infty} \Gamma_{t,v+\omega_0,\tau}(v\tau) A_t(\omega_0 + v) e^{itv} dZ(\omega_0 + v)$. Since Z is a process with orthogonal increments, we obtain

$$E\left[|U_t(\omega_0)|^2\right] = \tau \int_{-\infty}^{+\infty} |\Gamma_{t,v+\omega_0,\tau}(v\tau)|^2\, |A_t(\omega_0+v)|^2\, d\mu(\omega_0+v).$$

From the lemma 2.1, we have $\Gamma_{t,v+\omega_0,\tau}(v\tau) = \Gamma(v\tau) + r(t,\omega_0,\tau)$ where

$$|r(t,\omega_0,\tau)| \leq \frac{\varepsilon}{A_t(\omega_0+v)}. \tag{6}$$

Thus,

$$E\left[|U_t(\omega_0)|^2\right] = \tau \int_{-\infty}^{+\infty} |\Gamma(v\tau)|^2\, h_t(v+\omega_0) dv + I_1 + I_2 + I_3,$$

where

$$I_1 = \tau \int_{-\infty}^{+\infty} \Gamma(v\tau) r^*(t,\omega_0,\tau)\, |A_t(v+\omega_0)|^2\, d\mu(v+\omega_0).$$

$$I_2 = \tau \int_{-\infty}^{+\infty} \Gamma^*(v\tau) r(t,\omega_0,\tau)\, |A_t(v+\omega_0)|^2\, d\mu(v+\omega_0),$$

$$I_3 = \tau \int_{-\infty}^{+\infty} |r(t,\omega_0,\tau)|^2\, |A_t(v+\omega_0)|^2\, d\mu(v+\omega_0).$$

Using the inequality (6), and the fact that the measure μ is finite, we get $I_3 \leq \tau\varepsilon^2 \int_{-\infty}^{+\infty} d\mu(v) = O(\tau\varepsilon^2)$, and $|I_2| \leq \tau\varepsilon \int_{-\infty}^{+\infty} |\Gamma(v\tau)|\,|A_t(v+\omega_0)|\, d\mu(v+\omega_0)$. To show that $I_2 = O(\tau\varepsilon)$, it remains to be proven that the right side of the last integral is finite. Indeed, let the set Λ be defined as follows:

$$\Lambda = \{v; |\Gamma(v\tau)|\,|A_t(v+\omega_0)| \leq 1\}.$$

$\bar{\Lambda}$ being the complementary of Λ. Therefore,

$$\begin{aligned}\int_{-\infty}^{+\infty} |\Gamma(v\tau)|\,|A_t(v+\omega_0)|\, d\mu(v+\omega_0) &\leq \int_{\Lambda} d\mu(v+\omega_0) + \int_{-\infty}^{+\infty} |\Gamma(v\tau)|^2\, h_t(v+\omega_0) dv \\ &\leq \frac{1}{\tau} \sup_{v'} h_t(v') \int_{-\infty}^{+\infty} |\Gamma(v')|^2\, dv'.\end{aligned}$$

Thus, $|I_2| = O((\tau+1)\varepsilon)$. In the same way, we can show $|I_1| = O((\tau+1)\varepsilon)$. Therefore,

$$E\left[|U_t(\omega_0)|^2\right] = \tau \int_{-\infty}^{+\infty} |\Gamma(v\tau)|^2 h_t(v+\omega_0)dv + O((\tau+1)\varepsilon).$$

Let $v\tau = \omega$, we get

$$E\left[|U_t(\omega_0)|^2\right] = \sum_{j\in Z} \int_{2j\pi}^{2(j+1)\pi} |\Gamma(\omega)|^2 h_t(\frac{\omega}{\tau}+\omega_0)d\omega + O((\tau+1)\varepsilon)$$

Using the following change of variable $v = \omega - 2j\pi$, and the fact that $\Gamma(v)$ is 2π-periodic, we can write

$$E\left[|U_t(\omega_0)|^2\right] = \int_0^{2\pi} |\Gamma(v)|^2 \sum_{j\in Z} h_t(\frac{v+2j\pi}{\tau}+\omega_0)dv + O((\tau+1)\varepsilon).$$

Let j be an integer number such that $h_t(\frac{v+2j\pi}{\tau}+\omega_0) \neq 0$ for all $v \in [0, 2\pi]$. Thus, $\frac{2j\pi}{\tau}+\omega_0 \leq \frac{v+2j\pi}{\tau}+\omega_0 \leq \frac{2(j+1)\pi}{\tau}+\omega_0$, the integer j verifies $\frac{2(j+1)\pi}{\tau}+\omega_0 < \Omega$ and $-\Omega < \frac{2j\pi}{\tau}+\omega_0$. Thus, $\frac{(-\Omega-\omega_0)\tau}{2\pi} < j < \frac{(\Omega-\omega_0)\tau}{2\pi} - 1$. As $\tau < \inf\left\{\frac{2\pi}{\Omega+\omega_0}, \frac{4\pi}{\Omega-\omega_0}\right\}$, we have $\frac{(\Omega-\omega_0)\tau}{2\pi} - 1 < 1$ and $-1 < \frac{(-\Omega-\omega_0)\tau}{2\pi}$. Consequently, $j = 0$. Then

$$E\left[|U_t(\omega_0)|^2\right] = \int_0^{2\pi} |\Gamma(v)|^2 h_t(\frac{v+\tau\omega_0}{\tau})dv + O((\tau+1)\varepsilon).$$

Since $|\Gamma(v)|^2$ is pseudo δ-function with respect to the function $x \to h_t\left(\frac{x}{\tau}\right)$ in order ε, we deduce that $E|U_t(\omega_0)|^2 = h_t(\omega_0) + O((\tau+1)\varepsilon)$.

4. Smoothing periodogram

In order to give a consistent estimate of $h_t(\omega_0)$, we propose the smoothing estimator defined as follows:

$$\widehat{h}_{t,T'}(\omega_0) = \sum_{v=-m}^{m} w_{T',v} |U_{t-v}(\omega_0)|^2,$$

where $w_{T',v}$, is a positive weight function depending on the parameter T', and satisfying the following conditions:

a) $w_{T',v} = 0$, for all v such that $|v| > m$,

b) $w_{T',v} = w_{T',-v}$,

c) $\sum\limits_{v=-m}^{m} w_{T',v} = 1$,

d) $\sum\limits_{v=-m}^{m} w^2_{T',v} < \infty$.

e) Suppose that there exists a constant C such that $\lim\limits_{T'\to\infty}\left\{T' \sum\limits_{u=-m}^{m} |W_{T',u}|^2\right\} = C$,

where $W_{T',u} = \sum\limits_{v=-m}^{m} e^{-ivu} w_{T',v}$. The following theorem shows that $\widehat{h}_{t,T'}(\omega_0)$ is an estimator asymptotically unbiased of the average values of $h_t(\omega_0)$ in the neighborhood of the time-instant t.

Theorem 4.1 *Let* $\omega_0 \in]-\Omega, +\Omega[$, *then*

$$E\left[\widehat{h}_{t,T'}(\omega_0)\right] = \overline{h}_t(\omega_0) + O((\tau+1)\varepsilon)$$

where $\overline{h}_t(\omega_0) = \sum\limits_{v=-m}^{m} w_{T',v} h_{t-v}(\omega_0)$.

It is easy to prove this theorem by using the theorem 2.2. When the process is gaussian, the following theorem shows that the variance of $\widehat{h}_{t,T'}(\omega_0)$ converges to zero and the estimator $\widehat{h}_{t,T'}(\omega_0)$ converges to $\overline{h}_t(\omega_0)$ in mean square.

Theorem 4.2 *Suppose that the process X_t is gaussian. Let $\omega \in]-\Omega, +\Omega[$, then*

$$Var[\widehat{h}_{t,T'}(\omega)] = O(\frac{\tau^2}{T'}).$$

$$E\left|\widehat{h}_{t,T'}(\omega_0) - \overline{h}_t(\omega_0)\right|^2 = O(\frac{\tau^2}{T'}) + O\left((\tau+1)^2\varepsilon^2\right)$$

Proof We have

$$\begin{aligned} E\left\{|U_t(\omega)|^2 |U_s(\omega')|^2\right\} &= \tau^2 \sum_u \sum_v \sum_r \sum_z g_u g_v^* g_r g_z^* E\left[X_{t-u\tau} X_{t-v\tau}^* X_{t-r\tau} X_{t-z\tau}^*\right] \\ & \quad e^{-i\omega(t-u\tau)} e^{i\omega(t-v\tau)} e^{-i\omega'(s-r\tau)} e^{i\omega'(s-z\tau)} \end{aligned}$$

Since X_t is Gaussian we get

$$\begin{aligned} E\left[X_{t-u\tau} X_{t-v\tau}^* X_{t-r\tau} X_{t-z\tau}^*\right] &= R(t-u\tau, t-v\tau) R(s-r\tau, s-z\tau) \\ &+ R(t-u\tau, s-z\tau) R(t-v\tau, s-r\tau \\ &+ R(t-u\tau, s-r\tau) R(t-r\tau, s-z\tau) \end{aligned}$$

where

$$R(p,q) = \int e^{i\omega(p-q)} A_p(\omega) A_q^*(\omega) d\omega = E\left[X_p X_q^*\right]. \tag{7}$$

To simplify, we suppose that $d\mu(\omega) \equiv d\omega$. Using (7) and the integral representation, we obtain

$$E\left\{|U_t(\omega)|^2 |U_s(\omega')|^2\right\} = E\left[|U_t(\omega)|^2\right] E\left[|U_s(\omega')|^2\right] + S_1 + S_2. \tag{8}$$

where

$$\begin{aligned} S_1 &= \tau^2 \sum_{u,v,r,z} g_u g_v^* g_r g_z^* e^{-i\omega(t-u\tau)} e^{-i\omega'(s-r\tau)} e^{i\omega(t-v\tau)} e^{i\omega'(s-z\tau)} \\ &\times \int\int e^{i(t-u\tau)\theta} e^{-i(s-z\tau)\theta} A_{t-u\tau}(\theta) A_{s-z\tau}^*(\theta) e^{-i(t-v\tau)\phi} e^{i(s-r\tau)\phi} A_{t-v\tau}^*(\phi) A_{s-r\tau}(\phi) d\theta d\phi \\ S_2 &= \tau^2 \int\int \left\{\sum_u g_u e^{-iu(-\omega\tau+\theta\tau)} A_{t-u\tau}(\theta)\right\} \left\{\sum_v g_v^* e^{iv(-\omega\tau+\phi\tau)} A_{t-v\tau}^*(\phi)\right\} \\ & \left\{\sum_r g_r e^{-ir(-\omega'\tau+\phi\tau)} A_{s-r\tau}(\phi)\right\} \left\{\sum_z g_z^* e^{iz(-\omega'\tau+\theta\tau)} A_{s-z\tau}^*(\theta)\right\} e^{i\theta(t-s)} e^{i\phi(s-t)} d\theta d\phi. \end{aligned}$$

The lemma 2.1 and the definition of Γ, imply

$$\begin{aligned} S_1 &\simeq \tau^2 \int\int \Gamma(\theta\tau - \omega\tau) A_t(\theta) \Gamma^*(\phi\tau - \omega\tau) A_t^*(\phi) \Gamma(\phi\tau - \omega'\tau) A_s(\phi) \\ &\times \Gamma^*(\theta\tau - \omega'\tau) A_s^*(\theta) e^{i\theta(t-s)} e^{i\phi(s-t)} d\theta d\phi. \end{aligned}$$

Since g_u and $X(t)$ are real, we have

$$A_t(-\theta) = A_t^*(\theta), \qquad \Gamma(-\theta+\omega) = \Gamma^*(\theta-\omega).$$

Consider the function $B_{t,\tau}$ defined by $B_{t,\tau}(x) = A_t(\frac{x}{\tau})$ and putting $\phi = -\phi$, the last equality becomes

$$\begin{aligned} S_1 \simeq\ & \tau^2 \left\{ \int \Gamma(\theta\tau - \omega\tau)\Gamma^*(\theta\tau - \omega'\tau)B_{t,\tau}(\tau\theta)B_{s,\tau}^*(\tau\theta)e^{i\theta(t-s)}d\theta \right\} \\ & \left\{ \int \Gamma^*(\phi\tau + \omega'\tau)\Gamma(\phi\tau + \omega\tau)B_{s,\tau}^*(\tau\phi)B_{t,\tau}(\tau\phi)e^{-i\phi(s-t)}d\phi \right\}. \end{aligned}$$

Since $|\Gamma(\omega)|^2$ is pseudo δ-function with respect to $f_{t,\tau}(\omega) = h_t(\frac{\omega}{\tau}) = |A_t(\frac{\omega}{\tau})|^2$, it is reasonable to suppose that $\Gamma(\omega)$ is pseudo δ-function with respect to $A_t(\frac{\omega}{\tau})$ which is equal to $B_{t,\tau}(\omega)$. From this approximation, we have

$$\begin{aligned} S_1 \simeq \frac{1}{4}\tau^2 & \left\{ B_{t,\tau}(\omega\tau)B_{s,\tau}^*(\omega\tau) + B_{t,\tau}(\omega\prime\tau)B_{s,\tau}^*(\tau\omega\prime) \right\} \left\{ B_{t,\tau}^*(\tau\omega)B_{s,\tau}(\tau\omega) + B_{t,\tau}^*(\tau\omega\prime)B_{s,\tau}(\tau\omega') \right\} \\ & \int\int \Gamma(\theta\tau - \omega\tau)\Gamma^*(\theta\tau - \omega'\tau)\Gamma(\phi\tau + \omega\tau)\Gamma^*(\phi\tau + \omega'\tau)e^{-i(\theta+\phi)(s-t)}d\theta d\phi. \end{aligned} \tag{9}$$

In the same way, we give a similar expression for S_2. The covariance can be written as:

$$Cov\left[|U_t(\omega)|^2, |U_s(\omega'|^2\right] = S_1 + S_2$$

From (8), we deduct that

$$Cov\left[\hat{f}_t(\omega), \hat{f}_s(\omega')\right] = \sum_{u,v=-m}^{m} w_{T',u}w_{T',v}(S_1 + S_2) = R_1 + R_2$$

R_1 and R_2 correspond to the respective contributions of S_1 and S_2. From (9), R_1 becomes:

$$\begin{aligned} R_1 \simeq \frac{1}{4}\tau^2 \sum_{i=1}^{4} \int\int \sum_{v=-m}^{m}\sum_{u=-m}^{m} & w_{T',u}w_{T',v}b_{t-u}^{(i)}b_{s-v}^{(i)}e^{-iu(\theta+\phi)}e^{iv(\theta+\phi)}\Gamma(\theta\tau - \omega\tau) \\ & \Gamma^*(\theta\tau - \omega'\tau)\Gamma(\phi\tau + \omega\tau)\Gamma^*(\phi\tau + \omega'\tau)e^{-i(s-t)(\theta+\phi)}d\theta d\phi \end{aligned}$$

where $b_{t-u}^{(1)} = |B_{t-u,\tau}(\omega\tau)|^2$, $b_{t-u}^{(2)} = B_{t-u,\tau}(\omega\tau)B_{t-u,\tau}^*(\omega'\tau)$, $b_{t-u}^{(3)} = B_{t-u,\tau}(\omega'\tau)B_{t-u,\tau}(\omega\tau)$ and $b_{t-u}^{(4)} = |B_{t-u,\tau}(\omega'\tau)|^2$

$$\begin{aligned} R_1 \simeq\ & \frac{1}{4}\tau^2 \sum_{i=1}^{4} \int\int \Gamma(\theta\tau - \omega\tau)\Gamma^*(\theta\tau - \omega'\tau)\Gamma(\phi\tau + \omega\tau)\Gamma^*(\phi\tau + \omega'\tau)K_t^i(\theta+\phi) \\ & K_s^{i*}(\theta+\phi)e^{-i(s-t)(\theta+\phi)}d\theta d\phi. \end{aligned}$$

with $K_t^i(\theta) = \sum_u w_{T',u}b_{t-u}^{(i)}e^{-iu\theta}$. Then $w_{T',v}b_{t-v}^{(i)} = \sum_\theta K_t^i(\theta)e^{iv\theta}$. Let $H_{t,\tau}^i(\theta) = K_t^i(\frac{\theta}{\tau})$, we have

$$\begin{aligned} R_1 \simeq\ & \frac{1}{4}\tau^2 \sum_{i=1}^{4} \int\int \Gamma(\theta\tau - \omega\tau)\Gamma^*(\theta\tau - \omega'\tau)\Gamma(\phi\tau + \omega\tau)\Gamma^*(\phi\tau + \omega'\tau)H_{t,\tau}^i(\tau\theta + \tau\phi)H_{s,\tau}^{i*}(\tau\theta + \tau\phi) \\ & e^{-i(s-t)(\theta+\phi)}d\theta d\phi. \end{aligned}$$

where $H^i_{t,\tau}(\theta) = \sum_u w_{T',u} b^{(i)}_{t-u} e^{-iu\frac{\theta}{\tau}} = \sum_v w_{T',v\tau} b^{(i)}_{t-v\tau} e^{-iv\theta}$ Therefore $H^i_{t,\tau}(\theta)$ is the Fourier transform of the product of two functions: $S_1 : v \to w_{T',v\tau}$ and $S_2 : v \to b^{(i)}_{t-v\tau}$. As in Priestley [10], [13], suppose that Fourier transform of S_2 is pseudo δ-function with respect to $\Gamma(x)$. Choose T' such that Fourier transform of S_1 is pseudo δ-function with respect to $\Gamma(x)$. Consequently, $H^i_{t,\tau}(x) H^{i*}_{s,\tau}(x)$ is pseudo δ-function with respect to $\Gamma(x)$. We obtain

$$\int K^i_t(\theta) K^{i*}_s(\theta) d\theta = \int H^i_{t,\tau}(\tau\theta) H^{i*}_{s,\tau}(\tau\theta) d\theta \quad = \quad 2\pi \sum_{u=-m}^{m} w_{T',u}{}^2 b^{(i)}_{t-u} b^{(i)*}_{t-u}$$

Set $\widetilde{H^i}(\tau\theta) = \frac{H^i_{t,\tau}(\tau\theta) H^{i*}_{s,\tau}(\tau\theta)}{\int H^i_{t,\tau}(\tau\theta) H^{i*}_{s,\tau}(\tau\theta) d\theta}$. We can write

$$\begin{aligned} R_1 \quad &\simeq \quad \frac{\pi}{2}\tau^2 \left\{ \sum_{u=-m}^{m} (w_{T',u})^2 \right\} \left\{ \int |\Gamma(\phi\tau + \omega\tau)|^2 \, |\Gamma(\phi\tau + \omega'\tau)|^2 \, d\phi \right\} \\ &\times \quad \left\{ \sum_{i=1}^{4} \overline{b^{(i)}_t b^{(i)*}_s} \int e^{-i(s-t)(\phi)} \widetilde{H^i}(\tau\phi) d\phi \right\} \end{aligned}$$

For $s = t$ and $\omega = \omega'$ the expression R_1 becomes

$$R_1 \quad \simeq \quad \frac{\pi}{2}\tau^2 \left\{ \sum_{u=-m}^{m} (w_{T',u})^2 \right\} \left\{ \int |\Gamma(\phi\tau + \omega\tau)|^4 \, d\phi \right\} \sum_{i=1}^{4} \left| b^{(i)}_t \right|^2$$

$$R_1 \quad \simeq \quad \frac{\pi}{2}\tau^2 \left\{ \sum_{u=-m}^{m} (w_{T',u})^2 \right\} \left\{ \int |\Gamma(\phi\tau + \omega\tau)|^4 \, d\phi \right\} 4h_t(\omega)^4$$

Using Parseval equality $\sum (w_{T',u})^2 = \sum \left| W_{T',v} \right|^2$, we obtain

$$R_1 \simeq \frac{2\pi\tau^2}{T'} h_t(\omega)^4 \sum T'(W_{T',v})^2 \int |\Gamma(\phi\tau + \omega\tau)|^4 \, d\phi.$$

From the condition e) satisfied by the weight function $w_{T',v}$, we conclude that $R_1 = O\left(\frac{\tau^2}{T'}\right)$. In the same way we show that $R_2 = O\left(\frac{\tau^2}{T'}\right)$. Thus,

$$Var[\widehat{h}_{t,T'}(\omega)] = O\left(\frac{\tau^2}{T'}\right).$$

It is easy to show that $E\left|\widehat{h}_{t,T'}(\omega_0) - \overline{h}_t(\omega_0)\right|^2$ is equal to the sum of the square of the bias and the variance of the estimator.

5. Conclusion

We have proposed in this paper some results about the estimation of the evolutionary spectral density for non-stationary time-continuous processes, which were sampled at discrete instants. The approach is based on the technique used by Sabre ([?]) for stable processes combining estimates of evolutionary spectrum introduced by Priestley ([9]). This work could be applied to several cases when the process is non-stationary with continuous time as in examples of:

- the spatial and temporal movement of some planets in astronomy can be considered as a non-stationary continuous-time processes

- the study of chemicals on agricultural soil which are considered as a non-stationary processes with continuous-space-parameter.
- the study of the telephone networks to increase the flow through non-stationary signals with continuous time.

This work could be supplemented by the study of optimal smoothing parameters using cross validation methods that have been proven in the field.

5.1. Acknowledgments

I would like to thank the anonymous referees for their interest in this paper and their valuable comments and suggestions.

References

[1] Al-Shoshan, A. I, Chaparro, L. F. 1996 Identification of non-minimum phase systems using evolutionary spectral theory, it Signal Processing **55** 79-92

[2] Cabrera S.D., Flores B. C., Thomas G., and Vega J. 1993 Evolutionary Spectral Estimation Based on Adaptive Use of Weighted Norms *Proceeding of Advenced Signal Processing Algorithms, Architectures and Implementations SPIE* **2027** 168-79

[3] Caetano M., Rodet X. 2009 Evolutionary Spectral Envelope Morphing by Spectral Shape Descriptors *In Proc. ICMC*

[4] Kayhan A S, El-Jaroudi A and Chaparro L F 1994 Evolutionary Periodogram for Nonstationary Signals, *IEEE Transactions on signal processing* **42** 6 1527-36

[5] Lomb N R 1976 Least-squares frequency analysis of unequally spaced data *Stroph. Space Sc.* **39** 2 44762.

[6] Masry E 1978 Aliasing-free sampling: An Alternative conceptualization and its Application *IEEE Transactions on Information Theory* **24** 317-24

[7] Mélard G, Herteleer-de Schutter A 1988 Contributions to evolutionary spectral theory it Journal of time series analysis **10** 1 41-63

[8] Moghtaderi A, Takahara G and Thomson D J 2009 Evolutionary spectrum estimation for uniformly modulated processes with improved boundary performance, Acoustics, Speech, and Signal Processing *IEEE International Conference* 2993-96

[9] Priestley M B 1965 Evolutionary spectra and non-stationary processes *J. Roy. Statist. Soc.* Serie B **27** 204-37

[10] Priestley M B 1966 Design relations for non-stationary processes *J. Roy. Statist. Soc.* Serie B **28** 228-40

[11] Priestley M B 1967 Power spectral analysis of non-stationary random processes it J.Sound. Vib. **6** 86-97

[12] Priestley M B 1988 Spectral analysis and time series *Probability and Mathematical Statistics* (Academic Press)

[13] Sabre R 2000 Discrete estimation of spectral density for symmetric stable process *Statistica* **3** 497-520

[14] Sabre R 2002 Aliasing free for stable random field *The Egy. Statistical journal* **6** 1 53-75

[15] Scargle J D 1989 Studies in astronomical time series analysis. III. Fourier transforms, autocorrelation functions, and gross-correlation functions of unevenly spaced data *Astronom. J.* **343** 1 87487

[16] Shah S I, Chaparro L F, El-Jaroudi A and Furman J M 1998 Evolutionary Maximum Entropy Spectral Estimation and Heart Rate Variability Analysis *Multidimensional Systems and Signal Processing* **9** 4 453-58

[17] Tarcsznski A, Dongdong Q 2005 Optimal random sampling dor spectrum estimation in dasp applications *Int. J. Appl. Math. Comput. Sci.* **15** 4 46369

[18] Tezcan J 2003 Evolutionary Power Spectrum Estimation Using Harmonic Wavelets *Seismic Design and Analysis of Structures* 37-41

[19] Van Bellegem S and von Sachs R 2008 Locally adaptive estimation of evolutionary wavelet spectra *Ann. Statist.* **36** Number 4 1879-1924

[20] Xinyuan Liu, Dingwu He, Liuqing Pei and Suming Zhang 2009 Singularity detection of the rabbit electrocardiogram: An evolutionary spectral method *Progress in Natural Science* **19** 10 1317-26

Journal of Physics: Conference Series **423** (2013) 012015 doi:10.1088/1742-6596/423/1/012015

Biogas Laminar Burning Velocity and Flammability Characteristics in Spark Ignited Premix Combustion

Willyanto Anggono[1], ING Wardana[2], M. Lawes[3], K. J. Hughes[4], Slamet Wahyudi[5], Nurkholis Hamidi[6], Akihiro Hayakawa[7]

[1]Petra Christian University/ Mechanical Engineering Department, Surabaya, Indonesia.

[1,2,5,6]Brawijaya University/ Doctoral Program Mechanical Engineering Department, Malang, Indonesia.

[1,4]The University of Leeds/ School of Process, Environmental and Materials Engineering, Leeds, England.

[3]The University of Leeds/ School of Mechanical Engineering, Leeds, England.

[7]Kyushu University/Department of Mechanical Engineering, Fukuoka, Japan.

E-mail: willy@petra.ac.id

Abstract. Spherically expanding flames propagating at constant pressure were employed to determine the laminar burning velocity and flammability characteristics of biogas-air mixtures in premixed combustion to uncover the fundamental flame propagation characteristics of a new alternative and renewable fuel. The results are compared with those from a methane-air flame. Biogas is a sustainable and renewable fuel that is produced in digestion facilities. The composition of biogas discussed in this paper consists of 66.4% methane, 30.6% carbon dioxide and 3% nitrogen. Burning velocity was measured at various equivalence ratios (ϕ) using a photographic technique in a high pressure fan-stirred bomb, the initial condition being at room temperature and atmospheric pressure. The flame for methane–air mixtures propagates from ϕ=0.6 till ϕ=1.3. The flame at $\phi \geq 1.4$ does not propagate because the combustion reaction is quenched by the larger mass of fuel. At $\phi \leq 0.5$, it does not propagate as well since the heat of reaction is insufficient to burn the mixtures. The flame for biogas–air mixtures propagates in a narrower range, that is from ϕ=0.6 to ϕ=1.2. Different from the methane flame, the biogas flame does not propagate at $\phi \geq 1.3$ because the heat absorbed by inhibitors strengthens the quenching effect by the larger mass of fuel. As in the methane flame, the biogas flame at $\phi \leq 0.5$ does not propagate. This shows that the effect of inhibitors in extremely lean mixtures is small. Compared to a methane-air mixture, the flammability characteristic (flammable region) of biogas becomes narrower in the presence of inhibitors (carbon dioxide and nitrogen) and the presence of inhibitors causes a reduction in the laminar burning velocity. The inhibitor gases work more effectively at rich mixtures because the rich biogas-air mixtures have a higher fraction of carbon dioxide and nitrogen components compared to the lean biogas-air mixtures.

1. Introduction

Biogas as "Powergas" is an alternative fuel, it is a sustainable and renewable fuel that is produced in digestion facilities. It does not contribute to the increase in atmospheric carbon dioxide concentrations because it comes from an organic source with a short carbon cycle and is thus a green solution in the development of sustainable fuels [1,2]. The digestion facilities can be constructed quickly in a few days using unskilled labor [3]. Biogas contains 50–70% methane and 30–50% carbon dioxide, as well as small amounts of other gases and typically has a calorific value of 21–24 MJ/m^3 [4]. Based on chemical analysis, the composition of the biogas produced in East Java, Indonesia is 66.4% methane, 30.6% carbon dioxide and 3% nitrogen [1,2]. Methane is a flammable gas, whereas, nitrogen and carbon dioxide are inhibitors [5]. The Kyoto protocol was intended to reduce greenhouse gas emissions, and to further this objective, research into biogas combustion in stoves, engines and gas turbines has had good results [4,6-10]. However, the laminar burning velocity and flammability characteristic of biogas, being a fundamental characteristic of a fuel, have not been studied yet. Thus, the aim of this paper was to investigate its laminar burning and flammability characteristics.

Demands for replacing fossil fuels to reduce emissions, require an improved fundamental understanding of the combustion processes that occur within the internal combustion engine. An important characteristic is the burning velocity, which directly affects pressure development and is often expressed in terms of laminar burning velocity [1,2,11-16].

Flammability limits, give the proportion of combustible gases in a mixture, between which limits this mixture is flammable. Gas mixtures consisting of combustible, oxidizing, and inert gases are only flammable under certain conditions. The lower flammable limit describes the leanest mixture that still sustains a flame, i.e. the mixture with the smallest fraction of combustible gas, while the upper flammable limit gives the richest flammable mixture. The laminar burning velocity and flammability are the most important flame propagation characteristics in spark ignited premixed combustion and as the fundamental flame propagation characteristic of biogas require further study; this paper looks into this matter with view to a better understanding of a new alternative and renewable fuel. The results are compared with those from methane-air flames to emphasize the contrast between the burning velocity and flammability characteristics of methane-air mixtures and biogas-air mixtures.

2. Experimental Methods

The laminar burning velocity of biogas premixed combustion was measured in the Mk II high pressure fan-stirred combustion vessel at the Leeds University School of Mechanical Engineering as shown in Fig. 1. Initially, all the experiments in this paper were performed at room temperature, at atmospheric pressure and with mixtures of various equivalence ratios (ϕ= 0.5, 0.6, 0.8, 1.0, 1.2, 1.3). The bomb is a spherical stainless steel vessel of 380 mm diameter, with three pairs of orthogonal windows each of 150 mm diameter and equipped with four electrically driven fans to ensure good mixing [1,2,11-14]. The fuel-air mixtures in the Mk II combustion bomb were centrally ignited and flame propagation was recorded by high speed schlieren cine-photography using a Photosonics Phantom digital camera as shown in Fig. 2, operating at a rate of 2500 frames/s with a resolution of 768 x 768. The flame radius was calculated as that of a circle encompassing the same area as that enclosed by the schlieren imaged flame [1,2,11-14].

Figure 1. Mk2 Combustion Bomb.

Figure 2. High Speed Schlieren Cine-Photography

The laminar burning velocity for a spherically expanding flame can be deduced from the schlieren photographs, the stretched flame velocity (S_n) can be derived from the flame radius versus time data as: $S_n = dr_u/dt$, where r_u is the flame radius in the Schlieren photographs and t is the elapsed time from the spark ignition. The flame stretch rate α is defined as $\alpha = d(\ln A)/dt = (dA)/(A\,dt)$, where A is the area of the flame. In the case of a spherically propagating premixed flame, the flame stretch rate can be calculated by $\alpha = (2/r_u)(dr_u/dt)$. A linear relationship between flame speed and the total stretch exists, and this is quantified by burned gas of Markstein length, L_b, and is defined at the radius, r_u, such that: $S_n - S_s = L_b\ \alpha$, where S_s is the unstretched flame speed, and is obtained as an intercept value of S_n at $\alpha = 0$ in the plot of S_n against α. The gradient of the best straight line fit to the experimental data gives L_b. The unstretched laminar burning velocity, u_l, was deduced from S_s using $u_l = S_s\ (\rho_b/\rho_u)$, where ρ_b is the density of the burned gas mixtures and ρ_u is the density of the unburned gas mixtures [1,2,11-14].

3. Results and Discussion

Based on the experimental investigation, the biogas-air mixtures at the intermediate equivalence ratios (ϕ=0.6, 0.8, 1.0, 1.2) were observed to produce a propagating flame, whereas at the extreme equivalence ratios (ϕ=0.5 and 1.3) no propagating flames were observed. The images resulting from the spherical flame propagation within the combustion bomb are shown in Fig. 3.

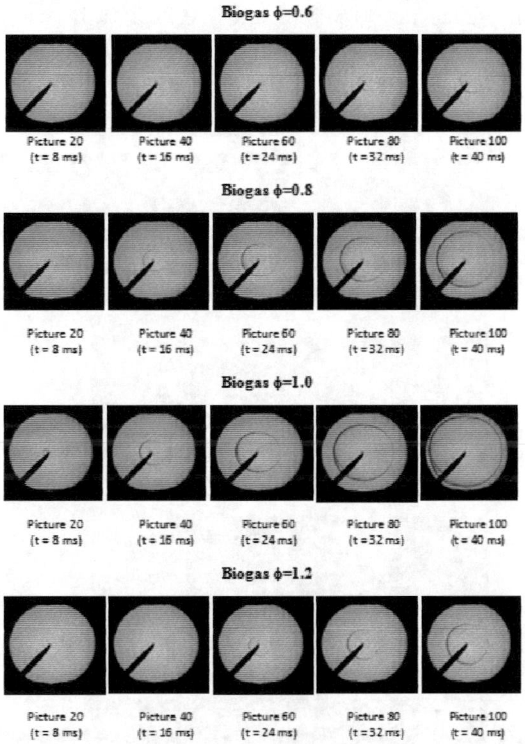

Figure 3. Flame Propagation Biogas-Air Mixtures at Various Equivalence Ratios.

The radius of the spherical flame propagation in Fig. 3 are presented in Fig. 4 as a function of elapsed time. Based on experiment results and the calculations as mentioned in the experimental method and previous studies [1,2,13], the laminar burning velocities of biogas-air mixtures in premixed combustion are 0.0743 m/s for lean (ϕ=0.6), 0.2086 m/s for lean (ϕ=0.8), 0.2638 m/s for stoichiometric (ϕ=1.0) and 0.1864 m/s for rich (ϕ=1.2) biogas-air mixtures, which are in agreement with previous studies [1,2].

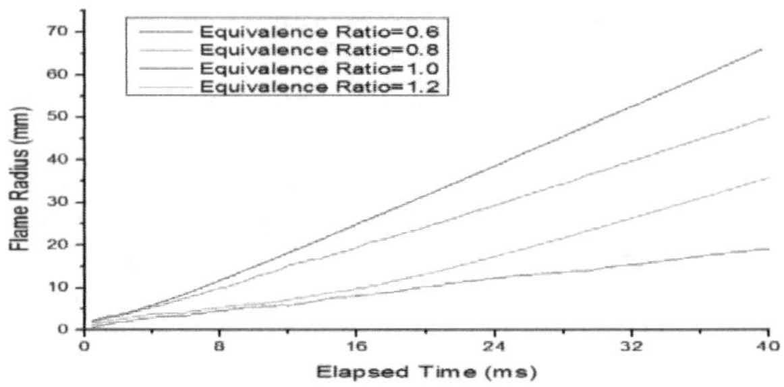

Figure 4. Flame Radius vs Elapsed Time of Biogas-Air Mixtures at Various Equivalence Ratios

For comparison, the laminar flame propagation of methane is also presented as shown in Fig. 5. The methane-air mixtures equivalence ratio (ϕ=0.6, 0.8, 1.0, 1.2, 1.3) were found to have produced a propagating flame, whereas for the extreme equivalence ratio (ϕ=0.5 and 1.4) no propagating flames were observed.

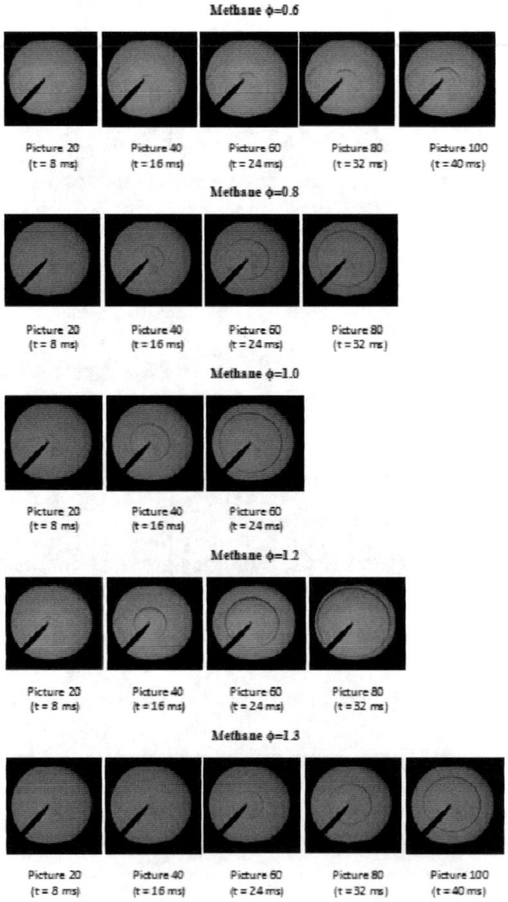

Figure 5. Flame Propagation Methane-Air Mixtures at Various Equivalence Ratios

Based on the experimental results and the same calculations and method for the laminar burning velocity of the various equivalence ratios of biogas-air mixtures, the laminar burning velocities of the methane-air mixtures were found. The laminar burning velocities were 0.0948 m/s for lean (ϕ=0.6), 0.2749 m/s for lean (ϕ=0.8), 0.3527 m/s for stochiometic (ϕ=1.0), 0.3082 m/s for rich (ϕ=1.2) and 0.2015 m/s for rich (ϕ=1.3) methane-air mixtures, which were in agreement with previous studies [1,2,13]. A summary of the results from biogas and methane are shown in Table 1 and Fig. 6.

Table 1. Laminar Burning Velocities of Methane and Biogas

ϕ	Methane Laminar burning velocity (m/s)	Biogas Laminar burning velocity (m/s)
0.5	No propagation	No propagation
0.6	0.0948	0.0743
0.8	0.2749	0.2086
1.0	0.3527	0.2638
1.2	0.3082	0.1864
1.3	0.2015	No propagation
1.4	No propagation	No propagation

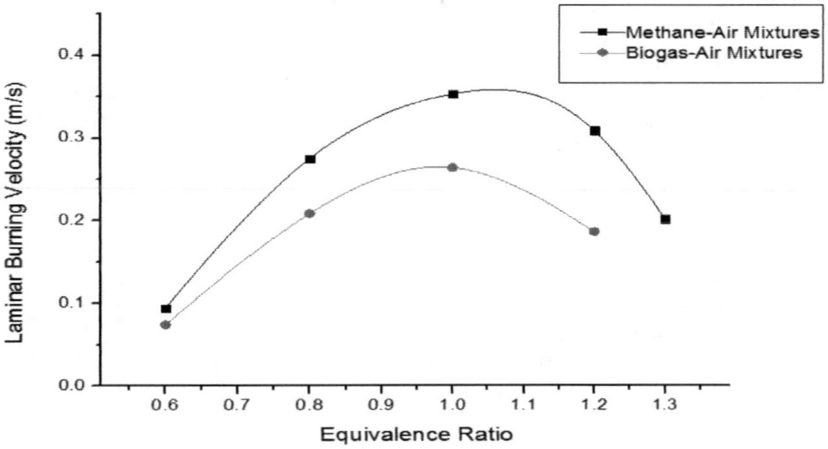

Figure 6. Laminar Burning Velocities of Methane-Air Mixtures and Biogas-Air Mixtures

From Table 1 and Fig. 6, it can be seen that laminar burning velocities of stoichiometric (ϕ=1) biogas-air mixtures and methane-air mixtures are higher than the lean and rich mixtures because the stoichiometric mixtures have just enough air for complete combustion of the available fuel [1,2]. As expected, because of the presence of carbon dioxide and nitrogen in the biogas, at the same equivalence ratio, the laminar burning velocity of biogas-air mixtures were lower than the laminar burning velocitiy of the methane-air mixtures. The carbon dioxide and nitrogen in the biogas are inhibitors that tended to decrease the laminar burning velocities [1,2,5,17].

The flames for the methane–air mixtures propagated from ϕ=0.6 till ϕ=1.3. But at $\phi \geq 1.4$ the flame did not propagate because the combustion reaction was quenched by the larger mass of fuel. At $\phi \leq 0.5$, the flame did not propagate either since reaction heat was insufficient to burn the mixtures. Flames for biogas–air mixtures propagated at narrower range, i.e. ϕ=0.6 to ϕ=1.2. In contrast, the biogas flame did not propagate at $\phi \geq 1.3$. This was due to the fact that the inhibitors in the fuel absorbed some of the heat from the combustion reaction. Therefore, the quenching effect in the larger mass of fuel was stronger. At $\phi \leq 0.5$, the same as in methane flame, the biogas flame did not propagate. This shows that the effect of inhibitors was very small at extreme lean mixtures. Compared to a methane-air mixture, the flammable region of biogas became narrower in the presence of inhibitors (carbon dioxide and nitrogen) and the presence of inhibitors caused a reduction in the laminar burning velocity for two reasons. Firstly, the dilution effect lead to a lower concentration of reactive species in the fuel-air mixture for a given equivalence ratio, which lead to a lower overall chemical reaction rate of bimolecular reactions in the fuel oxidation reaction process. Secondly, the presence of these inhibitor gases absorbed some of the heat generated, thus lowering the flame temperature which in turn will tend to reduce the overall rate of many of the chemical reactions within the fuel oxidation process. The inhibitor gases were more effective in rich mixtures because the rich biogas-air mixtures had a higher fraction of carbon dioxide and nitrogen components compared to lean biogas-air mixtures.

4. Conclusion

The flames for the methane–air mixtures from $\phi=0.6$ till $\phi=1.3$ propagated, whereas, the flames in mixtures of $\phi \geq 1.4$ did not propagate because the combustion reaction was quenched by the larger mass of fuel. The methane flame at equivalence ratio of $\phi \leq 0.5$ did not propagate either. This was due to the fact that the reaction heat was insufficient to burn the mixtures. The flames for biogas–air mixtures were propagated at narrower ranges, that is from $\phi=0.6$ to $\phi=1.2$. The biogas flame did not propagate for $\phi \geq 1.3$ because the inhibitors in the mixture absorbed some of the heat from the combustion reaction so that the quenching effect in the larger mass of fuel was stronger. As for the methane flame, biogas flame at $\phi \leq 0.5$ did not propagate which shows that the inhibitor effect was small in the lean mixtures. The laminar burning velocities of the biogas-air mixtures in the premix combustion were 0.0743 m/s for $\phi=0.6$, 0.2086 m/s for $\phi=0.8$, 0.2638 m/s for $\phi=1.0$ and 0.1864 m/s for $\phi=1.2$ biogas-air mixtures respectively. The inhibitor gases were more effective at rich mixtures because the rich biogas-air mixture had a higher fraction of the carbon dioxide and nitrogen components compared to the lean biogas-air mixtures. The flammability characteristic (flammable region) of biogas became narrower in the presence of inhibitors (carbon dioxide and nitrogen) and the presence of the inhibitors caused a reduction in the laminar burning velocity

Acknowledgments

Many Thanks to Direktorat Jendral Pendidikan Tinggi Kementrian Pendidikan Nasional Republik Indonesia, Universitas Brawijaya, Petra Christian University, Leeds University, Prof. M. Pourkashanian (Leeds University), Professor Derek Bradley (Leeds University), Professor Mike Pilling (Leeds University), and Bambang Pudjianto (BBA Gas) for their support during this research.

References

[1] Anggono. W., Wardana. I.N.G., Pourkashanian. M., Hughes. K.J., Lawes. M.,Wahyudi. S., Hamidi. N., and Hayakawa. A., 2012. Experimental and Numerical Simulation on Boigas Flame Propagation Characteristic in Spark Ignition Premixed Combustion, Proceeding The3rd International Conference on Engineering and ICT (ICEI2012) Melaka-Malaysia, Vol. 2, pp. 290-294.

[2] Anggono. W., Wardana. I.N.G., Lawes. M., Hughes. K.J., Wahyudi. S., Hamidi. N., 2012. Laminar Burning Characteristics of Biogas-Air Mixtures in Spark Ignited Premix Combustion. Journal of Applied Sciences Research.

[3] Lichtman. R., et. al., 1996. The Improved Biogas Systems Project: results and future work, Energy for Sustainable Development, Volume 3, Issue 4, pp. 28–42.

[4] Bond. T., and Templeton. M.R., 2011. History and future of domestic biogas plants in the developing world., International Journal Energy for Sustainable Development Volume 15, Issue 4, pp. 347–354.

[5] Ilminnafik. N., Hamidi. N., and Wardana. I.N.G., 2011, Behavior of flame propagation in LPG premixed combustion with carbon dioxide inhibitor, International Journal of academic research. Vol. 3. No. 2. March, 2011, Part III, pp. 705-708.

[6] Alwis. A.D., 2002. Biogas-A review of Sri Lanka's performance with a renewable energy technology, Energy for Sustainable Development, Volume 6, Issue 1, March 2002, Pp. 30–37.

[7] Lafay. Y., Taupin .B., Martins .G., Cabot .G., Renou .B., and Boukhalfa, 2007. Experiment study of biogas combustion using a gas turbine configuration. Experiments in Fluids: Springer 43, pp. 395-410.

[8] Nathan. S.S.,Mallikarjuna. J.M., and Ramesh. A., 2010. An experimental study of the biogas–diesel HCCI mode of engine operation, Energy Conversion and Management 51, pp. 1347–1353.

[9] Porpatham. E.,Ramesh. A., and Nagalingam. B., 2008. Investigation on the effect of concentration of methane in biogas when used as a fuel for a spark ignition engine, Fuel 87, pp.1651-1659.

[10] Sridhar G., Paul P.J. and Mukunda H.S. 2001. Biomass Derived Producer Gas as a Reciprocating Engine Fuel – An Experimental Analysis, Biomass & Bioenergy, Vol. 21, pp. 61-72.

[11] Bradley. D., Hicks. R.A., Lawes. M., Sheppard. C.G.W. and Wooley, R, 1998. The measurement of laminar burning velocities and Markstein numbers for isooctane-air and iso-octane-n-heptane-air mixtures at elevated temperatures and pressures in an explosion bomb, Combustion and Flame 115, pp. 126-144.

[12] Gillespie. L., Lawes. M., Sheppard. C.G.W., and Woolley. R., 2000. Aspects of laminar and turbulent burning velocity relevant to SI engines. SAE Paper Series 2000-01-0192.

[13] Gu. X.J., Haq. M.Z., Lawes. M., and Wooley. R., 2000. Laminar burning velocity and Markstein lengths of methane-air mixtures, Combustion and Flame 121, pp. 41-58.

[14] Serrano. C., Hernandez. J.J., Mandilas. C., Sheppard. C.G.W., and Woolley. R., 2008. Laminar burning behaviour of biomass gasification-derived producer gas, International Journal of Hydrogen Energy 33, pp. 851 – 862.

[15] Hermanns. R.T.E., 2007. Laminar Burning Velocities of Methane-Hydrogen-Air Mixtures, Doctoral Thesis Technische Universiteit Eindhoven, The Netherlands.

[16] Marshall. S.P., Taylor. S., Stone. C.R., Davies. T.J., and Cracknell. R.F., 2011. Laminar burning velocity measurements of liquid fuels at elevated pressures and temperatures with combustion residuals, Combustion and Flame 158, pp. 1920–1932.

[17] Ronney, P. D., 2001. Premixed-Gas Flames in Microgravity Combustion: Fires inFree Fall (H. Ross, Ed.), Academic Press, London, U.K.,pp. 35-82.

Journal of Physics: Conference Series **423** (2013) 012016 doi:10.1088/1742-6596/423/1/012016

Radiological properties of MAGIC normoxic polymer gel dosimetry

M Aljamal[1], A Zakaria[1], S Shamsuddin[2]

[1] Medical Radiation Dept., University Sains Malaysia, Kelantan, Malaysia.
[2] Biomedical Dept, University Sains Malaysia, Kelantan, Malaysia.

E-mail: aljamal81@gmail.com

Abstract. For a polymer gel dosimeter to be of use in radiation dosimetry, it should display water-equivalent radiological properties. In this study, the radiological properties of the MAGIC (Methacrylic and Ascorbic acid in Gelatin Initiated by Copper) normoxic polymer gels were investigated. The mass density (ρ) was determined based on Archimedes' principle. The weight fraction of elemental composition and the effective atomic number (Z_{eff}) were calculated. The electron density was also measured with 90° scattering angle at room temperature. The linear attenuation coefficient (μ) of unirradiated gel, irradiated gel, and water were determined using Am-241 based on narrow beam geometry. Monte Carlo simulation was used to calculate the depth doses response of MAGIC gel and water for 6MV photon beam. The weight fractions of elements composition of MAGIC gel were close to that for water. The mass density was found to be 1027 ± 2 kg m^{-3}, which is also very close to mass density of muscle tissue (1030 kg m^{-3}) and 2.7% higher than that of water. The electron density (ρ_e) and atomic number (Z_{eff}) were found to be 3.43×10^{29} e m^{-3} and 7.105, respectively. The electron density measured was 2.6% greater than that for water. The atomic number was very close to that for water. The prepared MAGIC gel was found to be water equivalent based on the study of element composition, mass density, electron density and atomic number. The linear attenuation coefficient of unirradiated gel was very close to that of water. The μ of irradiated gel was found to be linear with dose 2-40 Gy. The depth dose response for MAGIC gel from a 6 MV photon beam had a percentage dose difference to water of less than 1%. Therefore it satisfies the criteria to be a good polymer gel dosimeter for radiotherapy.

1. Introduction

Recent advances in radiotherapy treatment techniques such as conformal radiotherapy enable radiation treatments to be tailored to the tumor shape. This is advantageous as treatment outcomes can be enhanced due to a reduced absorbed dose to healthy tissue. However, these new treatments place great demands on dosimetry techniques due to the complex nature of the radiation dose distributions used. To completely verify these complex dose distributions before treatment starts, three-dimensional (3D) dosimetry system is needed. Dosimeters currently in use, such as ionization chambers and thermoluminescent (TLD) dosimeters, have limitations as they only measure the dose at a single point, and radiographic films only measure a 2D distribution. Various works have shown that polymer gel could be used as dosimeter to acquire 3D dose distribution [2, 3]. Gel dosimeters consist of a gel infused with radiation sensitive materials. Upon irradiation free radicals released during the radiolysis of the water within the gel initiate polymerization and cross-linking of the monomers. The amount of free radicals released is proportional to the dose received by the gel dosimeter and the resultant amount of polymer formed is therefore also proportional to dose until an upper limit is reached. Oxygen is an efficient scavenger of free radicals and must be removed from polymer gel dosimeters prior to irradiation or the polymerization process will be inhibited [4-6]. This results in the requirement for specialized equipment to prevent penetration of oxygen to polymer gel mixture and has been a disadvantage of using polymer gel dosimeters. Fong *et al.* [1] reported a polymer gel formulation which contains oxygen scavengers and can be manufactured in normal atmospheric conditions. This formulation consists of methacrylic acid, copper sulphate, ascorbic acid, hydroquinone, gelatin and water and is given the acronym MAGIC gel.

For a material to be used as radiation dosimetry, it should display water-equivalent radiological properties. The radiological properties including effective atomic number, number of electrons per gram, mass density and electron density should be same as radiological properties of water or tissue [7, 8]. Several studies have been done to investigate water equivalency of different polymer gel dosimetry for photon and electron beam dosimetry [1, 9-18]. MAGIC gel dosimeter has previously been investigated for its radiological interaction properties and radiological water equivalence [1, 10].

In the present work, we have evaluated the water equivalency and radiological attenuation properties of MAGIC gel by calculating the weight fraction of element composition, mass density, effective atomic number, number of electron per gram, electron density and the linear attenuation coefficient of unirradiated and irradiated MAGIC gel dosimetry. The depth doses for MAGIC gel and water were also calculated using Monte Carlo simulation.

2. MATERIALS AND METHODS

2.1. MAGIC gel preparation

The MAGIC gel was prepared based on the formulation proposed in the literature by Fong *et al.* [1]. The chemical materials used to produce the MAGIC gel and their concentrations are as follows: gelatin type A (300 bloom) (8%), ascorbic acid (0.0367%), copper sulphate (0.002%), methacrylic acid (8%) and HPLC grade distilled water (83.9%).

2.2 Density measurements

The density of the gel can be determined from Archimedes' principle [19]. A volumetric glass flask was filled with gel at room temperature (23°C). The density of the MAGIC gel was determined by the following equation:

$$\rho\, gel = \frac{m_{gel}}{V_{gel}}$$

Where ρ_{gel} is the density of the gel, m_{gel} is the mass of gel in the flask, V_{gel} is the volume of the flask filled with gel. The uncertainty in the mass density of the gel was calculated as the standard error in the mean value.

2.3 Element composition measurements

Since the formula of a compound expresses the ratio of the numbers of its constituent atoms, a formula also conveys information about the relative masses of the elements it contains. But in order to make this connection, we need to know the relative weight fraction of the different elements in each compound.

The chemical compounds used to produce the MAGIC gel were gelatin type A (300 bloom) ($C17H_{32}N_5O_6$), ascorbic acid ($C_6 H_8 O_6$), copper sulphate ($Cu SO_{4.}5H_2O$), methacrylic acid ($H_2C (CH_3) CO_2H$) and HPLC grade water. The weight fraction was determined for carbon (C), hydrogen (H), oxygen (O), nitrogen (N), copper (Cu II) and sulphur (S) in each compound above using the following equation:

$$Weight\ fraction = \frac{mass\ of\ each\ element\ in\ the\ whole\ compound}{mass\ of\ whole\ compound}$$

2.4. Effective atomic number (Z_{eff}) measurement
The effective atomic number (Z_{eff}) was calculated by summation of weight fractional proportion (w_k) of each atom in MAGIC gel multiplied by the atomic number (Z_k) of the atom as shown in the following equation [1]:

$$Z_{eff} = \sum_k w_k Z_k$$

2.5 Linear Attenuation Coefficient Measurements
For linear attenuation coefficient measurement, the polymer gel was poured into polystyrene spectrophotometry cuvettes with parallel flat sides and the total thickness of 1.2 cm. The gel in cuvettes were irradiated to different doses (2 Gy to 40 Gy) at a dose rate of 250 cGy per minute using a 6 MV photon beam with source to surface distance (SSD = 100 cm) from linear accelerator (Siemens Primus, USA).

The linear attenuation coefficient of unirradiated gel, irradiated gel, and de-ionised water were made separately at room temperature using narrow beam geometry (School of Physics, University Sains Malaysia). Measurements were carried out with a similar set-up used previously [20] as shown in figure 1. The radiation source used was Am-241 (Amersham, Sydney) with an activity of 1.67 GBq, which has a major photopeak at 59.5 keV [21]. Transmitted radiation was detected using XR-100T-CdTe detector (E G & G Ortec, Atlanta USA). The PX2T power supply and amplifier were used with coarse gain of 50 and unipolar pulse with shaping time of 3 μs. The live time used was 2000 s. The spectrum was analyzed on a personal computer with Aptec MCA Application Multichannel Analyser software (Canberra Industries, Meriden USA). The background counts were obtained without the Am-241 source. The Am-241 source was placed in the source holder. Then a lead collimator (1 mm diameter) was placed in the collimator holder and counts (I_0) without the sample was obtained. Radiation counts were then obtained with empty cuvettes in the radiation path to correct for the attenuation due to cuvette walls. Cuvette containing unirradiated gel was placed in the radiation path and the count was obtained. The irradiated gel of various doses was placed in the radiation path separately and radiation counts were obtained for each one. For the measurement of the linear attenuation coefficient of deionised water, radiation counts were obtained through cuvettes filled with deionised water. Background counts were subtracted from the recorded counts for each of the measurement above. Then the count of the empty cuvettes was subtracted from this resultant count to obtain I . The following equation was used to determine the linear attenuation coefficient of MAGIC gel:

$$I = I_0 e^{-\mu x}$$

Where I is the intensity of the beam, I_0 is the intensity of the beam incident on the material and x is the thickness of material through which the beam has passed.

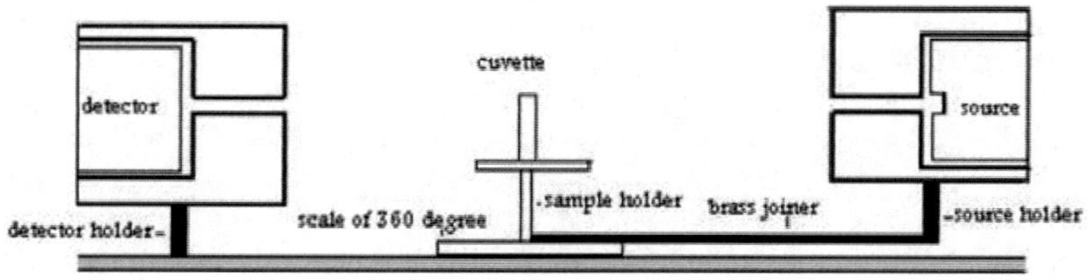

Figure 1. A narrow beam geometry used for linear attenuation coefficient measurements

In theory, Compton scattering takes place when a photon strikes an effectively free electron which is at rest. However, this situation is only an approximation when the binding energy of the electrons involved is small compared to the energy of the incident photons. If narrow beam geometry is adopted and a detector is placed at some angle from a radiation source in order to detect photons which are deflected by an object, the number of the scattered photons detected can be used to estimate the electron density of the object. The electron density was determined by:

$$\frac{[N]_{gel}}{[N]_{wat}} = \frac{[\rho_e]_{gel}}{[\rho_e]_{wat}}$$

where $[N]$ is the number of scattered photons detected. $(\rho_e)_{gel}$ and $(\rho_e)_{wat}$ are the electron density of gel and water, respectively. The electron density of water is $\rho_e = 3.34 \times 10^{29}$ e m $^{-3}$ which is usually used as the reference material [22]. The electron density of MAGIC gel was determined at room temperature using narrow beam geometry. The main components of the experimental set-up were a scanner, radiation source, detector and multi channel analyser (MCA).

The detector was moved 90° from the source. The 90° scattering angle was selected because the best resolution can be obtained at this angle [23]. The background counts were obtained without the Am-241 source. Cuvette containing unirradiated gel was placed in the center of sample holder and the radiation count $[N]_{gel}$ was obtained. Then in the same fashion the radiation counts $[N]_{wat}$ for water were obtained through cuvettes filled with deionised water. Background counts were subtracted from the recorded counts for each of the measurement above.

2.6 Monte Carlo depth dose simulation

In order to compare dose response of MAGIC gel and water to megavoltage energy, depth doses were calculated for 6MV photon beam using Monte Carlo simulation. The EGSnrc/BEAMnrc and EGSnrc/DOSXYZnrc codes were used to perform the dose calculation [24, 25]. The BEAMnrc software was used to simulate Primus Linear accelerator head geometry (Siemens Primus, USA) and to generate phase space file. PEGS4 (EGS preprocessor) was used to calculate cross sectional data by using the element composition and the mass density of MAGIC gel as the inputs to PEGS4 software. The electron cut-off energy (ECUT) was set to 0.7 MeV while the photon cut-off energy (PCUT) was set to 0.01 MeV [26]. The beam parameters used were a 100 cm SSD with a 10x10 cm^2 field size. A total of 800 million histories were run in the accelerator head calculations. A phase space was scored in a plane perpendicular to the beam axis at 100 cm distance from the target. Dose calculations in DOSXYZnrc were performed using a water phantom with the dimensions 20x20x15 cm^3 and the phantom was positioned at an SSD of 100 cm. For depth-dose calculations the voxel size was 0.5x0.5 cm^2 in the x and y plane and 0.2 cm in the longitudinal direction (z-depth direction).

3. Results and Discussion

A polymer gel dosimeter to be used in radiation dosimetry should display water-equivalent properties. The suitable method for evaluating water-equivalence is to measure the mass density, electron density and effective atomic number. The weight fraction of elemental composition obtained from this study for carbon, hydrogen, oxygen, nitrogen, sulfur and copper (II) in MAGIC gel is shown in table 1.

Table 1. The composition of MAGIC gel, water, human muscle tissue and fat (weight fractions denoted as w_k).

Material	w_C	w_H	w_O	w_N	w_S	$w_{Cu(ii)}$
MAGIC	0.0843	0.1067	0.7994	0.01392	2.56 x 10^{-6}	5.08 x 10^{-6}
Water	0.000	0.1111	0.8889	0.00	0.00	0.00
MAGIC[a]	0.075	0.1062	0.8021	0.0139	2.58 x 10^{-6}	5.08 x 10^{-6}
MAGIC[b]	0.086	0.1048	0.7987	0.0115	2.56 x 10^{-6}	5.09 x 10^{-6}
Muscle[a]	0.123	0.1020	0.7298	0.0350	0.00	0.00
Fat[b]	0.5732	0.1120	0.3031	0.0110	6.0 x 10^{-5}	0.00

[a] (Fong *et al.*, 2001)
[b] (Venning *et al.*,2005)

The weight fractions of elements composition given were close to that for water. From the literature, the weight fractions of elements composition calculated in this study were different from previous study done by Fong *et al.* [1] and Venning *et al.* [10]. This small difference can be attributed to the change in concentration of some main chemical compounds in MAGIC gel, where the quantity of water was increased and the amount of methacrylic acid was reduced. Hydroquinone was also excluded because there was still very small amount of hydroquinone (100-250 ppm) in the methacrylic acid to prevent auto-polymerization on the shelf.

The electron density and mass density are the most important issue that must be considered for any dosimeter to be used in external radiotherapy treatment [10]. The mass density (ρ), electron density (ρ_e) and effective atomic number (Z) of the MAGIC gel are given in table 2.

Table 2. Mass density, electron density, number of electron per gram and, effective atomic number for MAGIC gel, water, human muscle tissue and fat

Material	ρ (kg m^{-3})	ρ_e (× 10^{29} e m^{-3})	$\rho e/\rho$ (×10^{26} kg^{-1})	Z_{eff}
MAGIC	1027±2	3.43	3.348	7.10
MAGIC[a]	1060	3.51	3.310	7.07
MAGIC[b]	1037±5	3.44	3.323	7.30
Water[a]	1000	3.34	3.34	7.22
Muscle[a]	1030	3.44	3.31	6.93
Fat[b]	916	3.05	3.33	6.33

[a](Fong *et al.*, 2001)
[b](Venning *et al.*,2005)

The mass density was found to be 1027 ± 2 kg m^{-3}, which is also very close to mass density of muscle tissue (1030 kg m^{-3}) and 2.7% higher than that of water and this was attributed to its high gelatine and monomer concentration. The obtained mass density differs from the previous studies done by Fong *et al.* [1] (1060 kgm^{-3}), and Venning *et al.* [10] (1037 ± 5 kg m^{-3}). The observed differences in mass density are most likely due to the difference in preparation and measuring techniques. The electron density (ρ_e) and effective atomic number (Z_{eff}) were found to be 3.43 × 10^{29} e m^{-3} and 7.105, respectively. The electron density measured was 2.6% greater than that for water. The effective atomic number was close to that for water with small difference with the previous studies [1, 10]. This difference can be attributed to the change in the weight fraction of elemental composition.

Figure 2 shows the variation of linear attenuation coefficient (LAC) with absorbed dose based on the measurement using irradiated samples. Excellent correlation (r = 0.993) was found between linear attenuation coefficient and absorbed dose. It was found that the linear attenuation coefficient of unirradiated gel was very close to water. The measured linear attenuation coefficient (LAC) for unirradiated MAGIC gel and water was found to be 0.84 ± 0.04 cm^{-1} and 0.85 ± 0.02 cm^{-1}, respectively. There was a linear relationship between linear attenuation coefficient and absorbed dose for MAGIC gel up to 40 Gy. So an increase in absorbed dose of MAGIC polymer gel leads to an increase in linear attenuation coefficient. This increase in linear attenuation coefficient is linked to the change in density of the irradiated gel [10].

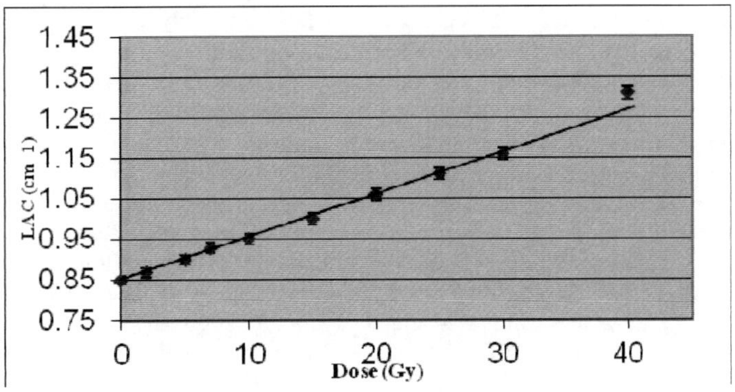

Figure 2. The variation of linear attenuation coefficient (LAC) with absorbed dose.

The depth dose curves for MAGIC gel and water that calculated using Monte Carlo simulation are shown in figure 3. The dose reduction with depth of MAGIC gel is very similar to that of water. The maximum difference of percentage depth dose of gel with that for water was found to be 1 %. Result obtained was consistent with the previous study done by Venning *et al.* [10].

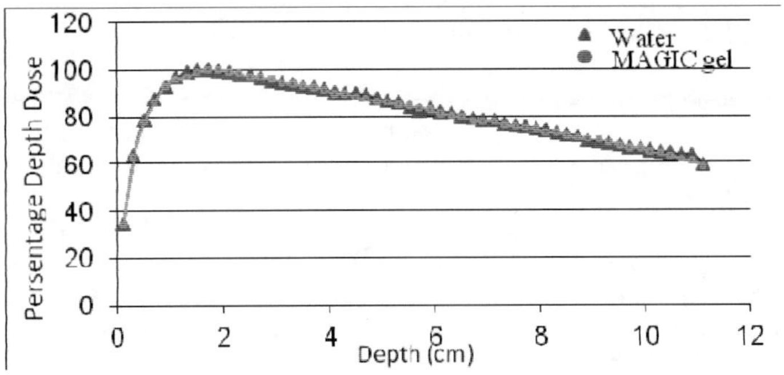

Figure 3. Monte Carlo calculated relative depth dose curves for MAGIC gel and water.

4. Conclusion

MAGIC normoxic polymer gel dosimeter has been investigated for its radiological water equivalence by investigating the element composition, mass density, electron density, number of electron per gram, effective atomic number, and linear attenuation coefficient of unirradiated gel. MAGIC gel can be considered a radiological water equivalent dosimeter since almost all of its radiological properties differ by less than 2.7% from that of water. The μ of irradiated gel was found to be linear with dose, up to 40 Gy. The Monte Carlo simulation for 6 MV photon beam was used to calculate the percentage depth dose for MAGIC gel and water and it was found that the dose reduction with depth of gel was very similar to that of water with difference less than 1%. It may be concluded that MAGIC gel can be considered water equivalent for depth dose measurements of 6 MV photon beam.

5. References

[1] Fong M, Keil C, Does D. and Gore C 2001 Polymer gels for magnetic resonance imaging of radiation dose distributions at normal room atmosphere. *Phys. Med. Biol.* **46** 3105-13.

[1] Venning J, Brindha B and Baldock C 2004 Preliminary study of a normoxic PAG gel dosimeter with tetrakis (hydroxymethyl) phosphonium chloride as an antioxidant. Third International Conference on Radiotherapy Gel Dosimetry. *J. Phys. :Conf. Ser.* **3** 155-158.

[2] Zahmatkesh H, Kousari R, Akhlaghpour S and Bagheri S 2004 "MRI gel dosimetry with methacrylic acid. Ascorbic acid. Hydroquinone and Copper in Aharose (MAGICA) gel". *Preliminary Proceedinges of DOSGEL.* Sep 13-16. Ghent. Belgium.

[3] De Deene Y, Reynaert N and De Wagter C 2001 On the accuracy of monomer/polymer gel dosimetry in the proximity of a high-dose-rate 192Ir source. *Phys. Med. Biol.* **46** 2801-25.

[4] Salomons J, Park S, McAuley B and Schreiner J 2002 Temperature increases associated with polymerization of irradiated PAG dosimeters. *Phys. Med. Biol.* **47** 1435-48.

[5] De Deene Y, Vergote K, Claeys C and De Wagter C 2006 The fundamental radiation properties of normoxic polymer gel dosimeters: a comparison between a methacrylic acid based gel and acrylamide based gels. *Phys. Med. Biol.* **51** 653-73.

[6] Khan F M 2010 *The Physics of Radiation Therapy* (Philadelphia, PA: Lippincott Williams & Wilkins)

[7] Baldock C, Deene Y D, Doran S, Ibbott G, Jirasek A, Lepage M, Mcauley K B, Oldham M and Schreiner L J 2010 Polymer gel dosimetry *Phys. Med. Biol.* **55** R1–63

[8] Pantelis E, Karlis A K, Kozicki M, Papagiannis P, Sakelliou L and Rosiak J M 2004 Polymer gel water equivalence and relative energy response with emphasis on low photon energy dosimetry in brachytherapy *Phys. Med. Biol.* **49** 3495–514

[9] Venning J, Nitschke N, Keall J & Baldock, C 2005 Radiological properties of normoxic polymer gel dosimeters. *Med. Phys.* **32** 1047-53.

[10] Sellakumar P, James E and Supe S 2007 Water equivalence of polymer gel dosimeters *Radiat. Phys. Chem.* **76** 1108–15.

[11] Brown S, Venning A, De Deene Y, Vial P, Oliver L, Adamovics J and Baldock C 2008 Radiological properties of the PRESAGE and PAGAT polymer dosimeters *Appl. Radiat. Isot.* **66** 1970–4.

[12] TaylorML, Franich D, Trapp J and Johnston N 2008 The effective atomic number of dosimetric gels *Australas. Phys. Eng. Sci. Med.* **31** 131–8.

[13] TaylorML, FranichD, Trapp J and Johnston N 2009 Electron interaction with gel dosimeters: effective atomic numbers for collisional, radiative and total interaction processes *Radiat. Res.* **171** 123–6.

[14] Gorjiara T, Hill R, Kim J-H, Kuncic Z, Adamovics J and Baldock C 2010a Study of dosimetric water equivalency of PRESAGER° for megavoltage and kilovoltage x-ray beams *J. Phys.: Conf. Ser.* **250** 012053

[15] Gorjiara T, Hill R, Kuncic Z, Adamovics J, Bosi S, Kim J H and Baldock C 2011 Investigation of radiological properties and water equivalency of PRESAGER° dosimeters *Med. Phys.* **38** 2265–74.

[16] Gorjiara T, Hill R, Kuncic Z and Baldock C 2010b Water equivalency evaluation of PRESAGER° dosimeters for dosimetry of Cs-137 and Ir-192 brachytherapy sources *J. Phys.: Conf. Ser.* **250** 012093.

[17] Gorjiara T, Hill R, Kuncic Z, Bosi S and Baldock C 2010c An evaluation of genipin gel as a water equivalent dosimeter for megavoltage electron beams and kilovoltage x-ray beams *J. Phys.: Conf. Ser.* **250** 012036.

[18] Smith F and Hashemi J 2006 *TA403.S663:2006 Foundations of materials science and engineering,* 4. McGraw-Hill.

[19] Trapp J V, Michael G, De Deene Y and Baldock C 2002 Attenuation of diagnostic energy photons by polymer gel dosimeters *Phys. Med. Biol.* **47** 4247 –58.

[20] Erdtmann G and Soyka W 1979 *The Gamma Rays of the Radioniclides, Tables for Applied Gamma Ray Spectroscopy. Topical Presentations in Nuclear Chemistry,* Verlag Chemie, New York, USA.

[21] Al-Bahri S and Spyrou M 1998 Electron density of normal and pathological breast tissues using a Compton scattering technique. *Appl. Radiat. Isot.* **49** 1677-84.

[22] Balogun, A 1999 Angular variation of scattering volume and its implications for Compton scattering tomography. *Appl. Radiat. Isot.* **50** 317-323.

[23] Rogers W 2006 Fifty years of Monte Carlo simulations for medical physics *Phys. Med. Biol.* **51** R287–301.

[24] Walters R, Kawrakow I and Rogers W 2009 DOSXYZnrc users manual *National Research Council of Canada Report PIRS-794 rev. B.*

[25] Sofie C, Isabelle G, Helen G, Jonas S, Stine K, Flemming K, Michelle H and Sven A° J Ba¨ck 2010 RapidArc treatment verification in 3D using polymer gel dosimetry and Monte Carlo simulation *Phys. Med. Biol.* **55** 4885–4898.

Journal of Physics: Conference Series **423** (2013) 012017 doi:10.1088/1742-6596/423/1/012017

Sinusoidal Beds as a Wave Reflector

Viska Noviantri

Department of Mathematics, Bina Nusantara University
West Jakarta, Indonesia

E-mail: viskanoviantri@yahoo.com

Abstract. In this paper, we study the relevance of sinusoidal beds as shoreline protection through the Bragg scattering mechanism. Here we take into account the presence of current and assume that the shore on the right of the sinusoidal beds can absorb wave completely. As well known, a relatively small amplitude of sinusoidal beds can reduce the amplitude of incident waves effectivelly, due to Bragg resonance. Bragg resonance will occur if the wavelength of the monochromatic wave is twice the wavelength of sinusoidal beds. We apply the multiple scale asymptotic expansion method to the linear Shallow Water Equation for sinusoidal beds. When there is current, two wave numbers lead to Bragg resonances. Otherwise, there will be only one wave number. A transmission and reflection coefficient are given and used to predict the amplitude reduction. The effect of sinusoidal beds to transmission and reflection waves are simulated analitycally. It shows that the sinusoidal beds can reduce the amplitude of incident wave.

1. Introduction

When a wave meets a different depth, it will scatter into a transmitted and reflected wave. Let us imagine an incident wave running above a flat bottom with a finite patch of sinusoidal disturbance. During its evolution, there occur many scattering processes. Assume that the shore on the right of the sinusoidal beds can absorb wave completely, then Bragg resonance occurs when the wavelength of incident wave is twice of the wavelength of the periodic bottom disturbance [2]. We find that a larger amplitude disturbance leads to larger reflected wave amplitude. This result explains that the long shore sandbar indeed can reduce the amplitude of incident wave. This motivates the idea of artificially constructing beds to protect a beach from large amplitude of incident waves such as a tidal waves and tsunami.

In this paper, we will take into account the effect of current by analytically. We apply the asymptotic expansion method for the linear shallow water equation for sinusoidal beds in the case when there is a current. The procedure is similar to those applied by C.C. Mei et. al. in [2, 3]. Similar method applied by Philip L.-F. Liu in [4] for water waves in a channel with corrugated boundaries. We obtain there are two wave numbers that may lead to Bragg resonance. These two wave numbers reduce to the Bragg resonance wave number in the no current case. Literatures such as [5] study the numerical method for these problems.

We use the data from Heathershaw experiment [1], to simulate the wave propagating over sinusoidal beds. The larger amplitude of sinusoidal beds and current; leads smaller amplitude of transmitted wave that hit the shore. The same results will occur when the case is subcritical and supercritical detuning.

2. Bragg resonance conditions with current

The aim is to find Bragg resonance condition for surface wave above sinusoidal beds in the case there is a current. The method used is similar with the one use by C.C. Mei in [2], which is multi scale asymptotic expansion.

Consider the SWE equation with current U as follows

$$\begin{cases} \eta_{\hat{t}} + \{(\eta + h)(U + u)\}_{\hat{x}} = 0 \\ u_{\hat{t}} + (U + u)u_{\hat{x}} + g(\eta + h)_{\hat{x}} = 0. \end{cases} \quad (1)$$

with $\eta(\hat{x},\hat{t})$ is surface displacement at position $\hat{x}$ and time $\hat{t}$, $u(\hat{x},\hat{t})$ is horizontal component of fluid particles, g is gravitation coefficient and a bottom topography

$$h(\hat{x}) = h_0(1 + \varepsilon E \sin K\hat{x}) \quad (2)$$

where h_0 is flat depth, $h_0\varepsilon E$ and K is amplitude and wave number of sinusoidal beds, respectively.

2.1. Riemann invariant form

Consider the linear equation of (1) for flat bottom case $h(x) = h_0$, which is

$$\begin{bmatrix} \eta_t \\ u_t \end{bmatrix} = \begin{bmatrix} -U & -h_0 \\ -g & -U \end{bmatrix} \begin{bmatrix} \eta_x \\ u_x \end{bmatrix} \quad (3)$$

In (3), equation for $\eta(\hat{x},\hat{t})$ and $u(\hat{x},\hat{t})$ are coupled. These solutions will obtain by writing them first in the following Riemann invariant form

$$\begin{bmatrix} G_{1t} \\ G_{2t} \end{bmatrix} = \begin{bmatrix} -U + \sqrt{gh_0} & 0 \\ 0 & -U - \sqrt{gh_0} \end{bmatrix} \begin{bmatrix} G_{1x} \\ G_{2x} \end{bmatrix} \quad (4)$$

with transformation

$$[G_1 \quad G_2]^T = P^{-1}[\eta \quad u]^T \quad (5)$$

where

$$P = \begin{bmatrix} -\sqrt{\dfrac{h_0}{g}} & \sqrt{\dfrac{h_0}{g}} \\ 1 & 1 \end{bmatrix} \quad (6)$$

Then, will get the following system of differential equation

$$\begin{bmatrix} G_{1tt} \\ G_{2tt} \end{bmatrix} = \begin{bmatrix} (-U + \sqrt{gh_0})^2 & 0 \\ 0 & (-U - \sqrt{gh_0})^2 \end{bmatrix} \begin{bmatrix} G_{1xx} \\ G_{2xx} \end{bmatrix} \quad (7)$$

Now, in equation (7), the equation for G_1 and G_2 are separated, and they have D'Alembert solution. Assume that each solution is monochromatic wave, then

$$\begin{bmatrix} G_1 \\ G_2 \end{bmatrix} = \begin{bmatrix} Ae^{ik_1x - i\omega t} + B^{ik_1x + i\omega t} \\ Ce^{ik_2x - i\omega t} + D^{ik_2x + i\omega t} \end{bmatrix} \quad (8)$$

Journal of Physics: Conference Series **423** (2013) 012017 doi:10.1088/1742-6596/423/1/012017

with disperse relation

$$\frac{\omega}{k_1} = \pm c_1, \quad \frac{\omega}{k_2} = \pm c_2, \tag{9}$$

where ω is the wave frequency, k_1 and k_2 are wave numbers, and $c_1 = \sqrt{gh_0} - U$ and $c_2 = \sqrt{gh_0} + U$ are phase velocity of each monochromatic wave.

2.2. Multiscale expansion method

We apply the multiple scale expansion method in order to find an analytical solution of (1), in the case of near resonance, which is valid for all $\hat{x}$ and $\hat{t} > 0$. And now, we introduce fast and slow variables in space and time as belows

$$x = \hat{x}, \bar{x} = \varepsilon\hat{x} \tag{10}$$

$$t = \hat{t}, \bar{t} = \varepsilon\hat{t}. \tag{11}$$

The relations between partial derivatives are

$$\partial_{\hat{t}} = \partial_t + \varepsilon\partial_{\bar{t}}, \quad \partial_{\hat{x}} = \partial_x + \varepsilon\partial_{\bar{x}}. \tag{12}$$

Next, we expand

$$\eta(x,\bar{x};t,\bar{t}) = \eta_0(x,\bar{x};t,\bar{t}) + \varepsilon\eta_1(x,\bar{x};t,\bar{t}) + ..., \tag{13}$$

$$u(x,\bar{x};t,\bar{t}) = u_0(x,\bar{x};t,\bar{t}) + \varepsilon u_1(x,\bar{x};t,\bar{t}) + ..., \tag{14}$$

with $\varepsilon > 0$ is a small parameter.

Substituting (13) and (14) into (1) gives us the following series of equations

$$O(1): \partial_{tt}\eta_0 - (U^2 + gh_0)\partial_{xx}\eta_0 - 2Uh_0\partial_{xx}u_0 = 0, \tag{15}$$

$$\begin{aligned} O(\varepsilon): \partial_{tt}\eta_1 - (U^2 + gh_0)\partial_{xx}\eta_1 - 2Uh_0\partial_{xx}u_1 = {} & 2\{\partial_{t\bar{t}}\eta_0 - (U^2 + gh_0)\partial_{x\bar{x}}\eta_0 - 2Uh_0\partial_{x\bar{x}}u_0\} \\ & + gh_0E\sin Kx\partial_{xx}\eta_0 + gh_0EK\cos Kx\partial_x\eta_0 \\ & + 2Uh_0E\sin Kx\partial_{xx}u_0 + 3Uh_0EK\cos Kx\partial_x u_0 \end{aligned} \tag{16}$$

Note that equation (15) is equivalent with (3) and hence the solutions obtained from inverse transform of (5). There are

$$\eta_0 = -\frac{1}{2}\sqrt{\frac{h_0}{g}}\left\{A(\bar{x},\bar{t})e^{i(k_1x-\omega t)} + B(\bar{x},\bar{t})e^{-i(k_1x+\omega t)} + c.c. - C(\bar{x},\bar{t})e^{i(k_2x-\omega t)} - D(\bar{x},\bar{t})e^{-i(k_2x+\omega t)} + c.c.\right\} \tag{17}$$

$$u_0 = \frac{1}{2}\left\{A(\bar{x},\bar{t})e^{i(k_1x-\omega t)} + B(\bar{x},\bar{t})e^{-i(k_1x+\omega t)} + c.c. + C(\bar{x},\bar{t})e^{i(k_2x-\omega t)} + D(\bar{x},\bar{t})e^{-i(k_2x+\omega t)} + c.c.\right\} \tag{18}$$

with $c.c$ denotes their complex conjugate. Note that η_0 and u_0 are superposition of waves with two wave numbers k_1 and k_2, which now their amplitudes are functions of $\bar{x}$ and $\bar{t}$. Here, $A(\bar{x},\bar{t})$ and $C(\bar{x},\bar{t})$ are amplitudes of right running monochromatic wave with wave number k_1 and k_2, respectively. The others, $B(\bar{x},\bar{t})$ and $D(\bar{x},\bar{t})$ are amplitudes of left running monochromatic wave with wave number k_1 and k_2, respectively. The amplitudes $A(\bar{x},\bar{t}), B(\bar{x},\bar{t})$, $C(\bar{x},\bar{t})$ and $D(\bar{x},\bar{t})$ are complex functions to be determined.

Next, we look for solution of the order-ε of equation (16) which are η_1 and u_1. Substituting (17) and (18) into the right hand side of (16) will yield exponent terms with wave number $\pm k_1, \pm k_2, \pm(K-k_1), \pm(K-k_2)$.

When $K = 2k_1$, the exponent terms on the r.h.s. with wave number $\pm k_1, K-k_1, -K+k_1$ have the same wave number with the natural mode $\exp i(k_1 x \pm \omega t)$. To avoid unbounded resonance of η_1 and u_1, we can simplify equate to zero the coefficients of those terms, and get the following equations

$$\begin{cases} A_{\bar{t}} + c_1 A_{\bar{x}} = \beta B \\ B_{\bar{t}} - c_1 B_{\bar{x}} = -\beta A. \end{cases} \tag{19}$$

with

$$\beta = \frac{c_0 E k_1}{2c_1}\left(\frac{c_0}{2} - 2U\right), \quad c_0 = \sqrt{g h_0}. \tag{20}$$

where $c_0 = \sqrt{g h_0}$. As a check, we put $U = 0$ in the equations (19), we get exactly the equations by C.C. Mei in [1].

Analogously when $K = 2k_2$, we avoid unbounded resonance of η_1 and u_1 caused by terms with wave number k_2, so we get

$$\begin{cases} C_{\bar{t}} + c_2 C_{\bar{x}} = \beta D \\ D_{\bar{t}} - c_2 D_{\bar{x}} = -\beta C. \end{cases} \tag{21}$$

3. Amplitude of transmission and reflection method

Next we study the relevance of sinusoidal bottom in reducing the amplitude of an incident monochromatic wave Imagine that we have a sinusoidal beds patch at 0 < x < L. A monochromatic wave coming from the left passes the sinusoidal beds and then propagates to the right until it hit the shore. In region x < 0, there is no interaction between the right and left propagating wave, so that the equation in this region is

$$c_1 A_{\bar{x}} + A_{\bar{t}} = 0$$
$$c_1^2 B_{\bar{x}} + B_{\bar{t}} = 0 \tag{22}$$

When the wave propagates above sinusoidal beds, there will be many scattering processes. The waves split into a transmitted and reflected wave. If the wave number of sinusoidal beds is twice of the wave number of incident wave, then interaction between the right and left propagating wave govern by equation (19).

System of equations (19) can be separated into equations for each $A(\bar{x},\bar{t})$ and $B(\bar{x},\bar{t})$

$$A_{\bar{t}\bar{t}} - c_1^2 A_{\bar{x}\bar{x}} + \beta^2 A = 0,$$
$$B_{\bar{t}\bar{t}} - c_1^2 B_{\bar{x}\bar{x}} + \beta^2 B = 0. \qquad (23)$$

Equations (23) known as the Klein-Gordon equations.

Assume that the shore on the right of the sinusoidal beds can absorb wave completely, then there is no reflected wave in $x > L$. Therefore, the right boundary condition is $B(\overline{L},\bar{t}) = 0$.

Let the incident wave is

$$\zeta = A(\bar{x},\bar{t})e^{ik_1x - i\omega t} \qquad (24)$$

where

$$A(\bar{x},\bar{t}) = A_0 e^{iK(\bar{x}-\omega\bar{t})}, \bar{x} < 0 \qquad (25)$$

In $0 < x < L$, assume that the solutions of (23) are

$$A(\bar{x},\bar{t}) = A_0 T(\bar{x}) e^{-iKc_0\bar{t}} \qquad (26)$$

$$B(\bar{x},\bar{t}) = A_0 R(\bar{x}) e^{-iKc_0\bar{t}} \qquad (27)$$

Here, $T(\bar{x})$ and $R(\bar{x})$ are transmission and reflection coefficient respectively. Note that the solutions must continue at $\bar{x} = 0$ and at $\bar{x} = L$.

Substituting (26) and (27) to (23) give the differential equation for $T(\bar{x})$ and $R(\bar{x})$. These solutions divide into three cases.

Case 1: Subcritical detuning
Subcritical detuning occurs when K = 2(k_1 +-εκ), where ε is small parameter and κ is constant. In this case, the transmission and reflection coefficient as

$$T(\bar{x}) = \frac{iQc_1 \cosh Q(\overline{L} - \bar{x}) + \alpha \sinh Q(\overline{L} - \bar{x})}{iQc_1 \cosh Q\overline{L} + \alpha \sinh Q\overline{L}} \qquad (28)$$

$$R(\bar{x}) = \frac{-\beta \sinh Q(\overline{L} - \bar{x})}{iQc_1 \cosh Q\overline{L} + \alpha \sinh Q\overline{L}} \qquad (29)$$

where $Qc_1 = \sqrt{\beta^2 - \alpha^2}$ and α=κc1

Case 2: Supercritical detuning
Supercritical detuning occurs when K = 2(k_1 + εκ), where ε is small parameter and κ is constant. In this case, the transmission and reflection coefficient as

$$T(\bar{x}) = \frac{Pc_1 \cos P(\bar{L}-\bar{x}) - i\alpha \sin P(\bar{L}-\bar{x})}{Pc_1 \cosh Q\bar{L} - i\alpha \sinh P\bar{L}} \tag{30}$$

$$R(\bar{x}) = \frac{i\beta \sin P(\bar{L}-\bar{x})}{Pc_1 \cosh P\bar{L} - i\alpha \sinh P\bar{L}} \tag{31}$$

where $Qc_1 = \sqrt{\beta^2 - \alpha^2}$ and α=κc1

Case 3: Perfect resonance
The transmission and reflection coefficient of perfect resonance ($K = 2k_1$) are

$$T(\bar{x}) = \frac{A}{A_0} = \frac{\cosh \frac{\beta(\bar{L}-\bar{x})}{c_1}}{\cosh \frac{\beta\bar{L}}{c_1}} \tag{32}$$

and

$$R(\bar{x}) = \frac{B}{A_0} = \frac{i\sinh \frac{\beta(\bar{L}-\bar{x})}{c_1}}{\cosh \frac{\beta\bar{L}}{c_1}} \tag{33}$$

simulation of Bragg resonance will make using the following data: flat depth $h_0 = 10$ m wave number of incident wave $k_1 = \pi$ /m, the sinusoidal beds patch has wave number $2k_1 = 2\pi$ /m, and length $L = 10$ m.

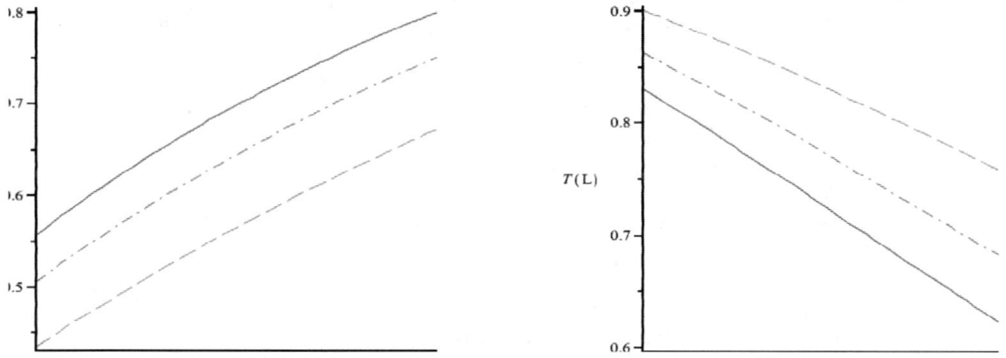

Figure 1. Left: $R(0)$ versus εE for some U, right: $T(L)$ versus εE for some U.

Sketch of $R(0)$ and $T(L)$ versus εE for some current U draw in Figure 1. We can see from them that the larger amplitude of sinusoidal beds leads larger amplitude of reflected wave at $x = 0$ and leads smaller amplitude of transmitted wave at $x = L$. Besides that, the larger current leads the larger transmitted wave amplitude at $x = L$ and leads the smaller reflected wave amplitude at $x = 0$.

Figure 2 shows the reflected wave amplitude in the middle of reflected wave amplitude in subcritical and supercritical detuning case when perfect resonance occurs.

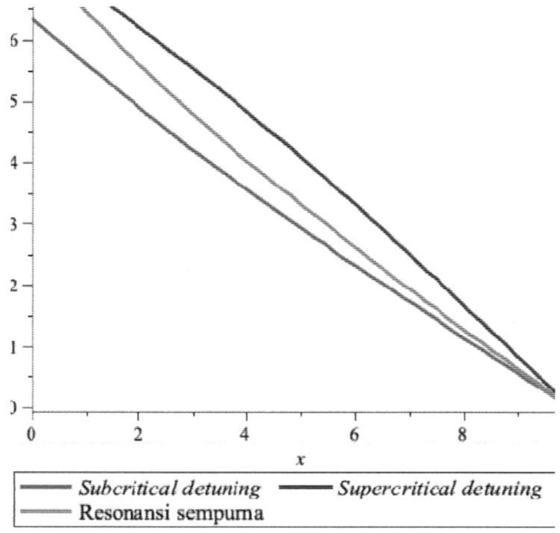

Figure 2. Comparison of reflected wave amplitude in perfect resonance, subcritical and supercritical detuning (εE=0.14 and U=0,5)

4. Conclusion

The solution of wave equation over sinusoidal-beds with current obtains by Riemann invariant form. A solution of them is superposition of monochromatic waves with two different wave numbers.

Sinusoidal beds may lead to Bragg resonance. When there is current, two wave numbers lead to Bragg resonances. Otherwise, there will be only one wave number. Bragg resonance occurs when the one of two different wavelength of incident wave is twice of the wavelength of the periodic bottom disturbance. The larger amplitude of sinusoidal beds leads larger amplitude of reflected wave at $x = 0$ and leads smaller amplitude of transmitted wave at $x = L$.

5. References

[1] A.D., Heathershaw 1982 Seabed-wave resonance and sandbar growth *Nature* 296 pp 343-345

[2] Chiang C. Mei 2004 Multiple Scattering by an Extended Region of In homogeneities *Lecture Notes* MIT

[3] Jie Yu And Chiang C. Mei 2000 Do Longshore Bars Shelter The Shore? J. Fluid Mech. 404 pp 251-268

[4] Philip L.-F. Liu 1987 Resonant reflection of water waves in a long channel with corrugated boundaries *J. Fluid Mech* 179 pg. 371-381

[5] Viska Noviantri and S.R.Pudjaprasetya 2010 The Relevance of Wavy Beds as Shoreline Protection *Proceedings of the 13th Asian Congress of Fluid Mechanics* Pg. 489-492

Journal of Physics: Conference Series **423** (2013) 012018 doi:10.1088/1742-6596/423/1/012018

Hierarchical Cluster Approach for Regionalization ofPeninsular Malaysia based on the Precipitation Amount

N H Ahmad[1] and I R Othman[2]and S M Deni[3]

[1,2,3]Department of statistics

Faculty of computer and mathematical sciences

UiTM Shah Alam Malaysia

Email : norleehainieahmad@gmail.com

Abstract: The rainfall series from 59 homogeneous stations in Peninsular Malaysia are used in this study. The series are indicated to be multifarious, with unpredictable fluctuations from year to year and from region to region. Therefore it is more appropriate to observe all stations on spatial scales corresponding to clusters instead of the individual series.Single site occurrence analysis is inadequate for regional planning. Hence, it is important to determine regional rainfall frequencies. Hierarchical cluster analysis with 11 different similarity measures and 7 different linkagesare used in this study to find homogeneous rainfall groups. Validity indices are evaluated to determine the best number of clusters in a rainfall data set. The study elucidatesthree homogeneous rainfall regionsnamely A, B and C withinPeninsular Malaysia, which correspond to areas with different natural features and rain pattern. Based on this research, a hierarchical clustering method is suggested in order to group individual rainfall series into a limited set of spatial clusters.

1. Introduction

A rainy day may be defined as a day that rains more than the threshold (in this study which is 1 mm), while the intensity of rainfall may be defined as the measure of the amount of rain that falls over time. According to Malaysian Meteorological Department, [19], the characteristic features of the climate of Malaysia are of uniformed temperature, high humidity and copious rainfall. Situated in the equatorial doldrums area, it is extremely rare to have a full day with completely clear sky even during periods of severe drought. Instead, it is also rare to have a stretch of a few days with completely no sunshine except during the northeast monsoon seasons. The seasonal wind flow patterns paired with the local topographic features determine the rainfall distribution patterns over the country.

During the northeast monsoon season, the exposed areas i.e. the east coast of Peninsular Malaysia experience heavy rainfalls. On the other hand, inland areas or areas which are sheltered by mountain ranges are relatively free from this influence. It is best to describe the rainfall distribution of the country according to seasons. Alternatively, lower number of rainy days is observed in the northwestern sector of the Peninsular where minimum precipitation occurs in that part of the country. By using method the coefficient of variability [1] proved that the annual rainfall variability at the eastern half is larger than at the western half. Besides, the researchdone by [2] in his case study of Sungai Kayu Ara, Damansara Selangor shows that greater rainfall volume fell on developed area, compared to suburban/rural area.

The nonparametric Mann-Kendall (MK) statistical test used by [3]to display the results for both southwest monsoon and northeast monsoon seasons indicate that there are significantly decreasing trends in the frequency of wet days during the extreme events for most of the stations inPeninsular Malaysia. However, a smaller number of significant trends were found for extreme intensity. In addition, [4]in their study seeks to identify the spatial patterns and time-variability of rainfall in Peninsular Malaysia on monthly, yearly and monsoon temporal scales using Shepard s interpolation scheme. The spatial variation analysis shows that the East Coast region, which received substantially higher amounts of rainfall during the northeast monsoon, has lower spatial rainfall variability and a more uniformed rainfall distribution than other regions. A larger range of the monthly spatial variation is observed in the West Coast region. Thus, rainfall variation within Peninsular Malaysia is complex, with unpredictable rainfall fluctuations from year to year and from region to region by comparison.

Quite a lot of methods are regularly used for the regionalization of hydrologic phenomena such as rainfall, streamflow and other components of the water cycle as to determine if an area has hydrologically homogeneous regions [5]. Multivariate techniques are powerful tools in classifying meteorological data such as rainfall. Principal component analysis, factor analysis, and different types of cluster analysis and support vector machine analysis have been used to classify daily rainfall patterns and their relationship to atmospheric circulation [6], flood and drought years [7], and streamflow drought [8]. The application of multivariate methods to classify hydrologic events on a national scale was suggested by [9] while L-moments and cluster analysis was applied by [10] to determine regional rainfall climates within conterminous United States, as he defined 104 homogeneous precipitation regions there. Clusters are seen to offer several advantages over geographical regions and they are interpreted in terms of basin characteristics through the use of a multivariate linear discriminant analysis as stated by [11].

There have been many studies relating to rainfall pattern of Malaysia. It provides beneficial information to various sectors including hydrological and water resources management.Nevertheless, if there are too many rain gauge stations need to be analyzed, time consumption is the utmost constraint. However, limited studies have been done to group spatial and temporal patterns to define homogeneous rainfall regions within Peninsular Malaysia in order toabridge hydrologic calculations and diminish masses of observations and variables for frequency analysis.

2. Rainfall Data

The annual rainfall data from 59homogeneous stations in [18] are obtained from Drainage and Irrigation Department and Malaysian Meteorological Department. Data are recorded from 1975 to 2010. There are 36 years of observation used in this study. Additionally, it is suggested by The World Meteorological Organization (WMO) to use 30-year rainfall periods for rainfall analysis. Table 1 shows the 59 stations in Peninsular Malaysia with their geographical coordinates while Figure 1 shows the map of Peninsular Malaysia along with the stations used in this study. Peninsular Malaysia consists of 13 states and the boundary of each state is also illustrated.

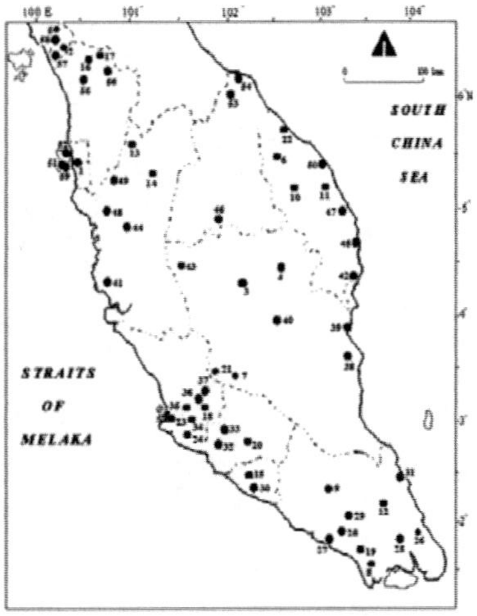

Figure 1: Location of rainfall gauge stations in Peninsular Malaysia

Table 1: The geographical coordinates of the rain gauge stations in Peninsular Malaysia

No.	District	Long	Lat	No.	District	Long	Lat
1	PINTU A.BAGAN,AIR ITAM	100.2	5.4	**31**	MERSING	103.8	2.5
2	ARAU	100.3	6.4	**32**	HOSPITAL PORT DICKSON	101.8	2.5
3	S. K. KG. AUR GADING	101.9	4.4	**33**	SETOR JPS SIKAMAT SEREMBAN	101.9	2.7
4	LDG. BATU KAWAN	100.4	5.3	**34**	PETALING JAYA	101.7	3.1
5	GUAR NANGKA	100.3	6.5	**35**	SUBANG	101.6	3.1
6	KLINIK BIDAN,JAMBU BONGKOK	103.4	4.9	**36**	EMPANGAN GENTING KELANG	101.7	3.2
7	JANDA BAIK	101.9	3.3	**37**	GOMBAK	101.7	3.3
8	JAM. SG. SIMPANGAN, JLN. EMPAT	102.1	2.4	**38**	RUMAH PAM PAHANG TUA,PEKAN	103.4	3.6
9	IBU BEKALAN KAHANG , KLUANG	103.6	2.2	**39**	KUANTAN	103.2	3.8
10	SEK. KEB. KEMASEK	103.5	4.4	**40**	RUMAH PAM PAYA KANGSAR	102.4	3.9
11	SEK. KEB. KG. JABI	102.6	5.7	**41**	SITIAWAN	100.7	4.2
12	LDG. KIAN HOE , KLUANG	103.3	2.0	**42**	JPS KEMAMAN	103.4	4.2
13	DISPENSARI KROH	101.0	5.7	**43**	LDG BOH	101.4	4.5
14	STN. PEMEREKSAAN HUTAN, LAWIN	101.1	5.3	**44**	IPOH	101.1	4.6
15	PEKAN MERLIMAU	102.4	2.2	**45**	SEK.MEN. SULTAN OMAR,DUNGUN	103.4	4.8
16	RUMAH PENJAGA JPS. PARIT NIBONG	100.5	5.1	**46**	GUA MUSANG	101.9	4.9
17	PENDANG	100.5	5.9	**47**	KG. MENERONG	103.1	4.9
18	JPS. WILAYAH PERSEKUTU	101.7	3.2	**48**	PUSAT KESIHATAN BT.KURAU	100.8	4.9
19	LDG. BENUT ,RENGAM	103.4	1.8	**49**	SELAMA	100.7	5.1
20	LDG. SG. SABALING	102.5	2.9	**50**	STOR JPS KUALA TRENGGANU	103.1	5.3
21	GENTING SEMPAH	101.8	3.4	**51**	KOLAM TAKONGAN AIR ITAM	100.3	5.4
22	KG. MERANG ,SETIU	102.9	5.5	**52**	KLINIK BKT. BENDERA	100.3	5.4
23	IBU BEKALAN SG. BERNAM	101.4	3.7	**53**	TO' UBAN	102.1	5.9
24	KG. SG. TUA	101.7	3.3	**54**	KOTA BHARU	102.3	6.2
25	SENAI	103.7	1.6	**55**	ALOR STAR	100.4	6.2
26	SEK.MEN.BKT BESAR DI KOTA TINGGI	103.7	1.8	**56**	AMPANG PEDU	100.8	6.2
27	SEK.MEN.INGGERIS BATU PAHAT	102.9	1.9	**57**	PADANG KATONG ,KANGAR	100.2	6.5
28	PINTU KAWALAN SEMBRONG	103.1	1.9	**58**	ABI KG. BAHRU	100.2	6.5
29	KLUANG	103.3	2.0	**59**	BAYAN LEPAS	100.3	5.3
30	MALACCA	102.3	2.3				

3. Methodology

There are three steps involved in determining the regionalization of Peninsular Malaysia based on the precipitation amount. The basic steps to develop clustering process in this study are presented in Figure 2.

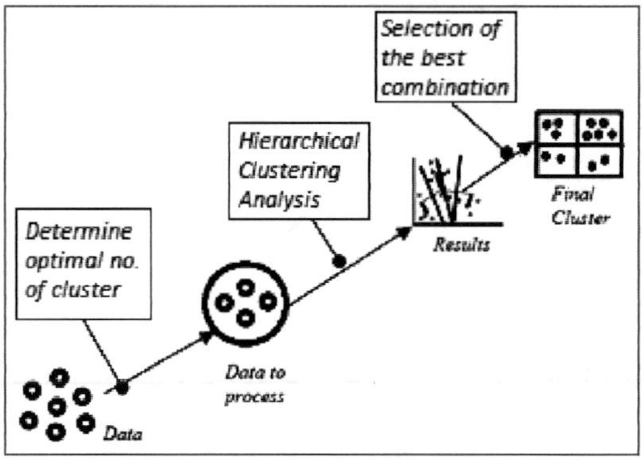

Figure 2: Steps of clustering process.

3.1. Determination the optimal number of cluster

Cluster analysis for rainfall precipitation involves the grouping of a given data set into groups (clusters) such that the data points in a cluster are highly similar in explaining the pattern of that particular area such as geographical, physical, statistical or stochastic features. A number of approaches can be used to do regionalization. Hierarchical cluster analysis and L-moments regional frequency analysis were both used in [5] study regarding regional precipitation climates of Iran. Multivariate linear discriminant analysis and hypothetical ungauged basins was applied by [11]in his study of drainage basins in Britain.

Cluster and principal component analysis were being employed by [9] to collect data from that of Sweden and found that cluster analysis is suitable to use on a national scale for a country with heterogeneous hydrological regimes. His research for rivers in southern Manitoba [9] Canadaincorporates both a basin similarity measure, imbedded in the clustering algorithm, and a regional homogeneity measure to evaluate station partitionings obtained from the clustering algorithm. Meanwhile, [13] used hierarchical clusters of rainfall distribution patterns in a Mediterranean region.

According to [14] hierarchical clustering techniques are proceeded by either a series of successive mergers or a series of successive divisions. *Agglomerative hierarchical methods* that are used in this study begin with the individual objects. Thus, there are initially as many clusters as that of object. Most similar objects are first grouped, and these initial groups are merged accordig to their similarities. Eventually, as the similarity decreases, all subgroups are fused into a single cluster.

Since there is no prior knowledge on number of clusters for regionalization of Peninsular Malaysia, eight different clustering validity indices were used to obtain the optimal number of clusters. The eight different clustering validity indices were Silhoutte index, Davies-Bouldin, Calinski-Harabasz Index, Dunn Index, C Index, Krzanowski-Lai Index, Hartigan Index and Weighted Inter-Intra Index.

The clustering validity indices applied hierarchical clustering with Euclidean distances and complete linkage where in this algorithm, it uses the largest distance between objects in the two clusters. The formula for Euclidean distance is: $d_{rs}^2 = \sum_{j=1}^{p} (x_{rj} - x_{sj})^2$

Besides, complete linkage is the best method for classification in this study. The complete linkageis written as follows:

$$d(r,s) = \max(dist(x_i, x_{sj})), i \in (1,\ldots,n), j \in (1,\ldots,n_s)$$

whererth and sth are rows of the data matrix, x that are denoted by $(x_{r1,}x_{r2,\ldots,}x_{rp})$and $(x_{s1,}x_{s2,\ldots,}x_{sp})$respectively. These two rows correspond to the observations on two objects for all P variables. The quantity d_{rs}^2 is referred to as the squared Euclidean distance. The Euclidean distance of dissimilarity is then used in the cluster techniques.Cluster Validity Analysis Platform (CVAP) package in Matlab® is used to obtain the optimal number of cluster for rainfall series in Peninsular Malaysia.

One of the most important issues in cluster analysis is the assessment of clustering results to find the segregation that best fits the data. The term optimal clustering defined as the outcome of running a clustering algorithm that best fits the inherent partitions of the data set [15]. The measure of validity of the clusters should be such that it will be able to impose an ordering of the clusters in terms of its goodness. Table 2 shows the summary of validity indices used in this study.

Table 2: Clustering Validity Indices

Method	Description
Silhoutte index	A larger Silhouette value indicates a better quality of a clustering result and largest silhouette indicates the optimal number of cluster.
Davies-Bouldin	A low value indicates good cluster structures and minimum value determines the optimal number of cluster.
Calinski-Harabasz Index	It is the pseudo F-statistic, which evaluates the clustering solution by looking at how similar the objects are within each cluster and how well the objects of different clusters are separated and maximum value indicates optimal number of cluster.
Dunn Index	Large values indicate the presence of compact and well-separated clusters and maximum value indicates optimal number of cluster.
C Index	Minimal C-index indicates optimal number of cluster.
Krzanowski-Lai Index	Maximum value indicates optimal number of cluster.
Hartigan Index	The estimated number of clusters is the smallest k. There will be no estimation if this condition is not satisfied. However, an elbow might indicate an optimal number of clusters.
Weighted Inter-Intra Index	Search forward (k=2, 3, 4 ...) and stop at the first down-tick of the index, which indicates optimal number of cluster.

3.2. Hierarchical Clustering

After the optimal number of cluster was obtained in the first section, hierarchical analysis was performed using 77 combinations of 11 different similarity measures and 7 different types of linkages in order to group the data into clusters. Table 3 shows the list of linkages used to create agglomerative hierarchical cluster tree as well as the list of similarity measures in Table 4. The similarity measures indicate the pairwise distance between pairs of objects. Table 3 and Table 4 were taken from the manual of Matlab® software because hierarchical clustering can be done easily using the function under cluster analysis provided by Matlab®.

Table 3 : List of linkages

Method	Description
Average	Unweighted average distance (UPGMA).
Centroid	Centroid distance (UPGMC). Y must contain Euclidean distances.
Complete	Furthest distance.
Median	Weighted center of mass distance (WPGMC). Y must contain Euclidean distances.
Single	Shortest distance. This is the default.
Ward	Inner squared distance (minimum variance algorithm). Y must contain Euclidean distances.
Weighted	Weighted average distance (WPGMA).

Table 4 : List of similarity measures

Metric	Description
Euclidean	Euclidean distance.
Seuclidean	Standardized Euclidean distance. Each coordinate in the sum of squares is inverse weighted by the sample variance of that coordinates.
Mahalanobis	Mahalanobis distance.
Cityblock	City block metric.
Minkowski	Minkowski metric.
Cosine	One minus the cosine of the included angle between points (treated as vectors).
Correlation	One minus the sample correlation between points (treated as sequences of values).
Spearman	One minus the sample Spearman's rank correlation between observations treated as sequences of values.
Hamming	Hamming distance, the percentage of coordinates that differ.
Jaccard	One minus the Jaccard coefficient, the percentage of nonzero coordinates that differ.
Chebychev	Chebychev distance (maximum coordinate difference).

3.3. Selection of the best combination

Previously, there are 77 combinations for hierarchical clustering. Thus, the best combination of both measure and linkage will be evaluated based on few factors such as Malaysia geographical, monsoon limitation and the even segregation of the stations in the cluster. Result will be discussed in the next section.

4. Result and discussion

Annual summation of rainfall series for Peninsular Malaysia is used to obtain the optimal number of cluster. The clustering process using Hierarchical complete clustering and Euclidean distance has been done in this study. This method is used because it suits the data well and widely used by the researcher. Figure 3 shows the results of 8 validity indices used to evaluate the clustering solutions. The optimal number of clusteris indicated by a square symbol.

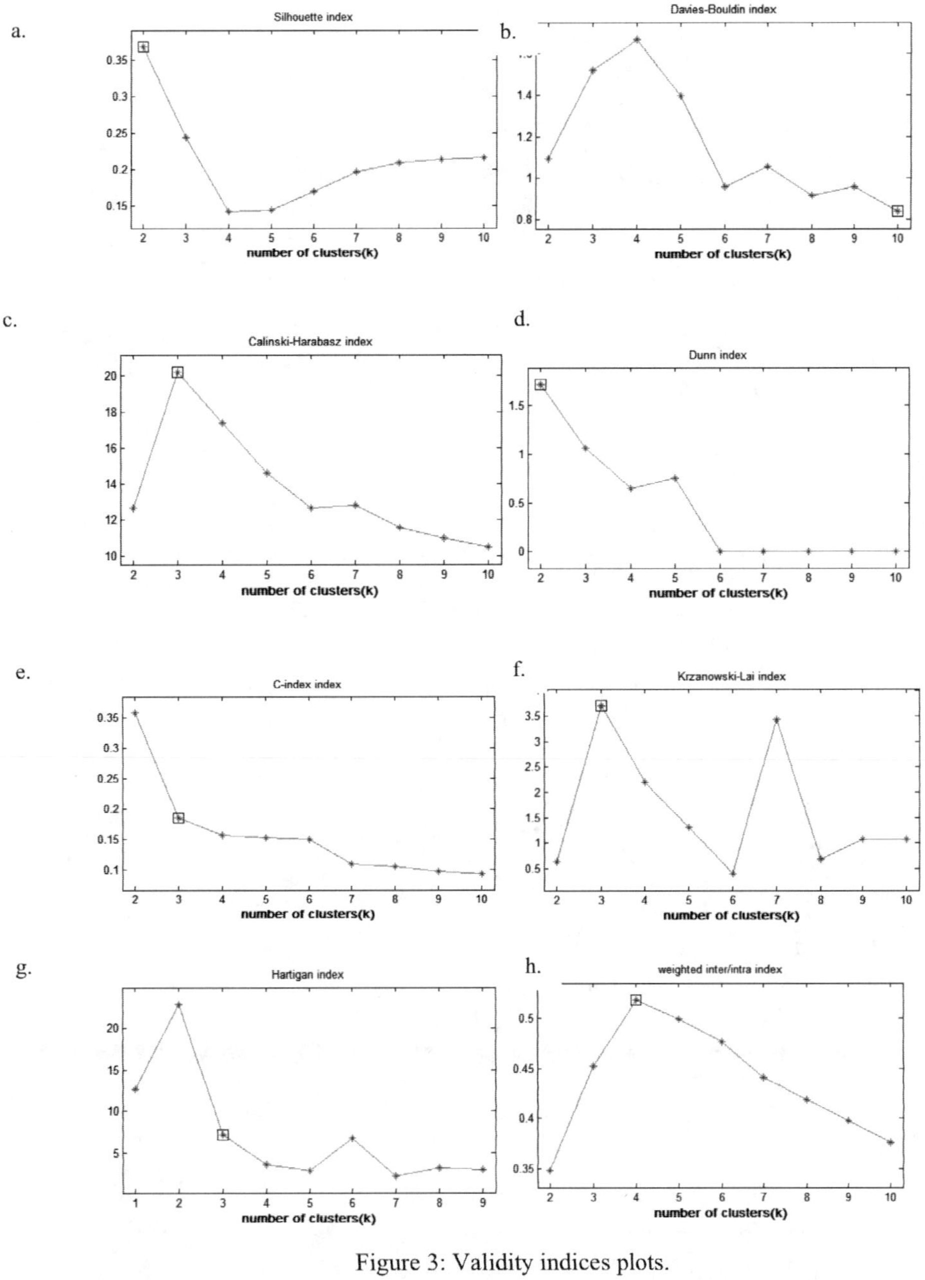

Figure 3: Validity indices plots.

Table 3 summarized the results and it indicates that the best number of cluster is three since majority or 50% of the validity indices show the same result.

Table 3: Optimal number of cluster (Euclidean)

	Method	Complete
a.	Silhoutte index	2
b.	Davies-Bouldin	10
c.	Calinski-Harabasz	3
d.	Dunn Index	2
e.	C Index	3
f.	Krzanowski-Lai Index	3
g.	Hartigan Index	3
h.	Weighted Inter-Intra Index	4

The final step is to allocate the rain gauge stations according to the appropriate cluster. By using hierarchical clustering,two best combinations of similarity measure and linkages were obtained. The first combination(i.e. Complete Cluster) is the combinationof complete linkage and correlation similarity measures. Complete cluster segregated Peninsular Malaysia into three parts which are top, middle and bottom. In the top cluster, there are 21 stations from northern and eastern part of Peninsular Malaysia followed by middle cluster that consists of 19 stations mostly from western part. Bottom cluster comprised the southern part of Peninsular Malaysia with 12 stations altogether.

The second combination (i.e. Weighted cluster) is the combination of weighted linkage and correlation similarity measures. Peninsular Malaysia is separated into three parts and basically it has the same basis with Complete cluster except for the number of stations in each part. In the top part, there are 25 stations, for middle there are 16stations and bottom part have 12 stations. Nevertheless, both figures have to eliminate 7 stations because of station mislocation. Figures 4(a) and Figure 4(b) illustrate the rain gauge stations over the Peninsular Malaysia according to the respective cluster. Red dots indicate the stations that are mislocated. In this study, we chose to remove these stations because special processes are needed to overcome mislocation problem.

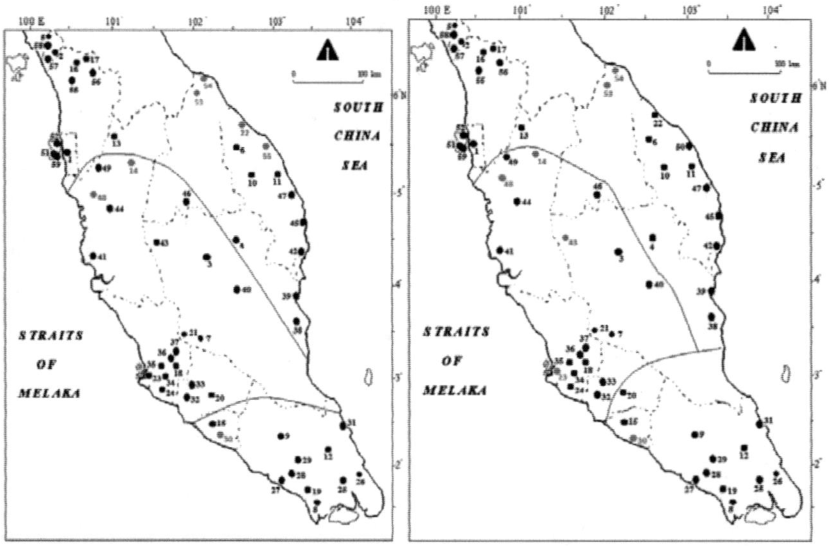

Figure 4: (a) Complete cluster (b) Weighted cluster

5. Conclusion

A lot of studies with pertain to rainfall pattern of Malaysia have been done. It provides beneficial information to various sectors including hydrological and water resources management.However, 59 rain gauge stations are too wide to be studied independently. Time and cost for the researcher will be the greatest restrictions to complete the task. Thus, this study has been done to group the rain gauge stations in Peninsular Malaysia into several clusters to reduce the masses of observation.

Three steps used in this study are the determination the optimal number of cluster, hierarchical clustering and selection of the best combination. The optimal number of cluster for Peninsular Malaysia is three. It representsPeninsular Malaysia as top, middle and bottom part, namely cluster A, B and C. In this study, the combination between complete linkage and correlation similarity measures is more suitable to fits the data set. This is based on the monsoon condition, the segregation of the stations in each cluster and its simplicity. The border of the cluster for this method is illustratedwellbecause it equally dividesPeninsular Malaysia as shown in Figure 5. The stations are also randomly distributed within the cluster and fairly dispersed for each cluster. Further analysis of this topic can be done and the seasonal climate pattern of Malaysia should be taken into account.

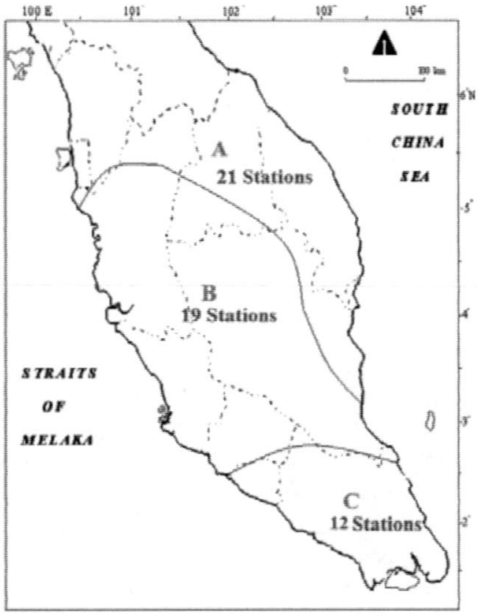

Figure 5: Final Cluster Map of Peninsular Malaysia

Acknowledgment

The authors are indebted to the staff of the Meteorological Department of Malaysia for providing the daily rainfall data for this study. They also register their sincere appreciation to the reviewers for their valuable suggestions and remarks which has improved the manuscript. This research is funded by Universiti Teknologi MARA (600-RMI/ST/FRGS 5/3/Fst (192/2010)).

References

[1] Camerlengo A L, N Saadon,A S Aziz and R Othman 1998 *Monthly Distribution of Precipitation in Peninsular Malaysia* Pertanika Journal Science & Technology 6(1): 59 - 70 (1998).

[2] Leong W C 2007 *Study On Malaysian Urban Rainfall-Runoff Characteristics: Case Study Of Sungai Kayu Ara, Damansara, Selangor*Msc. Civil Engineering, Universiti Sains Malaysia.

[3] Suhaila J, DeniS M, W Zawiah and Jemain AA 2010 *Trends In Peninsular Malaysia Rainfall Data During The Southwest Monsoon And Northeast Monsoon Seasons: 1975 2004* Sains Malaysiana 39(4)(2010): 533 542 .

[4] C LWong, R Venneker, S Uhlenbrook, A B M Jamil and Y. Zhoul 2009 *Variability of rainfall in Peninsular Malaysia*Hydrolorgy Earth System Sciences Discussion 6, 5471 5503.

[5] R Modarres2006 *Regional precipitation climates of Iran* Journal of Hydrology (NZ) 45 (1): 13-27, 2006.

[6] Romero R, Summer G, Ramis C and Genoves A 1999 *A classification of the atmospheric circulation patterns producing significant daily rainfall in the Spanish Mediterranean area*International Journal of Climatol **19**: 765 785

[7] Singh C V 1999 *Principal components of monsoon rainfall in normal, flood and drought years over India* International Journal of Climatology 19: 639-952.

[8] Stahl K., Demuth S 1999 *Methods for regional classification of streamflow drought series: Cluster analysis* Technical report to the ARIDE project, No.1

[9] Gottschalk L 1985 *Hydrological regionalization of Sweden* Hydrological Sciences Journal 30: 65-83.

[10] Guttman N B 1993 *The use of L-Moments in the determination of regional precipitation climates* Journal of Climate 6 (12): 2309-2325.

[11] WiltshireS E 1986 *Regional flood frequency analysis II: Multivariate classification of drainage basins in Britain* Hydrological Sciences - Journal - 31, 3, 9/1986.

[12] Donald H. Burn 1989 *Cluster Analysis As Applied To Regional Flood Frequency* Journals/ Water Resources Planningand Management, Vol. 115, No. 5, September, 1989.

[13] Ramos M C 2001 *Divisive and hierarchical clustering techniques to analyze variability of rainfall distribution patterns in a Mediterranean region* Journal of Hydrology 57: 123-138.

[14] Richard A Johnson and Dean W Wichern 1982 Applied Multivariate Statistical Analysis © 1982 by Prentice-Hall, Inc.,Eaglewood Cliffs, N.J 07632.

[15] Maria Halkidi, Yannis Batistakis And Michalis Vazirgiannis 2001 *On Clustering Validation Techniques*Journal of Intelligent Information Systems, 17:2/3, 107–145, 2001

[16] Kaufman L, Rousseuw P J 1990 *Finding groups in data: An introduction to cluster analysis* Wiley, New York, 344 pp.

[17] Deni S M, A A Jemain and Ibrahim K 2009 *The best probability models for dry and wet spells in Peninsular Malaysia during monsoon seasons*International Journal Of Climatology.

[18] N H Ahmad and Deni S M 2012 *Homogeneity Test on Daily Rainfall Series for Malaysia* Proceeding for 1st ISM International Statistical Conference 2012.

[19] F Yunus *Delineation of Climate Divisions for Peninsular Malaysia* Meteorological Officer (Report)of Meteorological Department of Malaysia

Digital Image Compression using Graph Coloring Quantization Based on Wavelet-SVD

Adiwijaya, M. Maharani, B. K. Dewi, F. A. Yulianto, B. Purnama

Telkom Institute of Technology,
Jl. Telekomunikasi no.1 Ters. Buah Batu, Bandung 40257, Indonesia

E-mail: adiwijaya@ittelkom.ac.id

Abstract. JPEG2000 is an example of standard of compression technique. Wavelet and scalar quantization are used in the JPEG2000 compression scheme. Despite the advantages of JPEG2000 based on wavelet has low computation, the wavelet has poor performance in low-correlated image. In this paper, we propose a graph coloring technique in the quantization process on image compression scheme based on Wavelet- Singular Value Decomposition. We show that the system provides good performance based on compression ratio and PSNR. The average compression ratio generated by the system stands between 50-60%, while the average PSNR stands between 40-80 dB.

1. Introduction

The general principal used in image processing is to reduce data duplication inside the image so the memory needed to reduce and represent data duplication inside the image can be smaller [6]. Compression based on the output given by this system is a lossy compression. Lossy compression is a type of compression where some information might be missing from output data but still tolerated by user. Compression model is divided in two forms which is compression and decompression. Forward transformation, quantization and entropy encoding are parts of compression process while backward transformation, dequantization and entropy decoding are parts of decompression process.

JPEG2000 is a standard compression technique with .jp2 format as its basic core system [3,7]. JPEG2000 using Wavelet as the transformation method. However, despite the advantage of wavelet transformation which has low computation, wavelet has poor performance in low-correlated image [4]. In this Paper, Wavelet - Singular Value Decomposition (SVD) domain will be used as the transformation method. SVD will be used to achieve a better performance for transformation process in low-correlated image, while Wavelet domain is used in high-correlated image [4]. Therefore, the transformation process is supposed to give better performance for both low-correlated image and high-correlated image. In the quantization process, graph-based quantization is used to cluster image component in frequency domain. The implementation of graph based quantization is based on coloring graph method. Coloring graph is a coloring dot from a graph method where adjacent dots have different colors while using colors as minimum as possible [1,2]. On previous research, it is known that graph-based quantization provides better result than vector quantization that is widely used.

2. The Proposed Method

The input image is a color image of 512x512 pixel with RGB color components. Then RGB components are transformed to a 512x512 pixel YCbCr color components. Down-sampling process is done to the chrominance components (Cb, Cr) until the current chrominance component pixel is half of before. The down-sampled color components size is 256x256, then each of color components is divided into 8x8 sub-blocks where each of sub-block consists of 64x64 pixel.

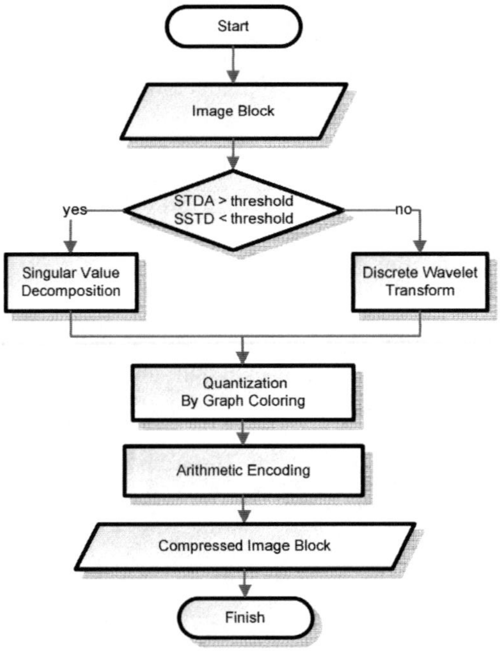

Figure 1. Flowchart of Proposed Scheme

Calculation of average of standard deviation (STDA) and standard deviation of standard deviation of each sub-block (SSTD) is necessary to determine which transformation method should be applied to related sub-block. The sub-block whose size is 64x64 is divided into another sub-block size of 8x8 pixel. For each of 8x8 size of sub-block from a 64x64 size of sub-block standard deviation is calculated. Then for each 64x64 size of sub-block the average of all standard deviation from 8x8 size sub-block from 64x64 size of sub-block related is calculated to get the STDA value and the standard deviation of all standard deviation from 8x8 size sub-block from 64x64 size of sub-block related is calculated to get the SSTD value.

High STDA values indicated that sub-block contains edges, low-correlated pixel, edges and low-correlated pixel or edges and high-correlated pixel (these can be sharp edges) [7]. Sharp edges should not be transformed using SVD because it needs larger codebook to reduce the given error [5]. That is why both STDA and SSTD threshold are needed to determine which transformation method should be used. SSTD value of sharp edges is usually much higher than low-correlated pixel or high-correlated pixel [5]. Every sub-block which has higher STDA value than STDA threshold and lower SSTD value than SSTD threshold should be transformed using SVD. And then, every sub-block which has lower STDA value than STDA threshold and higher SSTD value than SSTD threshold should be transformed using DWT.

We used 7 sample images in Table 1, which have different range of STDA to ensure that threshold value used in the system will be valid for all images. The DWT and SVD are computationally expensive, but the SVD is even more expensive [5]. Considering the fact that SVD is too much time consuming and running time in compression image is one of important matters, the sub-blocktransformed using SVD is a candidate sub-block for SVD which has higher STDA value than STDA threshold and lower SSTD value than SSTD threshold. Candidate sub-block for SVD is limited to sub-block which has STDA value in range 2/3 from average of all STDA and maximum STDA for each image. This should keep the running time of compression image system stay low. The calculation of STDA and SSTD is applied for luminance component (Y) only.

Table 1 Sample images

No	Image	Range STDA
1	Brick	24 – 48
2	Flowers	8 – 37
3	Barbara	5 – 32.4
4	Peppers	3.6 – 18.97
5	Crayon	0 – 21.9
6	Airplane	0.7 – 27.8
7	Sailboat	3.1 – 25.9

Testing with STDA threshold in range 15 – 50 is done for all sample images. From 7 sample images, there are 5 images which has sub-block that transformed using SVD. We provide them in Table 2.

Table 2 STDA and SSTD of sub-block that transformed using SVD

No	Image	STDA	SSTD
1	Barbara	32,03	13,17
		32,41	12,19
		29,34	9,78
		28,52	9,15
		29,89	2,25
		28,72	8,48
2	Airplane	26,81	19,68
		25,61	24,06
		27,78	14,19
		25,72	11,47
3	Sailboat	23,18	10,41
		22,46	11,97
		25,86	9,02
		24,09	15,29
4	Brick	45,16	19,21
		45,19	14,46
		44,92	16,96
		47,96	15,32
5	Flowers	33,69	15,56
		37,73	17,09

Based on test of 20 as STDA threshold and range 15 - 17 as SSTD threshold by considering given Peak Signal to Noise Ratio (PSNR) and compression ratio value, the threshold value applied in this compression image system is 20 as STDA threshold and 15 as SSTD threshold.

Every sub-block transformed using SVD is divided into 32x32 pixel and current amount of sub-block is twice than before. And then the results of SVD transformation are 3 matrices which is matrix left singular vector U, matrix singular value $\sum$ and right singular vector V^T for each sub-block. Two values from the lowest values of matrix singular value $\sum$ are assigned with 0. Because a good image representation can be achieved by taking only a few largest eigenvalues and corresponding eigenvectors [5].

Every luminance components of sub-block transformed using DWT is encoded using Daubechies 9/7 filter bank into 3 levels. The chrominance components are encoded into maximum 2 levels. Transformed sub-block using DWT results are 4 coefficient which is LL, LH, HL, and HH. For each sub-block, half of lowest level of HH coefficient is assigned with 0.

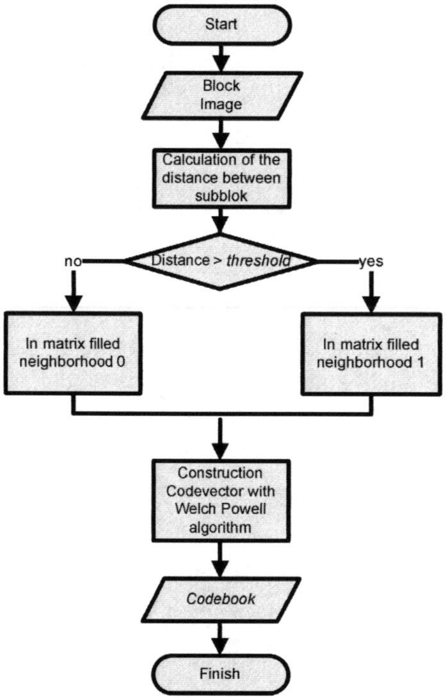

Figure 2. Flowchart of Quantization by Graph Coloring

Quantization process for transformation outputs from DWT and SVD is done separately. In Figure 2, euclidean distance is used to get distance value from all input vectors. All input vectors from transformation process are treated as dots. A distance threshold is used to determine adjacent relationship between dots. A pair of dots whose distance is larger than distance threshold has 1 as adjacent matrix value. A pair of dots whose distance is less than distance threshold has 0 as adjacent matrix value. After all values of adjacent matrix are assigned, coloring dots from a graph process is done using Welch Powell algorithm [2]. The amount of clusters representation as a graph quantization results is the amount of colors given by Welch Powell algorithm. Codeword values of each clusters is average distance value of all members of related cluster. Codebook is a group of codeword.

Outputs of DWT transformation are divided into two groups which is LL coefficient and group of other coefficients because LL coefficient is the most important coefficient which is related the most with image visual so the given error will be reduced by dividing group this way. Outputs of quantization of DWT coefficients are two codebooks. In Table 3, mean distance value of transformed sub-block using DWT based on chrominance and luminance components.

Table 3 Mean distance value of transformed sub-block using DWT

No	Image	Mean distance value	
		Luminance	Chrominance
1	Brick	6291.2	401.8
2	Flowers	4845	1009
3	Barbara	4457.8	516.6
4	Peppers	4393.4	907.8
5	Crayon	6575.9	1335.5
6	Airplane	3783.6	396.7
7	Sailboat	5266.8	665.8

Based on test range of 1000 - 3000 as luminance distance threshold and range of 100 - 300 as chrominance distance threshold with considering given Peak Signal to Noise Ratio (PSNR), compression ratio value and the amount of colors needed, the distance threshold value applied in this compression image system is 1000 as luminance distance threshold and 300 as chrominance distance threshold for DWT transformation output.

Outputs of SVD transformation are also divided into two groups which is singular matrix value $\sum$ and group of U and V matrix because singular matrix value $\sum$ is image base which is related the most with image visual so the given error will be reduced by dividing group this way. Outputs of quantization of SVD matrix are two codebooks. In Table 4, mean distance value of transformed sub-block using SVD based on chrominance and luminance components.Distance threshold used for SVD transformation output chrominance and luminance components is grouped into one distance threshold because of its less difference value. Based on test range of 0.6 – 1.1 as distance threshold value by considering given Peak Signal to Noise Ratio (PSNR), compression ratio value and the amount of colors needed, the distance threshold value applied in this compression image system is 0.9 for SVD transformation output.

Table 4 Mean distance value of transformed sub-block using DWT

No	Image	Mean distance value	
		Luminance	Chrominance
1	Brick	1.3966	1.3781
2	Airplane	1.3991	1.3835
3	Barbara	1.3999	1.3848
4	Sailboat	1.3991	1.3835

All of quantization results are arranged serially based on sub-block into a vector which will become the input of arithmetic coding or huffman coding. Transformed sub-block using SVD is marked by 0 as bit value and 1 as bit value if sub-block transformed using DWT. Codeword result of arithmetic coding has value in range 0 until 1. This range is given based on frequency probability of each symbols.

The compressed bit value from encoder output is decoded using arithmetic decoding so the result value is same as input value of arithmetic encoder. Dequantization process is one on one mapping process between codevector index as result of arithmetic decoding and codevector from codebook. The output vectors are arranged based on sub-block as input of backward transformation method grounded on bit mark of each sub-block after mapping process. If the bit mark is 0 the backward transformation process uses SVD. Then the eigenvectors and eigenvalues matrix are reconstructed into one matrix.

3. System Evaluation

In this section, we present the performance of system evaluation and compare with JPEG2000. We use compression ratio and PSNR as parameters of system evaluation. Compression ratio is the decreased percentage of current compressed size than original size [8].

$$Ratio = \left(1 - \left(\frac{size\ citra\ terkompresi}{size\ citra\ asli}\right)\right) * 100\% \quad (1)$$

PSNR is a comparison value between maximum value of reconstrued image and the original image (before compressed) in decibel unit (dB). Let A be an original image, B as the reconstructed image, *mn* (*m* x *n*) as image resolution of both of images and *max* as most possible maximum value of image pixel :

$$PSNR = 10\log_{10}\left(\frac{max^2 mn}{\sum_{i=1}^{m}\sum_{j=1}^{n}[A(i,j)-B(i,j)]^2}\right) \quad (2)$$

In this paper, we calculated two types of PSNR which is luminance PSNR and average PSNR. Luminance PSNR is a calculated PSNR for luminance component only. Average PSNR is average of calculated PSNR for luminance and chrominance PSNR. Luminance component is a color component which is most sensitive to human eyes.

Table 5 System performance compare with JPEG2000

No	Image	Ratio (%)	STDA		T		Luminance PSNR (dB)			Average PSNR (dB)			Time (Second)		
			Range	mean	D	S	PA	PH	J	PA	PH	J	PA	PH	J
1	Brick	55	24- 48	37,32	63	1	80.6	80,8	56.8	55.0	54,8	57.2	7	5	3
2	Flowers	54	8 – 38	18,4	64	-	87.5	87,1	58.4	57.5	57,2	58.0	7	4	3
3	Barbara	57	5 – 32	16,58	58	6	55.1	91,4	61.1	48.4	59,8	60.5	15	4	3
4	Peppers	53	4 – 19	10,9	64	-	48.7	104,1	59.4	45.1	63,6	58.8	7	4	3
5	Sailboat	57	3 – 26	15,17	61	3	50.9	97,1	55.6	44.9	59,9	56.4	9	4	3
6	Lena	57	2 – 23	9,99	62	2	46.6	107,1	59.2	45.5	65,4	58.9	7	4	3
7	Baboon	55	7 – 30	17,66	57	7	56.2	102,3	54.5	46.4	61,4	55.3	18	4	3
8	Pink Flowers	55	12 – 29	22,45	60	4	62.3	93,2	63.2	51.5	60,7	63.1	10	5	3
9	Flower	59	0 – 20	9,4	64	-	45.6	89,5	58.6	44.6	59,2	58.3	7	4	3
10	Airplane	68	1 – 28	11,64	61	3	46.4	69,8	59.1	46.1	52,9	59.1	8	4	3
Mean value							58.0	92,2	58.6	48.5	59,5	58.6	9.6	4,2	3

STDA = Average standard deviation of image sub-block
T = Transformation method used by system
D = Amount of sub-block transformed using DWT
S = Amount of sub-block transformed using SVD
PA = Proposed Scheme with Arithmetic Coding
PH = Proposed Scheme with Huffman Coding
J = JPEG2000

Based on conducted experiments, we have some conclusions as follow. The Luminace PSNR of reconstructed image quality of image compression scheme using proposed scheme is 48,5 dB (by Arithmetic coding) and 59,5 dB (by Huffman coding), while using JPEG2000 is 58,64 dB. The proposed scheme has higher PSNR than JPEG2000 because error generated by SVD for transforming low-correlated sub-block is smaller than the error generated by DWT to transform the sub-block.

Table 6 Performance of Image Reconstructed

Source	Recontructed with Arithmetic Coding	Recontructed with Huffman Coding	Recontructed with JPEG2000

	PSNR : 48.4	PSNR :59,8	PSNR :60.5

4. Conclusion

A graph coloring technique in the quantization process on image compression scheme based on Wavelet-SVD (Singular Value Decomposition) has a good performance. Based on compression ratio and PSNR, the system provides good performance compared with JPEG2000. The average compression ratio generated by the system stands between 50-60%, while the average PSNR stands between 40-80 dB. The proposed scheme has an advantage over the image with high average standard deviation value or image that have low degree of similarity between its sub-block because it produces a smaller error rate on dequantization process thus providing a higher PSNR value. Higher level of decomposition will yield a higher value of PSNR on reconstructed image to a certain threshold where the trade off with the decreasing compression ratio still felt balanced.

5. References

[1] Adiwijaya, Salman A N M, Suprijanto D, and Baskoro E T, 2011, A characterization of the corona product of a cycle with some graphs based on its *f*-chromatic index, *AIP Conf. Proc. 1450*, pp. 155-158

[2] Agnarrson G, dan Greenlaw R., 2007, *Graph Theory:Modeling, Applications,and Algorithms,* Prentice Hall, USA

[3] Bako C, 2004, JPEG 2000 Image Compression. *Analog Dialogue* **38**

[4] Ochoa H, dan Rao K R, 2007, A Discrete Wavelet Transform – Singular Value Decomposition System for Image Coding. *Journal of applied research and technology,***5:2** 121-136

[5] Ochoa H, dan Rao K R, 2003, A Hybrid DWT-SVD Image-Coding System (HDWTSVD) for Color Images.*Systemics, Cybernetics and Informatics***1:2** 64-69

[6] Pitas I., *Digital Image Processing Algorithms and Applications*, John Wiley & Sons, Inc, Canada, 2000.

[7] Rabbani, Majid dan Joshi, Rajan,An overview of the JPEG 2000 still image compression standard.*SignalProcessing: ImageCommunication* **17**(2002)3–48

[8] Sayood K, *Introduction to Data Compression*, San Fransisco, Morgan Kaufmann, USA, 2006

Constructing New Biorthogonal Wavelet Type which Matched for Extracting the Iris Image Features

R. Rizal Isnanto[1], Suhardjo[2], Adhi Susanto[3]

[1]Doctor Candidate of Electrical of Electrical Engineering UGM Yogyakarta, Indonesia and Lecturer of Engineering Faculty, UNDIP Semarang, Indonesia
[2] Professor in Medicine Faculty UGM Yogyakarta
[3] Professor in Electrical Engineering and Information Technology Department UGM Yogyakarta

e-mail: rizal_isnanto@undip.ac.id and rizal_isnanto@yahoo.com

Abstract. Some former research have been made for obtaining a new type of wavelet. In case of iris recognition using orthogonal or biorthogonal wavelets, it had been obtained that Haar filter is most suitable to recognize the iris image. However, designing the new wavelet should be done to find a most matched wavelet to extract the iris image features, for which we can easily apply it for identification, recognition, or authentication purposes. In this research, a new biorthogonal wavelet was designed based on Haar filter properties and Haar's orthogonality conditions. As result, it can be obtained a new biorthogonal 5/7 filter type wavelet which has a better than other types of wavelets, including Haar, to extract the iris image features based on its mean-squared error (MSE) and Euclidean distance parameters.

1. Introduction

1.1 Background

Humans as individuals, have unique characteristics and distinctive. These characteristics can be used to recognize or identify persons. This is known as biometric recognition. Iris is the part of the circle around eye pupil. Although iris has a relatively narrow region compared with entire area of the human body, iris has a very unique pattern, different in each individual and the pattern will remain stable. For those reasons, iris can be used as the basis for the recognition in biometrics.

Many algorithms have been applied as a method of iris recognition, such as PCA (Principal Component Analysis), ICA (Independent Component Analysis), Gabor-Wavelet algorithm [6], characterizing Key Local Variation, Laplace Pyramid, Gray Level Co-ocurrence Matrix (GLCM) [3] and others.

Wavelet transform as one method for analyzing textures has been used as iris image feature extractor. However, it is still limited applying the standard orthogonal wavelet types, for example: Haar, Daubechies, Coiflet, and Symlet. In this research, a new type of biorthogonal wavelet will be develop. Some parameters in Haar filter properties and Haar's orthogonality conditions will be applied. The length of filters are 5 (five) for decomposition and 7 (seven) for reconstruction, The notation used to come is Biorthogonal 5/7 for the upcoming filter.

Journal of Physics: Conference Series **423** (2013) 012020 doi:10.1088/1742-6596/423/1/012020

To extract the iris image features, two parameters used, i.e. mean-squared error (MSE) and Euclidean distance parameters.

1.2 Research Objective

The purpose of making this research is to develop a new type of biorthogonal wavelet. Some parameters in Haar filter properties and Haar's orthogonality conditions will be applied in the design steps. The length of filters are 5 (five) for decomposition and 7 (seven) for reconstruction, The notation used to come is Biorthogonal 5/7 for the upcoming filter. To extract the iris image features, two parameters used, i.e. mean-squared error (MSE) and Euclidean distance parameters.

1.3 Limitations

In order not to deviate far from the problems, then this study has problem limitations as follows:

a. Iris image used is the image from database of CASIA V1.0 – Portions of the research in this paper use the CASIA-IrisV1 collected by the Chinese Academy of Sciences' Institute of Automation (CASIA) – and iris images captured using Irdosoft 4.0 camera.
b. Research uses of Haar filter parameters and its orthogonality properties as the basis for reconstructing a new 5/7 biorthogonal wavelet. The reason of using Haar parameter is that in previous research, Haar wavelet is the best wavelet for recognizing iris images based on recognition rate [9],[11] as well as based on energy compaction [10].
c. Recognition method used is the normalized Euclidean distance method

2. Basis of Theory

2.1 Iris Eyes

Iris can serve as the basis for biometric systems. Each iris has a texture that is very detailed and unique to each person and remain stable for decades. The eye can not be altered through surgery without causing any damage to eyesight. Fig.1 shows the anatomy of the eye, and examples of human iris [2].

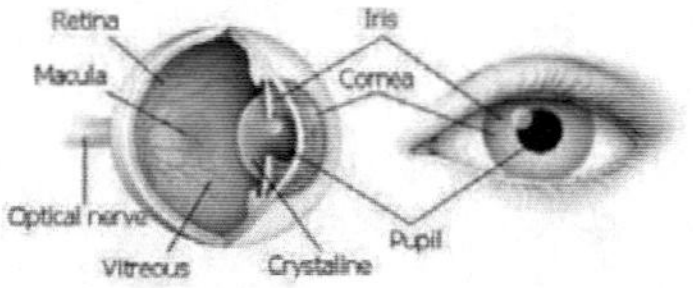

Figure 1. Anatomy of eyes and example of iris region.

The advantages of using the iris for reliable identification system are [2] as follows.

1. Iris is insulated and shielded from the outside environment.
2. In the iris, it is not possible to do some surgeries without causing defects in the eye.
3. Iris has a physiological response to light, which allows testing of the natural use of the possibility of fraud and faked eye lenses and so forth.

2.2 Wavelet Transform

Wavelets are mathematical functions that satisfy certain requirements are able to perform the decomposition of a function [13]. Hierarchically, wavelet is used to represent data or other functions. Wavelet can be used to describe a model or the original image into a mathematical function regardless of the shape of the model in the form of image, a curve or a plane. Wavelet transform is a function that converts the signals from region to region the frequency or time scale. The most appropriate wavelet transform used in image processing because there is not much information is lost during the reconstruction.

Wavelet is a base derived from a wavelet basis function which is also said to be a scaling or scaling function. Scaling function has properties which can be assembled from a number of copies that have been dilated, translated and scaled. This function is derived from the dilation equation, which is considered as the basis of wavelet theory.

2.3 Image Decomposition

Image processing using wavelet transform is done by filtering image using wavelet filter. Result of this filtering is 4 image subspaces from original image. These four subspaces are in wavelet domain. Four subspaces mentioned are lowpass-lowpass (LL), lowpass-highpass (LH), highpass-lowpass (HL), and highpass-highpass (HH). This process is called decomposition.

The decomposition can be continued using lowpass-lowpass (LL) as its input for getting next decomposition stages. Figure 4 shows a decomposition image from level 1 until level 3. Maximum level of decomposition in image processing using wavelet (except biorthogonal wavelets) can be formulated as follows [4].

$$\text{Level}_{\max} = \frac{\log(\text{data length}/(\text{filter length - 1}))}{\log(2)} \tag{1}$$

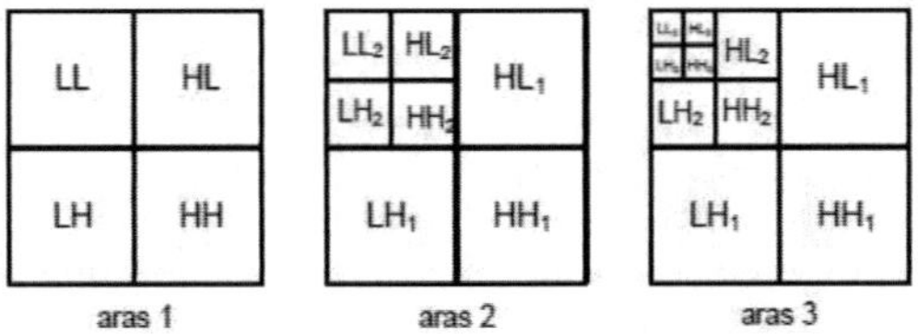

Figure 2. Diagram of image decomposition.

2.4 Normalized Euclidean Distance

After passing through feature extraction process and parameter values are obtained, then the next stage is calculating the nearest distance (Euclidean distance) of feature vector values of an image [14]. The closer the Euclidean distance, getting closer to a certain image. For example, values of feature vectors of input image $A_i = (A_1, A_2, \ldots\ldots, A_n)$ and values of feature vector of *j*-th image is $B_j = (B_{1j}, B_{2j}, \ldots, B_{nj})$, then Euclidean distance between values of feature vectors of input image and values of feature vector of *j*-th image can be expressed as:

$$D(A,B) = \sqrt{\sum_{i=0}^{n} \frac{(|A_i - B_i|)^2}{A_i}} \tag{2}$$

where:

$D(A,B)$ = Euclidean distance between iris A and B

A_i = Feature vector of iris A

B_i = Feature vector of iris B

n = vector length (sum of textural features) of vector A and vector B

2.5 Succesive Approxiamation or Cascade Algorithm

One approach to calculate $\varphi(t)$ and $\psi(t)$ is a form of succesive approximation that is used theoretically to prove existence and uniqueness of $\varphi(t)$ and can also be used to actually calculate them. The basic recursion equation that comes from multiresolurion formulation is

$$\varphi(t) = \sum_n h[n]\sqrt{2}\varphi^{(k)}(2t-n) \tag{3}$$

where $h(n)$ is the scaling coefficients and $\varphi(t)$ is the scaling function which satisfies this equation.

In order to solve the basic recursion equation (3), an iterative an algorithm that will generate successive approximations to $\psi(t)$ is applied [7]. If the algorithm converges to a fixed point, then that fixed point is a solution to (3).

The iterations are defined by

$$\varphi^{(k+1)}(t) = \sum_{n=0}^{N-1} h[n]\sqrt{2}\varphi^{(k)}(2t-n) \tag{4}$$

This iteration converges to $\varphi(t)$. From this $\varphi(t)$, then the wavelet function can be generated from

$$\psi(t) = \sum_{n=-\infty}^{\infty} g[n]\sqrt{2}\varphi^{(k)}(2t-n) \tag{5}$$

Because it applies the same operation over and over to the output of the previous application, it is known as the *cascade algorithm* [13].

2.6 Biorthogonal wavelet

A biorthogonal wavelet is a wavelet where the associated wavelet transform is invertible but not necessarily orthogonal. Designing biorthogonal wavelets allows more degrees of freedom than orthogonal wavelets. One additional degree of freedom is the possibility to construct symmetric wavelet functions.

In the biorthogonal case, there are two scaling functions $\varphi, \tilde{\varphi}$, which may generate different multiresolution analyses, and accordingly two different wavelet functions $\psi, \tilde{\psi}$. So the numbers M and N of coefficients in the scaling sequences $a, \tilde{a}$ may differ. The scaling sequences must satisfy the following biorthogonality condition [12].

$$\sum_{n\in Z}^{\infty} a_n \tilde{a}_{n+2m} = 2.\delta_{m,0} \tag{6}$$

Then the wavelet sequences can be determined as

$$b_n = (-1)^n \tilde{a}_{M-1-n}, n = 0,\ldots,M-1 \tag{7}$$

and

$$\tilde{b}_n = (-1)^n a_{M-1-n}, n = 0,\ldots,N-1 \tag{8}$$

3. Design of Research

In constructing a new 5/7 biorthogonal wavelet, chart to be followed is depicted in figure 3.

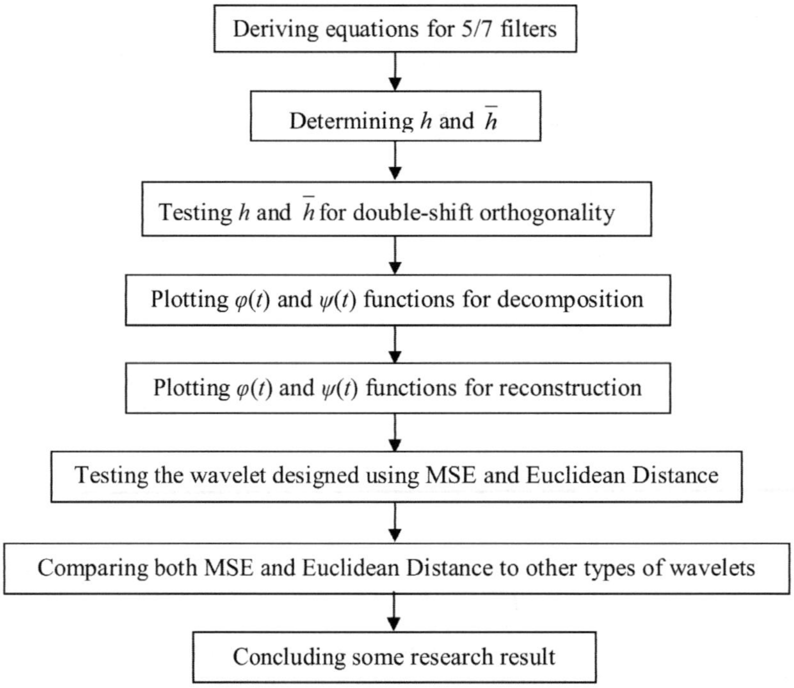

Figure 3. Chart of research design

From figure 3, it can be detailed as follows.

a. Deriving and then obtaining some equations for 5-tap filters as well as for 7-tap filters which satisfies orthogonality conditions.
b. Determining the components of h dan $\bar{h}$ which depicting filter components for decomposition and reconstruction, respectively.
c. Testing the results of point b in case of its double-shift orthogonality.
d. Plotting and depicting two graphics both scaling function as well as wavelet function for decomposition purpose using cascade algorithm applied up to 20 iterations.
e. Plotting and depicting two graphics both scaling function as well as wavelet function for reconstruction purpose using cascade algorithm applied up to 20 iterations.
f. Testing the wavelet designed using both MSE and Euclidean distance measures between reconstructed iris images after decomposition and its original images.
g. Comparing both MSE and Euclidean distance (results from point f) against other types of wavelets which had ever observed (i.e. Haar, db5, coif3, sym4, and bior2.4).
h. Concluding some results of points f and g to give research conclusions.

4. Designing the Biorthogonal Wavelet using 5/7 Filter

The meaning of 5/7 indicates that one filter which has length of 5 (five) for decomposition and 7 (seven) for reconstruction.

The reasons why we choose 5/7 biorthogonal filter are: biorthogonal wavelet bases maintain the symmetry for wavelets and scaling functions by relaxing the orthogonality constraint [15], while none of the orthogonal wavelet systems, except Haar, has symmetrical coefficients [8]. Also, the orthogonal wavelets filter and scaling filter must be of the same length, and the length must be even [7]. This

restriction has been relaxed for biorthogonal filters. Since we had done some research previously to construct Biorthogonal 2/6 and 3/5 filters [15] to represent short-medium length filters, then we choose Biorthogonal 5/7 as a representative of a relatively medium-medium filters with odd lengths both for analysis and synthesis. The categorization of short – medium – long refers to the research to observe the relationship between the averaged SNR and the length of the filters when orthogonal wavelets – Daubechies' wavelets, symlets, and coiflets have been used. From that research, a maximum SNR is achieved for wavelets which corresponding filter is with medium length [16]. Short filter range is 1 up to 4, medium filter range is 5 up to 12, and long filter range is which exceeds 12.
For example: we need 5/7 biorthogonal with symmetric scaling filters. The 5-tap filter then can be written as a modified form which is stated by Soman, *et al.* [8]:

$$H(z)=\sqrt{2}\left(\frac{1+z}{2}\right)^2\left(\frac{1-a}{2}z^{-1}+a+\frac{1-a}{2}z\right) \tag{9}$$

The first factor gives us two vanishing moments, while the second factor determines:

$$H(1)=\sqrt{2} \tag{10}$$

and

$$\left(\frac{1-a}{2}z^{-1}+a+\frac{1-a}{2}z\right)_{z=-1}\neq 0 \tag{11}$$

The resulting filter is a symmetric low pass filter which has 1 parameter with 5 taps.

By expanding and simplifying (11) for $H(z)$, we can get:

$$H(z)=\sqrt{2}\left(\left(\frac{1-a}{8}\right)z^{-2}+\left(\frac{1}{4}\right)z^{-1}+\left(\frac{1-a}{4}\right)+\left(\frac{1}{4}\right)z+\left(\frac{1-a}{8}\right)z^{2}\right) \tag{12}$$

or

$$h=\left(h_{-2},h_{-1},h_0,h_1,h_2\right)=\sqrt{2}\left(\frac{1-a}{8},\frac{1}{4},\frac{1-a}{4},\frac{1}{4},\frac{1-a}{8}\right) \tag{13}$$

Meanwhile, the second filter $\overline{H}(z)$ can be stated as:

$$\overline{H}(z)=\sqrt{2}\left(\frac{1+z}{2}\right)^2\left(b+\frac{c}{2}z+\frac{c}{2}z^{-1}+\frac{1-b-c}{2}z^{2}+\frac{1-b-c}{2}z^{-2}\right) \tag{14}$$

Also, by expanding and simplifying (14) for $\overline{H}(z)$, we can get:

$$\begin{aligned}\overline{H}(z)=\sqrt{2}\Bigg(&\left(\frac{1-b-c}{8}\right)z^{-3}+\left(\frac{2-2b-c}{8}\right)z^{-2}+\left(\frac{1+b+c}{8}\right)z^{-1}+\ldots\\&\ldots\left(\frac{2b+c}{4}\right)+\left(\frac{1+b+c}{8}\right)z+\left(\frac{2-2b-c}{8}\right)z^{2}+\left(\frac{1-b-c}{8}\right)z^{3}\Bigg)\end{aligned} \tag{15}$$

or

$$\begin{aligned}\overline{h}&=\left(\overline{h}_{-3},\overline{h}_{-2},\overline{h}_{-1},\overline{h}_0,\overline{h}_1,\overline{h}_2,\overline{h}_3\right)\\&=\sqrt{2}\left(\frac{1-b-c}{8},\frac{2-2b-c}{8},\frac{1+b+c}{8},\frac{2b+c}{4},\frac{1+b+c}{8},\frac{2-2b-c}{8},\frac{1-b-c}{8}\right)\end{aligned} \tag{16}$$

Double-shift orthogonality property yields the equations below:

$$h_0\bar{h}_3 + h_1\bar{h}_4 + h_2\bar{h}_5 + h_3\bar{h}_6 = \left(\frac{1-a}{4}\times\frac{2b+c}{4}\right)+\left(\frac{1}{2}\times\frac{1+b+c}{8}\right)+\left(\frac{1+a}{2}\times\frac{2-2b-c}{8}\right)+\left(\frac{1}{2}\times\frac{1-b-c}{8}\right)=0$$

$$\Leftrightarrow\quad -\frac{a}{2}-b-\frac{3c}{16}+\frac{ab}{2}+\frac{ac}{4}+1=0 \tag{17}$$

$$h_0\bar{h}_5 + h_1\bar{h}_6 = \left(\frac{1-a}{4}\times\frac{2-2b-c}{8}\right)+\left(\frac{1}{2}\times\frac{1-b-c}{8}\right)=0$$

$$\Leftrightarrow\quad \frac{a}{2}-ab-\frac{ac}{2}+1=0 \tag{18}$$

Solutions on (16) and (17) yield:

$$b=\frac{1}{2}\left(a-3+\frac{6}{a}\right) \tag{19}$$

$$c=-a+4-\frac{4}{a} \tag{20}$$

For $a = \frac{7}{4}$, then value of $b = \frac{61}{56}$ and $c = -\frac{1}{28}$. Therefore we can determine:

$$h=(h_{-2},h_{-1},h_0,h_1,h_2)=\left(\frac{-3\sqrt{2}}{32},\frac{\sqrt{2}}{4},\frac{11\sqrt{2}}{16},\frac{\sqrt{2}}{4},\frac{-3\sqrt{2}}{32}\right) \tag{21}$$

and

$$\bar{h}=(\bar{h}_{-3},\bar{h}_{-2},\bar{h}_{-1},\bar{h}_0,\bar{h}_1,\bar{h}_2,\bar{h}_3)=\left(\frac{-3\sqrt{2}}{448},\frac{-\sqrt{2}}{56},\frac{115\sqrt{2}}{448},\frac{15\sqrt{2}}{28},\frac{115\sqrt{2}}{448},\frac{-\sqrt{2}}{56},\frac{-3\sqrt{2}}{448}\right) \tag{22}$$

It can be shown that the normality condition is satisfied. That is, it can be seen, with no shift, multiplying the filter coefficients in corresponding position, and then summing all multiplied coefficients, the result is 1. This result can be explained easily in figure 4.

0	$\frac{-3\sqrt{2}}{32}$	$\frac{\sqrt{2}}{4}$	$\frac{11\sqrt{2}}{16}$	$\frac{\sqrt{2}}{4}$	$\frac{-3\sqrt{2}}{32}$	0
$\frac{-3\sqrt{2}}{448}$	$\frac{-\sqrt{2}}{56}$	$\frac{115\sqrt{2}}{448}$	$\frac{15\sqrt{2}}{28}$	$\frac{115\sqrt{2}}{448}$	$\frac{-\sqrt{2}}{56}$	$\frac{-3\sqrt{2}}{448}$

Figure 4. Filters coefficient of 5/7 Biorthogonal

Applying the cascade algorithm up to 20 iteration to form both scaling and wavelet functions for decomposition can be shown in figure 5. From this figure, we can see visually that (a) we can construct plots of both scaling and wavelet function for decomposition using successive approximations (or Cascade Algorithm) from 1 to 20 iterations. After 20 iterations applied, we can get smooth functions graphics, or we can say, after 20 iterations, buth scaling and wavelet functions converge to fixed points; (b) biorthogonal wavelet bases maintain the symmetrical property both for wavelets and scaling functions by relaxing the orthogonality constraint.

Meanwhile, Applying the cascade algorithm up to 20 iteration to form both scaling and wavelet functions for reconstruction can be shown in figure 6. Two conclusions on constructing both scaling and wavelet functions for decomposition, also concluded similarly when it is applied for reconstruction process, i.e. graphics displayed more smooth after 20 iterations and that we get the symmetrical wavelet and scaling functions after 20 iterations too.

4.1 Image Reconstruction Tests using MSE and Euclidean Distance

From the wavelet design previously, we have constructed a 5/7 biothogonal, where 5 (five) indicates the length of decomposition filter, that is N_d =5. while 7 (seven) indicates the length of reconstruction filter, that is N_r =7.

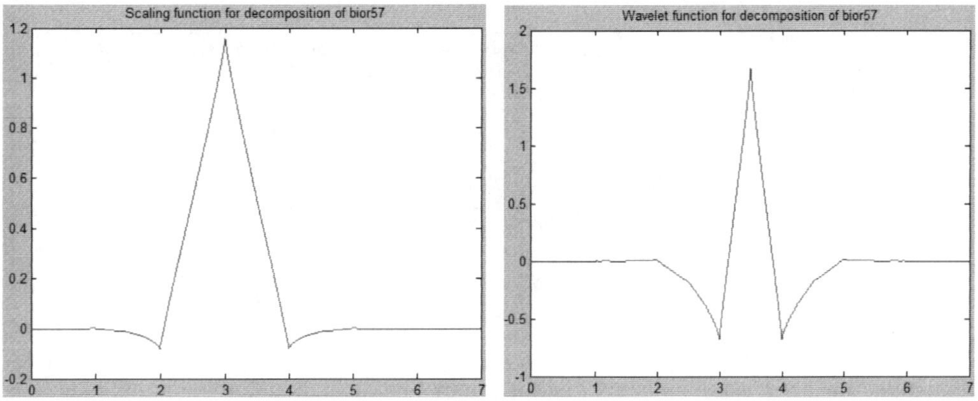

Figure 5. Scaling (left) and wavelet (right) functions of 5/7 biorthogonal filters for decomposition.

Then, orthogonal/biorthogonal wavelet types, i.e. Haar, db5, coif3, sym4, and bior2.4 which used in former research together with the new wavalet designed were tested. Some test was conducted to measure the performance based on mean-squared error (MSE) between reconstructed images of decomposized iris images and their original input images. In these tests, an unwrapped iris image from CASIA database, as well as from an image captured by Irdosoft 3.8 camera, as we can see in figure 7.

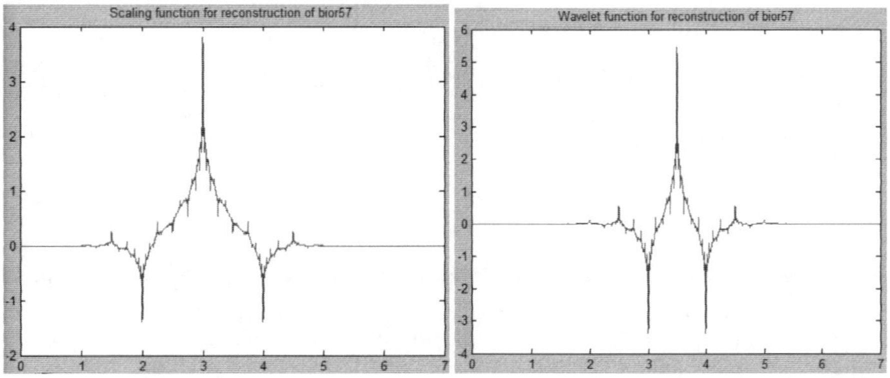

Figure 6. Scaling (left) and wavelet (right) functions of 5/7 biorthogonal filters for reconstruction.

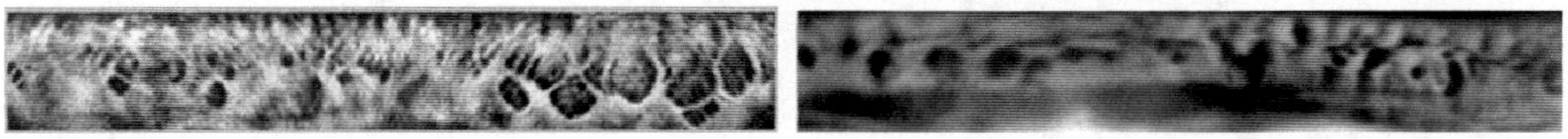

(a) example of CASIA image (b) example of iris captured by Irdosoft 4.0 camera

Figure 7. Example of iris images for Euclidean distance and MSE tests

Table 1 shows the results of Euclidean distance and MSE tests between reconstructed images and its original images which are shown in figure 7.

Table 1. Euclidean distance and MSE between reconstructed images and the original images.

No.	Wavelet types	CASIA database image		Image captured by Irdosoft 3.8	
		Euclidean distance	MSE	Euclidean distance	MSE
1	Haar	$1{,}1567 \times 10^{-13}$	$9{,}5565 \times 10^{-28}$	$9{,}5329 \times 10^{-14}$	$8{,}2615 \times 10^{-28}$
2	db5	$1{,}0832 \times 10^{-10}$	$8{,}3811 \times 10^{-22}$	$9{,}7398 \times 10^{-11}$	$8{,}6239 \times 10^{-22}$
3	coif3	$7{,}1285 \times 10^{-11}$	$3{,}6297 \times 10^{-22}$	$5{,}7406 \times 10^{-11}$	$2{,}9959 \times 10^{-22}$
4	sym4	$5{,}8662 \times 10^{-11}$	$2{,}4580 \times 10^{-22}$	$4{,}2551 \times 10^{-11}$	$1{,}6460 \times 10^{-22}$
5	bior2.4	$6{,}0709 \times 10^{-14}$	$2{,}6325 \times 10^{-28}$	$6{,}5122 \times 10^{-14}$	$3{,}8554 \times 10^{-28}$
6	bior5.7	$7{,}1054 \times 10^{-14}$	$5{,}0487 \times 10^{-28}$	$4{,}9228 \times 10^{-14}$	$2{,}2031 \times 10^{-28}$

*) new 5/7 biorthogonal wavelet

As we can see in Table 1, the new 5/7 biorthogonal wavelet (bior5.7) has the best result in reconstructing the decomposized image which is captured by Irdosoft 3.8 camera, when compared with other types of wavelet (Haar, db5, coif3, sym4, and bior2.4). We can notice this best result in two measures, i.e Euclidean distance as well as MSE, applying bior5.7 give us minimum Euclidean Distance and also minimum MSE. However, if applied to CASIA database image, bior5.7 give the good result compared with 4 (four) type of wavelets (Haar, db5, coif3, and sym4), although it only differs slicely below the bior2.4 which results in best performance, regarding in Euclidean distance and MSE.

5. Conclusion

From developing new wavelet, step-by-step until getting the results, it can be obtained some conclusions below.

1. A new type of wavelet has been found, i.e. bior5.7, which has properties: biorthogonal; filter length for decomposition and reconstruction, respectively, are $N_d = 5$ and $N_r = 7$.
2. The new 5/7 biorthogonal wavelet (bior5.7) has the best result in reconstructing the decomposized image which is captured by Irdosoft 3.8 camera, when compared with other types of wavelet (Haar, db5, coif3, sym4, and bior2.4). This can be seen in two measures, i.e Euclidean distance as well as MSE, applying bior5.7 give us minimum Euclidean Distance ($4{,}9228 \times 10^{-14}$) and also minimum MSE ($2{,}2031 \times 10^{-28}$).
3. When applied to CASIA database image, bior5.7 give the good result compared with 4 (four) type of wavelets (Haar, db5, coif3, and sym4), although it only differs slicely below the bior2.4 which results in best performance, regarding in Euclidean distance and MSE.

5.1 Recommendation and comments

From the research results, we would like recommend as well as giving comments as the herebelows.

1. For improving this manuscript in advance, the research can be continued to observe other wavelets and/or construct other type of wavelets, for which we can get the best-matched wavelet for iris extraction.
2. The new wavelet is expected to be used and to be a reference to next researchers that will explore the iris feature extraction for which the high rate of recognition can be achieved.
3. This research is highly expected to provide more detailed description on designing step to new wavelet which is matched for iris images. Furthermore, it can triggers other researchers to develop

the new wavelets for different biometrics systems, i.e. fingerprints, face, palmprint, and dynamic signatures.

References

[1] CASIA-IrisV1, http://biometrics.idealtest.org/

[2] Isnanto, R.R., T.S. Widodo, Suhardjo, and A. Susanto, 2012, "Increased Iris Recognition Rate by Observing Wavelet Decomposition Levels and Number of Images Stored", *Journal of Information and Communication Technologies (JICT), ISSN: 2047-3168, Vol. 2 Issue 5, 2012, pp. 1-7.*

[3] Kusuma A A, *"Pengenalan Iris Mata Menggunakan Pencirian Matriks Ko-Okurensi Aras Keabuan",* Skripsi S-1, Universitas Diponegoro, Semarang, 2009.

[4] Misiti M, Misiti Y, Oppenheim G, and Poggi J M, *Wavelet Toolbox For Use with Matlab*, User's Guide Version 3, The Math Works, Natick, MA, 2004.

[5] Mustofa, A., 2007, "Sistem Pengenalan Penutur dengan Metode Mel-Frequency Wrapping", *Jurnal Teknik Elektro, vol 7, No.2, September 2007, pp. 88-96.*

[6] Wijayanto W S, *Identifikasi Iris Mata dengan Tapis Gabor Wavelet dan Jaringan Syaraf Tiruan Learning Vector Quantization (LVQ),* Diponegoro University, Semarang, 2005.

[7] Burrus C S, Gopinath R A, and Guo H, 1998, *Introduction to Wavelets and Wavelet Transforms: a Primer*, Prentice-Hall, Upper Saddle River, New Jersey.

[8] Soman, K P, Ramachandran K I, and Resmi N G, 2010, *Inside into Wavelets: From Theory to Practice,* 3rd edition, PHI Learning Private Limited, New Delhi.

[9] Isnanto R R, Widodo T S, Suhardjo, and Susanto A, 2012, "Comparison Analysis on Recognition Rate of Iris using Haar and Daubechies Wavelet Tranforms", *International Journal of Information Technology and Network Application (IJITNA), ISSN: 2168-2178, Vol. 2 No. 1, pp. 21-27.*

[10] Isnanto R R, Widodo T S, Suhardjo, and Susanto A, 2012, "Wavelet Types Comparison for Extracting Iris Features Based on Energy Compaction", *Proceedings of The 3rd ICSIIT 2012 (International Conference on Soft Computing, Intelligent System, and Information Technology), ISBN: 978-602-97124-1-4, Bali, pp. 88-93.*

[11] Isnanto R R, Widodo T S, Suhardjo, and Susanto A, 2012, "Comparison Analysis on Recognition Rate of Iris using Haar and Symlet Wavelet Tranforms", *Proceedings of The 13th Seminar on Intelligent Technology and Its Applications, SITIA 2012, ISSN: 2252-8296, Surabaya, pp. 1-6.*

[12] Mallat S, 1999, *A Wavelet Tour of Signal Processing,* 2nd edition, Academic Press, London.

[13] Strang G and Nguyen T, 1996, *Wavelets and Filter Banks,* Wellesley-Cambridge Press, Wellesley, MA.

[14] Qi, X, 2003, "Object Contour Matching Using the Biorthogonal Wavelet Transform", *Proc. of the 5th International Conference on Computer Vision, Pattern Recognition, and Image Processing (CVPRIP'03)*, pp. 668-671, Cary, North Carolina, September 26-30, 2003

[15] Isnanto R R, 2012, *Texture-based Feature Extraction on Iris Image using Wavelets Transform,* Electrical Engineering and Information Technology Department, Faculty of Engineering, Gadjah Mada University, Indonesia.

[16] Nikolaev N and Gotchev A, 2000, "ECG Signal Denoising Using Wavelet Domain Wiener Filtering", *Proceeding of the European Association for Signal and Image Processing (EURASIP).*

Acknowledgments

Authors wishing to acknowledge assistance or encouragement from colleagues, they are: Mr. Rukun Santoso and Mr. Achmad Hidayatno who had gives some interesting views for insight understanding into wavelets. Also, authors have to thank to BOPTN FT Undip for their financial support for which this research can be done.

Empirical Analysis of Stochastic Volatility Model by Hybrid Monte Carlo Algorithm

Tetsuya Takaishi
Hiroshima University of Economics, Hiroshima 731-0192, JAPAN
E-mail: takaishi@hiroshima-u.ac.jp

Abstract. The stochastic volatility model is one of volatility models which infer latent volatility of asset returns. The Bayesian inference of the stochastic volatility (SV) model is performed by the hybrid Monte Carlo (HMC) algorithm which is superior to other Markov Chain Monte Carlo methods in sampling volatility variables. We perform the HMC simulations of the SV model for two liquid stock returns traded on the Tokyo Stock Exchange and measure the volatilities of those stock returns. Then we calculate the accuracy of the volatility measurement using the realized volatility as a proxy of the true volatility and compare the SV model with the GARCH model which is one of other volatility models. Using the accuracy calculated with the realized volatility we find that empirically the SV model performs better than the GARCH model.

1. Introduction

Statistical properties of asset price returns have been extensively studied both in finance and econophysics. It is well-known that asset price returns show fat-tailed distributions and volatility clustering which are now classified as stylized facts[1, 2]. Although in empirical finance the volatility is an important value to measure the risk it is quite difficult to extract the true volatility from asset price returns themselves. A popular technique to measure the volatility is to use volatility models which mimic volatility properties of asset returns such as volatility clustering. A famous volatility model is the generalized autoregressive conditional hetroskedasticity (GARCH) model[3, 4] where the volatility variable changes deterministically depending on the past squared value of the returns and past volatilities. There exist many extended versions of GARCH models, such as EGARCH[5], GJR[6], QGARCH[7, 8], GARCH-RE[9] models and many others, e.g. [10, 11], which are designed to increase the performance of the model.

The stochastic volatility (SV) model[12, 13] is another model which captures the properties of the volatility. Unlike the GARCH model, the volatility of the SV model changes stochastically in time. As a result the likelihood function of the SV model is given as a multiple integral of the volatility variables. Such an integral in general is not analytically calculable and thus the determination of the parameters of the SV model by the maximum likelihood method becomes extremely difficult. To overcome this difficulty the Markov Chain Monte Carlo (MCMC) method based on the Bayesian approach is proposed and developed[12]. In the MCMC method of the SV model one has to update not only the parameter variables but also the volatility ones under a joint probability distribution of the model parameters and the volatility variables. Although there exist various MCMC methods to perform the Bayesian program the performance of the

MCMC technique depends on each MCMC method. A local update scheme is usually not effective for the SV model which has an increasing number of volatility variables with the data size of return time series[14]. In order to improve the efficiency of the local update method the blocked scheme which updates several variables simultaneously is proposed[14, 15].

In this study we use the HMC algorithm[16] which is a global algorithm updating variables simultaneously. This global property of the HMC algorithm could be advantageous for updating volatility variables of the SV model. Actually it is shown that the HMC algorithm can decorrelate first enough the sampled data of the volatility variables[17, 18]. The HMC has been also applied to the parameter estimation of the GARCH model[19].

Using the HMC algorithm we perform the Bayesian inference of the SV model for two stock returns (Mitsubishi Co. and Panasonic Co.) traded on the Tokyo Stock Exchange and measure the volatility of those stock returns. In order to quantify the accuracy of the volatility estimated from the SV model we calculate a loss function using the realized volatility from the high-frequency intraday returns as a proxy of the true volatility. The realized volatility as a proxy of the true volatility has been used for ranking GARCH-type models[20]. In this study we also calculate the accuracy of the volatility from the GARCH model and compare the SV model and the GARCH model based on the criterion of the volatility accuracy.

2. Stochastic Volatility Model

The standard SV model[12, 13] is defined by

$$y_t = \sigma_t \epsilon_t = \exp(h_t/2)\epsilon_t, \tag{1}$$

$$h_t = \mu + \phi(h_{t-1} - \mu) + \eta_t, \tag{2}$$

where h_t is defined by $h_t = \ln \sigma_t^2$, and σ_t is called volatility. We also call h_t volatility variable. The error terms ϵ_t and η_t are taken from independent normal distributions $N(0,1)$ and $N(0,\sigma_\eta^2)$ respectively.

This model contains three parameters μ, ϕ and σ_η^2 which have to be inferred so that the model could match the asset return data y_t measured in the financial markets. Let θ be an abbreviation for μ, ϕ and σ_η^2, i.e. $\theta = (\mu, \phi, \sigma_\eta^2)$. Then the likelihood function $L(\theta)$ of the SV model can be written as

$$L(\theta) = \int \prod_{t=1}^{n} f(\epsilon_t|\sigma_t^2) f(h_t|\theta) dh_1 dh_2 ... dh_n, \tag{3}$$

where

$$f(\epsilon_t|\sigma_t^2) = \left(2\pi\sigma_t^2\right)^{-\frac{1}{2}} \exp\left(-\frac{y_t^2}{2\sigma_t^2}\right), \tag{4}$$

$$f(h_1|\theta) = \left(\frac{2\pi\sigma_\eta^2}{1-\phi^2}\right)^{-\frac{1}{2}} \exp\left(-\frac{[h_1-\mu]^2}{2\sigma_\eta^2/(1-\phi^2)}\right), \tag{5}$$

$$f(h_t|\theta) = \left(2\pi\sigma_\eta^2\right)^{-\frac{1}{2}} \exp\left(-\frac{[h_t - \mu - \phi(h_{t-1}-\mu)]^2}{2\sigma_\eta^2}\right). \tag{6}$$

Unlike the GARCH model where the maximum likelihood method should work, $L(\theta)$ of the SV model is constructed as a multiple integral of the volatility variables which makes the maximum likelihood estimation difficult. A possible estimation technique for the SV model is the Bayesian inference performed by the MCMC method. In the Bayesian inference model parameters are inferred as the expectation values given by

$$\langle \theta_{\rm i} \rangle = \int \theta_{\rm i} f(\theta|y) \Pi_{k=1}^{3} d\theta_k, \tag{7}$$

where $(\theta_1, \theta_2, \theta_3) = (\mu, \phi, \sigma_\eta^2)$ and $f(\theta|y)$ is the probability distribution of θ constructed by the likelihood function and the prior distribution of $\pi(\theta)$ as

$$f(\theta|y) \sim L(\theta)\pi(\theta). \tag{8}$$

In general eq.(7) can not be performed analytically and usually it is estimated numerically by the MCMC method. Since $L(\theta)$ also contains the integral of volatility variables h_t we have to update not only the model parameter θ but also the volatility variables h_t in the MCMC method. To update the model parameter θ we follow the standard update technique[12, 13].

The probability distribution of the volatility variables h_t is given by

$$P(h_t) \equiv P(h_1, h_2, ..., h_n) \sim \tag{9}$$
$$\exp\left(- \sum_{i=1}^{n} \{ \tfrac{h_i}{2} + \tfrac{\epsilon_i^2}{2} e^{-h_i} \} - \tfrac{[h_1-\mu]^2}{2\sigma_\eta^2/(1-\phi^2)} - \sum_{i=2}^{n} \tfrac{[h_i-\mu-\phi(h_{i-1}-\mu)]^2}{2\sigma_\eta^2} \right).$$

To update the volatility variables with this probability distribution we use the HMC algorithm.

3. Hybrid Monte Carlo Algorithm

The HMC algorithm is a global MCMC one which can update simultaneously all the variables of the model we consider. Originally the HMC algorithm is developed for the MCMC simulations of the lattice Quantum Chromo Dynamics (QCD) calculations[16, 21] where a very large scale simulation with a huge number of variables is inevitable. The basic idea of the HMC algorithm is a combination of molecular dynamics (MD) simulation and Metropolis accept/reject step. In the HMC algorithm we first introduce momentum variables p_i conjugate to h_i and define the Hamiltonian of the SV model[17, 18] as

$$H(p_t, h_t) = \sum_{i=1}^{n} \frac{1}{2} p_i^2 + \sum_{i=1}^{n} \{ \frac{h_i}{2} + \frac{\epsilon_i^2}{2} e^{-h_i} \} + \frac{[h_1 - \mu]^2}{2\sigma_\eta^2/(1-\phi^2)} + \sum_{i=2}^{n} \frac{[h_i - \mu - \phi(h_{i-1} - \mu)]^2}{2\sigma_\eta^2}. \tag{10}$$

Then we integrate the Hamilton's equations of motion,

$$\frac{dh_i}{d\tau} = \frac{\partial H}{\partial p_i}, \tag{11}$$
$$\frac{dp_i}{d\tau} = -\frac{\partial H}{\partial h_i}, \tag{12}$$

numerically by doing the MD simulation in the fictitious time τ. The simplest integrator for the MD simulation is the 2nd order leapfrog integrator[16] given by

$$\begin{aligned} h_i(\tau + \Delta\tau/2) &= h_i(\tau) + \frac{\Delta\tau}{2} p_i(\tau) \\ p_i(\tau + \Delta\tau) &= p_i(\tau) - \Delta\tau \frac{\partial H(p_t, h_t)}{\partial h_i} \\ h_i(\tau + \Delta\tau) &= h_i(\tau + \Delta\tau/2) + \frac{\Delta\tau}{2} p_i(\tau + \Delta\tau), \end{aligned} \tag{13}$$

where $\Delta\tau$ is a step size of the MC simulation. Although we adopt this 2nd order leapfrog integrator in this study we could also use the other improved integrators[22] or higher order integrators[23, 24].

After integrating the Hamilton's equations of motion up to some constant time, we obtain a set of new candidate variables (p', h'). These candidates are accepted with the Metropolis test[25] with the following probability,

$$P = \min\{1, \frac{\exp(-H(p', h'))}{\exp(-H(p, h))}\} = \min\{1, \exp(-\Delta H)\}, \tag{14}$$

where ΔH is the energy difference given by $\Delta H = H(p', h') - H(p, h)$. The acceptance rate of eq.(14) can be tuned by $\Delta\tau$. The high acceptance rate is not necessary for the HMC and it is shown that the optimum acceptance rate of the HMC with the 2nd order integrator is around 60%[23].

4. Realized Volatility

In this study we use the realized volatility as a proxy of the true volatility and use it for the accuracy calculation of the volatility inferred from the volatility models. When high-frequency intraday return data is available we can construct the realized volatility as a sum of squared intraday returns[26, 27, 28]. Let us assume that the logarithmic price process $\ln p(s)$ follows a continuous time stochastic diffusion,

$$d\ln p(s) = \tilde{\sigma}(s)dW(s), \tag{15}$$

where $W(s)$ stands for a standard Brownian motion and $\tilde{\sigma}(s)$ is a spot volatility at time s. Then the integrated volatility is defined by

$$\sigma_T^2(t) = \int_t^{t+T} \tilde{\sigma}(s)^2 ds, \tag{16}$$

where T stands for the interval to be integrated. If we consider daily volatility T should take one day. Since $\tilde{\sigma}(s)$ is latent and not available from market data, the value of eq.(16) can not be obtained directly.

Constructing n intraday returns from high-frequency data, the realized volatility RV_t is given by a sum of squared intraday returns,

$$RV_t = \sum_{i=1}^{n} r_{t+i\delta}^2, \tag{17}$$

where δ is a sampling time interval defined by $\delta = T/n$. Note that small sampling time interval corresponds to high sampling frequency.

If there is no microstructure noise[29, 30], RV_t goes to the integrated volatility of eq.(16) in the limit of $n \to \infty$. Empirically it is well-known that the realized volatility measure suffers from the microstructure noise and the distortion from the microstructure noise will be serious at very high-frequency. Such distortion can be depicted in the volatility signature plot[31]. To avoid a large distortion on the realized volatility and at the same time to maintain the accuracy of the realized volatility measure we have to employ a good sampling frequency. The optimum sampling frequency under the independent microstructure noise was derived and found to be between one and five minutes[32]

When we consider daily stock realized volatility we have to cope with non-trading hours issue. Usually high-frequency data are not available for the entire 24 hours. At the Tokyo stock

Table 1. Results estimated by the HMC algorithm. **SD** stands for Standard Deviation and **SE** stands for Statistical Error. The statistical errors are estimated by the jackknife method. τ_{int} is the autocorrelation time defined by $\tau_{int} = 1 + 2\sum_{t=1}^{\infty} ACF(t)$, where $ACF(t)$ is the autocorrelation function. We also show h_{10} as a representative one of the volatility variables h_t.

		ϕ	μ	σ_η^2	h_{10}
Mitsubishi Co.		0.987	-7.33	0.0186	-7.48
	SD	0.007	1.07	0.067	0.31
	SE	0.0005	0.007	0.0008	0.01
	τ_{int}	260(130)	1.3(1)	800(370)	30(4)
Panasonic Co.		0.975	-7.87	0.045	-7.57
	SD	0.012	0.50	0.017	0.36
	SE	0.0013	0.004	0.0025	0.01
	τ_{int}	410(250)	1.8(4)	800(480)	35(14)

exchange market domestic stocks are traded in the two trading sessions: (1)morning trading session (MS) 9:00-11:00. (2)afternoon trading session (AS) 12:30-15:00. The daily realized volatility calculated without including intraday returns during the non-traded periods can be underestimated.

An idea to circumvent the problem is advocated by Hansen and Lunde[33]. They introduced an adjustment factor which modifies the realized volatility so that the average of the realized volatility matches the variance of the daily returns. Let $(R_1, ..., R_N)$ be N daily returns. The adjustment factor c is given by

$$c = \frac{\sum_{t=1}^{N}(R_t - \bar{R})^2}{\sum_{t=1}^{N} RV_t}, \tag{18}$$

where $\bar{R}$ denotes the average of R_t. Here we call c the HL factor. Then using this factor the daily realized volatility is modified to cRV_t. In this study we calculate the realized volatility and the HL factor at several sampling frequencies and use the modified realized volatility cRV_t as a proxy of the true daily volatility.

5. Empirical Analysis

In this section we empirically study the SV model by applying it for asset returns traded on the Tokyo Stock Exchange. The empirical study is based on daily data of two stocks (Mitsubishi Co. and Panasonic Co.). These stocks are included in the list of the Topix core 30 index which collects liquid stocks of the Tokyo Stock Exchange. The sampling period of the stock data is June 3, 1996 to December 30, 2009. Figure 1 shows the daily close-to-close returns of the two stocks.

Using the daily close-to-close return data of Mitsubishi Co. and Panasonic Co. we perform the Bayesian inference by the MCMC method for the SV model. The first 10000 Monte Carlo samples are discarded and then 40000 samples are recorded for the analysis. The MCMC sampling of the volatility variables is performed by the HMC method. The MCMC sampling of the model parameters $(\mu, \phi, \sigma_\eta^2)$ is implemented by the standard MCMC method[13]. The values of the model parameters obtained by the MCMC method are summarized in table 1. We also list the results of h_{10} in the table as a representative one of volatility variables. It is found that the autocorrelation time of h_{10} is not big but around 30 which is much smaller than that of the Metropolis algorithm[18].

We measure the accuracy of the volatility from the SV model by calculating the following

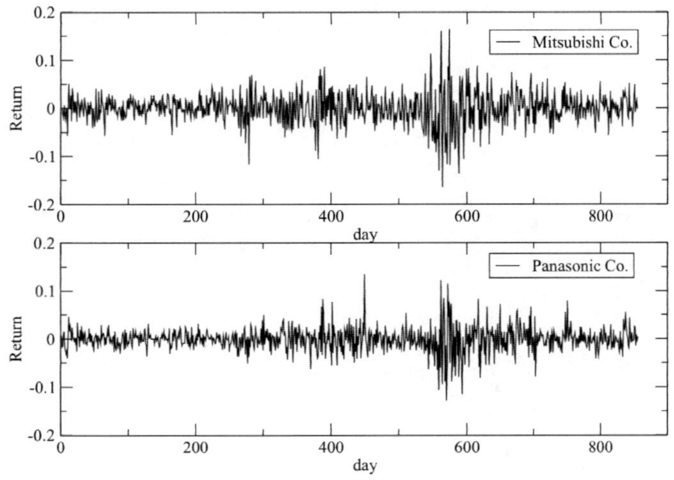

Figure 1. Daily close-to-close return time series of Mitsubishi Co. (top) and Panasonic Co. (bottom).

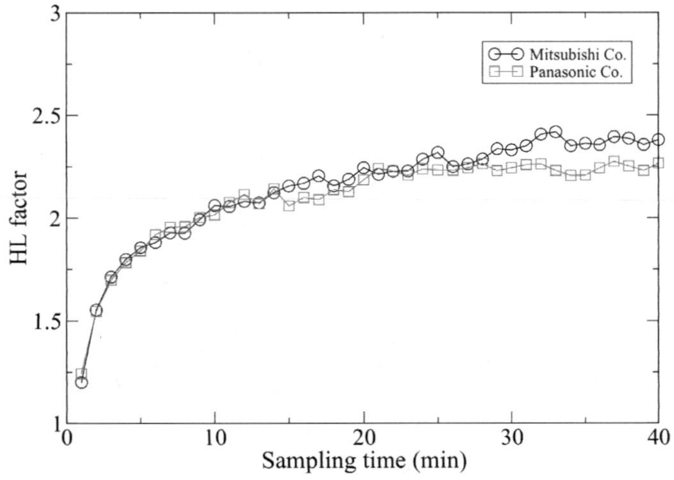

Figure 2. HL factor as a function of sampling time interval.

loss function.

$$RMSPE = \sqrt{\frac{1}{N}\sum_{t=1}^{N}\left(\frac{\bar{\sigma}_t^2 - cRV_t}{cRV_t}\right)^2}, \tag{19}$$

where RMSPE stands for the Root Mean Squared Percentage Error and $\bar{\sigma}_t^2$ is a volatility value at time t averaged over the volatility data sampled by the HMC.

The realized volatility is calculated using sampling time from 1-min. to 20-min. For each sampling time we also calculate the HL factor of eq.(18). Figure 2 shows the HL factors of Mitsubishi Co. and Panasonic Co as a function of sampling time interval. We find that both

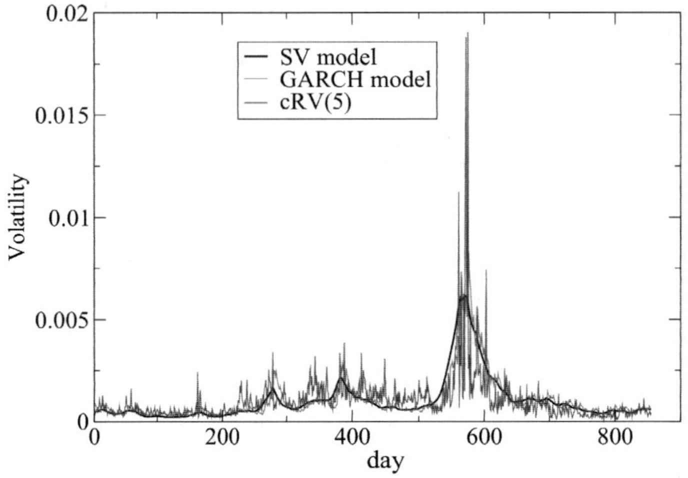

Figure 3. Volatility of Mitsubishi Co. obtained from the SV model, the GARCH model and the realized volatility ($cRV(5)$) at 5-min sampling frequency.

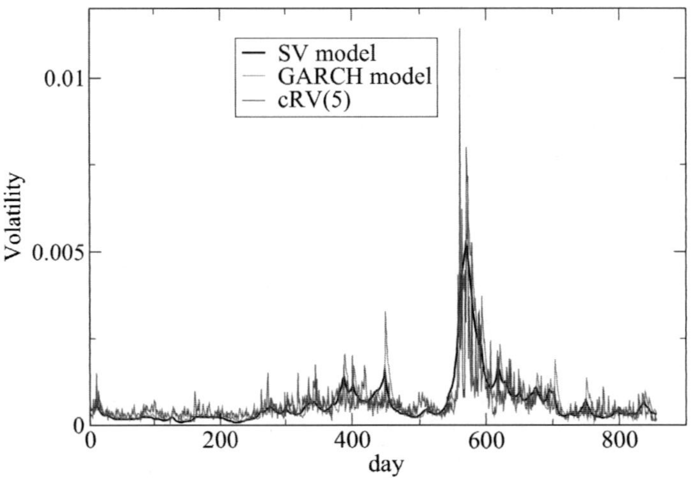

Figure 4. Volatility of Panasonic Co. obtained from the SV model, the GARCH model and the realized volatility ($cRV(5)$) at 5-min sampling frequency.

the HL factors from two stocks behave similarly and increase with sampling time interval. The smaller HL factors at small sampling time interval, i.e. at high sampling frequency are caused by the microstructure noise which artificially inflates the realized volatility. At bigger sampling time interval, i.e. at lower sampling frequency where the microstructure noise effect is negligible the HL factors take bigger than 2, which means that returns during non-trading hours moves as much as returns during trading hours. Especially it is found that overnight return change dominates in non-trading hours[34].

Using the HL factor c we scale the realized volatility RV_t to cRV_t. Figure 3-4 compare the scaled realized volatility cRV_t and the volatility from the SV model. In the figures we only

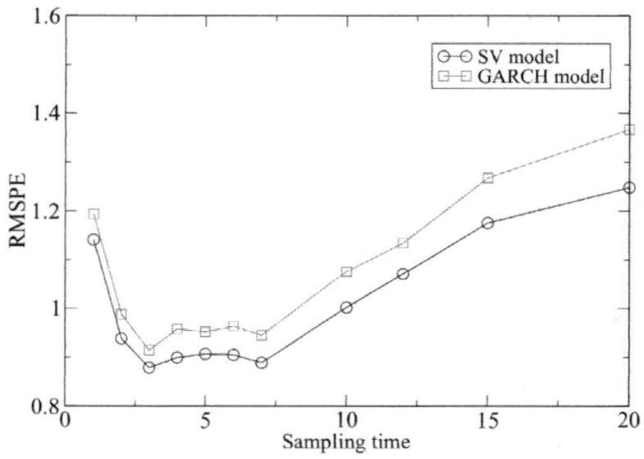

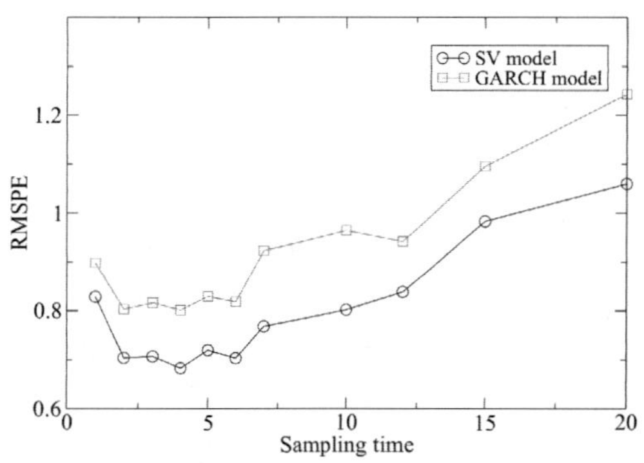

Figure 5. RMSPE of the SV model and the GARCH model for Mitsubishi Co. (left) and Panasonic Co. (right).

show the realized volatility constructed at 5-min sampling frequency as a representative one. The volatility estimated from the GARCH model is also shown in the figures. We used the GARCH(1,1) model with normal errors which is commonly used in empirical analysis of asset volatility. The MCMC method[35, 36, 37, 38] was also used for the volatility estimation of the GARCH model.

Figure 5 shows the RMSPE of the SV and the GARCH model. It is seen that the RMSPE of the SV model is smaller than that of the GARCH model which indicates that the SV model is superior to the GARCH model. It is also interesting to notice that the minimum of the RMSPE takes around 5-min and this observation is consistent with the optimum sampling time interval suggested in [32].

6. Conclusions

We applied the HMC algorithm to the Bayesian inference of the SV model and estimated the volatility of two stock returns (Mitsubishi Co. and Panasonic Co.) traded on the Tokyo Stock Exchange. We find that the volatility variables sampled by the HMC algorithm are well decorrelated compared to the Metopolis algorithm. In order to quantify the accuracy of the estimated volatility we calculated the RMSPE using the realized volatility modified by the HL factor as a proxy of the true volatility. Comparing the RMSPE of the SV model with that of the GARCH model we find that the volatility accuracy of the SV model is superior to that of the GARCH model. Therefore empirically the SV model is preferred to the GARCH model based on the criterion of the volatility accuracy.

Since in this study we only used the GARCH(1,1) model it might be interesting further to compare the SV model with other relevant GARCH-type models.

Acknowledgments.
Numerical calculations in this work were carried out at the Yukawa Institute Computer Facility and the facilities of the Institute of Statistical Mathematics. This work was supported by Grant-in-Aid for Scientific Research (C) (No.22500267).

References

[1] Mantegna R and Stanley H E, *Introduction to Econophysics* (Cambridge University Press, 1999).
[2] Cont R 2001 Empirical Properties of Asset Returns: Stylized Facts and Statistical Issues *Quantitative Finance* **1** 223–236

[3] Engle R.R 1982 Autoregressive Conditional Heteroskedasticity with Estimates of the Variance of the United Kingdom inflation *Econometrica* **60** 987–1007
[4] Bollerslev T 1986 Generalized Autoregressive Conditional Heteroskedasticity *Journal of Econometrics* **31** 307–327.
[5] Nelson D B 1991 Conditional Heteroskedasticity in Asset Returns: A New Approach *Econometrica* **59** 347–370
[6] Glosten L R, Jaganathan R and Runkle D E 1993 On the Relation Between the Expected Value and the Volatility of the Nominal Excess on Stocks *Journal of Finance* **48** 1779–1801
[7] Engle R F and Ng V 1993 Measuring and Testing the Impact of News on Volatility *Journal of Finance* **48** 1749–1778
[8] Sentana E 1995 Quadratic ARCH models *Review of Economic Studies* **62** 639–661
[9] Takasihi T and Chen T T 2012 Bayesian Inference of the GARCH model with Rational Errors *International Proceedings of Economics Development and Research* **29** 303–307
[10] Bollerslev T, Chou R Y and Kroner K F 1992 ARCH modeling in finance *Journal of Econometrics* **52** 5–59
[11] Xekalaki E and Degiannakis S 2010 *ARCH Models for Financial Applications* (John Wiley & Sons Ltd.)
[12] Jacquier E, Polson N G and Rossi P E 1994 Bayesian Analysis of Stochastic Volatility Models *Journal of Business & Economic Statistics* **12** 371–389
[13] Kim S, Shephard N and Chib S 1998 Stochastic Volatility: Likelihood Inference and Comparison with ARCH Models *Review of Economic Studies* **65** 361–393.
[14] Shephard N and Pitt M K 1997 Likelihood Analysis of Non-Gaussian Measurement Time Series *Biometrika* **84** 653–667
[15] Watanabe T and Omori Y 2004 A Multi-move Sampler for Estimating Non-Gaussian Time Series Models *Biometrika* **91** 246–248
[16] Duane S, Kennedy A D, Pendleton B J and Roweth D 1987 Hybrid Monte Carlo *Phys. Lett. B* **195** 216–222
[17] Takaishi T 2008 Financial Time Series Analysis of SV Model by Hybrid Monte Carlo *Lecture Notes in Computer Science* **5226** 929–936
[18] Takasihi T 2009 Bayesian Inference of Stochastic Volatility Model by Hybrid Monte Carlo *Journal of Circuits, Systems, and Computers* **18** 1381–1396
[19] Takaishi T 2006 Bayesian estimation of GARCH model by Hybrid Monte Carlo *Proceedings of the 9th Joint Conference on Information Sciences 2006*, CIEF-214 doi:10.2991/jcis.2006.159
[20] Hansen P R and Lunde A 2006 Consistent ranking of volatility models *Journal of Econometrics* **131** 97–121
[21] Ukawa A 1989 Lattice QCD Simulations Beyond the Quenched Approximation *Nucl. Phys. B (Proc. Suppl.)* **10** 66–145
[22] Takaishi T and de Forcrand Ph 2008 Testing and Tuning Symplectic Integrators for Hybrid Monte Carlo Algorithm in Lattice QCD *Phys. Rev. E* **73** 036706
[23] Takaishi T 2000 Choice of Integrators in the Hybrid Monte Carlo Algorithm *Comput. Phys. Commun.* **133** 6–17
[24] Takaishi T 2002 Higher Order Hybrid Monte Carlo at Finite Temperature *Phys. Lett. B* **540** 159–165
[25] Metropolis N, Rosenbluth A W, Rosenbluth M N, Teller A H and Teller E 1953 Equations of State Calculations by Fast Computing Machines *J. of Chem. Phys.* **21** 1087–1091
[26] Andersen T G and Bollerslev T 1998 Answering the Skeptics: Yes, Standard Volatility Models do Provide Accurate Forecasts *International Economic Review* **39** 885–905
[27] Andersen T G, Bollerslev T, Diebold F X and Labys P 2001 The distribution of realized exchange rate volatility *Journal of the American Statistical Association* **96** 42–55
[28] Andersen T G, Bollerslev T, Diebold F X and Ebens H 2001 The distribution of realized stock return volatility *Journal of Financial Economics* **61** 43–76
[29] Campbell J Y, Lo A W and MacKinlay A C 1997 *The Econometrics of Financial Markets* (Princeton University Press)
[30] Zhou B 1996 High-frequency data and volatility in foreign-exchange rates *Journal of Business & Economics Statistics* **14** 45–52
[31] Andersen T G, Bollerslev T, Diebold F X and Labys P 2000 Great Realization *Risk, March* 105–108
[32] Bandi F M and Russell J R 2008 Microstructure noise, realized variance, and optimal sampling *The Review of Economic Studies* **75** 339-369
[33] Hansen P R and Lunde A 2005 A forecast comparison of volatility models: does anything beat a GARCH(1,1)? *Journal of Applied Econometrics* **20** 873-889
[34] Takaishi T, Chen T T and Zheng Z 2012 Analysis of Realized Volatility in Two Trading Sessions of the Japanese Stock Market *Prog. Theor. Phys. Supplement* **194** 43-54
[35] Takaishi T 2009 An Adaptive Markov Chain Monte Carlo Method for GARCH Model *Lecture Notes of*

the Institute for Computer Sciences, Social Informatics and Telecommunications Engineering. Complex Sciences Vol. **5** 1424

[36] Takaishi T 2009 Bayesian Estimation of GARCH Model with an Adaptive Proposal Density *New Advances in Intelligent Decision Technologies, Studies in Computational Intelligence* Vol. **199** 635

[37] Takaishi T 2009 Bayesian Inference on QGARCH Model Using the Adaptive Construction Scheme *Proceedings of 8th IEEE/ACIS International Conference on Computer and Information Science* 525–529 doi:10.1109/ICIS.2009.173

[38] Takaishi T 2010 Bayesian inference with an adaptive proposal density for GARCH models *J. Phys.: Conf. Ser.* **221** 012011

Journal of Physics: Conference Series **423** (2013) 012022 doi:10.1088/1742-6596/423/1/012022

Conformally Flat Bulk Tilted Cosmological Model

Anita Bagora (Menaria)[1] **, Rakeshwar Purohit**[2]

[1]Department of Mathematics, Jaipur National University, Jaipur-302027, India

[2]*Department of Mathematics,M.L. Sukhadia University, Udaipur-313001,India*

Email:anita_bagora@yahoo.com

Abstract. Conformally flat plane symmetric bulk viscous fluid tilted cosmological model is investigated. We assume that $\zeta\theta = K$ (constant), where ζ is the coefficient of bulk viscosity and θ is the expansion in the model. It has been assumed that the supplementary condition between metric potential as $A = B^n$, where n is constant. Special model is also investigated in the absence of bulk viscosity. The physical and geometrical aspects of the model with singularities in the presence and absence of bulk viscosity are also discussed.

1. Introduction

Cosmological models are studied as one of the most important applications of general relativity. Einstein [1918] in his basic researches based on the spherically symmetric, isotropic and homogeneous model of universe but without red shift [1] and hence non-expanding, whereas de-Sitter gave the expanding static model which was empty [1]. The non-static models were propounded by Friedmann [1924], Lemaitre [1929], Robertson and Walker [1935] and started the process of searching for the various cosmological models which resembles the actual universe up to maximum extent [1]. Bondy and Gold [1948] initiated the steady state theory based upon the cosmological principle. At the present stage of evolution, the universe is spherically symmetric with homogeneous and isotropic matter distribution. But close to the big bang singularity, the assumptions of spherical symmetry or isotropy are not strictly valid and the researchers Randall [2], da Silva and Wang [3],Prakash[4],Pradhan and Pandey [5] etc. started considering plane symmetry. H. Takeno [6] defined the conformal curvature tensor $C_{ijk\ell}$ and the condition that a Riemannian space be conformally flat is given by vanishing of the conformal curvature tensor, i.e. $C_{ijk\ell} = 0$. A number of conformally flat, physically significant models are known like Schwarzschild interior solution. These were later on studied by Singh and Roy [7], Singh and Abdussattar [8],Roy and Bali [9], Pandey and Tiwari [10], Pradhan [11], Pradhan, Srivastava and Singh [12] .

A considerable interest is being shown towards the tilted universes. A spatially homogeneous universe is said to be tilted if the fluid velocity vector is non-orthogonal to the group orbits. Ellis and Baldwin [13] have shown that the actual universe is likely to be tilted one. King and Ellis [14], Ellis and King [15], Collins and Ellis [16] studied the general dynamics of tilted models. Various aspects of tilted cosmological models have been studied by Dunn and Tupper [17], Matravers *et al.* [18], Hewitt *et al.* [19], Horwood *et al.* [20], Coley and Tupper [21], Beesham [22], Bali and Sharma [23].

In most treatments of cosmology, cosmic fluid is considered as perfect fluid. However, bulk viscosity is expected to play an important role at certain stages of expanding universe. Various authors [24−28] have been shown that bulk viscosity leads to inflationary like solution and acts like a negative energy field in an expanding universe. Furthermore, there are several processes which are expected to give rise to viscous effects. These are the decoupling of neutrinos during the radiation era and the decoupling of radiation and matter during the recombination era. Bulk viscosity is associated with the GUT phase transition and string creation. A number of authors have discussed cosmological solutions with bulk viscosity in various context [29-33].

Tilted solutions containing perfect fluid have been obtained by Farnsworth [34] and Reboucas [35].Cosmological models with heat flow have been studied by various author viz. Novello and Reboucas [36], Ray [37], Reboucas and Lima [38], Roy and Banerjee [39-40].Singh *et al.* [41-42] have also investigated Bianchi type I homogeneous tilted cosmological models in different context. Bagora [43-44] obtained tilted Bianchi type I and III cosmological model for disordered radiation and stiff fluid distribution. Motivated by these studies, in this paper we propose to find conformally flat plane symmetric bulk viscous fluid tilted cosmological model. We assume that $\zeta\theta = K$ (constant), where ζ is the coefficient of bulk viscosity and θ is the expansion in the model. It has been assumed that the condition $A = B^n$, between metric potentials. Special model is also investigated in the absence of bulk viscosity. The physical and geometrical aspects of the model in the presence and absence of bulk viscosity are also discussed.

2. The Metric and Field Equations

We consider plane symmetric metric in the form

$$ds^2 = -dt^2 + A^2(dx^2 + dy^2) + B^2 dz^2 , \tag{1}$$

where A and B are functions of 't' alone.

The energy-momentum tensor for perfect fluid distribution with heat conduction given by Ellis [45] and for bulk viscosity given by Landau and Lifshitz [46] is given by

$$T_i^j = (\in + p)v_i v^j + p g_i^j + q_i v^j + v_i q^j - \zeta\theta(g_i^j + v_i v^j) , \tag{2}$$

together with

$$g_{ij} v_i v^j = -1 , \tag{3}$$

$$q_i q^j > 0 , \tag{4}$$

$$q_i v^i = 0 , \tag{5}$$

where p is the isotropic pressure, $\in$ the matter density and q_i the heat conduction vector orthogonal to v^i. The fluid flow vector has the components $\left(0, 0, \frac{\sinh\lambda}{B}, \cosh\lambda\right)$ satisfying (3) and λ is the tilt angle.

The Einstein field equation

$$R_i^j - \frac{1}{2} R g_i^j = -8\pi T_i^j , \text{ (units such that } c = G = 1)$$

For the line element (1) are

$$\frac{A_{44}}{A} + \frac{B_{44}}{B} + \frac{A_4 C_4}{AC} = -8\pi(p - K) , \tag{6}$$

$$\frac{A_4^2}{A^2} + \frac{2A_{44}}{A} = -8\pi\left[(\in + p)\sinh^2\lambda + p + 2q_3\frac{\sinh\lambda}{B} - K\cosh^2\lambda\right], \tag{7}$$

$$\frac{A_4^2}{A^2} + \frac{2A_4 B_4}{AB} = 8\pi\left[(\in + p)\cosh^2\lambda - p + 2q_3\frac{\sinh\lambda}{B} - K\sinh^2\lambda\right], \tag{8}$$

$$-8\pi\left[(\in + p)\, B\sinh\lambda\cosh\lambda + q_3\cosh\lambda + q_3\frac{\sinh^2\lambda}{\cosh\lambda} - KB\sinh\lambda\cosh\lambda\right] = 0 . \tag{9}$$

Here the suffix ‘4’ stands for ordinary differentiation with respect to cosmic time ‘t’ alone.

3. Solution of the Field Equations

Equations from (6)-(9) are four equations in six unknown A, B, $\in$, p, λ and q_3, therefore to determine the complete solution we require two more conditions.

Firstly, we assume that the space-time is conformally flat which leads to

$$C_{1212} = \frac{1}{3}\left[A^2 A_4^{\,2} - A^3 A_{44} - A^3 A_4 \frac{B_4}{B} + A^4 \frac{B_{44}}{B} \right] = 0 . \tag{10}$$

Secondly, we consider relation between metric potential as

$$A = B^n . \tag{11}$$

Also, we assume that $\zeta\theta = K$. (12)

The condition $\zeta\theta = K$ is due to the peculiar characteristic of the bulk viscosity.It acts like a negative energy field in an expanding universe (Johri and Sudharshan[47]) i.e $\zeta\theta = K$. According to that expansion is inversely proportional to bulk viscosity.

Equations (10) and (11) lead to

$$B^3 B_{44} + (n-1) B^2 B_4^{\,2} = 0 . \tag{13}$$

From equations (7), (8) and (11), we have

$$n\frac{B_{44}}{B} + (n^2 - 2n)\frac{B_4^{\,2}}{B^2} = -4\pi\left[(\in + p)\cosh 2\lambda + 4q_3 \frac{\sinh\lambda}{B} - K\cosh 2\lambda \right], \tag{14}$$

and

$$n\frac{B_{44}}{B} + 2n^2\left(\frac{B_4}{B}\right)^2 = 4\pi(\in - p + K) . \tag{15}$$

From equations (9) and (11), we have

$$\frac{-16\pi q_3 \sinh\lambda}{B} = 4\pi(\in + p + K)\sinh 2\lambda \tanh 2\lambda . \tag{16}$$

Again from equations (14) and (16), we have

$$n\frac{B_{44}}{B} + (n^2 - 2n)\frac{B_4^{\,2}}{B^2} = \frac{-4\pi(\in + p + K)}{\cosh 2\lambda} . \tag{17}$$

Equation (13) gives

$$B = \sqrt[n]{n(\ell t + m)} , \tag{18}$$

where ‘ℓ’and ‘m’ are constants of integration.

Equation (18) leads to

$$A^2 = n^2 (\ell t + m)^2 . \tag{19}$$

Hence the metric (1) reduces to

$$ds^2 = -dt^2 + n^2(\ell t + m)^2 (dx^2 + dy^2) + n^{2/n}(\ell t + m)^{2/n} dz^2 . \tag{20}$$

By using the suitable transformations the metric (20) becomes

$$ds^2 = \frac{-dT^2}{\ell^2} + n^2 T^2 (dX^2 + dY^2) + n^{2/n} T^{2/n} dZ^2 . \tag{21}$$

4. Some Physical and Geometrical Features

The density and pressure for the model (21) are given by

$$8\pi \in = \frac{8n^2 \pi K T^2 - 1}{n^2 T^2} , \tag{22}$$

Journal of Physics: Conference Series **423** (2013) 012022 doi:10.1088/1742-6596/423/1/012022

$$8\pi p = \frac{2n^2 + 2n - 1}{n^2 T^2}. \qquad (23)$$

The tilt angle λ is given by

$$\cosh\lambda = \sqrt{\frac{n^2 + 2n - 1 + 8\pi n^2 K T^2}{2n}}, \qquad (24)$$

$$\sinh\lambda = \sqrt{\frac{n^2 - 1 + 8\pi n^2 K T^2}{2n}}. \qquad (25)$$

The reality conditions

$\in + p > 0$, (ii) $\in + 3p > 0$, lead to

$$4n^2 \pi K T^2 - 1 + n^2 + n \geq 0. \qquad (26)$$

The scalar of expansion θ calculated for the flow vector v^i is given by

$$\theta = \frac{[8(1+3n)n^2 \pi K T^2 + 4n^4 + 10n^3 - 2n]}{nT\sqrt{2n\{n^2 + 2n - 1 + 8\pi n^2 K T^2\}}}. \qquad (27)$$

The components of fluid flow vector v^i and heat conduction vector q_i for the model (21) are given by

$$v^3 = \frac{1}{(nT)^{1/n}} \sqrt{\frac{n^2 - 1 + 8\pi n^2 K T^2}{2n}}, \qquad (28)$$

$$v^4 = \sqrt{\frac{n^2 + 2n - 1 + 8\pi n^2 K T^2}{2n}}, \qquad (29)$$

$$q_3 = \frac{-n^{1/n} T^{1/n} \{n^2 + 2n - 1 + 8\pi n^2 K T^2\}}{8\pi n^2 T^2} \sqrt{\frac{n^2 - 1 + 8\pi n^2 K T^2}{2n}}, \qquad (30)$$

$$q_4 = \frac{\{n^2 - 1 + 8\pi n^2 K T^2\}}{8\pi n^2 T^2} \sqrt{\frac{n^2 + 2n - 1 + 8\pi n^2 K T^2}{2n}}. \qquad (31)$$

The non-vanishing components of shear tensor (σ_{ij}) and rotation tensor (ω_{ij}) are given by

$$\sigma_{33} = \frac{n^{2/n} \{(1-n)(n^2 + n - 1) + 8\pi n^2 K T^2\}}{3n^2 T^{1-2/n}} \sqrt{\frac{n^2 + 2n - 1 + 8\pi n^2 K T^2}{2n}}, \qquad (32)$$

$$\sigma_{34} = \frac{-\{(1-n)(n^2 + n - 1) + 8\pi n^2 K T^2\}}{3n^{2-1/n} T^{1-1/n}} \sqrt{\frac{n^2 - 1 + 8\pi n^2 K T^2}{2n}}, \qquad (33)$$

$$\omega_{34} = \frac{-\{n^2 - n + 8\pi(n^2 + 2)n^2 K T^2\}}{n^{2-1/n} T^{1-1/n} \sqrt{2n(n^2 - 1 + 8\pi n^2 K T^2)}}. \qquad (34)$$

Thus

$$\sigma_{33} v^3 + \sigma_{34} v^4 = 0. \qquad (35)$$

Similarly $\omega_{11} v^1 + \omega_{14} v^4 = 0$. (36)

In the absence of bulk viscosity, the above mentioned quantities lead to

$$8\pi \in = \frac{-1}{n^2 T^2}, \qquad (37)$$

$$8\pi p = \frac{2n^2 + 2n - 1}{n^2 T^2}, \qquad (38)$$

$$\cosh\lambda = \sqrt{\frac{n^2 + 2n - 1}{2n}}, \qquad (39)$$

$$\sinh\lambda = \sqrt{\frac{n^2-1}{2n}}, \tag{40}$$

$$\theta = \frac{2[2n^3+5n^2-1]}{T\sqrt{2n(n^2+2n-1)}}, \tag{41}$$

$$v^3 = \frac{1}{(nT)^{1/n}}\sqrt{\frac{n^2-1}{2n}}, \tag{42}$$

$$v^4 = \sqrt{\frac{n^2+2n-1}{2n}}, \tag{43}$$

$$q_3 = \frac{-(n^2+2n-1)}{8\pi n^{2-1/n}T^{2-1/n}}\sqrt{\frac{n^2-1}{2n}}, \tag{44}$$

$$q_4 = \frac{(n^2-1)}{8\pi n^2T^2}\sqrt{\frac{n^2+2n-1}{2n}}, \tag{45}$$

$$\sigma_{33} = \frac{n^{2/n}(1-n)(n^2+n-1)}{3n^2T^{1-2/n}}\sqrt{\frac{n^2+2n-1}{2n}}, \tag{46}$$

$$\sigma_{34} = \frac{-(1-n)(n^2+n-1)}{3n^{2-1/n}T^{1-1/n}}\sqrt{\frac{n^2-1}{2n}}, \tag{47}$$

$$\omega_{34} = \frac{-(n-1)}{n^{2-1/n}T^{1-1/n}\sqrt{2(n^2-1)}}. \tag{48}$$

Thus

$$\sigma_{33}v^3 + \sigma_{34}v^4 = 0. \tag{49}$$

The physical significance of conditions (35), (36) and (49) are explained by Ellis[48]:The shear tensor (σ_{ij}) determines the distortion arising in the fluid flow, leaving the volume invariant. The direction of principal axis is unchanged by the distortion, but all other directions are changed. Thus we have $\sigma_{ij}v^j = 0$,

which leads to

$$\sigma_{33}v^3 + \sigma_{34}v^4 = 0 \quad (\because v_3 \neq 0, v_4 \neq 0)$$

Shear (σ) is given by

$$\sigma^2 = \frac{1}{2}\sigma_{ij}\sigma^{ij}. \text{Thus } \sigma^2 \geq 0 \text{ and } \sigma = 0 \Leftrightarrow \sigma_{ij} = 0.$$

The vorticity tensor (ω_{ij}) determines a rigid rotation of cluster of galaxies with respect to a local inertial rest frame. Thus, we have

$$\omega_{ij} = \eta_{ijk\ell}\omega^k\omega^\ell, \text{where } \eta_{ijk\ell} \text{ is pseudo tensor and } \omega^i = \frac{1}{2}\eta_{ijk\ell}v_j\omega_{k\ell}.$$

Thus $\omega_{ij}v^j = 0$.

This leads to

$$\omega_{33}v^3 + \omega_{34}v^4 = 0 \ (\because v_1 \neq 0, v_4 \neq 0)$$

The magnitude of ω_{ij} is ω and is defined as

$$\omega^2 = \frac{1}{2}\omega_{ij}\omega^{ij}.$$

Also $\omega = 0 \Leftrightarrow \omega_{ij} = 0$.

The rates of expansion H_i in the direction of x, y and z axes are given by

$$H_1 = \frac{1}{T}, \tag{50}$$

$$H_2 = \frac{1}{nT}, \tag{51}$$

$$H_3 = \frac{1}{nT}. \tag{52}$$

5. Conclusion

The model (21) represents a tilted model. The model starts with a big-bang at T=0 and the expansion in the model decreases as time increases. Also, $\sigma_{ij}v^j = 0$ and $\omega_{ij}v^j = 0$ are satisfied as $\sigma_{11}v^1 + \sigma_{14}v^4 = 0$ and The model has point type singularity at T=0 (MacCallum [49]).

In the absence of bulk viscosity, $\in \to \infty$ when $T \to 0$ and $\in \to 0$ when $T \to \infty$ therefore $\in$ is the decreasing function of time. The reality conditions $\in$+p>0, $\in$+3p>0 given by Ellis [50] are satisfied when $n^2>1$. The model is an expanding universe in which the lines of flow of matter are geodesic, shearing and rotating. The rate of expansion is decreasing function of time and tilt angle is function of time. The metric potentials are constant at t=1 i.e. the space time reduces to flat space time. For particularly n=1 this model becomes non-shearing with no rotation. So that the universe is not tilted for n>1.

References

[1] Karade T M, Khadekar G S and Bendre M S 2000 *Lectures on General Relativity*, Einstein Foundation International, chapters **7** and **8**
[2] Rendall Alan D 1995 *Gen. Rel. Grav.* **27** 213
[3] da Silva M F A, Wang A Z and Santos N O 1908 *Phy. Lett. A* **244** 462
[4] Prakash S 1980 *Indian J. Pure & Appl. Math.* **11(7)** 931
[5] Pradhan A and Pandey H R 2003 *Int. J. Mod. Phys. D* **12(5)** 941
[6] Takeno H 1966 The Theory of Spherically Symmetric Space-Times; revised edition, Research institute for Theoretical Physics, Hiroshima University
[7] Singh K P and Roy S R 1966 *Proc. Nat. Inst. Sci. India*, **32A** 223
[8] Singh K P and Abdusattar, *Gen. Rel. Grav.* 1974 **5(1)** 115
[9] Roy S R and Bali R 1978 *Indian J. Pure & Appl. Math.* **9** 871
[10] Pandey S N and Tiwari R 1981 *Indian J. Pure & Appl. Math.* **12** 261
[11] Pradhan A and Pandey O P 2002 *Spacetime & Substance* **4(14)** 169
[12] Pradhan A and Singh S K 2004 *Int. J. Mod. Phys. D* **13** 503
[13] Ellis G F R and Baldwin J E 1984 *Mon. Not. Roy. Astro Soc.* **206** 377
[14] King A R and Ellis G F R 1973 *Comm. Maths. Phys.* **31** 209
[15] Ellis G F R and King A R 1974 *Comm. Maths. Phys.* **38** 119
[16] Collins C B and Ellis G F R 1979 *Phys. Rep.* **56** 65

[17] Dunn K A and Tupper B O J 1978 *Astrophy. J.* **222(2)** 405
[18] Matravers D R, Madsen M S and Vogel D L 1985 *Astrophys. Space Sc.* **112** 193
[19] Hewitt C G, Bridson R and Wainwright J 2001 *Gen. Rel. Grav.* **33** 65
[20] Horwood†, Hancock‡ M J, The D and J Wainwright†2002 *Preprint gr-qc/0210031v2*
[21] Coley A A and Tupper B O J 1983 *Phys. Lett. A* **95** 357
[22] Beesham A 1986 *Astrophys. J.* **125** 99
[23] Bali R and Sharma K 2003 *Astrophys. Space Sc.* **283** 11
[24] Misner C W 1968 *Astrophys. J.* **151** 431
[25] Ellis G F R 1986 In *General Relativity and Cosmology*, ed. R.K. Sachs; Academic Press; New York
[26] Hu B L 1983 In *Advance in Astrophysics*, eds. L. J. Fung and R. Ruffini, (World Scientific, Singapore)
[27] Padmanabhan T and Chitre S M 1987 *Phys. Lett. A* **120** 433
[28] Johri V B and Sudarshan R 1989 *Proc. Int. Conf. on Mathematical Modelling in Science and Technology*,
[29] Srinath L S *et al* eds (World Scientific, Singapore)
[30] Grøn Ø 1990 *Astrophys. Space Sci.* **173** 191
[31] Pradhan A, Yadav V K and Chakrabarty I 2001 *Int J. Mod. Phys. D* **10** 339
[32] Chakrabarty I, Pradhan A and Saste N N 2001 *Int J. Mod. Phys. D* **10** 741
[33] Pradhan A and Aotemshi I 2002 *Int J. Mod. Phys. D* **11(6)** 1419
[34] Chimento L P, Jakubi A S and Pavon D 1999 *Class. Quantum Grav.* **16** 1625
[35] Singh G P, Ghosh S G and Beesham A 1997 *Aust. J. Phys.* **50(5)** 903
[36] Roy S R and Narain S 1981 *Indian J. Pure & Appl. Math.* **12(2)** 284
[37] Farnsworth D L 1967 *J. Maths. Phys.* **8(12)** 2315
[38] Reboucas M J 1982 *Nuovo. Cim. B* **67** 120
[39] Novello M and Reboucas M J 1978 *Astrophys. J.* **225** 719
[40] Ray D J 1980 *J. Math. Phys.* **21** 2797
[41] Reboucas M J and de Lima J A S 1982 *J. Math. Phys.* **23** 1547
[42] Roy S R and Banerjee S K 1988 *Astrophys. Space Sci.* **150(2)** 213
[43] Roy S R and Banerjee S K 1996 *Gen. Rel. Gravit.* **28**(1) 27
[44] Rathore G S, Bagora A and Gandhi S 2008 *Adv. Studies Theor. Phys.* **2(19)** 909
[45] Bagora A, Rathore G S and Gandhi S 2009 *Rom. J. Phys.* **54 (7-8)** 777
[46] Bagora A 2010 *EJTP* **4(14)** 373
[47] Bagora A 2009 *Astrophys. Space Sci.* **319** 155
[48] Ellis G F R 1971 *General Relativity and Cosmology*; ed. R.K. Sachs; Academic Press; New York, 116.
[49] Landau L D and Lifshitz E M 1963 *Fluid Mechanics*; Pergamon Press **6** 505
[50] Johri V B and Sudarshan R 1988 *Proc. Int. Conf. on Mathematical Modelling in Science and Technology*, World Scientific, Singapore
[51] Ellis G F R 1971 *General Relativity and Cosmology*; ed. R.K. Sachs; Academic Press; New York, 113
[52] MacCallum, M A H 1971 *Comm. Maths. Phys.* **20** 57.
[53] Ellis G F R 1971 General Relativity and Cosmology; ed. R.K. Sachs; Academic Press; New York, 117.

Journal of Physics: Conference Series **423** (2013) 012023 doi:10.1088/1742-6596/423/1/012023

Introducing the Dimensional Continuous Space-Time Theory

Luiz Cesar Martini

Electrical and Computing Engineering Faculty
University of Campinas - Brazil

martini@decom.fee.unicamp.br

Abstract. This article is an introduction to a new theory. The name of the theory is justified by the dimensional description of the continuous space-time of the matter, energy and empty space, that gathers all the real things that exists in the universe. The theory presents itself as the consolidation of the classical, quantum and relativity theories. A basic equation that describes the formation of the Universe, relating time, space, matter, energy and movement, is deduced. The four fundamentals physics constants, light speed in empty space, gravitational constant, Boltzmann's constant and Planck's constant and also the fundamentals particles mass, the electrical charges, the energies, the empty space and time are also obtained from this basic equation. This theory provides a new vision of the Big-Bang and how the galaxies, stars, black holes and planets were formed. Based on it, is possible to have a perfect comprehension of the duality between wave-particle, which is an intrinsic characteristic of the matter and energy. It will be possible to comprehend the formation of orbitals and get the equationing of atomics orbits. It presents a singular comprehension of the mass relativity, length and time. It is demonstrated that the continuous space-time is tridimensional, inelastic and temporally instantaneous, eliminating the possibility of spatial fold, slot space, worm hole, time travels and parallel universes. It is shown that many concepts, like dark matter and strong forces, that hypothetically keep the cohesion of the atomics nucleons, are without sense.

1. Introduction

In this article, the main equation of this theory is formally deduced, that is named the primitive equation of the continuous space-time. In order to give a first notion of the introduced concepts of this theory, imagine that any object occupying a specific space has dimensions that fill and define a volume in this space and this space cannot be occupied instantaneously by other object.

Imagine for example a stick in the space. This stick exists only because it has a potential energy that creates, defines and keeps its dimensions, determining a volume in the space.

In order to characterize the empiric concept of the existence of a volume occupying a determined space, it will be established the concept of a dimensional axis of potential energy, in which a elementary dimensional axis of potential energy occupying a space do not allows that other elementary axis of potential energy occupies the same space. This dimensional axis that constitutes an incremental axis, for example, of matter, that to exist physically needs a potential energy that creates and keeps it occupying a volume in the space. This first concept is introduced formally in the following [1].

2. Embryo of the Primitive Continuous Space-Time

In the last section, it will be established the rudiments of a theory that supposes the existence of a dimensional length that has a potential energy. This potential energy is stabilized from two opposed pairs of conjugated forces that have the same common origin.

Other In the last section, it will be established the rudiments of a theory that supposes the existence of a dimensional length that has a potential energy. This potential energy is stabilized from two opposed pairs of conjugated forces that have the same common origin [2].

It is defined a dimensional length α, whose end points are defined by the points P_1 and P_{-1}. At both points, there are two forces, one of contraction and the other of expansion, that are in equilibrium and maintain the potential at these points.

The conjugated forces of contraction f_{r_1} and $f_{r_{-1}}$ are named anti-space-time conjugated forces and the conjugated forces of expansion f_{s_1} and $f_{s_{-1}}$ are named space-time conjugated forces.

These two couples of conjugated forces $(f_{r_1};f_{r_{-1}})$ and $(f_{s_1};f_{s_{-1}})$ constitute one energy dimensional axis that sustains the elementary points P_1 and P_{-1}. This energy dimensional axis constituted by the elementary points P_1 and P_{-1} is the 0 dimensional basis of the space and it is going to be named as a elementary dimension of potential energy.

One elementary dimension of potential energy has a space field S, which energy is proportional to S^2, that is

$$E = kS^2 \tag{1}$$

where k is a constant that depends on the dimensional length α. The space field S on the other hand should be dependent of the dimensional length α too. imagine that the space field

should be an exponential function of dimension length α, if it is differentiated or integrated it must result in the same space field. Suppose an origin point in the space where many elementary axis of energy pass through. These axis and their energies are jointly independent, so they are mathematically orthogonal, that means that the total energy is the sum of the energies of each elementary axis. Then for two elementary axis of energy,

$$E(P_{1,-1}+P_{2,-2}) = E(P_{1,-1}) + E(P_{2,-2})$$

where $P_{1,-1}$ is the elementary axis whose extreme points are P_1 and P_{-1} and $P_{2,-2}$ is the elementary axis whose extreme points are P_2 and P_{-2}. Equivalently,

$$\begin{aligned} E_1 &= E(P_{1,-1}) = k\overrightarrow{S_1}\overrightarrow{S_1} = kS_1^2 \\ E_2 &= E(P_{2,-2}) = k\overrightarrow{S_2}\overrightarrow{S_2} = kS_2^2 \\ E_{1,2} &= E(P_{1,-1}+P_{2,-2}) = kS_1^2 + kS_2^2 \end{aligned} \tag{2}$$

It is defined a volume V that contains two elementary dimensional axis of potential energy E_1 and E_2, where the total energy is given by equation (2) and the space fields S_1 and S_2 are mathematically orthogonal. So, the volume V containing a set of m elementary dimensions is a euclidian space. As consequence, the addition of the energies E_1 and E_2 forms a one-dimensional mathematic space, that can be obtained from two zero-dimensional mathematic spaces, that has equal energies, $E_1 = E_2$.

In order to produce a two-dimensional mathematic space, it will start from two one-dimensional mathematic spaces. So it needs to add two more elementary dimension of potential energy. Now, considering a great number of elementary axis of potential energy that are uniformly distributed inside a volume V and constituted by elementary dimensions of length α, where the i-th elementary energy is given by $E_i = E_0$. Where E_0 is the elementary axis of potential energy, that is the same in all directions.

In order to establish an empiric formation law, it can firstly notice that the D-dimensional space of potential energy is formed by the addition of a power of two elementary energy dimensions.

Journal of Physics: Conference Series **423** (2013) 012023 doi:10.1088/1742-6596/423/1/012023

Starting from a zero-dimensional mathematic space of energy $E(0)$, that is $E(0)=E_1=E_0$.

The one-dimensional mathematical space of energy $E(1)$ is obtained by the addition of two zero-dimensional space of energy, that is $E(1)=E(0)+E(0)=E_1+E_2=2E_0$. The two-dimensional mathematical space of energy $E(2)$ is obtained by the addition of two one-dimensional mathematical space of energy, that is $E(2)=E(1)+E(1)=E_1+E_2+E_3+E_4=4E_0$

So, the D-dimensional mathematical space of energy $E(D)$ is given by:

$$E(D)=2^D E_0=e^{2\alpha_0 D}E_0 \tag{3}$$

So, D is defined as the dimension of a D-dimensional mathematical space of energy. In this way, it is found an elementary law that describes the total energy inside the volume V.

In the following, it is discovered that the total energy is dependent on the mathematical dimension D. Defining the space constant α_0 as

$$\alpha_0=\frac{1}{2}\ln 2 \tag{4}$$

The energy $E(D)$ represents the energy of a dimensional space containing a set of elementary dimensions of potential energy.

The space of energy has dimension equal to D constructed from the bases of segments of one-dimensional axis of energy, given by

$$E_{1b}=E(D)^{\frac{1}{D}}=2E_0^{\frac{1}{D}}=e^{2\alpha_0}E_0^{\frac{1}{D}} \tag{5}$$

So, the base zero-dimensional of one energy elementary dimension has energy, given by

$$E_{0b}=E_0^{\frac{1}{D}} \tag{6}$$

The (3) define an embryonal formula of the continuous space-time of the matter, energy and empty space that represents the totality of all real things in the Universe.

It can be demonstrated that

$$D=\infty \quad \text{for the time and for the empty space} \tag{7}$$

$$D=D_0=56 \quad \text{for the matter and for the energy} \tag{8}$$

On the other hand, conjugated forces of space-time, f_s, do no exist for the empty space, there are only conjugated forces of anti-space-time, f_r.

3. Introducing the Temporal Variation in the Length α

Starting from the embryonal equations (1) and (3), it is possible to find an expression for E_0. Since the elementary dimension of energy has a length α, setting $k=\alpha$ in (1). Since the space potential $S_{(0)}$ must be a function of α and an exponential law has to be obeyed, so doing.

$$E_0=kS_0^2=\alpha e^{2\alpha_0\alpha} \tag{9}$$

justified by the dimensional law 2^D, where by analogy it has been done $D=\alpha$ as elementary dimension.

In this theory, the universe is constituted only by dimensions whose unit of length is described in an absolute mathematical measurement system, that is perfectly characterized by the international measurement system (SI) [3]. It can be demonstrated that using the absolute mathematical measurement system, that

$$\alpha=\frac{1}{\pi} \tag{10}$$

Now appears the first question: How to introduce in the expression of E_0 space and anti-space representations and its temporal dependence? So, doing

$$E_0 = \alpha S_0^2 = \alpha S_{0s} S_{0r}$$
$$E_0 = \alpha e^{\alpha_0 \alpha} e^{\alpha_0 \alpha} = \alpha S_{0s} S_{0r} \tag{11}$$

where S_{0s} is a space component and S_{0r} is a anti-space component of the elementary energy E_0. So it has

$$S_{0s} = e^{\alpha 0 D_{1s}} e^{\alpha_0 L_{1s}} = e^{\alpha_0 \frac{\alpha}{2}} S_{1s} \tag{12}$$

where D_{1s} is a dimension of the stationary space given by $D_{1s} = L_{1s} = \dfrac{\alpha}{2}$ and L_{1s} is a dimensional length of the dimension D_{1s} of the stationary space.

As a consequence,

$$S_{0r} = e^{\alpha 0 D_{1r}} e^{\alpha_0 L_{1r}} = e^{\alpha_0 \frac{\alpha}{2}} S_{1r}$$
$$D_{1r} = L_{1r} = \frac{\alpha}{2} \tag{13}$$

Where S_{1s} is the elementary stationary component potential of the energy space and S_{1r} is the elementary stationary component potential of de anti space energy.

Now it is introduced a variation into the lengths L_{1s} and L_{1r} so it will be substituted respectively L_{1s} and L_{1r} by

$$L_{1s} = \alpha_s(t_s) = f_s(t_s) + \frac{\alpha}{2}$$
$$L_{1r} = \alpha_r(t_s) = f_r(t_s) + \frac{\alpha}{2} \tag{14}$$

With the harmonic fluctuations $\alpha_s(t_s)$ and $\alpha_r(t_s)$ having a complete temporal dependence of t_s in a most simple form. It is designed t_s as the subjective time or the time that rises to the future like the common concept that we have about the time.

Otherwise, the real time is a dimension that oscillates as a pendule with variable velocity. So it is defined a spatial fluctuation dependent of the time designed as space-time function given by $f_s(t_s) = \left(\alpha/2\right)\sin(\omega_0 t_s)$ where ω_0 must be determined.

4. Identification of Light Speed as a Universal Constant

On the other hand, before the Big-Bang E_0 can not have fluctuations because of the equilibrium between anti space and space time forces.

In this first moment because there is no correct notion of the time movement t_s, it is erroneously induced that $\alpha_r(t_s)$ is equal to $-\dfrac{\alpha}{2}\sin(\omega_0 t_s) + \dfrac{\alpha}{2}$.

So, before the Big Bang $E_0 = \alpha e^{\alpha_0 (D_{1s} + D_{1r})} e^{\alpha_0 \alpha_s(t_s)} e^{\alpha_0 \alpha_r(t_s)}$

Now it must focus to the physical and mathematical nature of S_{1s} and S_{1r}. The nature of the anti space and space forces are totaly dynamics, in this sense they only can be represented by the dynamics components of $\alpha_s(t_s)$ and $\alpha_r(t_s)$.

Then it is defined the elementary embryonal equations components of dynamics space S_s and of dynamics anti space S_r in the same sense of S_{1s} and S_{1r}, so that

$$\begin{aligned} S_{1s} &= e^{\alpha_0 D_{2s}} e^{\alpha_0 f_s(t_s)} = e^{\alpha_0 D_{2s}} S_s \\ S_{1r} &= e^{\alpha_0 D_{2r}} e^{\alpha_0 f_r(t_s)} = e^{\alpha_0 D_{2r}} S_r \end{aligned} \quad (15)$$

Then $D_{2s} = D_{2r} = \alpha/2$. So they have a dynamics components of the energies given by

$$\begin{aligned} S_{0s} &= e^{\alpha_0 D} e^{\alpha_0 (D_{1s} + D_{2s})} S_s \\ S_{0r} &= e^{\alpha_0 D} e^{\alpha_0 (D_{1r} + D_{2r})} S_r \end{aligned}$$

with

$$\begin{aligned} D_s &= D_{1s} + D_{2s} = \alpha = \frac{1}{\pi} \\ D_r &= D_{1r} + D_{2r} = \alpha = \frac{1}{\pi} \end{aligned}$$

where $D_s = D_r = \alpha = 1/\pi$ are the dimensions of dynamics space and anti-space components. Its correspondents lengths L_s and L_r are equal to α that is the peak to peak variation of $f_s(t_s)$ and $f_r(t_s)$ respectively.

$$\begin{aligned} E(D) &= \alpha e^{2\alpha_0 D} e^{\alpha_0 (D_s + D_r)} S_s S_r \\ &= \alpha e^{2\alpha_0 (D+\alpha)} S_s S_r \end{aligned} \quad (16)$$

Now the dimensional physical and mathematical space has dimension $D+\alpha$ and the quantities $\beta(D) = e^{2\alpha_0(D+\alpha)}$ are fundamental nature, dimensional function, that only depends of the dimension D.

This function $\beta(D)$ is physically and mathematically without dimension and to the mater and energy it has $D = D_0 = 56$. Then to the mater and energy, it is identified $\beta(D_0)$ with the square of the light velocity c in the empty space, expressed in meters per seconds.

In the equation (16) by doing $D = D_0$, the value c numeric equal to the light velocity in the empty space, is the phase time velocity of the wave time into the mater and energy.

$$c = e^{\alpha_0 (D_0 + \frac{1}{\pi})} \quad (17)$$

where $c = 2{,}997438563 \times 10^8$ m/s. The percentage difference between theoretical and practical values is $PD = -0{,}016$.

Now it is inquired that velocity has dimension, but this theory will proof that dimension time is exactly like space dimension, so meter by seconds is not dimensional.

This theory will proof that everything that exists into the universe is express in the unity of mathematical dimensional length α. So in a concise form it has

$$E(D_0) = \alpha c^2 S_s S_r \quad (18)$$

5. Determination of the Angular Frequency ω_0

Aiming to find a most complete equation that (16) represents, it will start admitting the principle that before the Big-Bang the continuous space time equation must not change with movement time. So the elementary cell of energy do not change with the time variation too.

The cell must be totaly static, without manifestation of temperature, without mass like it is known, or electric charge. The cell must be only an elementary cell of potential energy that keeps its dimensions stables. So, in this way, the composition of the continuous space time equation must be present the dynamics equations S_s and S_r representatives of the forces of space and anti space that mutually one neutralizes the other from a derivative process.

As a start point principle, it will consider that the time was always present and it, in fact, made a continuous and indefinite process derivative in the energy at the elementary cells.

Meanwhile, before the Big-Bang the continuous temporal derivative in the energy, does not promote fluctuations in this static energy.

So, initially it will admit that the derivative of the energy before the Big-Bang was zero and it will be making

$$\frac{dE(D_0)}{dt_s} = 0$$

In this way, the primitive Universe was totaly static in the zero absolute temperature. It was constituted only by elementary cells, completely stationaries, disposed side by side, totaly occupying the empty space and All Universe Energy was stored in this elementary cells as potential energy.

Each elementary cell contains 2^{D0} elementary dimensions of potential energy in a homogeneous spherical distribution of radius $\alpha / 2$.

Because the elementary cells has this homogenous distribution it can be proven that the potential gravitational between the cells is always constant, independent of the volume of the cells that is considered.

It is going to be choose the referential space-time function $f_s(t_s)$ to have a sinusoidal form and the referential anti space-time function $f_r(t_s)$ to be a cosine, that it will be justified later and is given by

$$f_r(t_s) = -\frac{\alpha}{2}\cos(\omega_0 t_s)$$

It will be postulate that the equation of action into the continuous space time is a continued derivative equation. For a while it will be doing this action equation before the Big-Bang to be equal zero

$$\frac{dE(D_0)}{dt_s} = \omega_0 \alpha_0 \alpha^2 \frac{\sqrt{2}}{2}\cos\left(\omega_0 t_s - \frac{\pi}{4}\right) c^2 S_s S_r = 0$$

The unique form of this expression to be zero is done $\omega_0 = 0$. But so where will be the temporal variation?.

6. Introduction of the Phase Variation

The question above will be solved by introducing a phase variation $\theta(t_s) = \omega_1 t_s$ where $\omega_1 = 2\pi \frac{V_f}{\lambda}$ and V_f is the phase velocity of the time wave, λ is the length of time wave produced by this phase velocity.

As in this moment the time will be considered into complete stationary space and not into movement matter, is possible to demonstrate that in this condition it must have $\lambda / v_f = 2\pi$

When it will be considered the matter in movement with velocity v, it will be demonstrated that

$$\omega_1 = \frac{1}{\sqrt{1 - \frac{v^2}{c^2}}}$$

that is the known Lorentz factor [4].

In this moment it will be assumed that ω_1 is a constant value to be found.

Aiming not to disturb the equation before the Big-bang, it is introduced two oppositive phases. So, before the Big-Bang, it is going to suppose initially that

$$f(t_s) = f_s(t_s + f_r(t_s) = \alpha\frac{\sqrt{2}}{2}\sin(\omega_0 t_s - \frac{\pi}{4} + \theta(t_s) - \theta(t_s))$$

In this first moment it will be supposed that space and anti space forces will be represented by the phases $\theta(t_s)$ and $-\theta(t_s)$.

What does it mean having into the space a sinusoid with zero frequency and $\theta(t_s)$ phase changing with the time?

It means that the amplitude of the sinusoid will be instantaneously the same in all universe and this amplitude has a variation with the time which amplitude is defined by the phase. This equation depends on the time and is producing zero derivative.

7. Alteration of Equilibrium Conditions

After the Big-Bang the $\frac{dE(D_0)}{dt_s}$ equation has to be changed.

The start of the Big-Bang was caused by a unbalance in a single elementary cell. With finality to promote a transformation in the equation, in this first moment $-\theta(t_s)$ will be eliminated. The introduction of two oppositive phase terms was resulting of a logical rationality, but it will be demonstrated that even before the Big-Bang, it was only necessary the phase $+\theta(t_s)$ and the phase $-\theta(t_s)$ was totaly unnecessary.

So to promote the changing in a energy differential equation it will eliminate $-\theta(t_s)$, then the transformation will give $f(t_s) = \alpha\frac{\sqrt{2}}{2}\sin\left(\omega_0 t_s - \frac{\pi}{4} + \theta(t_s)\right)$

Now the derivation of the energy equation is starting to produce a factor of energy reduction and dimensional contraction. The energy factor will form the mass and charge of the matter.

After the Big Bang it will postulate that to constitute the materia, the differential equation ${dE(d_0)}/{dt_s}$ will be separated in two components that are totaly tied and produces combined action, but having independent actuation.

Considering the product $P(t_s) = S_s S_r = e^{\alpha_0 f(t_s)}$. The mathematical derivative of the product is

$$P'(t_s) = \alpha_0 f'(t_s)P(t_s) = \alpha_0 f'(t_s)e^{\alpha_0 f(t_s)}$$

$$f'(t_s) = \frac{df(t_s)}{dt_s} = \frac{\sqrt{2}}{2}\cos(\omega_1 t_s - \frac{\pi}{4})$$

$$P'(t_s) = \alpha_0\frac{\sqrt{2}}{2}\cos(\omega_1 t_s - \frac{\pi}{4})e^{\alpha_0 f(t_s)}$$

Now it must attempt to the true physical derivative form of $P(t_s)$.

The time derivative actuation is produced over all functions in all equations. and must separate space and anti space.

So it is postulated that the derivative form physically must produce:

$$P_o' = \alpha_0 f_s'(t_s)e^{\alpha_0 f_s'(t_s)} + \alpha_0 f_r'(t_s)e^{\alpha_0 f_r'(t_s)}$$

Then it must have

$$E'(D_0) = \alpha c^2 P'_o = \alpha c^2 \left[\alpha_0 \frac{\alpha}{2} \right] \cos(\omega_1 t_s) e^{\alpha_0 f\_s(t_s)} + \alpha c^2 \left[\alpha_0 \frac{\alpha}{2} \right] \sin(\omega_1 t_s) e^{\alpha_0 f'_r(t_s)}$$

The term $\alpha_0 \alpha/2$ will be proved in the Theory that it is an incremental contribution to the mass and charge of the elementary particles.

Then to the space and anti-space components it will have respectively

$$E'_s(d_0) = \alpha c^2 \left[\alpha_0 \alpha/2 \right] \cos(\omega_1 t_s) e^{\alpha_0 f'_s(t_s)}$$

$$E'_r(d_0) = \alpha c^2 \left[\alpha_0 \alpha/2 \right] \sin(\omega_1 t_s) e^{\alpha_0 f'_r(t_s)}$$

$$f'_s(t_s) = \alpha/2 \cos(\omega_1 t_s)$$

$$f'_r(t_s) = \alpha/2\ sin(\omega_1 t_s)$$

Now it will try to reconstruct the primitive equation from its separation doing firstly

$$g(t_s) = E'_s(t_s) e^{\alpha_0 f'_r(t_s)} + E'_r(t_s) e^{\alpha_0 f'_s(t_s)} = \alpha c^2 \left[\alpha_0 \alpha/2 \right] \sqrt{2} cos(\omega_1 t_s - \frac{\pi}{4}) e^{\alpha_0 f'(t_s)}$$

So in the equation above has the additive term from a sinuous and cosine given by

$$h(t_s) = \sqrt{2} \cos(\omega_1 t_s - \frac{\pi}{4})$$

This term is not reconstructed, and its represents an other separation term between $E'_s(D)$ and $E'_r(D)$ and it can be seen that

$$h(t_s) = \sqrt{2} \frac{e^{j(\omega_1 t_s - \frac{\pi}{4})}}{2} + \sqrt{2} \frac{e^{-j(\omega_1 t_s - \frac{\pi}{4})}}{2}$$

To restore the original form of this term its substitution will be postulated by

$$q(t_s) = \sqrt{2} \frac{e^{j(\omega_1 t_s - \frac{\pi}{4})}}{\sqrt{2}} \frac{e^{-j(\omega_1 t_s - \frac{\pi}{4})}}{\sqrt{2}}$$

In the reconstruction the term $\left[\alpha_0 \alpha/2 \right]$ is obtained because it is resultant from the process of the separation by derivation.

Because it has introduced a complex time it must consider the time into $f_s(t_s)$ and $f_r(t_s)$ initially as a complex term, than it will have the reconstructed form

$$E'(D_0) = \alpha c^2 \sqrt{2} \frac{e^{j(\omega_1 t_s - \frac{\pi}{4})}}{\sqrt{2}} \frac{e^{-j(\omega_1 t_s - \frac{\pi}{4})}}{\sqrt{2}} e^{\alpha_0 f'(t_s)}$$

$$E'(D_0) = \alpha c^2 \sqrt{2} \frac{e^{j\omega_1 t_s}}{\sqrt{2}} \frac{e^{-j\omega_1 t_s}}{\sqrt{2}} e^{\alpha_0 \frac{\sqrt{2}}{2} \cos(j\omega_1 t_s - j\omega_1 t_s - \frac{\pi}{4})}$$

The complex coordinates that appears will be explained when the Relativity Theory is introduced [5]. The complex coordinate that is found is the temporal coordinate which is necessary to be complex to obtain a real phase of the time into the space.

8. Real Condition of the Invariability before the Big-Bang

Returning to consider again the derivative of the primitive energy equation.

It was made a preliminary analysis considering its mathematical derivative as been $zero$, however its derivative can not produce nullity of the continuous space time equation, but in sense that keep it unmodified.

So the continue derivative of the primitive continuous space time equation before the big bang has to produce itself, keeping its continuity.

So the real condition in a mathematical derivative definition before the Big-Bang must be $\frac{dE(D_0)}{d(jt_s)} = E(D_0)$, and do not zero.

To satisfy this condition it must eliminate one of two complex terms keeping only one of them. Studying the Relativity Theory the term $+j\omega_1 t_s$ must be conserved. So the following term will be eliminated. $\frac{e^{-j\omega_1 t_s}}{\sqrt{2}}$ and consequently $-j\omega_1 t_s$ in the $f'(t_s)$.

So in the primitive equation there is only a temporal complex coordinate $+j\omega_1 t_s$. Than it will have the primitive equation without derivative $E(D_0) = \alpha c^2 e^{j\omega_1 t_s} e^{\alpha_0 f(t_s)}$ where in this new condition $f(t_s) = j\frac{\alpha}{2} sin(\omega_1 t_s) - j\frac{\alpha}{2}\cos(\omega_1 t_s)$ and its mathematical derivative equation with respect to complex temporal coordinate jt_s, before the Big Bang, will be

$$\frac{dE(D_0)}{d(jt_s)} = \omega_1 \alpha c^2 e^{j\omega_1 t_s} e^{\alpha_0 f'(t_s)} + \alpha c^2 f'(t_s) e^{j\omega_1 t_s} e^{\alpha_0 f'(t_s)}$$

so that the derivative with respect to complex time coordinate jt_s produce the properly derivative energy of the continuous space time, so $\omega_1 = 1$ radian and $f'(t_s) = 0$ taking our attention to the term $f(t_s)$.

In this moment there is a question to solve: the continue derivation of $f_s(t_s)$ and $f_r(t_s)$ with respect to jt_s, must produce the nullity of the term $f'(t_s) = f_s'(t_s) + f_r'(t_s)$. How can it be possible?).

9. Characterization of the Dimensional Variation

Now the above question will be solved! When the complex component is introduced and keeping the sinuous and cosine components, the complex component $e^{j\omega_1 t_s}$, is a description of the time. So the sinuous and cosine components in $f(t_s)$, can not be the time representation, but the spatial dimensional actuation of the time on the dimension. So inside of the sinuous and cosine terms the time is not the subjective time but the rotating amplitude constant vector $\vec{t}_0$ which amplitude is the constant equal to 1.

Otherwise, the time rotating vector actuation on the dimensional components in sinuous and cosine depends respectively of t_{ss} and t_{sr} and do not directly of t_s.

It can be seen that

$$t_{ss} = jt_0 \sin(\omega_1 t_s) = j\sin(\omega_1 t_s)$$
$$t_{sr} = jt_0 \cos(\omega_1 t_s) = j\cos(\omega_1 t_s)$$

In the terms $f_s(t_s)$ and $f_r(t_s)$ it will be done respectively the transformations

$$j\omega_1 t_s = \omega_1 t_s \frac{t_{ss}}{\sin(\omega_1 t_s)} = \omega_1 t_s \frac{t_{sr}}{\cos(\omega_1 t_s)}$$

so it will have

$$f_s(t_{ss}) = \frac{\alpha}{2}\sin\left(\omega_1 t_s \frac{t_{ss}}{\sin(\omega_1 t_s)}\right)$$

$$f_r(t_{sr}) = -\frac{\alpha}{2}\cos\left(\omega_1 t_s \frac{t_{sr}}{\cos(\omega_1 t_s)}\right)$$

that results

$$f_s(t_{ss}) = \alpha/2 \; t_{ss}$$
$$f_r(t_{sr}) = -\alpha/2 \; t_{sr}$$

As seen above, $f_s(t_{ss})$ and $f_r(t_{sr})$ do not depends on ω_1. Utilizing the Relativity Theory [6] it can be proven that t_{ss} and t_{sr} do not depends of an angular variant frequency ω_1 and in the expression of t_{ss} and t_{sr}, the angular frequency is $\omega_1 = -j$.

So in reality the correct forms are the following.

$$\begin{aligned} t_{ss} &= \sin(t_s) \\ t_{sr} &= \cos(t_s) \end{aligned} \tag{20}$$

$$\begin{aligned} f_s(t_{ss}) &= \alpha/2 \; t_{ss} = \alpha/2 \sin(t_s) \\ f_r(t_{sr}) &= -\alpha/2 \; t_{sr} = -\alpha/2 \cos(t_s) \end{aligned} \tag{21}$$

And the derivative of $f_s(t_{ss})$ and $f_r(t_{sr})$ is not with respect to jt_s but is with respect to components t_{ss} and t_{sr} respectively.

In this condition the sum of the continue derivation of $f_s(t_{ss})$ and $f_r(t_{sr})$ will be producing

$$f'(t_{ss}, t_{sr}) = \frac{df_s(t_{ss})}{dt_{ss}} + \frac{df_r(t_{sr})}{dt_{sr}} = 0 \tag{22}$$

10. The General Form of the Primitive Equation in Static and Dynamics Forms

So after the definition of time functions $f_s(t_{ss})$ and $f_r(t_{sr})$, now is presented the equations of the continuous space-time before the Big-Bang [7] in the static form:.

$$E(D) = \alpha e^{2\alpha_0(D+\alpha)} e^{j\omega_1 t_s} e^{\alpha_0 f(t_{ss}, t_{sr})} \tag{23}$$

$\alpha_0 = 1/2 \ln 2$ space constant

$\alpha = 1/\pi$ dimensional length

$\omega_1 = 1$ angular frequency of the time before the Big-Bang

$f(t_{ss}, t_{sr}) = f_s(t_{ss}) + f_r(t_{sr})$ static time function

$t_{ss} = \sin(t_s)$ active time in the space component

$t_{sr} = \cos(t_s)$ active time in the anti-space component

t_s = subjective time

$f_s(t_{ss}) = \alpha/2 \; t_{ss}$ time function of the space component

$f_r(t_{sr}) = -\alpha/2 \; t_{sr}$ time function of the antispace component

$\alpha_0 = \frac{1}{2} \ln 2$ space constant

$v_f = e^{\alpha_0 (D+\alpha)}$ velocity phase of the time

$D = \infty$ to the empty space and the time

$D = D_0 = 56$ to the matter and energy

$S_s = E^{\alpha_0 f_s (t_{ss})}$ space component of the elementary static energy

$S_r = E^{\alpha_0 f_r (t_{sr})}$ anti-spacecomponent of the elementary static energy

Now is presented the equations of the continuous space-time before the Big-Bang [5] in the dynamic form:

$$E'(D) = \omega_1 \alpha e^{2\alpha_0 (D+\alpha)} e^{j\omega_1 t_s} e^{\alpha_0 f'(t_{ss}, t_{sr})} \tag{24}$$

$f'(t_{ss}, t_{sr}) = f_s'(t_{ss}) + f_r'(t_{sr})$ dynamic time function

$f_{s'}(t_{ss}) = \alpha/2$ time function of the space component in continuous derivative

$f_{r'}(t_{sr}) = -\alpha/2$ time function of the anti-space component in continuous derivative

$S_{s'} = E^{\alpha_0 f_s'(t_{ss})}$ space component of the elementary dynamic energy

$S_{r'} = E^{\alpha_0 f_r'(t_{sr})}$ anti-space component of the elementary dynamic energy

Then the primitive equation (24) of the continuous space time in its complete form in continuous derivation defined as Continuity equation of the continuous space time before the Big-Bang. After the Big-Bang everything starts to move and to the movement it has

$$\omega_1 = \frac{1}{\sqrt{1 - \frac{v^2}{c^2}}} \tag{25}$$

that is the angular frequency of the time in a matter moving with speed v.

So the elementary cells of potential energy of the primitive universe into the empty space totaly stationary, without any movement or other form of energy are express by the equations (23) and (24)in static equilibrium. These equations are general and can be worked in many ways, obtaining surprising results. when the angular frequency ω_1 is introduced, the number $D_0 = 56$ to the mater and energy, and $D = \infty$ to the time and empty space, the separation between space time and anti space time forces, the real time t instead of the subjective time t_s, and the basis of space and time..

11. Some Theoretical Results

The Equation (23) can be conveniently separated in three terms like.

$$E(D) = e^{2\alpha_0 (D+\alpha)} T S_s S_r \tag{26}$$

where is defined

$T = e^{j\omega_1 t_s}$ time equation

$S_s = e^{\alpha_0 f_s (t_{ss})}$ space equation

$S_r = e^{\alpha_0 f_r (t_{sr})}$ anti space equation

From the earlier equations, the space-time and the anti space-time equations are

$$C_s = e^{2\alpha_0(D+\alpha)}TS_s \quad \text{space-time equation} \tag{27}$$

$$C_r = e^{2\alpha_0(D+\alpha)}TS_r \quad \text{anti space-time equation} \tag{28}$$

After the Big-Bang the primitive equation of the continuous space time of the elementary cell of potential energy, was separated in two parts producing the equations of space time and anti space time spectrally tied and non separable into the matter, but in the energy there is only the space time component that is responsible by the wave propagation. The empty space has only the anti space component and the time has only the space component.

The empty space is characterized by its presence in distinct points. The time is characterized by the instantaneous actuation in all empty space. So only element of empty space is separated of another element of empty space, but the time is integralized on all empty space. The empty space, because it has only anti space component, it is physically totally static. Because the time has only space component and infinite space dimension it has an infinite phase velocity of propagation. So the time is a stationary wave and its instantaneous energy is simultaneous over all space.

The process of dimensional reduction where are engaged derivative and/or integral of the functions TS_s or TS_r will produce matter and energy, as resulting of this process. Now is presented some results of this theory:

$N = 7$ Number of atomics Shells

$$RT_1 = \frac{1}{2\pi\sqrt{2}} - \frac{1}{4\pi\sqrt{2}} \quad \text{Partial reduction of the time function}$$

$$RT_2 = \frac{1}{2\pi\sqrt{2}} \quad \text{Complete dimensional reduction of the time function}$$

$$M(0) = \frac{1}{2}\left(e^{\alpha_0\frac{\alpha}{2}} + e^{-\alpha_0\frac{\alpha}{2}}\right) \quad \text{Dimension alfactor of correction}$$

$$Z(0) = \frac{1}{2}\left(M(0) + \frac{1}{M(0)}\right) \quad \text{Dimensional factor correction}$$

where the factor $M(0)$ is associated to the space and the $\frac{1}{M(0)}$ factor is associated to the anti-space. The factor $Z(0)$ is associated to the conjugated actuation between space and anti space.

Light speed into the empty space

$$c = e^{\alpha_0(D_0+\frac{1}{\pi})} = 2{,}9974385630 \times 10^8 \text{ m/s} \tag{29}$$

$$\text{PD} = -0{,}016$$

where PD is the percentage difference.

Static proton mass

$$m'_p = \frac{\alpha_0}{\pi}(-\alpha_0^{D_0})\alpha_0^{RT_1}\frac{1}{M(0)}\text{kg} = 1{,}671795279 \times 10^{-27}\text{ kg} \tag{30}$$

$$\text{PD} = \text{-}0{,}049$$

Static neutron mass

$$m'_n = \frac{\alpha_0}{\pi}(-\alpha_0^{D_0})\alpha_0^{RT_1}Z(0)\text{kg} = 1{,}674341082 \times 10^{-27}\text{kg} \tag{31}$$

$$\text{PD} = -0{,}034$$

Boltzmann constant[8]

$$k = \frac{1}{\alpha_0 \frac{\alpha_0}{\pi} \alpha_0^{2RT_1 - RT_2}} e^{-D_0} = 1,380299907 \times 10^{-23}\,\text{J/K} \tag{32}$$

$$\text{PD} = -0,0262$$

Bohr Radius [8]

$$r_0 = \frac{1}{(m(8)-1)c} = \frac{1}{63c} = 5,295526678 \times 10^{-11}\,\text{m} \tag{33}$$

$$\text{PD} = 0,070$$

where

$$m(8) = \sum_{l=0}^{7} (2l+1) = 64$$

$$m(8) - 1 = D_0 + N$$

12. Conclusions

In this article, the equation called the primitive equation of the continuous space-time was formally deduced. This paper presents a completely new work that is a small part of a work quite extensive, resulting from years of intense research of the author. It was presented as an introduction, so it can be appreciated, analyzed and criticized. The conclusion will depend on future works all interconnected, and these conclusions may only be obtained after full publication of several other items that compound the theory.

13. References

[1] Gerald Mott. 2006 A REVIEW OF EINSTEIN'S THEORY OF SPECIAL RELATIVITY International Journal of Modern Physics 15.

[2] Hummel R. E. The Wave-Particle 2001 Duality Electronic Properties of Materials 6-13.

[3] Thompson A. and Taylor B. 2008 Guide for the use of the International System of Units NIST *Special Publication* **811**.

[4] Hegerfeldt, G.C 1972 The Lorentz transformations: Derivation of linearity and scale factor *Società Italiana di Fisica* 10 257-267

[5] U. E. Schröder. 1990 Physical and conceptual foundations of the theory of special relativity Relativity 2 11-58

[6] Moshe Carmeli. Special Relativity Theory Relativity 1 1-49

[7] Virginia Trimble 1996 Backgrounds and The Big Bang: Some Extracts From Their History, International Astronomical Union 168 9-16

[8] Weisstein E. Physical Constants at http://scienceworld.wolfram.com/physics/BohrRadius.html 2012.

Journal of Physics: Conference Series **423** (2013) 012024 doi:10.1088/1742-6596/423/1/012024

Parametric Control on Fractional-Order Response for Lü Chaotic System

K Moaddy[1], A. G. Radwan[2,3], K. N. Salama[4], S. Momani[5] and I. Hashim[1]

[1]School of Mathematical Science, Univirsiti Kebangsaan Malaysia, 43600, Malaysia
[2]NISC Research Center, Nile University, Cairo, Egypt
[3]Engineering Mathematics Dept., Faculty of Engineering, Cairo University, Egypt
[4]Electrical Engineering Department, (KAUST), Thuwal, Kingdome of Saudi Arabia
[5]Department of Mathematics, University of Jordan, 11942 Amman, Jordan

Corresponding author'se-mail:- moaddy@yahoo.com

Abstract. This paper discusses the influence of the fractional order parameter on conventional chaotic systems. These fractional-order parameters increase the system degree of freedom allowing it to enter new domains and thus it can be used as a control for such dynamical systems. This paper investigates the behaviour of the equally-fractional-order Lü chaotic system when changing the fractional-order parameter and determines the fractional-order ranges for chaotic behaviour. Five different parameter values and six fractional-order cases are discussed through this paper. Unlike the conventional parameters, as the fractional-order increases the system response begins with stability, passing by chaotic behaviour then reaches periodic response. As the system parameter α increases, a shift in the fractional order is required to maintain chaotic response.Therefore, the range of chaotic response can be expanded or minimized by controlling the fractional-order parameter. The non-standard finite difference method is used to solve the fractional-order Lü chaotic system numerically to validate these responses.

1. Introduction

Recently, many natural systems which dynamics are complex can be modelled as chaotic dynamical systems which are extremely sensitive to the initial conditions as weather. The fundamentals of the chaotic system and some of its characteristics in integer and fractional-order domain were discussed in [1]. In addition, the impact of the chaotic systems become seriously once many chaotic responses have been implemented practically and especially on the transistor-level [2,3] which is suitable in the VLSI technologies. Many chaotic systems have been introduced last few decades as Lorenz, Rössler, and Lu chaotic systems with many applications such as in control, communication, and encryption.

Integer calculus is considered a very narrow subset from fractional-calculus, which made it a must for dynamical systems to be generally re-discussed based on fractional-order differentiation due to the extra degrees of freedom, flexibility, and better characterization it offers During the last few decades, fractional calculus has become a powerful tool to describe the dynamics of complex systems which appear frequently in several branches of science and engineering. This generalization showed many new fundamentals which exist in the fractional-order domain and disappear in the integer-order case. One of the main goals for many applications is to discuss the effect of the fractional-order parameter on the system response where the results can be used in better description and control of these systems in many applications. The characterization of real dynamical systems has proved to be superior to the fundamentals of traditional calculus; this opened the door for the use offractional calculus in numerous applications in the field of viscoelasticity, robotics, feedback amplifiers, electrical circuits, control theory, electro analytical chemistry, fractional multi-poles, chemistry and biological sciences [4-14]. For example the stability analysis of fractional-order systems has been discussed in [6], the fractional order PID controller became ubiquitous in industry as shown in [9 - 11], The behaviour of the fractional-order chaotic systems has been discussed in [13], and the modelling of some complex dynamics in biological tissues using fractional calculus in [14, 15].

One of the main differences between the ordinary differential equation systems (integer order) and the corresponding fractional-order-differential equation systems is that, the definition of the fractional order system depends on the whole history of the responses instead of the finite memory in the integer case. Chaotic systems have a profound effect on their numerical solutions and are highly sensitive to time step sizes. Thus it is beneficial to find a reliable analytical tool to test its long-term accuracy and efficiency. Also hyper-chaotic systems have more complex dynamical behaviours because they are defined as chaotic systems with two positive Lyapunov exponents. In addition to the previous, since its discovery, chaotic synchronization has been applied in many different fields, including biological and physical systems, structural engineering, and ecological models [16].

A great deal of effort has been directed recently in an attempt to find robust and stable numerical and analytical methods for solving fractional equations. Such as the fractional difference method [17], the Adomian decomposition method [18, 19], the variational iteration method [20] and the Adams-Bashforth-Moulton method [21-22].

In this paper, the non-standard finite difference method (NSFD) [23-25] is implemented to give numerical solutions for various types of differential equations of integer and fractional orders [26-29]. NSFD has developed as an alternative method for solving a wide range of problems whose mathematical models involve algebraic, differential, biological, and chaotic systems. The technique has many advantages over the classical techniques, and provides an efficient numerical solution. The main aim of this paper is to study the proper fractional-order range in which the Lu system exhibits chaotic behaviour. Several cases are investigated for different orders and changing only a single system parameter. Stable, periodic and chaotic responses are shown for each system parameter but with different fractional order ranges.

2. Preliminaries and Notations

In this section we give some basic definitions and properties of the fractional calculus theory and non-standard discretization which will be used further in this paper.

2.1. *Grünwald-Letnikov approximation.*

We will beginwith the single fractional differential equation, [26]

$$D^{\alpha}x(t) = f\big(t, x(t)\big), T \geq t \geq 0,\ \ x(t_0) = x_0, \tag{1}$$

where $\alpha > 0$ and D^{α} denotes the fractional derivative, defined by

$$D^{\alpha}x(t) = J^{n-\alpha}D^{n}x(t), \tag{2}$$

where $n - 1 < \alpha \leq n, n \in N$ and J^{n} in the nth-order Riemann – Liouville integral operator defined as

$$J^{n}x(t) = \frac{1}{\Gamma(n)}\int_{0}^{t}(t-\tau)^{n-1}\,x(\tau)d\tau. \tag{3}$$

To apply Mickens scheme, we have chosen the Grünwald-Letnikov method of approximation for the one-dimensional fractional derivative as follows:

$$\lim_{h\to 0} h^{-\alpha} \sum_{j=0}^{N} (-1)^j \binom{\alpha}{j} x(t-jh), \tag{4}$$

where $N = t/h$ and t denotes the integer part of t andhis the step size. Therefore, equation (4) isdiscretized as follow

$$\sum_{j=0}^{n+1} c_j^{\alpha}\, x(t-jh) = f\big(t_n, x(t_n)\big), n = 1, 2, 3, \cdots, \tag{5}$$

where $tn = nh$ and c_j^{α} are the Grunwald-Letnikov coefficients defined as

$$c_j^{\alpha} = \left(1 - \frac{1+\alpha}{j}\right) c_{j-1}^{\alpha}, \text{and } c_0^{\alpha} = h^{-\alpha}, j = 1, 2, 3, \cdots. \tag{6}$$

2.2. *Non-standard discretization*

We seek to obtain the NSFD solution for a system of fractional differential equations of the form

$$D^{\alpha_1}(x) = f_1(x, y, z), \tag{7}$$

$$D^{\alpha_2}(y) = f_2(x, y, z), \tag{8}$$

$$D^{\alpha_3}(z) = f_3(x, y, z), \tag{9}$$

where $0 < \alpha \leq 1$, and initial conditions x_0, y_0, z_0.

Applying the NSFD scheme by replacing the step size h by a function $\varphi(h)$ and using the Grünwald-Letnikov discretization method, yields the following relations

$$x_{n+1} = \frac{\left(-\sum_{j=1}^{n+1} c_j^{\alpha_1} x_{n+1-j} + f_1(x_{n+1}, y_n, z_n)\right)}{c_0^{\alpha_1}}, \tag{10}$$

$$y_{n+1} = \frac{\left(-\sum_{j=1}^{n+1} c_j^{\alpha_2} y_{n+1-j} + f_2(x_{n+1}, y_{n+1}, z_n)\right)}{c_0^{\alpha_2}}, \tag{11}$$

$$z_{n+1} = \frac{\left(-\sum_{j=1}^{n+1} c_j^{\alpha_3} z_{n+1-j} + f_3(x_{n+1}, y_{n+1}, z_{n+1})\right)}{c_0^{\alpha_3}}, \tag{12}$$

where $c_0^{\alpha_i} = \varphi_i(h)^{-1}$are functions of the step size $h = \Delta t$, with the following properties:

$$\varphi_i(h) = h + O(h^2), \ \ where\, h \to 0. \tag{13}$$

Examples of functions $\varphi_i(h)$that satisfy (13) are:$h, sin\, h, sinh\, h, e^h - 1$, and in most applications, the general choice of $\varphi_i(h)$ is

$$\frac{1-e^{-R_i h}}{R}, \tag{14}$$

the function R_i can be chosen as the following form

$$R_i = max\left(\left|\frac{\partial f_i}{\partial x_i}\right|_{x=x_i}\right). \tag{15}$$

The nonlinear terms in system (10)-(12) can be in general being replaced by nonlocal discrete representations. For example,

$$y^2 \approx y_k y_{k+1}, \tag{16}$$

$$xy \approx 2xy - xy \rightarrow 2x_{n+1}y_n - x_{n+1}y_{n+1}, \tag{17}$$

where

$$h = \frac{T}{N}, t_n = nh, n = 0, 1, \ldots, N\epsilon Z^+. \tag{18}$$

3. NSFD Simulation Results

The Fractional order Lü system is the lowest-order chaotic system among all the chaotic systems [30]. The minimum effective dimension reported is 0.30. The system is given by

$$D^\alpha x = a(y - x), \tag{19}$$

$$D^\alpha y = by - xy\,, \tag{20}$$

$$D^\alpha z = xy - cz, \tag{21}$$

where $a, b,$ and c are the system parameters, (x, y, z) are the state variables, and α is the fractional order. Now, we apply the NSFD to obtain the numerical solution for the fractional-order Lü system. Using the Grünwald-Letnikov discretization method and applying NSFD scheme by replacing the step size h by a function $\varphi_i(h)$ and using the form in (16) and (17) for the nonlinear term xy the system (19) –(21) yields

$$\sum_{j=0}^{n+1} c_j^\alpha\, x(t - jh) = a\big(y(t_n) - x(t_n)\big), \tag{22}$$

$$\sum_{j=0}^{n+1} c_j^\alpha\, y(t - jh) = by(t_n)\,, -(2x(t_{n+1})y(t_n) - x(t_{n+1})y(t_{n+1})\,, \tag{23}$$

$$\sum_{j=0}^{n+1} c_j^\alpha\, z(t - jh) = 2x(t_{n+1})y(t_n), \tag{24}$$

where $x(t_0) = x_0,\ y(t_0) = y_0,$ and $z(t_0) = z_0$. Doing some algebraic manipulation to systems (22)-(24), yields the following relations

$$x(t_{n+1}) = \varphi^h \left(-\sum_{j=1}^{n+1} c_j^\alpha\, x(t - jh) a\big(y(t_n) - x(t_n)\big) \right), \tag{25}$$

$$y(t_{n+1}) = \frac{-\sum_{j=1}^{n+1} c_j^\alpha\, x(t - jh) + \big(b - 2x(t_{n+1})\big)y(t_n)}{c_0^\alpha - x(t_{n+1})}, \tag{26}$$

$$z(t_{n+1}) = \varphi^h \left(-\sum_{j=1}^{n+1} c_j^\alpha\, z(t - jh) + 2x(t_{n+1})y(t_n) - x(t_{n+1})y(t_{n+1}) - cz(t_n) \right), \tag{27}$$

where $c_0^\alpha = \varphi^{-h}$ and we chose $\varphi(h) = \sin h$ as a suitable function [26]. Conventionally when α = 1, the system has two equilibrium points at (0,0,0) and $(b, b, b^2/c)$ which depend on the parameters b and c only. The system exhibits chaotic behavior when the parameters set $(a, b, c) = (36.0, 28.0, 3.0)$. In the following simulations the effect of the parameter a; which doesn't affect the equilibrium points;in addition to different values of the fractional-order parameter α will be studied for more than thirty cases to determine the region where chaotic responses appear.

All the following simulations are performed using NSFD method using (25) up to (28) with $b = 28, c = 3$ and by varying the system parameter a and also the fractional-order parameter α.

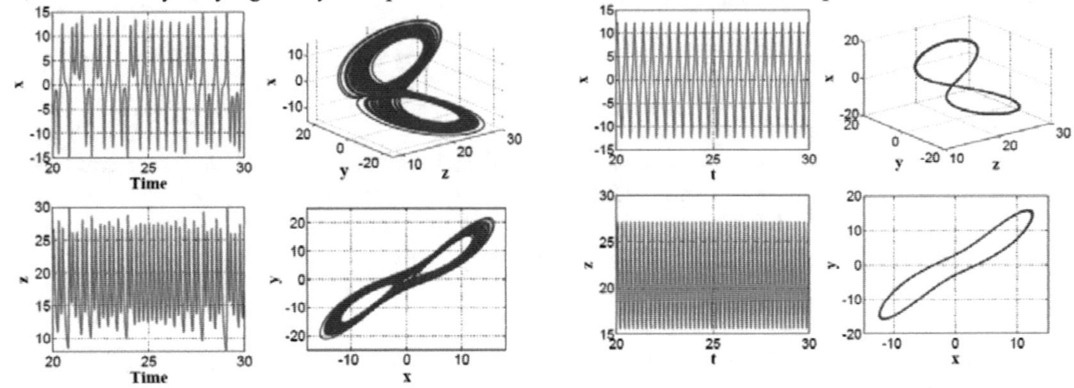

(a) when α = 0.75 (b) when α = 0.8

Figure 1. System response when $a = 19.5$

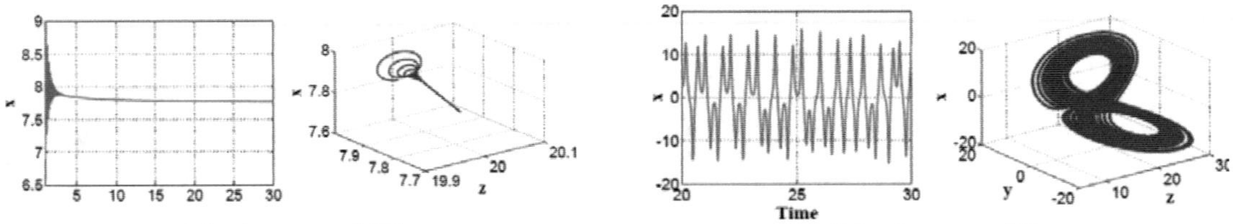

(a) when α = 0.75 (b) when α = 0.8

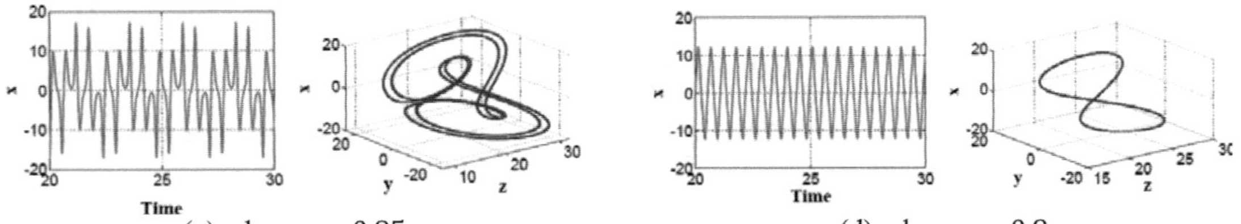

(c) when α = 0.85 (d) when α = 0.9

Figure 2. System response when $a = 22$

Figure 1 shows the system responses when $a = 19.5$ for two different fractional-orders. When α less than 0.75 the system displays stable response. However, as α increases to 0.75, the system behaves chaotically as shown in Fig.1(a). But for α = 0.8 and higher orders, the system response is periodic with period 1. Therefore, the system response will be chaotic around α = 0.75 with a small range.

As the system parameter a increases to 22 and when α < 1, the system response and x time waveform pass by stable, chaotic, high-period, and single-period responses when the fractional-order α equals to 0.75, 0.8, 0.85, and 0.9 cases as shown in Fig.2. Therefore, the range of the fractional-order α for chaotic response increases as the parameter a increases. In addition, when α = 0.85, the system is periodic with period 5 as shown in Fig. 2(c).

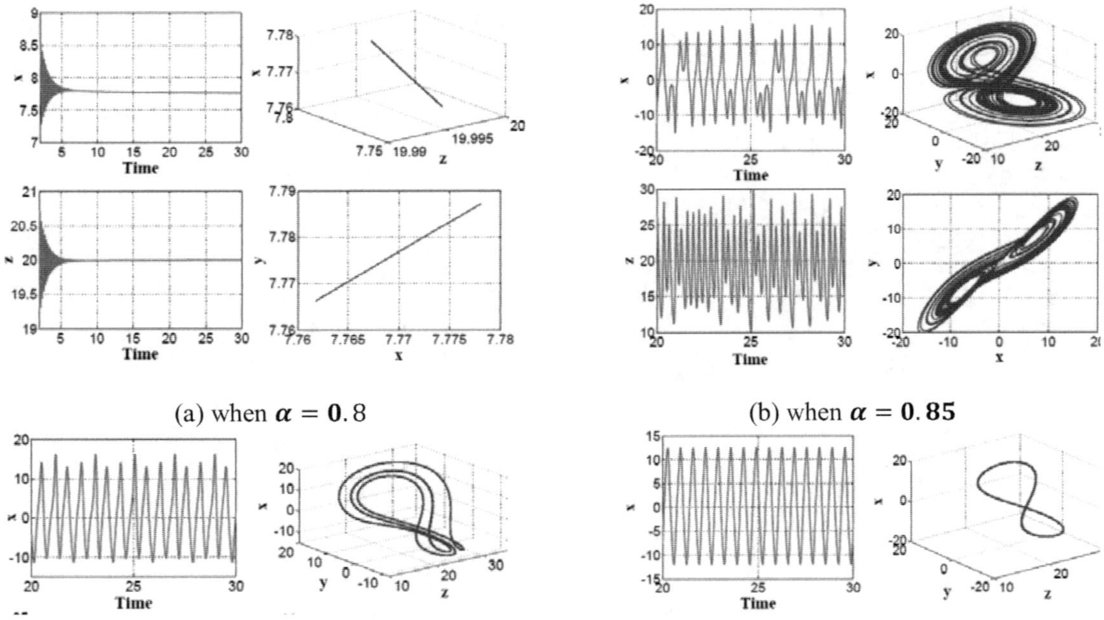

(a) when $\boldsymbol{\alpha} = \mathbf{0.8}$ (b) when $\boldsymbol{\alpha} = \mathbf{0.85}$

(c) when $\boldsymbol{\alpha} = \mathbf{0.95}$ (d) when $\boldsymbol{\alpha} = \mathbf{1.0}$

Figure 3. System response when $a = 25$

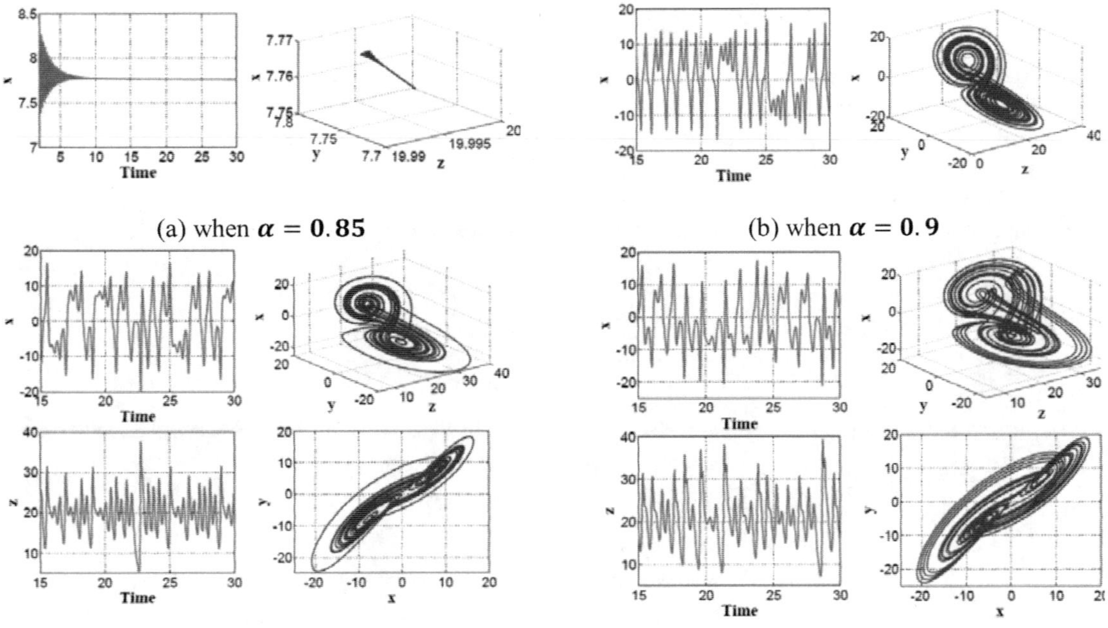

(a) when $\boldsymbol{\alpha} = \mathbf{0.85}$ (b) when $\boldsymbol{\alpha} = \mathbf{0.9}$

(c) when $\boldsymbol{\alpha} = \mathbf{0.95}$ (d) when $\boldsymbol{\alpha} = \mathbf{1.0}$

Figure 4. System response when $a = 30$

The responses when the system parameter aequals to 25 with different values of the fractional-order α are shown in Fig.3 where the range of chaotic responses increases and different periodic attractors

are obtained. Moreover, the range of the stable responses is shifted up and appears when $\alpha = 0.8$ with the same equilibrium value as the previous case when $a = 19.5$. The system shows chaotic responses and high periods when the fractional-order α belongs to the interval from 0.85 up to 0.95.
Similarly, as the system parameter a increases the chaotic response is shifted up as shown in Fig. 4 and figure 5 with the same equilibrium point. While when $a = 30$, the system becomes stable where $\alpha$is less than or equal to 0.85 and the chaotic response starts to appear in the range $[0.9, 1.0]$ as shown in Fig. 4. In this case the periodic response is expected to exist when the fractional-order $\alpha > 1$. When $a = 36$, the system will be stable up to $\alpha = 0.9$ and the chaotic response appears when $\alpha = 1.0$ which is the conventional case.

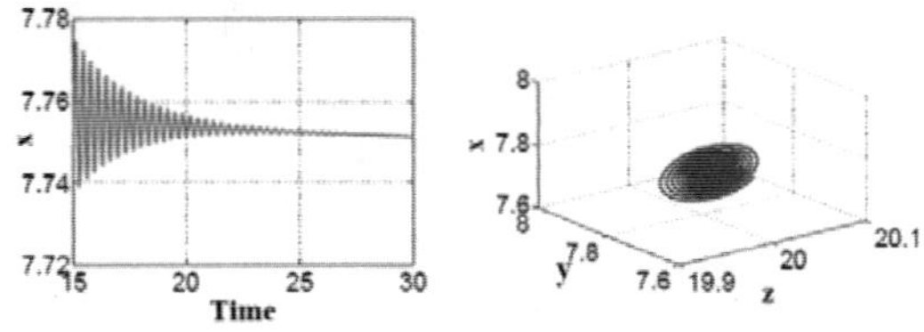

(a) when $\alpha = 0.9$

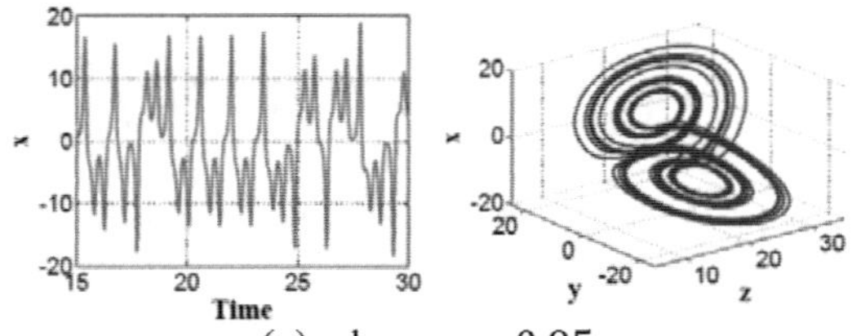

(a) when $\alpha = 0.95$

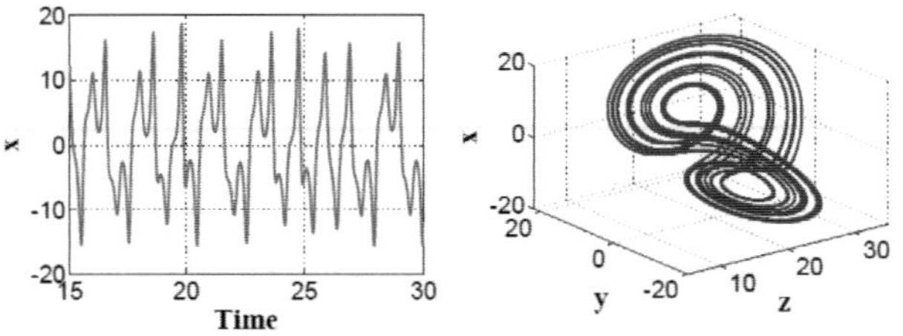

(a) when $\alpha = 1.0$

Figure 5. System response when $a = 36$

From the previous figures, we can conclude the results in Table 1, where the chaotic responses appear for a wide range of the system parameter abut in different ranges of the fractional-order parameterα. Therefore, as a increases the range of α for chaotic response increases and is shifted down. Moreover, it is expected that the Lü system can behave chaotically for larger values of $a > 36$ but with fractional-order $\alpha > 1$. In addition, as the range of α increases, more cases of high-periodic responses will appear.

Table 1. System performance versus the parameters

a	19.5	22	25	30	36
$\alpha < 7.5$	Stable	Stable	Stable	Stable	Stable
$\alpha = 7.5$	Chaotic	Stable	Stable	Stable	Stable
$\alpha = 0.8$	Period1	Chaotic	Stable	Stable	Stable
$\alpha = 0.85$	Period1	Period5	Chaotic	Stable	Stable
$\alpha = 0.9$	Period1	Period1	Chaotic	Chaotic	Stable
$\alpha = 0.95$	Period1	Period1	Period3	Chaotic	Chaotic
$\alpha = 1.0$	Period1	Period1	Period1	Chaotic	Chaotic

Figure 6 shows the response in the case when $\alpha = 0.8$ and with time varying parameter $a = 22 + 3sin(0.025\pi t)$. It is clear that at the beginning when a close to 22, the system response behave chaotically and as a increases close to 25 the system response changed into stable then back to chaotic in the middle cycle where a back close to 22. Moreover, when the system parameter a reaches its minimum value 19, the system become periodic as shown in the last interval of Fig.6. Therefore, this figure shows the continuous behaviour of the system when a changed smoothly with time.

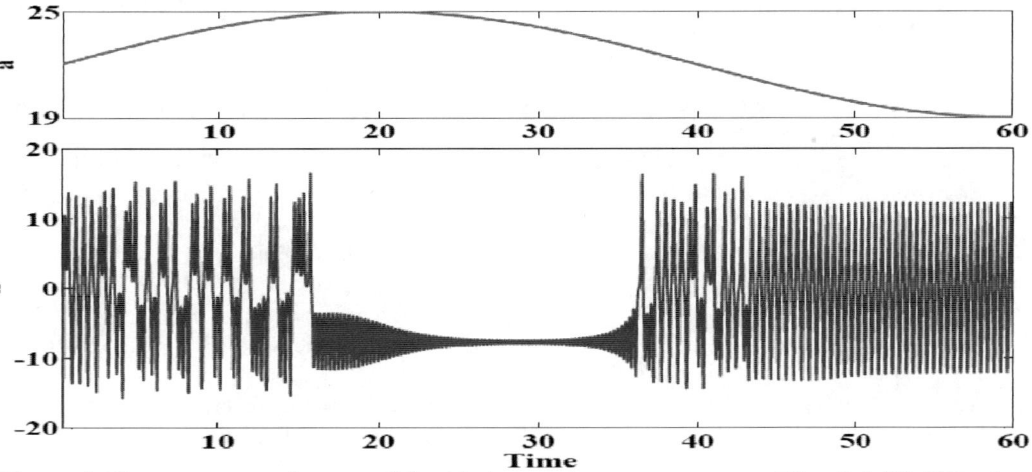

Figure 6. The x response when $\alpha = 0.8$ with time variant parameter $a = 22 + 3sin(0.025\pi t)$.

4. Conclusion

This paper studied the fractional-order Lü system with different system parameters and for equal fractional-orders. As the parameter a increases, the system behaviour changes and a higher fractional-order is needed to get similar response. Thus, the conventional integer system with stable responses can be chaotic forlower fractional-orders under the same circuit parameters. Moreover, for each fixed set of parameters, there is a range of fractional-ordersat which the response is stable, chaotic, and periodic. More than thirty cases have been investigated and simulated using the non-standard finite difference scheme to validate the smooth and gradual change in the system response as the system parameter a or the fractional-order parameter$\alpha$is changed smoothly with time.

References

[1] Zaslavsky M 2002 Chaos, fractional kinetics, and anomalous transport, *Phys. Rep.* **371** 461–580

[2] Radwan A G, Soliman A M and EL-sedeek A L 2003 An inductorless CMOS realization of Chua's circuit *Chaos, Sol. & Fract.* **18** 149–158

[3] Radwan A G, Soliman A M and EL-sedeek A L 2004 MOS realization of the modified Lorenz chaotic system *Chaos, Sol. & Fract.* **21** 553–561

[4] Baleanu D, Diethelm K, Scalas E and Trujillo J 2009 *Fractional Calculus Model and Numerical Methods* (World Scientific)

[5] Heymans N and Podlubny I 2005 Physical interpretation of initial conditions for fractional differential equations with Rieman-Liouville fractional derivatives *Rheol. Acta* **45** 765–771

[6] Radwan A G ,Moaddy K and Momani S 2011 Stability and nonstandard finite difference method of the generalized Chua's circuit *Comput. Math. Appl.* **62** 961–970

[7] Radwan A G, Shamim A and Salama K N 2011 Theory of Fractional-order Elements Based Impedance Matching Networks *IEEE Micro. Wire. Compo. Lett.* **21** 120–122

[8] Radwan A G, Soliman A M, Elwakil A S 2007 *Proc. Int. Conf on Microelectronics* (ICM) 91–94

[9] Li Y, Chen Y and Ahn H 2011 Fractional order iterative learning control for fractional-order linear systems, *Asian J. of Cont.* **13** 54–63

[10] Chen Y, Petras I and Xue D 2009 Fractional Order Control - A Tutorial, Proc. *American Control Conf.* 1397–1411

[11] Petras I and Bednarova D Control of fractional-order nonlinear systems 2011 *Acta Mech. Auto.* **5** 96–100
[12] Li CP and Deng WH 2007 Remarks on fractional derivatives *Appl. Math. Comput.* **187** 777–784.
[13] Podlubny I 1999 *Fractional differential equations* (New York: Academic Press)
[14] Magin R L 2010 Fractional calculus models of complex dynamics in biological tissues *Comput. Math. Appl.* **59** 1586–1593
[15] Moaddy K, Radwan A G, Salama K N, Momani S and Hashim I 2012 The Fractional-order modeling and synchronization of electrically coupled neurons system *Comput. Math. Appl.* **64** 3329–3339
[16] Kocarev L and Parlitz U 1995 General approach for chaotic synchronization with applications to communication, *Phys. Re. Lett.* **74** 5–28
[17] Oldham K and Spanier J 1974 *The fractional calculus* (New York: Academic Press)
[18] Momani S 2005 Analytic and approximate solutions of the space-and time-fractional telegraph equations *Appl. Math. Comput.* **170** 1126–1134
[19] Momani S 2005 Analytical approximate solution for fractional heat-like and wave-like equations with variable coefficients using the decomposition method, *Appl. Math. Comput.* **165** 459–472
[20] Odibat Z and Momani S 2009 The variational iteration method: An efficient scheme for handling fractional partial differential equations in fluid mechanics *Comput. Math. Appl.* **58** 2199–220.
[21] Diethelm K and Ford N J 2002 Analysis of fractional differential equations, *J. Math. Anal. Appl.* **265** 229–48
[22] Diethelm K, Ford N J and Freed A D 2004 Detailed error analysis for a fractional Adams method *Numr. Algo.* **26** 31–52
[23] Mickens R E Nonstandard Finite 1994 *Difference Models of Differential equations* (Singapore: World Scientific)
[24] Mickens R E 2000 *Applications of nonstandard finite difference schemes* (Singapore: World Scientific)
[25] Mickens R E 2005 Advances *in the applications of Nonstandard Finite Difference Schemes* (Singapore :World Scientific)
[26] Hussian G, Alnaser M and Momani S Nonstandard 2008 Discretization of fractional differential equations *Proce. of 8th seminar of differential equations and dynamical systems*(Isfahan, Iran)
[27] Moaddy K, Hashim I, Alomari A K and Momani S 2011 A new hybrid nonstandard finite difference- Adomian scheme for solution of nonlinear equations *Sains Malaysiana* **40** 515–519
[28] Moaddy K, Hashim I and Momani S 2011 Non-standard finite difference schemes for solving fractional-order Rössler chaotic and hyperchaotic systems *Comput. Math. Appl.* **62** 1068–1074
[29] Moaddy K, Momani S and Hashim I 2011 The nonstandard finite difference scheme for linear fractional PDEs in fluid mechanics *Comput. Math. Appl.* **61** 2209–2216
[30] Lü JG 2006 Chaotic dynamics of the fractional order Lü system and its synchronization. *Phys. Lett. A* **354** 05–11

Effect of internal tubes on the flow structures in gas-solid fluidized beds

J Gubis[1], **H R Norouzi**[2], **N Mostoufi**[2]*, **R Zarghami**[2]

[1]Department of Chemical Engineering, Institute of Chemical Technology, Prague, Czech Republic.
[2] Process Design and Simulation Research Centre, School of Chemical Engineering, College of Engineering, University of Tehran, P.O. Box 11155/4563, Tehran, Iran.

* Corresponding author: mostoufi@ut.ac.ir

Abstract. The influence of internal tubes present in the fluidized bed on its hydrodynamic behavior was examined in this study. One tube was completely immersed in the bed of 650 μm glass beads at 10 cm above the air distributor in the 15 cm inner diameter fluidized bed column of aspect ratio 1. The pressure fluctuations signal was collected over 300 s and was consequently analyzed by the both statistical and wavelet analyses. The results showed that the presence of tube within the bed suppresses the occurrence of macro-scale phenomena like formation of large bubbles and its stability and supports the meso-structures such smaller bubbles, voids and clusters. It leads to an earlier onset of bubbling flow regime as well as turbulent flow regime. The observed minimum bubbling velocities were compared with data in literature and were found in a good agreement with them.

Keywords. fluidized bed, pressure fluctuations, bed internals, hydrodynamic behavior

1. Introduction

The gas–solid flow in a fluidized bed is characterized by very distinct flow structures. To understand these complicated flow structures, accurate and reliable techniques must be developed to determine hydrodynamic characteristics of the fluidized bed. Many investigators have used pressure signals [1, 2] and their fluctuations to determine and distinguish between phenomena occurring within the bed like formation, movement, coalescence and break-up of bubbles, slugging etc. [3-6]. Various analysis methods of the pressure signals have been developed, for both time and frequency domains, such as wavelet analysis, Fourier transform, chaos analysis etc., to extract the necessary information about the hydrodynamic behavior of the fluidized bed, namely minimum fluidization velocity, regime transition velocities, bubble characteristics or to estimate the dominant phenomena in the bed [6-8].

Wavelet transform analysis was utilized for non-linear and transient time series data for a better resolution in the frequency domain [9]. For example, this method can be used to characterize regime transition from minimum fluidization to bubbling and from bubbling to turbulent fluidization [9].

Traditionally, statistical analyses have been used to determine hydrodynamic changes and fluidization quality of the bed [3, 5 11]. Statistical measures that can be used to analyze the signals include standard deviation, skewness, and kurtosis. Puncochar et al. [5] and Felipe and Rocha[1] found that standard deviation of pressure measurement is appropriate to estimate U_{mf} of Geldart A and B particles. Lee and Kim used the skewness and kurtosis of pressure fluctuations to determine the transition from bubbling to turbulent regime [12]. They found that the maximum point in the kurtosis and the minimum point in the skewness correspond to transition velocities. Since the fluidization process is very complex, any change of conditions will lead to change of the hydrodynamic behavior of the bed, e.g. by addition of a fixed obstacle into the bed for the sake of the enhancement of heat transfer area.

The main objective of this study was to recognize the effects of internals attached into the bed on the hydrodynamic characteristics of the fluidized bed. The pressure signals of the bed were recorded and subsequently analyzed using wavelet analysis tools and statistical analysis and the results were compared with literature data.

2. Experiments

The experimental setup is shown in Figure 1 schematically. The experiments were carried out in a gas-solid fluidized bed column made of a Plexiglas pipe of 15 cm inner diameter and 2 m height. The whole system was electrically grounded to minimize electrostatic effects. Air at room temperature entered the column through a perforated plate distributor with 0.75 mm holes on 8 mm triangular pitch. Air flow rate was controlled by a digital mass flow meter. The superficial air velocity range was 0 - 2 m/s. The height of the static bed was set to 15 cm (aspect ratio of 1). Glass bead particles with density of 2525 kg/m3 and 670 μm mean diameter were used in the experiments. A 1 -inch cylindrical tube was inserted through the whole column diameter, located at 10 cm above the distributor, thus it was completely immersed in the bed. Pressure fluctuations signal was collected over 300 seconds at the sampling frequency of 400 Hz. The frequency was considered to be high enough to avoid any information loss since the frequency of observed phenomena within the bed reaches 20 Hz at the maximum [13].

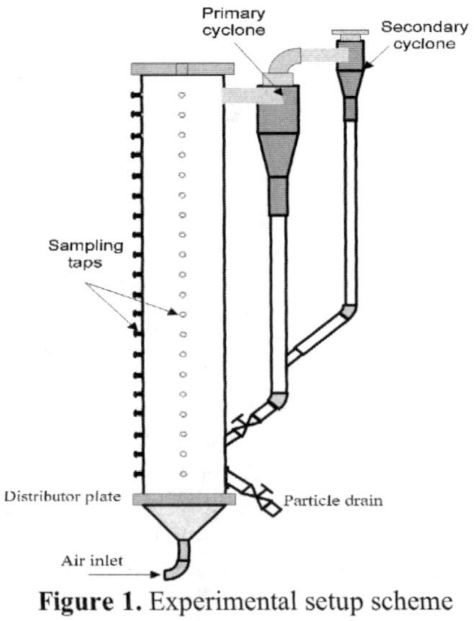

Figure 1. Experimental setup scheme

3. Methods of data analyses

The obtained differential pressure signals were analyzed by using both the wavelet analysis tools and statistical analysis (standard deviation, skewness and kurtosis). Using the wavelet analysis, the decomposition level was set to 8, thus the signals were decomposed into nine sub-signals, where approximation levels a1-a8 represent low frequency components and detailed coefficients D1-D8 represent high frequency components [14]. By recombining the sub-signals, three different scales belonging to different flow structures were extracted from the original signals [8]. The micro-scale, represented by the D1 to D4 sub-signals with frequency more than 10 Hz is related to particle interactions. The meso-scale is represented by the D5 and D6 sub-signals and corresponds to small bubbles, clusters and voids in the bed in frequency range 3-10 Hz. The macro scale represents large bubbles and bubble eruptions on the bed's surface and slugging and consists of D7, D8 and a8 sub-signals. After that the energies of the flow structures were computed by the following equations

$$E_i^a = \sum_{t=0}^{n} |a_i(t)|^2 \quad (1)$$

$$E_i^D = \sum_{t=0}^{n} |D_i(t)|^2 \quad (2)$$

where $q(t)$ and $D_i(t)$ represent the approximation and detail sub-signals. Finally, the total energy of the signal is expressed:

$$E = E_i^a + \sum_{i=0}^{i} E_i^D \quad (3)$$

Statistical analysis of the pressure signal was applied to justify consistency of results obtained from wavelet analysis. Standard deviation, skewness and kurtosis analyses were used. Disadvantages of the statistical analysis lie in the fact that it does not work with de-noised signal and does not allow to extract specific flow structure in the bed being predominate. The standard deviation is defined as:

$$\sigma = \sqrt{\frac{1}{n-1}\sum_{i=1}^{n}(x_i - \bar{x})^2} \quad (4)$$

Skewness, the measure of asymmetry of data distribution function is defined as:

$$S = \frac{\sum_{i=1}^{n}(x_i - \bar{x})^3}{(n-1)\sigma^3} \quad (5)$$

and kurtosis, the measure of the sharpness of the distribution function as:

$$K = \frac{\sum_{i=1}^{n}(x_i - \bar{x})^4}{(n-1)\sigma^4} \quad (6)$$

4. Results and discussion

The energies of different flow structures were computed and compared for the experimental setups containing either none or one tube. The energies of the pressure signals are depicted as a function of air velocity in the Figure 2. Obviously, the presence of an internal supports an earlier onset of bubbling regime and also the onset of turbulent flow regime. The minimum bubbling velocities, U_{mb}, were determined as 0.29 m/s (system with no internals) and as 0.26 m/s (system with one internal tube). It was found to be in a good agreement with the literature [15], which proposes the U_{mb} to be 0.286 m/s. The maximum in the energy demarks the onset of turbulent flow regime [13, 16]. While for the plain bed the maximum occurs at the velocity near to 1.9 m/s, it is shifted to 1.2 m/s by insertion of one tube.

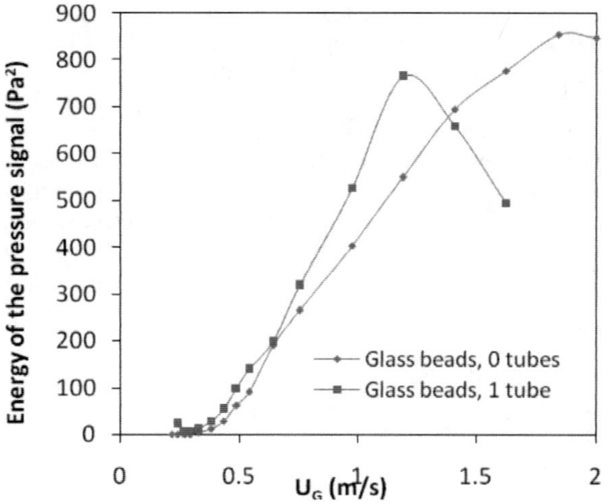

Figure 2. Change in the total energy vs air velocity in presence of a tube

The changes in energy of the flow structures in the bed caused by the insertion of an internal as a function of the air velocity are shown in Figure 3. The energy of the macro structures decreases with the insertion of tube while the energy of meso structures increases. It shows that the presence of the internal suppresses the formation and flow of large bubbles and on the other hand supports formation of smaller bubbles, voids and clusters. It was also observed visually in the bed. The energy of the micro structures remains mostly unchanged and can be considered as independent on the presence of the internals.

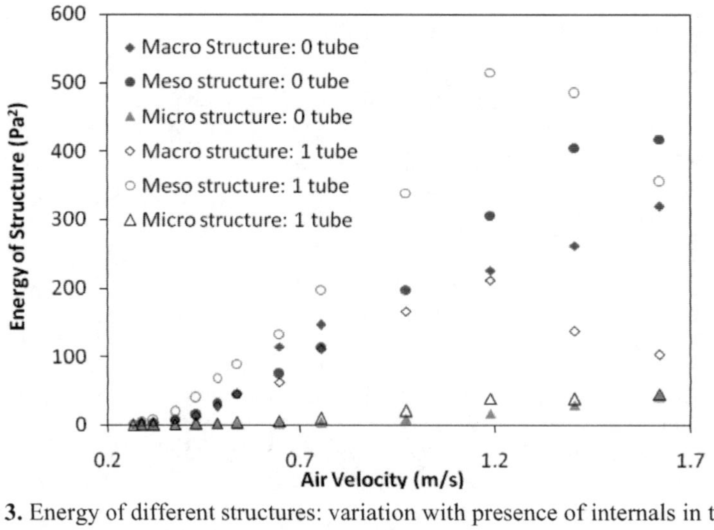

Figure 3. Energy of different structures: variation with presence of internals in the bed

The contribution of energy of different flow structures is shown in the Figure 4 for the experiments with no tubes present in the bed. In the velocities lower than the minimum fludization velocity the major percentage of energy belongs to the micro structures – particle-particle and particle-wall interactions. The major change in the value of contribution of the macro structures corresponds to the flow regime transition (onset of fluidization). From that point, the contribution of micro structures is almost negligible and the major percentage of energy is held by the macro and meso structures – formation, rise and break-up of the bubbles, voids and clusters.

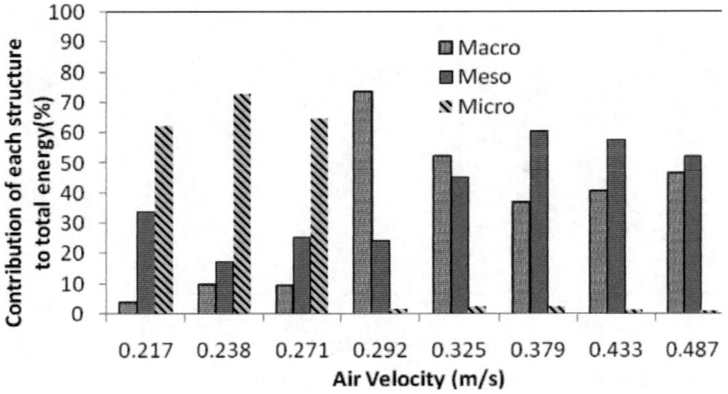

Figure 4. Contribution of energy of each structure to the total energy around minimum fluidization velocity

The comparison of results of both statistical and wavelet analyses is shown in Figure 5. There is conformity in the trends for experiments with one tube present in the bed. The trend lines reach their maximum, denoting the onset of turbulent flow regime in the bed, at the same air velocity in case of one tube inserted to the bed. However, this trend was not observed in case of bed without any tube. The earlier onset of turbulent in presence of internals cannot be therefore reliably justified and further investigations are needed.

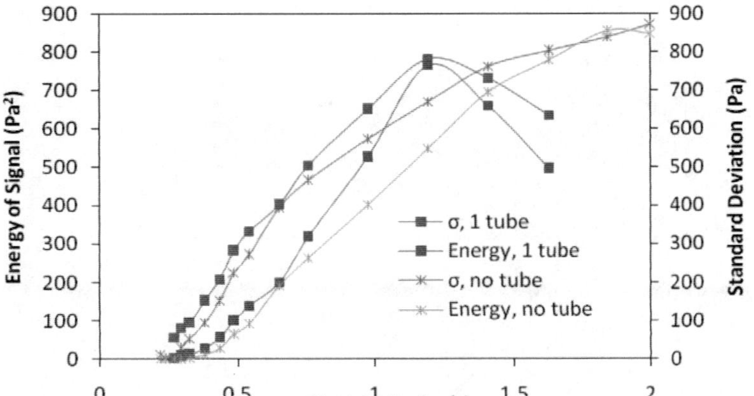

Figure 5. Comparison of the analysis methods

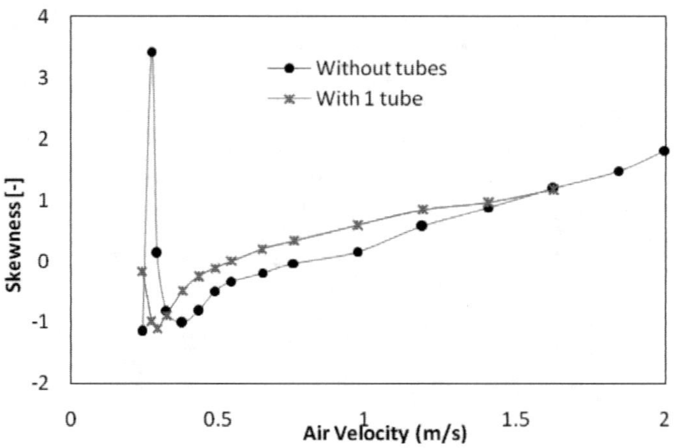

Figure 6. Skewness as a function of air velocity

The skewness as function of velocity is depicted in Figure 6. It shows the same influence of an internal on the bed hydrodynamic behavior – the shift of reaching the zero value of skewness corresponds to the transition between the structures in the bed [17], while the internal is present in the bed. In the Figure 7, the kurtosis as function of the air velocity shows same trend, when it reaches its minimum at lower velocities, which corresponds to the different flow structures in the bed.

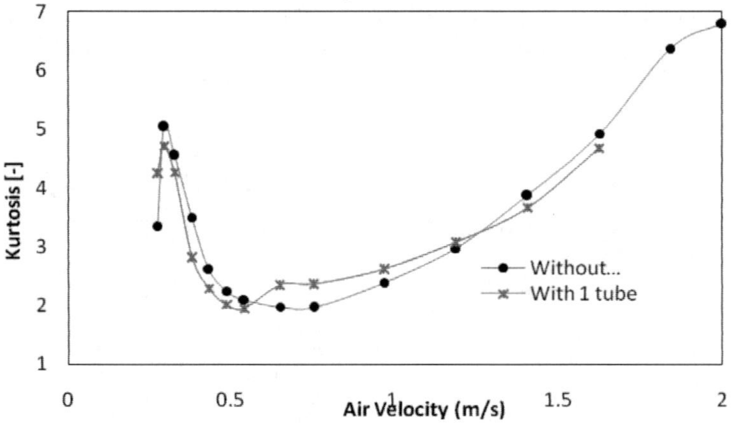

Figure 7. Kurtosis as function of the air velocity

5. Conclusions

The influence of internals present in the fluidized bed on the hydrodynamic behavior has been studied. The results show that the presence of tube within the bed suppresses the occurrence of macro-scale phenomena like formation of large bubbles and its stability and supports the meso-structures such smaller bubbles, voids and clusters. It leads to an earlier onset of bubbling flow regime as well as turbulent flow regime. The observed minimum bubbling velocities were compared with data in literature and were found in a good agreement with them. Future work will be focused on experiments with several tubes within the bed and experiments with particles with weaker electrostatic effects.

Acknowlegement
The authors would like to thank Ms. Mojgan Abbasi for her guides through this work.

References

[1] Felipe CAS and Rocha SCS 2007 *Powder Technol.* **174** 104 –113
[2] Bi H T 2007 *Chem. Eng. Sci.* **62** 3473 – 3493
[3] Hong SC, Jo BR, Doh DS and Choi CS 1990 *Powder Technol.* **60** 215–221
[4] Lu X and Li H 1999 *Chem. Eng. J.* **75** 113–119
[5] Puncochar M, Cermak J and Selucky K 1985 *Chem Eng Commun.* **35** 81 –87
[6] Shou MC and Leu LP 2005 *Chem. Eng. Res. Des.* **83** 478–491
[7] Johnsson F, Zijerveld RC, Schouten JC, van den Bleek CM and Leckner B 2000 *Int J. Multiphase Flow* **26** 663–715
[8] Ren J, Mao Q, LI J and Lin W 2001 *Chem. Eng. Sci.* **56** 981 –988
[9] Shou M C and Leu L P 2005 *Chem. Eng. Res. Des.* **83** 478–491
[10] Davies C E, Fenton K 1997 *IPENZ Trans.* **24** 12–20
[11] Yu T, Yang LP 2008 Chem. Eng. Sci. **63** 1950–1970
[12] Lee GS and Kim SD1988 J Chem Eng Jpn. **21** 515
[13] Abbasi M, Sotudeh-Gharebagh R, Mostoufi N, Zarghami R and Mahjoob M J 2009 *Powder Technol.* **196** 278–285
[14] Mallat S 1989 *IEEE Trans. Pattern Anal. Mach. Intell.* **11** 674 –693
[15] Wen CY and Yu YH 1966 *AIChE J.* **12** 610
[16] Abbasi M, Sotudeh-Gharebagh R, Mostoufi N and Mahjoob M J 2009 *AIChE J.* **56** 597 –603
[17] Zarghami, R 2009. *Conditional Monitoring of Fluidization Quality in Fluidized Beds*. Ph.D. Dissertation, University of Tehran.

Identification of Climate Change with Generalized Extreme Value (GEV) Distribution Approach

Anita Rahayu
Department of Statistics, Bina Nusantara University
Jalan KH. Syahdan No. 9 Palmerah, Jakarta 11480, Indonesia

E-mail: n33_tha@yahoo.co.id

Abstract. Some events are difficult to avoid and gives considerable influence to humans and the environment is extreme weather and climate change. Many of the problems that require knowledge about the behavior of extreme values and one of the methods used are the Extreme Value Theory (EVT). EVT used to draw up reliable systems in a variety of conditions, so as to minimize the risk of a major disaster. There are two methods for identifying extreme value, Block Maxima with Generalized Extreme Value (GEV) distribution approach and Peaks over Threshold (POT) with Generalized Pareto Distribution (GPD) approach. This research in Indramayu with January 1961-December 2003 period, the method used is Block Maxima with GEV distribution approach. The result showed that there is no climate change in Indramayu with January 1961-December 2003 period.

1. Introduction

As the vast archipelago country, Indonesia was strongly influenced by global climate change, therefore, Indonesia was able to predict the impacts of climate variability and change and rising sea levels. That ability is needed to perform the development in various aspects of life such as agriculture, fisheries, forestry, urban with infrastructure, health, coastal wetlands, and small islands. That ability is also required so Indonesia can prepare mitigation measures, anticipation, and adaptation to minimize the negative impacts of climate variability and change. One method that can be used to minimize that impact is Extreme Value Theory (EVT).There are two methods in identifying extreme value, Block Maxima with Generalized Extreme Value (GEV) distribution approach and Peaks over Threshold (POT) with Generalized Pareto Distribution (GPD) approach. Some studies have used the EVT method that is applied in various fields, including: economy [1, 2, 3, 4, 5, 6], telecommunications [7, 8, 9, 10], as well as the hydrology and climatology [11, 12, 13, 14, 15, 16, 17, 18, 19, 20].

As an island nation located in tropics, Indonesia is one of the country's most vulnerable to the threat and impact of extreme events. The purpose of doing this study is to apply EVT in identifying climate change in Indramayu.

2. Extreme Value Theory (EVT)

Extreme events that often occur in the areas of insurance, economics, climatology, hydrology, and telecommunications are indicated by the presence of an observation that is very high (maximum) and very low (minimum). The interesting thing is to determine the probability (maximum and minimum) of rare occurrence (tail distribution). One of the statistical methods used to study the behavior of the tail distribution is the Extreme Value Theory (EVT). There are two methods in identifying movements of the extreme value are taking the maximum value in a given period (called Block Maxima method) and retrieve the values that pass through a given threshold (called Peaks Over Threshold method) [21].

2.1 Block Maxima Method

One of the methods in identifying movements of the extreme value is Block Maxima, which identifies the extreme value based on the value of maximum observation data are grouped according to certain periods. In this method, observation data is divided into blocks in a certain period, example monthly, quarterly, semester, or year. Then for each block is determined the magnitude of observation data and the value is the maximum value of the extremes for each block and were used as the sample. Figure 1shows that the observations which were used as the sample is x_1, x_2, x_3, x_4, x_5.

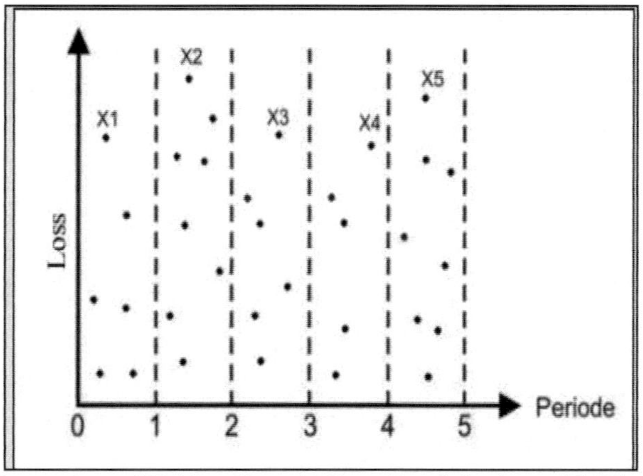

Figure 1. Illustration of Taking Sample Data Precipitation with Block Maxima

Generalized Extreme Value (GEV) distributionis a family of continuous distribution is built into the EVT to combine the Gumbel, Frechet, and Weibulldistribution, known as the extreme value distribution of type I, II, and III. On the GEV distribution, *X* is the random variable which has Probability Density Function (PDF) is as follows.

$$f(x;\mu,\sigma,\xi) = \begin{cases} \frac{1}{\sigma}\left\{1+\xi\left(\frac{x-\mu}{\sigma}\right)\right\}^{-\frac{1}{\xi}-1} \exp\left\{-\left(1+\xi\left(\frac{x-\mu}{\sigma}\right)\right)^{-\frac{1}{\xi}}\right\}, & \xi \neq 0 \\ \frac{1}{\sigma}\exp\left\{-\frac{x-\mu}{\sigma}\right\}\exp\left\{-\exp\left(-\frac{x-\mu}{\sigma}\right)\right\}, & \xi = 0 \end{cases} \quad (1)$$

Where a parameter of location is μ, a parameter of scale is σ, and a parameter of shape is ξ. Shape parameter ξ determines the behavior of the tail distribution. Distribution type defined with $\xi = 0$, $\xi > 0$, and $\xi < 0$ and can be likened to the Gumbel, Frechet, and Weibull distribution.

3. Parameter Estimation of Generalized Extreme Value (GEV) Distribution

Parameter estimation of Generalized Extreme Value (GEV) distribution can be done with Maximum Likelihood Estimation (MLE) using the following stages.

a. Take a random sample of n $x_1, x_2, \ldots, x_n$

b. Make likelihood function

$$L\left(\underset{\sim}{\theta} \middle| x_1, x_2, \ldots, x_n\right) = f\left(x_1, x_2, \ldots, x_n, \underset{\sim}{\theta}\right) = \prod_{i=1}^{n} f\left(x_i, \underset{\sim}{\theta}\right) \tag{2}$$

c. Likelihood function Utility by making ln from equation (2)

Qualified need $\dfrac{\partial \ln L\left(\underset{\sim}{\theta}\right)}{\partial \underset{\sim}{\theta}} = \underset{\sim}{0}$ so obtained $\hat{\underset{\sim}{\theta}}$

Qualified enough $\dfrac{\partial^2 \ln L\left(\underset{\sim}{\theta}\right)}{\partial \underset{\sim}{\theta} \partial \underset{\sim}{\theta}^T} = H\left(\underset{\sim}{\theta}\right)$ called hessian matrix

$\hat{\underset{\sim}{\theta}}$ Maximize $L\left(\underset{\sim}{\theta}\right)$ the definite positive terms $\boldsymbol{H}\left(\hat{\underset{\sim}{\theta}}\right)$

If the results obtained from the first derivative of the ln function likelihood against each of these parameters is not closed form then required further numerical analysis to complete the equation.

4. Return Level

The return level is the maximum value which expected to be reached within the period of time one k with period p, or in other words, in the k and p period, precipitation will reach the maximum value R_p^k one time [2]. The formula to estimate return levels are:

$$\hat{R}_p^k = \begin{cases} \hat{\mu} - \dfrac{\hat{\sigma}}{\hat{\xi}}\left\{1 - \left(-\ln\left(1 - \dfrac{1}{k}\right)\right)^{-\hat{\xi}}\right\}, & \xi \neq 0 \\ \hat{\mu} - \hat{\sigma} \ln\left\{\left(-\ln\left(1 - \dfrac{1}{k}\right)\right)\right\} & , \xi = 0 \end{cases} \tag{3}$$

5. Method

The data used in this research is dasarian rainfall data in Indramayu station during the period January 1961-December 2003.Thedata obtained from the Agency of Meteorology Climatology and Geophysics (BMKG).The steps of the research methods used are:

a. Create description of dasarian rainfall data in Indramayustation with 1961-2003 periods
b. Identifying the existence of a fat tail distribution
c. Separates the data into two periods, namely, the period I (1961-1990) and period II (1991-2003)
d. Identifying extreme values using Block Maxima method, which compiled the rainfall data based on block 3 monthly for each period, that is, December-January-February (DJF), March-April-May (MAM), June-July-August (JJA), and September-October-November (SON)
e. Identify the distribution pattern of dasarian rainfall data in 1961-2003
f. Identify the distribution pattern of rainfall data for each period
g. Perform parameter estimation for each period with MLE method, as well as create a confidence interval for each estimate parameters that have been obtained from the MLE method
h. Do hypothesis testing for each parameter estimation obtained from the MLE method using normal approach

i. Evaluating the suitability of the data distribution pattern of theoretical distribution pattern
j. Calculating the estimate return levels

6. Apply Extreme Value Theory (EVT) in Identifying Climate Change in Indramayu

Rainfall pattern in Indramayu station with 1961-2003 method is monsoon which shown with a plot of average monthly rainfall pattern resembles the letter U.

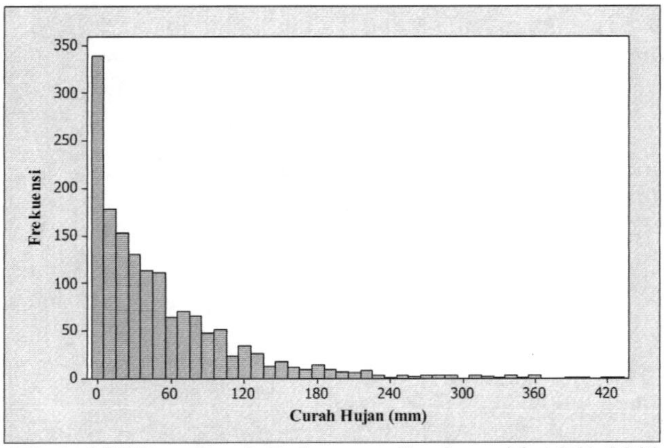

Figure 2. RainfallHistogram inIndramayu Station with 1961-2003 Period

Figure 2shows that the distribution pattern of dasarianrainfall data in Indramayu station with 1961-2003 period had tail distribution that goes down slowly when compared to a normal distribution, as a result of the extreme value of the opportunities will be greater than with a normal distribution modeling or commonly referred to as a fat-tailed distribution (heavy tail). With the indications on this research can use the EVT with extreme data retrieval using Block Maxima method.

Table 1.Parameter Estimation and Confidence Interval 95 % for GEV Distribution

Parameter	MLE Method	
	Period I (1961-1990)	**Period II (1991-2003)**
$\hat{\mu}$	96 [85; 108]	96 [77; 115]
$\hat{\sigma}$	63[54; 71]	66[52; 81]
$\hat{\xi}$	0,1 [-0,01; 0,6]	0,1 [-0,2; 0,8]

On Block Maxima method, dasarianrainfall data in Indramayustation with 1961-2003 periods are divided into 3 monthly blocks for each period. The block is the DJF, MAM, JJA, and SON. Then for each block is determined the magnitude of the maximum rainfall data and that data is extremes data for each block and were used as the sample. With the Block Maxima method obtained 172 extreme data in 1961-2003, with details of the 120 extreme data for the period I (1961-1990) and 52 extreme data for the period II (1991-2003).

The location parameter μ indicates the location of the point from data centralization, the scale parameter σ indicates the pattern of diversity data, and shape parameter ξ indicates the right end point behavior of functions there. Based on Table 1 visible that the value of the location parameter estimation μ obtained by the MLE method for period I and II with the centralization of data is 96 mm. Additionally the value of each parameter estimation obtained by the MLE method is lies in confidence interval of 95%.

After obtained a value for the parameter estimation $\mu, \sigma,$ and ξ, the next step is to do hypothesis testing in the period I and II for each parameter estimation obtained from the MLE method. The hypotheses used are:

- For μ parameter

  $H_0 : \mu = \mu_0$

  $H_1 : \mu \neq \mu_0$

  Test statistic: $Z_{hitung} = \dfrac{\hat{\mu} - \mu_0}{SE(\hat{\mu})}$

- For σ parameter

  $H_0 : \sigma = \sigma_0$

  $H_1 : \sigma \neq \sigma_0$

  Test statistic: $Z_{hitung} = \dfrac{\hat{\sigma} - \sigma_0}{SE(\hat{\sigma})}$

- For ξ parameter

  $H_0 : \xi = \xi_0$

  $H_1 : \xi \neq \xi_0$

  Test statistic: $Z_{hitung} = \dfrac{\hat{\xi} - \xi_0}{SE(\hat{\xi})}$

For the third hypothesis testing above, the decision reject H_0 if the value $|Z_{hitung}| > Z_{\alpha/2}$

Table 2. Hypothesis Testing with the Normal Approach

Period	μ Parameter		Z_{hitung}	$Z_{\alpha/2}$	Decision	σ Parameter		Z_{hitung}	$Z_{\alpha/2}$	Decision
	$\hat{\mu}$	μ_0				$\hat{\sigma}$	σ_0			
1	96	96	0	1,96	Accept H_0	63	64,5	-0,29	1,96	Accept H_0
(SE)	6,6					5,1				
2	96	96	0	1,96	Accept H_0	66	64,5	0,19	1,96	Accept H_0
(SE)	10,4					7,8				

Table 2 (Continued). Hypothesis Testing with the Normal Approach

Period	ξ Parameter		Z_{hitung}	$Z_{\alpha/2}$	Decision
	$\hat{\xi}$	ξ_0			
1	0,1	0,1	0	1,96	Accept H_0
(SE)	0,1				
2	0,1	0,1	0	1,96	Accept H_0
(SE)	0,1				

Based on Table 2 shows that the value of $|Z_{hitung}|$ for each parameter estimation μ, σ, and ξ in period I and II is less than the value of $Z_{\alpha/2}$ so the decision is accept H_0 and the meaning are $\mu = \mu_0$, $\sigma = \sigma_0$ and $\xi = \xi_0$. Because $\xi = \xi_0$ with $\xi_0 = 0$, so can defined that rainfall extreme data in Indramayu station with period I and II have Gumbel distribution. Therefore, there is no change in the distribution for period I and II and then it can be said that there is no climate change in Indramayu station with January 1960-December 2003 period.

Table 3. Return Level

Reperiod	Period I (1961-1990)		Period II (1991-2003)	
	Time	*Return Level*	Time	*Return Level*
2 block = 6 months	January 1991-Juni 1991	120 mm	January 2004-Juni 2004	120 mm
3 block = 9 months	January 1991-September 1991	157 mm	January 2004-September 2004	157 mm
4 block = 12 months	January 1991-Desember1991	182 mm	January 2004-Desember 2004	182 mm

Based on Table 3 to see that maximum rainfall expected on average can be exceeded once in a period of 6, 9, and 12months for period I was 120 mm; 157 mm; and 182 mm, while for the period II was 120 mm, 157 mm, and 182 mm.

7. Conclusion

Based on the results and discussion of the obtained some of the following conclusion:
There is no climate change in Indramayu station from January 1961-December 2003. This is shown by the absence of a change in distribution in period I (1961-1990) and period II (1991-2003) and the value of each parameter estimation obtained by the MLE method is lies in confidence interval of 95%.

8. References

[1] Diebold F X, Schuermann T and Stroughair J D 1998 *J.Risk Finance* **1** 30-36

[2] Gilli and Kellezi 2003 *An Application of Extreme Value Theory for Measuring Risk* (Switzerland: Geneva)

[3] Hastaryta R and Effendie A R 2006 Estimasivalue-at-risk denganpendekatanextreme value theory-generalized paretodistribution (Studikasus IHSG 1997-2004)

[4] Gander J P 2009 *J.Economics* **3** 1-41

[5] Georgescu V 2009 Measuring risk with extreme value theory *Thesis CASE (Center of Applied Statistics and Economics)* Humboldt University

[6] Gourier E, Farkas W and Abbate D 2009 *J.Operational Risk 4* **3** 3-26

[7] Alves M L F and Gomes M L 1996 Statistical choice of extreme value domains of attraction – a comparative analysis *Communication in statistics-theory and method* **25** 789-811

[8] Deane J H B, Johnstone G G and Ledford A W 1997 *J.Telecommunication* **33** 832-833

[9] Kang J, Hamilton B A and McLean 2000 *J.Telecommunication*

[10] Kalyani S and Giridhar K 2006 *J.Telecommunication and Computer Networks* **6** 2893-2898

[11] Hosking J R M, Wood E F and Wallis J R 1984 *J.Technometrics* **27** 251-261

[12] Coles S, Heffernan J and Tawn J 1999 *J.Applied Sciences* **2** 339-365

[13] Arshad M, Rasool M T and Ahmad I 2002 *J.Applied Sciences* **2** 774-776

[14] Katz R W, Parlange M B and Naveau P 2002 *Advances in Water Resources* **25** 1287–1304

[15] Klein T A M G and Können G P 2003 *J.Climate* **16** 3665–3680

[16] Li Y, Cai W and Campbell E P 2005 *J.Climate* **18** 852-863

[17] Gilleland E and Katz R W 2006 Analyzing seasonal to interannual extreme weather and climate variability with the extremes Toolkit (extRemes) *Based on the Recommendation of the 18th Conference on Climate Variability and Change, 86th American Meteorological Society (AMS) Annual Meeting* Atlanta-Georgia 29 January–2 February 2006

[18] Prang J D 2006 Sebarannilaiekstremterampatdalamfenomenacurahhujan*TesisJurusanStatistika* IPB-Bogor

[19] Diebolt J, Guillou A and Naveau P 2008 *J.Statistical* **6** 33-50

[20] Giles D E, Feng H and Godwin R T 2011 *Econometrics Working Paper* **35** 200-207

[21] McNeil A J 1999 *Extreme Value Theory for Risk Managers* Department Mathematic ETH Zen trum-Zurich

Journal of Physics: Conference Series **423** (2013) 012027 doi:10.1088/1742-6596/423/1/012027

Python to learn programming

A Bogdanchikov[1], M Zhaparov[1] and R Suliyev[1]

[1] Suleyman Demirel University, 1/1 Abylaikhan St., Kaskelen, Almaty, Kazakhstan

E-mail: meirambek.zhaparov@gmail.com

Abstract. Today we have a lot of programming languages that can realize our needs, but the most important question is how to teach programming to beginner students. In this paper we suggest using Python for this purpose, because it is a programming language that has neatly organized syntax and powerful tools to solve any task. Moreover it is very close to simple math thinking. Python is chosen as a primary programming language for freshmen in most of leading universities. Writing code in python is easy. In this paper we give some examples of program codes written in Java, C++ and Python language, and we make a comparison between them. Firstly, this paper proposes advantages of Python language in relation to C++ and JAVA. Then it shows the results of a comparison of short program codes written in three different languages, followed by a discussion on how students understand programming. Finally experimental results of students' success in programming courses are shown.

1. Introduction

Python programming language is most suitable as a first language to learn for newbie programmers, because it has powerful tools that reflect the way people think and the way they implement the code. [1] Moreover it minimizes extra keywords that are necessary to write syntactically correct program. This approach seems to be more productive than teaching C++ or Java languages, which have a lot of terms and elements that are related to the specifics of a language rather than to an algorithm realization. In addition, instructors in over a dozen universities, such as MIT, UC Berkeley, UC Davis, Sonoma State University, the University of Washington, the University of Waterloo, Luther College, and Swarthmore College, have used it for teaching the introductory programming course to the students of computer science department. [2]

Today, for a computer scientist, it is important to learn at least one programming language, because all innovations and technologies are based on thorough understanding of computers, operating system, software API or some hardware peripherals. All of which are created by programmers that use specific way of thinking. And to gain that way of thinking, one has to get used to one of the programming languages and become qualified in software development. [6]

For any person, who starts to learn programming, it is important to concentrate on programming concepts rather than on language specifics, because they may be different for various programming languages. But Python provides the highest level of programming. So the student does not have to think about memory management, which is unavoidable in C++, or class hierarchy, that is unavoidable in Java, or variable types and declarations, that exist in almost each programming language.

Journal of Physics: Conference Series **423** (2013) 012027 doi:10.1088/1742-6596/423/1/012027

2. Advantages

The easiest thing in Python is to write "Hello World" program. Most programming languages require writing a lot of specific methods, or functions, class or program declarations, etc. But Python gives ability to start programming without those requirements. You can check examples we gave in "Comparison" chapter.

Python is interpreted language, so with the use of command-line interpreter student can easily check how operators or functions work. [1] Python interpreter has built-in help module, which can significantly improve the process of understanding of different aspects of language.

In order to understand programming languages, freshmen (students which have no experience in programming) need to learn how to think like computer scientist, and that requires a great effort and a complete shift in their paradigm of thinking. [5] Implementation of Python code is easy enough, so that a person who passed a course of elementary mathematics will find such tools as 'variables' and 'functions' easy-to-use.

Whenever a programmer needs a prototype of software, Python, with its rich library, can be used. [3] Then the software may be rewritten in lower language if needed. Advantages of Python are significant, so using it as primary language to learn programming can highly affect the speed of learning computer science in general.

3. Comparison

Fangohr H in his paper compared C, MATLAB and Python [7], and here we are showing the differences between Python, JAVA and C++.

As first and very primitive example of simplicity of Python language, we will demonstrate the code of "*hello world*" program written in three different languages: Java, C++ and Python as shown in tables 1, 2 and 3 respectively.

Table 1. JAVA code of 'hello world' program

```
public class Hello{
    public static void main(String args[]){
        System.out.println("Hello, world!");
    }
}
```

Table 2. C++ code of 'hello world' program

```
#include <iostream>
using namespace std;
int main(){
     cout<<"Hello, world!"<<endl;
}
```

Table 3. Python code of 'hello world' program

```
print "Hello, world!"
```

So to explain the first program to a student, an instructor will need to explain many other useful, but unnecessary terms like *public static void main* or *using namespace std*, but in case of Python there is only single line of code and nothing else.

Another example is class declaration and inheritance. Tables 4, 5 and 6 display the code that is necessary to create small hierarchy of classes: *Person* and *Student* with constructors.

Table 4. JAVA code of Person hierarchy

```
public class Person{
      protected String name;
      protected String surname;
      public Person(String n, String s){
           this.name = n; this.surname = s;
     }
}
class Student extends Person{
      protected String group;
      public Student(String n, String s, String g){
           super(n,s);   this.group = g;
      }
}
```

Table 5. C++ code of Person hierarchy

```
#include <string>
using namespace std;
class Person
{
    public:
        Person(string &n, string &s);
    protected:
        string name, surname;
};
Person::Person(string &n, string &s)
{
     name = n; surname = s;
}
class Student: public Person
{
    public:
       Student(string &n, string &s, string &g);
    protected:
       string name, surname, group;
};
Student::Student(string &n, string &s, string &g):Person(n, s)
{
    group = g
}
```

Table 6. Python code of Person hierarchy

```
class Person:
    def __init__(self, name, surname):
        self.name = name; self.surname = surname
class Student(Person):
    def __init__(self, name, surname, group):
        Person.__init__(self, name, surname)
        self.group = group
```

4. Discussion

Python do not have security attributes (like: public/private/protected), so the program become simpler and shorted, more strict and understandable. Also Python is very dynamic so fields/attributes can be created on the fly, which cannot be done in JAVA or in C++. Polymorphism is nature of Python functions and class methods, unlike C++ with its virtual or non-virtual functions. Operator overloading gives extra power for Python objects, because it can be used for any natural expression, unlike JAVA's restricted syntax.

Programming indentation in Python plays a great role in program structure, because of that any program written in Python is easily read and understood. But in the case of JAVA or C++ every student is trying to make his\her program shorter, fitting it in one line, or trying to write every statement with same indentation, which makes the program difficult to be read and understood not only for the teacher but for the student himself.

Python includes wonderful algorithms in its native libraries. [3] So the student does not need to understand *long* arithmetic to make huge calculations. It already exists in native *long* data type in Python. There are good tools to sort, find, slice, and join any sequence of data. Sometimes they may not be so effective, like those which are written in C++ for a specific purpose. But for the beginners it is not the primary priority.

Python has its specific way to store and use variables, so the programmer no longer needs to define types of the variables, they will be defined by the values that variables store. Of course it is important to know how different types of variables are stored in memory, this can be discussed in lectures, but for the practice it is easier to skip the information about data types.

Also Python has adaptive and intuitive set of keywords and commands that impressively help students to learn programming.

5. Experiment

We compared results of midterm examinations of the course given in Java programming language (fall 2011) and the same course taught in Python programming language (fall 2012). Total midterm score of students, who took the course in Python language, increased by 16% in relation to previous year students, who took the same course in JAVA. Three groups of students (EN1-A-04, EN1-B-04 and EN1-C-04) were taken into consideration while comparison. Total of 76 students in three groups participated in the course "CS101 Algorithms & Programming (JAVA)" in fall 2011, and 65 students from first three groups are taken into account for course "CS101 Algorithms & Programming (Python)" in fall 2012. Results of their midterm marks are shown in Figure 1 below.

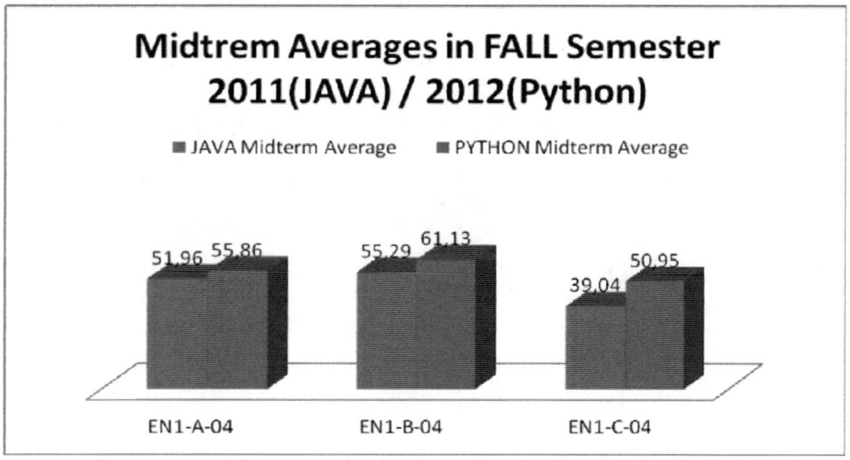

Figure 1. Midterm results for 2011(JAVA) / 2012 (Python)

6. Conclusion

In the conclusion, we would like to say that the initial understanding of programming is essential for a computer scientist. And because of complexities of advanced programming languages it appears that learning them is not so easy. First programming language imprints students' understanding and desire to learn programming and computer science in whole. So it is important to apply precise and adequate strategy to the process of learning the language.

Today some of leading universities either use Python to teach introduction to programming to their students or develop and teach their own simple compiler that can be easily understood by students. [2]

As a result we can say that students understand programming well with the use of Python. They have command line tool which allows them to test their theories immediately. Moreover there is a very good help tool in interpreter which always reminds about the structure of specific classes. Students enjoy simple *turtle* library [3] that provides very simple interface to draw on canvas. That helps them to understand simple statements like cycles and conditions. But the major result, of course, depends on students' desire to learn programming and computer science.

In experiment section we compared the results of midterm marks of the course taught using Java language with the results of the midterm marks of the same course taught using Python. We observed that results increased by 16%.

References

[1] Downey A, Elkner J and Meyers C 2008 *"`How to think like Computer Scientist` Learning with Python"*
[2] http://www.pythontutor.com/
[3] http://docs.python.org/3/library/
[4] Blank D, Kay JS, Marshall JB, O'Hara K and Russo M *"Calico: A Multi-Programming-Language, Multi-Context Framework Designed for Computer Science Education"*
[5] Anshul KJ, Manik S and Manu SG *"Algorithm Building and Learning Programming Languages Using a New Educational Paradigm"*
[6] Sanders ID and Langford S *"Students' Perceptions of Python as a First Programming Language at Wits"*
[7] Fangohr H *"A comparison of C, MATLAB, and python as teaching languages in engineering"*

Journal of Physics: Conference Series **423** (2013) 012028 doi:10.1088/1742-6596/423/1/012028

Computer Networks teaching by microlearning principles

Zhamanov A[1]

[1]Suleyman Demirel University, Ablayhan Street No 1/1, Kaskelen, Almaty, Kazakhstan, 040900

E-mail: zhamanov@gmail.com

Zhamapor M[2]

[2] Suleyman Demirel University, Ablayhan Street No 1/1, Kaskelen, Almaty, Kazakhstan, 040900

E-mail: meirambek.zhaparov@gmail.com

Abstract. In nowadays there are many kinds of problems in the system of higher education. One of them is: everyday students get very huge amount of information. All of them take very big amount of time for a student to understand lesson and it is very hard for him/her to do everything in time. This paper explains a study of work microlearning application for computer networks. It consists of general introduction and purpose.

1. Introduction

In learning and developing environments macro, meso and micro parts are used. Macro contains meso parts, meso contains micro parts (this presents the smallest part of them) and all of them construct together products. For example in computer science to make a big project (macro) is divided into subprojects (meso) and then each subproject contains its modules (micro). It is called "divide and conquer". This example shows that in learning it is important to begin from small parts to construct big parts. Microlearning is a new method of teaching which become more popular day by day.
According to Hug[1][2] there are 7 dimensions of microlearning.

- Time: Each topic can be touched at most 5-7 minutes. If topic is more complex it is divided into small parts. For example IPv4 topic is separated 4 small parts, represented in 4 microlessons below.
- Content: Each topic is represented in A4 page, where topic description, simple test or simple tasks are given.
- Curriculum: As shown in microlessons below, topics are divided into small subtopics.
- Form: Mostly each Computer Network protocol is taken simpler as can be, for example to explain IPv4 addresses types, can be given comparison with a university within many groups and some students in those groups.
- Process: Learning of Computer Networks is same as swimming; it needs a lot of practice. Because each topic contains simple test questions and tasks students can solve each of them in 2-10 minutes.

- Mediality: Microlearning in Computer Networks is prepared as book in PDF format with hyperlinks. It can be also used in cell phones because of PDF format.
- Learning type: In class there is a projector, where lecturer shows A4 page, then after explanation students' answers for test questions and do simple tasks.

To pass through all subjects it will be better to use literature in microlearning style which provides understanding of Computer Networks terminology in basic way by dividing information into small micro parts which have a log of graphics and pictures and not too many words for explanation. These examples show that in learning it is important to begin from small parts to construct big parts.

2. Microlearning in Computer Networks

Computer Networks have a lot of educational resources, such as Cisco Networking Academy [3] and world of network [4]. These resources are good for experts, but not for beginners, because people are afraid of things that they do not know.

Microlearning gives students opportunity to learn computer networks without fear.
In this section we demonstrate 4 microlessons.

This is topic. Lesson 1 – IPv4 Characteristics

Now begins explanation of this topic.

- Consists of 32 bits
- Divided into 4 octets
- Each octed consists of 8 bits (1byte)
- Represented in decimal form

Example of IPv4 address: 192.168.1.21
Computer works only with binary numeral system, to understand listed above IPv4 address computer converts it from decimal to binary.

Table 1. Shows decimal and binary representations of IPv4 address

	1st octed	2nd octed	3rd octed	4th octed
Decimal representation	192	168	1	21
Binary representation	11000000	10101000	00000001	00010101

Now we make a simple test to check and revise.

Test for 1st lesson:

1. How many octets in IPv4?
   a. 3
   b. 4
   c. 5
   d. 1
2. In which numeral system computer understand IPv4 address?
   a. Decimal
   b. Binary
   c. Hexadecimal
   d. Octal
3. How many bits take 3 octets?
   a. 1
   b. 8
   c. 18
   d. 24
4. In which numeral system IPv4 represented?
   a. Decimal
   b. Binary
   c. Hexadecimal
   d. Octal

By this way our lesson is finished.

Another example is given below.

Lesson 2 – Decimal to binary Convertation

- Divide by 2 technique
- Mapping technique

2.1.1. Divide by 2 technique
Example: $17_{(10)} = ?_{(2)}$
Figure 1 explains "divide by 2 technique" in process

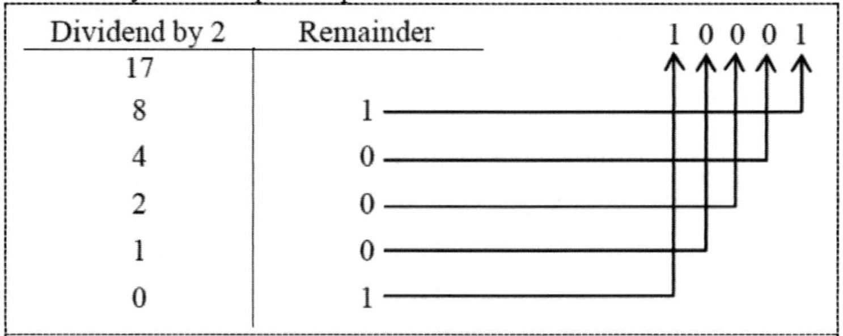

Figure 1. Division by 2 in process

Result: $17_{(10)} = 10001_{(2)}$

2.1.2. Mapping technique
Example: $114_{(10)} = ?_{(2)}$

Table 2. Describes Mapping process

Position(value)	64	32	16	8	4	2	1
Multiplication	*	*	*	*	*	*	*
Given binary stream	1	1	1	0	0	1	0
Result of Multiplication	64	32	16	0	0	2	0
Addition	64 +	32 +	16 +	0 +	0 +	2 +	0

Result = 114

2.1.3. Task for 2nd lesson:

1. Convert numbers by using Divide by 2 technique
   a. $14_{(10)} =$ _________ $_{(2)}$
   b. $67_{(10)} =$ _________ $_{(2)}$
   c. $78_{(10)} =$ _________ $_{(2)}$
2. Convert numbers by using Mapping technique
   a. $234_{(10)} =$ _________ $_{(2)}$
   b. $123_{(10)} =$ _________ $_{(2)}$
   c. $510_{(10)} =$ _________ $_{(2)}$

Lesson 3 – Types of IPv4 address

- Network address
- Host address
- Broadcast address

Group's and student's names in university are good examples of network and host addresses in computer networks. In Green group there are students A, B, C and in Red group can be same names of students and they are unique only in group. In figure 2 represented example of comparison network addresses, host addresses with university groups and students.

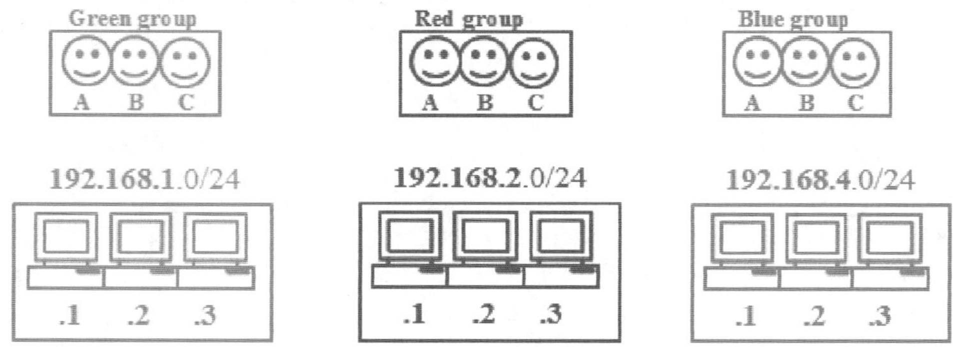

Figure 2. Groups and students in comparison with network addresses and host addresses

Network address
Network address is the common part of every networking device in same network (cannot be assigned to networking device).

Host address
Host address is the unique address of networking device in single network.

Broadcast address
Broadcast address is the common address for all networking devices in one network which is usually used to make request to all devices in same network (cannot be assigned to networking device).

Test for 3rd lesson:

1. What three types of IPv4 addresses explained in this lesson (choose three)?
   a. Multicast
   b. Broadcast
   c. Host
   d. Network
2. Which addresses cannot be assigned to networking devices (choose two)?
   a. Broadcast
   b. Host
   c. Network
   d. Super

Lesson 4 - Finding Network, Host and Broadcast addresses in IPv4

IPv4 Characteristics

- Consists of 32 bits
- Separated into Network and Host addresses by prefix
- Prefix defines number of bits of Network address starts from the left side
- Subnet mask shows how many bits belong to network address, defined by prefix
- Broadcast address is the last combination in host address portion

Example of IPv4 with prefix: 192.168.1.4/24

Table 3. IPv4 address and Subnet mask binary and decimal representation

	8bits	8bits	8bits	8bits	Prefix
Decimal of IPv4	192	168	1	4	/24
Binary of IPv4	11000000	10101000	00000001	00000100	
Decimal mask	255	255	255	0	
Binary mask	11111111	11111111	11111111	00000000	

Computer defines network address by IPv4 address "AND" Subnet mask

Table 4. Logical AND operation

A	B	A and B
0	0	0
0	1	0
1	0	0
1	1	1

Table 5. Separating Network address from Host address

	8bits	8bits	8bits	8bits	Prefix
IPv4	11000000	10101000	00000001	00000100	/24
Subnet mask	11111111	11111111	11111111	00000000	
IPv4∩Subnet mask	11000000	10101000	00000001	00000000	
Network address	192	168	1	0	
Host address	192	168	1	4	
Broadcast address	192	168	1	255	
		Network Address		Host	

Task for 4th lesson:
Find Network, Host and Broadcast addresses

A. 192.168.11.123/17
B. 10.12.54.34/28
C. 99.88.77.66/30

Table 6. Answer table for 4th lesson's task

	A	B	C
Network address			
Host address			
Broadcast address			
Usable addresses			

3. Conclusion

Microlearning gives a great opportunity to mankind to be more educated. Now every person unconsciously uses the microlearning approach while reading tutorials, forums, wiki, blogs and other resources. We started to teach students in our university by using microlearning method and we have good feedback from them. Now students provide more interest for subject and learn more material than in previous years. Our paper shows how to use microlearning principles in teaching computer networks. This study is new and needs more research in this area. We already established "Microlearning Research Center" in our university, in future we plan to develop it and provide Microlearning method to other universities to solve problem of higher education.

References

[1] Hug, Theo; Lindner, Martin; Bruck, Peter A. (2006): Microlearning: Emerging Concepts, Practices and Technologies after e-Learning. Proceedings of Microlearning 2005. Innsbruck: Innsbruck University Press, 2006.

[2] Micro Learning and Narration, Theo Hug, Innsbruck, Paper presented at the fourth Media in Transition conference, May 6-8, 2005, MIT, Cambridge (MA), USA

[3] cisco.netacad.net

[4] http://www.networkworld.com/

Journal of Physics: Conference Series **423** (2013) 012029 doi:10.1088/1742-6596/423/1/012029

Validation of ANUGA hydraulic model using exact solutions to shallow water wave problems

S Mungkasi[1,2] and S G Roberts[1]

[1] Mathematical Sciences Institute, The Australian National University, Canberra, Australia
[2] Department of Mathematics, Sanata Dharma University, Yogyakarta, Indonesia

E-mail: sudi.mungkasi@anu.edu.au, stephen.roberts@anu.edu.au

Abstract. ANUGA is an open source and free software developed by the Australian National University (ANU) and Geoscience Australia (GA). This software is a hydraulic numerical model used to solve the two-dimensional shallow water equations. The numerical method underlying it is a finite volume method. This paper presents some validation results of ANUGA with respect to exact solutions to shallow water flow problems. We identify the strengths of ANUGA and comment on future work that may be taken into account for ANUGA development.

1. Introduction

Shallow water flows have been studied mathematically since the ninteenth century. One available mathematical model is the shallow water equations derived by de Saint-Venant [5]. Based on the shallow water equations, a well-known solution to a dam break problem was derived by Ritter [16]. More comprehensive study of shallow water flows has been conducted by other authors, such as Stoker [19, 20].

Nowadays, the shallow water equations are widely applied to model river flows, tsunami inundations, and floods. However, the exact analytical solutions to the shallow water equations are available only for some specific cases. Numerical solutions are then desired as approximations of the exact solutions. Due to the applications of the shallow water equations, numerical packages (softwares) to solve these equations have been made available. Some of the softwares are commercial and some are free. One available free software is ANUGA. ANUGA is an open source software developed by the Australian National University (ANU) and Geoscience Australia (GA).

This paper validates ANUGA with respect to some exact analytical solutions to the shallow water equations. We identify the strengths of ANUGA when used to simulate shallow water flow problems. Comments on ANUGA shall also be provided for consideration on ANUGA development.

This paper is organised as follows. We provide a brief review of ANUGA software in Section 2. Some exact analytical solutions to the shallow water equations are collected in Section 3. In Section 4, we present numerical results of ANUGA experiments in comparison to the exact analytical solutions. Section 5 concludes this paper with some remarks.

2. ANUGA software

ANUGA is written in Python programming language for the User Interface together with C programming language for the expensive computation parts. Python gives a flexible interface, while C leads to fast computations. The combination of these two languages make ANUGA a robust software. This shall be demonstrated in the numerical experiments we present in Section 4.

The mathematical background underlying ANUGA is a finite volume method used to solve the two-dimensional shallow water equations. A description of it can be found in some documents, such as the ANUGA User Manual [18] and our previous work [11].

ANUGA has advantages for simulation of shallow water flows or waves. First, it can be used to simulate tsunami, dam break, flood inundations and other types of shallow flows. Second, it can handle wetting and drying processes reasonably. Third, it is able to accurately resolve discontinuous water surface and velocity profiles, such as shock waves or hydraulic jumps. See previous publications [1, 3, 4, 6, 7, 8, 15, 17, 22, 23] about ANUGA for more detailed explanations on these advantages.

However, ANUGA has some limitations due to the implemented mathematical model [6]. First, it is not able to simulate turbulence and breaking waves. (To simulate turbulence and breaking waves, the Navier-Stokes equations or the three-dimensional shallow water equations should be used.) Second, ANUGA does not cover sedimentations, erosions and debris-type flows. Third, for simulating tsunamis, a sea floor displacement is not handled by ANUGA. (A tsunami generation should be taken into account for another model, such as Method of Splitting Tsunamis (MOST) and the URS Corporation's Probabilistic Tsunami Hazard Analysis [15].)

3. Some exact analytical solutions to the shallow water equations

In this section, we collect four analytical solutions to shallow water flow (wave) problems. The first three are analytical solutions to problems in one dimension. The fourth is an analytical solution to a problem in two dimensions. All of the analytical solutions can be implemented to test the performance of numerical methods used to solve two-dimensional problems.

In this paper,we use the following notations. For two-dimensional problems:

- x, y represent the two-dimensional spatial variables,
- t represents the time variable,
- $u = u(x, y, t)$ denotes the water velocity in the x-direction, that is, the x-velocity,
- $v = v(x, y, t)$ denotes the water velocity in the y-direction, that is, the y-velocity,
- $h = h(x, y, t)$ denotes the water height (depth),
- $z = z(x, y)$ denotes the topography elevation (bed),
- $w = h + z$ is called the stage, and
- g is the acceleration due to gravity.

The momentum in the x-direction called the x-momentum is given by uh, and the momentum in the y-direction called the y-momentum is given by vh. The notations for one-dimensional problems are similar to those for two-dimensional problems, but note that variables y and v do not arise in one-dimensional problems.

3.1. Dam break involving a dry area in one dimension

In this subsection, we recall the analytical solution to a dam break problem in one dimension.

The initial condition is described as follows. A dam wall is located at points $x = x_0$. The topography is horizontal. Water height on the left of the dam wall is $h = h_1$, and on the right of the dam wall is $h = h_0$.

For the case with $x_0 = 0$, $h_1 > 0$, and $h_0 = 0$, the analytical solution to the dam break is as follows. After the dam wall is removed $(t > 0)$, the water height is

$$h(x,t) = \begin{cases} h_1 & \text{if } x \leq -t\sqrt{gh_1} \\ \frac{4}{9g}(\sqrt{gh_1} - \frac{x}{2t})^2 & \text{if } -t\sqrt{gh_1} < x \leq 2t\sqrt{gh_1} \\ 0 & \text{if } x \geq 2t\sqrt{gh_1}, \end{cases} \tag{1}$$

and the velocity is

$$u(x,t) = \begin{cases} 0 & \text{if } x \leq -t\sqrt{gh_1} \\ \frac{2}{3}(\sqrt{gh_1} + \frac{x}{t}) & \text{if } -t\sqrt{gh_1} < x \leq 2t\sqrt{gh_1} \\ 0 & \text{if } x \geq 2t\sqrt{gh_1}. \end{cases} \tag{2}$$

The derivation of this analytical solution can be found in some references, such as Ritter [16], Stoker [19, 20], Mangeney, Heinrich, and Roche [10], as well as Mungkasi and Roberts [12, 13].

3.2. Dam break involving a shock wave in one dimension

Recall the dam break problem stated in the previous subsection. A dam wall is located at points $x = x_0$. The topography is horizontal. Water height on the left of the dam wall is $h = h_1$, and on the right of the dam wall is $h = h_0$.

For $h_1 \geq h_0 > 0$, the solution to the dam break problem is

$$h(x) = \begin{cases} h_1 & \text{if } x \leq -t\sqrt{gh_1} \\ \frac{4}{9g}(\sqrt{gh_1} - \frac{x}{2t})^2 & \text{if } -t\sqrt{gh_1} < x \leq t(u_2 - \sqrt{gh_2}) \\ \frac{h_0}{2}\left(\sqrt{1 + \frac{8\dot{\xi}^2}{gh_0}} - 1\right) & \text{if } t(u_2 - \sqrt{gh_2}) < x < t\dot{\xi} \\ h_0 & \text{if } x \geq t\dot{\xi} \end{cases} \tag{3}$$

and

$$u(x) = \begin{cases} 0 & \text{if } x \leq -t\sqrt{gh_1} \\ \frac{2}{3}(\sqrt{gh_1} + \frac{x}{t}) & \text{if } -t\sqrt{gh_1} < x \leq t(u_2 - \sqrt{gh_2}) \\ \dot{\xi} - \frac{gh_0}{4\dot{\xi}}\left(1 + \sqrt{1 + \frac{8\dot{\xi}^2}{gh_0}}\right) & \text{if } t(u_2 - \sqrt{gh_2}) < x < t\dot{\xi} \\ 0 & \text{if } x \geq t\dot{\xi} \end{cases} \tag{4}$$

at any time $t > 0$, where $\dot{\xi}$ is the shock speed constant

$$\dot{\xi} = 2\sqrt{gh_1} + \frac{gh_0}{4\dot{\xi}}\left(1 + \sqrt{1 + \frac{8\dot{\xi}^2}{gh_0}}\right) - \left(2gh_0\sqrt{1 + \frac{8\dot{\xi}^2}{gh_0}} - 2gh_0\right)^{\frac{1}{2}}. \tag{5}$$

The derivation of this analytical solution can be found in some references, for example, Stoker [19, 20] and Mungkasi and Roberts [13].

3.3. Periodic wave on a sloping beach in one dimension

Consider a sloping beach with a positive constant slope. When the water is unperturbed, we have the dimensional setting: (i). the origin of the spatial domain is on the water surface, (ii). the horizontal distance of the water surface between the origin and the shoreline is L, (iii). the vertical distance (water depth or height) between the origin and the topography is h_0.

One available analytical solution is as follows. The stage is

$$w = -\frac{1}{2}u^2 + AJ_0\left(\frac{4\pi\sqrt{w+1-x}}{T}\right)\cos\left(\frac{2\pi\,(u+t)}{T}\right), \tag{6}$$

and the velocity is

$$u = -\frac{AJ_1\left(\frac{4\pi\sqrt{w+1-x}}{T}\right)}{\sqrt{w+1-x}}\sin\left(\frac{2\pi\,(u+t)}{T}\right). \tag{7}$$

Note that this analytical solution is dimensionless, where the original dimensional quantities have been scaled [2, 9, 14]. The scalings are as follow:

- horizontal distances are scaled by L,
- vertical distances by h_0,
- time by $L/\sqrt{gh_0}$, and
- velocity by $\sqrt{gh_0}$.

3.4. *Periodic wave on a paraboloid channel in two dimensions*

Now we consider a two-dimensional domain. We define parameters $D_0 = 1000$, $L = 2500$, $R_0 = 2000$ such that

$$\begin{aligned} A &= \frac{L^4 - R_0^4}{L^4 + R_0^4}, && (8)\\ \omega &= \frac{2}{L\sqrt{2gD_0}}, && (9)\\ T &= \frac{2\pi}{\omega} && (10)\end{aligned}$$

are the amplitude of oscillation at the origin, the angular frequency, and the period of oscillation. Furthermore, the topography is given by

$$z(x,y) = -D_0\left(1 - \frac{r^2}{L^2}\right), \tag{11}$$

where

$$r = \sqrt{x^2 + y^2}. \tag{12}$$

One available analytical solution is as follows. The water height is

$$h(x,y) = D_0\left[\frac{\sqrt{1-A^2}}{1 - A\cos(\omega t)} - 1 - \frac{r^2}{L^2}\left(\frac{1-A^2}{[1 - A\cos(\omega t)]^2} - 1\right)\right]. \tag{13}$$

In polar coordinates (r,θ), the velocity u_r in r-direction and the velocity u_θ in θ-direction are given by

$$u_r = \frac{\omega r A \sin(\omega\theta)}{2\,[1 - A\cos(\omega\theta)]^2}, \tag{14}$$

$$u_\theta = 0. \tag{15}$$

This analytical solution was derived by Thacker [21] and implemented by Yoon and Cho [24].

4. Computational experiments on problems in two-dimensional settings

In this section, we present our computational results. The results regard validations of ANUGA using exact solutions to shallow water flow (wave) problems. Note that we use the four analytical solutions presented in the previous section as the benchmarks. Some of those analytical solutions are for one-dimensional problems originally, but they can also be used to test two-dimensional numerical methods.

The numerical setting is as follows. The second order method is used. The spatial domain is discretised using the `rectangular_cross` function available in ANUGA. Quantities are measured in SI units. The acceleration due to the gravity is taken as $g = 9.81$. The friction term is set to be zero.

Four computational experiments are performed. Results for a planar dam break involving a dry area are presented in Subsection 4.1. In Subsection 4.2, results for a planar dam break involving a shock wave are provided. Subsection 4.3 presents results for a periodic wave on a sloping beach. Finally, results for a periodic wave on a paraboloid channel are summarised in Subsection 4.4.

4.1. Planar dam break involving a dry area

We consider the spatial domain $\{(x, y)|x \in [-60, 60], y \in [-10, 10]\}$. It is descretised into 100 by 20 rectangular-crosses, in which each rectangular cross has four uniform triangles. Therefore, we have 8000 triangles as the discretisation of the spatial domain. The dam wall is at $x = 0$. Still water of height 10 is on the left of the dam wall and dry area is on the right of the dam wall.

The numerical solutions of the stage are shown in Figures 1, 3, and 5 for time $t = 0, 1, 2$ respectively. The corresponding x-momenta are depicted in Figures 2, 4, and 6 respectively. From these figures, we see that the numerical solutions match well with the analytical solution. However, a small inaccuracy around wet/dry interface occurs.

4.2. Planar dam break involving a shock wave

Recall the spatial domain and its discretisation used in the previous subsection, but now with a different initial condition. Taking the initial condition as still water of height 10 on the left of the dam wall and of height 1 on the right of the dam wall, we have a shock wave which involves in the solution after dam wall is removed.

The numerical solutions of the stage are shown in Figures 7, 9, and 11 for time $t = 0, 1, 2$ respectively. The corresponding x-velocities are depicted in Figures 8, 10, and 12 respectively. From these figures, the numerical solutions match well with the analytical solution. In addition, the shock wave is resolved accurately. It is interesting to see that diffusion is more visible around the rarefaction than around the shock. It suggests that the limiter implemented in ANUGA works very well for the second order method that we test in resolving the shock.

4.3. Periodic wave on a sloping beach

We consider the spatial domain $\{(x, y)|x \in [0, 55000], y \in [0, 500]\}$. It is descretised into 550 by 5 rectangular-crosses, in which each rectangular cross has four uniform triangles. Therefore, we have 11000 triangles as the discretisation of the spatial domain.

We set some required parameters. The dimensional parameters are as follows:

- the reference length is $L = 5 \cdot 10^4$,
- the reference depth is $h_0 = 5 \cdot 10^2$,
- the period of oscillation is $T_p = 9 \cdot 10^2$, and
- the amplitude at origin is $a = 1.0$.

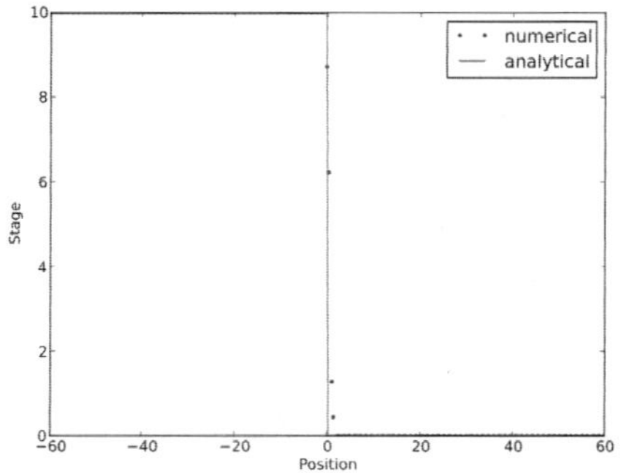

Figure 1. Stage of a cross section of the planar dam break involving a dry area at $t = 0$.

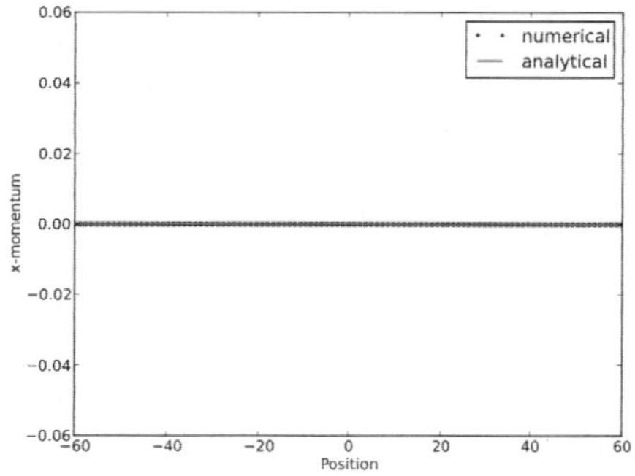

Figure 2. x-momentum of a cross section of the planar dam break involving a dry area at $t = 0$.

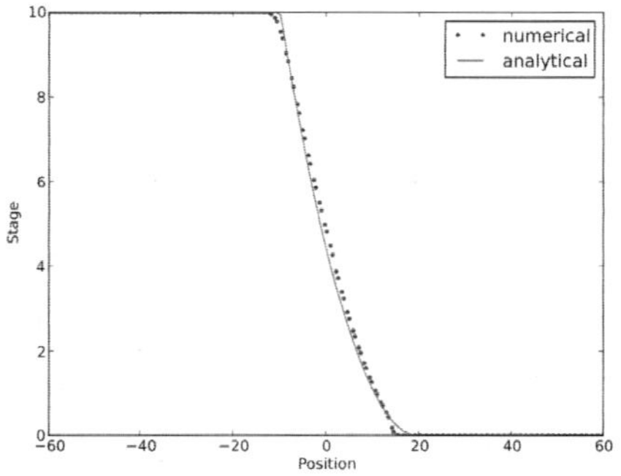

Figure 3. Stage of a cross section of the planar dam break involving a dry area at $t = 1$.

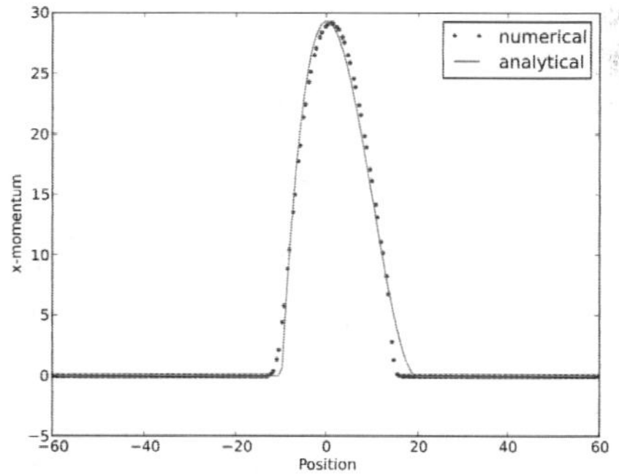

Figure 4. x-momentum of a cross section of the planar dam break involving a dry area at $t = 1$.

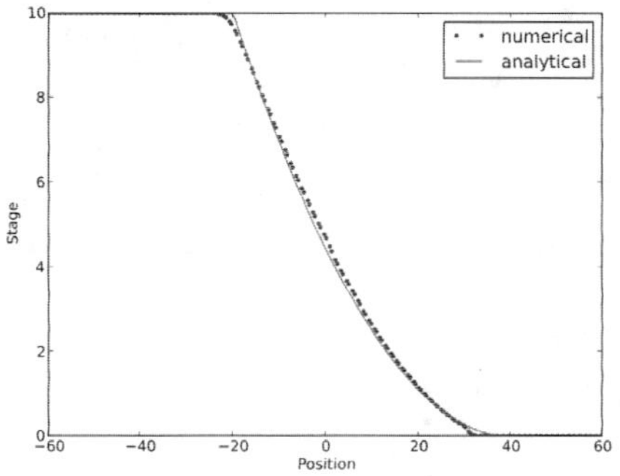

Figure 5. Stage of a cross section of the planar dam break involving a dry area at $t = 2$.

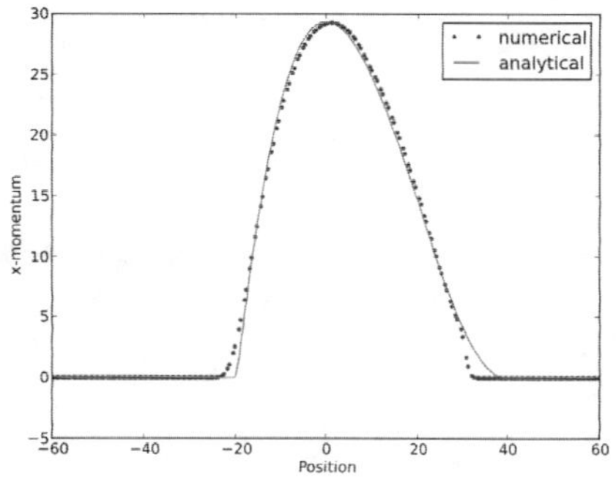

Figure 6. x-momentum of a cross section of the planar dam break involving a dry area at $t = 2$.

Journal of Physics: Conference Series **423** (2013) 012029 doi:10.1088/1742-6596/423/1/012029

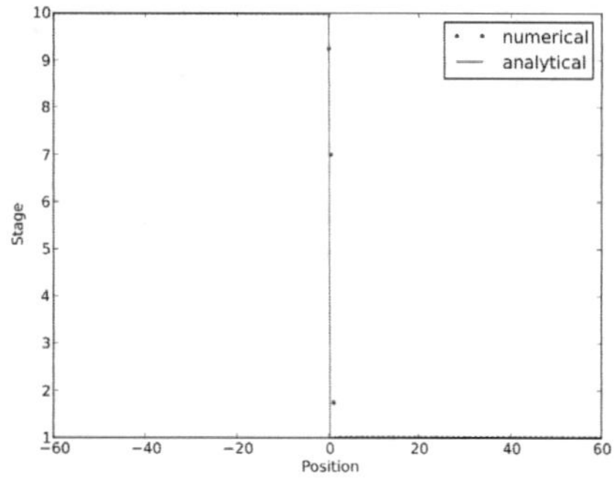

Figure 7. Stage of a cross section of the planar dam break involving a shock wave at $t = 0$.

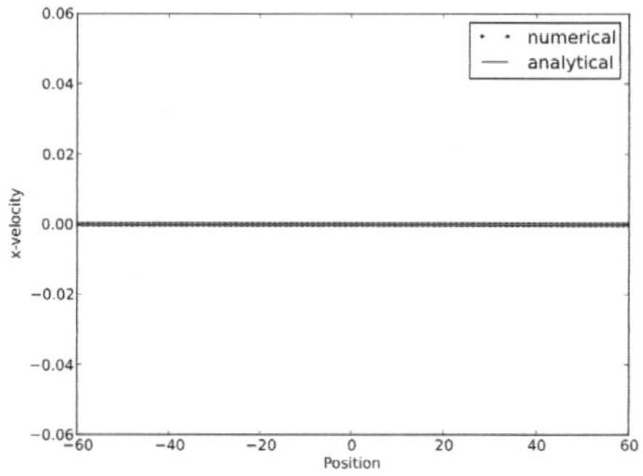

Figure 8. x-velocity of a cross section of the planar dam break involving a shock wave at $t = 0$.

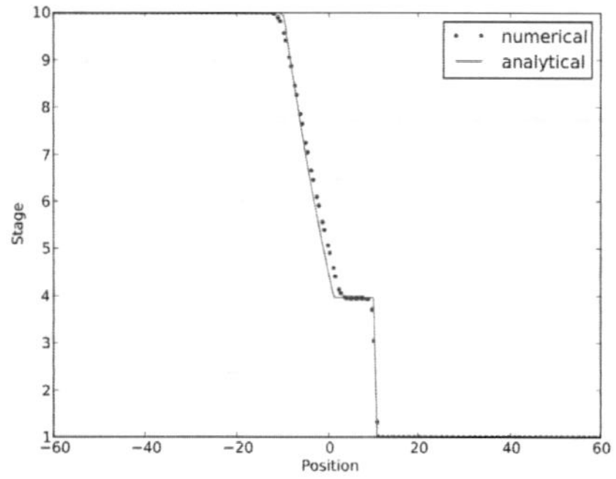

Figure 9. Stage of a cross section of the planar dam break involving a shock wave at $t = 1$.

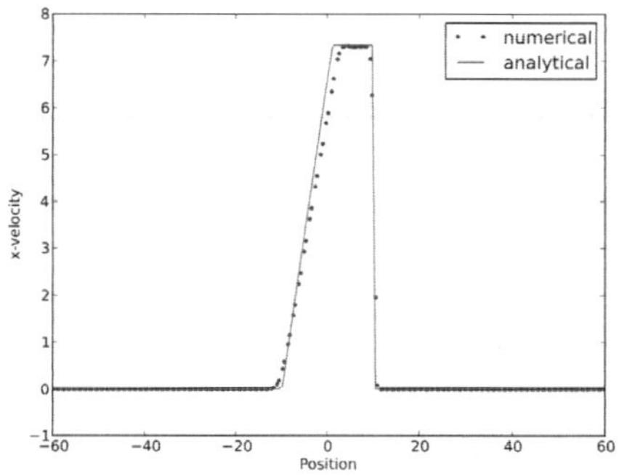

Figure 10. x-velocity of a cross section of the planar dam break involving a shock wave at $t = 1$.

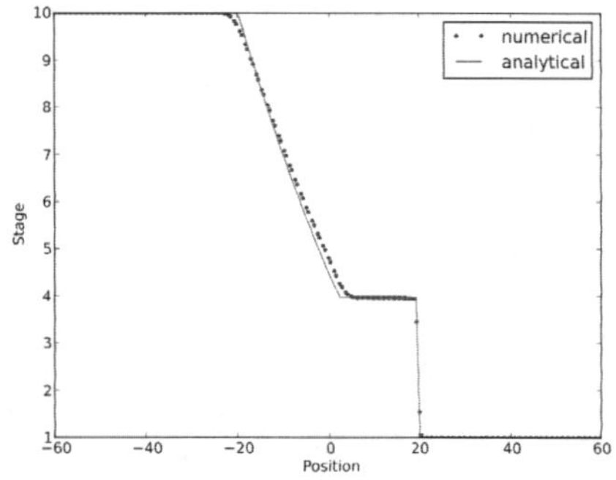

Figure 11. Stage of a cross section of the planar dam break involving a shock wave at $t = 2$.

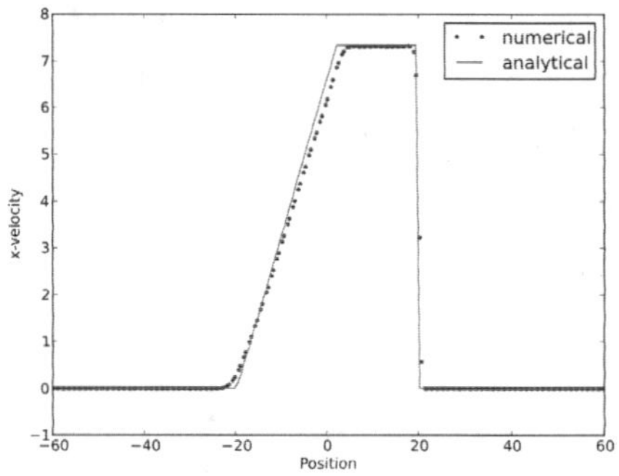

Figure 12. x-velocity of a cross section of the planar dam break involving a shock wave at $t = 2$.

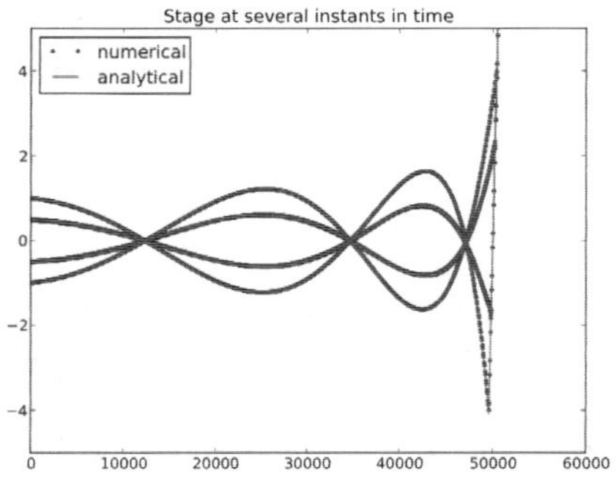

Figure 13. Stage of a cross section of the Carrier-Greenspan periodic wave at several instants in time ($t = kT_p/6$ where $k = 55, ..., 60$).

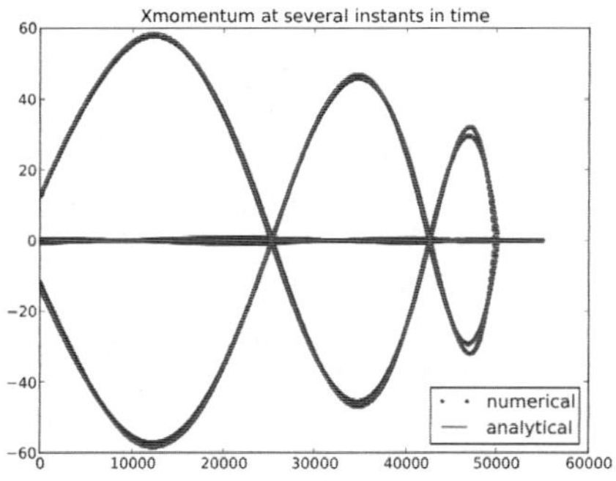

Figure 14. x-momentum of a cross section of the Carrier-Greenspan periodic wave at several instants in time ($t = kT_p/6$ where $k = 55, ..., 60$).

For the periodic wave on a sloping beach, we use the notation T_p for the dimensional period of oscillation and the notation T for the dimensionless one. The dimensionless parameters are found by some scalings as described in Subsection 3.3. We set the following dimensionless parameters:

- perturbation at origin
$$\epsilon = \frac{a}{h_0}, \tag{16}$$
- period of oscillation
$$T = \frac{T_P\sqrt{gh_0}}{L}, \tag{17}$$
- amplitude at shoreline
$$A = \frac{\epsilon}{J_0(4\pi/T)}, \tag{18}$$

where J_0 is the Bessel function of the first kind of order 0. See our previous work [14] for more detailed dimensional and dimensionless settings.

The numerical solutions for the stage and x-momentum are shown in Figures 13 and 14 respectively at time $t = kT_p/6$ where $k = 55, ..., 60$. These numerical solutions are found by considering the initial condition given by the analytical solution set up at time $t = 0$. Based on these results, ANUGA can handle long waves accurately. The errors for the stage and momentum are indeed seen to be negligible.

4.4. Periodic wave on a paraboloid channel

Now we consider the spatial domain $\{(x, y)|x \in [-4000, 4000], y \in [-4000, 4000]\}$. It is descretised into 100 by 100 rectangular-crosses, in which each rectangular cross has four uniform triangles. Therefore, we have $4 \cdot 10^5$ triangles as the discretisation of the spatial domain.

The initial condition is given by the analytical solution set up at time $t = 0$. The numerical solution of the stage at time $t = 3.75T$ is shown in Figure 15. The stage corresponding to the origin of the spatial domain with respect to time is depicted in Figure 16. The exact solution in this test is much harder to resolve than the previous test. The difficulty occurs due to large vertical variations of the water. The vertical variations are indeed much larger than the previous test. Even though the results shown in Figure 15 lead to small error, as the time evolves, the error grows as illustrated in Figure 16. The oscillation is damped gradually. This means that

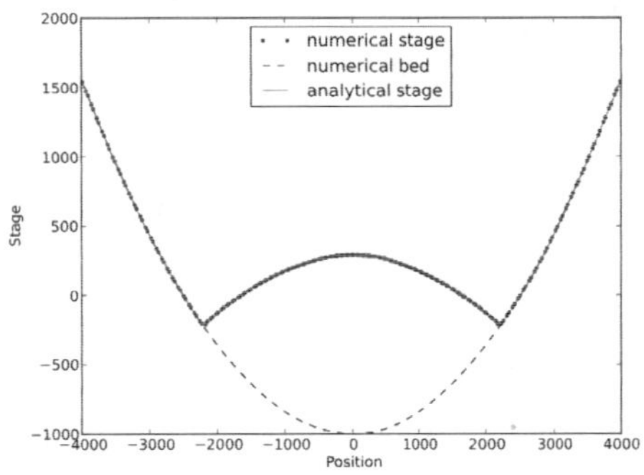

Figure 15. Stage of a cross section of the water oscillation on a paraboloid channel at time $t = 3.75T$.

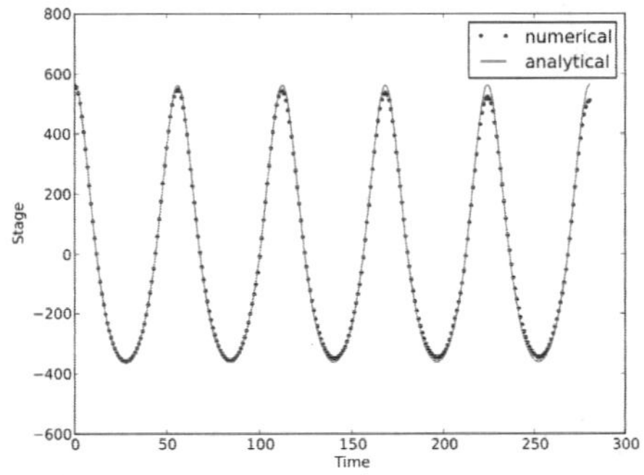

Figure 16. Stage corresponding to the origin of the spatial domain for time $t = 0, ..., 5T$.

the physical energy of this smooth flow decreases over time. One way to reduce the damping or the loss of the physical energy is by running the code with a large number of rectangular crosses (hence a large number of triangles for the spatial domain discretisation), but this leads to an expensive computation.

5. Conclusions

ANUGA has been found to be a stable and robust software to simulate shallow water flows or waves. It can handle wetting and drying process and solve problems with shock waves accurately. For a very long running time, however, it seems that there exists a loss of physical energy, even when the flow is smooth and frictionless. A possible future development of ANUGA is to minimise this loss of physical energy of a smooth and frictionless flow with a reasonably cheap computation.

Acknowledgments

This work was financially supported by The Australian National University (ANU). The authors thank Mathew Langford at the ANU for proofreading the earlier version of this paper.

References

[1] Baldock T E, Barnes M P, Guard P A, Hie T, Hanslow D, Ranasinghe R, Gray D, and Nielsen O 2007 Modelling tsunami inundation on coastlines with characteristic form *Proc. 16th Australasian Fluid Mechanics Conf. (AFMC)* 939–942 AFMS.

[2] Carrier G F and Greenspan HP 1958 Water waves of finite amplitude on a sloping beach *Journal of Fluid Mechanics* **4** 97–109.

[3] Cousins W J, Power W L, Destegul U Z, King A B, Trevethick R, Blong R, Weir B, and Miliauskas B 2009 Earthquake and tsunami losses from major earthquakes affecting the Wellington region *Proc. 2009 New Zealand Society for Earthquake Engineering Conf.* paper number 24 NZSEE.

[4] Davies G 2011 A well-balanced discretization for a shallow water inundation model *Proc. MODSIM 2011 Int. Congress on Modelling and Simulation* 2824–2830 MSSANZ.

[5] de Saint-Venant A J C B 1871 Theorie du mouvement non-permanent des eaux, avec application aux crues des rivieres et a l'introduction des marees dans leur lits *Comptes Rendus de l'Academie des Sciences, Paris* **73** 147–154, 237–240.

[6] Floth U, Vott A, May S M, Bruckner H, and Brockmuller S 2009 Geo-scientific evidence versus computer models of tsunami landfall in the Lefkada coastal zone (NW Greece) *Marburger Geographische Schriften* **145** 140-156 (Marburg: Selbstverlag).

[7] Jakeman J D, Bartzis N, Nielsen O, and Roberts S 2007 Inundation modelling of the December 2004 Indian Ocean tsunami *Proc. MODSIM 2007 Int. Congress on Modelling and Simulation* 1667–1673 MSSANZ.
[8] Jakeman J D, Nielsen O M, Putten K V, Mleczko R, Burbidge D, and Horspool N 2010 Towards spatially distributed quantitative assessment of tsunami inundation models *Ocean Dynamics* **60** 1115–1138.
[9] Johns B 1982 Numerical integration of the shallow water equations over a sloping shelf *International Journal for Numerical Methods in Fluids* **2** 253–261.
[10] Mangeney A, Heinrich P, and Roche R 2000 Analytical solution for testing debris avalanche numerical models *Pure and Applied Geophysics* **157** 1081–1096.
[11] Mungkasi S and Roberts S G 2011 A finite volume method for shallow water flows on triangular computational grids *Proc. 2011 Int. Conf. Advanced Computer Science and Information System (ICACSIS)* 79–84 IEEE.
[12] Mungkasi S and Roberts S G 2011 A new analytical solution for testing debris avalanche numerical models *ANZIAM Journal* **52** C349–C363.
[13] Mungkasi S and Roberts S G 2012 Analytical solutions involving shock waves for testing debris avalanche numerical models *Pure and Applied Geophysics* **169** 1847–1858.
[14] Mungkasi S and Roberts S G 2012 Approximations of the Carrier-Greenspan periodic solution to the shallow water wave equations for flows on a sloping beach *International Journal for Numerical Methods in Fluids* **69** 763–780.
[15] Nielsen O, Sexton J, Gray D, and Bartzis N 2006 Modelling answers tsunami questions *AusGeo News* **83** 1–5 Geoscience Australia.
[16] Ritter A 1892 Die fortpflanzung der wasserwellen *Zeitschrift des Vereines Deutscher Ingenieure* **36** 947-954.
[17] Roberts S G, Nielsen O M, and Jakeman J 2008 Simulation of tsunami and flash floods *In Bock H G et al (Eds) Modeling, Simulation and Optimization of Complex Processes* 489–498 (Berlin: Springer-Verlag).
[18] Roberts S, Nielsen O, Gray D, and Sexton J 2010 *ANUGA User Manual* Commonwealth of Australia (Geoscience Australia) and the Australian National University.
[19] Stoker J J 1948 The formation of breakers and bores *Communications on Pure and Applied Mathematics* **1** 1-87.
[20] Stoker J J 1957 *Water Waves: The Mathematical Theory with Application* (New York: Interscience Publishers).
[21] Thacker W C 1981 Some exact solutions to the nonlinear shallow water equations *Journal of Fluid Mechanics* **107** 499–508.
[22] Van Drie R, Milevski P, and Simon M 2011 Validation of a 2-D Hydraulic Model-ANUGA, to undertake hydrologic analysis *Proc. 34th Int. Association for Hydro-Environment Engineering and Research Congress 2011* 450–457 IAHR.
[23] Van Drie R, Simon M, and Schymitzek I 2008 2D Hydraulic modelling over a wide range of applications with ANUGA *Proc. 9th National Conf. on Hydraulics in Water Engineering* 23–26 Engineers Australia.
[24] Yoon S B and Cho J H 2001 Numerical simulation of coastal inundation over discontinuous topography *Water Engineering Research* **2** 75–87.

Study of phonon modes and elastic properties of $Sc_{36}Al_{24}Co_{20}Y_{20}$ and $Gd_{36}Al_{24}Co_{20}Y_{20}$ rare-earth bulk metallic glasses

P H Suthar[1], P N Gajjar[2], B Y Thakore[3], A R Jani[3]

[1]Department of Physics, C U Shah Science College, Ahmedabad 380 014, Gujarat, India
[2]Department of Physics, University School of Sciences, Gujarat University, Ahmedabad 380 009, Gujarat, India
[3] Department of Physics, Sardar Patel University, Vallabh Vidyanagar 388 120, Gujarat, India

E-mail:sutharpunit@rediffmail.com

Abstract. A phonon modes and elastic properties of two different rare-earth based bulk metallic glasses $Sc_{36}Al_{24}Co_{20}Y_{20}$ and $Gd_{36}Al_{24}Co_{20}Y_{20}$ are computed using Hubbard-Beeby approach and our well established model potential. The local field correlation functions due to Hartree (H), Taylor (T), Ichimaru and Utsumi (IU), Farid et al (F) and Sarkar Sen et al (S) are employed to investigate the influence of the screening effects on the vibrational dynamics of $Sc_{36}Al_{24}Co_{20}Y_{20}$ and $Gd_{36}Al_{24}Co_{20}Y_{20}$ bulk metallic glasses. The results for bulk modulus B_T, modulus of rigidity G, Poisson's ratio ξ, Young's modulus Y, Debye temperature Θ_D, propagation velocity of elastic waves and dispersion curves are reported. The computed elastic properties are found to be in good agreement with experimental and other available data.

1. Introduction

Recently formation of rare- earth based bulk metallic glassy (rare earth-Al-Co-Y) alloys are distinguished itself from other metallic glasses due to their better glasses forming ability. Moreover, rare-earth based bulk metallic glasses have attracted more attention for their unique physical properties. Such as mechanical properties, magnetic properties, elastic properties, thermo plastic behaviour and bond properties of material are one of the key issues for both scientific significance and practical application in condensed matter physics [1-4].

In regards of these Wang [4] has reported the development of series of the rare-earth based bulk metallic glasses with desirable properties based on the correlation. The elastic properties of metallic glass can be roughly predicted to guide the selection of alloying elements for developing rare-earth base bulk metallic glass (BMG) with higher glass forming ability (GFA). In the present paper, focus is given on the $Sc_{36}Al_{24}Co_{20}Y_{20}$ and $Gd_{36}Al_{24}Co_{20}Y_{20}$ rare-earth bulk metallic glasses [4] and the computation is carried out for phonon dispersion curve (PDC), bulk modulus B_T, Poisson's ratio ξ, modulus of rigidity G, Young's modulus Y and Debye temperature Θ_D [5-7]. Our well recognized bare-ion model potential of the following form is to describe bare ion interaction [8].

$$V_b(q)=\frac{-4\pi Ze^2}{\Omega_0 q^2}\left[\cos(qr_c)-\frac{\exp(-1)qr_c}{(1+q^2r_c^2)}\{\sin(qr_c)+qr_c\cos(qr_c)\}\right] \quad (1)$$

Here Z, Ω, **q**, e and r_c are the valency, atomic volume, wave vector, charge of electron and the parameter of the potential respectively. The five exchange and correlation function due to Hartree (H) (without exchange and correlation effect) [9], Taylor (T) [10], Ichimaru and Utsumi (IU) [11], Farid et al (F) [12] and Sarkar Sen et al (S) [13] are employed for the investigation of relative effect of local field correction function on PDC and elastic properties of $Sc_{36}Al_{24}Co_{20}Y_{20}$ and $Gd_{36}Al_{24}Co_{20}Y_{20}$ Bulk metallic glasses.

2. Theory

The effective interaction pair potential for the bulk metallic glass as a one metallic fluid, i.e, the concepts of effective atom [6, 7], can be written as [6, 7]

$$V_{eff}(r)=\left(\frac{Z_{eff}^2e^2}{r}\right)+\frac{\Omega_{0eff}}{\pi}\int F_{eff}(q)\left[\frac{\sin(qr)}{qr}\right]q^2dq \quad (2)$$

Here Z_{eff} and Ω_{0eff} are the effective valence and atomic volume of present system respectively. The energy wave number characteristics appearing in the equation (2) is written as [6, 7]

$$F_{eff}(q)=\left(\frac{-\Omega_{0eff}}{16\pi}\right)\left|V_b^{eff}(q)\right|\frac{\left[\varepsilon_H^{eff}(q)-1\right]}{\left\{1+\left[\varepsilon_H^{eff}(q)-1\right]\left[1-f_{eff}(q)\right]\right\}} \quad (3)$$

Here ,$V_b^{eff}(q)$ is effective bare ion potential it is given in equation (1), $\varepsilon_H^{eff}(q)$ is the Hartree dielectric response function and $f_{eff}(q)$ is the local field correction functions to introduce the exchange and correction effects. A quality which is equally important as the pair potential while studying a disorder system is pair correlation functions $g(r)$, which is computed theoretically from the effective pair potential.

In the Hubbard and Beeby [5] approach, the longitudinal and transverse phonon frequencies are computed using [5-7];

$$\omega_L^2(q)=\omega_E^2\left[1-\frac{3\sin(q\sigma)}{q\sigma}-\frac{6\cos(q\sigma)}{(q\sigma)^2}+\frac{6\sin(q\sigma)}{(q\sigma)^3}\right] \quad (4)$$

$$\omega_T^2(q)=\omega_E^2\left[1+\frac{3\cos(q\sigma)}{(q\sigma)^2}-\frac{6\sin(q\sigma)}{(q\sigma)^3}\right] \quad (5)$$

Here, ω_E represents the maximum phonon frequency and is given as

$$\omega_E=\frac{4\pi n_{eff}}{3M_{eff}}\int_0^\infty g(r)r^2V''_{eff}(r)dr \quad (6)$$

With

$$V''_{eff}(r) = \frac{4Z^2}{r^3} + \frac{\Omega_0}{\pi^2}\int_0^\infty F(q)q^2\left[\frac{2\sin(qr)}{qr^3} - \frac{2\cos(qr)}{r^2} - \frac{q\sin(qr)}{r}\right]dq\ , \tag{7}$$

and

$$g(r) = \exp\left[\left(\frac{-V''_{eff}(r)}{k_B T}\right) - 1\right] \tag{8}$$

Here, ρ, M, $g(r)$, Ω_0, and $f(q)$ be the number density, atomic mass, pair correlation function, atomic volume and energy wave number characteristic, respectively.

In the long-wavelength limit, the phonon dispersion curve shows an elastic behaviour. Hence, the longitudinal v_l and transverse v_t sound velocities are also calculated by

$$v_l = \omega_E\left(\frac{3\xi}{10}\right)^{1/2} \quad \text{and} \quad v_t = \omega_E\left(\frac{\xi}{10}\right)^{1/2}$$

Various elastic properties are then determined by the longitudinal and transverse phonon frequencies.

The bulk modulus B_T, Poisson's ratio ξ, modulus of rigidity G, Young's modulus Y and the Debye temperature θ_D are calculated using the expression below[6, 7],

$$B_T = \rho\left(v_l^2 - \frac{4}{3}v_t^2\right) \tag{10}$$

$$G = \rho v_t^2 \tag{11}$$

$$\xi = \frac{1-2(v_t^2/v_l^2)}{2-2(v_t^2/v_l^2)} \tag{12}$$

$$Y = 2G(\xi+1) \tag{13}$$

and

$$\theta_D = \frac{h}{k_B}\left[\left(\frac{9\rho}{4\pi}\right)^{1/3}\left(\frac{1}{v_l^3} + \frac{2}{v_t^3}\right)^{-1/3}\right] \tag{14}$$

Where, ρ is the isotropic density of the solid, 'h' is plank constant and k_B is the Boltzmann constant.

3. Results and Discussions

Table 1. Input Parameter and constants

Parameters	$Sc_{36}Al_{24}Co_{20}Y_{20}$ BMG	$Gd_{36}Al_{24}Co_{20}Y_{20}$ BMG
Z_{eff}	2.80	2.80
Ω_{eff} (au)^3	146.98	166.61
r_c^{eff} (au)	1.1846	1.2350
M_{eff} (amu)	52.228	92.652

The input parameters and constants required for the present study are shown in table 1. Our well established model potential [8] is used along with five different local field correction functions for the first time to generate pair potential for the $Sc_{36}Al_{24}Co_{20}Y_{20}$ and $Gd_{36}Al_{24}Co_{20}Y_{20}$ rare-earth bulk metallic glasses system. From figure 1 and figure 2 it is observed that position of the first minima is affected by the type of screening functions. The maximum depth in the pair potential in the present study is obtained due to S screening function. It is also noticed that the first zero for $V(r = r_0)$ due to five local field correction functions occurs at $r_0 \approx 12.52$Å and $r_0 \approx 14.84$Å for the $Sc_{36}Al_{24}Co_{20}Y_{20}$ and $Gd_{36}Al_{24}Co_{20}Y_{20}$ rare-earth bulk metallic glasses, respectively. The oscillatory nature is also seen even at large r-region.

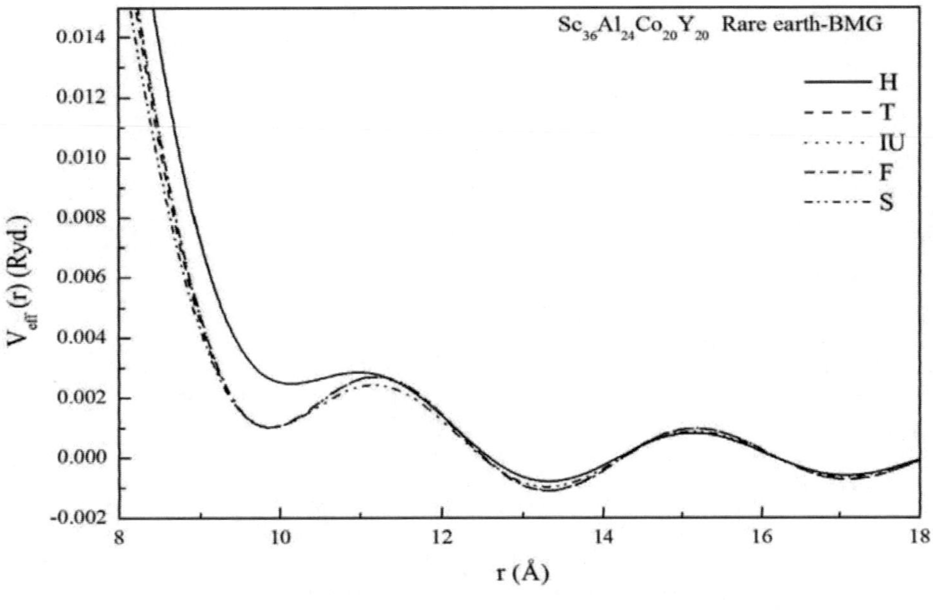

Figure 1. Effective pair potential of $Sc_{36}Al_{24}Co_{20}Y_{20}$ BMG.

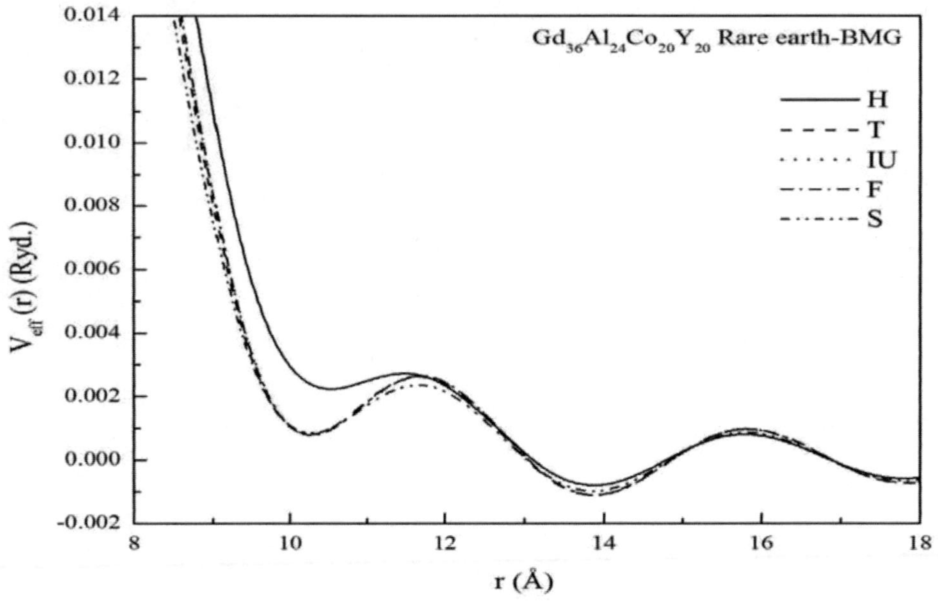

Figure 2. Effective pair potential of $Gd_{36}Al_{24}Co_{20}Y_{20}$ BMG.

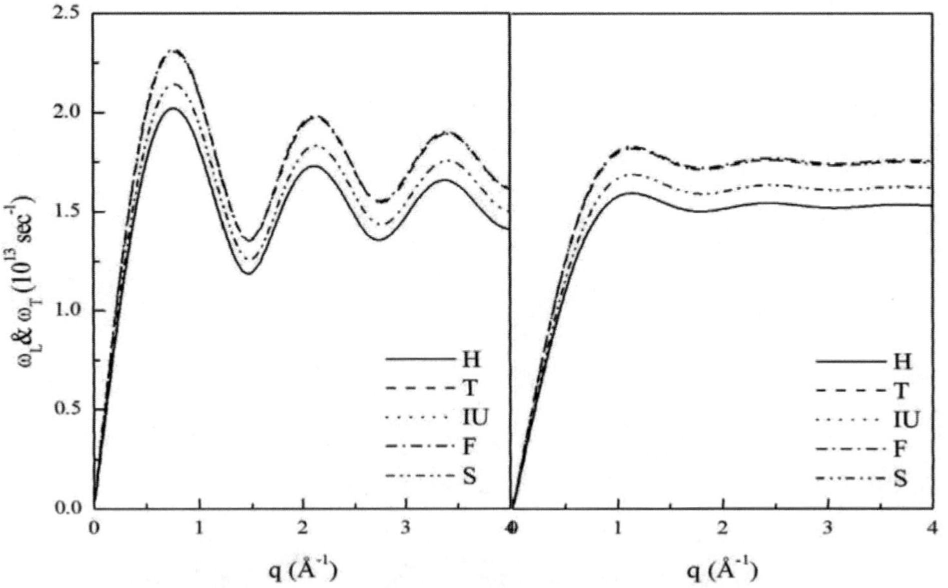

Figure 3. Longitudinal and transverses phonon modes of $Sc_{36}Al_{24}Co_{20}Y_{20}$ BMG

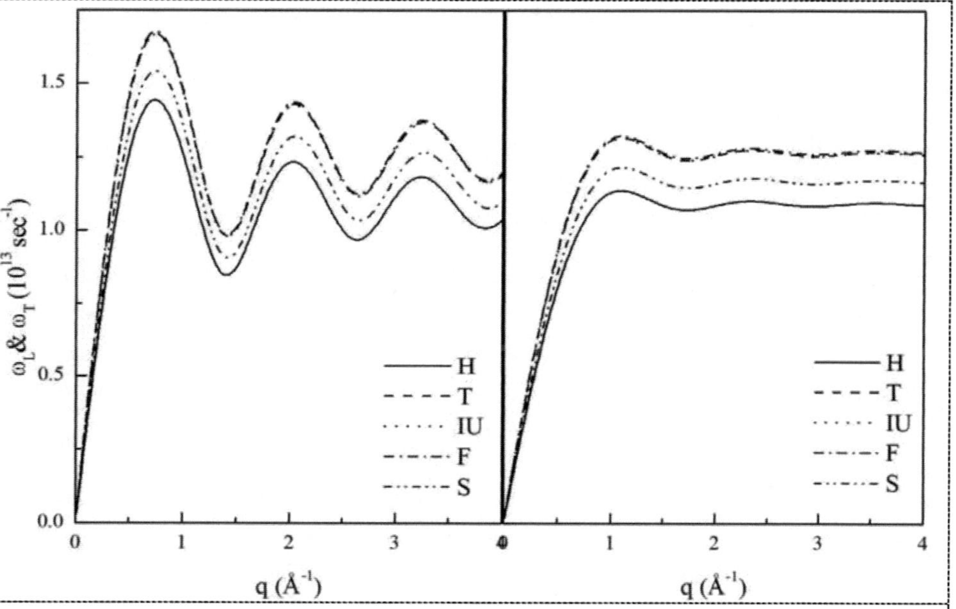

Figure 4. Longitudinal and transverses phonon modes of $Gd_{36}Al_{24}Co_{20}Y_{20}$ BMG

The computed results of the PDC generated using HB scheme along with five different screening functions are shown in figure 3 and figure 4 for the $Sc_{36}Al_{24}Co_{20}Y_{20}$ and $Gd_{36}Al_{24}Co_{20}Y_{20}$ rare-earth bulk metallic glasses, respectively. From figures 3-4, it is observed that the inclusion of local field correction function raises the phonon frequencies in both longitudinal as well as transverse branches. The first minimum in the longitudinal branch of $Sc_{36}Al_{24}Co_{20}Y_{20}$ BMG falls at $q \approx 1.41-1.42$ Å^{-1} and similarly the first minimum in the longitudinal branch of $Gd_{36}Al_{24}Co_{20}Y_{20}$ BMG fall at $q \approx 1.47-1.48$ Å^{-1}. In an absence of experimental data for structure factors, we can offer reverse comment that the structure factor S (**q**) will show it's at these q_s. It is also noticed that the related magnitude of first peak of PDC increasers but the position of first peak is not affected by different screening functions. Phonons of transverse branch undergo large thermal modulation due to anharmonicity of lattice vibrations in this branch. In the long wave length limit, the dispersion curves are linear and confirming characteristics of elastic wave.

Table 2. Elastic properties of $Sc_{36}Al_{24}Co_{20}Y_{20}$ BMG.

Properties	H [9]	T [10]	IU [11]	F [12]	S [13]	Exp. [4]
$\upsilon_L \times 10^5$ (cm s^{-2})	4.2442	4.8069	4.8155	4.8406	4.4569	--
$\upsilon_T \times 10^5$ (cm s^{-2})	2.4504	2.7753	2.7802	2.7947	2.5732	--
B_T (GPa)	43.982	56.418	56.620	57.212	48.502	77.5
G (GPa)	26.389	33.851	34.327	34.327	29.101	32.3
ξ	0.25	0.25	0.25	0.25	0.25	0.31
Y (GPa)	65.973	84.626	84.930	85.818	72.753	85.2
θ_D (K)	290.10	328.56	329.15	330.87	304.64	--

Table 3. Elastic properties of $Gd_{36}Al_{24}Co_{20}Y_{20}$ BMG.

Properties	H [9]	T [10]	IU [11]	F [12]	S [13]	Exp. [4]
$\upsilon_L \times 10^5$ (cm s^{-2})	3.1567	3.6292	3.6328	3.6547	3.3499	--
$\upsilon_T \times 10^5$ (cm s^{-2})	1.8225	2.0953	2.0974	2.1101	1.9340	--
B_T (GPa)	34.119	45.097	45.185	45.733	38.422	57.4
G (GPa)	20.471	27.058	27.111	27.490	23.053	23.6
ξ	0.25	0.25	0.25	0.25	0.25	0.319
Y (GPa)	51.178	67.646	67.778	68.599	57.632	62.2
θ_D (K)	206.94	237.92	238.15	239.59	219.60	--

Presently computed elastic properties for $Sc_{36}Al_{24}Co_{20}Y_{20}$ and $Gd_{36}Al_{24}Co_{20}Y_{20}$ rare-earth bulk metallic glasses are listed in table 2 and table 3, respectively. From these tables, one can see that the values are obtained due to T, IU and F screening functions are very close to one another as compared to the H and S screening functions. The experimental only data for bulk modulus 'B_T', modulus of rigidity 'G', Poisson's ration 'ξ' and Young modulus 'Y' are available. The comparison shows that presently computed results are lower for bulk modulus 'B_T' in both the bulk metallic glasses. The excellent is achieved for modulus of rigidity 'G' and Young modulus 'Y'. Since the unavailability of experimental or other theoretical results of collective excitations for these bulk metallic glasses, we are unable to compare the longitudinal and transverse phonon frequencies and Debye temperature of these bulk metallic glasses.

We have also computed maximum and minimum percentile deviation for the elastic properties of bulk modulus, modulus of rigidity and Young modulus with the available experimental data from table 2-3, respectively. In the case of bulk modulus, the computed maximum percentile deviation with respect to the available experimental data is found for H function and the values are 43.29% and 40.56% for $Sc_{36}Al_{24}Co_{20}Y_{20}$ and $Gd_{36}Al_{24}Co_{20}Y_{20}$ bulk metallic glasses, respectively and minimum deviation is found for F screening function and the values are 26.17% and 20.32% for $Sc_{36}Al_{24}Co_{20}Y_{20}$ and $Gd_{36}Al_{24}Co_{20}Y_{20}$ bulk metallic glasses, respectively. For modulus of rigidity, the maximum deviation is found for H function and the values are 18.30% and 13.25% for $Sc_{36}Al_{24}Co_{20}Y_{20}$ and $Gd_{36}Al_{24}Co_{20}Y_{20}$ bulk metallic glasses, respectively and minimum deviation is found for IU and S functions, the values are 6.27% and 2.52% for $Sc_{36}Al_{24}Co_{20}Y_{20}$ and $Gd_{36}Al_{24}Co_{20}Y_{20}$ bulk metallic glasses respectively. Similarly, for Young's modulus, the computed maximum percentile deviation is found for H function and the values are 22.56% and 13.52% for $Sc_{36}Al_{24}Co_{20}Y_{20}$ and $Gd_{36}Al_{24}Co_{20}Y_{20}$ bulk metallic glasses, respectively. Minimum deviation is found for F and S functions, the values are 0.45% and 7.34% for $Sc_{36}Al_{24}Co_{20}Y_{20}$ and $Gd_{36}Al_{24}Co_{20}Y_{20}$ bulk metallic glasses, respectively. Moreover, we have also calculated the relative influence of various local field correlation functions with respect to the static H-dielectric function. From table 2, the percentile influence for sound velocity for T, IU, F and S are resulted as 13.25%, 13.46%, 14.05% and 5.01% respectively. The percentile influence for bulk metallic glasses, modulus of rigidity, Young's modulus and Debye temperature for T, IU, F and S screening functions are computed as 28.27%, 28.73%, 30.08% and 10.27%, respectively. With respect to the highest influence of H- screening function is observed for F local field correlation function.

From table 3, we have calculated the percentile influence for sound velocity for T, IU, f and S screening functions are resulted as 14.96%, 15.08%, 15.78% and 6.12%, respectively. While percentile influence for bulk metallic glasses, modulus of rigidity, Young modulus and Debye temperature for T, IU, F and S screening functions are computed as 32.18%, 32.43%, 34.04% and 12.61% respectively. It is also observed that the F screening function gives the highest influence over H-screening function then other. However, the present findings are significant when comparison is carried out with the results of elastic constants, and it is found in line and supports the pseudopotential used in the present investigations. From this calculation it is observed that the presently computed data serve as guideline for further research in this direction.

4. Conclusion

We conclude here that the PDC generated from the HB approach reproduces satisfactorily general characteristics of dispersion curves for $Sc_{36}Al_{24}Co_{20}Y_{20}$ and $Gd_{36}Al_{24}Co_{20}Y_{20}$ bulk metallic glasses. The rare-earth bulk metallic glasses have not been investigated theoretically previously adopting model potential formalism and HB approach. The present local model potential depends on effective core radius which is used here to describe the electron-ion interaction. It is obvious that the present results are generated without any fitting procedure. This also confirms the applicability of the model potential in the aforesaid properties and supports the present approach.

Acknowledgement

The author PNG acknowledge the computer facilities developed under DST-FIST level-1 and financial support of DRS-SAP-I. BYT is thankful to UGC for providing financial support under major research project.

References

[1] Geer A L 1995 *Science* **267** 1947
[2] Wang W H, Dong C and Shek C H 2004 *Mater. Sci. Eng.* R **44** 45
[3] Jiang Q K, Zhang G Q, Yang L, Wang X D, Saksl K, Franz H, Wunderlich R, Fecht H and Jiang J Z 2007 *Acta Mater.* **55** 4409
[4] Wang W H 2006 *J. Appl. Phys.* **99** 093506
[5] Hubbard J and Beeby J L 1969 *J. Phys. C* **2** 556
[6] Thakore B Y, Gajjar P N and Jani A R 2000 *Bull. Mat. Sci.* **23** 5
[7] Thakore B Y, Dabhi H, Joshi M and Jani A R 2009 *Phys. Scr.* **79** 0250007
[8] Thakore P B, Gajjar P N and Jani A R 2004 *Communication in Phys.* **14** 15
[9] Harrison W A 1969 *Phys. Stat. Sol. (6)* **181** 1036
[10] Taylor R 1981 *J. Phys. F* **8** 1699
[11] Ichimaru S and Utsumi K 1981 *Phys. Rev.* B **24** 3220
[12] Farid B, Heine V, Engel G E and Robertson I S 1993 *Phys. Rev.* **48** 11602
[13] Sarkar A, Sen D S and Roy H D 1998 *Mod. Phys. Lett.* **12** 639

Journal of Physics: Conference Series **423** (2013) 012031 doi:10.1088/1742-6596/423/1/012031

Solution of Schrodinger equation for Three Dimensional Harmonics Oscillator plus Rosen-Morse Non-central potential using NU Method and Romanovski Polynomials

C Cari[1] and A Suparmi[2]
[1]Physics Department, Sebelas Maret University, Surakarta 57126
Ph:081548578367,
E-mail : carinln@yahoo.com
[2]Physics Department, Sebelas Maret University, Surakarta 57126
Ph:081548578367,
E-mail : suparmiuns@gmail.com

Abstract. The energy eigenvalues and eigenfunctions of Schrodinger equation for three dimensional harmonic oscillator potential plus Rosen-Morse non-central potential are investigated using NU method and Romanovski polynomial. The bound state energy eigenvalues are given in a closed form and corresponding radial wave functions are expressed in associated Laguerre polynomials while angular eigen functions are given in terms of Romanovski polynomials. The Rosen-Morse potential is considered to be a perturbation factor to the three dimensional harmonic oscillator potential that causes the increase of radial wave function amplitude and decrease of angular momentum length. Keywords: Schrodinger Equation, Three dimensional Harmonic Oscillator potential, Rosen-morse non-central potential, NU method, Romanovski Polynomials

1. Introduction

The exact analytical solutions of Schrodinger equations for some physical potentials are very essential since the knowledge of wavefunctions and energy contains all possible important information of the physical properties of quantum system. Recently, conciderably efforts have been paid to obtain the exact solution of the central and non-central potentials. There are only a few potentials for which the Schrodinger equation can be solved exactly.

In recent years, numerous studies have been made in analyzing the bound state of charged particle moving in a vector potential and a non-central scalar potential, such as an electron moving in a Coulomb field with simultaneously precence of Aharonov-Bohm field [1-3], or/and magnetic monopole [4], Makarov potential [5] or ring-shaped-oscillator potential [6-8], etc. In most of these studies, the eigenvalues and eigenfunctions are obtained using separation variables in spherical coordinate system. Very recently, supersymmetric quantum mechanics with the idea of shape invariance [8-9], factorization method [10-11], and Nikiforov-Uvarov method [12-13] are widely used to derive the energy spectrum and the wave function of a charge particle moving in non-central potential.

Very recently, an alternative method used to solve Schrodinger equation for a class of shape invariant potentials is NU method and Romanovski polynomials [14-16]. NU method was developed by Nikiforov-Uvarov [12] is based on solving the second order linear differential equations by reducing it to a generalized equation of hypergeometric type by a suitable change of variable. As NU method, finite Romanovski polynomials, which is a traditional method, consist of reducing Schrodinger equation by an appropriate change of the variable to that of very form of generalized hypergeometric equation [17]. The polynomial was discovered by Sir E.J.Routh [18] and rediscovered 45 years later by V. I. Romanovski [19]. The notion "finite" refers to the observation that, for any given set of parameters (i.e. in any potential) only a finite of polynomials appear orthogonal[20]. It seems that the two methods are very similar but they solve Schrodinger equation differently. NU method is applied wider than Romanovski polynomials.

In this paper we will attempt to solve the Schrodinger equation for three dimensional harmonic oscillator potential with simultaneously presence of Rosen-Morse non-central potential using NU method for radial Schrodinger equation and Romanovski polynomials for angular Schrodinger equation. Three dimensional harmonics oscillator is one of exactly solvable potential that used to describe the nuclei, atomic or molecular vibration. Non-central potential composed of spherical harmonics oscillator with square of inverse potential together with ring-shaped non-central potential, or double ring shaped potential have been investigated intensively by some authors [21-23]. The Rosen-Morse potential is trigonometric potential which was proposed by Rosen-Morse [24] in 1935 and was used to describe the diatomic molecular vibration. The approximate bound state solution for trigonometric Rosen-Morse potential have also been studied for *l*-state solution [15,25], and Coulombic Rosen Morse non-central potential, particularly for $\cot\theta$ part, has been investigated intensively [15].

This paper is organized as follows. In section 2, we review NU method and the finite Romanovski polynomials briefly. In section 3, we find the bound state energy solution and radial wave function expressed in associated Laguerre polynomials, and the angular wave functions in terms of Romanovski polynomials. 4. A brief conclusion is presented in section 5.

2. Review of formulas for NU method and finite Romanovski Polynomials

2.1. Review of Nikiforov-Uvarov Method

The one-dimensional Schrodinger equation of any shape invariant potential can be reduced into hypergeometric or confluent hypergeometric type differential equation by suitable variable transformation[12,26]. The hypergeometric type differential equation, which is solved using Nikiforov-Uvarov method, is presented as

$$\frac{\partial^2\Psi(s)}{\partial s^2}+\frac{\bar{\tau}(s)}{\sigma(s)}\frac{\partial\Psi(s)}{\sigma^2}+\frac{\bar{\sigma}(s)}{\sigma^2}\Psi(s)=0 \tag{1}$$

where $\sigma(s)$ and $\bar{\sigma}(s)$ are polynomials at most in the second order, and $\bar{\tau}(s)$ is first order polynomial. Equation (1) can be solved using separation of variable method which is expressed as

$$\Psi=\phi(s)y(s) \tag{2}$$

By inserting equation (2) into equation (1) we get hypergeometric type equation

$$\sigma\frac{\partial^2 y}{\partial s^2}+\tau\frac{\partial^2 y}{\partial s}+\lambda y=0 \tag{3}$$

and $\phi(s)$ is a logarithmic derivative whose solution obtained from condition

$$\frac{\phi'}{\phi}=\frac{\pi}{\sigma} \tag{4}$$

while the function $\pi(s)$ and the parameter λ are defined as

$$\pi=\left(\frac{\sigma'-\bar{\tau}}{2}\right)\pm\sqrt{\left(\frac{\sigma'-\bar{\tau}}{2}\right)^2-\bar{\sigma}+k\sigma} \tag{5}$$

$$\lambda=k+\pi' \tag{6}$$

Journal of Physics: Conference Series **423** (2013) 012031 doi:10.1088/1742-6596/423/1/012031

The value of k in equation (5) can be found from the condition that the expression under the square root of equation (5) must be square of polynomial which is mostly first degree polynomial and therefore the discriminate of the quadratic expression is zero. A new eigenvalue of equation (3) is

$$\lambda = \lambda_n = -n\tau' - \frac{n(n-1)}{2}\sigma'' \quad , \text{n} = 0,\ 1,\ 2 \tag{7}$$

where

$$\tau = \bar{\tau} + 2\pi \tag{8}$$

The new energy eigenvalue is obtained using equation (6) and (7).

To generate the energy eigenvalues and the corresponding eigenfunctions, the condition that$\tau' < 0$ is required. The solution of the second part of the wave function, $y_n(s)$, which is connected to Rodrigues relation,[8] is given as

$$y_n(z) = \frac{c_n}{w(z)} \frac{d^n}{dz^n}\left(\sigma^n(z) w(z)\right) \tag{9}$$

where C_n is normalization constant, and the weight function $\rho(s)$ must satisfies the condition

$$\frac{\partial(\sigma w)}{\partial s} = \tau(s) w(s) \tag{10}$$

The wave function of the system is therefore obtained from equation (4), (9) and (10).

2.2. Riview of finite Romanovski Polynomials

The hypergeometric type differential equation, which can be solved using finite Romanovski polynomials that was developed by Romanovski [19] is presented as in equation (3) in NU method where

$$\sigma(s) = as^2 + bs + c;\ \ \tau = fs + e \ \text{ and } \ \{-n(n-1) + 2n(1-p)\} = \lambda = \lambda_n \tag{11}$$

Equation (11) is described in the textbook by Nikiforov-Uvarov [12] where it is cast into self adjoint form and its weight function , $w(s)$, satisfies the so called Pearson differential equation

$$\frac{d(\sigma(s) w(s))}{ds} = \tau(s) w(s) \tag{12}$$

The weight function is obtained by solving the Pearson differential equation, that is

$$w(s) = \exp\left(\int \frac{(f-2a)s + (e-b)}{as^2 + bs + c} ds\right) \tag{13}$$

The corresponding polynomials are classified according to the weight function, and are built up from the Rodrigues representation as

$$y_n = \frac{1}{w(s)} \frac{d^n}{ds^n}\left\{(as^2 + bs + c)^n w(s)\right\} \tag{14}$$

For Romanovski polynomial, the values of parameters in equation (13) are:

$$a = 1,\ b = 0,\ c = 1,\ f = 2(1-p) \ \text{ and } \ e = q \text{ with } p > 0 \tag{15}$$

By inserting equation (15) into equation (13) we obtain the weight function

$$w(s) = \exp\left(\int \frac{(f-2a)s + (c-b)}{as^2 + bs + c}\right) = \exp\left(\int \frac{(2-2p-2)s + q}{s^2+1} ds\right)$$

$$w(s) = (1+s^2)^{-p} e^{q \tan^{-1}(s)} \tag{16}$$

The polynomial associated with equation (16) are named after Romanovski and will be denoted by $R_n^{(p,q)}(s)$. Due to the decrease of the weight function by s^{-2p} [18], integral of the type

$$\int_{-\infty}^{\infty} w^{(p,q)} R_n^{(p,q)}(s) R_{n'}^{(p,q)}(s) ds \tag{17}$$

will be convergent only if $n' + n < 2p - 1$ (18)

This means that only a finite number of Romanovski polynomials are orthogonal.

The differential equation satisfied by Romanovski Polynomial obtained by inserting equations (11) and (15) into equation (3) is

$$(1+s^2)\frac{\partial^2 R_n^{(p,q)}}{\partial s^2}+\{2s(-p+1)+q\}\frac{\partial R_n^{(p,q)}(s)}{\partial s}-\{n(n-1)+2n(1-p)\}R_n^{(p,q)}(s)=0 \quad (19)$$

where $y_n = R_n^{(p,q)}(s)$. The Schrodinger equation of the potential of interest will be reduced into the form which is similar to equation (19) by an appropriate transformation of variable, $r = f(s)$, and by introducing a new wave function which is given as

$$\Psi(r)= g_n(s) = (1+s^2)^{\frac{\beta}{2}} e^{\frac{-\alpha}{2}\tan^{-1}s} D_n^{(\beta,\alpha)}(s) \quad (20)$$

The eigen function in equation (20) is the solution of Schrodinger equation for potential interest where

$$D_n^{(\beta,\alpha)}(s) = R_n^{(p,q)}(s) \quad (21)$$

The Romanovski polynomials obtained from equations (14), 15 and (16) is expressed as

$$R_n^{(p,q)}(s) = D_n^{(\beta,\alpha)}(s) = \frac{1}{(1+s^2)^{-p}e^{q\tan^{-1}(s)}}\frac{d^n}{ds^n}\left\{(1+s^2)^n(1+s^2)^{-p}e^{q\tan^{-1}(s)}\right\} \quad (22)$$

3. Schrodinger equation of three dimensional harmonic oscillator (3D HO) plus Rosen-Morse potential

The three dimensional Schrodinger equation for 3D HO potential with simultaneously the presence of trigonometric Rosen-Morse non-central potensial is expressed as

$$-\frac{\hbar^2}{2M}\left\{\frac{1}{r^2}\frac{\partial}{\partial r}\left(r^2\frac{\partial}{\partial r}\right)+\frac{1}{r^2\sin\theta}\frac{\partial}{\partial\theta}\left(\sin\theta\frac{\partial}{\partial\theta}\right)+\frac{1}{r^2\sin^2\theta}\frac{\partial^2}{\partial\varphi^2}\right\}\Psi(r,\theta,\varphi)+\left\{\frac{M\omega^2r^2}{2}+\frac{\hbar^2}{2Mr^2}\left(\frac{\nu(\nu+1)}{\sin^2\theta}-2\mu\cot\theta\right)\right\}\Psi(r,\theta,\varphi)=E\Psi(r,\theta,\varphi)$$

(23)

Equation (23) is solved using variable separation method by setting $\psi(r,\theta,\varphi) = R(r)P(\theta)\phi(\varphi)$ so we obtain

$$\frac{1}{R}\frac{\partial}{\partial r}\left(r^2\frac{\partial}{\partial r}\right)-\frac{r^2}{\hbar^2}M^2\omega^2r^2+\frac{2Mr^2}{\hbar^2}E=-\frac{1}{P\sin\theta}\frac{\partial}{\partial\theta}\left(\sin\theta\frac{\partial}{\partial\theta}\right)-\frac{1}{\phi\sin^2\theta}\frac{\partial^2}{\partial\varphi^2}+\left(\frac{\nu(\nu+1)}{\sin^2\theta}-2\mu\cot\theta\right)=\lambda=\ell(\ell+1)$$

(24)

For azimuthal part we have

$$\frac{1}{\phi}\frac{\partial^2}{\partial\varphi^2}=-m^2 \quad (25)$$

therefore the solution of azimuthal part of wave function as usual is

$$\phi = A_m e^{im\varphi} \quad (26)$$

From equation (24) we obtain the radial and angular parts of Schrodinger equation given as

$$\frac{1}{R}\frac{\partial}{\partial r}\left(r^2\frac{\partial}{\partial r}\right)-\frac{r^2}{\hbar^2}M^2\omega^2r^2+\frac{2Mr^2}{\hbar^2}E=l(l+1) \quad (27)$$

$$-\frac{1}{P\sin\theta}\frac{\partial}{\partial\theta}\left(\sin\theta\frac{\partial}{\partial\theta}\right)+\frac{m^2}{\sin^2\theta}+\left(\frac{\nu(\nu+1)}{\sin^2\theta}-2\mu\cot\theta\right)=l(l+1) \quad (28)$$

From the solution of equation (27) will be obtained the energy spectrum of 3D HO potential and the radial part of wave function while from equation (28) will be obtained the angular wave function and the value of l which is the orbital quantum number.

3.1 Solution of radial Schrodinger equation

By setting $\frac{2M}{\hbar^2}E=\varepsilon^2 \quad \frac{M^2\omega^2}{\hbar^2}=\gamma^2$ and $R=\frac{\chi}{r}$ (29)

and inserting it into equation (27) we get

$$\frac{\partial^2\chi}{\partial r^2}+\left(-\gamma^2r^2+\varepsilon^2-\frac{l(l+1)}{r^2}\right)\chi=0 \tag{30}$$

Making a change variable $r^2 = x$ in equation (30) and change it into equation which is given as

$$x\frac{\partial^2\chi}{\partial x^2}+\frac{1}{2}\frac{\partial\chi}{\partial x}-\left(\frac{\gamma^2x^2-\varepsilon^2x+l(l+1)}{4x}\right)\chi=0 \tag{31}$$

By comparing equation (1) and (31) we have

$$\sigma=x,\qquad \vec{\tau}=\frac{1}{2}\qquad \vec{\sigma}=-\left\{\frac{l(l+1)+\gamma^2x^2-\varepsilon^2x}{4}\right\} \tag{32}$$

From equations (5) and (32) we get

$$\pi=\frac{1}{4}\pm\sqrt{\frac{1}{16}+\frac{l(l+1)+\gamma^2x^2-\varepsilon^2x}{4}+kx} \tag{33}$$

The value of k is obtained from the condition that quadratic expression under the square root in equation (33) has to be completely square of first degree of polynomial. From this condition equation (33) is rewritten as

$$\pi=\frac{1}{4}\pm\frac{\gamma}{2}\left\{x+\frac{4k-\varepsilon^2}{2\gamma^2}\right\} \tag{34}$$

and the discriminate of the quadratic expression under the square root that has to be zero is given as

$$\left(k-\frac{\varepsilon^2}{4}\right)^2-\left(\frac{\gamma^2}{4}\right)\left(l+\frac{1}{2}\right)^2=0 \tag{35}$$

From equation (35) we obtain the value of k as

$$k_1=\frac{\varepsilon^2}{4}+\frac{\gamma}{2}(l+\frac{1}{2})\qquad\text{or}\qquad k_2=\frac{\varepsilon^2}{4}-\frac{\gamma}{2}(l+\frac{1}{2}) \tag{36}$$

By imposing the condition that $\tau' < 0$ then from equation (34) and (35) we get

$$\pi_1=-\frac{\gamma x}{2}-\frac{l}{2}\ \text{for } k_1\qquad\text{or}\qquad \pi_2=-\frac{\gamma x}{2}+\frac{(l+1)}{2}\ \text{for } k_2 \tag{37}$$

By using equation (6), (8), (32) and (37) we obtain

$$\lambda_1=\frac{\varepsilon^2}{4}+\frac{\gamma}{2}\left(l-\frac{1}{2}\right)\ \text{for } k_1\qquad\text{or}\qquad \lambda_2=\frac{\varepsilon^2}{4}-\frac{\gamma}{2}\left(l+\frac{3}{2}\right)\ \text{for } k_2 \tag{38}$$

$$\tau_1=-\gamma x-\left(l-\frac{1}{2}\right)\ \text{for } k_1\qquad\text{or}\qquad \tau_2=-\gamma x+\left(l+\frac{3}{2}\right)\ \text{for } k_2 \tag{39}$$

From equations (7), (32) and (39) we obtain the same values of λ_n either for k_1 or k_2, that is

$$\lambda=\lambda_n=-n_r(-\gamma)=\gamma n_r \tag{40}$$

The energy eigenvalue obtained by equating equations (38) and (40) is given as

$$\frac{\varepsilon_1^2}{4}=\gamma\left(n_r-\left(\frac{l}{2}-\frac{1}{4}\right)\right)\ \text{for } k_1\qquad\text{or}\qquad \frac{\varepsilon_1^2}{4}=\gamma\left(n_r+\left(\frac{l}{2}+\frac{3}{4}\right)\right)\ \text{for } k_2 \tag{41}$$

To have physical meaning, the choice of the values are $k_2, \pi_2, \tau_2, \lambda_2$ and λ_{n2} therefore the energy spectrum of 3D HO plus Rosen-Morse non-central potential is obtained from equation (41) which is given as

$$E=\hbar\omega\left(2n_r+l+\frac{3}{2}\right) \tag{42}$$

where n_r is radial quantum number, l is orbital quantum number and its values depend on the parameters of Rosen-Morse non-central potential. The orbital quantum number obtained from the solution of angular Schrodinger equation in section 3.2 is expressed as

$$l = l' = \sqrt{\left(\sqrt{\nu(\nu+1)+m^2}+n_l+\frac{1}{2}\right)^2 - \frac{\mu^2}{\left(\sqrt{\nu(\nu+1)+m^2}+n_l+\frac{1}{2}\right)^2}} - \frac{1}{2} \tag{43}$$

We can see from equation (43) that for fixed values of principal quantum number, n, and radial quantum number n_r, the values of $l = l'$ is not fixed since it depends on the Rosen Morse's parameter.
The radial wave functions are calculated using equations (4), (9), (10), (37), and (39). The first part of the wave function obtained using equations (4), (32) and (37) is given as

$$\phi(r) = (x)^{\frac{(l'+1)}{2}} e^{-\frac{\gamma x}{2}} \tag{44}$$

and by using equations (10), (32), and (39) we get the weight function which is given as

$$w(x) = \rho(x) = x^{l'+\frac{1}{2}} e^{-\gamma x} \tag{45}$$

The second part of radial wave function obtained using equations (9) and (45) is given as

$$y_{n_r}(x) = \frac{c_{n_r}}{w(x)}\frac{d^{n_r}}{dx^{n_r}}\left(\sigma^{n_r}(x)w(x)\right) = \frac{c_{n_r}}{x^{l'+\frac{1}{2}}e^{-\gamma x}}\frac{d^{n_r}}{dx^{n_r}}\left(x^{l'+\frac{1}{2}+n}e^{-\gamma x}\right) \tag{46}$$

By setting $\beta x = z$ in equation (46), we get the second part of radial wave function and the total wave function in terms of associated Laguerre polynomials

$$y_{n_r}(x) = C_{n_r}(\gamma)^{l'+\frac{1}{2}} n_r! L_{n_r}^{l'+\frac{1}{2}}(z) \quad \text{and} \quad \chi_{n_r} = B_{n_r} z^{\frac{(l'+1)}{2}} e^{-\frac{z}{2}} L_{n_r}^{l'+\frac{1}{2}}(z) \tag{47}$$

The radial wave functions with and without the presence of Rosen-Morse potential and its related energy spectra for $n_l = n_r$ are listed in Table 1. The graph of radial wave functions, $R_{n_l m \nu\mu n_r}(r)$, for $n_r = n_l = 1$, $m=1$ and different values of ν and μ are shown in Figures 1 and 2. As shown in Table 1and Figures 1-2, the two terms of Rosen Morse potential, $\csc^2\theta$ and $\cot\theta$, give different effect on the radial wave function. The $\csc^2\theta$ term causes the increase of the energy and the radial wave function's amplitude, as shown in Figure1, while $\cot\theta$ term causes the decrease in energy and radial wave function's amplitude, as shown in Figure 2. The normalization factor B_n in equation (47) can be obtained from the normalization condition of radial wave function which is expressed as

$$\int_0^\infty \chi_{n_r'}(r)\chi_{n_r}(r)dr = \delta_{n_r' n_r} \tag{48}$$

By using equations (47) and (48) we get

$$B_{n_r} = \sqrt{\frac{2\sqrt{\gamma}n_r!}{\left(n_r + l' + \frac{1}{2}\right)!}} \tag{49}$$

By the absent of Rosen Morse potential the radial wave function becomes the wave function of 3D HO.

Table 1. The second part of and the radial wave functions for $n_l = n_r$,and the energy for various values of l'.

No	n_l	m	ν	μ	l'	y_{n_r}	$R_{n_l m \nu\mu n_r}(y)$	$E_{n_r}(\hbar\omega)$
1	2	1	0	0	3	$C_2(24.75)-11\gamma r^2+\gamma^2 r^4)$	$C_2 r^3 e^{-\frac{\gamma r^2}{2}}\{(24.75)-11\gamma r^2+\gamma^2 r^4)\}$	8.5

2	2	1	2	0	4.65	$C_2\{(43.97) - 14.3\gamma r^2 + \gamma^2 r$	$C_2 r^{4.65} e^{-\frac{\gamma r^2}{2}}\left\{\begin{matrix}(43.97) - 14.3\gamma r^2 \\ + \gamma^2 r^4)\end{matrix}\right\}$	10.15
3	2	1	0	2	2.95	$C_2\{(24..25) - 10.9\gamma r^2 + \gamma^2 r$	$C_2 r^{3.45} e^{-\frac{\gamma r^2}{2}}\left\{\begin{matrix}(24.25) - 10.9\gamma r^2 \\ + \gamma^2 r^4)\end{matrix}\right\}$	8.45
4	2	1	2	2	4.64	$C_2\{(43.84) - 14.3\gamma r^2 + \gamma^2 r$	$C_2 r^{4.64} e^{-\frac{\gamma r^2}{2}}\left\{\begin{matrix}(43.84) - 14.3\gamma r^2 \\ + \gamma^2 r^4)\end{matrix}\right\}$	10.14
5	3	0	0	0	3	$C_3\left\{\begin{matrix}(160.88) - (107.25)\gamma r^2 \\ + 19.5\gamma^2 r^4 - \gamma^3 r^6\end{matrix}\right\}$	$C_3 r^3 e^{-\frac{\gamma r^2}{2}}\left\{\begin{matrix}(160.88) - (107.25)\gamma r^2 \\ + 19.5\gamma^2 r^4 - \gamma^3 r^6\end{matrix}\right\}$	10.5
6	3	1	0	0	4	$C_3\left\{\begin{matrix}(268.13) - (146.25)\gamma r^2 \\ + 22.5\gamma^2 r^4 - \gamma^3 r^6\end{matrix}\right\}$	$C_3 r^4 e^{-\frac{\gamma r^2}{2}}\left\{\begin{matrix}(268.13) - (146.25)\gamma r^2 \\ + 22.5\gamma^2 r^4 - \gamma^3 r^6\end{matrix}\right.$	11.5
7	3	1	2	0	5.65	$C_3\left\{\begin{matrix}(533.19) - (223.72)\gamma r^2 \\ + 27.45\gamma^2 r^4 - \gamma^3 r^6\end{matrix}\right\}$	$C_3 r^{5.65} e^{-\frac{\gamma r^2}{2}}\left\{\begin{matrix}(533.19) - (223.72)\gamma r^2 \\ + 27.45\gamma^2 r^4 - \gamma^3 r^6\end{matrix}\right\}$	13.15
8	3	1	0	2	3.98	$C_3\left\{\begin{matrix}(265.62) - (145.41)\gamma r^2 \\ + 22.44\gamma^2 r^4 - \gamma^3 r^6\end{matrix}\right\}$	$C_3 r^{3.98} e^{-\frac{\gamma r^2}{2}}\left\{\begin{matrix}(265.62) - (145.41)\gamma r^2 \\ + 22.44\gamma^2 r^4 - \gamma^3 r^6\end{matrix}\right.$	11.48
9	3	1	2	2	5.64	$C_3\left\{\begin{matrix}(531.21) - (223.2)\gamma r^2 \\ + 27.42\gamma^2 r^4 - \gamma^3 r^6\end{matrix}\right\}$	$C_3 r^{5.64} e^{-\frac{\gamma r^2}{2}}\left\{\begin{matrix}(531.21) - (223.2)\gamma r^2 \\ + 27.42\gamma^2 r^4 - \gamma^3 r^6\end{matrix}\right\}$	13.14

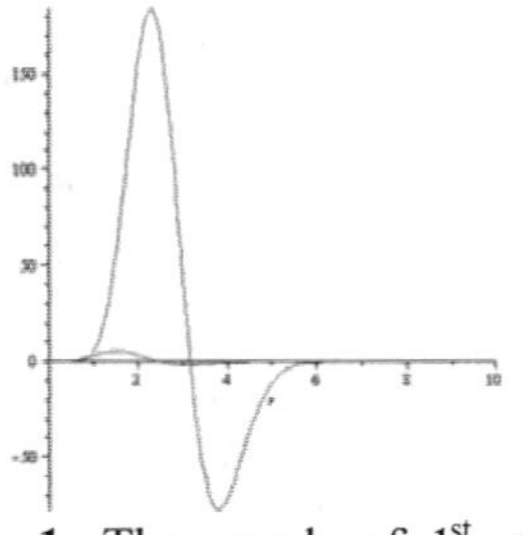

Figure 1. The graph of 1st state of Radial wave function for 3D HO- Rosen Morse system with $\csc^2\theta$ term

— $R_{11201}(r) = C_1 r^{3.65} e^{-\frac{\gamma r^2}{2}}(5.15 - \gamma r^2)$

— $R_{11601}(r) = C_1 r^{8.06} e^{-\frac{\gamma r^2}{2}}(9.56 - \gamma r^2)$

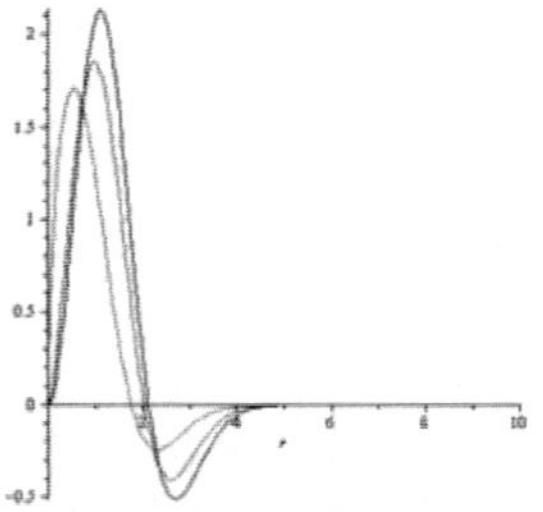

Figure 2. The graph of 1st state of Radial wave function for 3D HO- Rosen Morse system with $\cot\theta$ term:

— $R_{11021}(r) = C_1 r^{1.55} e^{-\frac{\gamma r^2}{2}}(4.05 - \gamma r^2)$

— $R_{11061}(r) = C_1 r^{1.95} e^{-\frac{\gamma r^2}{2}}(4.45 - \gamma r^2)$

— $R_{110101}(r) = C_1 r^{0.52} e^{-\frac{\gamma r^2}{2}}(3.02 - \gamma r^2)$

3.2 The solution of angular Schrodinger equation

By setting $P = \frac{Q}{\sqrt{\sin\theta}}$ in equation (28) then equation (28) becomes

$$\frac{d^2Q}{d\theta^2}-\left(\frac{\nu(\nu+1)+m^2-0.25}{\sin^2\theta}-2\mu\cot\theta\right)Q+\left(l(l+1)+\frac{1}{4}\right)Q=0 \tag{50}$$

To solve equation (50) we introduce a new variable $\cot\theta = s$ and equation (50) change into

$$\left(1+s^2\right)\frac{\partial^2 Q}{\partial s^2}+2s\frac{\partial Q}{\partial s}-\left\{\left(\nu(\nu+1)+m^2-\frac{1}{4}\right)-\frac{2\mu x}{\left(1+s^2\right)}-\frac{l(l+1)+\frac{1}{4}}{\left(1+s^2\right)}\right\}Q=0$$

(51)

Equation (51) is solved in terms of Romanovski polynomial by setting

$$Q(\theta)=g_n(s)=\left(1+s^2\right)^{\frac{\beta}{2}}e^{-\frac{\alpha}{2}\tan^{-1}}D_n^{(\beta,\alpha)}(s) \tag{52}$$

By inserting equation (52) into equation (51) we obtain

$$\left(1+s\right)\frac{\partial^2 D}{\partial s^2}+\left\{2s(\beta+1)-\alpha\right\}\frac{\partial D}{\partial s}-\left\{\frac{\beta s\alpha-\frac{\alpha}{4}+\beta^2-2\mu s-(l(l+1)+\frac{1}{4})}{1+s^2}+\nu(\nu+1)+m^2-\frac{1}{4}-\beta^2-\beta\right\}D=0 \tag{53}$$

Equation (53) reduces to differential equation satisfied by Romanovski polynomials

$$\left(1+s^2\right)\frac{\partial^2 D}{\partial s^2}+\left\{2s(\beta+1)-\alpha\right\}\frac{\partial D}{\partial s}-\left\{\nu(\nu+1)+m^2-\frac{1}{4}-\beta^2-\beta\right\}D=0 \tag{54}$$

when
$$\beta s\alpha-\frac{\alpha^2}{4}+\beta^2-2\mu s-\left\{l(l+1)+\frac{1}{4}\right\}=0 \tag{55}$$

By comparing equations (10) and (54) we obtain

$$(\beta+1)=(-p+1);\alpha=-q \text{ and } \nu(\nu+1)+m^2-\frac{1}{4}-\beta^2-\beta=n(n-1)+2n(1-p) \tag{56}$$

From equation (55) we have

$$-\frac{\alpha^2}{4}+\beta^2-\left\{l(l+1)+\frac{1}{4}\right\}=0;\quad \beta\alpha-2\mu=0 \tag{57}$$

By using equation (56) together with equation (57) and by imposing the condition that p>0 we get

$$\beta_{nl}=-\sqrt{\nu(\nu+1)+m^2}-n-\frac{1}{2}\qquad\text{and}\qquad \alpha_{nl}=-\frac{2\mu}{\sqrt{\nu(\nu+1)+m^2}+n+\frac{1}{2}} \tag{58}$$

and
$$l=l'=\sqrt{\left(\sqrt{\nu(\nu+1)+m^2}+n+\frac{1}{2}\right)^2-\frac{\mu^2}{\left(\sqrt{\nu(\nu+1)+m^2}+n+\frac{1}{2}\right)^2}}-\frac{1}{2} \tag{59}$$

The weight function obtained from equations (7), (56), (58) and (59) is given as

$$w^{(-\beta,-\alpha)}=\left(1+s^2\right)^{\beta_{nl}}e^{-\alpha_{nl}\tan^{-1}s} \tag{61}$$

Using equations (13) and (61) we obtain the Romanovski polynomials given as

$$R_n^{(-\beta,-\alpha)}(s)=\frac{1}{\left(1+s^2\right)^{\beta_{nl}}e^{-\alpha_{nl}\tan^{-1}s}}\frac{d^{n_l}}{ds^{n_l}}\left\{\left(1+s^2\right)^{\beta_{nl}+n_l}e^{-\alpha_{nl}\tan^{-1}s}\right\} \tag{62}$$

The angular wave function obtained from equations (11) and (62) is given as

$$Q_{nl}\left(\theta=\cot^{-1}s\right)=g_{nl}(s)=\left(1+s^2\right)^{\frac{\beta_{nl}}{2}}e^{\frac{-\alpha_{nl}}{2}\tan^{-1}s}R_{n_l}^{(-\beta,-\alpha)}(s) \tag{63}$$

The polar wave function obtained from equation (63) is given as

$$P_{n_l}=P_l^m(\theta)=\frac{Q_{n_l}(\theta)}{\sqrt{\sin\theta}}=\left(1+s^2\right)^{\frac{\beta_{n_l}}{2}+\frac{1}{4}}e^{\frac{-\alpha_{n_l}}{2}\tan^{-1}s}R_{n_l}^{(-\beta,-\alpha)}(s) \qquad (64)$$

Since the values of β_{n2} and α_{n2} are n-dependence therefore the orthogonal integral of Romanovski polynomial is not produced from the orthogonality integral of wave function thus the polynomial is infinity [18]. The orthogonality integral of the wave function expressed in equation (76) is given as

$$\int P_{n_l}(\theta)P_{n_l'}(\theta)\sin\theta d\theta=\delta_{n_l n_{l'}} \qquad (65)$$

The effect of Rosen Morse potential to the angular wave function is shown in Figures 3-5. Figures 3(a)-5(a) show the graph of three dimensional polar representations of the absolute value of $Y_{l'}^{m'}$ and Figures 3(b)-5(b) show the polar diagrams of $Y_{l'}^{m'}$ which is obtained from equations (26) and (64). Figure (3) shows the angular wave function with the absent of RM potential, Figure (4) shows $\csc^2\theta$ term causes the increase of the absolute value of the angular wave function, Figure (5) shows $\cot\theta$ term causes the decrease of the absolute value of the angular wave function in the interval of $0<\theta<\frac{\pi}{2}$ but increasing in the $\frac{\pi}{2}<\theta<\pi$ interval.

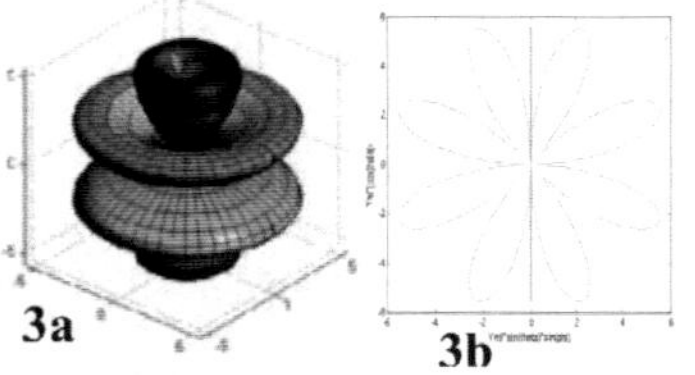

3a 3b

Figure 3(4a) shows graph of threee dimensional polar representations and (4b) polar diagram of $P_{3100}=\left|Y_4^1\right|=15(-4\cos^3\theta\sin\theta$ $(+3\cos\theta\sin^3\theta)$

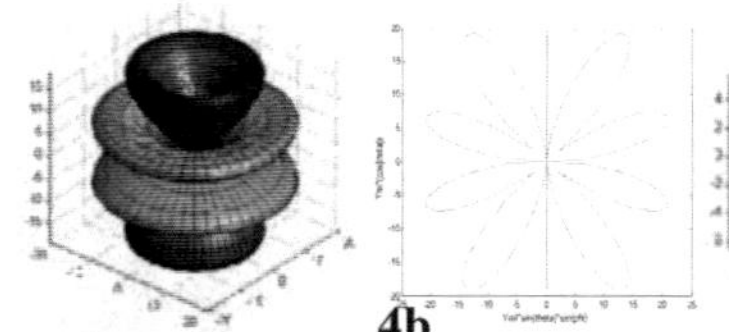

4a 4b

Figure 4: (4a) shows graph of three dimensional polar representations and (4b) polar diagram of $P_{3120}=\left|Y_{5.65}^{2.65}\right|=(\sin\theta)^{5.65}$ $(-381.72\cot^3\theta+156.87\cot\theta)$

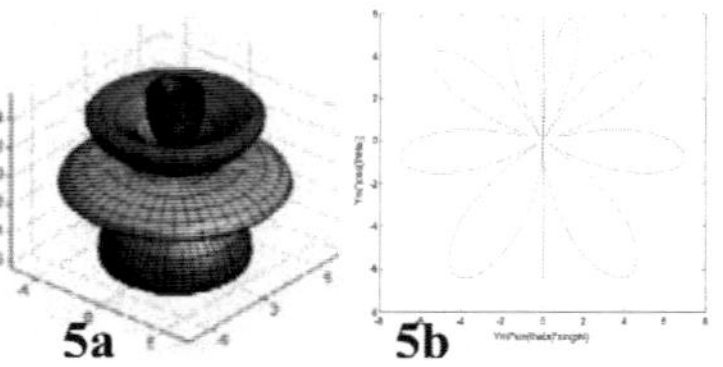

5a 5b

Figure 5. (5a) graph of three dimensional polar representations and (5b) polar diagram of $P_{3102}=\left|Y_{3.98}^{1}\right|=2\sin^4\theta$ $e^{0.45\tan^{-1}\cot\theta}(-60\cot^3\theta+$ $54\cot^2\theta+32.85\cot\theta-9.17)$

The total un-normalized wave function for the n level is given as

$$\psi(r,\theta,\varphi)=e^{im\varphi}\left\{C_{n_r}r^{-(l'+1)}e^{\frac{\gamma r^2}{2}}\frac{d^{n_r}}{dr^{n_r}}\left(r^{2l'+1+2n_r}e^{-\gamma r^2}\right)\right\}$$

$$\left[(\csc^2\theta)^{\frac{\beta_{n_l}}{2}+\frac{1}{4}}e^{\frac{\alpha_{n_l}}{2}\tan^{-1}(\cot\theta)}\frac{d^{n_l}}{d\cot\theta^{n_l}}\left\{(\csc^2\theta)^{\beta_{n_l}+n_l}e^{-\alpha_{n_l}\tan^{-1}(\cot\theta)}\right\}\right] \qquad (79)$$

By the absent of Rosen-Morse potential the wave function in equation (79) reduces to the three dimensional spherical harmonics oscillator wave function.

4. Conclusion

The Schrodinger equation for three dimensional harmonic oscillator potential with simultaneously the presence of Rosen-Morse non-central potential is solved approximately using NU method and Romanovski polynomial. The energy spectrum of non-central system is expressed in the closed form and is reduced to the energy spectrum for three dimensional harmonics oscillator potential by the absent of

Rosen-Morse potential. The presence of Rosen Morse non-central potential causes the energy of 3D HO increases. The effect of the two of potential terms to the energy is different. The $\csc^2\theta$ term (ν values) causes the increase of energy, but the $\cot\theta$ term (μ values) causes the energy of 3D HO decreases. In the same way with the energy, the radial wave function amplitude increases by the increase of ν values, but decreases by the increase of μ values. The approximate radial wave function is expressed in term of associated Laguerre polynomials and reduces to the wave function of three dimensional harmonics oscillator for the absent of Rosen-Morse potential. The angular wave functions are expressed in terms of Romanovski polynomials, and reduce to associated Legendre polynomials by the absent of trigonometric Rosen-Morse non-central potential. The $\cot\theta$ term (μ values) causes the length of the vector of angular function decreases and is shifted to the smaller values of polar angle, the larger is the value of μ the faster is the decrease of the vector length and the polar angle for small values of n_l, but this effect reduces when the values of n_l get larger.

References

[1] V M Villalba 1994 *Phys. Lett.* A **193** 218-222
[2] M Kibbler and C Campiogotto 1993 *Phys. Lett.* A 181 1
[3] Y Aharonov and D Bohm 1959 Phys. Rev. 115 485
[4] L V Hoang, L X Hai, L I Komarov and T S Romanova 1992 *J. Phys.* A 25 6461
[5] M Movahedi and M Hamzavi 2011 Int. J. Phys. Sci. **6** 4 89
[6] B Gonul and I Zorba 2000 *Phys. Lett.* A **269** 83
[7] H X Quan, L Guang, W Z Min, N L Bin, and M Yan 2010 *Com.Theor. Phys*. (China) **53** 242
[8] H Salehi, 2011, SciRes App. Math. 2 999-1004 doi:10.4236/am.2011.28138
[9] J C Sheng, Z Ying, Z X Lin, S LTian 2001 *Commun .theor. Phys*. **36** 641-646
[10] J Sadeghi 2007 *Act.Phys.Polonica*, A **112** 1 23
[11] J Sadeghi and B Pourhassan 2008 *EJTP* **5** 17 193
[12] A F Nikiforov and U B Uvarov 1988 *Special Function in Mathematical Physics*, Birkhausa, Basel
[13] F Yasuk, C Berkdemir, and A Berkdemir, and C Onem 2005 *Phys. Scr*. **71** 472
[14] M Movahedi and M Hamzavi 2001 *Int. J. Phys. Sci.* **6** 4 891-896
[15] D E Alvarez -Castillo, C B Compean, M Kirchbach 2011 arXiv: 1105.1354v1[quant-ph] 1-10
[16] D E Alvarez -Castillo and M Kirbach 2007 Rev. Mex. Fis. E53.2 143-154
[17] W Koepf and M Masjed-Jamei 2006 *Integral Transform and Special Functions* **17** 559
[18] E J Routh 1884 *Proc. London Math. Soc.* **16** 245
[19] V I Romanovski *1929 Compt. Rend. Ac. Sci. Paris* **188** 1023
[20] A Raposo, H J Weber, D E Alvarez-Castillo, and M Kirchbach *2007 C. Eur. Phys. J.* **5** 253
[21] M C Zhang 2009 *Int. J. Theor. Phys*. **48** 2625-2632.
[22] M Aktas, 2007 arXiv: 0701063v1 [quant-ph] 1-14
[23] M R Shojaei and A A Rajabi 2011 *Int. J. Phys. Sci.* **6** 33 7441- 7446 DOI: 10.5897/IJPS11.653,
[24] N Rosen and P M Morse 1932 *Phys. Rev* **42** 2
[25] D Agboola, Pram 2011 *J.Phys* **76** 6 875-885
[26] S M Ikhdair 2012 arXiv: 1104.0302v2[quant-ph]

Acknowledgement

This work is partially supported by Hibah Pascasarjana Sebelas Maret University grant
No : 2340/UN 27.10/PG/2012

Green synthesis of silver nanoparticles using latex extract of *Thevetia peruviana*: a novel approach towards poisonous plant utilization

N Nyoman Rupiasih[1,3,4], Avinash Aher[2], Suresh Gosavi[2] and P B Vidyasagar[2]
[1] Department of Physics, Faculty of Mathematics and Natural Sciences, Udayana University, Kampus Bukit Jimbaran, Badung, Bali 80362, Indonesia
[2] Department of Physics, University of Pune, Ganeshkhind, Pune-411007, India

E-mail: rupiasih69@yahoo.com

Abstract. The development of green synthesis of nanoparticles has received increasing attention due to ease of preparation, less chemical handling, and eco-friendly. In the present study, silver nanoparticles (AgNPs) were successfully synthesized from $AgNO_3$ through a simple green synthesis route using the latex of *Thevetia peruviana* as reducing as well as capping agent. Synthesized silver nanoparticles were characterized using UV-vis, FTIR, SEM-EDS, XRD and HRTEM. UV-vis spectra showed absorption at 570 nm with shoulder at 395 nm. FTIR spectra analysis confirmed the functional groups involved in the AgNPs formation. X-ray diffraction pattern of AgNPs exhibited the diffraction at $2\theta = 34.48°$, 40.12° and 45.39° corresponding to (222), (400), and (420) Braggs reflection planes of silver nanoparticles. TEM analysis showed the particles size is spherical in nature with its size distribution between 10-30 nm. The HRTEM image showed 2.59Å interplaner spacing.

Keywords: Green synthesis, *Thevetia peruviana*, Latex, Silver nitrate, Silver nanoparticles

1. Introduction

The field of nanotechnology is one of the most active areas of research in recent materials science. Nanoparticles are often referred to as clusters, nanospheres, nanorods and nanocups of the size ranges from 1 to 100 nm which exhibit completely new or improved properties based on size-distribution, and morphology [1-3].

Nanoparticles with controlled size and composition are of fundamental and technological interest as they provide solutions to technological and environmental challenges in the areas of solar energy conversion, catalysis, medicine, and water treatment. Global warming and climate change have induced a worldwide awareness and efforts to reduce generated hazardous wastes. Thus, "green"

[3]Author to whom any correspondence should be addressed: rupiasih69@yahoo.com; [4] Grant: the Ministry of Education and Culture Republic of Indonesia, No.: 0542/023.04.1.01/00/2011 date 5 June 2011 the 2nd DIPA Revision

chemistry and chemical processes are progressively being integrated in science and industry for sustainable development [4-5]. One of them is synthesis of nanoparticles by biological process which implemented to develop safe, cost-effective and environmentally friendly technologies [4].

A great deal of effort has been put into the biosynthesis of inorganic material, especially metal nanoparticles using microorganisms and plants [1-2, 6]. Use of plant extract for the synthesis of nanoparticles could be advantageous over other environmental friendly biological processes by eliminating the elaborate process of maintaining cell cultures [4, 6-7]. Gardea-Torresdey and co-workers first reported the formation of gold and silver nanoparticles by living plants [8-9]. Their synthetic protocol by plant extract or biomass exemplifies the promising application of the green synthesis of metal nanoparticles. Very recently green silver nanoparticles have been synthesized using various natural products like green tea (*Camellia sinensis*), neem (*Azadirachta indica*) leaf broth, natural rubber, starch, aloe vera plant extract, lemongrass leaves extract, etc [7, 10]. The shape and size of the nanoparticles synthesized using plants can be controlled and modulated by changing the pH [2]. The reduction of silver ions (Ag^{+}) in aqueous solution generally yields colloidal silver with particle diameters of several nanometers.

Medicinal herbs are the local heritage with global importance. It has curative properties due to the presence of various complex chemical substances of different composition, which are found as secondary plant metabolite in one or more parts of the plants [4-5, 7, 11]. These plant metabolites according to their composition are grouped as alkaloids, glycosides, corticosteroids, essential oils, etc.

Thevetia peruviana is an ever-green ornamental dicotyledonous plant that belongs to *Apocyanaceae* family. It is commonly found in the tropics and sub-tropics where as it is native to Central and South America [12]. The whole plant contains a milky juice (white latex) containing a compound called thevetin that is used as a heart stimulant but in its natural form is extremely poisonous. It's poisonous owing to the presence of cardiac glycosides or cardiac toxins, which acts directly in the heart [12-14]. Ingestion of these plant parts could lead to death [12-13]. It is cultivated as ornamental tree standards in gardens and parks in temperate climates [15]. In the West Indies half a leaf is known as an emetic and purgative. In Java Indian immigrants are reported to smoke using the dried leaves. Tincture of bark is also emetic and purgative and has been used as a febrifuge, but tests with bark and fruit extracts have proved inactive against avian malaria. In Indonesia it is reported used as a fish-poison [16].

Studies have indicated that biomolecules like protein, phenols, and flavonoids not only play a role in reducing the ions to the nanosize, but also play an important role in the capping of the nanoparticles [17]. The reduction of Ag^{+} ions by combinations of biomolecules found in these extracts such as vitamins, enzymes/proteins, organic acids such as citrates, amino acids, and polysaccharides [17–18] is environmentally benign, yet chemically complex.

The present paper demonstrate method for the synthesis of silver nanoparticles by the reduction of aqueous silver ions using latex from *Thevetia peruviana,* a poisonous as well as medicinal herbs plant [15]. It resulted in the extracellular formation of AgNPs at room temperature which was further harvested by centrifugation and simple natural drying evaporation.

2. Materials and methods

2.1. Preparation of latex extracts

Fresh fruits of *Thevetia peruviana* were collected, washed thoroughly 2-3 times with tap water, continued with distilled water, and used for extraction. Milky white latex was obtained by cutting the green stems of the fruits. 100 ml of distilled water was added with 100 µl of fresh latex extract, mixed gently and filtered through Whatman No.1 filter paper. The solution was stored at 4°C for further used.

2.2. Synthesis of silver nanoparticles
A 50 µl of aqueous solution of 1 M silver nitrate ($AgNO_3$) was added into 10 ml of extract solution for reduction of silver nitrate into Ag^+ ions and kept at room temperature. After around 6 hours, the colour of the mixture turned into light purple and continuously changed to purple-brown indicating the formation of silver nanoparticles.

2.3. Characterization techniques
UV-Vis absorption spectra were measured using JASCO V-670 UV-Vis spectrophotometer in the wavelength range 200-800 nm. The progress of the reaction between metal ions and the latex extract solution were monitored in extract solution at different reaction time between 0-52 hours. Elemental analysis was done using Energy Dispersive X-ray (EDS) JEOL JSM 6360A. Crystalline nature of silver nanoparticles was examined by X-ray diffraction analysis using X-ray diffractometer (BRUKER AXS D8) with a Cu-Kα (λ = 1.54 Å) in the range 20-80^0. Morphology and size of silver nanoparticles were investigated using transmission electron microscopy using TEM-TECNAI G^2 20U-TWIN instrument. FTIR measurements were done using JASCO (FT/IR-6100) spectrophotometer in the range 400-4000 nm.

3. Results and discussion
Figure 1 shows the UV-Vis absorption spectra of silver nanoparticles (AgNPs) synthesized using latex extract of *Thevetia peruviana* at different time interval. It shows the characteristic of surface plasmon resonance (SPR) absorption band at around 570 nm for AgNPs (purple-brown solution). It is observed that the intensity of SPR bands increases as the reaction time progresses from 0-44 h and after 44 h of reaction time the SPR intensity nearly constant indicating that the reaction process is almost completed. The observed absorption spectra are broad in nature with a tail at longer wavelength and shoulder at around 395 nm indicating the formation of polydispers AgNPs. Figure 1 also shows the red shifted band from 510 nm to 570 nm occurred with increasing reaction time from 0-52 h.

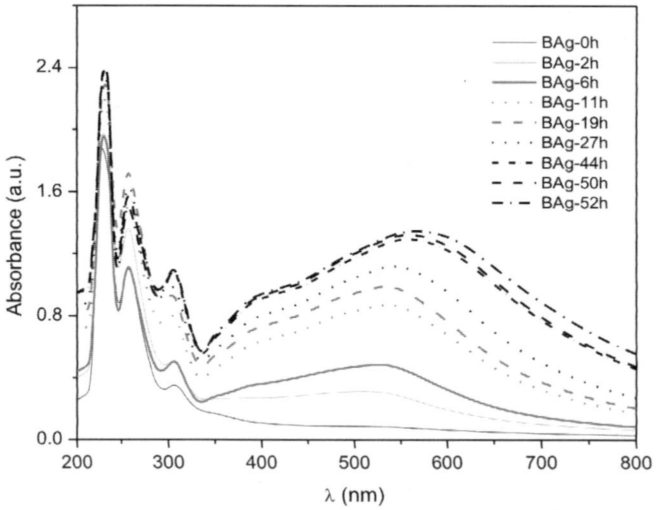

Figure 1. UV-vis absorption spectra of silver nanoparticles synthesized at room temperature using latex extract of *Thevetia peruviana* at different time durations: 0-52 h.

The strong absorption peak at around 230 nm indicates the presence of several organic compounds which are known to interact with silver ions. An absorption band at around 260 nm is attributed to the aromatic amino acids of proteins. It is well known that the absorption band at 260 nm arises due to electronic excitations in tryptophan and tyrosine residues in the proteins. This observation indicates the release of proteins into solution by latex of *Thevetia peruviana* and suggests a possible mechanism for the reduction of the metal ions present in the solution.

Fig.2 shows representative EDX profile for synthesized silver nanoparticles. The energy dispersive X-ray analysis (EDS) reveals strong signal in the silver region and confirms the formation of silver nanoparticles. There is also a strong signal for Si, which due to Si (111) wafers that has used as substrate to prepare thin film. Other elemental signals are recorded possibly due to elements from enzymes or proteins present within the latex of *Thevetia peruviana.*

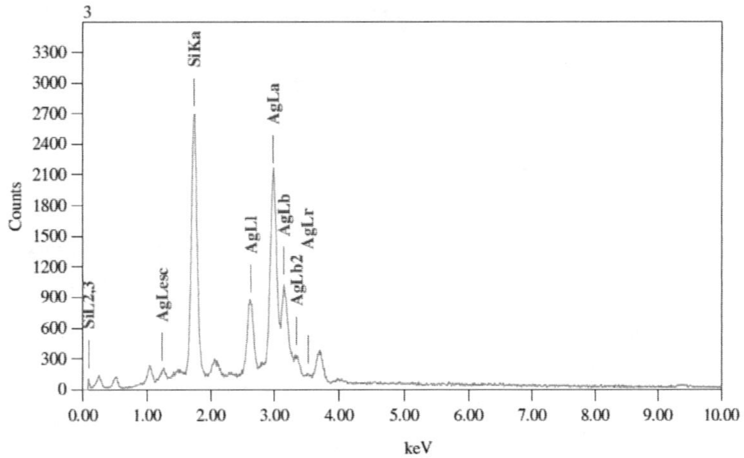

Figure 2. EDS spectrum for silver nanoparticles synthesized at room temperature using latex extract of *Thevetia peruviana*.

A thorough study on the literature of *Thevetia peruviana* reveals that the major components of the plant are protein, phenols, oil, mome inositol, Ethyl palmitate, Ethyl linolenate, and lesser amount of glycosides with the major component is thevetin. The presence of theveridoside, digitoxigenin, cerberin, peruvoside and theveside has been established in the plant [14]. It has been reported by Yao Zhou et al. [18] that the carbonyl groups from the amino acid residues and peptides of proteins have a strong affinity to bind metals. So that protein can act as encapsulating agent and thus protect the nanoparticles from agglomeration. For this phenomenal, FTIR measurements were carried out to identify the possible biomolecules responsible for the reduction of the Ag^+ ions and the capping of the bioreduced silver nanoparticles synthesized by the latex extract of *Thevetia peruviana*. The solution after complete reduction of Ag^+ was centrifuged at 15000 rpm for 10 minutes to isolate the silver nanoparticles free from proteins or other compounds present in the solution. The representative spectra of nanoparticles obtained (figure 3) manifest absorption bands located at 3426, 1640, 1390, 1338, 1104, and 1030 cm^{-1}. The absorption peak at around 1030 cm^{-1} can be assigned as absorption peaks of -C-O-C- or -C-O-. The peak at around 1640 cm^{-1} is assigned to the amide I bonds of proteins. The bonds or functional groups such as -C-O-C-, -C-O- and -C=C- derived from heterocyclic compounds and the amide I bond derived from the proteins which are present in the extract are the capping ligands of the nanoparticles [1, 7, 10, 17]. The band at 1390 and 1104 cm^{-1}, as well as a strong band at 1338 cm^{-1} indicating presence of amine groups, as would be expected due to plant-origin of the samples. The peak at around 3426 cm^{-1} was characteristic of -N-H stretching of amide I band [17].

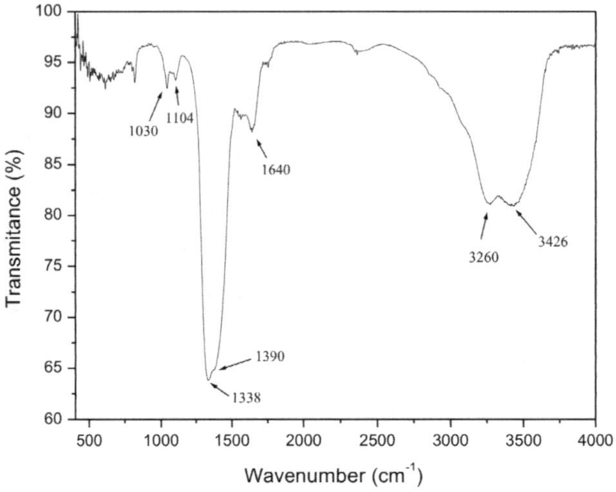

Figure 3. FTIR spectra of silver nanoparticles synthesized at room temperature using latex extract of *Thevetia peruviana*.

Figure 4 shows that the XRD patterns of natural dried silver nanoparticles synthesized using latex of *Thevetia peruviana*. A number of Bragg reflections with 2θ values of 34.48^0, 40.12^0, and 45.39^0 sets of lattice planes are observed which may be indexed to the (222), (400), and (420) faces of silver respectively. XRD pattern thus clearly illustrates that the silver nanoparticles formed in this present synthesis are crystalline in nature and having Face Cantered Cubic crystal structure.

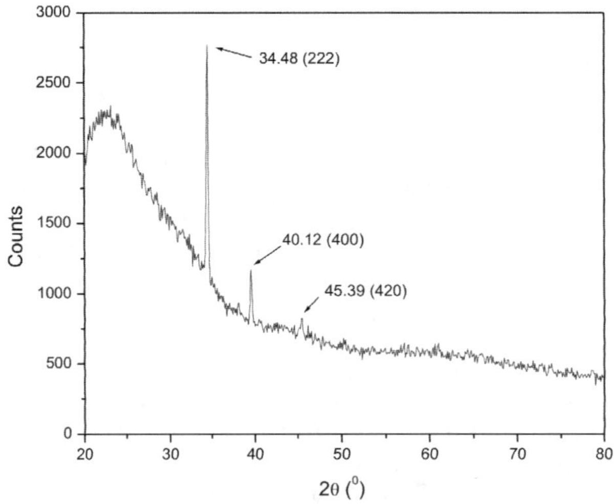

Figure 4. XRD patterns of silver nanoparticles synthesized at room temperature using latex extract of *Thevetia peruviana*.

Journal of Physics: Conference Series **423** (2013) 012032 doi:10.1088/1742-6596/423/1/012032

Figure 5 shows the TEM micrograph of the Ag nanoparticles synthesized by latex of *Thevetia peruviana*. Figure 5a is the representative TEM image of synthesized silver nanoparticles. The morphology of AgNPs is found to be predominantly spherical. The nanoparticles are polydispers having two dominant size distributions ranges from 10-20 nm and 20-30 nm as shown in figure 5b. The figure 5c shows HRTEM image of silver nanoparticle. The fringe spacing is found to be 2.59Å. These results were in agreement with those reported by Kiruba Daniel et al. [1], Harekrishna Bar et al. [7] and Naheed Ahmad et al. [17].

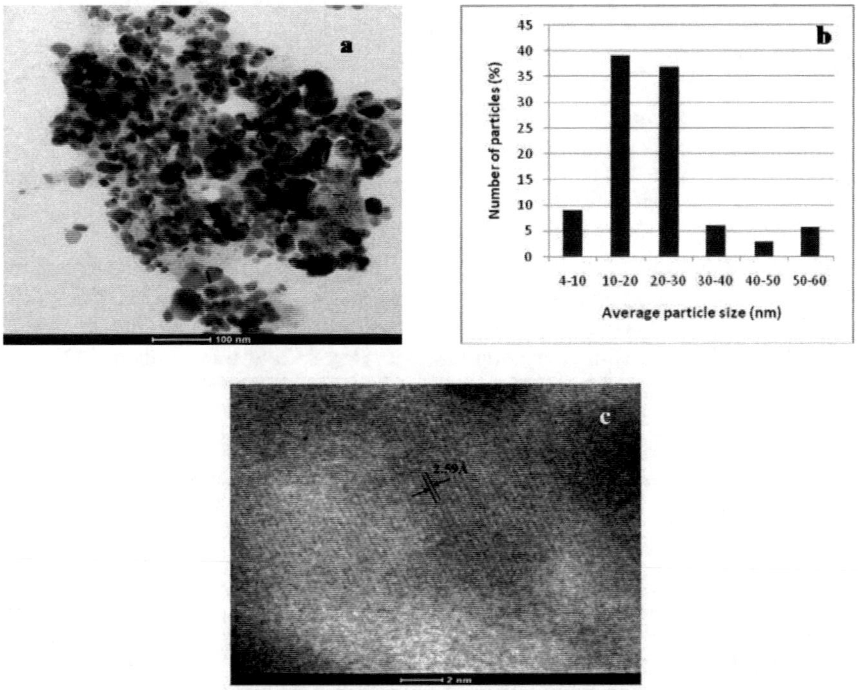

Figure 5. a) Representative TEM image of silver nanoparticles synthesized at room temperature using latex extract of *Thevetia peruviana*. b) Histogram of particle size distribution and C) High Resolution Transmission electron microscopy (HRTEM) image of silver nanoparticles.

4. Conclusion

The rapid synthesis of silver nanoparticles using latex extract of *Thevetia peruviana* has been demonstrated. Present green synthesis showed that the environment friendly and renewable latex of *Thevetia peruviana* could be used as an effective capping as well as reducing agent for the synthesis of silver nanoparticles. This eco-friendly route for Synthesis of silver nanoparticles is a challenging alternative to chemical synthesis.

Acknowledgments

The author thanks the Higher Education Ministry (DIKTI), Jakarta, Indonesia for "Abroad SAME Program 2012 Funding" and to Dept. of Physics, University of Pune, India as the university based of research conducted. The author also thanks to Miss Jyotsana Dixit, Biophysics Lab. Dept. of Physics, University of Pune, India for valuable helps.

References:

[1] Kiruba Daniel S C G, Kumar R, Sathish V, Sivakumar M, Sunitha S and Anitha Sironmani T 2011 Green Synthesis (Ocimum tenuiflorum) of Silver Nanoparticles and Toxicity Studies in Zebra Fish (Danio rerio) Model *International Journal of NanoScience and Nanotechnology* **2** (2) 103-117

[2] Sangiliyandi Gurunathana, Kalimuthu Kalishwaralala, Ramanathan Vaidyanathana, Venkataraman Deepaka, Sureshbabu Ram Kumar Pandiana, Jeyaraj Muniyandia, Nellaiah Hariharana, Soo Hyun Eom 2009 Biosynthesis, purification and characterization of silver nanoparticles using Escherichia coli *Colloids and Surfaces B: Biointerfaces* **74** 328-335

[3] Guangquan Li, Dan He, Yongqing Qian, Buyuan Guan, Song Gao, Yan Cui, Koji Yokoyama and Li Wang 2012 Fungus-Mediated Green Synthesis of Silver Nanoparticles Using *Aspergillus terreus Int. J. Mol. Sci.* ***13*** 466-476

[4] Asmita J Gavhane1, Padmanabhan P, Suresh P Kamble and Suresh N Jangle 2012 Synthesis of silver nanoparticles using extract of Neem leaf and triphala and evaluation of their Antimicrobial activities *Int. J. Pharm. Bio. Sci.* **3**(3) 88-100

[5] Geethalakshmi R and Sarada D V L 2010 Synthesis of plant-mediated silver nanoparticles using *Trianthema decandra* extract and evaluation of their anti microbial activities International *Journal of Engineering Science and Technology* **2**(5) 970-975

[6] Ponarulselvam S, Panneerselvam C, Murugan K, Aarthi N, Kalimuthu K and Thangamani S 2012 Synthesis of silver nanoparticles using leaves of Catharanthus roseus Linn. G. Don and their antiplasmodial activities *Asian Pacific Journal of Tropical Biomedicine* 574-580

[7] Harekrishna Bar, Dipak Kr Bhui, Gobinda P Sahoo, Priyanka Sarkar, Sankar P De, Ajay Misra 2009 Green synthesis of silver nanoparticles using latex of *Jatropha curcas Colloids and Surfaces A*: Physicochem. Eng. Aspects **339** 134-139

[8] Gardea-Torresdey J L, Parsons J G, Dokken K, Peralta-Videa J, Troiani H E, Santiago P and Jose-Yacaman M 2002 Formation and growth of Au nanoparticles inside live Alfalfa plants. *Nano Lett* **2** 397-401

[9] Gardea-Torresdey J L, Gomez E, Peralta-Videa E, Parsons J G, Troiani H E and Jose-Yacaman M 2003 Alfalfa sprouts: a natural source for the synthesis of silver nanoparticles. *Langmuir* **19** 1357-1361

[10] Manish Hudlikar, Shreeram Joglekar and Mayur Dhaygude Kisan Kodam 2012 Latex-mediated synthesis of ZnS nanoparticles: green synthesis approach *J Nanopart Res* **14** 865

[11] Venkataraman Deepak, Kalimuthu Kalishwaralal, Sureshbabu Ram Kumar Pandian and Sangiliyandi Gurunathan 2011 An Insight into the Bacterial Biogenesis of Silver Nanoparticles, Industrial Production and Scale-up *Rai M and Duran N (eds.), Metal Nanoparticles in Microbiology* (Berlin Heidelberg: Springer-Verlag) chapter 2 pp 173-214

[12] Usman L A, Oluwaniyi O O, Ibiyemi S A, Muhammad N O and Ameen O M 2009 The potential of oleander (thevetia peruviana) in African agricultural and industrial development: a case study of nigeria *Journal of Applied Biosciences* **24** 1477-1487

[13] Mathuravalli K and Eswara Lakshmi R 2012 Analysis of phytochemical components and anti-microbial activity of the toxic plant -*Thevetia Peruviana Indian J. Innovations Dev.* **1** (2) 97-101

[14] Sangodare R S A, Agbaji A S, Dakare M A, Usman Y O, Magomya A, Paul E D, Ibraheem A S, Etim N A, Ibrahim A, Aribido O and Shikeola O 2012 Investigation of the Chemical Constituent of Extracts of *Thevetia peruviana* Seed Using GC-MS and FT-IR *International Journal of Food Nutrition and Safety* **2**(1) 27-36

[15] Garima Zibbu and Amla Batra 2011 *Thevetia peruviana* (Pers.) Schum. : A plant with enormous therapeutic potential *Journal of Pharmacy Research* **4**(12) 4461-4464

[16] Burkill HM 1985 Entry for Thevetia Neriifolia Juss. [Family Apocynaceae] The Useful Plants of West Tropical Africa, *Jstor Plant Science* Vol **1** Http://Plants.Jstor.Org/Upwta/1_411 (Accessed On 09-10-2012)

[17] Naheed Ahmad, Seema Sharma, Singh V N, Shamsi S F, Anjum Fatma and Mehta B R 2011 Biosynthesis of Silver Nanoparticles from Desmodium triflorum: A Novel Approach TowardsWeed Utilization *Biotechnology Research International* **2011** 8 pages
[18] Yao Zhou, Wenshuang Lin, Jiale Huang, Wenta Wang, Yixian Gao, Liqin Lin, Qingbiao Li, Ling Lin and Mingming Du 2010 Biosynthesis of Gold Nanoparticles by Foliar Broths: Roles of Biocompounds and Other Attributes of the Extracts *Nanoscale Res Lett* **5** 1351-1359

Journal of Physics: Conference Series **423** (2013) 012033 doi:10.1088/1742-6596/423/1/012033

Reconsideration of Deformation Potential Constants in Nitride Semiconductors with Wurtzite Structure

S Emura and K Shirai

The Institute of Scientific and Industrial Research, Osaka University, Mihoga-oka 8-1, Ibaraki, Osaka. 567-0047, Japan

E-mail: emura@sanken.osaka-u.ac.jp

Abstract. The deformation potential in nitride semiconductors with wurtzite structure is theoretically re-examined. The ligand field potential term including the trigonal part is definitely dealt in the model. The alternative approach to the expression for the deformation potential constants is tried in the framework of invariant regime. The basic expressions derived by perturbation theory are given, where the deformation potential constants are defined for the stresses instead of the strains, because this definition is much more straightforward for experimentalist.

1. Introduction

Examining the responsiveness of materials to external perturbations such as electric, magnetic fields and stress, it is often required in device design. In multi-quantum well systems of semiconductors, the interfacial stresses are inevitably generated, affecting the various physical properties of the semiconductors such as the band gap. The effects are especially pronounced in semiconductors with a wurtzite structure because three valence band states lie in close proximity.

The effects of strain on the band structures of semiconductor have been under investigation for half a century. Bir and Pikus [1] developed a general expression for these effects in semiconductors including those with wurtzite structure, which belongs to the space group P_{63mc}. Their energy matrix is widely used in a variety of situations, and they defined six deformation potential constant (DPC)s in wurtzite crystals, two (D_{c1} and D_{c2}) for the conduction band and four (D_1 D_2, D_3 and D_4) for the valence band. For a hexagonal GaN under the stresses, the ligand field splitting parameter Δ_1 may play a particular role on the alignment of three valence band states because the spin-orbit and the ligand field splitting energies are both small (< 20 meV) [2], leading to an easy rearrangement of the three valence bands under stress.

In the hexagonal GaN, the experimentally reported DPCs, especially for D_3 and D_4, are largely scattered [3 – 5]. This large discrepancy among various reports mostly comes from the difference of the definition of DPC used in the analysis of the experimental results. The prototype of DPCs is one defined in the invariant category [1]. In invariant method by Bir and Pikus, the total Hamiltonian (electronic part + lattice part and the interaction part between the both) is invariant under the external or internal stress (strain) application, even if the crystal symmetry is lowered by the applied stress $f_{\Gamma\gamma}$ belonging to the irreducible representation Γ and the component γ. Therefore, the stress Hamiltonian term $f_{\Gamma\gamma}Q_{\Gamma\gamma}$, where Q is a normal coordinate, must be transformed into total symmetry in the crystal symmetry. This condition is satisfied only in the case that the both $f_{\Gamma\gamma}$ and $Q_{\Gamma\gamma}$ transform under the same irreducible representation. Stress (Strain) is the second rank symmetry tensor, and the six different symmetry tensors, *xx, yy, zz, xy, yz, and zx*, should be reduced into adequate irreducible tensors in the invariant method, as well known. The irreducible second-rank symmetry tensors in C_{6v} group are *xx + yy, zz, xy, xx – yy, yz, and xz* (see the second rank basis of C_{6v} group in the table of irreducible representation [6]). On analyzing the experimental data in the invariant category, the applied stresses (strains) such as *xx, yy,* and *zz* must be developed into the irreducible second-rank symmetry tensors. Recent papers sometimes employ the unreduced stress (strain) tensors, which span Cartesian coordinate, to obtain the values of the DPCs from the observed data. These DPCs are the different definition from the DPCs used in the invariant method. Thus, the reported values of the DPCs are scattered. In the framework that the conduction band is formed in an s orbital of III group and the valence band is composed in 2p orbital of nitrogen ion, the relationship of the cubic approximation, $D_3 = -2D_4$ [1], is absolutely right. The slight gap from $D_3 = -2D_4$ may be caused in mixing of other orbital into the relevant valence band state.

In this contribution, we present the correction of the Bir and Pikus matrix in the next section, and furthermore, try to show a revised expression for the DPCs of wurtzite symmetry, following Henry, Schnatterly, and Slichter treatment [7 – 9]. The basic expression will be given.

2. Correction of Bir and Pikus expression

We definitely deal with the ligand field splitting term [6, 10], which arises by mean of the lowered (wurtzite) symmetry from cubic (zinc blende) structure. It is worth paying attention that the trace of the energy matrix for the energy splitting term must be zero, namely a center of gravity of the energy is not shifted on the energy splitting brought about lower symmetry. The equivalent operator ($J_z^2 - 2/3$) [3, 10] for the *p*-like basis should be used to calculate the matrix elements instead of J_z^2, which is employed by Bir and Pikus, leading to the different band alignment as shown in Fig.1. It should be noticed that the conduction band is also shifted, because $J_z^2 \varphi_c(\boldsymbol{r}) = 0$ but $(J_z^2 - 2/3)\, \varphi_c(\boldsymbol{r}) \neq 0$, where $\varphi_c(\boldsymbol{r})$ is an electron wavefunction in the conduction band.

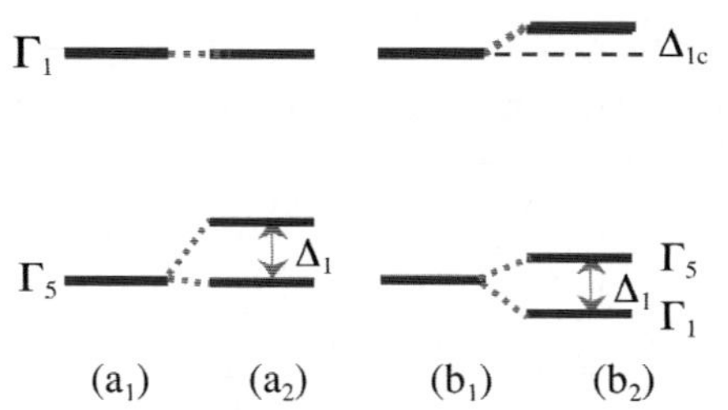

Figure 1 Band regime. (a) is predicted by Bir and Pikus scheme, and (b) the present scheme. The suffix 2 indicates that the ligand field splitting term is included. Γ_1 in the upper part indicates the conduction band symmetry, and that in the lower part the valence band symmetry, which is split off by the trigonal ligand field. Δ_1 is the ligand field splitting energy.

This correction significantly influences the manner of the determination of the DPCs. Considering the method of determining the DPCs from the viewpoint of experimentalists, the Bir and Pikus scheme (see Fig. 1(a)) can independently provide the DPCs $D_1 + D_{c1}$ (= C_1) and $D_2 + D_{c2}$ (= C_2), where D_{c1} and D_{c2} are the DPCs in the conduction band, and which correspond to the totally symmetric deformation, from the hydrostatic pressure and stress dependence of the transition energy from the lowest valence band Γ_1 to the conduction band, because they take no trigonal ligand field part in the Γ_1 (p_z) valence band into account. In our corrected scheme (see Fig. 1(b)), the transition energy involves the energy lowered by the trigonal component of the ligand field potential in the band $\Gamma_1(p_z)$. Therefore, C_1 and C_2 cannot be uniquely determined by the hydrostatic and the stress experiments. Here, the notations of the DPCs are similar to those of Chuang and Chang [11].

In accordance with the above description, we develop a strain matrix that is corrected by definitely including the ligand field splitting parameter Δ_1 into the strain Hamiltonian. We extract only the terms relevant to the strain effects for the valence band to avoid complexity. The total ligand field splitting energy is regarded as Δ_1 (=<$\varphi_v(\boldsymbol{r})|V_r(\boldsymbol{r})|\varphi_v(\boldsymbol{r})$>, where $V_r(\boldsymbol{r})$ is the potential in a trigonal portion, and $\varphi_v(\boldsymbol{r})$ is the wavefunction of the valence band. The partial ligand field splitting energy of $\frac{1}{3}\Delta_1$ is allotted to the two-fold degenerated band $\Gamma_5(p_x, p_y)$ and the $-\frac{2}{3}\Delta_1$ to the band $\Gamma_1(p_z)$. Hence, the energy gravity is held. Spin-orbit interaction is ignored, because the stress energy is much higher than that of the spin-orbit interaction at an interface of a hetero-structure or a multi-layer composed of materials with different lattice constants.

$$\begin{bmatrix} A & V & W \\ V & B & W \\ W & W & C \end{bmatrix} \quad (1)$$

$$\mathbf{A} = \Delta_0 + (1/3)\Delta_1 + D_1'(zz)\boldsymbol{\varepsilon}_{zz} + D_2'(xx+yy)\boldsymbol{\varepsilon}_{xx+yy} + D_5(xx-yy)\boldsymbol{\varepsilon}_{xx-yy},$$

$$\mathbf{B} = \Delta_0 + (1/3)\Delta_1 + D_1'(zz)\boldsymbol{\varepsilon}_{zz} + D_2'(xx+yy)\boldsymbol{\varepsilon}_{xx+yy} - D_5(xx-yy)\boldsymbol{\varepsilon}_{xx-yy},$$

$$\mathbf{C} = \Delta_0 - (2/3)\Delta_1 + D_1''(zz)\boldsymbol{\varepsilon}_{zz} + D_2''(xx+yy)\boldsymbol{\varepsilon}_{xx+yy},$$

$$\mathbf{V} = D_5\boldsymbol{\varepsilon}_{xy},$$

$$\mathbf{W} = D_6\boldsymbol{\varepsilon}_{yz} \text{ or } D_6\boldsymbol{\varepsilon}_{zx}, \quad (2)$$

where $D_1'(zz) = D_1 + (1/3)D_3$, $D_2'(xx+yy) = D_2 + (1/3)D_4$, $D_1''(zz) = D_1 - (2/3)D_3$, and $D_2''(xx+yy) = D_2 - (2/3)D_4$. Δ_0 (=<$\varphi_v(\boldsymbol{r})|V_{cubic}(\boldsymbol{r})|\varphi_v(\boldsymbol{r})$>. $V_{cubic}(\boldsymbol{r})$ is the potential energy of the cubic portion of the potential. Furthermore, the following relations stand,

$$D_1'(zz) - D_1''(zz) = D_3, \quad (3)$$

$$D_2'(xx+yy) - D_2''(xx+yy) = D_4. \quad (4)$$

The notation of $\varepsilon_{\Gamma\gamma}$ implies an irreducible strain. The suffixes of the strains and the functions in the parentheses of D's such as $(xx+yy)$ indicate the γth basis functions belonging to irreducible representation Γ to clarify the physical meaning of strain much more. For the conduction band, four different DPCs ($D_{c1}(zz)$ and $D_{c2}(xx+yy)$ for Δ_0, and $D_{c3}(zz)$ and $D_{c4}(xx+yy)$ for Δ_{1c}) should be assigned because Δ_{1c} = <$\varphi_c(\boldsymbol{r})|V_r|\varphi_c(\boldsymbol{r})$> $\neq 0$. However, as these four DPCs cannot be determined from the stress experimental results, we can settle four DPCs into two as $D_c(zz)$ and $D_c(xx+yy)$ without any practical problems.

3. Alternative approach to deformation potential of wurtzite symmetry

In the same era, the alternative approach in the invariant category is developed by Henry, Schnatterly, and Slichter (HSS) [7 - 9], who treated cubic symmetry case and showed more understandable physical comprise. We apply HSS treatment to obtain the expression of the DPCs for wurtzite symmetry. Here, we will discuss an alternative approach to the deformation potentials, based on the treatment by HSS moment method [7 - 9]. We introduce into the one-electron Hamiltonian a system of normal coordinates $\boldsymbol{Q}_{\Gamma\gamma}$, which relate to the lattice normal modes belonging to irreducible representations $\Gamma\gamma$ and span the eigen-spaces of the phonon modes. It should pay attention to the fact that Hooke's law, $f_{\Gamma\gamma} \propto \Delta\boldsymbol{Q}_{\Gamma\gamma}$, makes ends meet directly in this coordination system. This External stresses do not directly affect the electronic state because a stress Hamiltonian contains no electron coordinate $\boldsymbol{r}$. On the other hand, they deform the crystal lattice, and this transformation affects specific lattice normal modes depending on the stress symmetry and deviates the potential $V(r, \boldsymbol{Q}_{\Gamma\gamma})$. The deformation of the potential causes alteration of the electronic states through an electron – phonon interaction. We take the extreme condition under which the alteration of the band energy by an applied or internal stress is much stronger than the splitting energy of the spin–orbital interaction, and deal only with the Γ point of the Brillouin zone. To make the role of the phonon clearer, we develop the potential using the normal coordinates $\boldsymbol{Q}_{\Gamma\gamma}$ to the first order.

$$V(r, \boldsymbol{Q}_{\Gamma\gamma}) = V(r, \boldsymbol{Q}_{\Gamma\gamma} = \boldsymbol{0}) + \Sigma \partial V(r, \boldsymbol{Q}_{\Gamma\gamma})/\partial \boldsymbol{Q}_{\Gamma\gamma} \mid_{\boldsymbol{Q}_{\Gamma\gamma}=0} \boldsymbol{Q}_{\Gamma\gamma}. \tag{5}$$

$\boldsymbol{Q}_{\Gamma\gamma} = \boldsymbol{0}$ implies the equilibrium positions of the crystal lattice. It is full worth noting that $\partial V(\boldsymbol{r}, \boldsymbol{Q}_{\Gamma\gamma})/\partial \boldsymbol{Q}_{\Gamma\gamma}$ must be a Hamiltonian operator in real space, and is of the same symmetry as the γth basis function of Γ (invariant requisition). The second term of Eq. (5) represents the change of the potential energy of electron as a result of the lattice deformation. Now, we add an external stress Hamiltonian $\mathscr{H}_{\rm stress}(\boldsymbol{Q}_{\Gamma\gamma}) = - \Sigma f_{\Gamma\gamma}\boldsymbol{Q}_{\Gamma\gamma}$ to Eq. (5), where $f_{\Gamma\gamma}$ must be a force of with the same symmetry as the lattice normal mode. This compulsion is also an invariant requisition. The lattice normal mode Hamiltonian $\mathscr{H}_{\rm L}(\boldsymbol{Q}) = \Sigma\, [P^2/(2M) + \frac{1}{2}M\omega^2\boldsymbol{Q}_{\Gamma\gamma}^2]$ will be modified by the external stresses, where P is momentum, M mass, and ω frequency of the lattice normal mode. The external stresses cause rearrangement of the equilibrium position of the lattice normal mode as follows,

$$\tfrac{1}{2}M\omega^2\boldsymbol{Q}_{\Gamma\gamma}^{\ 2} - f_{\Gamma\gamma}\boldsymbol{Q}_{\Gamma\gamma} = \tfrac{1}{2}M\omega^2(\boldsymbol{Q}_{\Gamma\gamma} - f_{\Gamma\gamma}/M\omega^2)^2 - \tfrac{1}{2}f_{\Gamma\gamma}^{\ 2}/(M\omega^2). \tag{6}$$

The new equilibrium position of the lattice mode under stress is given as

$$\boldsymbol{Q}_{\Gamma\gamma}' = \boldsymbol{Q}_{\Gamma\gamma} - f_{\Gamma\gamma}/M\omega^2 = \boldsymbol{Q}_{\Gamma\gamma} - \Delta\boldsymbol{Q}_{\Gamma\gamma},\ (\Delta\boldsymbol{Q}_{\Gamma\gamma} = f_{\Gamma\gamma}/M\omega^2), \tag{7}$$

$\Delta\boldsymbol{Q}_{\Gamma\gamma}$ means a strain transformed under an irreducible representation $\Gamma\gamma$. Substituting Eq. (7) into Eq. (5),

$$V(r, \boldsymbol{Q}_{\Gamma\gamma}') = V(r, \boldsymbol{Q}_{\Gamma\gamma}') + \Sigma \partial V(r, \boldsymbol{Q}_{\Gamma\gamma}')/\partial \boldsymbol{Q}_{\Gamma\gamma}' \mid_{\boldsymbol{Q}_{\Gamma\gamma}'=0} \Delta\boldsymbol{Q}_{\Gamma\gamma}. \tag{8}$$

Here, we take the coordinate origin at the new equilibrium position. The stress Hamiltonian perturbing the electronic state can be rewritten using the new coordinate $\boldsymbol{Q}_{\Gamma\gamma}'$ and the irreducible strain amplitude $\Delta\boldsymbol{Q}_{\Gamma\gamma}$.

$$\mathscr{H}_{\rm stress}(\boldsymbol{Q}_{\Gamma\gamma}') = \Sigma \delta V'(r, \boldsymbol{Q}_{\Gamma\gamma}')\Delta\boldsymbol{Q}_{\Gamma\gamma}, \tag{9}$$

where $\delta V'(\boldsymbol{r}, \boldsymbol{Q}_{\Gamma\gamma}') = \partial V(r, \boldsymbol{Q}_{\Gamma\gamma}')/\partial \boldsymbol{Q}_{\Gamma\gamma}' \mid_{\boldsymbol{Q}_{\Gamma\gamma}'=0}$. (10)

Hence, these descriptions clarify that the external stresses modify the lattice equilibrium positions, resulting in the change of electronic states through the linear term Eq. (5) of the electron-phonon interaction (see Refs. [7 – 9] for details).

Particular attention that $\Delta\boldsymbol{Q}_{\Gamma\gamma} = \boldsymbol{\varepsilon}_{\Gamma\gamma}$ is the irreducible strain labeled "$\Gamma\gamma$", which transforms in the same form as the stress $f_{\Gamma\gamma}$ under C_{6v} symmetry operations should be paid. Thus, the stress matrix element can be represented as $<\chi_v(\boldsymbol{r}) \mid \mathscr{H}_{stress}(\boldsymbol{Q}_{\Gamma\gamma}) \mid \chi_v(\boldsymbol{r})>$. Assuming a linear relation between stress and strain, $f_{\Gamma\gamma} \propto \Delta\boldsymbol{Q}_{\Gamma\gamma}$ (Hooke's law),

$$<\chi_v(\boldsymbol{r}) \mid \Sigma\delta V'(\boldsymbol{r}, \boldsymbol{Q}_{\Gamma\gamma}') \mid \chi_v(\boldsymbol{r})>\Delta\boldsymbol{Q}_{\Gamma} = \Sigma\, K_{\Gamma} <\chi_v(\boldsymbol{r}) \mid \delta V'(\boldsymbol{r}, \boldsymbol{Q}_{\Gamma\gamma}') \mid \chi_v(\boldsymbol{r})>f_{\Gamma\gamma}, \tag{11}$$

where $\chi_v(\boldsymbol{r})$ represents a p-like wave function of the valence band, and transforms according to the irreducible representation A_1 or E_1 under C_{6v} symmetry operations. $\mathbf{K}_{\Gamma}$ is a proportional constant between an external stress ($f_{\Gamma\gamma}$) and the resulting strain ($\Delta\boldsymbol{Q}_{\Gamma}$). It should again be emphasized that in normal coordinate space, strains are linearly connected to stresses by Hooke's law.

There are six ways to externally squeeze a crystal. For hexagonal symmetry P_{63mc}, these are classified into four types in accordance with reducing the second rank tensors under P_{63mc} symmetry operations as follows,

$$\{A_1 + E_1\} \otimes \{A_1 + E_1\}_{symmetry} = A_1^a(zz) + A_1^b(xx+yy) + E_1(xz, yz) + E_2(xx\text{-}yy, xy). \tag{12}$$

Three anti-symmetric second-rank tensors are removed, because the stresses (strains) are symmetric. The functions in the parentheses indicate the basis functions transforming under each irreducible representation. Each stress type must be assigned to one deformation potential constant. Hence, there should be four different deformation potential constants in each band for hexagonal symmetry P_{63mc} at the maximum. By substituting Eq. (12) into Eq. (11), we can define the new DPCs in a form directly responsible to stresses. For $\Gamma_5(E_1)$ band,

$$<\chi_v(\boldsymbol{r}) \mid \Sigma\delta V'(\boldsymbol{r}, \boldsymbol{Q}_{\Gamma\gamma}') \mid \chi_u(\boldsymbol{r})>f_{\Gamma\gamma} = \Sigma D_{\Gamma}\, (\Theta_{\Gamma\gamma})_{vu}\, f_{\Gamma\gamma}, \tag{13}$$

where $\chi_v(\boldsymbol{r})$ is an electron wavefunction, and for the valence band of nitride semiconductors, v (u) is a p_x or p_y orbital (vice versa), and $(\Theta_{\Gamma\gamma})_{vu}$ is a tensor matrix element [7] obeying the following relations.

$$\Sigma\, (\Theta_{\Gamma\gamma})_{vu} \cdot (\Theta_{\Gamma'\gamma'})_{uv} = A_{\Gamma}\delta_{\Gamma\Gamma'}\delta_{\gamma\gamma'}. \tag{14}$$

A_{Γ} is a constant depending on Γ only. The sum runs over v and u, and

$$\Sigma\, (\Theta_{\Gamma\gamma})_{vu} \cdot (\Theta_{\Gamma\gamma})_{uw} = \delta_{vw}, \tag{15}$$

where the sum also runs over v and Γ.

For $\Gamma_1(z)$ ($A_1(z)$) band, the same expression as Eq. (13) stands, but independently. As a cross-strain (shear strain) terms,

$$D_{cross}(zx) = D_{ctoss}(yz) = <\chi_v(p_z) \mid \delta V'(zx) \mid \chi_v(p_x)>. \tag{16}$$

Any external stresses must be decomposed into the above irreducible stress components to analyze observed data. For analysis of the experimental observations, the concrete representation will be exhibited elsewhere.

For the case of lattice mismatch in hetero-structure, the band alignment is significantly influenced by the internal stresses caused by the lattice mismatch.

4. Summary

Bir and Pikus's strain matrix for semiconductors with wurtzite structure (hexagonal symmetry) was re-examined, and corrected by properly adopting the trigonal part of the ligand field term. Introducing a normal coordinate system, an alternative approach to the deformation potential is tried in the framework of invariant regime. The basic expressions of the DPCs for the valence band of nitride semiconductors are given. This definition connecting to not strain but stress is formalised, makes its physical situation more evident, and is much more straightforward for experimentalist.

References

[1] Bir G L, and Pikus G E 1974 *Symmetry and Strained-Induced Effects in Semiconductors* translated from Russian by P. Shelnitz (New York and Toronto, John Wiley & Sons, Inc.), and references therein
[2] BanalR. G, Funato M and Kawakami Y 2009, *Phys. Rev.* B**79** 121308(R)
[3] Peng H Y, McCluskey M D, Gupta Y M, Kneissl M and Johnson N M 2005 *Phys. Rev.* B**71** 115207
[4] Ishii R, Kaneta A, Funato M and Kawakami Y 2010, *Phys. Rev.* B**81** 155202
[5] Wagner J M and Bechstedt F 2002 Phys. Rev. B**66** 115202
[6] Griffith J S 1962 *The Irreducible Tensor Method for Molecular Symmetry Group* (New York, Prentice-Hall)
[7] Henry R C H, Schnatterly S E and Slichter C P, 1965 *Phys. Rev.* **137** 130
[8] Schnatterly S E, 1965 *Phys. Rev.* **140** A1364
[9] Henry R C H and Schnatterly S E 1968 *Physics of Color Center* edited by Fowler W B (New York and London, Academic Press) chap. 6.
[10] Kamimura H, Sugano S and Tanabe Y 1988 *Ligand Field Theory and Its Application* (Tokyo, Syokabo) (in Japanese).
[11] Chuang S L and Chang C S 1996 *Phys. Rev.* B**54** 2491

Journal of Physics: Conference Series **423** (2013) 012034 doi:10.1088/1742-6596/423/1/012034

Effectiveness of Pavement Management System and its Effects to the Closing of Final Account in Construction Project in Malaysia

Zarabizan Zakaria[1], Syuhaida Ismail[1], Aminah Md Yusof[2]
[1]RAZAK School of Engineering and Advanced Technology, University of Technology Malaysia, 54100 Kuala Lumpur Malaysia
[2]Faculty of Civil Engineering, University of Technology Malaysia, 81310 UTM, Johor, Malaysia

E-mail: zarabizan2@live.utm.my

Abstract. Federal roads maintenance needs a systematic and effective mechanism to ensure that the roads are in good condition and provide comfort to the road user. In implementing effective maintenance, budget is main the factor limiting this endeavor. Thus Public Works Department (PWD) Malaysia used Highway Development and Management (HDM-4) System to help the management of PWD Malaysia in determining the location and length of the road to be repaired according to the priority based on its analysis. For that purpose, PWD Malaysia has applied Pavement Management System (PMS) which utilizes HDM-4 as the analysis engine to conduct technical and economic analysis in generating annual work programs for pavement maintenance. As a result, a lot of feedback and comment have been received from Supervisory and Roads Maintenance Unit (UPPJ) Zonal on the accuracy of the system output and problems that arise in the closing of final account. Therefore, the objective of this paper is to evaluate current system accuracy in terms of generating the annual work program for periodic pavement maintenance, to identify factors contributing to the system inaccuracy in selecting the location and length of roads that require for treatment and to propose improvement measures for the system accuracy. The factors affecting the closing of final account caused by result received from the pavement management system are also defined. The scope of this paper is on the existing HDM-4 System which cover four states specifically Perlis, Selangor, Kelantan and Johor which is analysed via the work program output data for the purpose of evaluating the system accuracy. The method used in this paper includes case study, interview, discussion and analysis of the HDM-4 System output data. This paper has identified work history not updated and the analysis is not using the current data as factors contributing to the system accuracy. From the result of this paper, it is found that HDM-4's system accuracy used by PWD Malaysia attains average 65 per cent only and had not achieved level that had been set by PWD Malaysia namely 80 per cent. Hence, this paper has revealed the causes of the occurrances in the pavement management system in construction project in Malaysia and investigated the consequences of the late payments and final account problems confronted by contractors in Malaysia, which eventually proposed strategic actions that could be taken by the contractors in securing their payments.

1. Introduction

The developments of infrastructure especially in road and highway network are very important to generate the country economic excellence growth. In line with the rapid development that has been taken thus far, it is clearly evidence that the government aspiration is to develop Malaysia as a developed country in the year 2020. It is essential to give the priority for infrastructure development. Thus, the Government of Malaysia has entrusted Ministry of Work Malaysia (MWM) to materialise the aspiration. MWM thru the agency of Public Works Department (PWD) has planned to forge these integrated developments. PWD also manages and maintains the existing Federal Road to ensure the good road network system is always safe, efficient and comfortable to the road users. The government also reserves and spends a huge budget for construction of the new road and maintains the existing road. In the Ninth Malaysia Plan (9MP), the government has budgeted the amount of RM 2.5 billion for road maintenance. In ensuring the smoothness of the Federal Road maintenance, government has decided the mechanism for systematic and well-planned road maintenance management.

Nevertheless, in implementing effective maintenance, budget is observed by this paper as the major factor limiting this effort. Thus, PWD has used HDM-4 System to help the management of PWD to determine the location and length of roads to be repaired according to the priority based on its analysis. For that purpose, PWD Malaysia has applied PMS which utilizes HDM-4 as the analysis engine to conduct technical and economic analysis in generating annual work programs for pavement periodic maintenance.

It is widely recognized that a good road infrastructure of a country is a pre-requisite to the development of a nation. Besides constructing new roads in Malaysia, Road Authorities are now aware of the needs to maintain and upkeep the existing roads in hand. With the large road network currently at about 17,500 km for Federal Roads and about 61,000 km for the state roads in Malaysia, roads maintenance budget accounts for sizeable proportion of public expenditures and is currently at about RM 580 million a year. If road user costs or vehicle operating costs (VOC) are taken into considerations, the total expenditure by road transport sector is even greater. Minimising the total road transport costs is the approach that is now widely accepted by most road authorities around the globe. To undertake such task in minimizing the road transport cost, a few strategies have been adopted in Malaysia. One of the most common approaches is a creation of a Pavement Management System (PMS) where the collection of data has been regarded as one of the most important components of the PMS. The results of PMS will be then transferred to the final account to show the effectiveness of allocated provision by the government.

2.0 Literature Review

2.1 Federal Road Maintenance Management in Peninsular Malaysia

In the year 2001, the Government of Malaysia has privatized the federal road maintenance in Peninsular Malaysia to three (3) concession companies and divided the nation into the zone as decided by PWD Malaysia. List of the company with the selected zone for the maintenance works are as shown in Table 1.

Table 1. Maintenance zone and the concessionaire

No	Zone	State	Consession Company
1.	North	Perlis, Kedah, Pulau Pinang & Perak;	Belati Wangsa Sdn. Bhd.
2.	Middle / East	Selangor, Pahang, Terengganu & Kelantan;	Roadcare Sdn. Bhd.
3.	South	Negeri Sembilan, Melaka & Johor	Selia Selenggara Sdn. Bhd.

The concession company which has been awarded to maintain the federal road have to complete their scope of work as instructed by the PWD Malaysia or as agreed in the contractual agreement signed by both PWD Malaysia and the concessionaire. The main scopes of work for the concessionaire are in routine maintenance, periodic pavement maintenance, periodic non pavement maintenance and emergency work. In these programmes, the maintenance of the road is only stressed for the periodic pavement maintenance only. This is because, every year PWD will comeout with the Periodic Pavement Maintenance programme for the Federal Road in Peninsular Malaysia undertaken by the concessionare. The lists of the programme location for the Periodic Pavement Maintenance are produced by RAMS thru HDM-4 system analysis which have been used by PWD Malaysia. Based on the Road Statistic Report 2005, the overall total length of the paved federal road in Peninsular Malaysia is 12,958.58 km excluding East Malaysia and Labuan. The report is produced by PWD Malaysia.

2.2 Pavement Management System (PMS)

According to Mulholland (1989) Pavement Management System (PMS) can be identified as a method of data collections, analysis and improve engineer to decide and design optimum resources for maintenance, repair and rebuild pavement. It is also defined by this paper as a set of activity coordination in optimising civil value of money by providing an economic, safe and comfortable road.

On the other hand, IKRAM (1993) defines PMS as coordination process and a set of comprehensive activities control to maintain pavement by using optimum available resources. In other words, PMS is a systematic process which objectively functions as a pavement quality controller and maintenance programme as a result of situation, limited reallocation and economical optimum within cost reduce and optimum agency maintenance cost. Elements included in PMS are as below:

- Major and minor road inventory to be updated once in two years
- Pavement assessment for all route by using International Roughness Index (IRI) and updated once in 2 years
- Pavement progress history
- Identification of all pavement section which has to be repaired
- Reallocation of budget to repair pavement for the latest period or 2 coming years
- Effect of reallocation of pavement budget for the upcoming years

PMS is designed to provide work framework structure to record route network database systematically. Main objective of PMS is to assist decision maker in the development of effective cost strategy and provide existing maintenance budget to be prioritized where the balance of budget will be used for other non-prioritized roads. Objectives of Pavement Management System are to determine optimum maintenance policy, determine maintenance budget post time, determine the best strategy under sub-optimum budget and assess of Progress Scheme Capacity as well as determine early remedial design and assessment of Axel Load Policy. According to Safry (2000), there are four processes involved in pavement management. The processes are policy, planning, programming, preparation and operation as discussed in the following sub-topics.

2.2.1 Policy

It is a basic foundation or regulation made by PWD Malaysia to determine financial budget in achieving road standard in certain time based. From the existing budget, PWD Malaysia has decided to collect data once in three years for each zone and accuracy of work programme produced by this system should achieve 80% and above.

2.2.2 Planning

Planning includes overall road analysis system which requires preparation for short term and long term strategy to estimate the development, restoration and maintenance cost of the road under variety budget and economy scenario. Forecasting can be done on the condition of the road network in terms of Pavement Performance Index under variaties of budget. When a need budget and service target archive have been fixed, a standard work schedule should be formed where it will show the best treatment for different level of damages.

2.2.3 Programming

This includes the preparation of yearly road work program, the selected road for maintenance or the selected road to be up graded and analyed. The Cost Analysis should be taken for applicability in terms of economy for each work to be done. Programming activity will produce estimation each year expenses under fixed budget for each type of different road and for each road section.

2.2.4 Preparation

Preparation is a short term planning where road planning package is to be applied. At this stage, the design should be completed and detailed, encompasses the detailed list of quantity and costing with instruction of work and contract. Detailed specification and cost will be prepared and the benefit cost analysis will be done to make sure the applicability.

2.2.5 Operation

Operation includes organization of operation and the result for the management operation to be done every day or week which the schedule encompasses of work programmed supervision in terms of worker, tool and material, completion work record and the use of this information and controlling. Each function represents consecutive activity called cycle of Pavement Management System as shown in Figure 1.

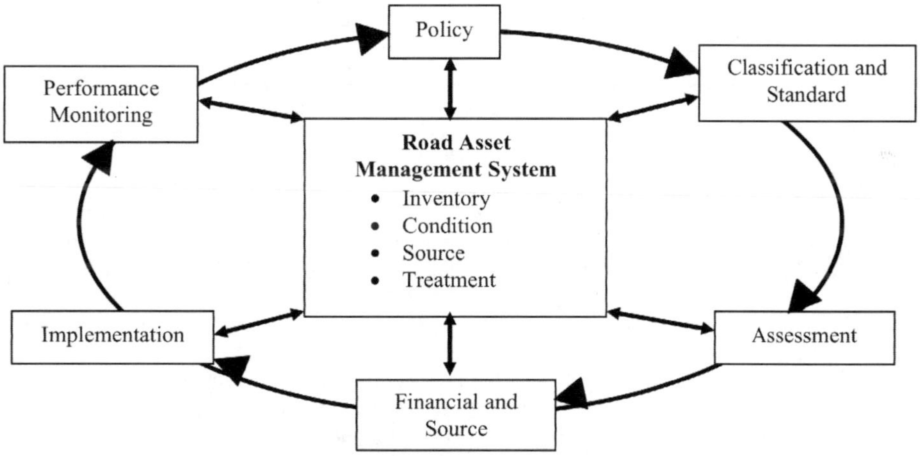

Figure 1. Process involved in pavement management

2.3 HDM-4 as a Mechanism System for Pavement Management System

HDM-4 provides an established system for analysis of road management and an alternative investment. This system can be applied in other matters like road management to concentrate on preventive work in maintenance road network, make a programme for construction work/rehabilitation/road care, make an estimation fund needed and allocation of budget, indicate the road network performance in providing safety driving and comfortable to users, assess/evaluate project and study the effectiveness of the policy implementation. The framework of HDM-4 is based on pavement life cycle analysis concept. It is used to forecast the matters on pavement life cycle: declining road, road work effect, road used effect and socio-economic as well as environmental impact.

When the road is built, road pavement will be affected by several factors such as vehicle load, environmental weathered and effect from irrigation which is supported sufficiently. Road pavement decline rate will give impact to standard maintenance which is used to pavement failure such as crack, raveling, puddle and etc. For long term pavement life cycle, it is depending on standard maintenance used for that particular road. Figure 2 shows the flow of pavement performance which is delegated by driven quality and always been measured as *International Roughness Index (IRI)*. From IRI graph versus time as shown in Figure 2, it is see that road pavement will decline year by year if not properly treated. Standard maintenance excepted by PWD is IRI 3.5 and above. If declining road scale has reached IRI 3.5 and above, pavement should be treated. Delay in treatment will cause road pavement get worst and treatment cost will increase. Figure 2 also shows that IRI will reduce after treatment is undertaken.

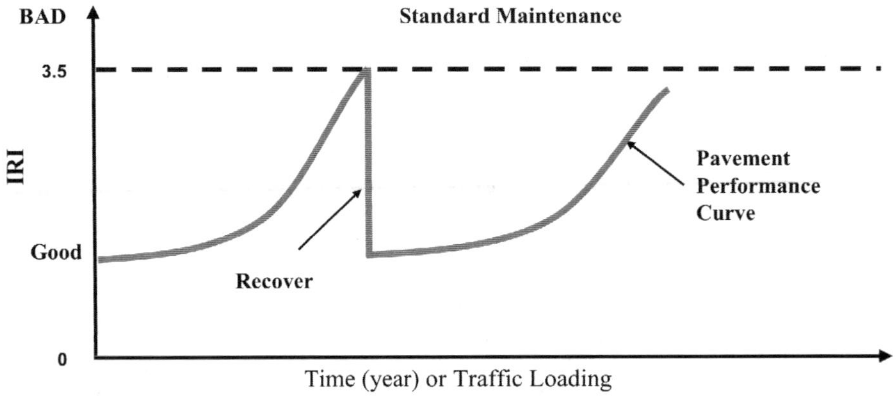

Figure 2. Graph Roughness versus Time

Pavement management process can be assumed as a function of pavement management activity cycle which highlights the concept and work frame where HDM-4 can fulfil pavement management. According to Kerali (1999), there are three main items in management concentrate suitable for HDM-4 which are:

a. **Project Analysis**
   i. Assessments for one or more road projects or choice in investigation.
   ii. Concentration in road network analysis or section with choice of treatment method by users, related to project cost and benefit every year.
   iii. Economic indicators are based on choice of differences in investigation.

b. **Programming Analysis**
   i. Used to prepare annual programs for road network where choice in investigate is identified based on limited reallocation of source.
   ii. Road network analyzed based on section and cost estimate will produce for road maintenance and financial requirement for each section for each year.

c. **Strategy Analysis**
   i. Used to analyze road network that is selected to provide medium and long term forecast planning for financial requirement for road development and maintenance under different budget scenario.
   ii. Road network is characterized according to road length in different category and can be explained based on parameters such as road class, pavement type, road situation and traffic flow.

iii. Forecast will be produced for financial requirement for medium and long term period which is within 5 to 15 years.

3.0 Final Account

Whilst there is no universally written answer to what final account in construction industry means, because there are rare studies on this subject, the term final account has often been with and in existence in the construction contracts in Malaysia for more than 30 years. Although there is no specific definition, the term final account is taken in the construction contracts simply as a mechanism for dealing with the final contract sum or final payment due to the contractor after the Final Measurement Period (typical building contract) or after the Maintenance / Defects Liability Period (typical civil engineering contract). Sometimes, it also refers to the calculation and agreement of the final construction cost between the employer and contractor and incorporates a fair valuation of the works carried out [2]. The cost at tender will only reflect the cost of fulfilling the client s original brief and will not contain allowances for any subsequent omissions from or additions to the design on which the bills of quantities were prepared [4].

Therefore at the final account stage, the contract sum will include such items as day-work charges, adjustments in accordance with variation orders, and so on, and hence in some ways a clearer picture of the total cost is presented [4]. In the contractual view, upon completion of the contract requirements, the contractor is entitled to final payment including any retained sums [10]. The effect of this is that the original contract sum will be changed for a variety of reasons in pursuance to the provisions within the contractual conditions. The final contract amount includes all additions, alternations and deductions resulting from project changes. More often than not, variations and claims are inherent in construction projects because uncertainties lead, invariably, to the need for adjustments [3]. Therefore, the assessment and agreement of this final contract sum, i.e. the final account, is usually of the utmost importance to both the client and contractor [2]. Hence, delays of final accounts closing can be resolved at the early stage if the documents are kept properly and project manager has better understanding of critical success/failure factors in the construction project [12].

4.0 Research Methodology

4.1 Introduction

HDM-4 has been used since 2002 by PWD Malaysia to organize their yearly periodic routine maintenance work. The work programme used has to be very accurate and precise in order to get the correct location and the length of the road that to be taken care of. So far, there was no research that can be used/followed as a based procedure to ensure the preciseness and the accurateness of the system. In this paper, the new system that can be used to justify the accurateness and preciseness of the system will be discussed in the next sub topic.

4.2 Purpose of Paper

Every paper has their own purpose and this paper is to ensure the effectiveness and accurateness of HDM-4 system so that every yearly periodic routine maintenance work programme performed by this system will be successful and precise. Via this paper, HDM-4 output, i.e list of the work programme will be taken and compared with the actual needs that are approved by UPPJ zone. For those location lists that did not exactly follow the site requirement, the replacement location will be proposed by UPPJ zone.

4.3 Paper: Sources and Tools

To make this HDM-4 system effectively generates the periodic routine maintenance work programme, the followings are the tools and sources needed.

Journal of Physics: Conference Series **423** (2013) 012034 doi:10.1088/1742-6596/423/1/012034

4.3.1 RAMS Data Base (RAMS DB)

Database needed to analyse all HDM-4 data includes road inventory, road pavement, pavement structure and its traffic. As for road inventory, data that are taken into consideration are pavements type, road lengths and its width, road geometric situation and type and width of the road border. For the road inventory, information that are taken into considerations are roughness, cracked, siltness, pothole and the depth of its textures. While for road structure, it will consider the road pavement life-time, its thickness of its asphalt, its base and road sub-base as well as Falling Weight Deflectometer (FWD) Central Deflection.

4.3.2 HDM-4 Software

Basic raw data from RAMS DB will be used as the main input into HDM-4. These data will be analyzed by HDM-4 system based on the category either its state or federal road, year where they starts its analysis, year they ended the anaysis, yearly budget approved and standard stage of maintenance. HDM-4 system will then start its analysis for the road conditions and its pavement by taking into consideration all the parameters set. The results of these analysis then will be used and set as its primary/based work programme to be followed for their annual routine pavement maintenance at the site location. All the work programme lists proposed by HDM-4 will be monitored by UPPJ zone officers.

4.0 Data Analysis Approach

This HDM-4 has the same workflow as what has been implemented in the yearly periodic routine maintenance work programme by PWD Malaysia. However, there are some differences where the list of work programme produced/proposed will be analyzed for its accurateness. The workflow for this accuracy calculation for HDM-4 system is shown in Figure 3.

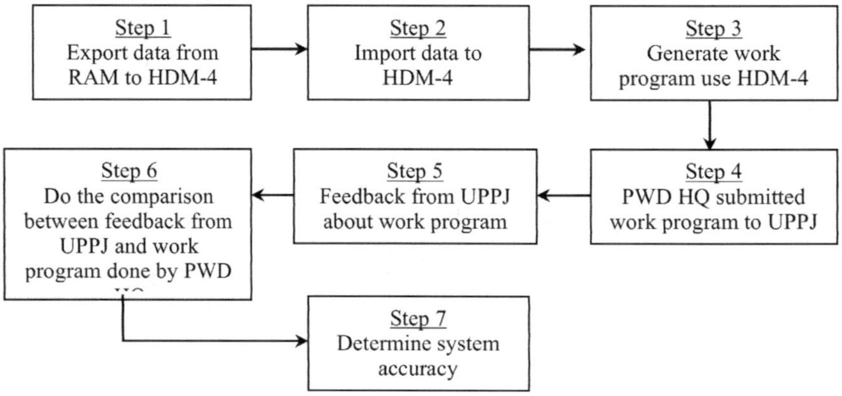

Figure 3. Workflow for accuracy calculation for HDM-4 system.

4.1 Data Obtained

In order to get the data, the first step is exporting the data from RAMS DB to HDM-4 system. These have been explained well in Topic 2 where all data in RAMS DB have been taken from IKRAM s data collection through their yearly Road Data Collection Programme. Certain informations and parameters needed have to be set up during the implementation of the data export process. In this paper, data that are to be studied are the sample data that have been collected from four states representing each zone. The selected states are Perlis (North zone), Selangor (Middle zone), Kelantan (East Zone) and Johor (South Zone).

4.2 Generation of Work Programme

The data taken from RAMS DB for road and pavements have to be imported into HDM-4. The two imported files are programme file and road network file. Programme analysis will be performed in HDM-4 to produce 2007 work programme based on all the set up parameters. As the results, HDM-4 will produce a proposed list of work programme for the periodic road maintenance programme for that state or road.

4.3 Work Programme Accurateness Checker
Periodic road maintenance programme that is produced by HDM-4 system will be then forwarded to UPPJ zone to have it checked, decided and approved the work implementation by looking at the proposed job scope and approved cyling budget. Both inspections and verifications by UPPJ zone will then lead to which implementation to be chosen based on the site condition and also the maximum approved budget. Based on the proposed work programme by HDM-4 system, UPPJ zone will cancel those locations which are still in good conditions and do not need the maintenance or those locations that exceed the given budget. Final list will then be forwarded to the Road Maintenance Unit, HQ PWD which will follow and use this programme as their baseline for periodic road maintenance programme and hence this paper will be the most accurate analysis ever implemented.

5.0 Accurateness Work Programme Analysis
In general, the list programme for the annual pavement maintenance that would be implemented will list all the location that needs the maintenance produced by HDM-4 and accepted by UPPJ zone together with the proposal of replaced location given by UPPJ zone after considering which location maintenance is proposed by HDM-4 but do not need to do on site. Figure 4 shows representation of final periodic routine maintenance work programme breakdown.

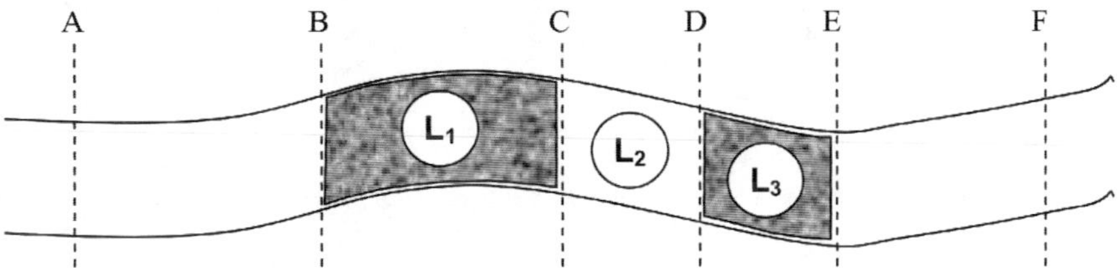

legend :

A – F Lengths of analyzed road

B – D Location proposed by HDM-4

(L_1) Road location from B C proposed by HDM-4 and approved by UPPJ

(L_2) Road location from C D proposed by HDM-4 and disagreed by UPPJ

(L_3) Road location from D E is a replacement proposed by UPPJ

Figure 4. Final periodic routine maintenance work programme breakdown.

Based on Figure 4, list of the periodic road maintenance programme produced by HDM-4 system or list periodic road maintenance programme proposed by UPPJ zone can be summarized as below:

a. L1 : Lengths of road on the locations that need maintenance in 2008 which were proposed by HDM-4 system and approved by UPPJ zone
b. L2 : Lengths of road on the locations that need maintenance in 2008 which were proposed by HDM-4 system but not approved by UPPJ zone.

c. L3 : Lists of road that need maintenance proposed by UPPJ in 2008 but were not approved by HDM-4 system.

Based on the discussion with the officer incharged for HDM-4 system in Malaysian HQ PWD and IKRAM a formula has been formulated to calculate the accurateness of this system based on the location and also the lengths proposed. Formula is as follows:

$$\textbf{System accuracy (\%)} = \frac{\textbf{L1}}{\textbf{(L1 + L2 + L3)}} \textbf{ x 100 \%} \qquad \textbf{(1)}$$

$$= \textbf{X \%}$$

Based on this system accuracy calculation, the implementor can see the effectiveness of the system whether it reaches the goal target by comparing with the policy that has been fixed at 80 per cent above. If this is not reachable, factors that contribute to the inaccuracy of the system will be studied to ensure future improvements and new input or ideas for effectiveness of the system.

5.1 Analysis Result HDM-4's Output of Paper Location

From responses received from UPPJ, HDM-4 output's analysis perpetrated upon paper location is as follows:

Table 2. Final location summary list

No	State	Location in Total	Location Agreed by HDM-4 and UPPJ	Location Suggested by HDM-4, Rejected by UPPJ	Location Replaced by UPPJ
1.	Perlis	21	15	6	0
2.	Selangor	33	21	12	9
3.	Kelantan	17	11	6	5
4.	Johor	50	33	17	9

Analysis result HDM-4's output, periodic maintenance proposal location list pavement in Perlis state of year 2007 is amounted to 21 locations as indicated in Table 2. Of this list, 15 locations were agreed by UPPJ zone while six location have been rejected by UPPJ zone based on the needs at the site. Nevertheless, it is found that no replacement list lodged by UPPJ zone of Perlis state.

On the other hand, HDM-4's analysis of Selangor state shows that periodic maintenance proposal location list pavement in Selangor state of year 2007 is amounted to 33 locations as indicated in Table 2. Of this list 21 locations were agreed by UPPJ zone while 12 locations have been rejected by UPPJ zone based on the needs at the site. UPPJ zone has also put nine different locations alternately on location subtrahend from HDM-4's proposal.

HDM-4's analysis of Kelantan state demonstrates that periodic maintenance proposal location list pavement in Kelantan state of year 2007 is amounted to 17 locations as indicated in Table 2. Of this list, 11 locations ware agreed by UPPJ zone while six locations have been rejected by UPPJ zone based on the need at the site. UPPJ zone has also put five different locations alternately on location subtrahend from HDM-4's proposal.

HDM-4's analysis of Johore state shows that periodic maintenance proposal location list pavement in Johore state of year 2007 is amounted to 50 locations as indicated in Table 2. Of this list 33 locations ware agreed by UPPJ zone while 17 locations have been rejected by UPPJ bases on the needs at the site. UPPJ zone has also put nine different location alternately on location subtrahend from HDM-4's proposal.

5.2 System Accuracy Analysis

HDM-4 output's result of analysis such as in Table 2 and responses of UPPJ, it has been categorised again such as in Table 3. Accurate analysis work programme is issued by HDM-4's system which has been counted using Equation 1. The result is summarized in Table 3 and the calculation method shown as follows :

Table 3. Retrieval result HDM-4 system accuracy s calculation

State	Precise system (%)	Notes
Perlis	69	No achieved on target
Selangor	62	No achieved on target
Kelantan	62	No achieved on target
Johor	66	No achieved on target
Average	65	No achieved on target

6.0 Summary and Conclusion

Through this paper some conclusions have been identified. Whereas for recommendation unit, it is aimed to propose to PWD Malaysia that further paper should be done to boost more HDM-4 s system s effectiveness. HDM-4 output s accuracy is an important element proposal in the process of producing periodic programme work pavement in PWD Malaysia. This is because it involves limited maintenance allocation and long road network number to be organized. From the result of this paper, it is found that HDM-4 s system accuracy used by PWD Malaysia attains average 65 percent only and had not achieved level that had been set by PWD Malaysia namely 80 percent. Accuracy assessment revenue summary periodic maintenance work programme pavement is suggested by HDM s system. Periodic maintenance work programme proposal pavement is issued by HDM-4 s system and it is very closely connected with data and information input included in generation current system programme carried out.

On the other hand, results obtained will also give effect to the allocation provisions for the implementation of PMS. The results HDM-4 s system accuracy attains average 65 percent only and had not achieved level that had been set by PWD Malaysia namely 80 percent, thus will affect to the final account process. This results will give impact to PWD in managing PMS roads in Malaysia. Therefore, introducing payment department for every contractors incuding UPPJ Zone to report any late payments and final account problems is needed to formed. This department would be responsible in keeping records of every client with pavement management record and late payment history. Punishments would be given to them if they fail closing the final account as stipulated in contracts. It shall be reminded that time is the essence and this department shall be able to settle the problems as soon as possible.

From this paper, somes factor that contribute to periodic maintenance work programme proposal inaccuracy pavement are identified. Besides, the results of the effectiveness of pavement management system that will affect the final account are also concluded. To help road management party in PWD Malaysia in maintaining federal road in Malaysia with more effectiveness and effeciency, several matters should be given attention to increase this HDM-4 s system s effectiveness. As such, from conclusion above-mentioned, it is clear that this paper has achieved its objective. We believe that the information of this paper can be immense help the construction players (clients, contractors and consultants) and academicians. The construction players can better understand and make efforts to reduce the incidences of inefficiency management of PMS. In future paper need to be carried out on HDM4's system's effectiveness, with paper proposal accuracy treatment that proposed by HDM4 system.

7.0. Acknowledgements

The authors would like to express their sincere gratitude to University of Technology Malaysia for providing the financial support for this paper to be published. The authors are also very grateful to anonymous reviewers for their valuable comments to this paper. This paper is financed by the Grant for Research University (GUP) Tier 1 of University of Technology Malaysia, Johor under Cost Centre No. Q.K.130000.7140.00H47.

8.0 References

[1] Ahmad S K and Azmi A E 2007 *Road evolution In Malaysia* (Kuala Lumpur: PWD)
[2] Ashworth A & Hogg K 2002 *Willis s Practice & Procedure for the Quantity Surveyor* (United Kingdom: Blackwell Science Ltd)
[3] Baloi D., Price A. 2003 Modelling global risk factors affecting construction cost performance, *International Journal of Project Management* **21** 261-9
[4] Cartlidge D P 1973 *Cost planning and building economics* (London: Hutchinson Educational Ltd)
[5] PWD Malaysia 2002 *Final report for development of a new pavement management system for federal roads in Malaysia* (Kuala Lumpur: PWD)
[6] Kerali H R 1999 *Overview of the new HDM-4 system present in XX1st World Road Congress in Kuala Lumpur*
[7] Kumpulan IKRAM Sdn. Bhd. and PWD Malaysia 2001 *Contract proposal for development of a new pavement management system for federal roads in Malaysia* (Kuala Lumpur: IKRAM Publication)
[8] Mulholland P J 1989 *Pavement management system (PMS) for local government state of the art report - Research report* (Autralian: Australian Road Research Board)
[9] Safry 2000 *Development of HDM-4 based on pavement management system for federal roads in Malaysia* (Kuala Lumpur: PWD)
[10] Stokes M 1997 Construction law in contractor s language, 3nd Edition (United States: R.R. Donnelley & Sons Company)
[11] World Road Association (PIARC) 2003 *HDM-4 Information and support centre online at http:/HDM-4.piarc.org, 2003*
[12] Zarabizan Zakaria, Syuhaida Ismail and Aminah Md Yusof 2012 Cause and impact of dispute and delay the closing of final account in Malaysia construction industry, *Journal of Southeast Asian Research* **2012** 12

Experimental study on the influence of the porosity of parallel plate stack on the temperature decrease of a thermoacoustic refrigerator

Ikhsan Setiawan[1,2], Agung Bambang Setio Utomo[1], Mitrayana[1], Masafumi Katsuta[2], Makoto Nohtomi[2]

[1]Physics Department, Faculty of Mathematics and Natural Sciences, Gadjah Mada University, Sekip Utara, Yogyakarta 55281, Indonesia
[2]Graduate School of Environment and Energy Engineering, Waseda University, Nishi-tomita 1011, Honjo-city, Saitama pref., Japan

E-mail: ikhsan_s@ugm.ac.id

Abstract. Thermoacoustic refrigerators are cooling devices which are environmentally friendly because they don't use hazardous gases like chlorofuorocarbons (CFCs) or hydrofuorocarbons (HFCs) but rather air or inert gases as working medium. They apply sound wave with high intensity to pump heat from the cold to hot the regions through a stack in a resonator tube. One of the important parameters of thermoacoustic refrigerators is the porosity (blockage ratio) of stack which is a fraction of cross sectional area of the resonator unblocked for the gas movement by the stack. This paper describes an experimental study on how the porosity of parallel plate stack affects the temperature decrease of a thermoacoustic refrigerator. The porosity of parallel plate stack is specified by the thickness of plates and the spacing between plates. We measured the maximum temperature decreases of thermacoustic refrigerator using stacks with various porosities in the range of 0.5 - 0.85, with plate spacing from 0.5 mm to 1.5 mm and plate thicknesses 0.3 mm, 0.4 mm, and 0.5 mm. The measurements were done with two resonators with length of 0.8 m and 1.0 m, with air at atmospheric pressure and room temperature, correspond to thermal penetration depths (δ_κ) of 0.26 mm and 0.29 mm, respectively. It was found that there is an optimum porosity which gives the largest temperature decreases, and there is a tendency that the optimum porosity shifts to a larger value and the temperature decrease become larger when we used a stack with thinner plates. On the other hand, the study on the dependence of the temperature decrease on the plate thickness and the plate spacing reveals more useful information than that on the stack porosity itself. We found that stack with thinner plates tends to give larger temperature decrease, and the plate spacing of around $4\delta_\kappa$ leads to the largest temperature decrease.

1. Introduction

Thermoacoustic refrigerators are environmentally friendly cooling devices since they use air or inert gases as working medium instead of chlorofluorocarbons (CFCs) and hydro-fluorocarbons (HFCs) as in conventional coolers. The relatively simple design with no or only a little moving parts and the abundance of the working gas make them can be built and maintained in a low cost budget and serve as an alternative and promising cooling technology in the future compared to the conventional systems.

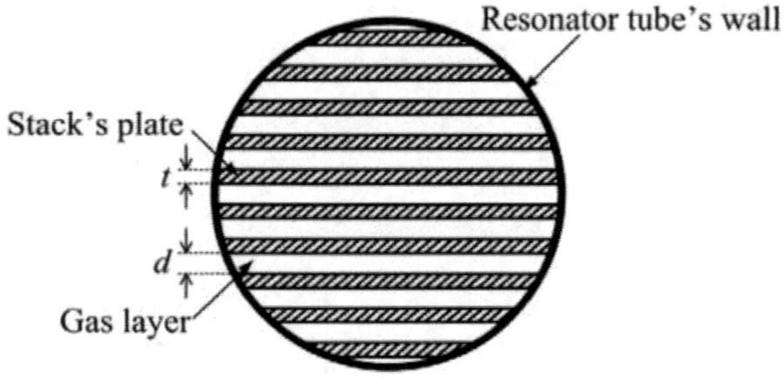

Figure 1. Schematic illustration of the cross-section of parallel plate stack, t is the plate thickness, and d is the plate spacing.

Thermoacoustic refrigerators mainly consist of a sound driver (e.g. loudspeaker), a resonator tube, a stack (porous medium), and working gas. The sound driver provides a high power standing sound wave in the working gas inside the resonator tube. A stack is typically placed near the pressure anti-node of the standing wave. The interaction between the expanding and compressing gas with the stack material is resulting in heat transport axially in the stack in the direction from the pressure node to the pressure anti-node. More detail descriptions on the working principles of thermoacoustic refrigerators can be found elsewhere.[1][2][3]

One of the important parts of thermoacoustic refrigerator is stack which is a kind of heat exchanger and placed in a resonator tube. Generally, stack consists of a large number of closely spaced surfaces aligned parallel to the length of the resonator tube. Tijani *et al* [4] studied the optimal spacing of stack plates for thermoacoustic refrigeration They found that a plate spacing of four times of the thermal penetration depth leads (δ_κ) to the lowest temperature, while the best performance was given by a $3\delta_\kappa$ plate spacing. Kim *et al* [5] investigated the optimum position of a stack in a thermoacoustic heat pump with various operating frequency in the range of 150 Hz - 300 Hz. They concluded that the stack positions for the most efficient operation of the thermoacoustic heat pump are at the maximum position (up-stream pumping) and the minimum position (down-stream pumping) of the curve of the temperature different vs position at the resonance frequency. Setiawan *et al* [6] studied experimentally the effects of the length and location of stack on the temperature decrease of a thermoacoustic cooler. For their thermoacoustic apparatus, they found the optimum stack length was 10 cm and the optimum stack location was at 10 cm or 0.125 normalized distance of stack center to the closed end of resonator. The impact of gas blockage on the performance of a thermoacoustic refrigerator in the relation to the existence of heat exchanger in the both end of stack has been inspected by Akhavanbaza *et al* [7] Their results show that the heat transfer decreases with an increase in the gas blockage, the relationship between gas blockage fraction and the temperature difference across the stack is linear, and a heat exchanger with larger thermal contact area (higher blockage fraction) increases the heat exchange between the heat exchanger fluid and the stack, but reduces the cooling power.

In this paper, we describe an experimental study on the influence of the porosity of parallel plate stack on the temperature decrease of a thermoacoustic cooler. The porosity, also called as blockage ratio (B), is the fraction of the cross sectional area of the resonator unblocked for the gas movement by the stack. For parallel plate stack, the porosity depends on the plates thickness and plate spacing, and expressed by [8]

$$B = \frac{d}{d+t} \tag{1}$$

where d is the plate spacing and t is the plate thickness as depicted in Fig. 1. When the plate spacing is too wide, then it could make the heat transport between gas and plates don't effectively occur, whereas when the plates spacing is too narrow, then the gas particles motion

Journal of Physics: Conference Series **423** (2013) 012035 doi:10.1088/1742-6596/423/1/012035

will be more hampered by viscous effect. The optimal plate spacing usually compare to the thermal penetration depth (δ_κ) of the gas in the stack. The thermal penetration depth is a measure of the distance that heat can diffuse in through a substance during time $1/\pi f$, and can be expressed by [3][4]

$$\delta_\kappa = \sqrt{\frac{\kappa}{\pi f \rho c_p}} \tag{2}$$

where κ is the thermal conductivity, ρ is the density, c_p is the isobaric specific heat of the gas, and f is the resonance frequency of sound in the resonator. On the other hand, the plate thickness determines not only the cross sectional blockage area of the stack, which blocks gas for moving back and forth, but also the thermal contact surface area between the gas and stack plates involved in the heat exchange process.

2. Experiment

Fig. 2 shows the schematic diagram of the experimental set-up. The thermoacoustic cooler in this experiment is a standing wave type and mainly consists of a loudspeaker, a resonator tube, a stack, and air as the working gas. A digital function generator (model GFG 8016G) was used to generate an audio sinusoidal signal. This signal was then amplified by a 350 W audio amplifier before fed into a 12" 400 W loudspeaker. The input voltage and current of the loudspeaker were measured by a digital voltmeter (V) and an ampere-meter (A), respectively. The loudspeaker together with its box is coupled to the resonator tube, so that they serve as a quarter-wavelength resonator system. We used free air at the atmospheric pressure and room temperature as the working gas that fills the resonator tube. In this experiment, we used two resonator tubes with different length (L), i.e. 80 cm and 100 cm, for which the estimated resonance frequencies were around 107 Hz and 86 Hz, respectively. A stack is placed near the closed end of the resonator so that the center of stack is located at a distance of around $L/5$ from the closed end of resonators. Both resonator tube and the stack housing are made of 1-1/4" PVC pipes.

In order to study the impact of the porosity of stack on the temperature decrease of the thermoacoustic refrigerator, we prepared fifteen plate parallel stacks with various porosities. The plates were made of mica plastic sheets with three kinds of thickness (t), i.e. 0.3 mm, 0.4 mm, and 0.5 mm. To provide gap (plate spacing, d) between the plates, we used nylon fishing lines which are glued onto the plates along the stack axis with distances of 1 cm. We used five different diameters of the nylon fishing lines, i.e. 0.5 mm, 0.85 mm, 1.0 mm, 1.2 mm, and 1.5 mm. This nylon fishing lines are ignored in the calculation of the stack porosities. The length

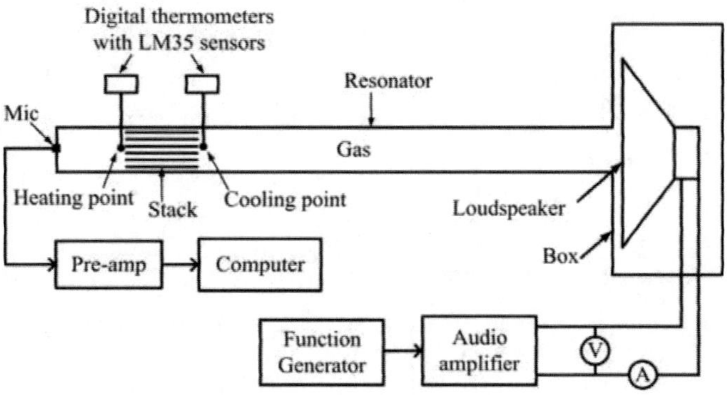

Figure 2. The schematic diagram of the experimental set-up.

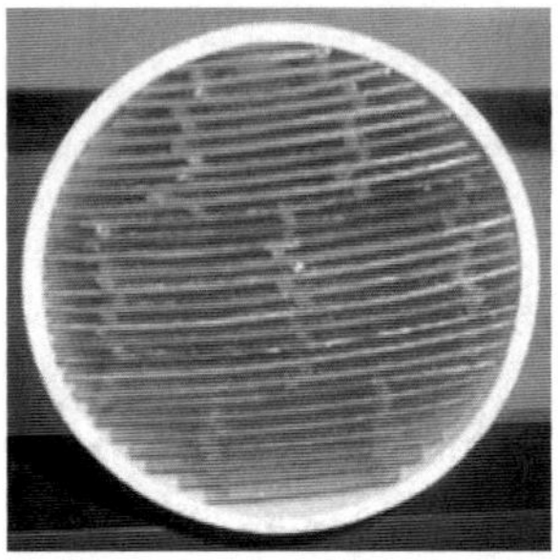

Figure 3. The cross section of a stack inside a 1-1/4 inches PVC pipe which is used in this experiment. The plate thickness $t = 0.5$ mm and the nylon diameter (plate spacing) of $d = 1.0$ mm.

of stacks are 10 cm chosen based on our previous result [8] and the diameter is 3.5 cm. Figure 3 shows the cross section of one of the stacks we used in this experiment, in which the plate thickness is 0.5 mm and plate spacing is 1.0 mm. The stacks are not so good built that the plates are slightly curved.

Experiments were done by operating the thermoacoustic cooler at its resonance frequencies. A small microphone is attached at the closed end of resonator to catch the sound and send the signal to a computer via a sound card. The resonance condition was recognized from the real time spectrum of the sound displayed by Oscilloscope 2.51 software on the computer's monitor. The high intensity sound from the loudspeaker provides work to pump heat from the cold region to the hot region (from right side to left side of the stack in Fig. 2). To study the influence of stack porosity on the temperature decrease and temperature difference of the thermoacoustic cooler, we measured the temperatures at the cooling point and the heating point the the both ends of the stack (see Fig. 2) for various stacks with different porosities, with various combination of plate thicknesses and plate spacings of the stacks. We used two digital thermometers with LM35 sensors to measure the temperatures at the heating and cooling points.

3. Results and Discussion

Figure 4 shows one of measurement results of the temperatures of the heating and cooling points for the the first 15 minutes operation of the thermoacoustic refrigerator. In this case, we used resonator with 80 cm in length and the stack with 0.3 mm plate thickness and 1.0 mm plate spacing. The temperatures of both measurement points after 15 minutes were relatively constant (not shown in the figure). The initial temperatures of both measurement points were 30.2 °C, and then we can see that the temperature of heating point was rising up to 43 °C and that of cooling point was decreasing down to 21.5 °C after the first 15 minutes. It means that the temperature increase (ΔT_H) was 12.8 °C, the temperature decrease (ΔT_C) was 8.7 °C, and the

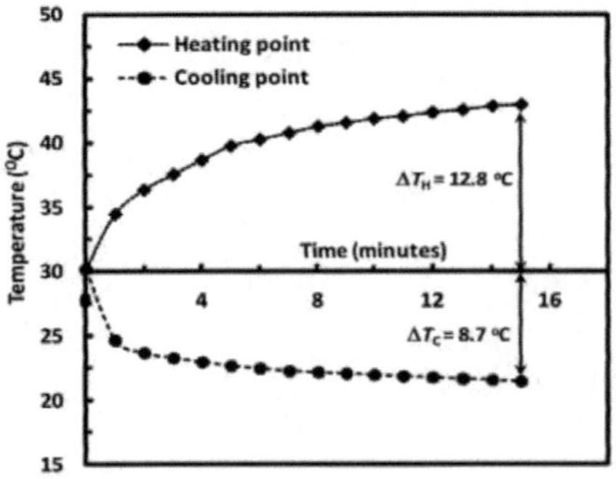

Figure 4. Temperatures of the heating and cooling points for the the first 15 minutes operation of the thermoacoustic refrigerator. In this case, $L = 80$ cm, $t = 0.3$ mm, and $d = 1.0$ mm. The dash lines are to guide the eye.

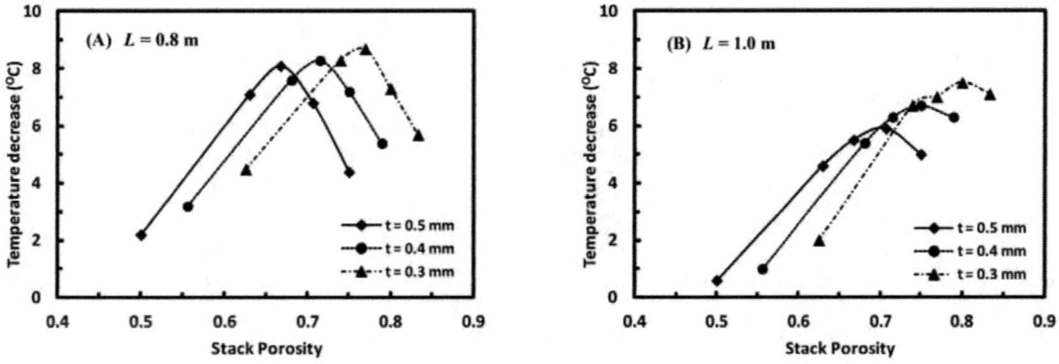

Figure 5. The dependence of temperature decrease on the stack porosity for (A) $L = 80$ cm and (B) $L = 100$ cm. The lines are to guide the eye.

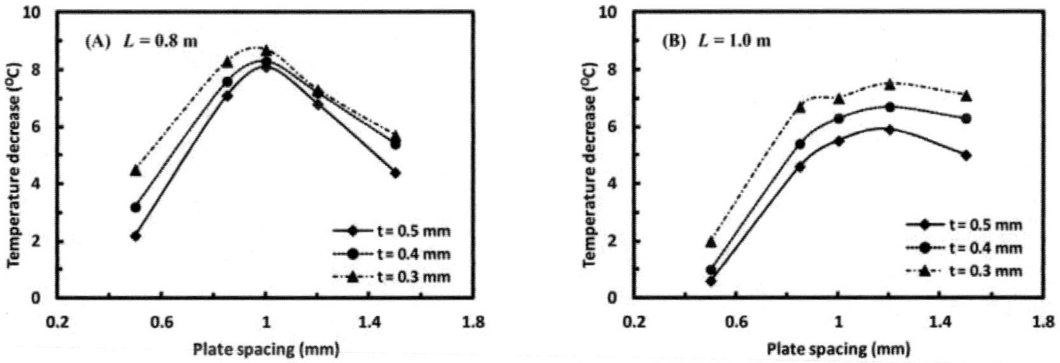

Figure 6. The dependence of temperature decrease on the plate spacing for (A) $L = 80$ cm and (B) $L = 100$ cm. The lines are to guide the eye.

temperature difference of 21.5 °C between the both ends of stack was established after that time period.

The measurement results of the maximum temperature decrease for various porosities of the stack are shown in Fig. 5 for $L = 80$ cm (A) and $L = 100$ cm (B). It can be seen in each figure that, for a given plate thickness, there is an optimum porosity which gives the largest temperature decreases. Moreover, there is a tendency that the optimum porosity shifts to a larger value and the temperature decrease become larger when we used a stack with thinner plates.

In order to have further analysis, we look the experimental data in a different way as shown in Fig. 6 in where the temperature decreases are plotted versus the plate spacing of the stacks instead of versus the stack porosity as in Fig. 5. We can see clearly that for a given plate thickness there is an optimum plate spacing which gives the largest temperature decrease, and the optimum values of plate spacing are the same for the same resonators length (L), regardless of the plate thickness. In this case, the optimum plate spacings are 1.0 mm for $L = 80$ cm and 1.2 mm for $L = 100$ cm. In addition, it can more clearly be seen in Fig. 6 that, for a given plate spacing, the stack with thinner plate yields a larger temperature decrease.

The dependence of temperature decrease on the plate thickness can be understood as follow.

The stack with thinner plate has not only more gas channels and hence larger thermal contact surface area between gas and stack material, but also smaller cross sectional blockage area. These conditions make the stack has more effective heat transport from the cold end to the hot end of the stack which in turn is resulting in a larger temperature decrease at the cooling point. On the other hand, the influence of plate spacing on the temperature decrease is related to both the viscous effect which is suffered by the moving gas and the thermal penetration depth (δ_κ) of the gas in the stack as has been described in the the previous section. By using Eq. (2) and the values of air properties at the experimental conditions (1 atm, 303 K), the thermal penetration depths were calculated approximately 0.26 mm and 0.29 mm for resonators with length of 80 cm and 100 cm, respectively. When we compare the optimum plate spacings to these thermal penetration depths, we can find that the optimum plate spacings are $3.8\delta_\kappa$ and $4.1\delta_\kappa$, respectively. These facts agree with the previous results obtained by Tijani *et al* [4] that plate spacing of $4\delta_\kappa$ leads to the lowest temperature in a thermoacoustic refrigerator.

By comparing the data presentation in Figs 5 and 6 and the analysis, it can be said that the data presentation in Fig.6 reveals more useful information than those in Fig. 5. This is natural because the porosity of a plate parallel stack is specified by the plate thickness and plate spacing of the stack. Therefore, we should consider the two parameters i.e plate thickness and plate spacing rather than the stack porosity only to characterize a plate parallel stack more properly. In addition, the result of this study suggests us to use thinner plate and around $4\delta_\kappa$ plate spacing for plate parallel stack to obtain a lower temperature decrease in thermoacoustic coolers.

4. Conclussion

An experimental study on the influence of the porosity of parallel plate stack on the temperature decrease of a thermoacoustic refrigerator has been successfully done. The range of porosity was 0.5 - 0.85, with plate spacing from 0.5 mm to 1.5 mm and plate thicknesses of 0.3 mm, 0.4 mm, and 0.5 mm. Two resonators with length of 0.8 m and 1.0 m were used, with air at atmospheric pressure and room temperature, correspond to thermal penetration depths (δ_κ) of 0.26 mm and 0.29 mm, respectively. It was found that there is an optimum porosity which gives the largest temperature decrease, and there is a tendency that the optimum porosity shifts to a larger value and the temperature decrease become larger when we use a stack with thinner plates.

Because the porosity of a plate parallel stack is specified by the plate thickness and plate spacing, the study on the dependence of the temperature decrease on the plate thickness and the plate spacing reveals more useful information than that on the stack porosity itself. It was found that the use of stack with thinner plate tends to result in larger temperature decrease, and the plate spacing around $4\delta_\kappa$ leads the largest temperature decrease, regardless of the plate thickness.

Acknowledgments

This research was financially supported by the Competitive Grant XVI Research Project, Gadjah Mada University, Indonesia.

[1] Swift G W 1988 *J. Acoust. Soc. Am.* **84** 1145
[2] Swift G W 2002 *Thermoacoustics: A Unifying Perspective for Some Engines and Refrigerators*, Acoustical Society of America, New York
[3] Russel D A and Weibull P 2002 *Am. J. Phys.* **70** 1231
[4] Tijani M E H, Zeegers J C H and de-Waele A T A M 2002 *J. Acoust. Soc. Am.* **112** 128
[5] Kim Y T and Kim M G 2000 *J. Kor. Phys. Soc.* **36** 279
[6] Setiawan I, Setio-Utomo A B, Nohtomi M and Katsuta M 2012, *Proc. 3rd Int. Conf. on Physics* (Yogyakarta: Physics Dept. Gadjah Mada University)
[7] Akhavanbazaz M, Siddiqui M H K and Bhat R B 2007 *Exp. Therm. Fluid Sci.* **32** 231
[8] Tijani M E H, Zeegers J C H and de-Waele A T A M 2002 *Cryogenics* **42** 59

ScieTech 2013 IOP Publishing
Journal of Physics: Conference Series **423** (2013) 012036 doi:10.1088/1742-6596/423/1/012036

Charge transport mechanism and low temperature phase transitions in KIO_3

M M Abdel Kader(*), F El-Kabbany, H M Naguib and W M Gamal
Physics Department, Faculty of Science, Cairo University, Giza, Egypt

Email: m_m_abdelkader@yahoo.com

Abstract. Our report deals with the measurement of some electrical properties, namely the ac conductivity $\sigma(\omega, T)$ and the complex dielectric permittivity $\varepsilon^*(\omega,T)$ in the temperature interval 95K < T < 280K and at some selected frequencies (0.7kHz – 20kHz) for polycrystalline samples of potassium iodate KIO_3 using a computerized RLC meter. The improper character of the ferroelectricity over the mentioned temperature range has been achieved by recording the ferroelectric hysteresis loops. The temperature dependence of each electrical parameter reveals that the compound undergoes two phase transitions at T≈ 258K and at T≈110K. The frequency dependent conductivity seems to be in accordance with the power law $\sigma(\omega,T)\alpha\omega^{S(T)}$ and the trend of temperature dependence of the frequency exponent s (0< s < 1) suggests that the quantum mechanical tunneling (QMT) model is the main mechanism of the charge transport. Comparison with the behavior of the NH_4IO_3 in the same temperature range was outlined.

1.Introduction

Potassium iodate KIO_3 is a member of the monovalent metal iodate series of the general molecular formula MIO_3 where M stands for Li, Cs, Na, K… and/or NH_4. This series of compounds attracted the attention of investigators due to their interesting properties. For example, the improper ferroelectric hydrogen-bonded NH_4IO_3 undergoes a ferroelectric phase transition at ≈ 368K [1-3]. Moreover, since the compound behaves as a proton conductor, therefore it has some practical applications such as chemical sensors, electrochromic displays and supercapacitors [3-4]. Furthermore, $LiIO_3$ undergo successive phase transitions and the high temperature phase behaves as a superionic conductor [5-6]. Properties such as piezoelectric, pyroelectric and non-linear optical effects are common for some members of this series [7-11].

For KIO_3, which is the subject of this article, it is known that most of the physical properties of the crystals of this compound such as, (dc) dielectric constant [12] piezoelectric and pyroelectric effects [13,14], electro-optical behavior [13], non-linear optical properties [7,14], , ferroelectricity

[15-16] in addition to nuclear quadruple resonance (NQR) [12] and Raman scattering experiment [17], have been investigated previously. Later on, some of these properties were discussed [18]. Due to the increasing interest of these properties and the expected applications of such a material in laser technology, it is logical to re-examine these properties using more advanced techniques. For example, a somewhat recent work dealing with the temperature dependence for piezoelectric and ferroelastic properties, using resonance – antiresonance method [19], has been reported. The data indicate that the compound undergoes successive five phase transitions (I-II, II-III, III-IV, IV-V, V-VI) at the transition temperatures: T_1 = 485K, T_2 = 345.5K, T_3 = 258K, T_4= 113K and T_5 = 33K respectively. Similar behavior was observed earlier [12] with somewhat slight different in the value of the transition temperature for some phases. The anomaly that has been observed around 150°C (423K) by Crane [14], and by others [12,19] and more recently at (428±2)K by the present authors [20] has been already explained as a change in electrical conduction between extrinsic and intrinsic mechanisms [20]. It is not related to any phase transition [19,20].

Among other things, KIO_3 single crystal exhibits noble non-linear optical properties in all phases [7]. Furthermore, its non-linear optical coefficient in the room temperature phase (phase III), is the largest between monovalent metal iodates (MIO_3) [7,21] and is also greater than those of KH_2PO_4 (KDP), BaB_2O_4 (BBO) and $LiBrO_5$ (LBO), which are known as double frequency materials in the ultraviolet wavelength [22]. Although the presence of ferroelectric domains and twins at room temperature (phase III) prevents or at least hinders practical applications, yet using the poling technique and specially after detwenning and domain removal, it can be used for fabrication of non-linear optical devices [22].

Another point of interest is that the static piezoelectric constants of the single crystals of KIO_3 in phase III are about 50 times greater than those of α-quartz [23]. Due to this large piezoelectric effect, KIO_3 is a promising candidate for the fabrication of piezoelectric sensors [24]. Furthermore, the thermal characterization including specific heat and thermal decomposition in the temperature interval 260K< T < 600K were investigated [25] using Perkin – Elmr DSC.

To clarify the dynamic properties of the high temperature phase transition of KIO_3 crystals, the temperature dependence of low frequency (soft) optical modes was measured by Raman scattering [24]. More recently, detailed polarized Raman study were performed on KIO_3 single crystals over 78.5-553K [26]. The improper character of ferroelectricity of the compound was suggested by Ahart et al. [24], supported by Liu et al. [26] and confirmed by Abdel-Kader et al. [20]

From the accurate measurements of some mentioned properties as a function of temperature[15, 17, 19,20, 22-25], it has been reported that, the KIO_3 undergoes a ferroelectric phase transition around 486K and several phase transitions as summarized in Table 1. The Table shows also the symmetry of some phases of KIO_3

Regarding the crystal structure, it is known that the structure of phase III using the single crystal x-ray diffraction technique, was determined by Crane [14] who also summarized the results of studies on KIO_3 until 1971 [14]. Furthermore, based on x-ray powder diffraction data, Hamid [27] suggested the space group of phase I as R3m which has not been confirmed later. He also suggested the symmetry of phase II and determined the structure of Phase I from intensities taken by photographic method. It is known that, single crystals x-ray diffraction technique, has been faced with the some problems due to the existence of ferroelectric and/or ferroelastic domains [27].

Again single crystal technique has been used by Kalinin et al [28] and has been faced with the same problems namely the tendency towards twinning and polymorphism [28].

To overcome all these problems, recently Kasatani et al. [29] determined, the crystal structure of phase I at 530K using high energy X-ray technique and the MEM/Rietveld analysis. On the other hand, the high resolution neutron powder diffraction has been employed for determination of the crystal structure of phase I at 523K [30], and also phase III at 300K [31a], at 100K and 10K [31b].

We have found that it is of interest to measure the ac conductivity and dielectric permittivity in the low temperature range 95K < T < 280K on polycrystalline samples. This technique is absent in literature for the title compound. Another reason for the use of this technique in the present work is the dependence of the physico – chemical properties of this compound on the method of its recrystallization. For example, different NQR ^{127}I spectra were found for crystals grown from water solution with or without HIO_3 [32]. Moreover, for crystallization of KIO_3 from an aqueous solution containing HIO_3 and KIO_3 with a ratio of HIO_3/KIO_3 greater than 7.4%, crystals of KIO_3 and $KH(IO_3)_2$ are formed [23].

According to the best of our knowledge, no data on "electrical transport mechanism" in the investigated low temperature region was reported. Also the motivation behind the present work, is to compare our data with that of the very similar compound namely NH_4IO_3 [33].

In order to discuss the different models for conduction mechanism, the starting point is that, the total measured ac conductivity $\sigma_{ac(total)}$ at a given angular frequency (ω), can be separated into dc and ac components namely:

$$\sigma_{ac(total)} = \sigma_{dc} + \sigma_{(ac)} \qquad (1)$$

The ac component $\sigma(\omega)$ is a function of both frequency (ω) and temperature (T) and may be written as $\sigma(\omega,T)$ whereas the dc component is a function of temperature only. According of Jonscher [34] $\sigma(\omega) \propto \omega^{s(T)}$, therefore

$$\sigma(\omega) = A(T)\omega^{s(T)} \qquad (2)$$

This is the well known power law [34-42] and has been so widely observed for highly disordered materials and hence it has come to be known as a universal dynamic response.

The pre-factor A (T) is a constant and represents the degree of polarizability [37] and the frequency exponent s ($0 < s < 1$) describes the interaction (mainly of an electrostatic type) between the mobile ions with lattice around them or in more details, and according to the many body interaction model [34,37], the interaction between all dipoles participating in the polarization process is characterized by the parameter s. A unit value of s implies a pure Debye type interaction. The value of s decreases with the increase of interaction [37,40]. The thermal behavior of the ac conductivity follows Arrhenius relation in most cases.

$$\sigma T = \sigma_o \exp(-E/kT) \qquad (3)$$

where E is the activation energy for the process, k is the Boltzmann's constant and σ_o is constant.

Furthermore, it is known that the complex dielectric permittivity $\varepsilon^*(\omega)$ can be divided into the real part $\varepsilon'(\omega)$ and the imaginary part $\varepsilon''(\omega)$ according to the equation:

$$\varepsilon^*(\omega) = \varepsilon'(\omega) - j\varepsilon''(\omega) \qquad (4)$$

(ε' and ε'' are frequency-dependent and represent the charging and loss currents respectively) [43].

Journal of Physics: Conference Series **423** (2013) 012036 doi:10.1088/1742-6596/423/1/012036

2. Experimental procedures

2.1. Sample preparation. The material used in the present work was manufactured by the BDH Chemical Company Ltd. For electrical and/or ferroelectric measurements, a suitable amount of the fine polycrystalline powder of the compound, sufficient to prepare 5–6 pellets was further grained under 2μm (micronized). The aim of this technique is to remove the effect of grain size and also to minimize the effect of porosity.

A microanalytical balance, type Sartorius, was used to achieve the equality of masses between all samples before compressing, (the mass of each sample was≈ 0.6 gm). These samples were pressed under the same pressure (of about $2*10^8 N/m^2$) so we have practically identical pellets of thickness ≈ 2mm and of diameter ≈ 1cm.

Good electrical contact was attained by painting the opposite faces of each pellet with air drying conducting silver paste. Before any measurements, the samples were inserted into a dessicator over night to remove any humidity. A sample holder with brass electrodes was specially designed to fit the present measurements.

2.2. Characterization techniques. To demonstrate and to check the presence of ferroelectricity at different temperatures, the circuit shown in 'figure 1a', was designed and manufactured in the Science Technical Center, Cairo University to fit the present measurement. The digital storage oscilloscope type Instek GDS-820S, provided with a computerized camera as connected in the designated circuit shown was used to record ferroelectric hysteresis loops at different temperatures. Calibration, using triglycine sulphate (TGS), was carried out just before any actual run.

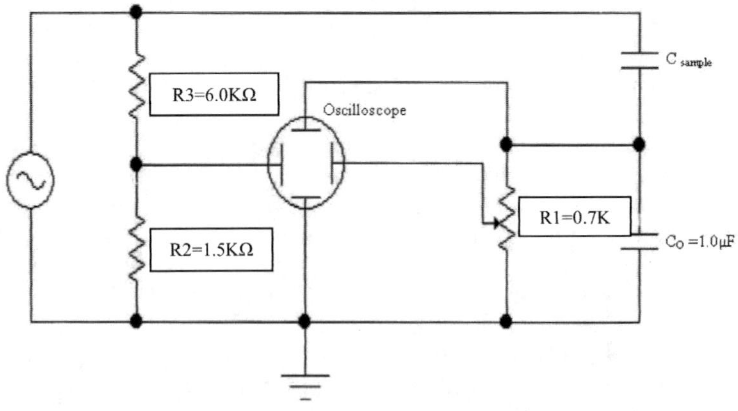

Figure 1a. The designated circuit for detecting ferroelectric hysteresis loops.

A computerized LCR bridge type Philips PM6304 was used to determine the ac conductivity and the dielectric permittivity at some selected frequencies. The bridge measures precisely the values of R, C, Z, Q… However, in the present work the values of R and C are converted into conductivity and permittivity using the dimensions of the pellet and a simple program, since the bridge is interfaced to computer. The dc conductivity was measured by using an electrometer, type Keithly 614. Data were collected on at least three virgin samples and the results were found to be quite consistent and reproducible. All measurements were performed during heating run. The heating rate was 0.5K/min and the experimental error of temperature is better than +1%. All measurements were performed at the thermal equilibrium. Calibration was done on a standard sample before any actual measurements.

3. Results

3.1. Ferroelectricity. The ferroelectric hysteresis loops (D-E loops) at some selected temperatures in the range 95K < T < 280K have been displayed and recorded on the monitor of the digital oscilloscope. Two of these loops are shown in 'figure 1b', together with D-E loops recorded at high temperature up to 490K where the compound is in the paraelectric phase. It seems likely that, the saturation in the D-E loops is not complete. This problem will be discussed later on (in the discussion section). The existence of this type of D-E loops is an indication that the compound behaves as an improper ferroelectric material.

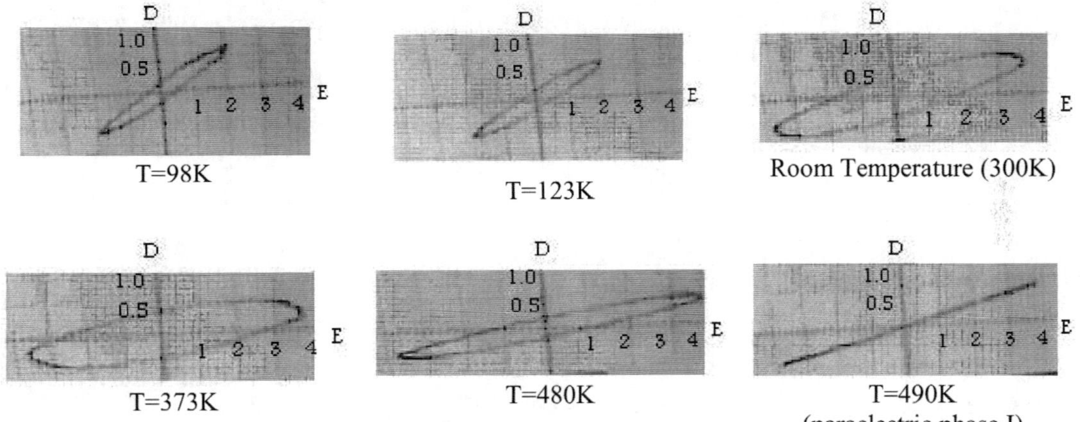

Figure 1b. The D-E hysteresis loops for•KIO_3. (Arb. Scale)

3.2. Dielectric Permittivity. The temperature dependence (95K<T<280K) of the real part ε' of the dielectric permittivity determined at some selected frequencies (0.7 kHz-20 kHz) is shown in 'figure 2'.

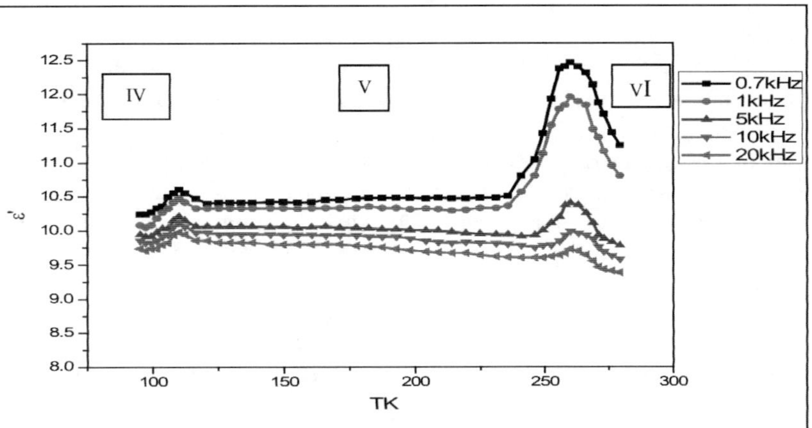

Figure 2. Temperature dependence of the real part (ε') of the dielectric permittivity measured at different frequencies for KIO_3.

In general, the value of ε' depends on both the frequency (f) and temperature (T). For a given frequency, each curve is characterized by the existence of two anomalies: a pronounced peak in the vicinity of $T_3 \approx 258K$ and a small peak (anomaly) at $T_4 \approx 110K$. Between the two transition regions, the value of ε' is almost temperature independent. It is known that, The compound undergoes a ferroelectric phase transition of an improper character (I-II) at $T_1 \sim (486\pm1)K$, a change in the conduction mechanism at (428±2)K and a structural phase transition (II-III) at $T_2 \sim 345K$.

The variation of ε", the imaginary part of the dielectric permittivity, as a function of temperature, 95K<T< 280K is shown in figure 3a'. The frequency range (1kHz-10kHz) has been only considered for the purposes of clarity.
Apparently, the general trend of this plot is similar to that of ε'-T graph, 'figure 2', except that the value of ε" at a given frequency and temperature is much reduced as compared with the corresponding value of ε', 'figure 2'.

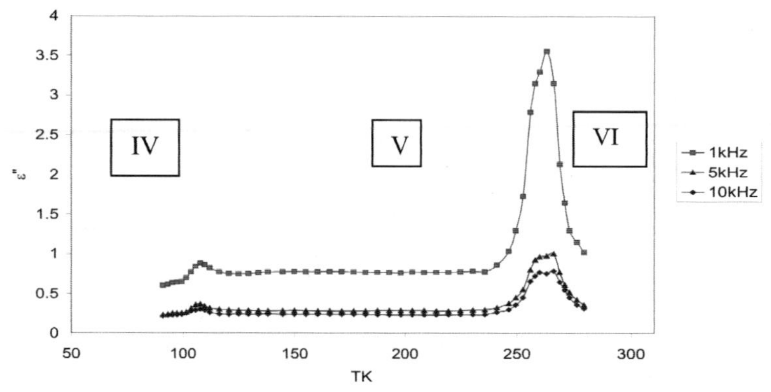

Figure 3a. Temperature dependence of the imaginary part (ε") of the dielectric permittivity measured at different frequencies for KIO_3.

'Figure 3b' shows the frequency dependence of the imaginary ε" plotted as ε" vs lnf at some selected temperatures. The plot indicates that the dispersion increases with increasing temperature and decreasing frequency which is in agreement with the typical behavior of the dielectric materials.

The variation of the (dielectric loss), tanδ = ε'/ ε" as a function of temperature (95K<T<280K) measured at some selected frequencies is shown in 'figure 4'. The general trend of the plot indicates also the existence of two phase transitions at the above mentioned temperatures and are reflected as two peaks.

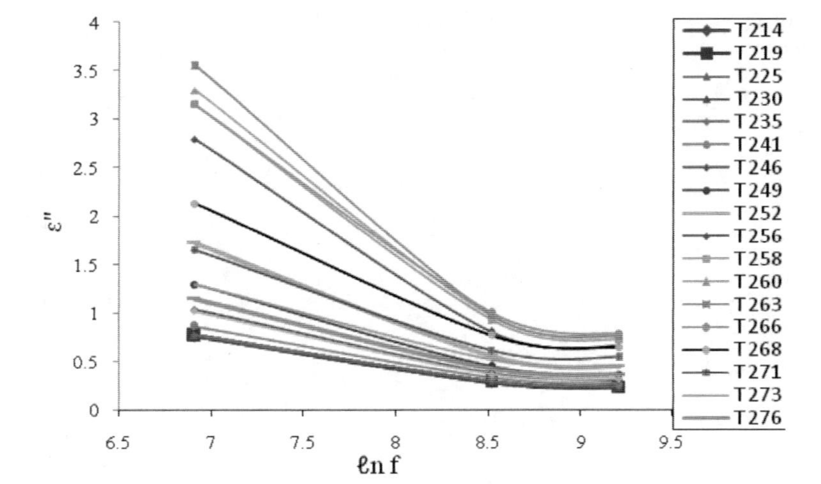

Figure 3b. Relation between ε'' *versus* ℓnf at different temperatures for KIO_3.

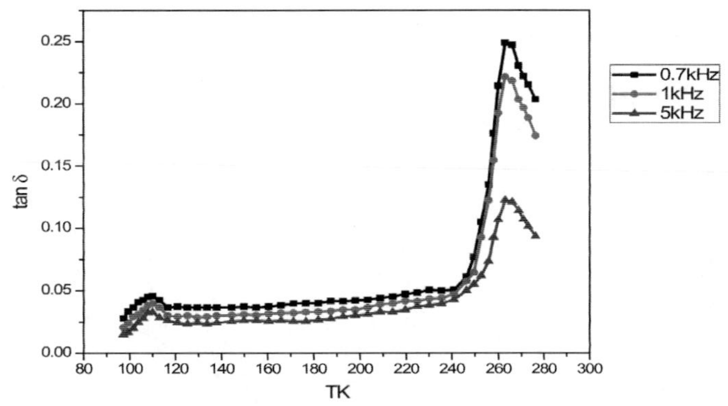

Figure 4. Temperature dependence of tanδ measured at different frequencies for•KIO_3.

3.3. Electrical conductivity.
Obviously the conductivity depends on both, the temperature and the frequency.

3.3.1 Temperature dependence. The ac conductivity $\sigma_{(\omega)}$ as function of temperature, 95K<T<280K and frequency 0.7kHz - 20 kHz, plotted as ℓn(σT) versus 1/T is shown in 'figure 5'. The behavior of σ seems to be in accordance of Arrhenius relation. For a given frequency, and as the temperature increases from about 95K, the first broad peak is observed at258K. The other phase transition is found at ≈ 110K. Between the two transition regions the behavior of σ obeys Arrhenius relation.

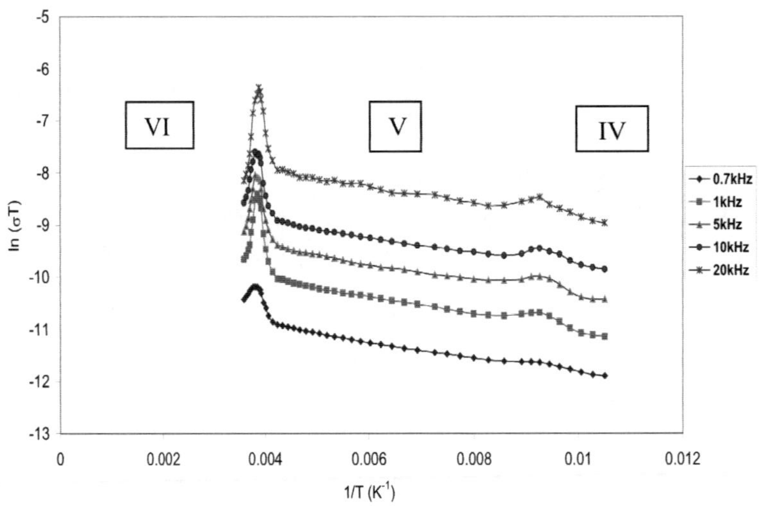

Figure 5.Variation of (ℓnσ T *versus*1/T) at different frequencies for KIO_3.

3.3.2 Frequency dependence. The variation of ℓnσ versus ℓnω at different temperatures is shown in 'figure 6' .

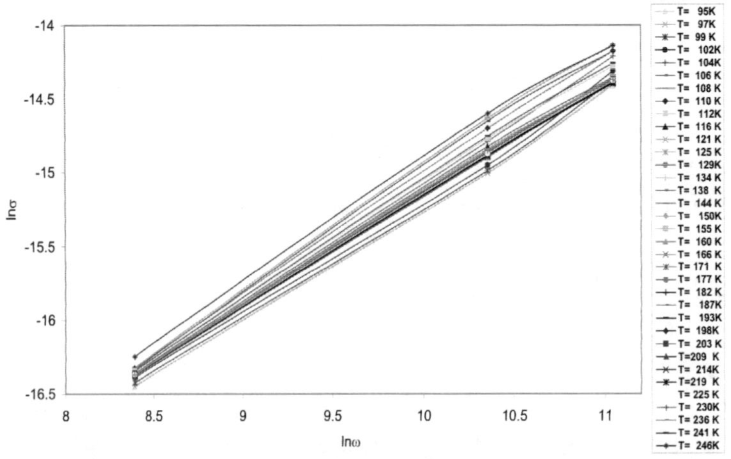

Figure 6. Double logarithmic plot: (ℓnσ *versus* ℓnω) for KIO_3.

The plot is consists of a series of straight lines of different slopes. The slope of each line gives the value of s, where $s = (\frac{\partial \ln \sigma}{\partial \ln \omega})_{T=const.}$ at that temperature.

The temperature dependence of s is shown in 'figure 7'. Apparently, s is practically temperature independent except for around the phase transition regions.

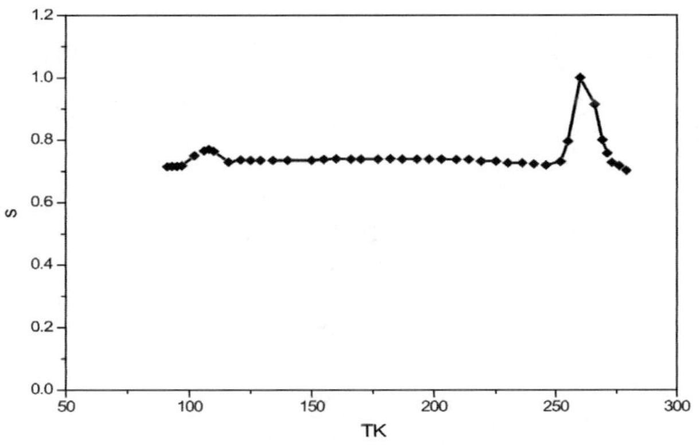

Figure 7. Temperature dependence of the frequency exponent (s) for KIO_3.

'Figure 8' shows the variation of the pre-exponential factor (A) with temperature, plotted as ℓn A vs T. The behavior of the plot indicates also the existence of two phase transitions at the same mentioned temperatures (258K and 110K) which are reflected in the graph shown as "inverted peaks"

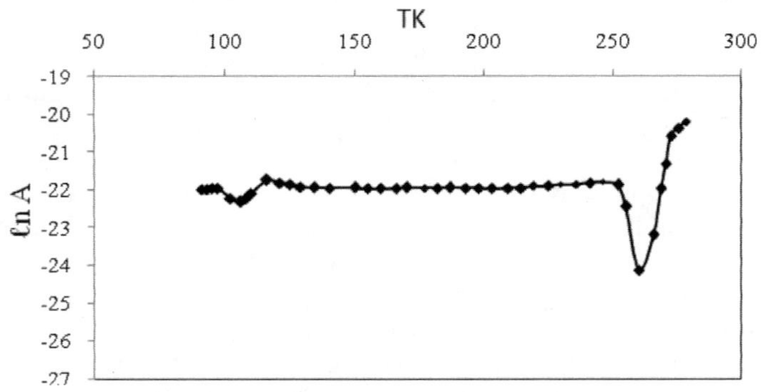

Figure 8. Variation of ℓn A versus T for KIO_3.

Journal of Physics: Conference Series **423** (2013) 012036 doi:10.1088/1742-6596/423/1/012036

4. Discussion.

The values of the transition temperatures T_3 (III- IV) = 258K and T_4 (IV –V) = 110K of the present work was found to be consistent with those found by previous workers (Table 1a and Table 1b). For example, from his previous investigation of the ferroelectricity in KIO_3 single crystal, Salje [17] observed two phase transitions in the low- temperature region at≈ -15°C (258K) and at -163°C (110K).

Furthermore, Ivanov et al [15] studied the temperature dependence of the optical axes angle 2v, the birefringence in the z-cut $\delta n_{(z)}$ and the spontaneous polarization axes P_{sf} in a wide temperature range and they observed that at T = -15°C (258K) all three curves [(2v (T), $\delta n_{(z)}$(T) and P_{sf} (T)] exhibit only breaks and thus they suggested that this transition is to be of a second order. Moreover, although Ivanov et al [15] did not investigate the phase transition (IV – V) in detail, yet they observed a regular quantitative change of the domain pattern which confirms the existence of a transition with a lowering of point symmetry. Moreover, they demonstrated the existence of ferroelectricity in the compound at the liquid nitrogen temperature -195°C (78K) by observing the D-E hysteresis loops which was found to be clear at that temperature and thus they concluded that phase V behaves as a ferroelectric which is in a good agreement with the present work and also with the work of Herlach [12], where the crystal remains ferroelectric below 485K [12]. In somewhat recent work Maeda et al., [19] demonstrated the existence of low temperature phase transitions at T_3= -15°C (258K) and T_4 = -160°C (113K). A similar behavior has been observed for NH_4IO_3 where the compound undergoes a ferroelectric phase transition at ≈ 368K and behaves as a ferroelectric material below 368K down to liquid nitrogen temperature [33].

Among other things, although some investigators have been reported on the low-temperature phase transitions in KIO_3, through the study of the temperature dependence of some parameters, yet, the only crystal structure of the low temperature is that determined by Lucas [31b]. The data do not show significant change, on the atomic level, relative to that of the room temperature, i.e. no strong evidence for low temperature structural phase transitions. As we mentioned above, the temperature dependence of some properties show detectable change at ≈ 258K and at ≈ 110K. This inconsistence is not a major problem, since the structural phase transitions that involve only minor changes in atomic positions are not easy to detect [44a]. Potassium chlorate $KClO_3$ represents an example of a crystal with a reported phase transition with very small structural change where the atomic arrangement and coordination above and below the phase transition are identical [44a]. The same conclusion was also given by Brooker et al [44b]. From Raman studies of the phase transition in $KClO_3$, they observed that the difference in the vibrational spectra of the two phases was very small and hence the transition (at 545K) involved only mainor rearrangement of the ClO_3 ion. Similar saturation may exist for the KIO_3 since the two compounds (KIO_3 and $KClO_3$) are members of halate series of the general molecular formula MXO_3 where (X=Cl, Br and I).
From the crystal structure point of view [29, 30, 31], the I and O atoms exist in the crystals as IO_3 molecules rather than the IO_6 complex. In other words unlike the typical perovskite structure of ABO_3 ferroelectrics where six O atoms form O_6 octahedron surrounding, a B atom each I atom in KIO_3 has three nearest O atoms forming an IO_3 molecule [26].
At this stage, it is necessary to mention that although there is a good agreement between different authors that the compound undergoes two phase transitions in the temperature interval 95K < T < 280K, yet there is a discrepancy concerning the nature and/or the exact mechanism of these transitions. Some authors consider these transitions to be associated with the orientational glass transitions [19]. Other authors [44b] proposed that these transitions might be associated with the structural changes of domain which is generally independent on atomic arrangements.

The present work tries to construct a bridge or link between different visions. The motion and/or the change of the structure of ferroelectric domains with temperature was first suggested by Byrom and Lucas [30]. On the other hand, and in a similar way to the role of ClO_3^- as the main cause of the structural phase transition in the $KClO_3$ compound, one cannot neglect the effect of the reorientational motion of the IO_3^-, since, as we mentioned above, the spatial correlation between IO_3^- decreases with increasing temperature [29]. Thus the thermal energy at each transition temperature may be enough to cause the change of the structure of ferroelectric domains and at the same time initiate (stimulate) the orientational motion of the IO_3^- and hence a very slight rearrangement of the molecule with a very minor change in the atomic position in a way similar to that observed in the $KClO_3$ [44a,b].

It is of interest to mention here that even for the structural phase transition of the KIO_3 that have been observed at ~ 487 K which is associated with the ferroelectric phase transition, the two X-ray diffraction patterns above and below the mentioned transition temperature are very similar [30]. In our vision, the exact mechanism of the low temperature phase transition is a future problem and requires the use of the high energy X-ray (synchrotron) powder (and not single) technique which is beyond the scope of the present work.

For the problems of unsaturated D-E hysteresis loops, 'figure 1'it is known that the compound behaves as an "improper ferroelectric material" [20, 24-26] and not as (usual) ferroelectrics. The characteristic features of the improper ferroelectric are [45-46]:

i) a very slight change of dielectric constant near Tc

ii) a low spontaneous polarization

Logically the spontaneous polarization is not the order parameter and not proportional to it as in the case of proper ferroelectrics. The factor ii) is responsible for the unsaturation in the D-E hysteresis loops. The case of KIO_3 is very similar to the improper ferroelectric NH_4IO_3 where the D-E hysteresis loops in the ferroelectric phase are also unsaturated [33, 47].

Regarding the charge transport mechanism in different phases, it is known that the three main processes that contribute significantly to the ac conductivity [36, 48-49] are the quantum mechanical tunneling QMT, the correlated barrier hopping (CBH), and the small polaron tunneling (SPT).

For QMT model, the value of s is given by

$$s = 1 - \frac{4}{\ln\left(\frac{\upsilon_{ph}}{\omega}\right)} = 1 - \frac{4}{\ln\left(\frac{1}{\omega\tau}\right)} \tag{5}$$

whereas for (CBH) model, we have

$$s = 1 - \frac{6kT}{E_0 - \ln\left(\frac{\upsilon_{ph}}{\omega}\right)} = 1 - \frac{6kT}{E_0 - \ln\left(\frac{1}{\omega\tau}\right)} \tag{6}$$

and finally for the (SPT),

$$s = 1 - \frac{4}{\ln\left(\frac{1}{\omega\tau_p}\right) - w_H / kT} \tag{7}$$

The symbols in equations (5), (6) and (7) take their conventional meanings where k is Boltzmann's constant, E_0 is the optical energy band, υ_{ph} is the frequency of the phonon. τ_p is the relaxation time of the polaron (of the order of 10^{-13} s) and w_H is the activation energy involved in the electron transfer process between a pair of states. Moreover, the value of s in equation (5) is temperature independent, whereas in equation (6) s decreases with increasing the temperature and is less dependent on the frequency. For SPT, equation (7) predicts that, s increases as T increases.

Referring to 'figure 7', where s is plotted as a function of temperature, it is clear that the value of s is practically temperature independent i.e. in accordance with equation (5). Thus, the QMT model is the most likely one in this range of temperature. Moreover, if we put υ_{ph} (~ 10^{-13} sec.$^{-1}$ and ω~ 10^4 rad/sec.) [36] into equation (5), then the value of s ~ 0.76, which is in a good agreement with our experimental value, 'figure 7'.

Furthermore, for QMT model, the value of $\sigma(\omega)$ as given by Auston and Mott [50] and Efros [51] is

$$\sigma(\omega) = \frac{\eta}{3} e^2 kT[N(E_f)(T)]^2 \alpha^{-5} \omega \left[\ln(\frac{\upsilon_{ph}}{\omega}) \right]^4 \quad (8)$$

Where: e is the charge of electron,
T is the absolute temperature,
$N(E_f)$ is the density of energy states near the Fermi level,
α is the electron wave function decay constant 1Å [33],
υ_{ph} is the phonon frequency, and η is a constant.

Equation (8) predicted that, there is a linear relation between ω and $\frac{\sigma(\omega)}{\left[\ln(\frac{\upsilon_{ph}}{\omega})\right]^4}$ such a relation is presented in 'figure 9', which again confirm equation (8) where a series of almost straight lines of different slopes are observed. Thus the QMT model is the best one that describes the conduction mechanism in the investigated temperature region.

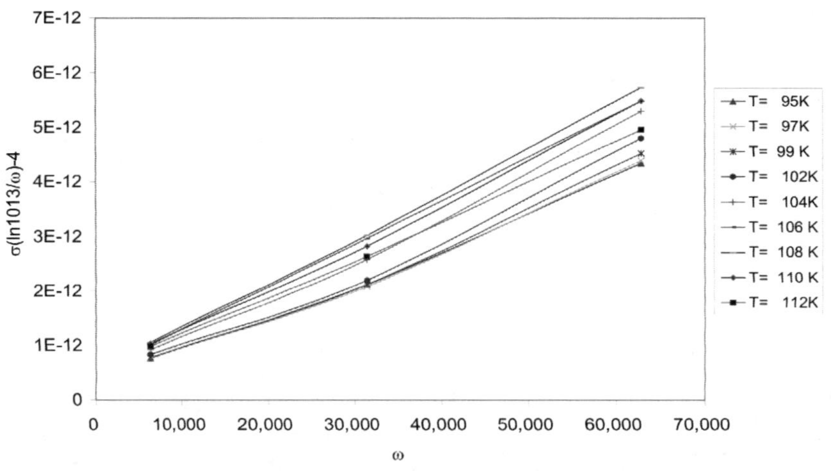

Figure 9. Variation of $\sigma(\omega)/[\ell n(\upsilon ph/\omega)]^4$ versus ω at different temperatures for KIO_3.•

5. Conclusion. The present data indicate that in the investigated low temperature region 95K < T < 280K, the compound remains ferroelectric (of an improper character) where the ferroelectric hysteresis loops, of quite insufficient saturation, are observed clearly over the mentioned temperature intervals. This is also the case for the similar compound NH_4IO_3 where it undergoes a ferroelectric phase transition of an improper character at ≈ 368K and remains ferroelectric down to liquid nitrogen temperature. For the compound KIO_3, the behavior of the temperature dependences for all electrical parameters confirm and support each other and indicate clearly that compound undergoes two structural phase transitions at $T_3 \approx 258K$ and at $T_4 \approx 110K$ over the mentioned temperature range. It is known that the iodine and oxygen atoms in the present compound exist as IO_3 rather than as IO_6 complex and the spatial correlation between IO_3 becomes weak with increasing temperature. The thermal energy (kT) at the transition temperature is enough to cause the rearrangement of IO_3 molecules in a way similar to the case of $KClO_3$. The thermal energy also causes an observable change in the structure of ferroelectric domain.

Regarding the electrical conduction, in the investigated low temperature region, the QMT model seems to be the main mechanism for the charge transport. This is to be expected, since this model usually exists in the low temperature range.

The low and high phase transitions as well as the change in conduction mechanism at (428±2)K are shown in the following scheme. The values of the transition temperatures are taken from ref. [20] and the present work except for transition at T = 33K is taken from ref. [19].

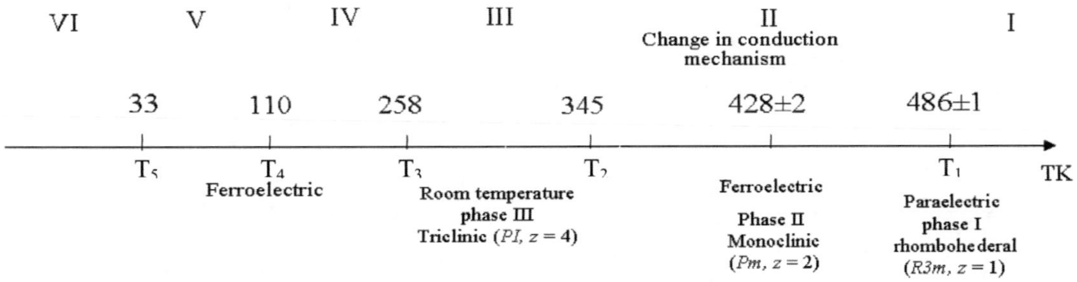

Table 1a. Summary of Phase Transitions in KIO_3

Phase Transition	Symmetry	Transition Temperature (TK)	Technique or property	Reference
I-II	I Rhombohedera l (*R3m, Z*=1). [20] Rhombohedera l (*R3, Z*=1). [30] Rhombohedera l ($R\bar{3}m$, Z=2). [18] Rhombohedera l (*R3m, Z*=2). [17]	485K (486±1)K 487K	Raman Spectroscopy Elastic and piezoelectric Detewenning and domain removing procedure Ferroelectricity and optical properties. Ac conductivity Thermal properties	[24] [19] [22] [15] [20] [25]
II-III	II Monoclinic (*Cm, Z*=8). [18] Monoclinic (*Pm, Z*=4). [17]	345K 345.5K 345.6K 343K	Raman Spectroscopy Ac conductivity Thermal properties Ferroelectricity and optical properties. Elastic and piezoelectric Elastic properties Detewenning and domain removing procedure	[24] [20] [25] [15] [19] [23] [22]
III-IV	III Triclinic (*P1, Z*=4). [31a] Triclinic (*P1, Z*=4). [17] Triclinic (*P1, Z*=4). [18]	258K	Ac conductivity Raman Spectroscopy Elastic and piezoelectric	Present work [17] [19]
IV-V	_	110K 113K	Ac conductivity Raman Spectroscopy Elastic and piezoelectric	Present work [17] [19]
V-VI	_	33K	Elastic and piezoelectric	[19]

Table 1b: Summary of the crystal structure at some selected temperatures for the compound KIO3

Temperature	Crystal Data	Technique	Reference
530K	Rhombohedral, R3m, Z = 1	High energy X-ray powder diffraction MEM/Reitveld analysis	[29]
523K	Rhombohedral, R3, Z = 1	Neuton powder diffraction technique (Rietveld method)	[30]
(Room Temperature)	Triclinic symmetry	Single crystal X-ray diffraction	[28]
523K	Triclinic, P1, Z = 4	Neutron powder diffraction technique	[31a]
300K	Triclinic, P1, Z = 4	(Rietveld method)	[31b]

References

[1] Barabash A, Gavrilko T, Eshimov K, Puchkovskaya G and Shanchuk A 1999 *J. Molec. Struc.* **145** 511-512.
[2] Bismayer U, Salje E and Viswanathan K 1979 *Phase Transitions* **1** 35.
[3] Abdel-Kader M M, El Kabbany F, Naguib H and Gamal W M 2008 *Phase Transitions* **81** 83.
[4] Colomban P. 1992 *Proton Conductors*, (Solids, Membranes and Gels Materials and Devices) Part I P.1 (Cambridge University Press, Cambridge,).
[5] Crettez J M, Righi A, Galez C, Bourson P and Mareira R L, 1998 *Sol. Stat. Comm.* **105** 481.
[6] Gaffar M A and Abu El-Fadl J A 1999 *J. Phys. Chem. Solid* **60** 1633.
[7] Bergman J G, Boyd Jr G D, Ashiken A and S-Kurtz 1969 *J. Appl. Phys.* **40** 2860.
[8] Keve E T, Abrahams S C and Berustein J L 1971 *J. Chem. Phys.* **54** 2556.
[9] Kurtz S K and Perry T T 1965 *App. Phys. Lett.* **12** 156.
[10] Crane G R, Bergman J C and Glass A M 1969 *J. Am. Ceramic Soc.* **52** 655.
[11] Helg H. and Gruncher H.1970 *J. Phys.Soc. Jap. Supl.* **28** 169.
[12] Herlach F, Helv. 1961 *Helv. Phys. Acta* **34** 305.
[13] Salje E, 1972 *Z. Krishtallagr.* **136** 135.
[14] Crane G R 1972 *J. Appl. Cryst.* **5** 360.
[15] Ivanov N R, Shuvalov L A and Chikgladze O A 1973 *Phys. Letters* **45A** 437.
[16] Salje E 1971 *Z. Krishtallagr.* **134** 107.
[17] Salje E 1974 *Z. Krishtallagr.* **139** 317.
[18] Landolt Bornstein 1982 Vol. **111/16b,** Springer 269.
[19] Maeda M, Takagi M and Suzuki I 2000 *J. Phys. Soc. Jpn.* **69** 267 .
[20] Abdel Kader M M, El-Kabbany F, Naguib H M and Gamal W M 2008 *Phase Transitions* **81** 29.
[21] Filimonov A A, Survotov L G, Paklomov V S, and Sonin A S 1965 *Sov. Phys. Cryst.* **10** 202.
[22] Yin X- and Lü M K 1992 *App. Phys. Lett.* **60** 2849.
[23] Haussühl S, Jyang W and Mengkai L Ü 1995 *Cryst. Res. Technology* **30** 535.
[24] Ahart M, Kojam S, Takagi M, Maeda M and Suzuki I 1998 *Jpn, J. Appl. Phys.* **37** 5687.
[25] Loiacono G M and Jacco J G, 1985 *Mater. Letters* **4** 27.
[26] Liu L, Wu R Q, Ni Z H, Shen Z X and Feng Y P 2006 *J. Phys. Conf. Ser.* **28** 105 and references there in.
[27] Hamid S A 1973 *Z. Kristallogr.* **137** 412.
[28] Kalinin V R, Llyukhin V V and Belov. N V 1978 Sov. *Phys. Dokl.* **12** 1978
[29] Kasatani H, Aoyagi S, Kuroiwa Y, Yagi K, Katayama R and Terauchi H 2003 *Nucl. Instr. Meth.* **B 199** 49.
[30] Byrom P G and Lucas B W 1987 *Acta Cryst.* **C43** 1651.
[31a] Lucas B W 1984 *Acta Cryst.* **C40** 1989.
[31b] Lucas B W 1985 *Acta Cryst.* **C 41** 1388.
[32] Bajsa D F, Barabash A I and Vertegel I G 1981 *Ukr. Fiz.Zhurnal* **10** 1745.
[33] Abdel Kader M M, El-Kabbany F, Naguib H M and Gamal W M 2012 *Phase Transitions*
[34] Jonsher A K 1983 *Dielectric relaxation in solids* (Chelsea Dielectric press). London, p. 223.
[35] Dyre J C 1988 *J. Appl. Phys.* **64** 2456.
[36] Elliott S R 1987 *Adv. Phys.* **36** 135.
[37] Krthik C and Varma K B R 2006 *J. Phys Chem. Sol.* **67** 2437.
[38] Louati B, Gargouri M, Guidara K and Mhiri T 2005 *J. Phys. Chem. Sol.* **66** 762.
[39] Dyre J C, T.B. Schroder, *Rev. Mod. Phys.* 72 (2000) 873.
[40] Chen R H, Chen-Chief Yen, SHern C S and Fukami T 2006 *Sol. State Ion.* **177** 2857.

[41] Lee W K, Liu J F and Nowick A S 1991 *Phys. Rev. Lett.* **67** 1559.
[42] Moawad H M M, Jain H and El-Mallawany R 2009 *J. Phys. Chem. Sol.* **70** 224.
[43] Al Refaie S N, El-Rayyan H S B 1992 *J. Mater. Sci. Lett.* **11** 958.
[44a] Ramachandran N and Lonappan M A 1957 *Acta Cryst.* **10** 281.
[44b] Brooker M H and Shapter J G 1989 *J. Phys. Chem. Solids* **50** 1087.
[45] Blinc R and Zekz B "*Soft modes in ferroelectrics and antiferroelctrics*" (Amsterdam, North-Holand Pub.co) P. 15 and P. 45.
[46] Astrukov B and Levanyuk A P, "*Ferroelectric Phenomena in crystals"* (Berlin. Springer) P. 74.

[47] Oka T, Mitsui T and Sawada S 1976 J. Phys. Soc. Jpn. **40** 913.
[48] Gamati F, Fattoum A, Bohli N, Dhaoui W and Mohamed A B 2007 *J. Phys.: Condens. Matter* **19** 36203.
[49] Long A R 1982 Adv. Phys. **31** 553.
[50] Austin I G and Mott N F 1969 *Adv. Phys.* **18** 41.
[51] Efros A L 1981 *Phillosophical Magazine* **43** 829.

Journal of Physics: Conference Series **423** (2013) 012037 doi:10.1088/1742-6596/423/1/012037

Modeling the Determinants Influencing the Diffusion of Mobile Internet

Saleh Alwahaishi[1] **and Václav Snášel**
Department of Computer Science, FEECS, VŠB- Technical University of Ostrava, 17. listopadu 15, 708 33 Ostrava-Poruba, Czech Republic

E-mail: salehw@vsb.cz

Abstract Understanding individual acceptance and use of Information and Communication Technology (ICT) is one of the most mature streams of information systems research. In Information Technology and Information System research, numerous theories are used to understand users' adoption of new technologies. Various models were developed including the Innovation Diffusion Theory, Theory of Reasoned Action, Theory of Planned Behavior, Technology Acceptance Model, and recently, the Unified Theory of Acceptance and Use of Technology. This research composes a new hybrid theoretical framework to identify the factors affecting the acceptance and use of Mobile Internet -as an ICT application- in a consumer context. The proposed model incorporates eight constructs: Performance Expectancy (PE), Effort Expectancy (EE), Facilitating Conditions (FC), Social Influences (SI), Perceived Value (PV), Perceived Playfulness (PP), Attention Focus (AF), and Behavioral intention (BI). Individual differences—namely, age, gender, education, income, and experience are moderating the effects of these constructs on behavioral intention and technology use.

1. Introduction

Information and communications technology or information and communication technology (ICT), is often used as an extended synonym for Information Technology (IT), but is a more specific term that stresses the role of unified communications and the integration of telecommunications, computers as well as necessary enterprise software, middleware, storage, and audio-visual systems, which enable users to access, store, transmit, and manipulate information [1]. ICT is an interdisciplinary area of research driven and shaped by the fast development of computing, communication, and Internet-related technologies, which have a great impact on our societies and daily lives. Over the last few decades there has been an increase in ICT research, which has changed and shaped the way societies and organizations operate and produce their goods and services. It is not only the generation of new technology but also, and perhaps even to a higher extent, its diffusion throughout the economy which affects productivity growth at the macro-level.

Technology adoption was defined as the use, or acceptance of a new technology, or new product [2]. Moreover, understanding individual acceptance and use of information technology is one of the most mature streams of information systems research. The Internet and mobile technology, the two most dynamic technological forces in modern information and communication technologies are converging into one ubiquitous mobile Internet service, which will change our way of both doing business and dealing with our daily routine activities. There is no doubt that the mobile Internet service is moving toward the new generation on which enables mobile users to enjoy a variety of new and upgraded multimedia mobile services.

According to a recent study by on global mobile data traffic forecast, Smartphones represent only 12 percent of total global handsets in use today, but they represent over 82 percent of total global handset traffic [3]. Moreover, the number of mobile phone subscriptions reached almost 6,000 million at the end of 2011, representing a penetration rate of 86.7 percent worldwide and 78.8 percent in developing countries [4]. Hence, during the period from Sep. 2011 to Sep. 2012, on an average around 265,000 applications were registered in the US App Store [5], and 140,000 application were registered in Android Market during same period [6].

2. ICT Acceptance and Use

As the use of information and communication technology (ICT) expands globally, there is need for further research into cultural aspects and implications of ICT. The acceptance of Information Technology (IT) has become a fundamental part of the research plan for most organizations [7]. A better understanding of the factors contributing to the acceptance or rejection of information technology is the first step toward the solution of the problem.

User acceptance is often the pivotal factor and a central focus of Information Systems (IS) implementation research in determining the success or failure of an information technology product [8, 9, 10, 11, 7]. Availability of information technology does not necessarily lead to its acceptance. Most information system failures result from a lack of user acceptance rather than poor quality of the system [12, 7, 11].

Previous research into user acceptance of information technology has mainly concentrated on users' attitudes toward acceptance while neglecting the role of social norms. It was also noticed that few IT characteristics were researched and these were not approached in a coherent manner [9, 10, 7, 11]. Thus, it was recognized that the study would need to consider a broad range of IT characteristics and investigate the normative side of the equation besides that of attitudes toward usage.

In the technologically developed world, IT adoption is faced by barriers, such as the lack of top management support, poor quality IS design and inadequately motivated and capable users [13]. In the developing world, the same barriers appear to be often impenetrable [14, 15]. In addition, problems found in developing counties are attributed to a lack of national infrastructure [16], capital resources, or government policies set in place to prevent technology transfer [17].Although there are isolated reports of countries where sufficient resources and government support exist, the technology has failed to be effectively transferred [17]. While the uses of IT are varied, the common tie of technology use in the developing counties is one of limited diffusion [17].

3. Review of Technology Acceptance Theories

In ICT research, numerous theories are used to understand users' adoption of new technologies. Researchers have attempted to predict and explain user behavior across many IS and IT domains, seeking to investigate and develop theory as to how to improve usage and examine what inhibits usage and intention to use the technology [18]. To develop the conceptual framework for our model it is useful to draw comparisons between the various theories. The theories based on intention of ICT adoption such as Technology Acceptance Model –TAM- [20,21] and Theory of Planned Behavior –TPB- [18] have shown that the adoption and usage of an IT system is eventually determined by the users' personal beliefs and attitudes toward the technology. Other models such as IDT state that user's perception of the characteristics of an innovation is more significant [22].

3.1 Innovation Diffusion Theory (IDT)
Innovation Diffusion Theory (IDT) notes that relative advantage, complexity, compatibility, trialability and observability predict user adoption [23]. Rogers defined an innovation as an idea or practice that is perceived as new by the adopting organization [22]. Braun argued that Rogers Innovation Diffusion Theory (IDT) analyzed the process of diffusion, and mapped the impact of a combination of social, economic, and technical forces on that process [24]. There is a general agreement among researchers that IDT is a suitable and valid theory for examining the process of adoption.

3.2 Theory of Reasoned Action (TRA)
The theory of reasoned action is a widely studied model from social psychology which is concerned with the determinants of consciously intended behaviors [25, 26]. The foundation of the TRA conceptual framework is provided by the distinction between beliefs, attitudes, intentions, and behaviors. The major concern of the conceptual framework, however, is with the relations between these variables, as shown in figure 1.

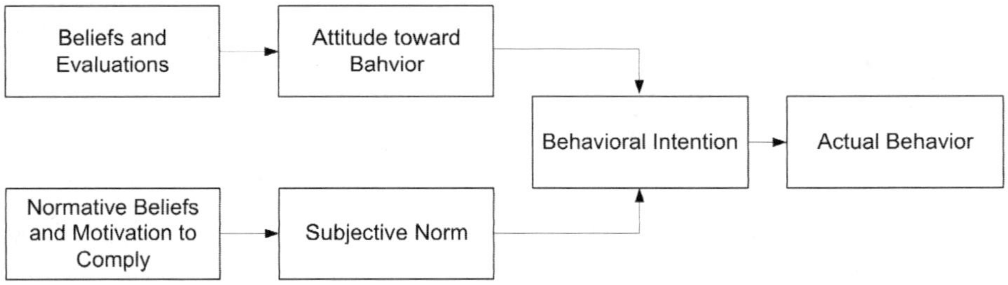

Figure 1 Theory of Reasoned Action [26].

According to TRA, a person's performance of a specified behavior is determined by his or her behavioral intention to perform the behavior, and behavioral intention is jointly determined by the person's attitude and subjective norms concerning the behavior in question.

3.3 Theory of Planned Behavior (TPB)
The theory of planned behavior goes beyond the theory of reasoned action (TRA) and incorporates a further construction, specifically perceived behavior control (PBC); this accounts for those situations where control over the target behavior is not fully volitional [27]. TPB is considered as to be among the more influential of the theories in predicting and explaining behavior [28]. Various studies showed the applicability of TPB to various domains, and verified the ability of this theory in providing a valuable framework to explain and predict the accepting of new information technology [29].

3.4 Technology Acceptance Model (TAM)
Originally introduced by Fred Davis as early as in the 1980s, the Technology Acceptance Model (TAM) sought to measure the willingness of people to accept and adopt new information technology innovations of that era, such as the electronic mail systems [20]. The model had two main determinants

which explained IT adoption: Perceived Usefulness and Perceived Ease of Use. In his work, Davis defined them as "*the degree to which a person believes that using a particular system would enhance his or her job performance*" and "*the degree to which a person believes that using a particular system would be free of effort*", respectively [20]. Contrary to his hypothesis, Davis reported that the relationship between perceived usefulness and adoption was significantly stronger than that of between perceived ease of use and adoption [20]. Furthermore, he noted that perceived ease of use might even precede perceived usefulness, suggesting the existence of a causal relationship instead of the independence of the determinants.

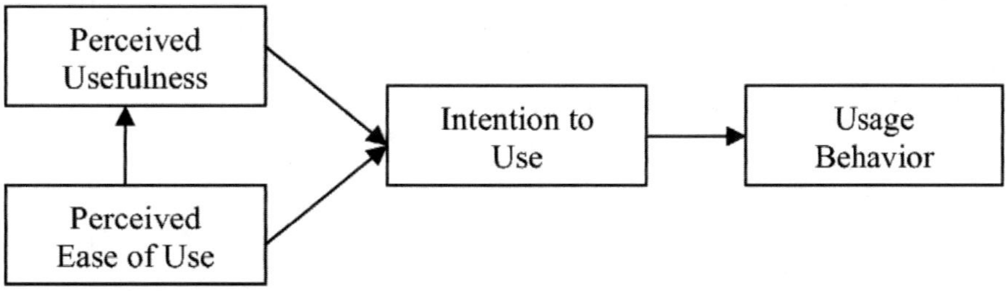

Figure 2 The earliest Technology Acceptance Model [20].

Figure 2 above suggests an interpretation that respondents tend to consider the usefulness of a new system before making a decision to use it. However, the easier the system is perceived to be, the more useful it becomes in the minds of the people, thus improving the overall perception and leading to increased usage. Still, there is a certain limitation to how usage is measured in the study, as Davis duly stated [20].

3.5 Unified Theory of Acceptance and Use of Technology (UTAUT)

UTAUT was developed by Venkatesh et al. to predict user adoption of an information technology [18]. UTAUT integrated eight theories, including the TAM, IDT, the theory of reasoned action (TRA), the motivational model, the theory of planned behavior (TPB), a model combining the TAM and TPB, the model of PC utilization and social cognitive theory (SCT). With empirical analysis, Venkatesh et al. [18] found that performance expectancy; effort expectancy, social influence and facilitating conditions are the main factors determining user adoption. Among them, performance expectancy is similar to perceived usefulness and relative advantage. Effort expectancy is similar to perceived ease of use and complexity. Social influence is similar to subjective norm. Since its inception, UTAUT has been used to explain user adoption of a variety of information technologies, including location-based services [19], mobile technologies [30], mobile banking [31], Internet banking, and health information technologies [32].

As can be seen from figure 3, the UTAUT streamlined the social influence concepts presented in TAM2 and moved some of the elements such as experience and voluntariness of use into background variables (moderating effects).

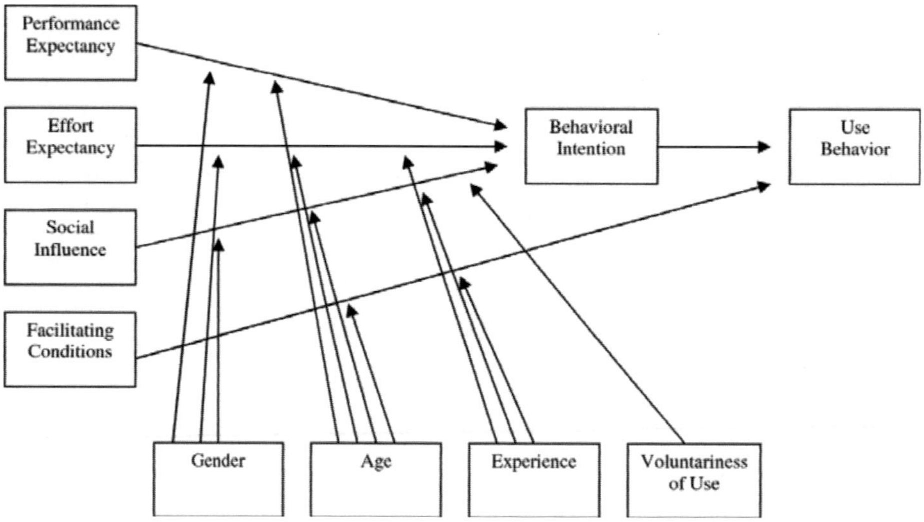

Figure 3 Unified Theory of Acceptance and Use of Technology [18].

4. Proposed Model

Due to the relatively low adoption rate of ICT, extant research has paid much attention to prior work [18, 33, 34] when identifying the factors affecting mobile user behavior.

Performance Expectancy reflects the perceived utility associated with using mobile Internet. Mobile Internet frees users from temporal and spatial limitations, and enables them to acquire information or services at anytime from anywhere. This can improve users' living and working performance and efficiency.

Effort Expectancy reflects the perceived difficulty of using mobile Internet. The constraints of mobile terminals such as small screens and inconvenient input have made it relatively difficult for users to search for information on mobile Internet.

Social Influence reflects the effect of referees' opinion on individual user behavior [34]. According to social influence theory, users tend to comply with other important referees' opinions [35]. Thus when others who are important to a user recommend him or her to use mobile Internet, he or she may follow their suggestions.

Facilitating Conditions mean that users have the resources and knowledge necessary to use mobile Internet. Users need to bear the costs of using mobile Internet, such as communication fees and service fees.

Perceived Value is defined as the consumer's overall assessment of the utility of a product based on perceptions of what is received and what is given.

Perceived Playfulness reflects the pleasure and enjoyment associated with using technology. Perceived enjoyment is an intrinsic motivation that emphasizes the usage process, whereas perceived usefulness is an extrinsic motivation that emphasizes the outcome [36]. In addition, perceived enjoyment may also facilitate continuance usage.

Attention Focus reflects a user's immersion when using ICT such as mobile Internet. Mobile users often perform multiple tasks on movement, such as listening to music and accessing mobile Internet. Thus their attention allocated to mobile Internet may be limited. If they cannot focus their attention, their experience may be affected [34].

Behavioral Intention is defined as a person's perceived likelihood or "subjective probability that he or she will engage in a given behavior" (Committee on Communication for Behavior Change in the 21st Century, 2002, p. 31).With increasing experience, consumers have more opportunities to reinforce their habit because they have more time to encounter the cues and perform the associated behavior [37].

Figure 4 presents the research model including the seven constructs and the behavioral intentions to accept and use the mobile Internet as well as gender, age, education, experience and income as moderator variables.

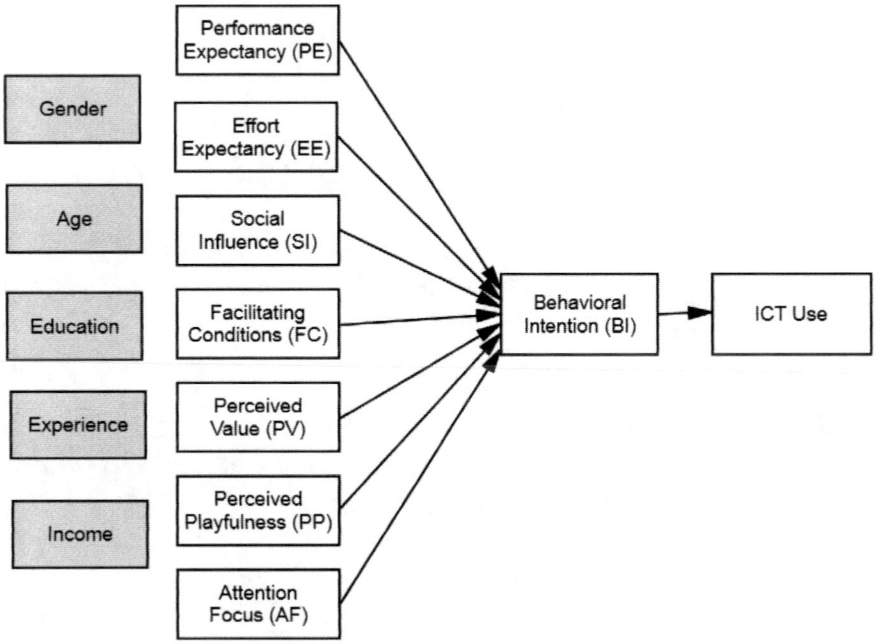

Figure 4 The Proposed Research Model.

5. Conclusion

This research clearly shows the need for a thorough understanding of user attitudes and preferences towards mobile Internet acceptance. As more technological innovations are introduced in rapid succession and an increased number of those innovations are failing, profound insights in the determinants towards adoption and use become more important. The research problem domain was discussed and the researcher's assumptions of the influential effects on mobile Internet acceptance and use are being explored. Also, the research justifications and contribution to the body of knowledge were briefly discussed. The framework presented within this article sets the goal to contribute to technology acceptance research by providing a comprehensive model, although it consists of a limited number of adoption and use determinants.

Journal of Physics: Conference Series **423** (2013) 012037 doi:10.1088/1742-6596/423/1/012037

References

[1] Wikipedia 2012 Information and communications technology.

[2] Agarwal, R. 2000 Individual acceptance of information technologies. Framing the domains of IT management: Projecting the future through the past 85-104.

[3] Cisco 2012 Visual Networking Index: Global Mobile Data Traffic Forecast Update 2011–2016.

[4] ITU 2012 Mobile industry growth forecast.

[5] 148Apps 2012 Count of Active Applications in the App Store.

[6] AndroLib 2012 Distribution of Apps and Games in Android Market.

[7] Igbaria, M. 1993 OMEGA Int. *J. of Mgmt Sci.* **21** 1 73-90.

[8] Swanson, E.B. 1988 Information System Implementation: Bridging the Gap Between Design and Utilization, Homewood, III-IRWIN, Inc.

[9] Davis, F.D., Bagozzi, R.P. and Warshaw, P.R. 1989 *Management Science* **35** 982- 1003.

[10] Thompson, RL., Higgins, CA and Howell, J.M. *MIS Quarterl* **15** 125-143.

[11] Davis, F.D. 1993 *Int. Journal of Machine studies* **38** 475-487.

[12] Torkzadeh, G. and Angulo, I.E. 1992 *Behaviour & Information Technology,* **11** 99-108.

[13] Kwon, T.H., Zmud, R.W. 1987 *Critical Issues in Information Systems Research* 227-51

[14] Danowitz, A., Y. Nassef, and S.E. Goodman. 1995 *Communications of the ACM*, **38** 23-28.

[15] Knight, C. H., and R. J. Dewhurst. 1994. J. Dairy Res. 441–449.

[16] Odedra,M., Bennett, M. , Goodman, S. , and Lawrie, M. 1993 *Communications of the ACM* **36** 25-29.

[17] Goodman, S.E and Green, J.D 1992 *Communication of ACM* **35** 21-25

[18] Venkatesh, V., Morris, M. G., Davis, G. B., and Davis, F. D. 2003 *MIS Quarterly* **27** 425-478.

[19] Xu, H. and Gupta, S. 2009 *Electronic Markets* **19** 137–149.

[20] Davis, F.D. 1989 *MIS Quarterly* **13** 319-340.

[21] Venkatesh, V. and Davis, F.D 2000 *Management Science* **46** 186-204.

[22] Rogers, E.M. 1995 *Diffusion of Innovations* (4th ed.). New York: The Free Press.

[23] Rogers, E. M. 1983 *Diffusion of Innovations* (3rd ed.). New York: The Free Press.

[24] Braun, P. 2004 *Information Technology & Tourism*, **6**

[25] Ajzen, I. and Fishbein, M. 1980 *Understanding Attitudes and Predicting Social Behavior*, Englewood Cliffs, NI: Prentice-Hall.

[26] Fishbein, M. and Ajzen, I. 1975 Belief, Attitude, Intention and Behavior: An Introduction to Theory and Research, Reading, MA: Addison-Wesley.

[27] Ajzen, I. 1985 *From cognition to behavior* 11-39.

[28] Sheppard, B., Hartwick, J. and Warshaw, P. 1988 *Journal of Consumer Research* **15** 325-343.

[29] Hung, S.Y., C.M. Chang, and T.J. Yu 2006 *Government Information Quarterly* **23** 97-122.

[30] Park, J., Yang, S. and Lehto, X. 2007 *Journal of Electronic Commerce Research* **8** 196–206.

[31] Zhou, T., Lu, Y. and Wang, B. 2010 *Computers in Human Behavior* **26** 760–767.

[32] Kijsanayotin, B., Pannarunothai, S. and Speedie, S.M. 2009 *International Journal of Medical Informatics* **78** 404–416.

[33] Venkatesh, V., Thong, J.Y.L., and Xin, X. 2012 *MIS Quarterly* **36** 157-178.

[34] Zhou, T., 2011 *Information Development,* **27** 207-218

[35] Bagozzi, R.P. and Lee, K.-H. 2002 *Social Psychology Quarterly* **65** 226–247.

[36] Davis, F.D., Bagozzi, R.P. and Warshaw, P.R. 1992 *Journal of Applied Social Psychology* **22** 1111–1132.

[37] Kim, S. S. and Malhotra, N. K. 2005 *Management Science* **51** 741-755.

Journal of Physics: Conference Series **423** (2013) 012038 doi:10.1088/1742-6596/423/1/012038

Low Power Transmitter for Wireless Capsule Endoscope

D X Lioe [1,2], S Shafie[1,4], H Ramiah[3], N Sulaiman[1], I A Halin[1]

[1] Department of Electrical and Electronic Engineering, Faculty of Engineering, Universiti Putra Malaysia, 43300 Serdang, Malaysia
[2] School of Engineering, Asia Pacific University of Technology and Innovation, Technology Park Malaysia, Bukit Jalil, 57000 Kuala Lumpur, Malaysia
[3] Department Electrical Engineering, University of Malaya, 50603 Kuala Lumpur, Malaysia
[4] Institute of Advanced Technology, Universiti Putra Malaysia, 43300 Serdang, Malaysia

Email: suhaidi@eng.upm.edu.my

Abstract. This paper presents the transmitter circuit designed for the application of wireless capsule endoscope to overcome the limitation of conventional endoscope. The design is performed using CMOS 0.13 μm technology. The transmitter is designed to operate at centre frequency of 433.92 MHz, which is one of the ISM band. Active mixer and ring oscillator made up the transmitter and it consumes 1.57 mA of current using a supply voltage of 1.2 V, brings the dc power consumption of the transmitter to be 1.88 mW. Data rate of 3.5 Mbps ensure it can transmit high quality medical imaging.

1. Introduction

The evolution of wireless communication and circuit integration has brought medical science and devices to a new dimension. Conventional endoscope for gastrointestinal diseases diagnoses which are uncomfortable for patients, have certain limitations in performing diagnosis or treatment. It is not feasible for the conventional endoscope, in terms of travelling through a long and convoluted small intestine. Wireless capsule endoscopy addresses the disadvantages of the conventional wired endoscopes where it can reach the small intestine and achieve diagnosis without discomforting the patients as well as easily travel through the digestion tract [1], [2].

For the implementation of a capsule endoscope, several important characteristics need to be taken into account, particularly the power consumption of the system as the system is a battery operated device. The power supply must be able to sustain for approximately 8 hours, for a full cycle of digestion until the capsule reaches the bowel. Apart from power consumption, in order to have a highly reliable diagnosis the image quality cannot be tolerated, and for a good quality of image inevitably the data rate has to be high enough to transfer the image wirelessly to the receiver wound around the patients' body. Size of the capsule should be minimized to avoid uncomfortable experience for patients in swallowing the capsule. The characteristics mentioned complicate the design of the wireless capsule endoscope and it is very much a challenge in realizing the system.

2. Design Consideration

Designing the system aims at optimizing each and every one of the crucial characteristics, basically with the trade off of power consumption versus performance, where the performance includes image size, bit depth, frame rates, which sums up to be the data rate. The balance of both criteria of performance and power consumption has to be reached so that a quality capsule endoscope system is designed.

2.1. Power consumption

The most essential aspect which gives the capsule endoscope to keep functioning is the power consumption. Considering it is a battery operated device, power management of the system must be well planned. Overall dc power consumption should not exceed 16 mW. Dividing the power consumption into functional area of lighting, camera etc, the power consumption of the transmitter of the system has to be within 7 mW. [3]

2.2. Carrier frequency

Carrier frequency of the system needs to be optimum for the performance of the system. Many standards of frequency have been utilized for data transmission such as Zigbee, Wireless Local Area Network etc [4]-[5]. Transmitting images wirelessly through human body causes loss in the signal, such as radiation loss, attenuation loss, reflection loss etc. Frequency range where the total loss are minimal is within 400 MHz to 600 MHz [6]. This frequency range covers several standards such as Medical Implant Communication Services (MICS) 402-405 MHz and Industrial, Scientific and Medical (ISM) 433.05 – 434.79 MHz. Another standard close to the range is Wireless Medical Telemetry Services (WMTS) 608 – 614 MHz. MICS is allocated by Federal Communications Commission (FCC) and are commonly utilized in implantable medical devices [7]. However, this standard offers low data rate of up to 500 kbps, which is not sufficient to meet the high data rate requirements to transmit high quality images. WMTS is only covered in the United States and it has inadequate spectrum thus not appropriate for this application. ISM band is widely available and there is no restriction on the bandwidth as well as modulation technique, and hence suitable to be used for wireless capsule endoscope.

2.3. Quality of imaging

Resolution, bit depth, frame rates and data rates all combines to determine the quality of the capsule endoscopy imaging, which essentially is the main objective of diagnosis. Poor quality of imaging would bring to uncertainty of diagnosis and leads to time and financial loss. Resolution of the image should be large enough to be examined by physicians without doubt. The proposed resolution for the system is Quarter Video Graphics Array (QVGA) where the pixel size is 320 x 240 pixels. Frame rate of the camera is set to be five frames per second (fps) in order not to miss any intestinal area in concerned, contrast to 2fps used in other work [8]. Flashes of lighting have to be synchronous with the frame rate in order to keep the power consumption low. Bit depth is another important attributes as difference in colour in intestine leads to discovery of diseases. 8 bits of colour is used in the design of the system to produce high quality imaging. Coming together with the mentioned characteristics, the data rate needed would be 320 x 240 x 5 x 8, which is equivalent to 3 Mbps. The transmitter of the system is designed to cater for data rate of 3.5 Mbps, compare to existing system of 1 -2 Mbps [9], [10].

3. Proposed Transmitter

Wireless capsule endoscope architecture consists of LEDs for lighting, image sensor, battery, transmitter and antenna, as shown in Figure 1. The building block in interest is the transmitter where it accounts for large percentage of the total power consumption and determines the quality of transmission. The proposed transmitter uses an active mixer and a ring oscillator to transmit the signal obtained from image sensor. Ring oscillator is used instead of LC voltage-controlled oscillator due to several reasons, one of which is the start-up time of ring oscillator is faster than that of the LC oscillator. Apart from that, ring oscillator offers lower power consumption and due to the inductorless architecture, requires smaller die area and better circuit integration [11].

Figure 2 shows the transmitter architecture which consists of a mixer and a ring oscillator. The mixer basically up converts the signal from the image sensor with the centre oscillation frequency of 433.92 MHz.

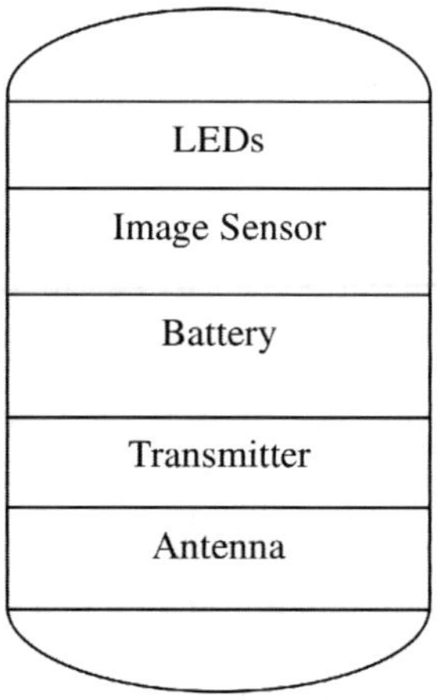

Figure 1. Simplified overall system of the capsule endoscope

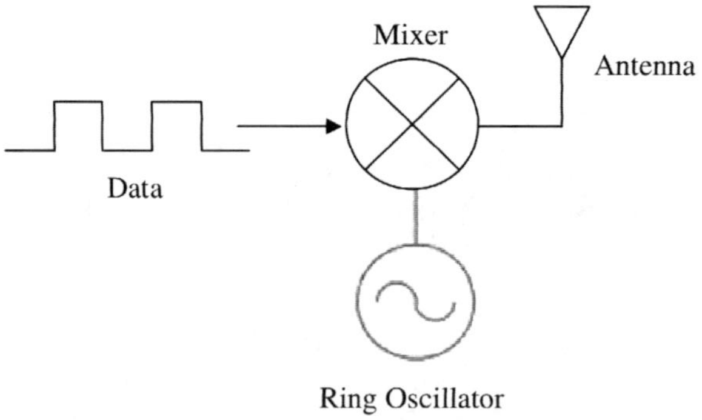

Figure 2. Transmitter architecture

4. Results

The transmitter circuit has been designed and simulated using 0.13 um CMOS technology. The layout of the active mixer and ring oscillator is as shown in Figure 3 with die area within 1 mm x 1 mm. The small area is important for achieving highly integrated circuit with optimized performance. The choice of using mixer and ring oscillator which are without inductors greatly contributes to the minimization of die area.

The combination of mixer and ring oscillator consumes 1.57 mA of current with voltage supply of 1.2 V which translates to power consumption of 1.88 mW. The resulted power consumption is way below the targeted power performance of 7.7 mW. The achievement of low power consumption enables complete diagnosis without interruption of possible power failure. Apart from that, the reduction of power consumption can be utilized for other blocks of the system, which consequently able to boost the overall performance of the system. Figure 4 shows the simulated output waveform of the transmitter. Performance of the designed transmitter is shown in Table 1.

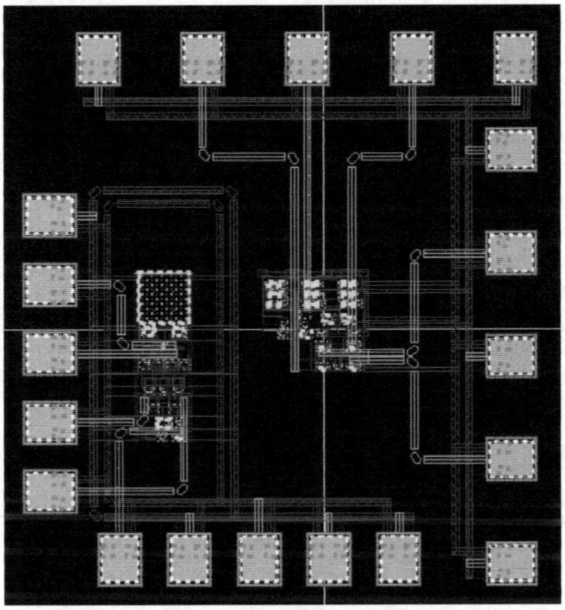

Figure 3. Layout of the designed mixer and ring oscillator

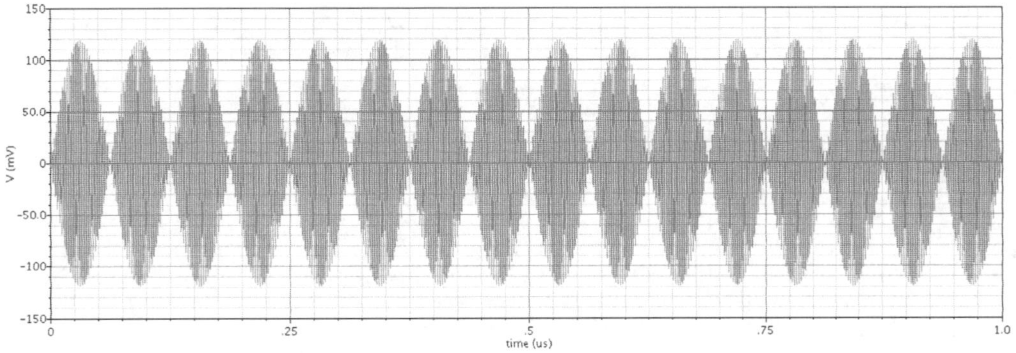

Figure 4. Simulated output waveform of the transmitter

Table 1. Performance table of designed transmitter

Technology	CMOS 0.13 μm
Voltage supply	1.2 V
Current Consumption	1.57 mA
Power Consumption	1.88 mW
Carrier Frequency	433.92 MHz
Data Rate	3.5 Mbps
Die Area	$1mm^2$

5. Conclusion

Low power transmitter for wireless capsule endoscope has been designed which achieves the purpose of low power consumption and high data rate for high quality medical imaging. It offers higher frame rates by having higher data rates of 3.5 Mbps than other works done and better imaging quality which could improves the diagnosis accuracy and acceptance among surgeons and patients. The low power design of 1.88 mW ensures the capsule endoscope works without power failure for the whole diagnosis, as well as enables more voltage headroom to other parts of capsule endoscope system.

The transmitter designed will be integrated with CMOS image sensor and other peripheral circuitries in order to develop a complete prototype of wireless capsule endoscope. Performance testing will be performed and image recovery experiment will be carried out using the prototype capsule endoscope.

References

[1] G. Iddan, G. Meron, A. Glukhovsky, P. Swain, "Wireless Capsule Endoscopy", Nature, vol.405, p417, 2000.

[2] Lee.Y.Y, Amry A.R., Hassan S., "Persistent Obscure Gastrointestinal Bleeding After Jejunal Resection of a Bleeding Jejunal Angiodysplasia in a Lady with Multiple Medical Problems", Journal of Medical Sciences Research, vol.2, pp. 51-53, 2007.

[3] S. Diao; Y. Zheng; Y. Gao; CH. Heng; M. Je, "A 7.2mW 15Mbps ASK CMOS Transmitter for Ingestible Capsule Endoscope", IEEE Asia Pacific Conference on Circuit and System, 2010, pp. 512 – 515

[4] G. H. Tan; R. M. Sidek; H. Ramiah; W. K. Chong, "Design of ultra-low voltage 0.5V CMOS current bleeding mixer", IEICE Electronics Express, Vol. 9 (2012) No. 11 pp. 990-997

[5] Miaou, Shaou-Gang; Jeng, Shiann-Shiun; Tsung, Chen-Wan; Hsiao, Chih-Hong; Lin, Tah-Yeong, "Transmitting Capsule Endoscope Images With Wireless Lan And Smart Antenna Systems", vol. 18, no. 05, pp. 246-254, 2006

[6] Kihyun Kim; Sumin Yun; Sungho Lee; Sangwook Nam; Young Joong Yoon; Changyul Cheon; , "A Design of a High-Speed and High-Efficiency Capsule Endoscopy System," Biomedical Engineering, IEEE Transactions on , vol.59, no.4, pp.1005-1011, April 2012

[7] FCC Rules and Regulations, "MICS Band Plan", Part 95, Jan. 2003

[8] Itoh, S.; Kawahito, S.; Terakawa, S., "A 2.6mW 2fps QVGA CMOS One-chip Wireless Camera with Digital Image Transmission Function for Capsule Endoscopes",. Proceedings of the 2006 IEEE International Symposium on Circuits and Systems, ISCAS 2006 (2006), Page(s):4 pp. 3353 – 3356.

[9] B. Chi, J. Yao, S. Han, X. Xie, G. Li, Z. Wang, "Low power transceiver analog front-end circuit

for bidirectional high data rate wireless telemetry in medical endoscopy applications", IEEE Transaction On Biomedical Engineering, Vol. 54, No. 7, 2007, pp. 1291-1299.

[10] J. Thoné, et al., Design of a 2 Mbps FSK near-field transmitter for wireless capsule endoscopy, Sens. Actuators A: Phys. (2009)

[11] T. Miyazaki , M. Hashimoto and H. Onodera "A performance comparison of PLLs for clock generation using ring oscillator VCO and LC oscillator in a digital CMOS process", Proc. IEEE Asia South Pacific Design Automation, pp.545 -546 2004

ScieTech 2013 IOP Publishing
Journal of Physics: Conference Series **423** (2013) 012039 doi:10.1088/1742-6596/423/1/012039

Approximate Solution Of Schrodinger Equation For Eckart Potential Combined With Trigonormetric Poschl-Teller Non-Central Potential Using Romanovski Polynomials

A. Suparmi[1], C. Cari[2], and J. Handhika[3]
[1]Physics Department, Sebelas Maret University, Surakarta 57126
Ph:081578669977 Fax (0271) 632450,
[2]Physics Department, Sebelas Maret University, Surakarta 57126
Ph:081548578367,

E-mail : carinln@yahoo.com, suparmiuns@gmail.com

Abstract. The approximate analytical solution of Schrodinger equation for Eckart potential combined with trigonometric Poschl-Teller noncentral potential is investigated using Romanovski polynomial. The approximate bound state energy eigenvalues are given in the close form, the corresponding approximate radial eigenfunctions is formulated in term of Romanovski polynomials, and the angular wave function is also expressed in term Romanovski polynomial. The effect of the presence of trigonometric Poschl-Teller potential changes the state of angular wave function level.

1. Introduction

The exact analytical solutions of Schrodinger equations for some physical potentials are very essential since they provide the important information of the quantum system. Recently, conciderably efforts have been paid to obtain the exact solution of the non-central potentials. These potentials include ring-shaped oscillator, Modified Kratzer, Coulombic ring-shaped, and ring-shaped non-spherical harmonic oscillator [1-4]. The bound state energy spectra of these potentials have been investigated by various techniques such as factorization methods, [5-7] supersymmetric quantum mechanics, [8-11] hypergeometric type equation,[12-14] Nikiforov-Uvarov method, [15-17] etc. The exact solution of the radial part of the Schrodinger equations for some potentials such as Eckart, Manning Rosen, Hulthen, modified Poschl-Teller potentials, are obtained if the the angular momentum $l=0$. However, for $l \neq 0$, the radial part of Schrodinger equation can only be solved approximately for different suitable approximation scheme. One of the suitable approximation scheme is conventionally proposed by Greene and Aldrich [18-19].

In this paper we will attempt to solve the Schrodinger equation for Eckart potential combined with Poschl-Teller non-central potential in term of finite Romanovski polynomials. The Romanovski polynomials consist of reducing Schrodinger equation by an appropriate change of the variable to form of generalized hypergeometric equation[11]. The polynomial was discovered by Sir E.J.Routh [20] and by V. I. Romanovski [21]. The notion "finite" refers to the observation that, for any given set of parameters (i.e. in any potential) only a finite of polynomials appear orthogonal [17,23].

Eckart potential and trigonometric Poschl-Teller potentials play the essential roles in interatomic and intermolecular forces [22-25] and can be applied to describe molecular vibrations. This potential is exactly solvable or quasi – exactly solvable and their bound state solutions have been reported.[26]. The Eckart potential combined with trigonometric Poschl-Teller non-central potential is separable potential therefore it is solved using variable separation method and has potential application in describing the molecular vibration and intermolecular forces for $l \neq 0$.

This paper is organized as follows. In section 1.1, we review the finite Romanovski polynomial briefly. In section 1.2, we find the bound state energy solution in term of new orbital quantum number and radial and also angular wave functions. A brief results and discussion is presented in section four and conclusion in section two.

2. Review of Finite Romanovski Polynomials

The generalized hypergeometric equation is given as [15]

$$\sigma(s)\frac{d^2 y_n}{ds^2} + \tau(s)\frac{dy_n(s)}{dx} + \lambda_n y_n(s) = 0 \tag{1}$$

where $\sigma(s) = as^2 + bs + c \quad \tau = ds + e$ and $\lambda_n = -\{n(n-1) + 2n(1-p)\}$ (2)

Equation (1) is described in the textbook by Nikiforov-Uvarov [15] where it is cast into self adjoint form and its weight function, *w(x)*, satisfies Pearson differential equation which is expressed as

$$\frac{d(\sigma(s)w(s))}{ds} = \tau(s)w(s) \tag{3}$$

The weight function obtained by solving the Pearson differential equation is

$$w(s) = \exp\left(\int \frac{(d-2a)s + (e-b)}{as^2 + bs + c} ds\right) \tag{4}$$

The corresponding polynomials to the weight function equation (4) are built up from the Rodrigues representation that is given as

$$y_n = \frac{1}{w(s)}\frac{d^n}{ds^n}\left((as^2 + bs + c)w(s)\right) \tag{5}$$

For Romanovski polynomial, the values of parameters in equation (4) are:

$$a = 1, b = 0, c = 1\ d = 2(1-p) \text{ and } e = q \text{ with p} > 0 \tag{6}$$

By inserting equation (6) into equation (4) we obtain the weight function as

$$w(s) = \left(1 + s^2\right)^{-p} e^{q\tan^{-1}(s)} \tag{7}$$

This weight function was first reported by Routh and then by Romanovski. The polynomials associated with equation (7) are Romanovski polynomials and denoted by $R_n^{(p,q)}(s)$. As long the weight function decreases by x^{-2p}, integral of the type

$$\int_{-\infty}^{\infty} w^{(p,q)} R_m^{(p,q)}(s) R_{m'}^{(p,q)}(s) ds \tag{8}$$

will be convergent only if

$$m' + m < 2p - 1 \tag{9}$$

This means that there are only a finite number of Romanovski polynomials that are orthogonal. The differential equation satisfied by Romanovski Polynomial is

The differential equation satisfied by Romanovski Polynomial obtained by inserting equations (2) and (6) into equation (1) is

$$\left(1 + s^2\right)\frac{\partial^2 R_n^{(p,q)}(s)}{\partial s^2} + \{2s(-p+1) + q\}\frac{\partial R_n^{(p,q)}(s)}{\partial s} - \{n(n-1) + 2n(1-p)\}R_n^{(p,q)}(s) = 0 \tag{10}$$

where $y_n = R_n^{(p,q)}(s)$. The Schrodinger equation of the potential of interest will be reduced into the form which is similar to equation (10) by an appropriate transformation of variable, *r = f(s),* and by introducing a new wave function which is given as

$$\psi(r) = g_n(s) = \left(1 + s^2\right)^{\frac{\beta}{2}} e^{-\frac{\alpha}{2}\tan^{-1}(s)} D_n^{(\beta,\alpha)}(s) \tag{11}$$

The eigen function in equation (11) is the solution of Schrodinger equation for potential interest where

$$D_n^{(\beta,\alpha)}(s) = R_n^{(\beta,\alpha)}(s) \tag{12}$$

The Romanovski polynomials obtained from Rodrigues formula with the weight function in equation (5) is expressed as

$$R_n^{(\beta,\alpha)}(s) = D_n^{(\beta,\alpha)}(s) = \frac{1}{\left(1+s^2\right)^{-p} e^{q\tan^{-1}(s)}} \frac{d^n}{ds^n}\left\{\left(1+s^2\right)^n \left(1+s^2\right)^{-p} e^{q\tan^{-1}(s)}\right\} \tag{13}$$

If the wave function of the nth level in equation (11) is rewritten as [27-28]

$$\psi_n(r) = \frac{1}{\sqrt{df(s)/ds}}\left(1+s^2\right)^{\frac{-p}{2}} e^{\frac{q}{2}\tan^{-1}(s)} R_n^{(p,q)}(s) \tag{14}$$

then the orthogonality integral of the wave functions expressed in equation (14) gives rise to orthogonality integral of the finite Romanovski polynomials, that is

$$\int_0^\infty \psi_n(r)\psi_{n'}(r)dr = \int_{-\infty}^{\infty} w^{(p,q)} R_n^{(p,q)}(s) R_{n'}^{(p,q)}(s)ds \tag{15}$$

In this case the values of p and q are not n-dependence where n is the degree of polynomials. However, if equation (9) and (15) are not fulfilled then the Romanovski polynomials is infinity [27-29].

3. Solution of Schrodinger equation for Ekckart potential combined by trigonometric Poschl-Teller non-central potential Using Romanovski Polynomials

The non-central potential which constructed by Eckart potential and trigonometric Poschl-Teller non-cental potential is given as

$$V(r,\theta) = \frac{\hbar^2}{2Ma^2}\left(V_0 \frac{e^{-r/a}}{(1-e^{-r/a})^2} - V_1 \frac{1+e^{-r/a}}{1-e^{-r/a}}\right) + \frac{\hbar^2}{2Mr^2}\left(\frac{\kappa(\kappa-1)}{\sin^2\theta} + \frac{\eta(\eta-1)}{\cos^2\theta}\right) \tag{16}$$

with V_0 and V_1 describe the depth of the potential well and are positives, $V_1 > V_0$, a is a positive parameter which to control the width of the potential well, $0 < (r/a) < \infty, \kappa > 1, \eta > 1$.

The three dimensional time-independent Schrodinger equation for Eckart potential combined with trigonometric Poschl-Teller non-central potential is

$$-\frac{\hbar^2}{2M}\left\{\frac{1}{r^2}\frac{\partial}{\partial r}\left(r^2\frac{\partial}{\partial r}\right) + \frac{1}{r^2\sin\theta}\frac{\partial}{\partial\theta}\left(\sin\theta\frac{\partial}{\partial\theta}\right) + \frac{1}{r^2\sin^2\theta}\frac{\partial^2}{\partial\varphi^2}\right\}\psi(r,\theta,\varphi) +$$

$$\frac{\hbar^2}{2Ma^2}\left(V_0 \frac{e^{-r/a}}{(1-e^{-r/a})^2} - V_1 \frac{1+e^{-r/a}}{1-e^{-r/a}}\right)\psi(r,\theta,\varphi) + \frac{\hbar^2}{2Mr^2}\left(\frac{\kappa(\kappa-1)}{\sin^2\theta} + \frac{\eta(\eta-1)}{\cos^2\theta}\right)\psi(r,\theta,\varphi) = E\psi(r,\theta,\varphi) \tag{17}$$

Equation (17) is solved using variable separation method so we get radial, polar and azimuthal parts of Schrodinger equation as following:

$$\frac{1}{R}\frac{\partial}{\partial r}\left(r^2\frac{\partial R}{\partial r}\right) - \frac{r^2}{a^2}\left(V_0 \frac{e^{-r/a}}{(1-e^{-r/a})^2} - V_1 \frac{1+e^{-r/a}}{1-e^{-r/a}}\right) + \frac{2Mr^2}{\hbar^2}E = \lambda = l(l+1) \tag{18}$$

$$\left\{-\frac{1}{P\sin\theta}\frac{\partial}{\partial\theta}\left(\sin\theta\frac{\partial P}{\partial\theta}\right) - \frac{1}{\Phi\sin^2\theta}\frac{\partial^2\Phi}{\partial\varphi^2}\right\} + \left(\frac{\kappa(\kappa-1)}{\sin^2\theta} + \frac{\eta(\eta-1)}{\cos^2\theta}\right) = \lambda \tag{19}$$

$$\frac{1}{\Phi}\frac{\partial^2\Phi}{\partial\varphi^2} = -m^2 \tag{20}$$

The azimuthal part of wave function obtained from equation (19) is given, as usual, as

$$\Phi = A_m e^{im\varphi} \tag{21}$$

The radial and polar parts of the Schrodinger equations are solved using Romanovski Polynomials.

3.1 Solution of radial part of Schrodinger equation.

By substitution $\frac{2M}{\hbar^2}E = -\varepsilon^2$ and $R = \frac{\chi(r)}{r}$ in equation (14) we get

$$\frac{\partial^2\chi(r)}{\partial r^2} - \frac{1}{a^2}\left(V_0\frac{e^{-r/a}}{(1-e^{-r/a})^2} - V_1\frac{1+e^{-r/a}}{1-e^{-r/a}}\right)\chi(r) - \varepsilon^2\chi(r) - \frac{l(l+1)}{r^2}\chi(r) = 0 \tag{22}$$

For $\frac{r}{a} <<< 1$ the approximation of the centrifugal term [18-19] in equation (22) is given as $\frac{1}{r^2} \cong \frac{1}{4a^2}\left(d_0 + \frac{1}{\sinh^2(r/2a)}\right)$ with $d_0 = 1/12$. In term of hyperbolic functions, equation (22) is re written as

$$\frac{d^2\chi(r)}{dr^2} - \frac{l(l+1)}{4a^2}\left(d_0 + \frac{1}{\sinh^2(r/2a)}\right)\chi(r) - \frac{1}{a^2}\left(\frac{V_0}{4\sinh^2(r/2a)} - V_1\coth(r/2a)\right)\chi(r) - \varepsilon^2\chi(r) = 0 \tag{23}$$

and by making an appropriate change of variable, $r = f(x) = 2a\coth^{-1}(ix)$ in equation (23), we get

$$(1+x^2)\frac{d^2\chi}{dx^2} + 2x\frac{d\chi}{dx} - \left\{V_0 + l(l+1) + \frac{4V_1 ix}{(1+x^2)} - \frac{l(l+1)d_0 + 4\varepsilon^2 a^2}{(1+x^2)}\right\}\chi = 0 \tag{24}$$

To solve equation (24) in terms of Romanovski polynomial, equation (7) suggests the substitution in equation (24) as [29]

$$\chi(f(x)) = g_n(x) = (1+x^2)^{\frac{\beta}{2}} e^{\frac{-\alpha}{2}\tan^{-1}x} D_n^{(\beta,\alpha)}(x) \tag{25}$$

where $1 < ix < \infty$

By inserting equation (25) into equation (24) we obtain

$$(1+x^2)\frac{\partial^2 D}{\partial x^2} + \{2x(\beta+1) - \alpha\}\frac{\partial D}{\partial x} - \left\{\frac{\beta\alpha x - \frac{\alpha^2}{4} + \beta^2 + 4V_1 ix - (l(l+1)d_0 + 4\varepsilon^2 a^2)}{1+x^2} + V_0 + l(l+1) - \beta^2 - \beta\right\}D = 0 \tag{26}$$

Equation (26) reduces to differential equation that satisfied by Romanovski polynomials if the numerator of $\frac{1}{1+x^2}$ is set to be zero, that are

$$-\frac{\alpha^2}{4} + \beta^2 - (l(l+1)d_0 + 4\varepsilon^2 a^2) = 0 \text{ and } \beta\alpha + 4V_1 i = 0 \tag{27}$$

and then equation (26) becomes

$$(1+x^2)\frac{\partial^2 D}{\partial x^2} + \{2x(\beta+1) - \alpha\}\frac{\partial D}{\partial x} - \{V_0 + l(l+1) - \beta^2 - \beta\}D = 0 \tag{28}$$

By comparing the parameters between equations (10) and (28) we obtain the following relation:

$$V_0 + l(l+1) - \beta^2 - \beta = n(n-1) + 2n(1-p) \quad ; \quad 2(\beta+1) = 2(-p+1) \text{ and } \alpha = -q \tag{29}$$

To have more physical meaning the value of β obtained from equations (29) is

$$\beta = \beta_n = -\sqrt{V_0 + \left(l + \frac{1}{2}\right)^2} - n - \frac{1}{2} \tag{30}$$

and from equations (27) and (30) we obtain

$$\alpha^2 = -2(l(+1)d_0 + 4\varepsilon^2 a^2) \pm 2\sqrt{\{l(l+1)d_0 + 4\varepsilon^2 a^2\}^2 - 16V_1^2} \tag{31}$$

and
$$\alpha=\alpha_n=-\frac{4V_1 i}{\sqrt{V_0+\left(l+\frac{1}{2}\right)^2}+n+\frac{1}{2}} \tag{32}$$

From equations (31) and (32) we obtain the energy spectrum of the system given as

$$E_n=-\frac{\hbar^2}{2M}\left\{\frac{\left(\sqrt{V_0+\left(l+\frac{1}{2}\right)^2}+n+\frac{1}{2}\right)^2}{4a^2}+\frac{V_1^2}{a^2\left(\sqrt{V_0+\left(l+\frac{1}{2}\right)^2}+n+\frac{1}{2}\right)^2}-\frac{l(l+1)d_0}{4a^2}\right\} \tag{33}$$

The energy spectra in equation (33) is in agreement with the result obtained using NU method.
The energy spectra of Eckart potential obtained using NU method for Eckart plus Hulthen potential is given as

$$E_n=-\frac{\hbar^2}{2M}\left\{\frac{\left(-\sqrt{V_0+\left(l+\frac{1}{2}\right)^2}-n-\frac{1}{2}\right)^2}{4a^2}+\frac{V_1^2}{a^2\left(-\sqrt{V_0+\left(l+\frac{1}{2}\right)^2}-n-\frac{1}{2}\right)^2}\right\} \tag{34}$$

The energy obtained using NU method given in equation (34) is a little different with the energy calculated using Romanovski polynomials in equation (33). This difference is caused by the different approximation for the centrifugal term which is approximated by $\frac{1}{r^2}\cong\frac{1}{4a^2}\left(\frac{1}{\sinh^2(r/2a)}\right)$ for NU method.

To determine the wave function, equations (30) and (32) are inserted into equations (7) and (11) so that we obtain the weight function *w(x)* and the Romanovski polynomials $R_n^{(-\beta,-\alpha)}(x)$ as

$$w^{(-\beta,-\alpha)}=\left(1+x^2\right)^{\left(\sqrt{V_0+\left(l+\frac{1}{2}\right)^2}+n+\frac{1}{2}\right)}\exp\left(-\frac{4iV_1}{\sqrt{V_0+\left(l+\frac{1}{2}\right)^2}+n+\frac{1}{2}}\tan^{-1}(x)\right) \tag{35}$$

and
$$D_n^{(\beta,\alpha)}(x)=R_n^{(-\beta,-\alpha)}(x)=\frac{1}{\left(1+x^2\right)^{\beta_n}e^{-\alpha_n\tan^{-1}(x)}}\frac{d^n}{dx^n}\left\{\left(1+x^2\right)^{\beta_n+n}e^{-\alpha_n\tan^{-1}(x)}\right\} \tag{36}$$

where β_n and α_n are expressed in equations (30) and (32). As a result, the wave function of the nth level is given by

$$\chi(f(x))=g_n(x)=\left(1+x^2\right)^{\frac{\beta_n}{2}}e^{\frac{-\alpha_n}{2}\tan^{-1}x}R_n^{(-\beta,-\alpha)}(x) \tag{37}$$

By using trigonometric-hyperbolic functions relation we have

$$e^{-\frac{\alpha_n}{2}\tan^{-1}x}=e^{-\frac{\alpha_n}{2}\tan^{-1}(-i\coth(r/2a))}=\left(\frac{1+\coth(r/2a)}{1-\coth(r/2a)}\right)^{\frac{i\alpha_n}{4}} \tag{38}$$

By inserting equation (38) into equation (37) we get the radial wave function given as

$$\chi_n(r)=g_n(r)=\left(1-\coth(r/2a)\right)^{\frac{\beta_n}{2}-\frac{i\alpha}{4}}\left(1+\coth(r/2a)\right)^{\frac{\beta_n}{2}+\frac{i\alpha_n}{4}}R_n^{(-\beta,-\alpha)}\left(-i\coth(r/2a)\right) \tag{39}$$

$$R_n^{(-\beta,-\alpha)}(r)=\frac{i^n}{(1-\coth(r/2a))^{\beta_n-\frac{i\alpha_n}{2}}(1+\coth(r/2a))^{\beta_n+\frac{i\alpha_n}{2}}}\frac{d^n}{d(\coth(r/2a))^n}\left\{(1-\coth(r/2a))^{\beta_n-\frac{i\alpha_n}{2}+n}(1+\coth(r/2a))^{\beta_n+\frac{i\alpha}{2}+n}\right\} \tag{40}$$

The radial wave function for Eckart potential with centrifugal term obtained from the solution of Eckart plus Hulthen potential with centrifugal term using NU method is given as

$$\chi_n = \left(\frac{1-\coth(r/2a)}{2}\right)^{\frac{p}{2}-\frac{V_1}{p}} \left(\frac{1+\coth(r/2a)}{2}\right)^{\frac{p}{2}+\frac{V_1}{p}} y_n\left(\frac{1-\coth r/2a}{2}\right) \quad (41)$$

$$_n y(s) = \frac{{}_nC\, 2^n (-1)^n}{(1-\coth(r/2a))^{p-\frac{2V_1}{p}} (1+\cosh(r/2a))^{p+\frac{2V_1}{p}}} \frac{d^n}{d(\coth(r/2a))^n} \left\{ (1-\coth(r/2a))^{p-\frac{2V_1}{p}+n} (1+\coth(r/2a))^{p+\frac{2V_1}{p}+n} \right\} \quad (42)$$

with
$$p = -\left(\sqrt{V_0(l+\frac{1}{2})^2} + n + \frac{1}{2}\right) \quad (43)$$

Equations (40), (41), (42), and (43) show that the radial wave functions for Eckart potential with centrifugal term obtained using Romanovski polynomial and NU method are the same.

Since the β_n and α_n parameters, expressed in equations (30) and (32), are n-dependence then the orthogonality of the wave functions may not produce to the orthogonality integral of the polynomials[29]. The orthogonality integral of the wave functions is given as

$$\int_0^\infty \chi_n(r)\chi_{n'}(r)dr = \delta_{nn'} \quad (44)$$

By carrying out the differentiations of equation (36), we find the lowest four Romanovski polynomials given as

$$R_0^{(-\beta_0,-\alpha_0)}(x) = 0 \quad (45a)$$

$$R_1^{(-\beta_1,-\alpha_1)}(x) = (\beta_1+1)2x - \alpha_1 \quad (45b)$$

$$R_2^{(-\beta_2,-\alpha_2)}(x) = 2(\beta_2+2)(2\beta_2+3)x^2 - 2\alpha_2(2\beta_2+3)x + \alpha_2^2 + 2\beta_2 + 4 \quad (45c)$$

$$R_3^{(-\beta_3,-\alpha_3)}(x) = 4x^3(\beta_3+3)(2\beta_3+5)(\beta_3+2) - 6a_3x^2(2\beta_3+5)(\beta_3+2) + 2x(6\beta_3^2+3\alpha_3^2\beta_3+28\beta_3+6\alpha_3^2+34) - 2\alpha_3(2\beta_3+5) - \alpha(\alpha^2+2\beta+6) \quad (45d)$$

where β_n and α_n are expressed in equations (30) and (32). The lowest four degrees of radial wave functions for arbitrary values of l are calculated by using equations (39) and (45).

3.2 The solution of angular Schrodinger equation for Eckart potential combined with non central Poschl-Teller potential.

To solve the polar Schrodinger equation expressed in equation (19), we set the angular wave function as

$$P = \frac{Q(\theta)}{\sqrt{\sin\theta}}, \frac{\partial P}{\partial\theta} = \frac{\frac{dQ}{d\theta}}{\sqrt{\sin\theta}} - \frac{1}{2}\frac{\cos\theta Q}{\sqrt{(\sin\theta)^3}} \quad (46)$$

where $Q(\theta)$ is the new angular wave function. By inserting equation (46) into equation (19) we obtain

$$\frac{d^2Q}{d\theta^2} - \left\{\frac{\kappa(\kappa-1)+m^2-(1/4)}{\sin^2\theta} + \frac{\eta(\eta-1)}{\cos^2\theta}\right\}Q + (l(l+1)+(1/4))Q = 0 \quad (47)$$

Equation (47) to be solved using Romanovski polynomials, therefore we have to substitute the variable θ and introduce a new wave function such that equation (47) reduces to general hypergemetric type equation expressed in equation (1) or into second order differential equation of Romanovski polynomials expressed in equation (10). By making a change of variable in equation (47), $\cos 2\theta = is$ then equation (47) becomes

$$(1+s^2)\frac{\partial^2 Q}{\partial s^2} + s\frac{dQ}{ds} + \left[\frac{\kappa(\kappa-1)+m^2-\frac{1}{4}+\eta(\eta-1)}{2(1+s^2)} + \frac{\kappa(\kappa-1)+m^2-\frac{1}{4}-\eta(\eta-1)}{2(1+s^2)}is - \frac{(l+\frac{1}{2})^2}{4}\right]Q = 0 \quad (48)$$

Equation (48) will be reduced into differential equation of Romanovski polynomial by setting

$$Q(\theta) = g_n(s) = (1+s^2)^{\frac{\beta}{2}} e^{-\frac{\alpha}{2}\tan^{-1}s} D_n^{(\beta,\alpha)}(s) \tag{49}$$

By inserting equation (49) into equation (48) we get

$$(1+s^2)\frac{\partial^2 D}{\partial s^2} + \{s(2\beta+1) - \alpha\}\frac{\partial D}{\partial s} - \left\{ \frac{2\beta\alpha - s\alpha - \frac{\alpha^2}{2} + 2\beta^2 - 2\beta - \left\{\kappa(\kappa-1) + m^2 - \frac{1}{4} - \eta(\eta-1)\right\} is - \left\{\kappa(\kappa-1) + m^2 - \frac{1}{4} + \eta(\eta-1)\right\}}{2(1+s^2)} + \frac{(l+\frac{1}{2})^2}{4} - \beta^2 \right\} D = 0 \tag{50a}$$

Equation (50a) reduces to the differential equation satisfied by Romanovski polynomials given as

$$(1+s^2)\frac{\partial^2 D}{\partial s^2} + \left\{ s(2\beta+1) - \alpha \right\}\frac{\partial D}{\partial s} - \left\{ \frac{(l+\frac{1}{2})^2}{4} - \beta^2 \right\} D = 0 \tag{50b}$$

when the numerator whose its denominator $2(1+s^2)$ is set to be zero, that are

$$-\left\{\kappa(\kappa-1) + m^2 - \frac{1}{4} + \eta(\eta-1)\right\} - \frac{\alpha^2}{2} + 2\beta^2 - 2\beta = 0 \tag{51}$$

$$-\left\{\kappa(\kappa-1) + m^2 - \frac{1}{4} - \eta(\eta-1)\right\} i + 2\beta\alpha - \alpha = 0 \tag{52}$$

By comparing equations (50b) and (10) we obtain

$$(2\beta+1) = 2(-p+1); \; a = -q \text{ and } \frac{\left(l+\frac{1}{2}\right)^2}{4} - \beta^2 = n(n-1) + 2n(1-p) \tag{54}$$

By using equations (51) and (52) and setting

$$\left(\kappa - \frac{1}{2}\right)^2 + m^2 - \frac{1}{4} = \kappa(\kappa-1) + m^2 = F, (\eta - \frac{1}{2})^2 = G \tag{55}$$

we obtain $\alpha = \pm i(\sqrt{F} \mp \sqrt{G})$ and $\left(\beta - \frac{1}{2}\right) = \frac{(F-G)}{\pm 2i(\sqrt{F} \mp \sqrt{G})} i$ (56)

To have physical meaning, the proper choice of the values of α and β in equation (56) are

$$\alpha = i\left(\sqrt{\kappa(\kappa-1)+m^2} + \left(\eta - \frac{1}{2}\right)\right) \text{ and } \beta = \frac{\left(\sqrt{F} + \sqrt{G}\right)+1}{2} = \frac{\sqrt{\kappa(\kappa-1)+m^2} + (\eta + \frac{1}{2})}{2} \tag{57}$$

From equations (54) and (57) we have

$$l = \left(\sqrt{(\kappa - \frac{1}{2})^2 + m^2 - \frac{1}{4}} + \eta + 2n \right) \tag{58}$$

Equation (58) shows that the values of l depend on the potential parameters, κ and η, and the degree of the Romanovski polynomial, *n*=*η*. The weight function obtained from equations (7) and (57) is given as

$$w^{(p,q)} = w^{(-\beta+\frac{1}{2},-\alpha)} = (1+s^2)^{\frac{\sqrt{F}+\sqrt{G}}{2}} e^{-i(\sqrt{F}-\sqrt{G})\tan^{-1}s} \tag{59}$$

The Romanovski polynomials are obtained by using equations (13) and (61) as

$$R_n^{(-\beta+\frac{1}{2},-\alpha)}(s) = \frac{1}{(1+s^2)^{\frac{\sqrt{F}+\sqrt{G}}{2}} e^{-i(\sqrt{F}-\sqrt{G})\tan^{-1}s}} \frac{d^n}{ds^n}\left\{\left((1+s^2)^{\frac{\sqrt{F}+\sqrt{G}}{2}+n} e^{-i(\sqrt{F}-\sqrt{G})\tan^{-1}s}\right)\right\} \tag{60}$$

and the angular wave functions obtained from equations (11) and (60) is given as

$$Q(\theta) = g_n(s) = (1+s^2)^{\frac{(\sqrt{F}+\sqrt{G})+1}{4}} e^{-\frac{i}{2}(\sqrt{F}-\sqrt{G})\tan^{-1}s} R_n^{(-\beta+\frac{1}{2},-\alpha)}(s) \tag{61}$$

The angular eigenfunction obtained by using equations (46) and (61) is given as

$$P_{nm\kappa\eta}(\theta) = \frac{Q_{nm\kappa\eta}}{\sqrt{\sin\theta}} = \frac{1}{\sqrt{\sin\theta}}\sqrt{(1+s^2)^{\frac{\sqrt{F}+\sqrt{G}+1}{2}} e^{-i(\sqrt{F}-\sqrt{G})\tan^{-1}s}} R_n^{(-\beta+\frac{1}{2},-\alpha)}(s) \tag{62a}$$

$$\text{or } P_{\eta m\kappa\eta}(\theta) = \frac{1}{\sqrt{\sin\theta}}(1-\cos 2\theta)^{\frac{\sqrt{F}}{2}+\frac{1}{4}}(1+\cos 2\theta)^{\frac{\sqrt{G}}{2}+\frac{1}{4}} R_{n_l}^{(-\beta+\frac{1}{2},-\alpha)}(-i\cos 2\theta) \tag{62b}$$

The orthogonality integral of the angular wave function obtained from equation (62) is given as

$$\int_0^{\frac{\pi}{2}} P_n^*(\theta)P_{n'}(\theta)\sin\theta d\theta = \int_{-1}^{1}(1+s^2)^{\frac{(\sqrt{F}+\sqrt{G})}{2}} e^{-\frac{i}{2}(\sqrt{F}-\sqrt{G})\tan^{-1}s} R_n^{(-\beta+\frac{1}{2},-\alpha)}(s) R_{n'}^{(-\beta+\frac{1}{2},-\alpha)}(s)ds \tag{63}$$

The orthogonality of Romanovski polynomials is produced from the orthogonality of wave function but equation (63) is not convergent [29-30] since

$$n+n' < 2p-1 \text{ or } n < -\frac{\sqrt{(\kappa-\frac{1}{2})^2+m^2-\frac{1}{4}+\eta+\frac{1}{2}}}{2} \tag{64}$$

Construction of Romanovski polynomials. The first four Romanovski polynomials obtained from equation(60) are given as:

$$R_0^{(-\beta+\frac{1}{2},-\alpha)}(s) = 1 \tag{65}$$

$$R_1^{(-\beta+\frac{1}{2},-\alpha)}(s) = (\sqrt{F}+\sqrt{G}+2)s - i(\sqrt{F}-\sqrt{G}) \tag{66}$$

$$R_2^{(-\beta+\frac{1}{2},-\alpha)}(s) = \left[(\sqrt{F}+\sqrt{G}+4)(\sqrt{F}+\sqrt{G}+3)s^2 - 2is(\sqrt{F}-\sqrt{G})(\sqrt{F}+\sqrt{G}+3) - (\sqrt{F}-\sqrt{G})^2 + (\sqrt{F}+\sqrt{G}+4)\right] \tag{67}$$

$$R_3^{(-\beta+\frac{1}{2},-\alpha)}(s) = \begin{bmatrix} (\sqrt{F}+\sqrt{G}+6)(\sqrt{F}+\sqrt{G}+5)((\sqrt{F}+\sqrt{G})+4)s^3 - 3is^2(\sqrt{F}-\sqrt{G})(\sqrt{F}+\sqrt{G}+5)(\sqrt{F}+\sqrt{G}+4) \\ -3s\{(\sqrt{F}+\sqrt{G}+4)(\sqrt{F}-\sqrt{G})^2 - (\sqrt{F}+\sqrt{G}+6)\} + i\{(\sqrt{F}-\sqrt{G})^2 - (3\sqrt{F}+3\sqrt{G}+16)\}(\sqrt{F}-\sqrt{G}) \end{bmatrix} \tag{68}$$

If there is no the presence of trigonometric Poschl-Teller potential, where $\kappa = 0$ and $\eta = 0$ then the angular wave function reduces to associated Legendre polynomials and the orbital quantum number equation (36) becomes $l = m + 2n$. However the associated Legendre polynomial obtained from this non-central potential are only those polynomials whose values of l and m are differed by even numbers since $l = m + 2n$. The effect of the presence of Poschl-Teller non-central potential to spherical harmonics is illustrated using the three dimensional representation and the polar diagram of the absolute value of un-normalized angular wave functions obtained from equations (55), (62), and (68). The 3D representations and polar diagram of $|Y_l^m|$ visualized using Mat Lab 7 are shown in Figure 1 for $n_l = 3, \kappa = 0, \eta = 0, l = 6, m = 0$, Figure 2 for

$n_l = 3, \kappa = 2, \eta = 0, l = 7.4, m = 0$ and Figure 3 for $n_l = 3, \kappa = 0, \eta = 2, l = 8, m = 0$. By comparing Figures 2 and 3 with Figure 1, it is concluded that there is a state change in angular wave function. Therefore it may be concluded that the number of the degeneracy of the system changes. The un-normalized angular wave functions illustrated in Figures 1-3 is in agreement with the result calculated using NU method [31]. By putting the new value of the orbital quantum number expressed in equation (68), the energy of Eckart potential combined with trigonometric Poschl-Teller non-central potential is rewritten as

$$E_{n_l n_r} = -\frac{\hbar^2}{2M}\left\{ \frac{V_1^2}{a^2\left[\sqrt{V_o + \left(\sqrt{\kappa(\kappa-1)+m^2} + \eta + 2n_l + \frac{1}{2}\right)^2} + n_r + \frac{1}{2}\right]^2} + \frac{\left[\sqrt{V_o + \left(\sqrt{\kappa(\kappa-1)+m^2} + \eta + 2n_l + \frac{1}{2}\right)^2} + n_r + \frac{1}{2}\right]^2}{4a^2} - \frac{l(l+1)d_0}{4a^2} \right\} \tag{69}$$

where n_l is a new orbital quantum number and its values are non-negative integer, while n_r is radial quantum number and is nonnegative integer. From equation (65) we can calculate the energy for special case, for Eckart potential, we set $\kappa = \eta = m = n_l = 0$, therefore the energy spectrum of Eckart potential is

$$E_{n_r} = -\frac{\hbar^2}{2M}\left\{ \frac{V_1^2}{a^2\left[\sqrt{V_0 + \left(\frac{1}{2}\right)^2} + n_r + \frac{1}{2}\right]^2} + \frac{\left[\sqrt{V_0 + \left(\frac{1}{2}\right)^2} + n_r + \frac{1}{2}\right]^2}{4a^2} - \frac{l(l+1)d_0}{4a^2} \right\} \tag{70}$$

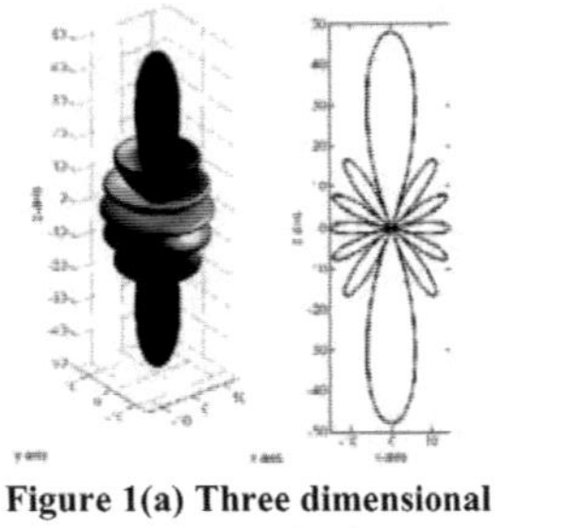

Figure 1(a) Three dimensional representation of $|Y_6^0|$ and (b) its polar diagram of $|Y_6^0| = \{86.63\cos^3 2\theta + 23.63\cos^2 2\theta - 55.13\cos 2\theta - 7.13\}$

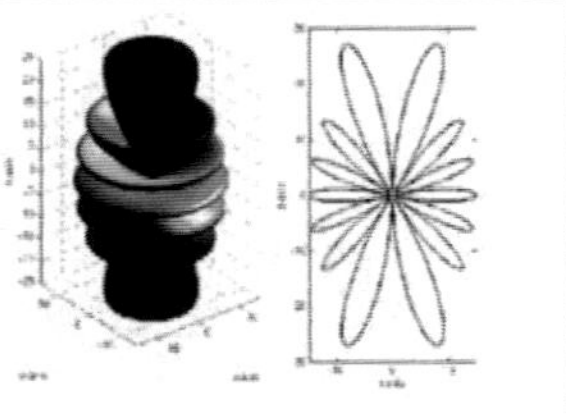

Figure 2(a) Three dimensional representation of $|Y_{7.4}^0|$ and (b) its polar diagram of $|Y_{7.4}^0| = \{199.48\cos^3 2\theta + 164.79\cos^2 2\theta - 48.36\cos 2\theta - 28.67\}\sin^{1.4}\theta$

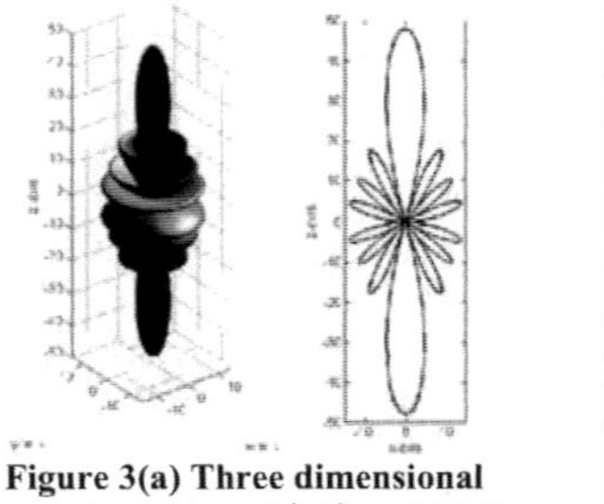

Figure 3(a) Three dimensional representation of $|Y_8^0|$ and (b) its polar diagram of $|Y_8^0| = \{268.13\cos^3 2\theta - 160.88\cos^2 2\theta - 86.63\cos 2\theta + 27.38\}\cos^2\theta$

The total un-normalized wave function of the system obtained from equations (39) and (65b) is given as

$$\psi(r,\theta,\varphi) = \left(1-\coth(r/2a)\right)^{\frac{\beta_{nr}}{2}-\frac{i\alpha_{nr}}{4}}\left(1+\coth(r/2a)\right)^{\frac{\beta_{nr}}{2}+\frac{i\alpha_{nr}}{4}} R_{n_r}^{(-\beta_{nr},-\alpha_{nr})}\left(-i\coth(r/2a)\right)$$

$$\times\frac{1}{\sqrt{\sin\theta}}\left(1-\cos 2\theta\right)^{\frac{\sqrt{F}}{2}+\frac{1}{4}}\left(1+\cos 2\theta\right)^{\frac{\sqrt{G}}{2}+\frac{1}{4}} R_{n_l}^{(-\beta+\frac{1}{2},-\alpha)}\left(-i\cos 2\theta\right)e^{im\varphi} \tag{71}$$

4. Results and Discussion

The energy spectra for Eckart plus trigonometric Poschl-Teller non-central potential reduce to energy spectra of Eckart potential plus centrifugal term by setting the Poschl-Teller potential 's parameter, $\kappa = \eta = m = 0$ in equation (69). The ground state wave function of Eckart potential is produced from radial wave function for $n_r = 0$ and $l = 0$. The radial wave function for Eckart potential with centrifugal term obtained using Romanovski polynomials is in agreement with the result obtained using NU method. The effect of Poschl-Teller non-central potential is dominantly coming from the $\sec^2\theta$ term which changes the state of the system, as shown in Figure 3, while the effect of $\cos ec^2\theta$ is similar with the effect of quantum magnetic number, m, as shown in equation (58). The energy eigenvalue decreases by the increase of the values of κ or/ and η. The angular wave function of Eckart potential plus trigonometric Poschl-Teller non-central potential obtained using Romanovski polynomials is in agreement with the result calculated using NU method. The associated Legendre polynomials is produced from angular wave function by setting $\kappa = \eta = 0$. This potential can be applied in an optical media, particularly as a model for graded refractive index [32].

5. Conclusion

The three dimensional Schrodinger equation for Eckart plus trigonometric Poschl-Teller non-central potential is solved approximately using Romanovski polynomials with suitable coordinate transformation. The energy spectrum is obtained in the closed form, the radial wave function is solved approximately and given in term of Romanovski polynomials. The angular wave function is expressed in term of Romanovski polynomials. The presence of Poschl-Teller non-central potential causes the change of the angular state wave function and the decrease of the energy eigenvalue.

References

[1] B Gonul and I Zorba 2000 *Phys. Lett.* A **269** 83.
[2] F Yasuk, C Berkdemir, and A Berkdemir 2005 *J. Phys. A:Math. Gen.* **38** 6579.
[3] C Berkdemir, A Berkdemir, and J Han 2006 *Chem. Phys. Let.* **417** 4 –6 326
[4] C Y Chen and S H Dong 2005 *Phys. Lett. A* **335** 37
[5] L Infeld and T D Hull 1951 *Rev. Mod. Phys.* **23**
[6] L E Gendenhtein 1983 *JETP Lett.* **38** 356
[7] F Cooper, J N Ginocchio and A Khare 1987 *Phys. Rev. D* **36** 2458
[8] E Witten 1981 *Nucl. Phys. B* **188** 513
[9] R Dutt, A Khare and U P Sukhatme 1988 *Am. J. Phys.* **56** 163
[10] A Comptet, A D Bandrauk, and D K Campbell 1985 *Phys. Lett. B*150 159
[11] H Goudarzi and V Vahidi 2011 *Adv. Studies Theor. Phys.* **5** 10 469
[12] N Cotfas, *Shape Invariant type Operators with Application to Quantum Mechanics,* math-ph /0603032; http//fpcm5.fizica.unibuc.ro/ncotfas
[13] G A Natanzon 1971 *Vestnik Leningrad. Univ.* 10:22
[14] F Yasuk, C Berkdemir, and R Berkdemir 2005 *J. Phys. A: Math. Gen*,**38** 6579
[15] A F Nikiforov and U B Uvarov 1988 *Special Function in Mathematical Physics*, Birkhausa, Basel
[16] A N Ikot and L E Akpabio 2010 *App. Phys. Res* **2** 2 202-208
[17] H Goudarzi and V Vahidi 2011 *Adv. Studies Theor. Phys.* **5** 10 469-476
[18] R I Greene and C Aldrich 1976 *Phys. Rev.* A **14** 2363
[19] Y Xu, S He and C S Jia 2010 *Phys.Scr.***81** 045001/
[20] E J Routh 1884 Proc. *London Math. Soc.* **16** 245
[21] V I Romanovski 1929 *Compt. Rend. Ac. Sci. Paris* **188** 1023
[22] C Eckart 1930 *Phys. Rev.* **35** 1303
[23] G Poschl and E Teller 1933 *Zeit.Phys.* **83** 149
[24] A N Ikot, L E Akpabio, and J A Obu 2011 *JVR* **6** 1 1-13
[25] A S Halberg 2011 *Int. J. Maths and Mathematical sciences* 1-9 DOI10.1155/2011/358198
[28] A N Ikot, A D Antia, L E Akpabio, and J A Obu 2011 *JVR* **6** 2 65-76
[29] D E Alvarenz-Castillo and M Kirchbach 2007 Rev. Mex.Fis. **E53** 2 143-154
[30] D E Alvarez-Castillo, C B Compean, and M Kirchbach 2011 arXiv:1105.1354v1[quant-ph]
[31] C Cari and A Suparmi 2012 *IOSR JAP* **2** 3 13-23
[32] N Dalarsson, 2005 Proc. 1st Int. Workshop, Belgrade, Serbia and Montenegro.

Acknowledgement
This work is partially supported by DIPA BLU Post Graduate Program Sebelas Maret University grant No. 2340/UN27.10/PG/2012

Journal of Physics: Conference Series **423** (2013) 012040 doi:10.1088/1742-6596/423/1/012040

Calculation of diffusion activation energy by thermodynamic model

P Wu[1], Y Zeng[2], H M Jin[2]

[1]Singapore University of Technology and Design, Singapore 138682, Singapore.
[2]Institute of High Performance Computing, Agency for Science, Technology and Research, 1 Fusionopolis Way, #16-16 Connexis, Singapore 138632, Singapore

Email: wuping@sutd.edu.sg

Abstract. Although self diffusion activation energy of metal is known to be strongly correlated to its melting point and heat of formation, prediction of diffusion activation energy remains a challenge and somehow an art. On the basis of the well developed quasi-chemical model, we demonstrate a general frame work to estimate surface/interface diffusion activation energy using the concentration and formation energy of the interface chemical bonds. Furthermore, this frame work reproduces all available diffusion activation energy data of Si on metals. We not only provide a practical method in predicting diffusion activation energy but also offer new physical insights about the interplay between chemical kinetics and thermodynamics.

1. **Introduction**

Diffusion activation energy data are frequently required in microstructure and phase transition research (e.g., in the glass formation process) where complicated chemical and structure parameters are involved [1-3]. However, diffusion activation energy data are scarce and the common techniques of experiment and computer simulation to derive these data are either not practical or time consuming due to wide range of changes of systems in terms of chemistry and microstructures [4, 5]. Therefore correlation methods were widely employed. Beke studied the dependence of diffusion to temperature and found that the diffusion coefficients usually obey the Arrhenius law [6]. Brown and Ashbby showed the rate of diffusion at the melting point is roughly a constant for a given crystal structure [7]. In addition, melting points show strong correlation to enthalpy of formation of alloys or compounds [8]. Any guidelines derived from these different approximation methods that leading to the prediction of diffusion activation energy will advance the diffusion research in general, although the nature of chemical bond formation is largely ignored in most of the previous correlation type of models. Recent advance in chemical thermodynamic calculations offers ready descriptions of detailed solution chemistry for systems of complicated chemistry as well as for multi-component and multiphase microstructures [9]. A direct link between the thermodynamics and diffusion activation will not only generate practical prediction method but also improve current theoretical understanding of chemical kinetics in general.

We demonstrated previously [10] that interface chemistry of metal and semiconductor can be qualified by quasi-chemical model (a solution chemistry model) [11, 12]; both the heat of formation of metal silicides and the eutectic temperatures of system consisted of the metal and semiconductors can be directly presented by the quasi-chemical model. Therefore, rather than a simple data correlation, our approach approximates the chemical bonding process, on the basis of the well developed quasi-chemical model, by calculating concentrations of different type of bonds based on the bond formation energy and solution chemical concentrations. This way, we maintain the rough physics of interface while developing correlation models.

In chemical catalysis research, factors that control chemical reaction paths and the corresponding potential energy surfaces [13, 14] are well studied and a simple linear correlation between the transition-state (TS) energy and the final-state (FS) energy of the respective elementary steps, namely, the Brønsted-Evans-Polanyi (BEP) relationship has been elucidated and applied in the analysis of surface elementary reactions. The BEP equation directly relates the change in activation energy of the reaction (a kinetic parameter) to the corresponding change of the reaction energy (a thermodynamic parameter). Liu and Hu [15] advanced the BEP method by using first principles calculations to systematically study a number of typical catalytic reactions on transition metal surfaces, and postulated that the reaction barrier can be decomposed using bond energies of all the associated reactant – surfaces and reactant – reactant pairs (or bonds). Their work directly linked the reaction barrier to the energetic of all sorts of involved chemical bonds and found that the dissociation reaction barrier is determined by the chemisorption energy at the final reaction state, in consistent with the BEP relationship.

These studies suggest that the bond energies of reactants and products may play a critical role in determining the reaction barrier and diffusion activation energy. In statistical thermodynamics, all equilibrium properties are derived by using partition functions of all energetic micro-configurations, therefore, a transition state (on an energy path way for diffusion) has already been counted and discriminated by the partition functions. We aim to pick up the micro-configuration(s) that corresponding to the transition state by reverse engineering the thermodynamic solution. Therefore, the quasi-chemical (QC) model which describes the energy and concentration of bonds of all stable and metastable micro-configurations is applied in this work. We first demonstrate such a reverse engineering technique (on QC) using the heat of formation and eutectic temperature data, and then diffusion activation energy. The resultant formulae can be used for prediction of activation energy merely from electro-negativities of the constituent elements.

2. Quasi-chemical calculation

In this section, the relationship of reaction bond energy and mole fraction of the reactant pairs of a binary system is described by the quasi-chemical model [11, 12]. The numerical results of the relationship associated with various initial component compositions are presented.

An A-B binary system contains three types of nearest-neighbor pairs, namely, A-A, B-B, and A-B. The relative amounts of the three types of pairs are determined by the energy change associated with the formation of two A-B pairs from one A-A and one B-B pair, as described in reaction (1).

$$[\text{A-A}] + [\text{B-B}] = 2[\text{A-B}] \quad (1)$$

Let X_{AA}, X_{BB} and X_{AB} be the mole fractions for A-A, B-B and A-B pairs, and X_A and X_B are the initial mole fractions of components A and B, respectively. When the two components are mixed, A-B pairs are formed at the expense of A-A and B-B pairs. It follows from the mass balance that:

$$2X_A = 2X_{AA} + X_{AB} \quad (2)$$

$$2X_B = 2X_{BB} + X_{AB} \quad (3)$$

The molar enthalpy change of reaction (1) is denoted by ω, the net pair bond energy. The enthalpy of mixing, ΔH, is then given as following:

$$\Delta H = \left(\frac{X_{AB}}{2}\right)\omega \tag{4}$$

A "quasi-chemical equilibrium constant" for reaction (1) is obtained as

$$\frac{X_{AB}^2}{X_{AA}X_{BB}} = 4e^{\frac{-\omega}{RT}} \tag{5}$$

Substituting Eqs. (2-3) into Eq. (5) yields Eq. (6) where it indicates that the mole fraction A-B bond, X_{AB}, is the function of the net pair bond energy ω, the initial component mole fraction X_A and the temperature T. R denotes the gas constant.

$$\frac{X_{AB}^2}{(2X_A - X_{AB})(2-2X_A - X_{AB})} = e^{\frac{-\omega}{RT}} \tag{6}$$

The analytical solution of X_{AB} cannot be obtained from Eq. (6). Thus we perform the numerical study by plotting ($-\omega$) versus logarithm X_{AB} given $T = 300$ K and $\omega = -25 \sim 0$ kJ mole^{-1}. The results are shown in Figure 1 with the initial component mole fraction X_A = 0.5, 0.6, 0.7, 0.8 and 0.9, respectively. It is noted that due to the symmetry in the composition of component A and B, the results of $X_A = 0.1$ and $X_A = 0.9$ are equal. Similarly, the result of $X_A = 0.2$ is same as that of $X_A = 0.8$, and so forth.

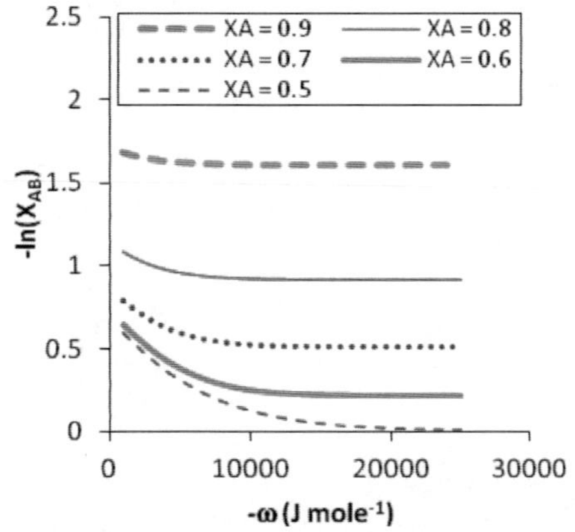

Figure 1. Dependence of net pair bond energy ω and the mole fraction of A-B bond X_{AB}.

3. Results

On the basis of quasi-chemical model, we examine four published data sets of formation heats, eutectic temperatures, and diffusion activation energies. Since the ω values for these systems are not available, we use the elemental electro-negativity (χ) to replace ω [10]. In particular, the squared difference of electro-negativities of two constituent elements $(\Delta\chi)^2$ is used. It is noted that $(\Delta\chi)^2$ is proportional to $-\omega$, as derived from the well-known semi-empirical Miedema's model [16]. In this work, the Martynov-Batsanov electro-negativities (eV$^{1/2}$) [17] are used. In the study of the following four systems, $(\Delta\chi)^2$ is chosen as the abscissa, in consistent with the result of quasi-chemical model (Figure 1).

We first demonstrate the reverse engineering approach using the formation heats of silicides formed on the interfaces of transition metals and n-type silicon [10]. In order to obtain comparable result, the ordinate is formulated as the logarithmic ($\Delta H_f/\Delta\chi^2$) where ΔH_f is the formation heat. Since the temperature is the same for this set of data which leads to a same constant for each data point, the temperature is not taken into account. The resulting trend of $(\Delta\chi)^2$ verses ($\Delta H_f/\Delta\chi^2$) is shown in Figure 2. It is also found that the dependences of X_{AB} and ω generated from QC model with a range of initial composition X_A as shown Figure 1 can be re-scaled to well match the trend of the experimental data. Beyond this composition range, it is impossible to reproduce the trend. As an illustration, the re-scaled dependence with $X_A = 0.6$ (bold line) is plotted in Figure 2. In comparison with the results of quasi-chemical model (Figure 1), $\ln(\Delta H_f/\Delta\chi^2)$ is equivalent to $-\ln(X_{AB})$, which implies a strong correlation between formation heat and the A-B bond properties (concentration and energy) at micro-configurations about X_A=0.6.

The second system deals with the eutectic temperature of binary systems of transition metal silicides [10], since it is widely accepted for a strong correlation between diffusion and temperature [7]. In the same fashion of Figure 2, the ordinate is formulated as the logarithmic ($T_{eu}/\Delta\chi^2$) where T_{eu} is the eutectic temperature. The result is shown in Figure 3. Similarly as in Figure 2, the QC dependence with $X_A = 0.6$ (bold line) is re-scaled to fit this set of experimental data. Furthermore, since $\ln(T_{eu}/\Delta\chi^2)$ is equivalent to $-\ln(X_{AB})$, it implies a strong correlation between the eutectic temperature and the A-B bond properties at micro-configurations about X_A=0.6.

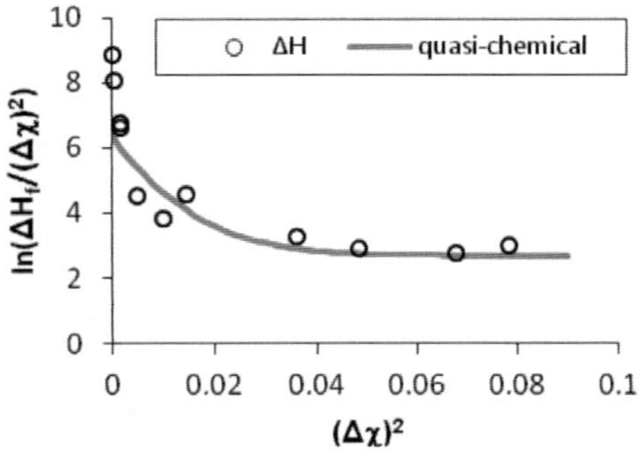

Figure 2. Dependence of $\ln(\Delta H_f/\Delta\chi^2)$ and $(\Delta\chi)^2$ for the formation heats of transition–metal and silicon.

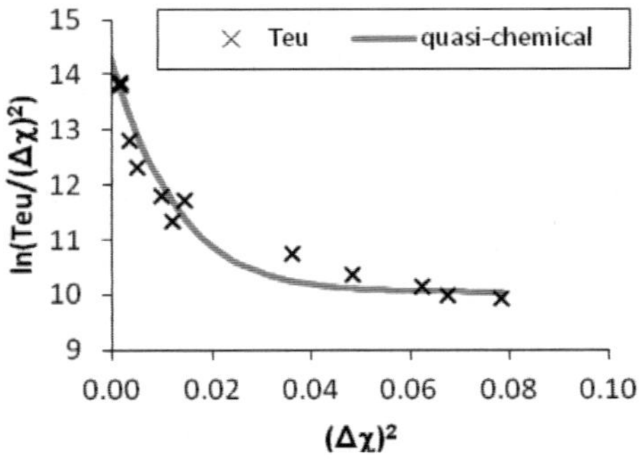

Figure 3. Dependence of $\ln(T_{eu}/\Delta\chi^2)$ and $(\Delta\chi)^2$ for the eutectic temperatures of transition–metal silicides.

Inspired by the above success, we further study the diffusion of Si into metals by the reverse engineering technique. The diffusion activation energies and the corresponding average temperatures are extracted from [18] and summarized in Table 1. The ordinate is chosen as $\ln(\frac{Q}{RT(\Delta\chi)^2})$ where Q is the activation energy, R the gas constant, and T the average temperature. The resulting trend of $(\Delta\chi)^2$ verses $\ln(\frac{Q}{RT(\Delta\chi)^2})$ is shown in Figure 4. Similarly, the QC dependence with $X_A = 0.6$ (bold line) is re-scaled to match this set of experimental data. Compared to Figure 1, $\ln(\frac{Q}{RT(\Delta\chi)^2})$ is equivalent to $-\ln(X_{AB})$, implying strong correlation between the activation energy and the A-B bonding properties.

Finally, we study the reaction barrier (activation energy, ΔE) in the diffusion of O, S, N on (111) surfaces of various metals (Ni, Re, Ag, Pd, Pt, Ru, Cu) [13]. Using the data listed in Table 2, we calculate $\ln(\Delta E/\Delta\chi^2)$ and plot it against $(\Delta\chi)^2$ in Figure 5. The term (RT) is omitted as the temperatures are the same for this set of data. Again, the QC dependence with $X_A = 0.6$ (bold line) is re-scaled to reproduce the experiment trend. Similarly, a strong correlation between the activation energy and the microscopic A-B bonding is demonstrated.

Table 1. Diffusion activation energies of Si into metals and the electro-negativities of metals.

Metal	Average Temp. °C	Activation energy cal $mole^{-1}$	Metal Electro-negativity $eV^{1/2}$
Ti	1050	39700	1.86
Zr	1100	55750	1.7
V	1100	61200	2.22
Nb	1100	48000	2.03
Ta	1000	34600	1.94
Cr	1000	22760	2
Mo	1350	78000	1.94
W	1000	63000	1.79
Fe	1000	22030	1.67

Table 2. Energies of initial and transition states in the diffusion of O, S, N on (111) surfaces of various metals (Data read from Ref.[13]), and electro-negativities of each system.

	Energy, eV		Electro-negativity, $eV^{1/2}$	
System	Initial State	Transition State	Non-Metal	Metal
C/Ru	-6.69	-6.12	2.37	1.97
C/Pt	-6.27	-5.38	2.37	1.91
S/Re	-5.69	-5.21	2.65	2.06
S/Ru	-5.64	-5.08	2.65	1.97
N/Ru+	-5.55	-4.73	2.85	1.97
O/Ru+	-5.40	-4.80	3.32	1.97
O/Ru	-5.07	-4.34	3.32	1.97
N/Ru	-5.06	-4.54	2.85	1.97
S/Ni	-5.02	-4.64	2.65	1.76
O/Ru-	-4.79	-4.35	3.32	1.97
S/Pt	-4.74	-4.18	2.65	1.91
N/Ru-	-4.59	-3.98	2.85	1.97
C/Cu	-4.25	-3.73	2.37	1.08
S/Cu	-4.14	-3.89	2.65	1.08
O/Pt+	-4.08	-3.47	3.32	1.91
C/Cu-	-4.02	-3.56	2.37	1.08
O/Pt	-3.63	-3.16	3.32	1.91
O/Pd	-3.54	-3.06	3.32	2.08
O/Ag	-3.06	-2.71	3.32	1.07
H/Pt	-2.58	-2.37	2.1	1.91

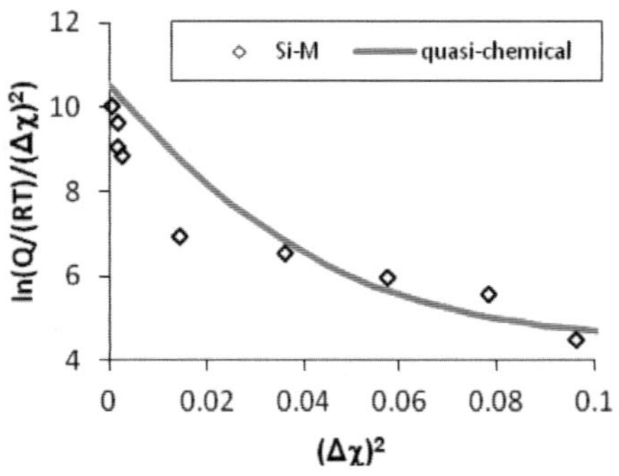

Figure 4. Dependence of $\ln(Q / RT(\Delta\chi)^2)$ and $(\Delta\chi)^2$ for the diffusion activation energies of Si diffusion into metals.

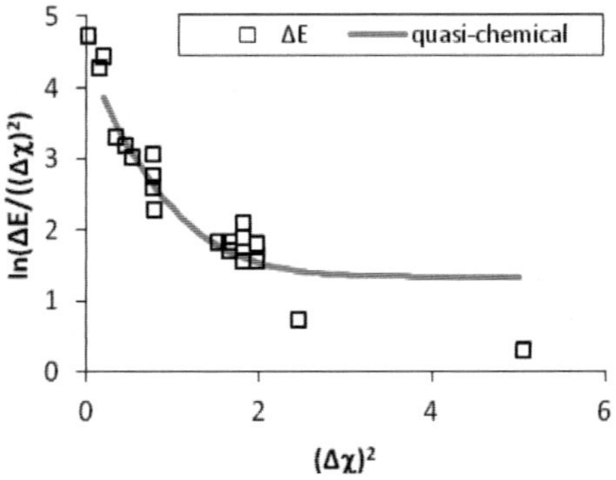

Figure 5. Dependence of $\ln(\Delta E/\Delta\chi^2)$ and $(\Delta\chi)^2$ for the reaction barriers of diffusion of O, S, N on (111) surfaces of various metals.

4. Conclusions

We demonstrated, on the basis of quasi-chemical model, a general frame work to formulate diffusion activation energy from solution thermodynamics. Although in this study our approach reproduced diffusion activation energy for relative simple systems, in principle similar calculations may be carried out for multi-component and multiphase systems because the needed bond fraction and bond energy data can be derived from quasi-chemical model. The method offers opportunities to further understand the interplay between chemical kinetics and thermodynamics.

References

[1] Gill E K and Morrison J A 1963 *Annu. Rev. Phy. Chem.* **14** 205
[2] Kang M, Zhang M X, Liu F and Zhu M 2009 *Mater. Trans.* **50** 123
[3] Wang Z, Li Y and Adams J B 2000 *Surf. Sci.* **450** 51
[4] Rouxel T 2011 *J. Chem. Phys.* **135** 184501
[5] Bletskan D I 2006 *Chalcogenide Lett.* **3** 81
[6] Beke D, Geszti T and Erdelyi G 1977 *Z. Metallkde* **68** 444
[7] Brown A M and Ashby M F 1980 *Acta Metall.* **28** 1085
[8] Murti Y V G S, Popson G A and Laskar A L 1990 *Phys. Status Solidi A* **120** K133
[9] Hecht U, Granasy L, Pusztai T, et al. 2004 *Mat. Sci. Eng. R* **46** 1
[10] Wu P and Zeng Y 2010 *J. Mater. Chem.* **20** 10345
[11] Wu P 1992 *Ph. D. Thesis*, University of Montreal
[12] Wu P and Pelton A D 1992 *J. Alloy Compd.* **179** 259
[13] Nilekar A U, Greeley J and Mavrikakis M 2006 *Angew. Chem. Int. Edit.* **45** 7046
[14] Santen R A van, Neurock M and Shetty S G 2010 *Chem. Rev.* ***110*** 2005
[15] Liu Z-P and Hu P 2001 *J. Chem. Phys.* **114** 8244
[16] Miedema A R 1992 *Physica B* **182** 1
[17] Villars P and Hulliger F 1987 *J. Less-Common Met.* **132** 289
[18] Samsonov G V and Vinitskii I M 1980 *Handbook of refractory compounds*, translated from Russian by Kenneth Shaw, IFI/Plenum, New York, pp 220

Journal of Physics: Conference Series **423** (2013) 012041 doi:10.1088/1742-6596/423/1/012041

Shear Strength of Single Lap Joint Aluminium-Thermoplastic Natural Rubber (Al-TPNR) Laminated Composite

M Z Muzakkar[1,3], S Ahmad[1], M A Yarmo[2], A Jalar[1], M Bijarimi[1,4]

[1] School of Applied Physics, Faculty of Science and Technology, Universiti Kebangsaan Malaysia, 43000 Bangi, Selangor, Malaysia
[2] School of Chemical Sciences and Food Technology, Faculty of Science and Technology, Universiti Kebangsaan Malaysia, 43000 Bangi, Selangor, Malaysia
[3] Department of Chemistry, Faculty of Mathematics and Science, Universitas Haluoleo, Kendari 93232, Indonesia
[4] Faculty of Chemical and Natural Resources Engineering, Universiti Malaysia Pahang, 26300 Kuantan, Malaysia

E-mail: zakir1703@yahoo.com

Abstract. In this work, we studied the effect of surface treatment on the aluminium surface and a coupling agent to improve adhesion between aluminium with organic polymer. Thermoplastic natural rubber (TPNR) matrix was prepared by melt blending of natural rubber (NR), liquid natural rubber (LNR) compatibilizer, linear low density polyethylene (LLDPE) and polyethylene grafted maleic anhydride (PE-g-MAH). The PEgMAH concentration used was varied from 0% - 25%. In addition, the aluminium surface was pre-treated with 3-glycidoxy propyl trimethoxy silane (3-GPS) to enhance the mechanical properties of laminated composite. It was found that the shear strength of single lap joint Al-TPNR laminated composite showing an increasing trend as a function of PE-g-MAH contents for the 3-GPS surface treated aluminium. Moreover, the scanning electron microscope (SEM) revealed that the strength improvement was associated with the chemical state of the compound involved.

1. Introduction

Recently, a lot of attention has been focused on the development of polymer matrix composite with metals (sandwich/laminate composite). These composites are used in a wide range of applications such as aerospace, automotive, implants, electronic, packaging, defense and medical. As compared to homogenous metal sheets, the metal-polymer laminated composite have many unique properties in terms of low density and weight, high bending strength as well as good sound, vibration and damping characteristics. [1-5].

The preparation of suitable metal-polymer laminate requires many practical considerations. It has been established that one of the most important considerations is the pretreatment process of metal surface either by mechanical, chemical, electrochemical and coupling agent [6-11,14,15]. Usually polar materials with good strength are selected as a polymer matrix. Functional groups like C=O, OH, COOH, NH_3, etc., make a polar polymer with inherent surface strength better than the non-polar polymers. The non-polar polymers can be surface modified by surface oxidizing, chemical etching,

[1] To whom any correspondence should be addressed.

surface grafting, flame treatment, electron beam, microwave irradiation, etc. [1,3,10,11]. Apart from these treatments, the adhesion between polymeric material with the primary metal structures also requires some attention. It is well-known that the polyethylene exhibits insufficient adhesive bond strength due to low surface energy. This property drawback has been solved by chemical, thermal, mechanical and electrical treatments. However, the dry treatment method is more suitable for industrial applications. Previous works have shown the effect of maleic anhydride coupling agent [4,10,11] by imparting polarity to a polyolefin surface that enables an improved adhesive bonding with aluminium and pretreated aluminium surface [6-8]. In the present work, we investigated the potential use of polyethylene grafted maleic anhydride (PE-g-MAH) coupling agent in the thermoplastic natural rubber (TPNR) LLDPE/NR/LNR matrix for Al-TPNR laminated composite. In the addition, the 3-GPS used as pretreatment to the surface of ground aluminium sheet in order to improve the shear strength of single lap joint aluminium-TPNR laminated composite.

2. Experimental

2.1. Material

Natural rubber (NR) was obtained from the Rubber Research Institute of Malaysia (RRIM). Liquid natural rubber (LNR) was prepared in our laboratory by photochemical oxidation of natural rubber (NR) [13]. Linear low density polyethylene (LLDPE) from Exxon Mobil Chemical Manufacturing Singapore was used. The metal used is a pure aluminium sheet with specifications; thickness 1.5 mm, serial number AA1100 (Al = 99.00%; Si & Fe = 0.05%, Mn = 0.05%; Zn = 0.10%; Cu = 0.05-0.2% and other content = 0.05%). Other materials used are polyethylene-grafted-maleic anhydride, PE-g-MAH (melting point = 105 °C; d = 0.925 g/mL) and silane coupling agent of 3-glycidoxy-propyl-trimethoxy-silane (3-GPS), $[NH_2(CH_2)_3Si(OCH_3)_3]$, both are manufactured in Aldrich Chemistry USA.

2.2. Preparation of Al-TPNR laminated composite

Thermoplastic natural rubber (TPNR); LLDPE, NR, LNR (80 : 15 : 5) and PE-g-MAH were blended by using an internal mixer, ThermoHaake, at 140 °C, rotor speed 50 $r.min^{-1}$ and mixing time of 12 minutes. Various concentrations of PE-g-MAH (0%, 5%, 10%, 15%, 20% and 25%) were incorporated into the compounds. The matrix is made of thin layers using a compression molding at a temperature of 160 °C, pressure 0.5 MPa for 15 min. Aluminium sheets were cut to single lap shear tensile test, following to the standard of ASTM D 1002-72 for the sandwich structure. The aluminium sheets were boiled in distilled water for 30 min, cleaned with acetone at room temperature, and then ground with sandpaper of P800A. After that, the aluminium sheets were cleaned again with acetone at room temperature and then dried in the ambient air. Several sheets of aluminium were treated by dipping into a solution of 3-GPS 1% (in ethanol), for 5 minutes, then stirring continuously for another 30 minutes. The sample matrix was then placed between two aluminium sheets and compressed to 10 minutes under the pressure of 0.5MPa, at 160 °C temperature.

2.3. Characterization of Al-TPNR laminated composite

Tensile properties were measured using Testometric M500-50CT tester with a crosshead speed of 1 mm/min. Fracture surface morphology of single lap shear tensile samples was carried out by using LEO 1450VP scanning electron microscope (SEM). Various elements, functional groups and bonds formed on the top surface of the aluminium were investigated using X-ray photoelectron spectroscopy (XPS) AXIS Ultra DLD and spectrum 400 FTIR with ATR method analysis.

3. Result and discussion

3.1. Lap Shear Strength

The shear strength of the single lap joint of laminate composite was performed to analyze the effect of surface roughness, polymer type and metal surface treatment to the bond strength between metal and polymer (interface adhesion mechanism). It has been widely postulated that the creation of covalent bonds at the interface is sufficient for creating viable adhesive strength in adhesion-related applications [3,11,14].

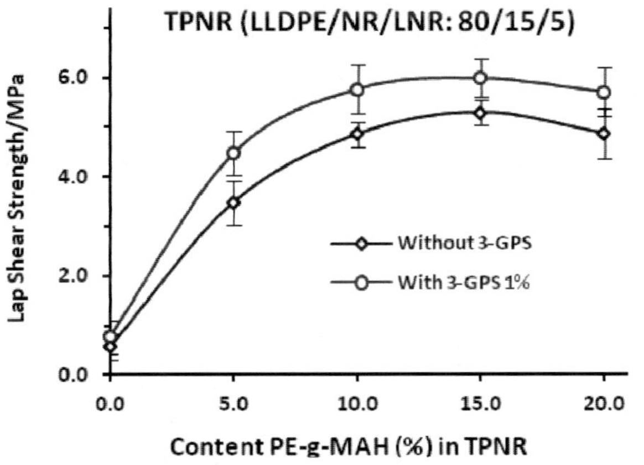

Figure 1. The effect of the PE-g-MAH content (%) on the shear strength of the single lap joint of Al-TPNR laminated composite: (–o–) with pretreatment 3-GPS 1% solution, (–◊–) without pretreatment by 3-GPS solution.

Figure 1 shows the shear strength of single lap joint Al-TPNR laminated composite with TPNR plus PE-g-MAH and the coupling agent, 3-GPS, which was added as a pretreatment on the surface of the aluminium sheet. With the addition of 15% PE-g-MAH into the TPNR, the lap shear strength increased from 0.58 MPa to 5.98 MPa as compared to TPNR system only. This remarkable increment of 931% was further enhanced for the laminate composite treated with 3-GPS 1% coupling agent. The lap shear strength was further increased to 13% against the untreated laminate composite. This could be contributed by the physical interactions, such as the van der Waals interaction between TPNR and surfactants of the pretreated aluminium sheets as well as the mechanical interlocking between TPNR and surface roughness of the aluminium sheets at the interface. With the addition of grafting maleic anhydride into polyolefin and coupling agent on the aluminium surface, the increase in the lap shear strength should be ascribed to the contribution of chemical interactions at the interface [3,5,9,11,15]. The contribution includes the chemical interactions and physical strength at the interface.

3.2. XPS analysis

An XPS spectrometer of the aluminium surface after the pretreatment procedure is shown in the Figure 2(a) and 2(b) respectively. Al 2p spectra after pretreatment showed the formation of Al-OOH, Al-Si bond (Al_2SiO_5), while C 1s showed formation of CO, COC, COH and COOH on the surface of aluminum. The formation of -OH functional groups on the surface of the aluminium, resulted in physical interaction, such as the van der Waals interaction between TPNR or PE-g-MAH and surfactants (3-GPS) of the pretreated aluminium sheets at the interface. This has contributed to the interfacial adhesive strength improvement if chemical interactions at the interface existed [4,6,10,14,15].

Journal of Physics: Conference Series **423** (2013) 012041 doi:10.1088/1742-6596/423/1/012041

3.3. FTIR spectroscopy analysis

Figure 3 shows FTIR spectra of LLDPE, PE-g-MAH, NR, TPNR (LLDPE/NR/LNR) and TPNR plus PE-g-MAH (LLDPE/NR/LNR/PE-g-MAH). The functional groups that are related to the chemical state of Al-TPNR laminated composite interfaces such as a C=O can be characterized in the frequency 1709 cm^{-1} (for PE-g-MAH), 1645 cm^{-1} (for system LLDPE/NR/LNR) and –OH in the frequency 3394 cm^{-1} respectively. The appearance of functional groups C=O and –OH was believed to have formed during mixing as a result of material oxidation. These functional groups are important in the physical and chemical interactions in the bonding interface of Al-TPNR laminated composite [3,4,10].

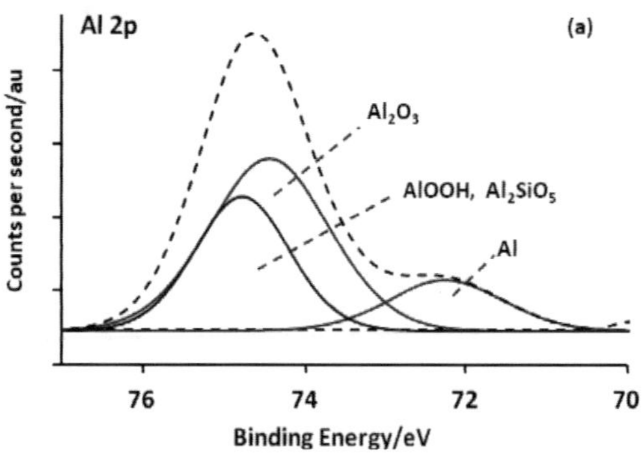

Figure 2a. XPS spectra Al 2p of on the aluminium surface with 3-GPS, separated polymer surface Al-TPNR laminated composite.

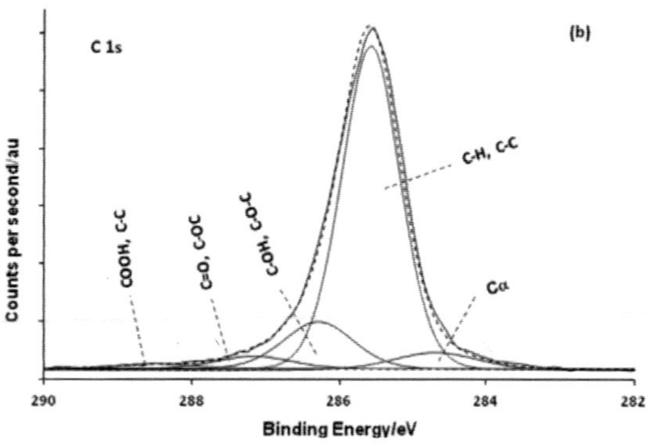

Figure 2b. XPS spectra C 1s of on the aluminium surface with 3-GPS, separated polymer surface Al-TPNR laminated composite

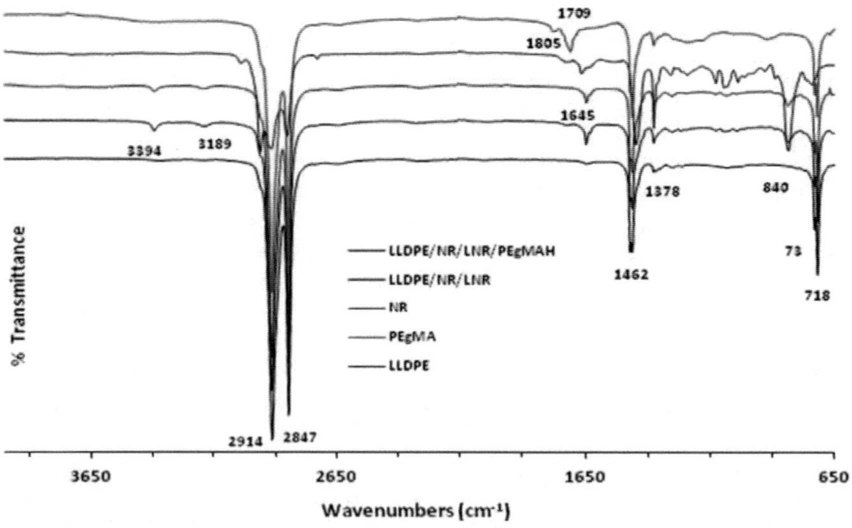

Figure 3. FTIR spectra for LLDPE, PEgMAH, NR, LLDPE/NR/LNR and LLDPE/NR/LNR/PEgMAH.

3.4. SEM analysis

Figure 4 (a, b) shows the SEM micrographs of the lap shear fractured aluminium surfaces without PE-g-MAH in TPNR matrix and coupling agent, 3-GPS. It can be seen in Figure 4 (a) that almost all aluminum surfaces are clearly visible and almost no inherent TPNR. In contrast, when the PE-g-MAH was added and aluminum surfaces pretreated with 3-GPS, it appears the fracture surface of aluminum remains as a strong matrix inherent as shown in Figure 4 (b). Chen et al [3] reported that pretreatment aluminum surface with the addition of coupling agent and maleic anhydride grafted into the plastic has enhanced the adhesion strength between polymer matrices with metal; hence an improved composite laminate could be obtained [10,15].

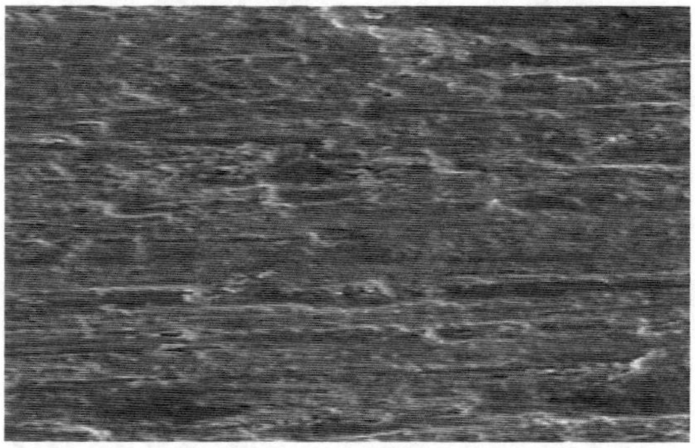

Figure 4(a). SEM images of lap shear fractured surface of the adhering aluminum sheet without 3-GPS and PE-g-MAH.

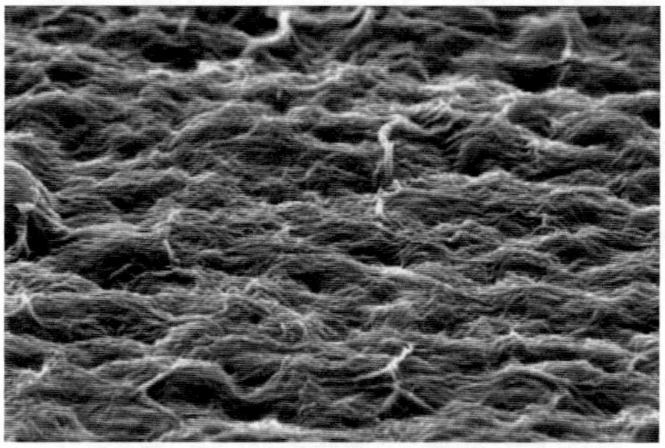

Figure 4(b). SEM images of lap shear fractured surface of adhered aluminium sheet with 3-GPS and PE-g-MAH 1 wt% added in TPNR matrix.

4. Conclusions

The shear strength of single lap joint Al-TPNR laminated composite increased by 931% when the PE-g-MAH was added at 15% into the TPNR. With the 3-GPS 1% coupling agent on the treated aluminium surface, the lap shear strength was further increased by another 13% as compared to the untreated laminate composite. These phenomena could be associated with the chemical interaction between the AlOOH, COOH, CO, COC, COH and Al-Si (Al_2SiO_5) as confirm by the XPS, FTIR and SEM analysis.

Acknowledgements

The authors would like to thank the Ministry of Education and Culture, Directorate Jenderal Higher Education (DIKTI) Republic of Indonesia and Universiti Kebangsaan Malaysia for sponsoring this research project (UKM-DLP-2012-029).

References

[1] Wolf R A 2006 Enercon Industries Corporation. Improving adhesion performance between low surface tension composite and dissimilar substrates. *ANTEC.* 1256–59.

[2] Weiss M, Dingle M E, Rolfe B F and Hodgson P D 2007 The Influence of Temperature on the Forming Behavior of Metal/Polymer Laminates in Sheet Metal Forming. *J. Eng. Mat. Tech.* **129** 530–37.

[3] Chen M A, Li H Z, Zhang X M 2007 Improvement of shear strength of aluminium-polypropylene lap joints by grafting maleic anhydride onto polypropylene. *Inter. J. Adhesion and Adhesives*. **27** 175–87.

[4] Ochoa P C, Vaidya U 2011 Mechanisms of interfacial adhesion in metal–polymer composites – Effect of chemical treatment. *Composites: Part A.* **42** 906–15

[5] Parsa M H, Ahkami S N, Ettehad M 2010 Experimental and finite element study on the spring back of double curved aluminum/polypropylene/aluminum sandwich sheet. *Mat. & Design.* **31** 4174–83.

[6] Abel M L, Joannic R, Fayosb M, Lafontaineb E, Shawc S J, Watts J F 2006 Effect of solvent nature on the interaction of γ-glycidoxy propyl trimethoxy silane on oxidized aluminium surface: A study by solution chemistry and surface analysis. *Inter. J. Adhesion and Adhesives.* **26** 16–27.

[7] Prolongo S G, Urena A 2009 Effect of surface pre-treatment on the adhesive strength of epoxy–aluminium joints. *Inter. J. Adhesion and Adhesives.* **29** 23–31.

[8] Sinmazçelik T, Avcu E, Bora M Ö, Çoban O 2011 A review: Fibre metal laminates, background, bonding types and applied test methods. *Mater. and Design.* **32** 3671–85.

[9] Bhowmik S, Chaki T K, Ray S, Hoffman F, Dorn L 2004 Experimental investigation into the effect of DC glow discharge pretreatment of HDPE on tensile lap shear strength. *Inter. J. Adhesion and Adhesives.* **24** 461–470.

[10] Chen M A, Xie X, Zhang X M 2009 Interactions of BTESPT silane and maleic anhydride grafted polypropylene with epoxy and application to improve adhesive durability between epoxy and aluminium sheet. *Prog. Org. Coat.* **66** 40–51.

[11] Gao H, Xie Y, Wang Q 2012 Grafting effects of polypropylene/polyethylene blends with maleic anhydride on the properties of the resulting wood–plastic composites. *Composites Part A: Appl. Sci. Manufac.* **43** 150–157.

[12] Wapner K, Stratmann M, Grundmeier G 2007 Structure and stability of adhesion promoting aminopropyl phosphonate layers at polymer/aluminium oxide interfaces. *Inter. J. Adhesion and Adhesives.* **28** 59–70.

[13] Seng L Y, Ahmad S, Rasid R, Noum S Y E, Hock Y C, Tarawneh M A 2011 Effects of Liquid Natural Rubber (LNR) on the Mechanical Properties of LNR Toughened Epoxy Composite. *Sains Malaysiana.* **40** 7 679–683

[14] Carradò A, Faerber J, Niemeyer S, Ziegmann G, Palkowski H 2011 Metal/polymer/metal hybrid systems: Towards potential formability applications. *Comp. Struc.* **93** 715 – 21.

[15] Xie Y, Hill C A S, Xiao Z, Militz H, Mai C 2010 Silane coupling agents used for natural fiber/polymer composites: A review. *Composites: Part A.* **41** 806–819.

Journal of Physics: Conference Series **423** (2013) 012042 doi:10.1088/1742-6596/423/1/012042

Wettability Effect of PECVD-SiO_x Films on Poly(lactic acid) Induced by Oxygen Plasma on Protein Adsorption and Cell Attachment

S Sarapirom[1], J S Lee[2], S B Jin[2], D H Song[3], L D Yu[1,4], J G Han[2] and C Chaiwong[1,4]

[1] Plasma and Beam Physics Research Facility, Department of Physics and Materials Science, Faculty of Science, Chiang Mai University, Chiang Mai 50200, Thailand

[2] Institute for Plasma-Nano Materials, Center of Advanced Plasma Surface Technology, Sungkyunkwan University, 300 ChunChun-dong, Jangan-gu, Suwon 440-746, Korea

[3] Department and Research Institute of Dental Biomaterials and Bioengineering, College of Dentistry, Yonsei University, 250 Seongsanno, Seodaemun-gu, Seoul 120-752, Korea

[4] Thailand Center of Excellence in Physics, Commission on Higher Education, 328 Si Ayutthaya Rd., Bangkok 10400, Thailand

E-mail: cchwng@gmail.com

Abstract. Surface wettability is an important property of biomaterials. Silicon oxide films have a wide range of applications due to a range of the properties such as the mechanical strength and surface wettability. This paper reports effect of the surface wettability of silicon oxide (SiO_x) films on protein adsorption and cell attachment and proliferation. SiO_x films were deposited onto poly(lactic acid) (PLA) substrate using plasma enhanced chemical vapor deposition (PECVD). Octamethylcyclotetrasiloxane (OMCTS:$Si_4O_4C_8H_{24}$) was used as a precursor with O_2 as a carrier gas. After deposition, the films were treated with O_2-plasma to adapt wettability. It was found that O_2-plasma enhanced the wettability of the films without changing the film thickness, while made the surface morphology slightly smoother. The polar component increased after O_2-plasma treatment as observed in the contact angle measurements. The surface energy of the films was calculated by means of the Owens-Wendt method to resolve the contributions of polar and dispersive components. The chemical structure was characterized using attenuated total reflectance-Fourier transform infrared (ATR-FTIR) spectroscopy. The films were dense with a high Si-network structure. The reduced carbon content (-CH_n, Si-CH_3) and increased hydrogen content (-OH) of the O_2-plasma treated SiO_x films led to the polar components enhancing the SiO_x wettability. Adsorption of bovine serum albumin (BSA) on the films was investigated by using x-ray photoelectron spectroscopy (XPS). More BSA was adsorbed onto the O_2-plasma treated SiO_x films. Attachment and proliferation of MC3T3-E1 mouse pre-osteoblasts and L929 mouse fibroblasts cells on the SiO_x films were evaluated via MTT assay. The cells were attached more to the untreated SiO_x films but proliferated more on the surface of the O_2-plasma treated SiO_x films depending on the cell types.

1. Introduction

Biomaterials have been important in transplantation medicine, tissue engineering, and bioartificial tissues. For advanced applications, they should not only be biocompatible but also specifically designed for peculiar cellular response, such as cell attachment, proliferation, and differentiation. Common interaction between material and biological systems occurs in a wide range of applications. While biomaterial is in contact with a biological environment, protein adsorption first takes place before other processes such as platelet or blood cell adhesion, and then cells contact with a protein layer on the material surface. Consequently, protein-surface interaction is critical to understanding for control and design of biomaterials.

There are several methods to modify the surfaces properties of biomaterials [1, 2], and each way depends on the objectives. For example, adhesive proteins (fibronectin, vitronectin or collagen) are immobilized onto different wettability surface for improving cells attachment or promoting specific interaction for biosensor/microarray [3-5], and bovine serum albumin is immobilized on hydrophobic and hydrophilic surface for blocking of non-pacific proteins [6]. Moreover, thin film coating on the materials can lead to excellent mechanical properties and biocompatibility of biomaterials. For instance, titanium dioxide improved cells-surface interaction [7], and silicon oxide films promoted selective adhesion of protein or cells to materials [8, 9].

Thin films of silicon-organic compounds are used in a range of applications. The characteristics of the films can be tailored by changing deposition process parameters. Polymer-like films are applied in optoelectronic devices, and protective coatings for different substrates. Quartz-like films can be used as diffusion barriers [10, 11]. In recent years, there is a growing interest in developing silicon oxide films for use in biomedical applications such as bioelectronics, and biofunctional materials. Plasma enhanced chemical vapor deposition (PECVD) is widely used to deposit silicon oxide films on polymeric biomaterials [12] because it can adjust the mechanical properties with the deposition parameters. Oxygen (O_2) plasma was used to adapt wettability of SiO_x films because O_2 plasma enhanced higher O_2-containing functionalities and more drastic morphology change leading to proteins and cells attachment [13, 14].

The adsorption of proteins in the blood plasma on the biomaterial surface is, in general, a dynamic phenomenon since there is a competition in the adsorption process of such proteins. Albumin (Alb) is the most abundant protein in blood plasma and able to bind with other molecules. It is usually employed as a model protein to be studied in protein-surface interactions. In an aqueous environment, proteins will tend to adopt both a hydrophobic and a hydrophilic surface. Proteins can be irreversibly bound to a hydrophobic surface through the dehydration of the interface and undergoes conformation changes. According to the Vroman effect, low-molecular-weight protein such as Alb is first adsorbed on the surface then higher molecular weight proteins, such as fibronectin [15]. Thus, if Alb is adsorbed with irreversible bonds, the cell adhesion protein can follow and enhance cell attachment and proliferation.

In this work, we were interested in and thus studied effect of wettability of silicon oxide films on protein adsorption, cell attachment and proliferation. Post-deposition O_2-plasma treatment was applied to modify the surface wettability of SiO_x films which were deposited on polymer using PECVD. The research results would be beneficial to certain biomedicine application potentials such as polymeric cell-culture dishes and biosensors whose cell attachment and proliferation properties could be modified by depositing SiO_x thin films.

2. Experiment

2.1. Materials preparation and modification

Poly(lactic acid) (PLA) membranes of (thickness 50 μm), obtained from the School of Mechanical Engineering, Institute of Engineering, Suranaree University of Technology, Thailand were used as the substrate. The membranes were cut into 1 cm × 4 cm strips.

SiO_x films were deposited on the PLA membranes by PECVD in Sungkyunkwan University, Korea using octamethylcyclotetrasiloxane (OMCTS: $Si_4O_4C_8H_{24}$) with oxygen as a carrier gas. The details of the deposition has been described elsewhere [12]. In brief, the membrane substrate samples were put on a glass slide sample holder in the PECVD chamber and the chamber was evacuated to a base pressure of 0.01 Pa. To remove the residues and increase films adhesion, the membranes were first sputtered using O_2-plasma at 3.3 Pa by a bottom electrode with a radio frequency (RF) power of 60 W for 20 seconds. After that, SiO_x films were deposited at an OMCTS vapor pressure of 3.3 Pa by a top electrode with an RF power of 120 W and a bottom electrode with an RF power of 60 W for 40 seconds. Then, for increasing the hydrophilicity, a part of the SiO_x films were further treated using O_2-plasma at 3.3 Pa and a bottom electrode with an RF power of 60 W for 60 seconds.

2.2. Surface characterization

The thickness of the deposited film was measured using an Alpha-Step IQ Surface Profiler (KLA-Tencor, USA). The film surface morphology was observed using atomic force microscopy (AFM) (Thermo Microscope Autoprobe CP-Research, USA). The AFM images were acquired in the contact mode using silicon tips with a scan rate of 1 Hz in air. The images were analyzed to measure the root mean square (Rms) and the surface-peak-to-valley roughness (Rp-v).

The static contact angles were measured with 2 µl of deionized water and diiodomethane (CH_2I_2), respectively, at room temperature. The image of the fluid drops was captured and analyzed by using the image analysis software. The surface energy was calculated using the Owens-Wendt equations [16, 17], which are expressed as follows:

$$\gamma_W\left(1+\cos\theta_W\right)=2\sqrt{\gamma_S^d\gamma_W^d}+2\sqrt{\gamma_S^p\gamma_W^p} \quad (1)$$

$$\gamma_D\left(1+\cos\theta_D\right)=2\sqrt{\gamma_S^d\gamma_D^d}+2\sqrt{\gamma_S^p\gamma_D^p} \quad (2)$$

where γ is the surface energy, θ is the contact angle, and the superscripts, d and p, represent the dispersive component and the polar component of the surface energy, respectively. The Subscript S, D, and W indicate solid, diiodomethane and water, erspectively. For water, γ_W = 72.8 mJ/m^2, γ_W^d = 22.1 mJ/m^2, and γ_W^p = 50.7 mJ/m^2. For diiodomethane, γ_D = 50.8 mJ/m^2, γ_D^p = 44.1 mJ/m^2, and γ_D^d = 6.7 mJ/m^2 [18].

The chemical structure of the film was characterized using attenuated total reflectance-Fourier transform infrared (ATR-FTIR) spectroscopy (Bruker-Optics) with germanium crystal. Each spectrum was obtained using an average of 64 scans in the range of 400-4000 cm^{-1} with a resolution of 4 cm^{-1}.

2.3. Protein adsorption analysis

The bovine serum albumin (BSA)(A2153) was purchased from Sigma-Aldrich Corporation, Germany. Strips of the films were cut into 1 cm × 1 cm and placed into wells of a 4-well plate. The protein was immobilized onto the films by adding 1 ml of BSA solution (0.5% W/V in distilled water) into each well and then incubated for 20 min on a rocking shaker at room temperature. After incubation, the films were washed in distilled water for 20 min, first soaking and then rinsing 3 times. The films were dried in a new plate and left at room temperature for 12 h. Some films were analyzed for the chemical structure using x-ray photoelectron spectroscopy (XPS) (Shimadzu, Japan) [19, 20]. Survey spectra of XPS were acquired from 0-1200 eV, with a pass energy of 80 eV and step size of 1 eV. The core level spectra (high-resolution spectra) were obtained with a pass energy of 20 eV and step size of 0.1 eV. Elemental compositions were calculated from peak areas obtained from the survey spectra. All XPS peaks were referenced to a C_{1s} signal at a binding energy of 284.6 eV, corresponding to the C-C and C-H bonds in hydrocarbons. The peaks were deconvoluted into Gaussian components to gain the insight into the bonding.

2.4. Cell culture

Fibroblast cells are normally used as a model of cell attachment and proliferation since they are the most common cells of connective tissues in animals and able to grow in a high rate. In this investigation, mouse fibroblast cells (L929) and mouse pre-osteoblast cells (MC3T3-E1) were used to study cell cultivation on the surfaces of the as-deposited SiO_x and O_2-plasma treated SiO_x films. Use of the fibroblast cells was aimed at artificial tissue test and use of the osteoblast cells was aimed at scaffold test. The cell attachment and proliferation were evaluated using MTT assay. The films were first sterilized under ultraviolet (UV) light for 1 h (both faces). The cells were diluted to 5×10^3 cells/ml. The culture medium was modified with 10% fetal bovine serum and 1% antibiotic antimycotic solution (WelGENE, Korea). The cells were cultured at 37°C in a 5%-humidified CO_2-atmosphere incubator. Then cell seeding films were measured with a fluorescence spectrometer at wavelengths of 544 and 590 nm. In these experiments, 3 replicates were used for calculating the percentage of each experiment. The cell attachment was observed in 1 day and cell proliferations were observed in 3 and 7 days.

3. Results and discussion

3.1. Surface characterization

The thickness of the as-deposited SiO_x films was 48.2±0.5 nm and that of the O_2-plasma treated SiO_x films was 47.8±0.8 nm as measured by the surface profiler. The film thickness was decreased by O_2-plasma treatment insignificantly, indicating that the O_2-plasma treatment did not have noticeable effect on changing the thickness of SiO_x films. Surface morphology of the films was observed by AFM with a scanning area of 3μm × 3μm. Roughness data were obtained from different regions on each sample. Figure1 shows the AFM images of the as-deposited SiO_x film and the O_2-plasma treated SiO_x film. The O_2-plasma treatment decreased the surface roughness (root-mean-square, R_{rms}) of the films from 26 nm to 19 nm due to plasma etching effect.

The material surface energy consists of polar and non-polar components. The polar component of surface energy comprises all other interactions due to the non-London forces. Polar molecules interact with dipole forces and hydrogen bonds. The dispersive or non-polar components of surface energy result from molecular interactions due to the London forces [21]. Figure 2a shows the results of the contact angle measurements. The water contact angle of the as-deposited SiO_x films was 57.0 degree and after O_2-plasma treatment the water contact angle obtained was significantly decreased down to 2.7 degrees, showing a significant increase in the hydrophilicity of the film. Figure 2b shows the components of the surface energy of the films. The polar components of the surface energy increased after O_2-plasma treatment. The total surface energy was higher after the O_2-plasma treatments, demonstrating that the surface of SiO_x became more hydrophilic. The enhanced hydrophilicity of the O_2-plasma treated SiO_x films was owing to increased polarity, as there were more polar components in the total surface energy. This could result from the incorporation of polar groups such as hydroxyl group on the surface.

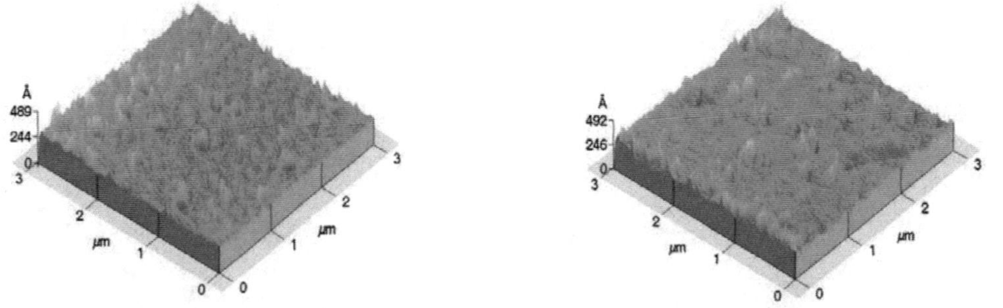

(a) (b)

Figure 1. AFM images of the surface morphology of (a) as-deposited SiO_x and (b) O_2-plasma treated SiO_x films.

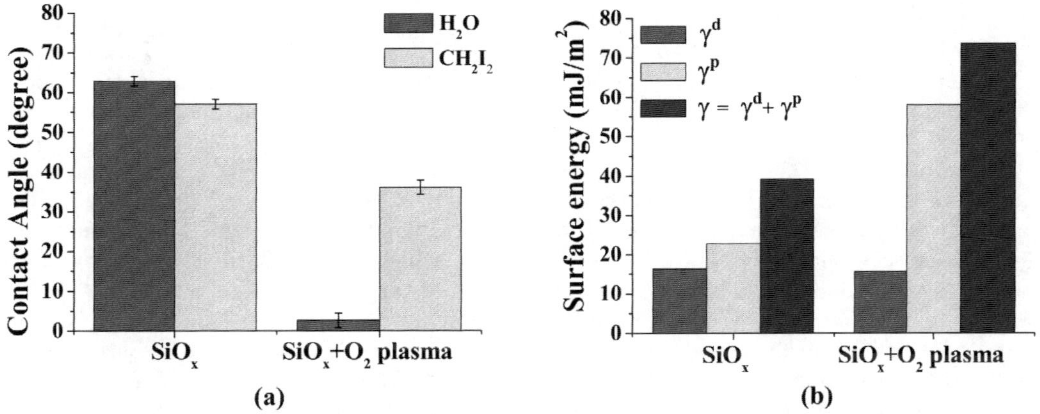

Figure 2. Surface properties of the SiO_x films. (a) Contact angle of water (H_2O) and diiodomethane (CH_2I_2) on the as-deposited SiO_x film and O_2-plasma treated SiO_x film. (b) Surface energy of the as-deposited SiO_x film and O_2-plasma treated SiO_x film, including dispersive components : γ_S^d and polar components : γ_S^p.

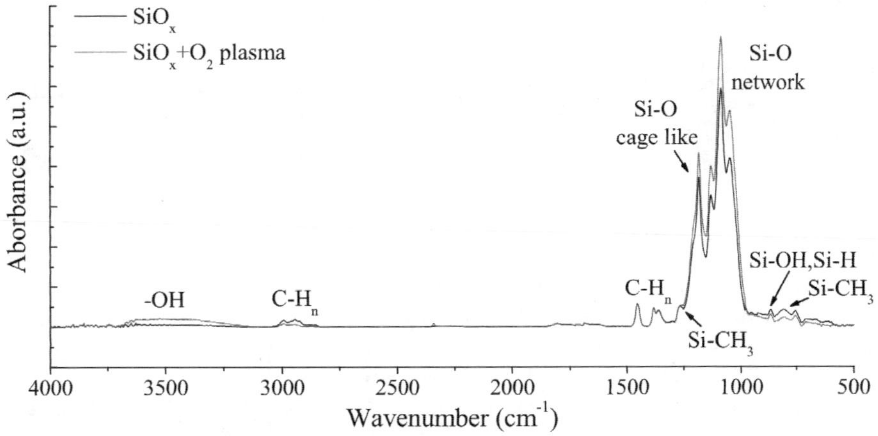

Figure 3. ATR- FTIR spectra of the as-deposited SiO_x and O_2-plasma treated SiO_x films.

The ATR-FTIR spectra of the as-deposited SiO_x film and O_2-plasma treated SiO_x film are show in Figure 3. The absorbance spectra of OMCTS were collected from 500-1350 cm^{-1} and compared with the spectral data of the known structures. In the PECVD of the films, a high ion current density enhanced the film density and hardness [10]. In this case, the Si-O network structure and Si-O cage-like structure showed very strong absorption bands in the range of 960-1250 cm^{-1}. The Si-O structure comprised Si-O-Si (1050 cm^{-1}), ring link Si-O-C (1085 cm^{-1}), open link Si-O-C (1128 cm^{-1}), and cage link Si-O-C (1180 cm^{-1}). The alkyl groups as $Si-CH_3$ and $-CH_n$ were found in the small region. The peaks at 754, 810, and 1268 cm^{-1} could be assigned to the Si-C stretching and the $-CH_3$ rocking modes as $Si-(CH_3)$ and $Si-(CH_3)_2$ [22]. The small peak at 874 cm^{-1} could be assigned to Si-OH and Si-H which indicated the incorporation of some moisture into the oxide films [23] which were related to the broad peak of -OH bond between 3150 and 3600 cm^{-1}. The peak between 2830 and 3025 cm^{-1} of the $-CH_n$ stretching bond consisted of $-CH_2$ and $-CH_3$. The $-CH_n$ peaks at 1360, 1379, and 1449 cm^{-1} could be assigned to $-CH_2$, $-CH_3$ and $–CH_4$, respectively [22]. It was found that after O_2-plasma treatment, the $-CH_n$ stretching bond and the $Si-CH_3$ bond decreased, whereas the -OH stretching bond increased. This implied that the O_2-plasma treatment reduced the carbon content of Si-O films, and hydrogen from moisture was incorporated into the oxide films and contributed to the polar surface to enhance the SiO_x wettability.

3.2. Cell culture

Cell attachment and proliferation of mouse fibroblast cells (L929) and mouse pre-osteoblast cells (MC3T3-E1) were observed in 1 day and 3-7 days, respectively, and the results are summarized in Figure 4. It showed both types of the cells attached on the as-deposited SiO_x films slightly more than on the O_2-plasma treated SiO_x films due to the lower BSA adsorption on the untreated films. The low BSA adsorptions resulted in the cell adhesion proteins more adsorbed on to the film surface, and thus the cells could have more attachment. But, both types of the cells proliferated on the O_2-plasma treated SiO_x films more than on the untreated as-deposited SiO_x films. Mouse pre-osteoblast cells proliferated slower than mouse fibroblast cells. The results indicated that the cell attachment depended on the protein adsorption on the surfaces, while the cell proliferation depended on the surface wettability.

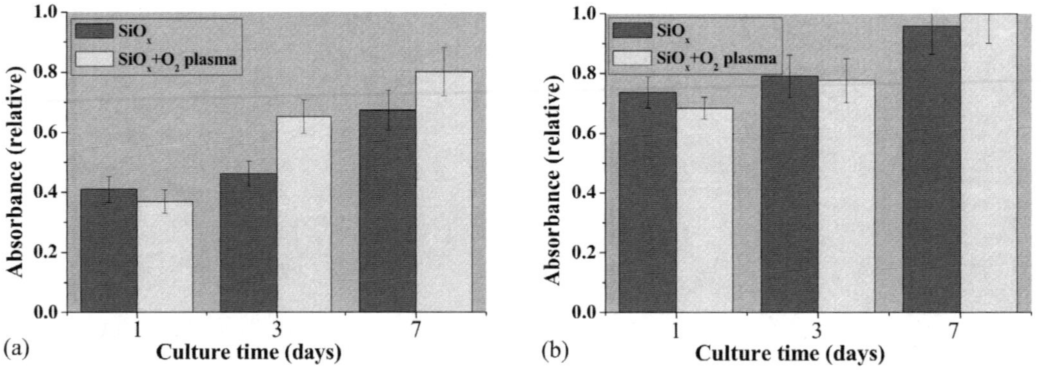

Figure 4. Relative absorbance of (a) L929 mouse fibroblast cells and (b) MC3T3-E1 mouse pre-osteoblat cells on the SiO_x and O_2-plasma treated SiO_x films measured by MTT assay after 1, 3, and 7 days, respectively.

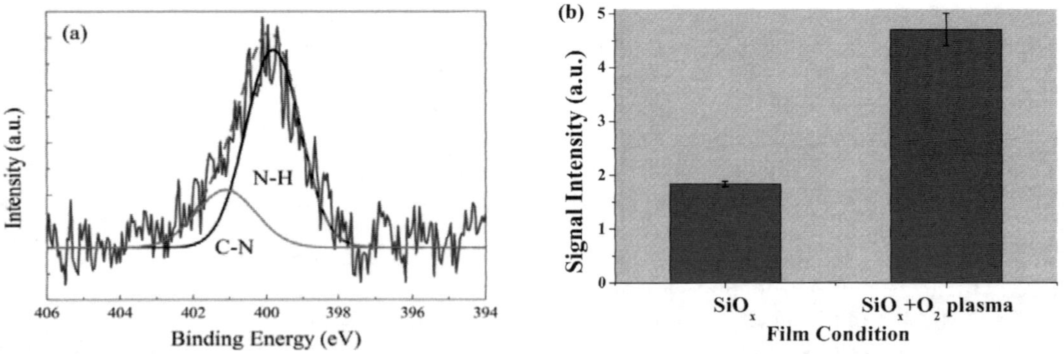

Figure 5. The concentration of the nitrogen bonds on BSA adsorbed on films SiO_x only and SiO_x treated with O_2-plasma measured by XPS. (a) An example of the original XPS spectrum with deconvoluted components. (b) The calculated N-signal intensity indicating the N concentration in the films.

3.3. Protein adsorption

Protein adsorption is a very complex process, depending on environment. There are three major factors: protein, surface and solution, and each factor depends on its properties. As several factors are present, it is quite difficult to focus on only a single factor. However, it may say correctly that the key factor of protein adsorption is the conformation change of proteins or peptides on substrate [24]. This is influenced by kinetic and thermodynamic considerations.

Applications of protein adsorption are almost related to blood plasma proteins. Alb is the most abundant protein in blood with a high concentration of 45 mg/ml, a molecular weight of 66 kDa, and an isoelectric point of 5.5 which has a negative charge in blood of pH 7.4. It is a globular protein in a soft protein type, which has a low internal stability in aqueous solution.

Generally, soft proteins adsorb on a hydrophobic surface rather than a hydrophilic surface, because the proteins tend to conserve their native structure on hydrophilic surface [25-28]. When biomaterials come into an aqueous solution, a water shell will form on the surface in a microsecond. Dehydration of the surface occurs in the next step, and the water structure forms hydrogen bonds with hydrophilic surface. On the hydrophilic surface, Alb will adsorb on the water layer with less tightly bond namely reversible bond. On the hydrophobic surface, water shell cannot form on the surface and Alb adsorbs on a water-free contact layer due to the hydrophobic effect with irreversible bond [29].

There have been many studies on controlling adsorption/desorption proteins to surface with the surface wettability. This study was focused on bovine serum albumin (BSA) adsorption onto SiO_x films with certain wettability. BSA was used to represent mammalian albumin. Some characteristics of BSA adsorption were investigated by XPS. Nitrogen peaks are caused by the presence of the amino acid sequence of BSA molecules. In the XPS spectrum the N1s peak was deconvoluted into two peaks, assigned to the N-H and C-N groups [30]. Figure 5 shows the concentration of the nitrogen bonds on BSA adsorbed on the SiO_x films and O-plasma treated SiO_x films measured by XPS. It is seen that the O-plasma-treated SiO_x films had more nitrogen than the untreated films, indicating the former absorbing more protein than the latter due to a lower energy band gap. The energy band gap has been rarely taken into account for a material surface factor of protein adsorption. Gandhiraman [31] studied the fibrinogen adsorption on $SiO_xC_yH_z$, TiOx and SiO-TiO films with the surface factors of wettability, roughness and energy band gap. They found that fibrinogen adsorption was the highest on the low energy band gap.

4. Conclusion

SiO_x films were deposited on PLA substrate using the PECVD technique and further treated with O_2-plasma to investigate the film surface wettability effect on the protein adsorption and cell attachment and proliferation. The O_2-plasma significantly enhanced the surface wettability of the SiO_x films without changing the thickness of the films but smoothing the surface morphology. Increase in the OH bond was responsible for the enhancement of the wettability due to hydrogen replacement and incorporation into the oxide films to contribute to the polar surface. BSA protein was more adsorbed on the hydrophilic SiO_x films treated by O_2-plasma with the low energy band gap, resulting in less cell attachment but cell-type-dependent cell proliferation, increased for the L929 mouse fibroblast cells but almost no changes for the MC3T3-E1 mouse pre-osteoblat cells. The results provide some hints for designing certain applications of SiO_x films on polymers in biomedicine.

Acknowledgements

This work was supported by the Institute for Plasma-Nano Materials, Center of Advanced Plasma Surface Technology, Sungkyunkwan University, Korea, the Department and Research Institute of Dental Biomaterials and Bioengineering, College of Dentistry, Yonsei University, Korea, the Thailand Center of Excellence in Physics, and the Graduate School of Chiang Mai University. We wish to thank Adcharaporn Boonma of the School of Mechanical Engineering, Institute of Engineering, Suranaree University of Technology for PLA films and Si-Eun Kim for assistance in cell culture.

References

[1] Chen H, Yuan L, Song W, Wu Z and Li D 2008 *Prog Polym Sci.* **33** 1059-87
[2] Goddard J M and Hotchkiss J H 2007 *Prog Polym Sci.* **32** 725
[3] Groth T and Altankov G 1996 *J Biomat Sci-Polym E.* **7** 305
[4] Yu Q, Zhang Y, Wang H, Brash J and Chen H 2011 *Acta Biomater.* **7** 1550-57
[5] Yang J, Bei J and Wang S 2002 *Biomaterials.* **23** 2607-14
[6] Jeyachandran Y L, Mielczarski J A, Mielczarski E and Rai B 2010 *J Colloid Interf Sci.* **341** 142
[7] Ryu G H, Yang W S, Roh H W, Lee I S, Kim J K, Lee G H, Lee D H, Park B J, Lee M S and Park J C 2005 *Surf Coat Tech.* **193** 64
[8] Huang C, Sun Y M, Tsai C Y, Wu S Y, Chang S C and Chang Y C 2012 *Surf Coat Tech.*
[9] Hayakawa T, Yoshinari M and Nemoto K 2004 *Biomaterials.* **25** 127
[10] Jin S B, Choi Y S, Kim Y J, Choi I S and Han J G 2010 *Surf Coat Tech.* **205** S139-43
[11] Chaiwong C, Rachtanapun P, Sarapirom S and Boonyawan D 2012 *Surf Coat Tech.*
[12] Jin S B, Choi Y S, Choi I S and Han J G 2011 *Thin Solid Films.* **519** 6763-68
[13] Satriano C, Marletta G and Kasemo B 2008 *Surf Interface Anal.* **40** 656
[14] Mangindaan D, Yared I, Kurniawan H, Sheu J R and Wang M J 2012 *J Biomed Mater Res A.* **100A** 3177-88
[15] Noh H and Vogler E A 2007 *Biomaterials.* **28** 422
[16] Rudawska A and Jacniacka E 2009 *Int J Adhes Adhes.* **29** 457
[17] Żenkiewicz M 2007 *JAMME.* **24** 145
[18] Jian Yang J B, Shenguo Wang 2002 *Biomaterials.* **23** 2607-14
[19] Vanea E and Simon V 2011 *Applied Surface Science.* **257** 2346-52
[20] Gruian C, Vanea E, Simon S and Simon V 2012 *BBA-Protein Proteomic.* **1824** 881
[21] Bacakova L, Filova E, Parizek M, Ruml T and Svorcik V 2011 *Biotechno Adv.* **29** 767
[22] Kim C Y, Navamathavan R, Lee H J and Choi C K 2008 *Surf Coat Tech.* **202** 5688-92
[23] Matilainen A, Britun N, Jin S B and Han J G 2010 *Surf Coat Tech.* **205** S300-04
[24] Mu Y 2011 *Phys Rev E.* **84** 031906
[25] D'Sa R A, Burke G A and Meenan B J 2010 *Acta Biomater.* **6** 2609-20
[26] Cole M A, Voelcker N H, Thissen H and Griesser H J 2009 *Biomaterials.* **30** 1827-50
[27] Vogler E A 2012 *Biomaterials.* **33** 1201-37
[28] Azioune A, Siroti F, Tanguy J, Jouini M, Chehimi M M, Miksa B and Slomkowski S 2005 *Electrochim Acta.* **50** 1661-67
[29] Werner C, Eichhorn K J, Grundke K, Simon F, Grählert W and Jacobasch H J 1999 *Colloid Surface A.* **156** 17
[30] Indest T, Laine J, Kleinschek K S and Zemljič L F 2010 *Colloid Surface A.* **360** 219
[31] Gandhiraman R P 2007 *PECVD silicon and titanium based coatings to enhance the biocompatibility of blood contacting biomedical devices* Dublin City University (Dublin City) 110

A study of El Niño-Southern oscillation impacts to the South China Sea region using ground-based GPS receiver

Wayan Suparta[1], Ahmad Iskandar[1], Mandeep Singh Jit Singh[1], Mohd. Alauddin Mohd. Ali[1], Baharudin Yatim[1], Fredolin Tangang[2]

[1]Institute of Space Science (ANGKASA), Universiti Kebangsaan Malaysia, 43600 Bangi, Selangor Darul Ehsan, Malaysia

[2]Research Centre for Tropical Climate Change System (IKLIM), Faculty of Science and Technology, Universiti Kebangsaan Malaysia, 43600 Bangi, Selangor Darul Ehsan, Malaysia

Email : wayan@ukm.my

Abstract. We observe an ENSO activity by using ground-based GPS receiver as an effort to study the effects of global warming and climate change in the tropical region. The precipitable water vapor (PWV) derived from Global Positioning System (GPS) meteorology in line with the sea surface temperature anomaly (SSTa) is used to indicate their response on ENSO activities. The PWV data used in this study was taken from the station at Universiti Malaysia Sabah, Kota Kinabalu (UMSK) over 2011, together with NTUS station (in the Singapore), PIMO (in Philippines) and BAKO (in Indonesia) are also compared. The relationship between PWV and SSTa at all stations on weekly basis exhibited modest with correlation coefficients between -0.30 and -0.78 significantly at the 99% confidence level. The negative correlation indicates that during a La Niña phase, the PWV is increased when the sea surface temperatures getting cold causes warm air mass in the central Pacific moved to west Pacific. The increased of PWV causes the GPS signals will be getting slower.

1. Introduction

One of the most significant advances in the natural climate variability in recent years that raised concern around the world is the cycle of El Niño-Southern Oscillation (ENSO). ENSO is a complex phenomenon that results from interaction between ocean and atmosphere in the tropical Pacific Ocean, which consisted of warm (El Niño) and cool (La Niña) phases. Based on the Oceanic Niño Index (ONI), the warm phase is defined with sea surface temperature anomaly (SSTa) greater than 0.5°C, and cool phase is less than -0.5°C. An ENSO cycle is occur with aperiodic scale from 2 to 9 years (Bjerknes, 1969) depending on its location of occurrence. An ENSO phenomenon can be monitored based on the differences between sea level pressure anomalies (SLPa) at Tahiti and Darwin (Ropelewski and Jones, 1987). Other indices to monitor ENSO are the sea surface temperature (SST) in the Pacific Ocean region from central to eastern (Rasmusson and Carpenter, 1982; Reynolds and

Smith 1995), and the zonal gradient of precipitation in the equatorial Pacific (Curtis and Adler, 2000). Although ENSO originates in the tropical Pacific, it affects human life and environment through global climate and weather events such as drought/flooding, landslides and tropical storms. Some aspects of the ENSO mechanisms that affect the South China Sea region are still not well understood.

In this paper, we propose a ground-based Global Positioning System (GPS) technique to monitor ENSO activities. As the GPS signals can be used for geophysical and atmospheric studies; so they have been employed to estimate the atmosphere parameters, such as precipitable water vapor (PWV) (e.g., Bevis *et al.*, 1994; Suparta *et al.*, 2008). For the observation, the PWV data derived from GPS measurements at four selected GPS stations were used. In addition, the sea surface temperature anomaly (SSTa) data associated with the PWV response was used to characterize the ENSO activity. A correlation analysis was employed to indicate PWV responded to ENSO events.

2. Metodology

2.1. Data and Location

The main base of observation in this study is use GPS data located at Universiti Malaysia Sabah, Kota Kinabalu Sabah (UMSK), Malaysia. The other GPS data were taken from the Nanyang Technological University (NTUS) in Singapore, the Manila Observatory (PIMO) in Philippines and Bakosurtanal (BAKO) in Indonesia for comparison. GPS data other than UMSK were downloaded from the Scripps Orbit and Permanent Array Center (SOPAC) site. At the same time, PWV data taken from Radiosonde (RS PWV) were used to compare the GPS PWV. The location of all GPS stations used in this study is depicted in Figure 1. UMSK station is located at the geographical coordinate 6.03°N latitude, 116.12°E longitude and ellipsoidal height of 63.49 m. The geographical coordinates for other stations is summarized in Table 1.

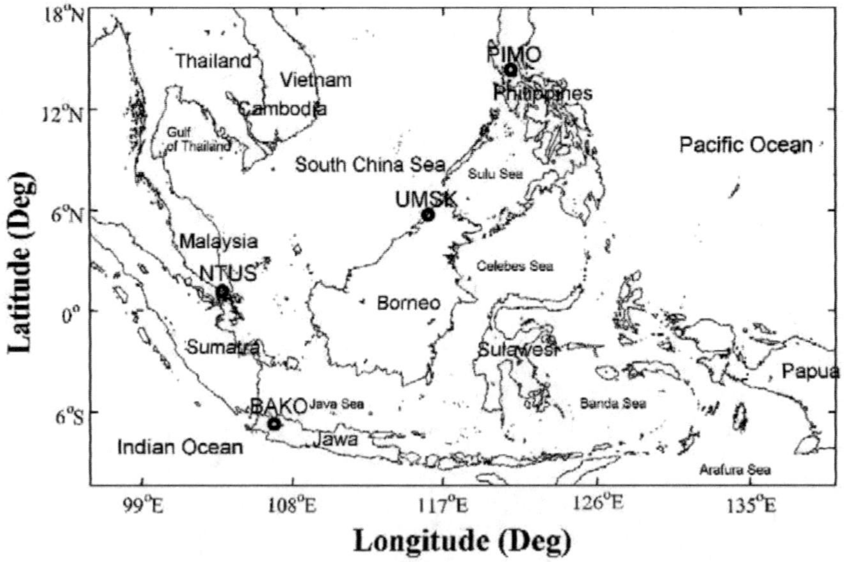

Figure 1. Location of GPS stations used in this study.

Table 1. Geographical coordinates and instruments set up of GPS receivers.

Station (Country)	Latitude (Deg)	Longitude (Deg)	Height (m)	Type of GPS receiver (year)	Cut off elevation-angle (Deg)
UMSK (Malaysia)	6.03 N	116.12 E	63.49	Trimble NetR8 (2011)	13
NTUS (Singapore)	1.35 N	103.68 E	75.38	LEICA GRX1200GGPRO (2007)	0
PIMO (Philippines)	14.64 N	121.08 E	95.53	ASHTECH UZ-12 (2009)	4
BAKO (Indonesia)	6.49 S	106.85 E	158.2	LEICA GRX1200+GNSS (2010)	0

2.2. Data Processing

To calculate the PWV, GPS data must be combined with meteorological data like pressure (*P* in hPa), temperature (*T* in °C) and relative humidity (*H* in percent) to remove the errors during GPS transmission to the ground. The meteorological data were obtained from the meteorology station which same location with GPS station, or downloaded from another source like weather underground site. Figure 1 shows the GPS signal propagation to a receiver on the ground that has been interacting with sea surface. As seen in the figure, when the GPS signal propagates through the Earth s atmosphere, it is affected by the variability of the refractive index of the ionosphere and troposphere. The excess delay of the signal causes bending of the signal and the total delay along the slant path can be determined (Suparta, 2012). In contrast, the total tropospheric delay (ZTD) in the neutral atmosphere, which comprised of zenith hydrostatic delay (ZHD) and zenith wet delay (ZWD) can be calculated based on the improved Modified Hopfield model. The ZHD was calculated using the Saastamoinen model. A Vienna mapping function (VMF1) was employed to reduce the atmospheric bias in the ZTD estimation (Suparta *et al.*, 2011). The ZWD was computed by subtracting the ZHD from the ZTD. The ZWD was then transformed into an estimate PWV by employing the surface temperature measured at a particular site. The total PWV (in mm) from a receiver position to the top of the atmosphere was calculated based on the formula proposed by Bevis *et al.* (1994). Details of PWV determination by means GPS observations for this study can be found in the paper of Suparta *et al.* (2008).

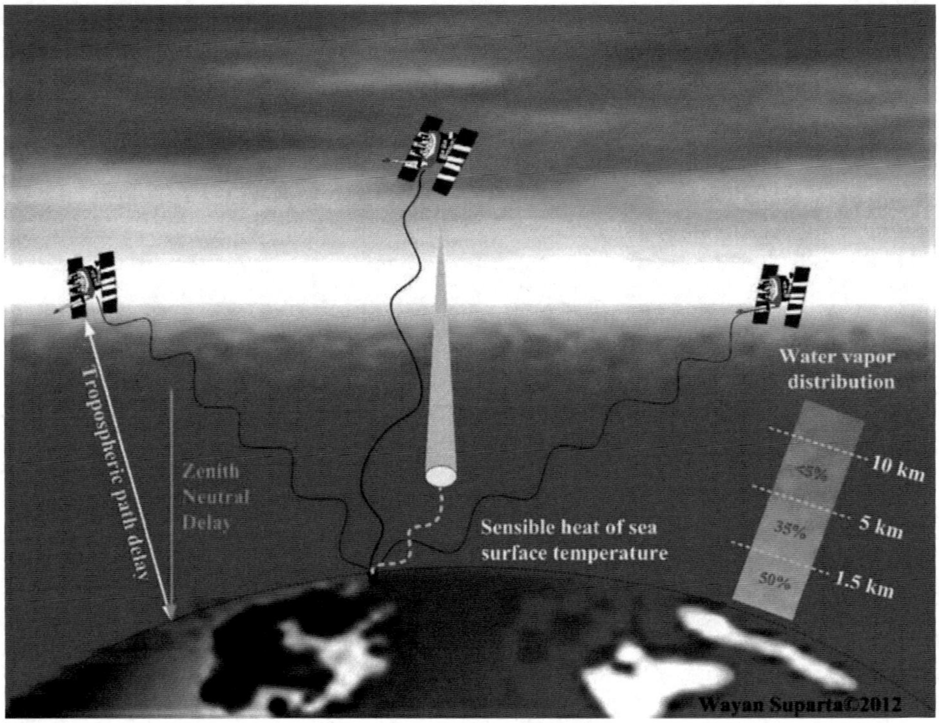

Figure 2. Propagation of GPS signals to a receiver on the ground that covers the sea surface influences.

To process and analyze the above parameters, we used the tropospheric water vapor program (*TroWav,* a set of Matlab codes) developed by Suparta *et al*. (2008), Suparta (2010) and Suparta *et al*. (2011). A schematic representation of the *TroWav* algorithm for processing the PWV is depicted in Figure 3. After cleaning and equalize the size matrix of both GPS and the surface meteorological data, they were processed to obtain PWV. Then to relate the ENSO activity with response to the PWV changes uses the SSTa Oceanic Niño Index (ONI) in pathways of Niño 3.4 and Niño 4 regions. The SSTa data were taken from the National Oceanic and Atmospheric Administration (NOAA). All the data were analyzed on a weekly basis based on GPS week because the SSTa data only provided by NOAA in a weekly.

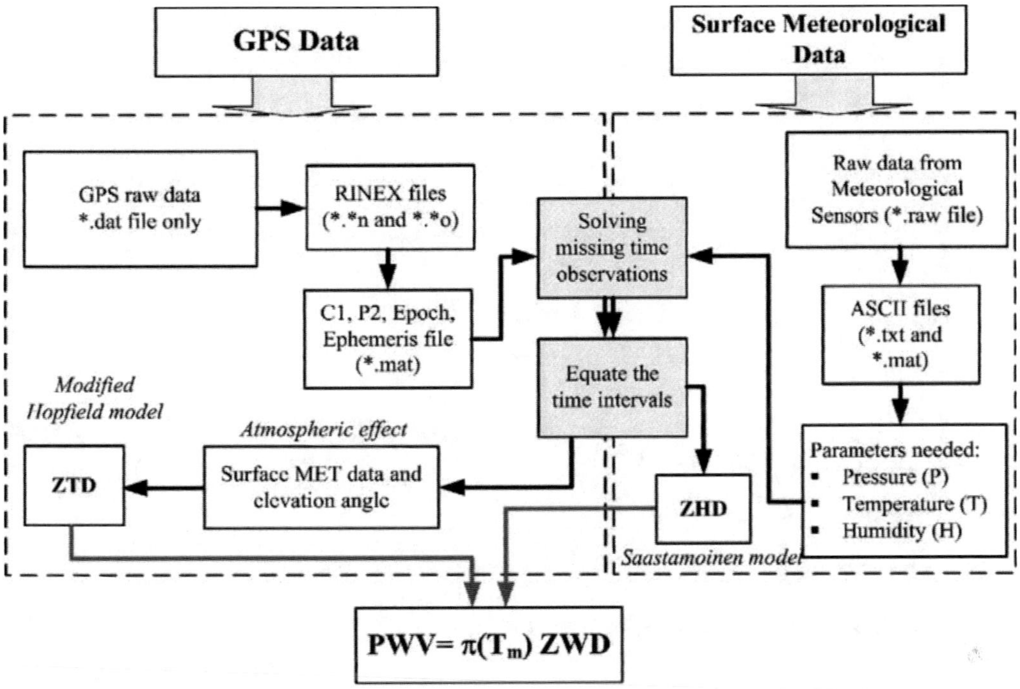

Figure 3. A schematic representation of the *TroWav* algorithm for processing the PWV adapted from Suparta (2010).

3. Result and Discussion

3.1. PWV and SSTa variability

The weekly PWV variability from the GPS and RS measurements is presented in Figure 4. Although the pattern of PWV at each station exhibited changeable, their variation showed a similar trend. For UMSK and NTUS stations, the PWV value is recorded between 45 mm and 55 mm. For PIMO station, the PWV was ranged between 35 mm and 50 mm, while BAKO station was between 35 mm and 45 mm. Table 2 gives the summary of PWV value between GPS and RS. From the table, the PWV value for GPS was ranged between 40 mm and 50 mm (45.76 mm on average) with a STD of 2.66 mm, while for RS was ranged between 37 mm and approximately 60 mm (50.61 mm on average) with a STD of 4.96 mm. The different variability in PWV at each station is due to many factors, including latitude coordinate, topography and the distance between GPS station with the sea. For example, the nearest distance between the stations with the sea will obtained higher variability of water vapor, and conversely, the more distant will lower the PWV variability. This influence gives the indication that the sea is the greatest evaporation, and PWV will move later to mainland by the wind. The further moves ashore, PWV will be reduced because some of the water vapor falls as rains.

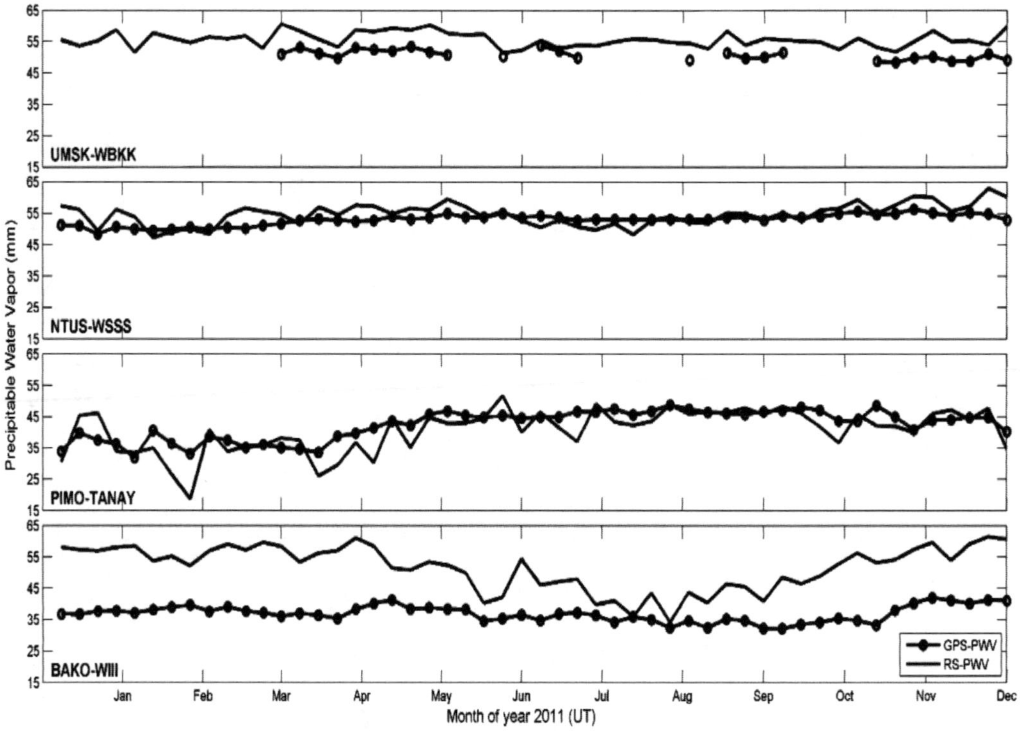

Figure 4. A weekly PWV variability from GPS and RS over the 2011 at eight selected stations.

Table 2. The statistical value of PWV for GPS and RS data for 2011.

Station	Min	Mean	Max	STD
UMSK (G)	48.48	50.81	53.66	1.56
NTUS (G)	48.30	52.97	56.39	1.78
PIMO (G)	31.80	42.39	48.80	4.78
BAKO (G)	32.21	36.90	42.02	2.53
WBKK (R)	51.39	55.66	60.72	2.36
WSSS (R)	47. 18	54.61	63.03	3.45
TANAY (R)	18.67	40.51	51.74	6.92
WIII (R)	34.22	51.68	61.47	7.11
Average	**39.03**	**48.19**	**54.72**	**3.81**

G and R are for GPS and Radiosonde, respectively

SSTa is a good indicator to represent an ENSO activity. The SSTa variability for Niño 3.4 and Niño 4 regions for the year 2011 is demonstrated in Figure 5. In general, only La Niña phase taken place in the West Pacific region during 2011 based on NOAA definition. The phase is active from January to April and the middle of July to December, while the other month is in normal phase. For the period of January to April, the intensity of La Niña was strong to weak condition with SSTa minimum of -1.8°C for Niño 3.4 region, while the middle of July period to December was weak to strong condition with SSTa minimum of -1.6°C for Niño 4 region.

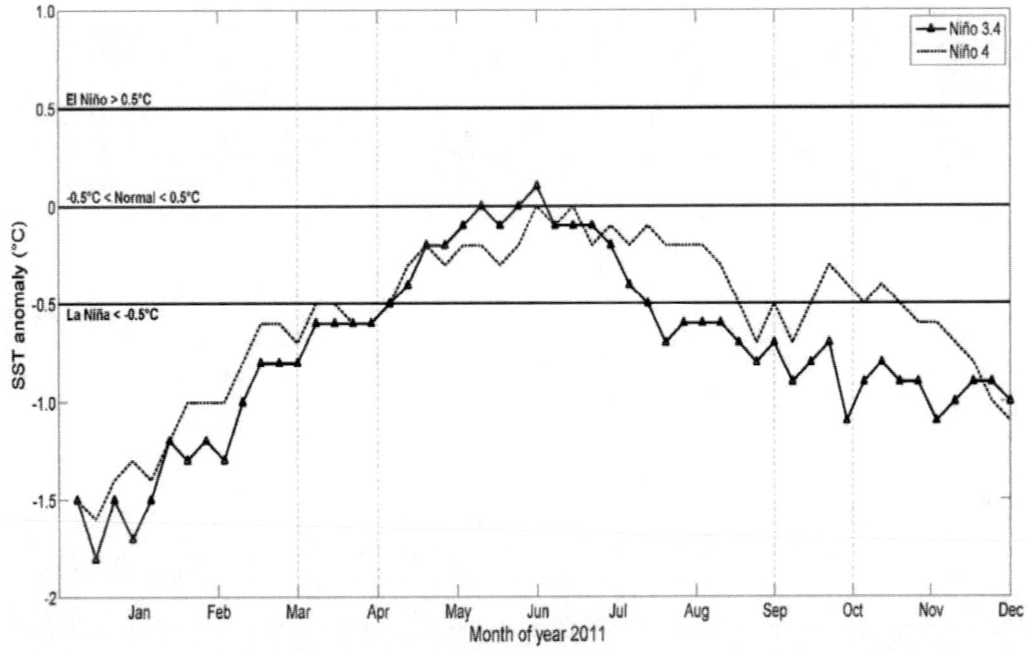

Figure 5. SSTa variability for Niño 3.4 and Niño 4 regions for the year 2011

3.2. Relationship between GPS PWV and SSTa

Figure 6 shows the relationship between GPS PWV and SSTa for a La Niña case (SSTa below -0.5°C) of Niño 4 region. Note that due to incomplete GPS data at UMSK station, the PWV from Radiosonde was used. Based on the correlation analysis, the relationship between the two variables at all stations shows modest correlation with correlation coefficients between -0.30 and -0.78 significantly at the 99% confidence level. For the Niño 3.4 region (not shown), the correlation coefficients for NTUS, PIMO, WBKK and BAKO stations are -0.58, -0.73, -0.67 and -0.66, respectively. The better relationship given by Niño 4 region than 3.4 region is due to the Niño 4 region is more western and near with the location of study (GPS station). More than that, the correlation coefficient obtained is negative, although the relation trend is positive. This indication during the cold phase can be explained as follows. Cooler air cannot hold much water vapor in the atmosphere because the density of water molecules easily depressed and falls as rain. The variation of water vapor is a function of the temperature. Moist climates tend to have a lower diurnal range in temperature. Because of latent heat, condensational cooling is at a maximum and there is a little water vapor to trap longwave radiation during the night. Therefore, during heavy rainfall, the temperature is cool. Generally speaking, water vapor in the atmosphere will be reduced by heavy rain.

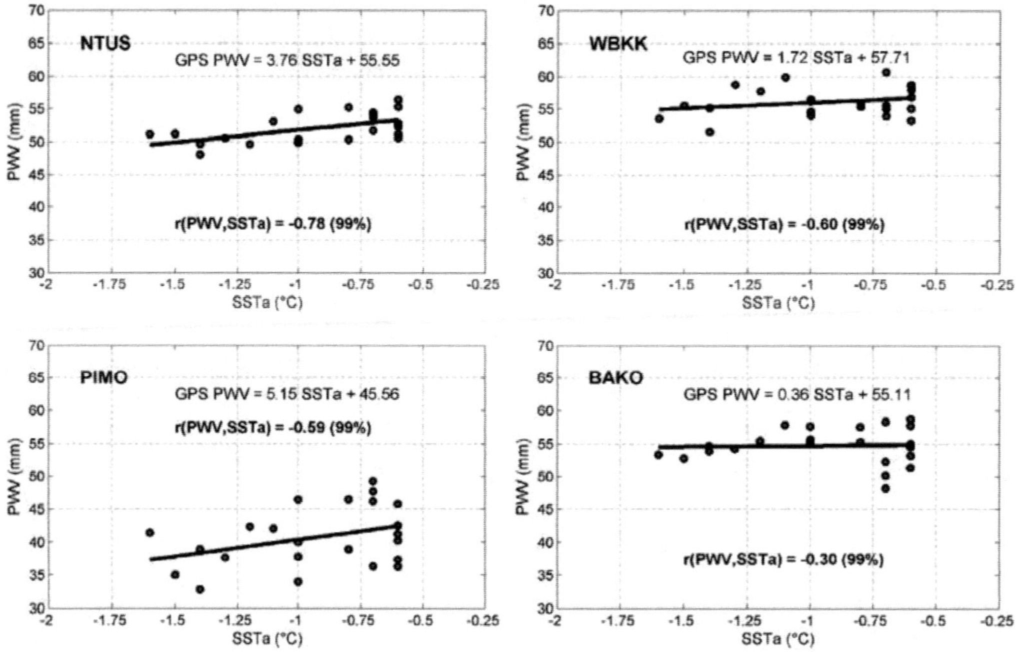

Figure 6. Scatterplot for relationship between GPS PWV and SSTa below -0.5°C (La Niña case) for Niño 4 region.

4. Conclusion

The impact of ENSO events such as La Niña over the South China Sea region to the variation in the GPS PWV during the period of 2011 has successfully addressed. Analyses of association between PWV and SSTa below -0.5°C (La Niña phase) found low-to-moderate negative correlations at four selected GPS stations. The negative correlation indicates that during the La Niña phase, the PWV is increased that causes warm air mass in the central Pacific Ocean moved to west Pacific. The increased of PWV causes the GPS signals will be getting slower because of ocean-atmosphere interactions. This implies that La Niña events that associated with unusually wet conditions (higher rainfalls) have an impact to the western Pacific Ocean as well as South China Sea region, which seem directly through the changes of the water vapor amount in the atmosphere. The La Niña effects with increased PWV will bring a consequence to increase the rainfall that can generate flooding and severe storms. As a conclusion, an ENSO activity can be studied from the atmospheric path delay of GPS signals through the response of PWV variability, and we suggest that this tool be used on monitoring platforms.

Acknowledgments

This research is funded by the Ministry of Higher Education Malaysia (MOHE) under grants, LRGS/TD/2011/UKM/PG/01 and UKM-LL-08-FGRS0212-2010. The authors would like to express their gratitude to the School of Science and Technology, Universiti Malaysia Sabah (UMS) for maintenance the GPS receiver and the meteorological systems. We would like to thank the National Oceanic and Atmospheric Administration (NOAA) for archiving the SST data and University of Wyoming for Radiosonde data.

References

[1] Bjerknes J 1969 Atmospheric teleconnections from the equatorial pacific *Monthly Weather Review* **97**(3) 163-172

[2] Ropelewski CF and PD Jones 1987 An extension of the Tahiti-Darwin Southern Oscillation Index *Monthly Weather Review* **115** 2161-2165

[3] Rasmusson EM and TH Carpenter 1982 Variations in tropical sea surface temperature and surface wind fields associated with the Southern Oscillation/El Niño *Monthly Weather Review* **110** 354-384

[4] Reynolds RW and TM Smith 1995 A high-resolution global sea surface temperature climatology *Journal of Climate* **8** 1571-1583

[5] Curtis S and R Adler 2000 ENSO Indices based on patterns of satellite-derived precipitation *Journal of Climate* **13** 2786-2793

[6] Bevis M, Businger S, Chiswel S, Herring TA, Anthes RA, Rocken C and Ware RH 1994 GPS Meteorology: mapping zenith wet delay onto precipitable water *Journal of Applied Meteorology* **33** 379-386

[7] Suparta W, Abdul Rashid ZA, Mohd Ali MA, Yatim B and Fraser GJ 2008 Observation of Antarctic precipitable water vapor and its response to the solar activity based on GPS sensing *Journal of Atmospheric and Solar Terrestrial Physics* **70** 1419-1447

[8] Suparta W, Jit Singh MS, Mohd. Ali MA, Yatim B and Mohd Yatim AN 2011 GPS water vapor monitoring and TroWav updated for ENSO studies *Proceeding of the 2011 International Conference on Instrumentation, Communication, Information Technology and Biomedical Engineering (ICICI-BME 2011)* pp 35-39

[9] Suparta W 2010 Using a Global Positioning System to estimate precipitable water vapor in Antarctica *Polar Geography* **33**(1) 63-79

[10] Suparta W 2012 Detection of a katabatic wind event with GPS meteorology measurements at Scott Base Antarctica *Meteorology and Atmospheric Physics* **115**(3-4) 153-162

Journal of Physics: Conference Series **423** (2013) 012044 doi:10.1088/1742-6596/423/1/012044

Application of microlearning technique and Twitter for educational purposes.

B H Aitchanov
Professor, PhD in CS, Academic of International Informatization Academy
KazNTU, Kazakhstan

A B Satabaldiyev
PhD student, Suleyman Demirel University, Kazakhstan
E-mail: askar.satabaldiyev@gmail.com

K N Latuta
PhD student, Suleyman Demirel University, Kazakhstan
Email: clatuta@gmail.com

Abstract. The current paper reviews the usage of social resource such as Twitter in microlearning technique for educational purposes. The problem is that most of instructors are unaware that with the help of social networks the students' productivity can increase. The research is applied on CS205 Advanced Programming in C++ course at Suleyman Demirel University (Kazakhstan). The collected results show that in a modern world of emerging mobile technologies, we are as educators should improve the way of teaching by adding electronically supported learning methods. In this study, the significance of microlearning technique is proposed.

1. Introduction

Our world changes very quickly. This also affects our work and private life. One should admit the fact that everything around us constantly changes and at the same time should adapt to continually decreasing time intervals. The ability to learn and the speed of learning thereby become an elementary factor for individual quality of life

Today people consider how to systematically use micro-sharing, made possible using tools like Twitter to connect with others. It allows to reach people's desktops, laptops, and devices already in pockets without any dependency on local email servers or a phone tree. Because this is the beginning of the mobile era, where people will be able to communicate, work, share information; make easier their lives by using mobile devices in their pockets. [1]

As it is mentioned above, Twitter is a public micro-sharing network used all over the world which has become an integral part of people's own professional practice or personal point of view. They use it to connect, share, and discover information far beyond any other networks. As a consequence, users finally start to understand that Twitter can be used for educational field as a micro-learning.

Twitter with its 140 character limitation is perfect for delivering short bytes of learning content – and for creating a community of learners around that content.

The main idea is following:

- A separate Twitter account was set up for this activity to send out the tweets to followers
- An example tweet looked like this

Table 1. Example of a tweet

Introduction to Java. First steps. JDK installation. http://javafever.com/corejava/jdk-install-guide.html

- Each tweet consists of:
    - a fact – an engaging statement that makes the learner to click through the link("teaser")
    - a link – supporting short resource – a web page, image, video,...
- Each tweet can be scheduled. There are number of tools to do this. (Scheduling tweets, ...)

Ubiquitous technology can be usefully applied for microlearning because it can reach users throughout the day, when they have idle time. Users can look through and revise subscribed course's data while spending time in public transport, waiting in line, or in the brief transition periods between activities. Brief interactions allow users to chip away at a larger learning goal and may serve a priming role by repeatedly bringing the learning task to their attention; users may then be more mentally prepared to take advantage of richer learning opportunities, such as those that occur naturally during actual social events [2].

How much time does an average human-being spend in Twitter? How much time do we waste, how much do we use for self-development? According to statistics of International Network for Social Network Analysis (INSNA) our young generation lives in virtual world. They always stay connected. The aim of this research is to combine what they like and what they need. As far as Twitter is an online network service with elements of microlearning, it enables users to learn anywhere and anytime.

2. Methods

There is a CS205 Advanced Programming in C++ course in Suleyman Demirel University, which is mostly oriented on sophomore students. As an experimental tool Twitter was chosen to teach Advanced Programming in C++ with microlearning technique.

The main idea of connecting twitter with programming languages was to improve knowledge of students and prove that social network can be very useful in skillful hands of educators. As a learning method, microlearning was chosen.

Microlearning deals with relatively small learning units and short-term learning activities. Microlearning often interacts with microcontent, which generally occurs either in e-learning media or in microcontent structures such as weblog postings (Mosel 2005).

Microlearning can be a hypothesis about the time needed to solve a learning task, for example answering to a specific question, memorizing an information unit, or finding a needed resource (Masie 2006). The time range for solving a learning task can cover a span from few seconds (e.g. in mobile learning) up to 15 minutes or more. There is some relation to the term microteaching, which is an established practice in teacher education .

In order to collect data for this research, we interviewed sophomore students from different groups and faculties using questionnaires, which consisted of 7 questions.

The most important questions are specified below:

- Do you use twitter?
  a) Yes b) No
- What do you want to learn using microlearning approach?
  a) Foreign languages b) Programming languages
  c) Science (chemistry, biology, e.t.c) d) Other
- How much of your time would you spend for educational purposes while implementing Twitter and microlearning technique?
  a) 10-20 minutes b) 1 hour c) 2-3 hours d) more than 3 hours

The results of this questionnaire are provided below:

- Total number of students, who participated, is 77.
- Number of students, who use twitter is 32, don't use – 45
- Only 9 students want to learn programming language using Twitter.

The question concerned with time has shown the most interesting results:

- 19 students want to spend 10-20 minutes, 36 students - 1 hour and 16 students - 2 hours per day to learn using social media.

According to the results of this questionnaire, students are unwilling to learn programming language using Twitter or they don't know how to use social media for educational purposes.

Course materials were processed and compressed into more brief and comprehensive data. Later this data was uploaded to instructor's website and other web resources such as YouTube. As a next step, we prepared a list of tweets which corresponded to earlier prepared syllabus for Advanced Programming in C++ course. Each tweet consisted of a brief description and link to resource (See Table 1), where more detailed information is stored. In order to have tweets sent on specific time and date according to the syllabus, we used tweet scheduler to keep the data updated and sent regularly without instructor's control. Students follow Twitter account, created for this experimental course, which enables getting messages and notifications anywhere, anytime on their mobile phones. This keeps students informed about which topic should be studied, revised and rehearsed.

3. Results

At the end of the term, we made a new questionnaire with 4 main questions for students, who took this course:

- How much time do you spend using Twitter per day?
  a) 10-20 min b) 1 hour c) >1 hour d) 0 min.
- Did you enjoy application of Twitter for Advanced programming in C++ course?
  a) Yes b) No c) Other
- What are your suggestions to improve experimental course?
- How did Twitter and microlearning technique affect your performance? Explain, why?

Results are given below as shown in figure 1 and figure 2:

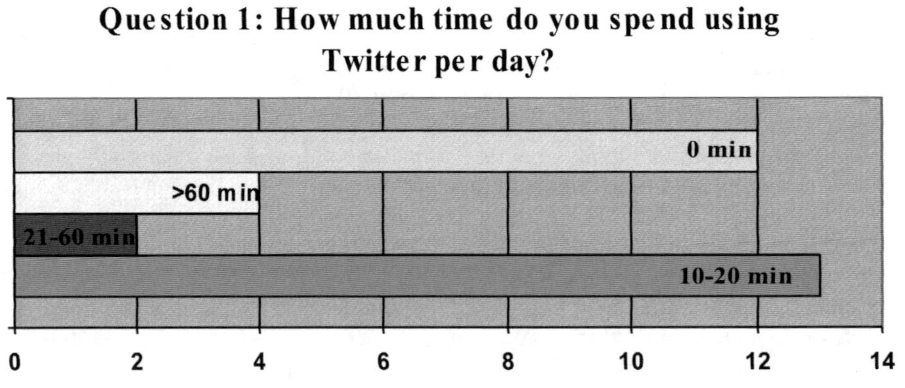

Figure 1. Results of answers for question 1

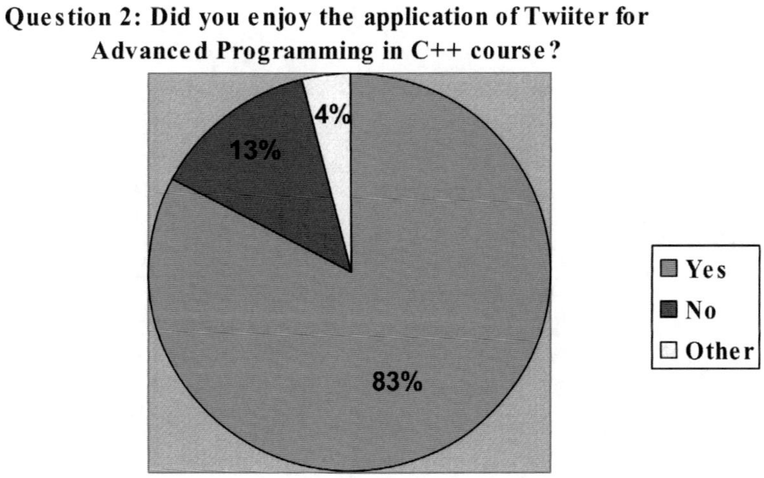

Figure 2. Percentage of answers for question 2

Also in suggestions and critics

- Tweet more often
- Supplementary materials should be used
- Students suggested using this technique to learn other spheres.

The positive sides of this survey that were mentioned by students are:

- Easily download material provided within tweeted links
- Keep tracking of topics that student need
- Communication with collaborators (help of other students)
- Do not spend time for searching information

4. Conclusions

Students were able to access study material outside university. This data was compressed and sliced into small chunks of information. Students subscribed for mobile notifications, that allowed receiving tweets anytime and anywhere, which is the main principle of micro-learning.

Moreover, information is nested on students' phones and Twitter accounts, which they can easily access via Internet.

The mobile technology is spreading very fast. Each second the tons of bytes of information are created. There is a need for learning technique, such as micro-learning. This paper demonstrates a simple example how to use Twitter and microlearning technique for educational purposes, but enterprises can use this approach as well. There is a necessity for researching this field of study, which is not yet complete. Many sections are under work and will be expanded in future.

5. References

[1] Corner M 2008 *Enterprise Micro-learning* (New York: Pistachio)
[2] Beaudin J S, Intille S S, Morris M E 2011 *MicroLearning on a Mobile Device* (MIT Media Lab)
[3] Clarey J 2009 *Microlearning presentation at LearnTrends* (LearnTrends)
[4] Lindner M 2007 *What is Microlearning?* (Austria, Research Studios Austria)
[5] Hug T, Lindner M, Bruck P A 2005 *Microcontent is Everywhere* (Eds.)

Oligostilbenoids from *Vatica pauciflora* and the Oxidative Effect on Chang Cells

Aisyah Salihah Kamarozaman[1], Jalifah Latip[2], Yana M Syah[3], Norfadilah Rajab[4] and Anuar Jaloh[4]

[1] Centre of Foundation Studies, MARA University of Technology, Puncak Alam Campus, 42300 Bandar Puncak Alam, Selangor, Malaysia
[2] Department of Chemical Sciences and Food Technology, Faculty of Sciences and Technology, National University of Malaysia, 43600 Bangi, Selangor, Malaysia
[3] Department of Chemistry, Faculty of Mathematics and Natural Sciences, Bandung Institute of Technology, Jln. Ganesha 10, 40132 Bandung, Indonesia
[4] Department of Biomedical Science, Faculty of Allied Health Science, National University of Malaysia, Jalan Raja Muda Abdul Aziz, 50300 Kuala Lumpur, Malaysia

Email: aisyahsalihah@yahoo.co.uk

ABSTRACT

Phytochemical studies on the woods and twigs of *Vatica pauciflora* (Dipterocarpaceae) have been conducted. The woods and twigs of *V. pauciflora* were extracted in methanol and ethanol respectively with Soxhlet apparatus. The crude methanol extracts from the woods were semi-purified by vacuum liquid chromatography to give 14 fractions (VMB01-VMB14). The VMB10 was further purified by radial chromatography to give pure dimmer stilbenoid. The same technique has been employed on the crude ethanol extracts from the twigs in which one pure fraction, VER612 was obtained. Based on the UV, IR, NMR and mass spectral data, the pure compounds were characterized as ε-viniferin and vaticanol G respectively. An oxidative damage test which was done on Chang liver cells has shown that ε-viniferin has a potential to be a chemopreventive agent to protect liver cells from oxidative damage.

Keywords: *Vatica pauciflora*, ε-viniferin, vaticanol G and chemopreventive.

1.0 Introduction

Vatica pauciflora which is known as *resak paya* or *resak laru* belongs to the family of Dipterocarpaceae [2,12]. This species has been recorded from the Malay Peninsula and the south of Sumatra [1,12]. Economically, *V. pauciflora* is one of the most species used as *laru* to prevent fermentation of toddy and its woods are used for temporary construction [2,12].

Journal of Physics: Conference Series **423** (2013) 012045 doi:10.1088/1742-6596/423/1/012045

The previous study on the stem bark of this species has successfully isolated five new oligostilbenoids which are paucifloral A, B and C, isovaticanol B and C; three new oligostilbene glucosides which are paucifloroside A, B and C; together with 17 known oligostilbenoids and bergenin [6]. Subsequent work done by the same researchers found other oligostilbenoids namely paucifloral D, E and F [8].

Apart from that, oligostilbenoids that have been isolated from *V. pauciflora* showed biological activities against murine leukemia P-388 cells [4,11], colon cancer cell line SW480 and leukemia HL60 cells [7].

Studies had shown that resveratrol, which is the basic skeleton of oligostilbenoids exhibited chemoprevention effect and cancer therapeutics on cultured cells systems. Resveratrol acted as an antioxidant, antimutagen and inducing drug metabolism enzyme phase II (anti-initiation activity) and detoxification carcinogen, anti-inflamatory effect as well as stopping the function of cyclooxygenase and hydroperoxide [3,9].

Our work on the woods and the twigs of *V. pauciflora* has successfully isolated two pure compounds which were characterized as ε-viniferin and vaticanol G. As part of our continuing study on the chemical and biological activities of *V.pauciflora*, we investigated the effect of oligostilbenoids in protecting Chang liver cells from oxidative damage which has been induced by hydrogen peroxide.

2.0 Experimental

2.1 Plant Material

V. pauciflora was collected from Gunung Panti, Pontian, Johor (voucher no. WYP 15, Herbarium, UKM Bangi). Dried and ground woods (400 g) and twigs (800 g) of *V. pauciflora* were extracted with methanol and ethanol respectively by using Soxhlet apparatus.

2.2 Extraction and Isolation

The crude methanol extract (16.73 g) was fractionated by Vacuum Liquid Chromatography (VLC) with a mixture of hexane-ethyl acetate in increasing polarity to give 14 fractions (VMB01-14). Fraction VMB10, which was a polar fraction was further purified by radial chromatography to give a pure compound labeled as VMB1004 (210 mg).

The crude ethanol extract (24.4 g) from the twigs of *V. pauciflora* was semi-purified by VLC with a mixture of hexane-ethyl acetate in increasing polarity to give 7 fractions (VER1-VER7). Further purification on VER6 by radial chromatography had given 12 combined fractions (VER601-VER612) which one of the fractions is a pure compound labeled as VER612 (20 mg).

The structure of VMB1004 and VER612 were elucidated by means of infrared (IR), ultraviolet-visible (UV-VIS), mass (MS) and one and two dimensioned nuclear magnetic resonance spectroscopy (NMR), optical rotation analysis and direct comparison with literature data [5,10] . Chemical shift, δ, ppm recorded in CD_3COCD_3.

2.3 Bioassay Evaluation

Hydrogen peroxide (H_2O_2) is introduced into Chang cells for about 30 minutes to induce oxidative damage to the cell. The media contain of H_2O_2 was discarded and replaced by the media containing treatment from pure compounds obtained from the isolation. The concentration of the treatment is at 0 μM, 50 μM and 100 μM. Then, the cells were incubated for 24 hours. MTT assay was done on the cells and the viability was obtained by comparing the optical density (OD) of the treated cell and treatment with media only.

3.0 Results and Discussions

ε-viniferin **(1)** was obtained as brownish viscous oil. HREIMS spectrum recorded the molecular ion $[M]^+$ peak at *m/z* 454 indicated a molecular formula of $C_{28}H_{22}O_6$, corresponding to a dimmer stilbenoid. The UV spectrum (λ_{max} 206, 248, 343 nm in MeOH) showed a characteristic of an oligostilbene.

The ^1H NMR spectrum of **1** showed the presence of four pairs of *ortho*-coupled aromatic protons at δ_H 6.73, 6.83, 7.18, 7.20 and two sets of *meta*-coupled aromatic protons at δ_H 6.23, 6.32, 6.72. There were doublet signals at δ_H 4.47, 5.42 which represents the protons at the dihydrobenzofuran ring. A pair of doublet signals at δ_H 6.71, 6.92 indicated the presence of olefinic protons.

The ^{13}C-APT NMR spectrum exhibited twenty – eight signals which consists of six oxyaryl (δ_c 158.3, 158.3, 159.6, 159.9, 159.9, 162.4), two aliphatic (δ_c 57.1, 93.9), eighteen aromatic (δ_c 96.8, 102.1, 104.1, 106.9, 106.9, 116.1, 116.1, 116.3, 116.3, 119.8, 127.9, 127.9, 128.7, 128.7, 129.8, 133.7, 136.3, 147.4) and two olefinic carbons (δ_c 123.4, 130.0).

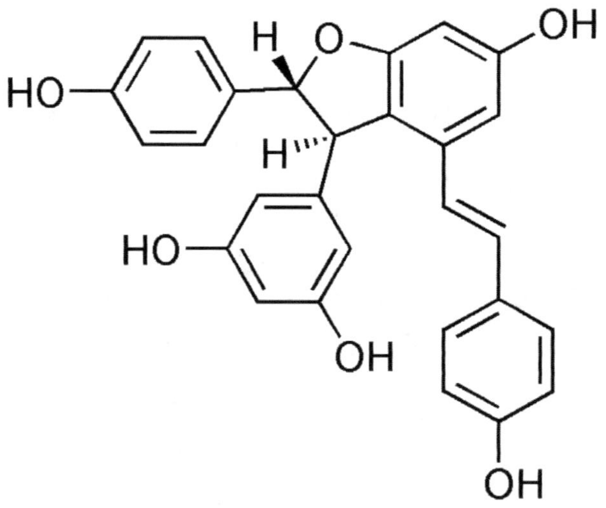

Figure 1. Structure of ε-viniferin (**1**)

Vaticanol G **(2)** obtained as brownish viscous oil. Negative ion FAB-MS: $[M-H]^-$ at *m/z* 679 indicated a molecular formula of $C_{42}H_{32}O_9$, corresponding to a trimer stilbenoid. The UV spectrum (λ_{max} 204, 282 nm in MeOH) showed the presence of aromatic and phenolic system. When the UV spectrum was recorded in the presence of NaOH, there was no batochromic shift occur at λ_{max} 282 which indicated that this molecule did not contain stilbene skeleton.

The ^{1}H NMR spectrum of **2** showed three pairs of broad singlet at δ_H 7.13, 6.66, 6.45, 6.42, 5.97 and 5.92 which represent two 1,4-disubstituted aromatic ring. The presence of 1,2,4-trisubstituted, 1,3,5-trisubstituted, 1,2,3,5-tetrasubstituted and 1,2,3,5,6-pentasubstituted aromatic rings were shown by the signals at δ_H 6.45, 6.17, 6.08, 6.05, 5.99, 5.93, 5.74 and 5.64. Other signals shown in the spectrum are the signals of methyne protons which are a broad singlet at δ_H 4.07, two pairs of doublet signals at δ_H 4.86, 4.60, 4.52 and 3.49 and a double doublet signal at δ_H 3.83.

The ^{13}C-APT NMR spectrum exhibited thirty – seven signals which represent fourty – two carbon atoms. There are nine oxyaryl (δ_c 159.2, 159.2, 158.6, 156.6, 156.0, 155.1, 154.9, 153.3, 153.1), ten quaternary (δ_c 147.8, 147.1, 142.0, 141.9, 140.0, 137.2, 129.3, 126.1, 121.9, 117.6), seventeen aromatic (δ_c 135.1, 130.7, 130.3, 130.3, 127.6, 119.5, 116.2, 114.8, 114.8, 114.3, 112.9, 111.5, 106.5, 106.5, 102.1, 101.7, 101.1) and six aliphatic (δ_c 63.1, 57.2, 57.1, 53.9, 50.4, 42.8) carbons.

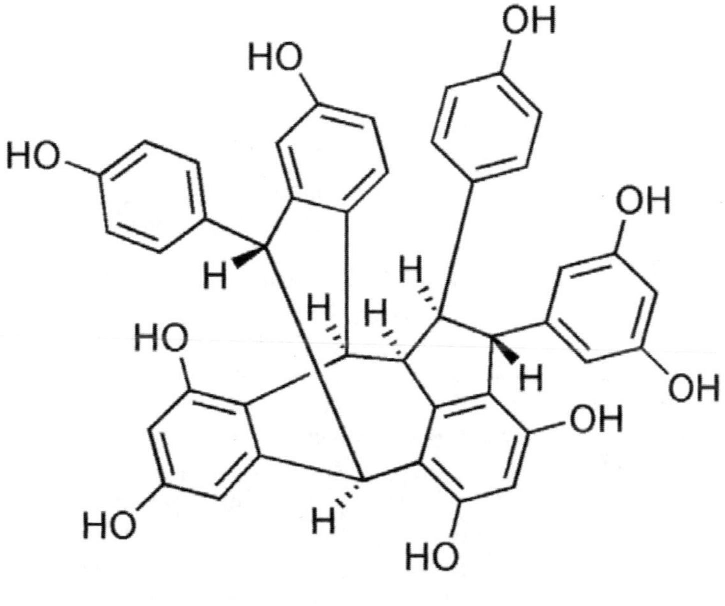

Figure 2. Structure of vaticanol G (**2**)

An oxidative damage test on compounds **1-2** was performed to determine the ability of these compounds in protecting Chang cells from oxidative damage induced by hydrogen peroxide (H_2O_2). The percentage of cell viability before the treatment with ε-viniferin **(1)** is 78.3% while the viability after the treatment at 50 μM and 100 μM is 106.9% and 111.0% respectively. Meanwhile, the percentage of cells treated with vaticanol G **(2)** at 50 μM and 100 μM is 85.4% and 84.4% respectively. Before the treatment, the cell viability is 76.6%. The proliferation of Chang cells had shown a significant increase after the treatment with ε-viniferin **(1)** while the treatment with vaticanol G **(2)** did not show any significant increase. This significant increase in Chang cells treated ε-viniferin **(1)** shown that the dimmer stilbenoid has the ability to protect the cells from oxidative damage. High antioxidant capacity in ε-viniferin **(1)** is one of the factors that contribute to the significant protection of oxidative damage in Chang cells.

4. Conclusions

The wood of *V. pauciflora* found to contain a dimmer stilbenoid, ε-viniferin which is the precursor of oligostilbenoids while the isolation from the twigs of *V. pauciflora* has successfully obtained vaticanol G which is a trimer stilbenoid. The major compound ε-viniferin has the potential to be applied as chemopreventive agent due to its significant ability to protect liver cells from oxidative damage.

5.0 Acknowledgements

Greatest appreciation to the Ministry of Science, Technology and Innovation, Malaysia for the financial support (02-01-02-SF0197) as well as Centre of Foundation Studies, UiTM Puncak Alam.

6.0 References

[1] Desch H E 1941 Dipterocarp Timbers of The Malay Peninsula. *Malayan Forest Records No. 14* Kuala Lumpur
[2] Foxworthy F W 1932 Diptercarpaceae of the Malay Peninsula *Malayan Forest Records No. 10* Singapore
[3] Fremont L 2000 Biological Effects of Resveratrol *Life Sciences* **66**(8) 663-673
[4] Haryoto S, Lia D Juliawaty, Yana M Syah, Sjamsul A Achmad, Euis H Hakim, Jalifah L and Ikram M Said 2008 Oligostilbenoids from *Shorea gibbosa* and their cytotoxic properties against P-388 cells *J. of Natural Medicine* **62** 195-198
[5] Ito T, Tanaka T, Nakaya K-I, Iinuma M, Takahashi Y, Naganawa H, Ohyama M, Nakanishi Y, Bastow K F and Lee K-H 2001 A novel bridged stilbenoid trimer and four highly condensed stilbenoid oligomers in *Vatica rassak Tetrahedron* **57** 7309-21
[6] Ito T, Tanaka T, Iinuma M, Iliya I, Nakaya K I, Ali Z, Takahashi Y, Sawa R, Shirataki Y, Murata J and Darnaedi D 2003 New resveratrol oligomers in the stem bark of *Vatica pauciflora Tetrahedron* **59**(28) 5347-63
[7] Ito T, Akao Y, Yi H, Ohguchi K, Matsumoto K, Tanaka T, Iinuma M and Nozawa Y 2003 Antitumor effect of resveratrol oligomers against human cancer cell lines and the molecular mechanism of apoptosis induced by vaticanol C *Carcinogenesis* **24**(9) 1489-97
[8] Ito T, Tanaka T, Iinuma M, Nakaya K, Takahashi Y, Sawa R, Murata J and Darnaedi D 2004 Three new resveratrol oligomers from the stem bark of *Vatica pauciflora J. Nat. Prod.* **67**(6) 932-937
[9] Jang M and Pezzuto J M 1998 Effects of resveratrol on 12-*O*-tetradecanoylphorbol-13-acetate-induced oxidative events and gene expression in mouse skin *Cancer Letters* **134**(1) 81-89
[10] Li W-W, Ding L S, Li B G and Chen Y Z 1996 Oligostilbenes from *Vitis heyneana Phytochemistry* **42**(4) 1163-65
[11] Muhtadi, Euis H Hakim, Lia D Juliawaty, Yana M Syah, Sjamsul A Achmad, Jalifah L and Emilio L Ghisalberti 2006 Cytotoxic resveratrol oligomers from the tree bark of *Dipterocarpus hasseltii Fitoterapia* **77** 550-555
[12] Symington C F 1974 *Foresters' Manual of Dipterocarps* Kuala Lumpur: Universiti Malaya

Journal of Physics: Conference Series **423** (2013) 012046 doi:10.1088/1742-6596/423/1/012046

Simplified bond-hyperpolarizability model of second-harmonic-generation in Si(111): theory and experiment

Hendradi Hardhienata[1]

Center for surface and nanoanalytics (ZONA), Johannes Kepler University, Altenbergerstr. 69, 4040 Linz, Austria

Andrii Prylepa

Christian Doppler laboratory for microscopic and spectroscopic material characterization, Johannes Kepler University, Altenbergerstr. 69, 4040 Linz, Austria
Center for surface and nanoanalytics (ZONA), Johannes Kepler University, Altenbergerstr. 69, 4040 Linz, Austria

David Stifter

Christian Doppler laboratory for microscopic and spectroscopic material characterization, Johannes Kepler University, Altenbergerstr. 69, 4040 Linz, Austria
Center for surface and nanoanalytics (ZONA), Johannes Kepler University, Altenbergerstr. 69, 4040 Linz, Austria

Kurt Hingerl

Center for surface and nanoanalytics (ZONA), Johannes Kepler University, Altenbergerstr. 69, 4040 Linz, Austria

E-mail: `kurt.hingerl@jku.at`

Abstract. Second-harmonic-generation (SHG) in centrosymmetric material such as Si(111) is usually understood either from the phenomenology theory or more recently using the simplified bond-hyperpolarizability model (SBHM) [G. D. Powell, J. F. Wang, and D. E. Aspnes, Phys. Rev. B 65, 205320/1 (2002)]. Although SBHM is derived from a classical point of view, it has the advantage over the former that it gives especially for lower symmetry systems a clear physical picture and a more efficient explanation of how nonlinearity is generated. In this paper we provide a step-by-step description of the SBHM in Si(111) for the linear and second harmonic case. We present a SHG experiment of Si(111) and show how it can be modelled by summing up the contribution of the fields produced by the anharmonic motion of the charges along the bonds.

[1] on leave from Theoretical Physics Division, Bogor Agricultural University, Jl. Meranti S, Darmaga, Indonesia

Journal of Physics: Conference Series **423** (2013) 012046 doi:10.1088/1742-6596/423/1/012046

1. Introduction

Nonlinear optics is one of the most versatile tools to investigate surface properties, especially for a material class which posseses inversion symmetry. Serious theoretical efforts to explain nonlinearity in such systems has significantly progressed especially since the pioneering work of Bloembergen *et.al.* [1]. A nonlinear optical probe has several advantages over other surface probes because the material damage and contamination associated with charged particle beams is eliminated, all pressure ranges are accessible, insulators can be studied without the problem of charging effects, and buried interfaces are accessible owing to the large penetration depth of the optical radiation [2]. In centrosymmetric material such as silicon (Si) the bulk and surface nonlinear contributions are comparable [3] and changes in the surface properties such as surface deposition will significantly alter the nonlinear intensity profile [4].

Because centrosymmetric materials is important to present day technology (e.g. silicon semiconductor, silicon thin film sensors) understanding the surface property and the physical mechanism behind is critical, with much effort both experimental and theoretical having been invented. Perhaps, the most notable development in understanding the creation of an outgoing 2ω wave or second-harmonic-generation (SHG) from centrosymmetric material surfaces and interfaces in the viewpoint of the phenomenological theory was performed by the group of Sipe *et.al.* [5] and further by Luepke *et.al.* [6]. The later group, calculated the nonlinear susceptibility for vicinal Si(111) by fitting a fourier series to reproduce the four polarization SHG intensity (p-incoming p-outgoing, ps, sp, and ss) obtained from their experiments. Unfortunately, already for a single atom with a relatively low symmetry such as silicon, the phenomenological theory requires many -seemingly non independent- variables to be fitted thus blurring the physical insights and mechanism of SHG inside the material.

As a response to this inefficiency, Aspnes *et.al.* [7] developed an ingenious method termed the simplified bond-hyperpolarizability model (SBHM) based on the somehow forgotten classical Ewald-Oseen extinction theory [8,9]. The theorem states that the electric field transmission and reflection can be obtained by direct superposition of dipoles inside the matherial rather than from macroscopic boundary calculation [10]. SBHM extends this classical view to cope with nonlinear optics by assuming that the SHG signal originates from the anharmonic motions of the charges along the bonds. Suprisingly, if the nonlinear field is far smaller than the linear field the calculation turns out to be much simpler because it does not lead to into an integral equation hence requiring self consistency check [11]. In this case it is possible to differentiate between the incoming driving fundamental field (ω) and the outgoing SHG field (2ω). As a result, the steps to obtain the far field contribution from the anharmonic bond radiation can be performed independently. Moreover, experimental data, e.g.Luepke's Si(111) experiment, can be very well fitted using SBHM by only a handful parameters rather than the cumbersome fitting of various fresnel parameters using phenomenology theory.

2. Basic Theory

An excellent discussion of how to calculate the far field intensity for centrosymmtric material such as Si(111) and Si(001) by assuming charge motion along their bonds already exists for vicinal Si sytems [6,12] and the interested reader is therefore suggested to refer to these papers for further information. However, for the sake of clarity, we will -although only briefly- show how SHG can be understood in the framework of the SBHM.

In classical electrodynamics one can obtain the far field expression as a result of dipole (or oscillating charge) radiation. The vector potential $\vec{A}(\vec{r},t)$, which satisfy the Lorenz gauge [13]:

$$\nabla \cdot \vec{A} = -\frac{1}{c}\frac{\partial \phi}{\partial t} \tag{1}$$

can be expressed in the form:

$$\vec{A}(\vec{r},t) = \frac{\mu_0}{4\pi} \int dt' d^3 dt' \vec{r'} \frac{\vec{j}(\vec{r'}, t' - \frac{|\vec{r}-\vec{r'}|}{c} - t)}{\left|\vec{r} - \vec{r'}\right|} \tag{2}$$

Assuming that the localized current $j(r)$ and charge density $\rho(r)$ as well as the vector potential $\vec{A}(\vec{r},t)$ can be written in harmonic form we can write eq.(2) in the far field approximation as:

$$\vec{A}(\vec{r'},t) = \frac{\mu_0}{4\pi}\frac{e^{ikr}}{r} \int d^3\vec{r'}\vec{j}(\vec{r'})e^{-i\omega t} = -\frac{\mu_0}{4\pi}\frac{e^{ikr}}{r} \int d^3\vec{r'}\vec{r'} \left|\nabla \cdot \vec{j}(\vec{r'})\right| \tag{3}$$

Using the continuity equation:

$$\nabla \cdot \vec{j}(\vec{r'}) = i\omega\rho(\vec{r}) \tag{4}$$

we have for the spatial dependence vector potential the expression:

$$\vec{A}(\vec{r}) = -\frac{i\omega\mu_0}{4\pi}\frac{e^{ikr}}{r}\vec{p} \tag{5}$$

where $\vec{p}$ is the dipole moment:

$$\vec{p} = \int d^3\vec{r'}\vec{r'}\rho(\vec{r}) \tag{6}$$

The expressions for the fields according to the defined vector potential are:

$$\vec{E}(\vec{r}) = \sqrt{\frac{\mu_0}{\epsilon_0}}\frac{i}{k}\nabla \times \vec{H}(\vec{r}) \tag{7}$$

$$\vec{H}(\vec{r}) = \frac{1}{\mu_0}\nabla \times \vec{A}(\vec{r}) \tag{8}$$

The far field can thus be calculated by inserting eq.(5) in eq.(8) and then inserting eq.(8) in eq.(7). With the help of the identity $(\hat{n} \times \vec{p}) \times \hat{n} = \vec{p} - (\hat{n}\hat{n}) \cdot \vec{p}$ we have:

$$\vec{E}(\vec{r}) = \frac{ke^{ikr}}{4\pi\epsilon_0 r}\left[(\overline{I} - \hat{n}\hat{n}) \cdot \vec{p}\right] \tag{9}$$

here $\overline{I}$ is the identity tensor. The next step will be finding an expression for the classical microscopic dipole per unit volume (polarization) $\vec{p}$. In this work we will use eq. (9) to calculate the reflected intensity of the linear and second-harmonic-generation.

3. SBHM for Si(111)

In the Maxwell classical approach all the microscopic details in the inhomogenous nonlinear wave equation are included and hidden in the electric susceptibility. In the classical optic approach this susceptibilty can be estimated by constructing a Lorentz oscillator model with damping to obtain the nonlinear susceptibility terms [14]. It is therefore not suprising -but ingenious- to use a similar approach microscopically, by starting from a force equation to find expressions for the polarization. Furthermore it has been shown that even with the simplification that the nonlinear polarization source only occur along the bond direction, the second-harmonic far field from the anharmonic charge radiation is in very good agreement with experimental result for vicinal Si(111). Here, we briefly follow the approach given by Ref. [7].

For a single Si(111) atom there are four bonds which we denote by $\vec{b}_1, \vec{b}_2, \vec{b}_3$ and $\vec{b}_4$. In the bulk, the orientation of the bonds can be expressed in cartesian coordinates as:

$$b_1 = \begin{pmatrix} 0 \\ 0 \\ 1 \end{pmatrix} \quad b_2 = \begin{pmatrix} \cos\left(\beta - \frac{\pi}{2}\right) \\ 0 \\ -\sin(\beta - \frac{\pi}{2}) \end{pmatrix}$$

$$b_3 = \begin{pmatrix} -\cos\left(\beta - \frac{\pi}{2}\right)\sin(\frac{\pi}{6}) \\ \cos\left(\beta - \frac{\pi}{2}\right)\cos(\frac{\pi}{6}) \\ -\sin(\beta - \frac{\pi}{2}) \end{pmatrix} \quad b_4 = \begin{pmatrix} -\cos\left(\beta - \frac{\pi}{2}\right)\sin(\frac{\pi}{6}) \\ -\cos\left(\beta - \frac{\pi}{2}\right)\cos(\frac{\pi}{6}) \\ -\sin(\beta - \frac{\pi}{2}) \end{pmatrix} \tag{10}$$

with $\beta = 109.47°$ is the angle between two bonds. Fig.1 depicts a graphical image of the bond direction and the definition of the coordinate. Note that the four Si(111) bonds are drawn as the red lines.

Although the orientation is valid for the bulk, the even nonlinear harmonic dipole contribution is generally considered small because it is forbidden by inversion symmetry or parity argument. However we believe that the incoming field decay effects due to absorption may break the symmetry and produce dipolar effects even inside the bulk (spatial dispersion). At the interface, a dipole contribution is allowed because of symmetry breaking, but the orientation of the bonds may not be arranged as homogenously as in the bulk. Nonetheless, we follow the assumption in Ref. [7] that, based on statistical averaging the majority of the bond vector at the interface can be treated similar as within the bulk, thus the bond model of the bulk also holds for the interface.

To model the reflected intensity vs rotation along the azimuthal direction (rotation in the

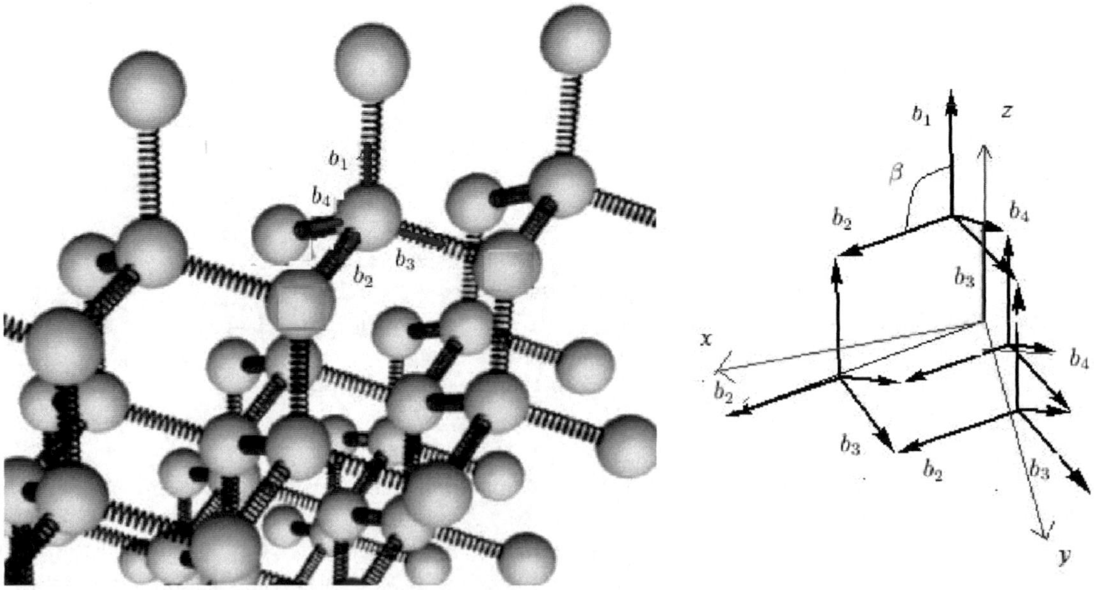

Figure 1. Sketch of Si lattice ((111) surface) and definition of the bond direction in cartesian coordinates.

x-y plane) we allow the bonds as a function of rotation via:

$$b_j(\phi) = b_j \cdot R_\phi = b_j \cdot \begin{pmatrix} \cos\phi & -\sin\phi & 0 \\ \sin\phi & \cos\phi & 0 \\ 0 & 0 & 1 \end{pmatrix} \tag{11}$$

where the subscript j runs from 1 to 4.

Having defined the bond vectors, we now analyze the motion of the charges. The force equation which describes this anharmonic motion along the bonds can be written in the form [7] taking the equilibrium position as zero:

$$F = q_j \vec{E}\hat{b_j} e^{-i\omega t} - \kappa_1 x - \kappa_2 x^2 - b\dot{x} = m\ddot{x} \tag{12}$$

where q, m, x are the electron charge, mass, and its displacement from equilibrium, respectively and κ_1 and κ_2 are the harmonic and anharmonic spring constants, and the term $-b\dot{x}$ is the common frictional loss in oscillation. Solving for $\triangle x_1$ and $\triangle x_2$ by using the assumption that x can be written as $x = x_0 + \triangle x_1 e^{-i\omega t} \triangle x_2 e^{-i2\omega t}$ gives for the lowest order of approximation:

$$\triangle x_1 = \frac{\vec{E} \cdot \hat{b_j}}{\kappa_1 - m\omega^2 - ib\omega}$$

$$\triangle x_2 = \frac{\kappa_2}{\kappa_1 - 4m\omega^2 - ib2\omega} \left(\frac{\vec{E} \cdot \hat{b_j}}{\kappa_1 - m\omega^2 - ib\omega} \right)^2 \tag{13}$$

therefore the linear polarization produced by each bond b_j is

$$p_{1j} = q_j \triangle x_1 = \alpha_{1j} \left(\hat{b}_j \cdot \vec{E} \right) \tag{14}$$

whereas we have for the SHG the nonlinear polarization:

$$p_{2j} = q_j \triangle x_2 = \alpha_{2j} \left(\hat{b}_j \cdot E \right)^2 \tag{15}$$

where α_1 and α_2 are the microscopic polarizability and second order hyperpolarizability given by:

$$\alpha_{1j} = \frac{1}{\kappa_1 - m\omega^2 - ib\omega}$$

$$\alpha_{2j} = \frac{\kappa_2}{\kappa_1 - 4m\omega^2 - ib2\omega} \left(\frac{1}{\kappa_1 - m\omega^2 - ib\omega} \right)^2 \tag{16}$$

In Si(111) surface, $\alpha_{j=1}$ is denoted as the "up" polarizability/hyperpolarizability α_u whereas the three remaining have the same value due to symmetry and are called the "down" polarizability/hyperpolarizability α_d. The total linear and second harmonic polarization produced by all the four bonds in Si(111) considering azimuthal rotation are thus :

$$\vec{P_1} = \frac{1}{V}\left(p_{11} + p_{12} + p_{13} + p_{14}\right) = \frac{1}{V}\sum_{j=1}^{4} \alpha_{1j} \left(\hat{b}_j(\phi) \cdot \vec{E} \right) \tag{17}$$

and

$$\vec{P_2} = \frac{1}{V}\left(p_{21} + p_{22} + p_{23} + p_{24}\right) = \frac{1}{V}\sum_{j=1}^{4} \alpha_{2j} \left(\hat{b}_j(\phi) \cdot \vec{E} \right)^2 \tag{18}$$

In the following chapters, we will show how the SBHM can reproduce experimental results of second-harmonic-generation in Si(111).

4. SHG Experiment of Si(111)

The setup for the azimuthal SHG measurements is shown in Fig. 2. As a source of a radiation a compact femtosecond fiber laser system, working at a wavelength of 1560 nm and producing pulses with duration of 86 fs and with a repetition rate of 80 MHz, was used. After a variable attenuator for adjusting the power, a half-wave plate (HWP) determines the state of polarisation for the incident beam, which is then filtered by a band-pass filter (F), transmitting only the fundamental wavelength. The incidence radiation is focused onto the sample by the objective lens L1 with a minimal focal waist down to 10 μm.

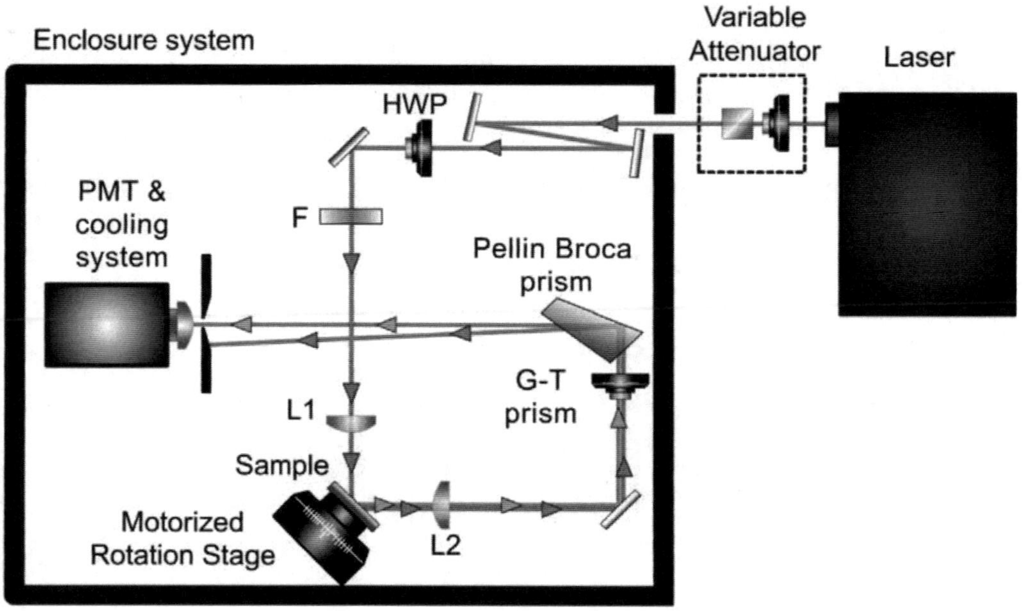

Figure 2. Schematic sketch of the experimental setup for SHG measurements (Abbreviations: HWP: half-wave plate, F:filter, L:lens, G-T:Glan-Taylor polarizing prism, PMT:photomultiplier).

The incidence angle for the radiation was set to 45°. The sample itself is mounted on a motorized rotation stage. SH and fundamental radiation from the sample are collected by the lens L2 and are directed to a rotatable Glan-Taylor calcite polarizer that selects the desired polarisation direction. A Pellin Broca prism and a slit are used for spatially separating the SH radiation, which is focused onto a cooled photomultiplier tube (Hamamatsu R911P). For the acquisition of the SHG signal photon counting was used, with a dark count rate of only 30 photons per minute.

As a sample, a Si(111) wafer with native oxide was used. In order to obtain a detectable SHG signal and not to damage the surface of the sample, the beam was slightly defocused, with the fiber laser operating at full power (350 mW).

5. Si(111) SHG experimental results and SBHM simulations

Second harmonic generation from a Si(111) face was obtained experimentally by plotting the intensity for various azimuthal angles (Si(111) rotated by 360° in the x-y plane). All four intensity profiles (pp, ps, sp, ss) are aquired with the results depicted in Fig. 3(b). To obtain the total field using SBHM we insert expressions for the total polarization in eq. (17) and eq. (18) into the far field formula eq. (9), with $\hat{n} = \widehat{k}$, where $\widehat{k}$ is the outgoing wave vector (the unit vector in the direction of the observer) and is given by $\hat{k} = -\cos\theta_0\hat{x} + \sin\theta_0\hat{z}$. The linear and second harmonic far field can now be written as:

$$\vec{E}_{\omega g}(r) = \frac{ke^{ikr}}{4\pi\epsilon_0 rV}\left[(\overline{I} - \hat{k}\hat{k}) \cdot \sum_{j=1}^{4}\alpha_{1j}\left(\hat{b}_j(\phi) \cdot \vec{E}_g\right)\right] \quad (19)$$

$$\vec{E}_{2\omega g}(r) = \frac{ke^{ikr}}{4\pi\epsilon_0 rV}\left[(\overline{I} - \hat{k}\hat{k}) \cdot \sum_{j=1}^{4}\alpha_{2j}\left(\hat{b}_j(\phi) \cdot \vec{E}_g\right)^2\right] \quad (20)$$

The term outside the paranthesis includes only constants and because SHG experiments are usually performed in arbitrary units the important terms are those given inside the paranthesis. Here g is the symbol for the polarization of the incoming wave and can be either p or s.

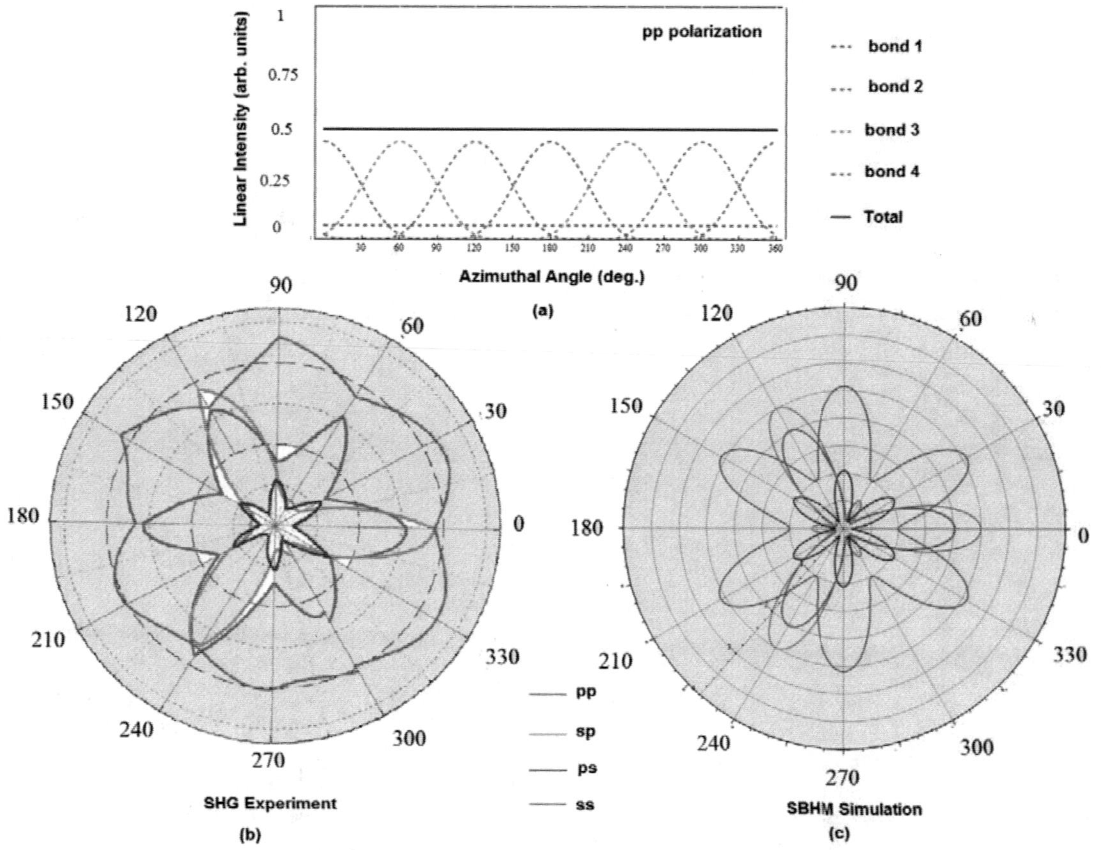

Figure 3. Si(111) linear and SHG intensity profiles vs azimuthal rotation in the x-y plane for various polarizations in arbitrary units of intensity. (a) The total linear pp-intensity yields a constant value independent of the azimutal rotation. (b) SHG intensity obtained from experiment and (c) SHG intensity simulation results using SBHM with α_u=α_d=1.

We next consider first eq. (17) and evaluate the intensity for four polarization direction: pp, ps, sp, and ss. The first letter in pp denotes the incoming and the second the outgoing wave. The far field from all four bonds in Si(111) for each polarization can be obtained using eq.(10). For the linear case we have:

$$\begin{aligned}
\vec{E}_{\omega,pp} &= \frac{1}{2}\left(\cos\left(\theta_{\ddot{o}}\right)\hat{x} + \sin\left(\theta_{\ddot{o}}\right)\hat{z}\right)\left[\left(-3\alpha_d\cos\left(\theta_i\right)\cos\left(\theta_o\right)\sin^2\left(\beta\right) + 3\alpha_d\cos\left(2\beta\right)\sin\left(\theta_i\right)\sin\left(\theta_o\right)\right)\right] \\
\vec{E}_{\omega,sp} &= 0 \\
\vec{E}_{\omega,sp} &= 0 \\
\vec{E}_{\omega,ss} &= \frac{3}{2}\alpha_d\sin^2\left(\beta\right)\hat{y}
\end{aligned} \tag{21}$$

An interesting result from this calculation is that the total linear fields for all polarization is independent of the azimuthal rotation ϕ, eventhough the field produced by each of the bond may depend on ϕ. This can be seen in Fig. 3(a) where we take the pp-case as an example. Similar results are obtained for the ss-polarization. Therefore, orientation of the bonds due to rotation in the x-y plane does not effect the reflected linear intensity.

More interesting results are obtained if we consider the second harmonic case. The total nonlinear polarization for all polarization modes are -in contrary to the linear case- a function of the azimuthal rotation of the crystal ϕ. Fig. 3(b) shows our experimental result. The pp, ps, and ss-polarization (red line) showsa 6-fold dependence, with peaks at every 60° interval. The ps and ss-polarization have the same azimuthal behaviour with a peak shift of 30° for the pp-polarization. The ss-polarization, notably has a high (DC) offset value and broad peak full-wave-half-maximum (FWHM), but peaks still being detected. Also the height of the pp and sp-polarization peaks are quite similar with the sp polarization having a 3-fold symmetry.

Remarkably, this seemingly complicated experimental result can be well explained using the SBHM, more explicitly by applying eq. (19) with the result depicted in Fig. 3(c). The formulas for the 4-polarization SHG far field from a Si(111) structure can be seen in Ref. [7]:

$$\begin{aligned}
\vec{E}_{2\omega,sp} &= \frac{3}{4}\alpha_d\sin^2\beta\left[\cos\theta_o\cos 3\phi\sin\beta - 2\cos\beta\sin\theta_o\right]\left[\hat{x}\cos\theta + \hat{z}\sin\theta\right] \\
\vec{E}_{2\omega,ps} &= \frac{3}{4}\alpha_d\sin^3\beta\cos^2\theta_i\sin 3\phi\,\hat{y} \\
\vec{E}_{2\omega,ss} &= -\frac{3}{4}\alpha_d\sin^3\beta\sin 3\phi\,\hat{y}
\end{aligned} \tag{22}$$

with a slightly different result obtained for the pp-polarization:

$$\begin{aligned}
\vec{E}_{2\omega,pp} = &\left(\alpha_u + 3\alpha_d\cos^3\beta\right)\sin^2\theta_i\sin\theta_o - 3\alpha_d\cos\beta\cos\theta_i\cos\theta_o\sin^2\beta\sin\theta_i \\
&+\frac{3}{4}\alpha_d\cos^2\theta_i\sin^2\beta(\cos\theta_o\cos 3\phi\sin\beta + 2\cos\beta\sin\theta_o)
\end{aligned} \tag{23}$$

By setting an arbitrary DC offsets which might be related to experimental noise, the model showed good agreement with the experiment, as is evident from the reproducable symmetry pattern with correct azimuthal position. In the simulation we have -for the sake of simplicity- set the values for the hyperpolarizabilities equal to unity and use for the SiO-Si interface incoming and outgoing angle of 29.5° (via snellius). The 3-fold sp symmetry (green line) is perfectly matching the experimental result as well as the ps-polarization blue line) with the correct relative peak difference between them (30°).

The intensity profiles can be easily explained as an effect of anharmonic charge radiation along the bond direction. One can easily check this by calculating the field produced by the individual bond contribution. For example, the 6-fold symmetry from the *ss*-polarization can be explained by each down bond contributing twice to the *s*-fundamental driving field when rotated by 360°. However, it has to be stated that for the *pp*-polarization SBHM gives a 3-fold symmetry and is in contradiction with the 6-fold symmetry given by experiment. We attribute this difference due to possible bulk dipole contribution because inside the bulk the incident angle is closer to the normal, and here SBHM also predicts a sixfold symmetry. This bulk dipole is still controversial and requires symmetry breaking in the form of a decaying field (absorption). Nevertheless, our simulation has shown that SHG measurement with SBHM can be used to investigate bond orientation at the surface of centrosymmetric materials in a more simple fashion without going to the complicated susceptibility tensor analysis (e.g. group theory).

6. Summary

Using the simplified bond hyperpolarizability model we show that the reflected linear intensity for all 4-polarization modes from a Si(111) structure is independent of the azimuthal rotation on the x-y plane. For the second harmonic intensity, SBHM gives a good agreement with experiment, sucessfully predicting the correct azimuthal profiles and symmetries of the surfaces except for the *pp*-polarization which we believe requires SBHM to be extended to cope with bulk effects. This shows that SBHM can be used as a simple model in predicting the bond orientation of a Si(111) wafer at the surface/interface and reconfirms second harmonic generation measurements as a sensitive non destructive method to investigate surface structures in material with inversion symmetry.

Acknowledegements: The authors would like to thank financial support by the Austrian Federal Ministry of Economy, the Austrian Family and Youth and the Austrian National Foundation for Research, Technology and Development.

References

[1] N. Bloembergen, R. K. Chang, S. S. Jha, and C. H. Lee, Phys. Rev. 174, 813 (1968)..
[2] J. F. McGilp, Journal of Physics D: Appl. Phys. 29, 1812 (20071996).
[3] P. Guyot-Sionnest and Y. R. Shen, Phys. Rev. B 38, 7985 (1988).
[4] B. Gokce, D. E. D. B. Dougherty, and K. Gundogdu, J. Vac. Sci. Technol. A 30 (2012).
[5] V. Mizrahi and J. E. Sipe, J. Op. Soc. Am. B. 5, 233402 (1988).
[6] G. Luepke, D. J. Bottomley, and H. M. van Driel, J. Opt. Soc. Am. B 11, 33 (1994).
[7] G. D. Powell, J.-F. Wang, and D. E. Aspnes, Phys. Rev. B 65, 205320 (2002).
[8] P.Ewald, Ann. Phys. 49, 1 (1916).
[9] C. W. Oseen, Ann. Phys. 48, 1 (1990).
[10] Fearn, James, and Milonni] H. Fearn, D. F. V. James, and P. W. Milonni, Am. J. Phys. 64, 986 (1996).
[11] D. Aspnes, Phys. Status Solidi B 2 247, 1873 (2010).
[12] J. Kwon, M. C. Downer, and B. S. Mendoza, Phys. Rev. B 73, 195330 (2006).
[13] J. D. Jackson, Classical Electrodynamics, 3rd. Ed p. 411 (1998).
[14] R. W. Boyd, Nonlinear Optics (Academic Press, 2003), ISBN 0-12-121682-9.

Journal of Physics: Conference Series **423** (2013) 012047 doi:10.1088/1742-6596/423/1/012047

Synthesis and characterization of InGaAs nanowires grown by MOCVD

D Gustiono[1,2], E Wibowo[3] and Z Othaman[1]

[1] Ibnu Sina Institute for Fundamental Science Studies, Universiti Teknologi Malaysia, 81310 UTM Skudai, Johor, Malaysia.

[2] Agency for Assessment and Application of Technology, BPPT second building Jakarta 10340, Indonesia

[3] Department of Physics, Faculty of Mathematic and Science, Universitas Negeri Semarang, Semarang, Indonesia.

E-mail : gustiono@ibnusina.utm.my

Abstract. Semiconductor nanowires have been intensively investigated in order to study their unique fundamental and application properties that develop at the nano-scale. One of main problems in the growth of III-V semiconductor nanowire is uniformity both of in dimension and composition of chemical elements. We synthesized InGaAs nanowire on GaAs (111) substrate at 400 °C – 480 °C temperatures for 30 minutes using MOCVD. The nanowires grow perpendicular to the substrate via direct impinging mechanism and they have hexagonal shape with diameter of 80-150 nm. Dimension of nanowire, length and diameter, increase with increases of growth temperature. Formations of tapering could be controlled with growth at lower temperature.

1. Introduction

Semiconductor nanowires attracted significant attention due to their potential applications as building blocks for future nanoscale circuits and optoelectronic devices. Semiconductor nanowire InGaAs is one of the most marvelous III–V ternary compounds offer amazing properties for applications in long wavelength optical transmission and integrated photonics applications [1,2]. The nanowire is also considered to be one of the fascinating materials suitable for transistor channels for large scale device applications due to its low electron effective mass [3,4]. Performance of nanowires depend on their physical and chemical properties, such as crystallinity, microstructure and chemical composition need to be uniform and well controlled [5,6].

Last decade, semiconductor nanowire with different compositions have been successfully grown using seed particles via vapor–liquid–solid (VLS), vapor–solid–solid (VSS), and solid–liquid– solid (SLS) mechanisms based on self-organized growth. [7-13]. In particular, III–V nanowire seed-particles assisted epitaxial growth using metal organic chemical vapor deposition (MOCVD) offers the ability to control the diameter and chemical composition [14]. The nanowire growth process, using gold as seed particles, promotes either VLS or VSS mechanism [14-16]. Initially, the gold nanoparticles on the semiconductor substrate surface form either a liquid (in case of VLS) or solid–

alloy (in case of VSS) with the group III species (source atoms) nanowire growth occurs with gold nanoparticles at the tip of each nanowire due to the source material precipitates out at the nanoparticle–semiconductor interface [7-9]. The diameter of NW is controlled by the size of the seed particle and its length is determined by the growth time and conditions [17].

In this work, we describe the growth of InGaAs nanowires by MOCVD. Cylindrical-hexagonal nanowire with diameter of 80-150 nm were grown perpendicular to the substrate. Dimension of nanowire, length and diameter, is going to increase with increases of growth temperature. Energy dispersive X-ray spectroscopy (EDX) analysis and transmission electron microscopy (TEM) observation of InGaAs nanowire is also reported.

2. Experimental procedures

We used GaAs(111) as substrate for the growth of nanowires and using MOCVD equipment as a technique of growth. The MOCVD chamber was maintained at low pressure 0.1 atm. Electronic grade hydrogen gas was used as the carrier for the ternary atoms to the growth chamber. Trimethylindium (TMIn), trimethylgallium(TMGa) and arsine (AsH_3) chemicals were used as precursors. GaAs with (111) energetically favorable growth direction was chosen as the substrate due to its low surface energy [2,19]. The substrate was immersed in 0.1% poly-L-lysine (PLL) solution for 3 min prior to the nanowires growth on the substrate. The negatively charged Au particles are attracted to the positively charged PLL layer and get embedded onto the substrate surface. The substrate was then cleaned with de-ionized water and subsequently blow-dried with nitrogen (N_2). It was then treated with 50% gold colloid solution (30 nm diameter gold particle) for 30 seconds. The substrate was placed in MOCVD chamber after gold seed particles were deposited on substrate surface. It was then heated up to 600 °C under constant partial pressure of AsH3 gas for 10 minutes to eliminate surface contamination and then cooled down to the desired growth temperature. When the growth temperature was reached, TMIn, TMGa, and AsH_3 were injected into the chamber. The flow rate of TMIn, TMGa, and AsH_3 were set at 100 sccm, 2 sccm and 32 sccm, respectively. The growth time was set at 30 minutes, growth temperature at 400 °C, indium mole fraction (x) at 0.41, and V/III ratio at 10. Field emission-scanning electron microscopy (FE-SEM) and TEM have been used to investigate the morphology of nanowires. The chemical composition of nanowires has been investigated using EDX analysis.

3. Results and discussions

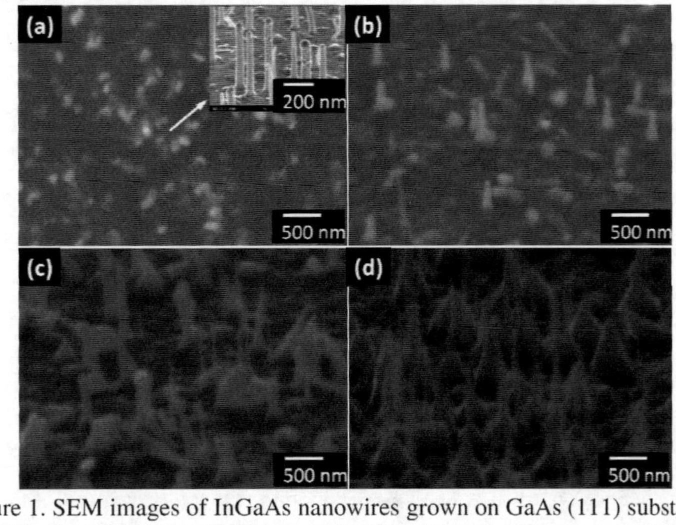

Figure 1. SEM images of InGaAs nanowires grown on GaAs (111) substarte at 400 °C (a), 420°C (b), 450°C (c) and 480°C (d) temperatures using MOCVD.

Figure 1 shows SEM images of InGaAs nanowires grown on GaAs (111) substrate from 400 °C to 480 °C temperatures for 30 minutes. InGaAs nanowires grow in multiple orientations, where orientations toward <111> are dominant. The nanowires have cylindrical shape with length about 0.2-1.5 μm and diameter about 80-150 nm as seen in figure 1(a). Nanowires grow by direct impinging mechanism via VSS mechanism where source atoms of precursors directly fall onto gold seed particles to form a liquid solution and partially molten state of an alloy [18]. A gold seed particle serves as a preferential site for the decomposition of the source atom via absorption and diffusion mechanism. Dimension of the nanowire increase with increases temperatures as shown in figure 1(b), figure 1(c) and figure 1(d). Increasing of growth temperature may give additional energy to source atoms or precursors, therefore precursor atoms fall onto the substrate surface obtain extra energy to move through substrate surface via diffusion mechanism. Since the seed particles act as the preferential sites for the decomposition of the source atoms which lead to nucleation. Source atoms from the substrate surface tend to move towards the seed particles and contribute to the growth of the nanowires via diffusion mechanism and induce tapering, whereby the top of the nanowire is highly small compared to the base one as seen in figure 1(d).

In addition, it can be observed in figure 1 that the InGaAs nanowires grown were of different diameters and lengths. The distribution of diameter and surface density of nanowires was due to the annealing effect (temperature and time) on the distribution of gold particles on the surface of substrate before the growth precursors were injected. Since they can obtain some extra energy, the adjacent gold particles could combine together to form the larger size through the Ostwald ripening mechanism [19]. On the other hand, gold particles that did not combine together retain their sizes. Besides, the incorporation of significant amount of precursor into the seed particle is another way to change the size of seed particle. Consequently, seed-particle-assisted nanowires grown have different diameters. Hence, Ostwald ripening and incorporation of growth precursor both can change the size of the gold seed particle. Due to the presence of these mechanisms, some nanowires were grown with diameter around six times larger than the diameter of gold particle. Morever, nanowires with smaller diameter grow shorter than the larger ones due to different growth rate. The classical analysis of Givargizov concluded that narrower wires grow more slowly due to Gibbs–Thomson effect [20]. Other studies, however, show the opposite, narrow wires growing more rapidly [21].

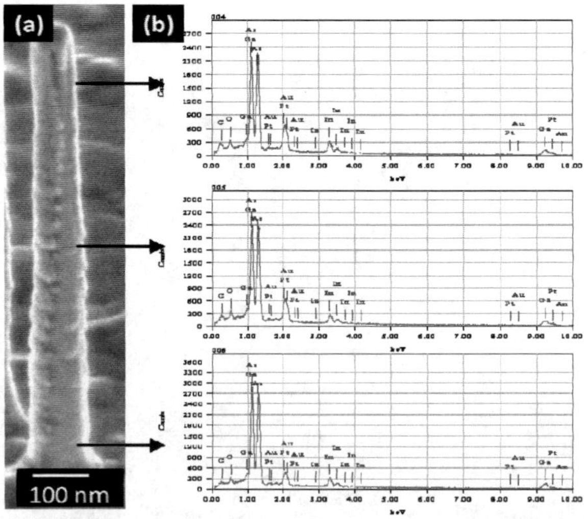

Figure 2. (a) Typical of SEM image of InGaAs nanowire grown at 400 °C temperature on GaAs (111) substrate using MOCVD. (b) EDX analysis of top, mid and bottom regions of the InGaAs nanowire.

Figure 2 show a typical SEM image of InGaAs nanowire grown on GaAs (111) substrate at 400 °C temperature for 30 minutes (a). The body of nanowires are formed by stacking of hexagonal shape, where stacking mode is caused by planar defects as twinning as seen in figure 2(a). EDX analysis has been used to investigate the chemical composition of individual InGaAs nanowire from various detection positions: tip, middle and bottom as shown in figure 2(b). Gold was still detected but uncounted even on the tip of the nanowire. This is likely due to very small amount of gold compared to the other elements of the nanowire. This phenomenon also proves that the increasing size of seed particle was mainly contributed by incorporation of growth precursor into the seed particle rather than Ostwald ripening mechanism. Therefore, the composition of seed particle is more likely from gold–precursor alloy nanoparticles.

Figure 3 shows one set of TEM images of InGaAs nanowire grown on GaAs (111) substrate at 420 °C temperature for 30 minutes at low and high resolutions. Figure 3(a) confirms again about the morphology of nanowire revealed dark region at top part of nanowire is gold seed particles and mid till bottom parts have tapered. Movement and diffusion of substrate atomics toward the seed particles induce tapering. TEM image also confirm the validity of the absorption, diffusion and precipitation process during direct impinging mechanism. As a result, gold– precursor alloy nanoparticles can only be found at the tip of NW, not elsewhere along the nanowires. Besides, TEM images provide evidence of the occurrence of Ostwald ripening and incorporation of growth precursor into the seed particle. The sidewalls of the nanowires are extremely flat, except for at the twin boundary where constriction can be seen clearly. The constriction is suggested accommodates mismatches at the twin boundary as shown in figure 3(b). Lattice space or range of atomic planes (110) was determined about 0.3433 nm and it was also determined that crystal structure of the nanowire is typical zinc-blende as shown in figure 1(c).

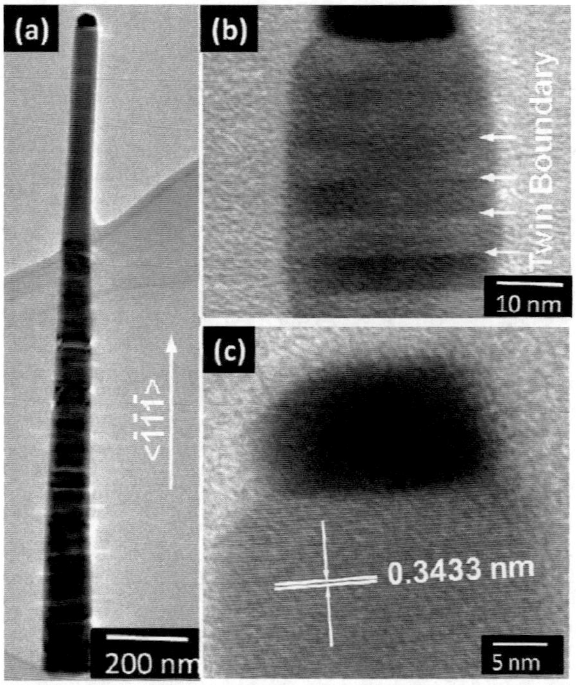

Figure 3. Typical of TEM images of InGaAs nanowire grown at 420°C temperature on GaAs (111) substrate using MOCVD were observed at low, low and high resolutions.

4. Conclusions

Semiconductor nanowire InGaAs has been grown at low growth temperature using MOCVD. The InGaAs nanowire has cylindrical-hexagonal shape grow perpendicular to the substrate via VSS direct impinging mechanism. The bodies of nanowire have planar defect and their crystal structure is typical zinc-blende. The In/Ga ratio on the tip, middle and bottom of nanowires were slightly different but tends to have uniform chemical composition. The growth temperature give effect of increases of length and diameter of nanowires and at lower temperature, tapering could be controlled.

Acknowledgments

We thank to the Ministry of Higher Education of Malaysia for the financial support via RU grants with Vote No: 78339 and 02J85.

References

[1] D. M. Cornet and R. R. La Pierre 2007 *Nanotechnology*18, 385305
[2] Y. Kim, H. J. Joyce, Q. Gao, H. H. Tan, C. Jagadish, M. Paladagu, J. Zou and A. A. Suvorova, T. Sato, Y. Kobayashi, J. Motohisa, S. Hara and T. Fukui 2008 *J. Cryst. Growth* 310, 5111
[3] T. Sato, J. Motohisa, J. Noborisaka, S. Hara and T. Fukui 2008 *J. Cryst. Growth* 310, 2359
[4] T. Sato, Y. Kobayashi, J. Motohisa, S. Hara and T. Fukui 2008 *J. Cryst. Growth* 310, 5111
[5] M. E. Messing, K. Hillerich, J. Johansson, K. Deppert and K. A. Dick 2009 Gold Bull. 42, 172
[6] H. J. Joyce, Y. Kim, Q. Gao, H. H. Tan and C. Jagadish 2007 *Proc. 2nd IEEE Int. Conf. Nano/ Micro Engineered and Molecular Systems*, Bangkok, Thailand.
[7] S. Kodambaka, J. Tersoff, M. C. Reuter and F. M. Ross 2009 *Phys. Rev. Lett.* 96, 096105
[8] D. Jishiashvili, V. Kapaklis, X. Devaux, C. Politis, E. Ketelia, N. Makhatadze, V. Gobronidze and Z. Shoilashvili 2009 *Adv. Sci. Lett.* 2, 40.
[9] E. J. Schwalbach and P. W. Voorhees 2008 *Nano Lett.* 8, 3739
[10] H. Adhikari, A. F. Marshall, C. E. D. Chidsey and P. C. McIntyre 2006 *Nano Lett.* 6, 318
[11] J. L. Lensch-Falk, E. R. Hemesath, F. J. Lopez and L. J. Lauhon 2007 *Am. Chem. Soc.* 129, 10670
[12] M. Omari, N. Kouklin, G. Lu, J. Chen and M. Gajdardziska-Josifovska 2008 *Nanotechnology* 19, 105301
[13] K. W. Kolasinski 2006 *Solid State and Mater. Sci.* 10, 182
[14] H. J. Joyce, Q. Gao, Y. Kim, H. H. Tan and C. Jagadish 2007 *Proc. 20th Annual Meeting of the Lasers and Electro-Optics Society*, Lake Buena Vista, Florida, USA
[15] K. A. Dick, K. Deppert, L. Samuelson and W. Seifert 2007 *J. Cryst. Growth* 298, 631
[16] H. J. Joyce, Y. Kim, Q. Gao, H. H. Tan and C. Jagadish 2006 *Proc. Int. Conf. Nanoscience and Nanotechnology*, Brisbane, Australia
[17] L. Samuelson, C. Thelander, M. T. Bj¨ork, M. Borgstr¨om, K. Deppert, K. A. Dick, A. E. Hansen, T. M˚artensson, N. Panev, A. I. Persson, W. Seifert, N. Sk¨old, M. W. Larsson and L. R. Wallenberg 2004 *Physica E* 25, 313
[18] X. Duan and C. M. Lieber 2004 Adv. Mater. 12, 4
[19] G. Cao, 2003 *Nanostructures and Nanomaterials: Synthesis, Properties and Applications* (Imperial College Press, London).
[20] E. I. Givargizov, 1975 *J. Cryst. Growth* 31, 20.
[21] K. A. Dick, D. Knut, L. Samuelson and W. Seifert 2006 *J. Cryst. Growth* 297, 326

Journal of Physics: Conference Series **423** (2013) 012048 doi:10.1088/1742-6596/423/1/012048

Characteristics of Leachate at Sukawinatan Landfill, Palembang, Indonesia

Eka Sri Yusmartini[1] , Dedi Setiabudidaya[2], Ridwan[3], Marsi[4] and Faizal[5]
[1]PhD Candidate, Postgraduate Program, Sriwijaya University, Indonesia
[2]Department of Physics, Faculty of Mathematics and Science, Sriwijaya University, Indonesia
[3]Nuclear Industrial Materials Technology Centre, Serpong, Indonesia
[4]Faculty of Agriculture, Sriwijaya University, Indonesia
[5]Department of Chemical Engineering, Faculty of Engineering, Sriwijaya University, Indonesia
E-mail: eka.yusmartini@gmail.com

Abstract. Landfill (TPA) Sukawinatan Palembang is an open dumping system which covers an area of 25 hectares. This system may bring an environmental damage to the surrounding area because it does not provide leachate treatment. Leachate is the landfill waste that dissolves many compounds that contain pollutants from both organic substances and heavy metal origin. This paper presents the results of laboratory analysis on samples of leachate as well as shallow groundwater from the surrounding area. The results were compared to established quality standards to evaluate whether the leachate has influenced the quality of the shallow groundwater in the surrounding area. The results show that there are some indications that the quality of groundwater has been polluted by the leachate of both organic substances and heavy metals produced by the Sukawinatan landfill.

1. Introduction

Palembang as the second largest city on the island of Sumatra is geographically located at the position of $104^0 37'$- $104^0 52'$ east longitude and 2^0 52 -$3^0 05$ ' south latitude. The total area of Palembang city of approximately 40,000Ha, which is divided into16 counties and 107subdistrict. Palembang is sited on swampy lowlands-marshes with tidal height between 3.5 meters to 4.12 meters above sea level. Currently, the solid waste of Palembang area is dumped in two landfills, namely Karya Jaya and Sukawinatan. This study focused on the Sukawinatan landfill, located in Sukarame District. The landfill covers an area of 25 hectares and began its operation in April 1994 with a design life of about 15 years. The operation of the landfill is managed by Department of Hygiene and Palembang City Cemetery.

Best practice for solid waste treatment in Palembang city is presented by Eddy Santana Putra [1]. One method is to reduce the volume of the solid waste that goes to landfill is by composting technique in household, offices, and schools. Despite this effort, the volume of waste that goes to Sukawinatan landfill is still large. The waste that goes to this landfill during the year of 2010 to 2011 is increasing by 111.463 pounds/day. The total volume of solid waste disposed to this landfill in 2009 was 150,900,928 pounds, increased to164,166,986 pounds in 2010 and then increased to187,703,406 pounds in 2011. The increased amount of solid wastes clearly adding to the burden on the landfill Sukawinatan if not treated properly, and thus will impact the use of the landfill it self.

The solid waste in the landfill is not only made up of solid components, but also liquid waste which contains chemical substances, both organic and inorganic, as well as a number of pathogenic bacteria, called leachate[2]. According to [3,4,5,6], leachate is the liquid that are likely to contain a large amount of organic contaminants, the COD(chemicaloxygen demand), BOD(biochemical oxygen demand), ammonia, hydrocarbons suspended solids, concentrations of heavy metals and inorganic salt. Leachate is also rich in phenols, nitrogen and phosphorus. If the leachate is no ttreated, it can be disposed into the environment and subsequently be apotential source of contamination to soil and groundwater, when absorbed into the soil, hence; it can lead to pollution of the waters[6, 7,8,9,10]. Influence of leachate contamination on groundwater is a major drawback of the method of open disposal of solid waste, although the advantage of this method is simple procedure and low operational costs. Given the effects and dangers of leachate discharged directly to the environtment may affect the water body, lowering the quality of surface and ground water, and also affect the ecosystem, the characteristic leachate generated in the landfill Sukawinatan need to be evaluated.

2. Composition and Characterization of Leachate

The quality and quantity of leachate is highly variable and is directly related to fluctuations of rainfall amount, composition/ characteristics of the waste, age, and landfill operational patterns[11]. Landfill leachate composition varies greatly depending on the season, leachate collection system and landfill age. The main pollutants in the leachate is organic matter and ammonia, where in if the landfill age increases, the concentration of organic matter (COD) in leachate decreased from 1800 mg/L in the second year to 610 mg /L in the sixth year [12]. Increase in the concentration of ammonia nitrogen from 98 mg to 364 mg $N_{NH4/l}$ also occur as the age of the landfill increases. Fluctuations of the other parameters (phosphorus, chloride, calcium, magnesium, sulfate, dissolved solids, heavy metals, also depends on the season of the year (seasonal variation) and the age of the landfill. Reported data also showed that that despite the landfill is still in the early age, some parameters such as pH value high (average 7.84, COD concentration is low (<2000 mg COD / L), the ratio of BOD / COD is low (<0.4), and the concentrations of heavy metals is low as well, so it can be indicated that the characterization is influenced by the conditions in the landfill methagonic at the beginning of the monitoring period.

Characterization of leachate from the landfill is generally represented by the COD (Chemical Oxygen Demand), total organic carbon (TOC), Biochemical Oxygen Demand (BOD), the ratio of BOD / COD, pH, suspended solids (SS), ammonium nitrogen (NH3-N), Total Kjldahl Nitrogen (TKN), the number of bacteria, turbidity and heavy metal values [13,14,15]. Concentration of leachate parameters change according to the age of the leachate. The relationship of leachate concentration with age can be seen in Table 1.

Table 1. Concentration of some leachate constituents at different phase (Kostova, 2006)

Lechate Constituen	Transition phase (0- 5 years)	Acid formation phase (5-10 years)	Methane Fermentation (10-20 years)	Final maturation phase(>20)
BOD	100-11000	1000-5700	100-3500	4-120
COD	500-22000	1500-71000	150-10000	30-900
TOC	100-3000	500-28000	50-2200	70-260
Ammonia	0-190	30-3000	6-430	6-430
NO_2-N	0.1-500	0.1-20	0.1-1.5	0.5-0.6
TDS	2500-14000	4000-55000	1100-6400	1460-4640

In general, if the landfill becomes older, the biological decomposition of waste deposited turned out to be longer, which consists of two sub-stages and methagonic acid. Leachate from different stages contain different elements, so people tend to be acid leachate because of volatile fatty acids. The content of heavy metals in leachate increase in acidic condition because the low pH can cause the metal to be easily dissolved. On the other hand, the metal content comes down on the fermentation and maturation phases, where the pH increases to neutral.

3. Quality of landfill leachate in Sukawinatan

The Sukawinatan Landfill Waste is an open dumping system. The increasing volume of waste obviously need better handling, especially the handling of liquid waste. The leachate is formed by external water ingress into the landfill waste, dissolve and rinse the dissolved materials including organic and inorganic compounds as the results of decomposition process [17].

The laboratory analysis of leachate samples were conducted at the Department of Hygiene and Funeral Palembang. Sampling of leachate was taken on February 21, 2011(sample A) and on7 November 2011 (sample B). The results and their comparison with quality standards can be seen in Table 2. Based on Table 2, parameters that do not meet quality standards according the South Sumatra Governor Regulation No. 18 of 2012 on the sample A is the content of ammonia(NH3-N). Ammonia levels that exceed the quality standards can impair respiratory tissues and cause odor in the water. COD and BOD values in sample A is still below the value of quality standards. Similarly, the content of heavy metals such as Chromium, Mn, Zn and Cu is acceptable. Sample B has a variety of different values such as BOD, COD, ammonia, chloride which exceeded the quality standard. The concentration of COD and BOD in both samples A and B match to those reported in the literature for sanitary landfill that reach an age of 10 – 20 year. In this case, the sanitary landfill is entering methane phase with the concentration BOD between 300 and 3500, and COD between 50 and 10000. As for sample A, the metal contents in sample B are acceptable. The low concentrations of metals in soil is influenced by open dumping leachate pH values, but it also plays a role in rainwater flushing on landfill waste [18].

Table 2. The results of laboratory analysis on samples of leachate

Parameter	Sample A	Sample B	Max. Limit	Test Methode
pH	7.45	6.88	6-9	SNI 06-6989.11-2004
Temperature	29.3	28.0	40	-
TSS	44	45	400	SNI 06-6989.3-2004
COD	291.1	354.8	300	SNI 06-6989.3-2004
BOD_5	145.7	218.1	150	SNI 06-2503-1991
NH_3-N	65.00	11.324	5	Salisilat Treatment
Nitrit	0.012	0.117	3	SNI 06-6989.9-2004
Nitrate	93.6	3.7	30	SPECTOFOTOMETRI
Sulfide	199	0.085	0.1	SPECTOFOTOMETRI
Cyanide	0.011	0.017	0.5	SPECTOFOTOMETRI
Chromium(Cr^{6+})	0,056	0.049	0.5	SPECTOFOTOMETRI
Chlorida	162.5	174.6	2	SPECTOFOTOMETRI
Fe	1.6	1.8	10	SPECTOFOTOMETRI
Cu	0.05	0.04	3	SPECTOFOTOMETRI
Zn	0.05	0.06	3	SPECTOFOTOMETRI
Mn	<0.2	0.4	5	SPECTOFOTOMETRI

4. Quality of shallow groundwater in the surrounding area

Residential areas surrounding the landfill Sukawinatan is highly populated with about 645 families living in a number of households. Well water is widely used by people around the settlement for their daily needs. Proximity of residential and landfill sites, besides causing ill-equipped landscape aesthetics also cause bad odor. In addition,the possibility of contamination of well water is also quite high. The results of laboratory analysis of water monitoring wells and water wells conducted by the Sanitation Department Palembang are presented in Table 3.

Based on Health Minister Regulation No.907 of 2002, the maximum water concentration of chromium in monitoring wells or household well is 0.012 mg/L and 0.017 mg/L. The levels of chromium in the leachate was detected in sample A is 0.056 mg/L and the B sample 0.049 mg/L. Even though the concentration of chromium in the leachate still meets the standard by Governor Regulation No. 8 of 2012, the level chromium in well water does not meet quality standards based Health Minister Regulation No.907. Complex Chromium(CrVI) as dichromate ions are adsorbed by soil can seep into the ground and then headed into the groundwater[18]. Permeation is dependent on soil pH. In neutral or alkaline solution of Cr(VI) is stable and easy to be absorbed[17]. In addition, the spread of chromium in groundwater depends on the sorption characteristics of the soil, including clay content, iron oxide and the amount of organic matter[18].

Table 3. The results of the analysis of well water

Parameter	Unit	Monitoring Well	Household Well	Maximum Limit
Chromium(Cr^{6+})	mg/L	0.012	0.017	0.005
Cu	mg/L	1.14	0.02	2
Cyanide	mg/L	0.012	0.005	0.007
Nitrate	mg/L	5.6	7.1	50
Nitrit	mg/L	0.025	0.007	3

Levels of Cu in the water monitoring wells reaching 1.14 mg / L and while the level in the household wells only 0.02 mg / L. This level is of Cu still acceptable according to Health Minister Regulation. Equally, the content of Cu in the leachate samples are still acceptable. Levels of nitrate and nitrite in both water monitoring wells and the household wells are still below the threshold of quality standards, as well as the leachate samples. The value of cyanide in leachate samples A and B respectively are 0.011 mg / L and 0.017 mg / L. In well water, the concentrations of cynida are 0.012 mg / L and 0.005 mg / L.

The concentration of cyanide in both leachate samples and monitoring wells and the household well are still below the limit stated in the quality standards. However, careful treatment is needed because high concentration of cyanide is harmful to human which may affect respiratory tissues, resulting in asphyxia, chronictoxicity and irritation will cause malaise. The concentration of nitrate in drinking water is also not expected because it is dangerous to human. It can cause methemoglobinemia or blue baby, which is a disease that affects children due to the oxygen-carrying capacity of the blood in the interference by nitrate.

5. Conclusions and Suggestion

The quality of leachate from two samples being analyzed in this study differs in terms of COD and BOD concentrations which may be related to the amount of rainfall. In this case, sample A was taken during dry season while sample B during rainy season. High ammonia concentrations can lead to disruption of water bodies such as wells or shallow groundwater. Although the metal content and effluent from the landfill still meets the applicable quality standards, leachate from the landfill is still a major source of environmental pollution on heavy metals. The presence of heavy metals and also the content of nitrate and nitrite ions in water or shallow ground water monitoring wells and the household wells need to be examined though concentration is still below the threshold standard.

To prevent and reduce effluent of leachate in landfill waste (landfill) Sukawinatan Palembang, leachate management systems need to be made prior to discharge into the environment. Management system must consider the characteristics of the leachate management system. The system should be selected that the process can eliminate or reduce the flow of contaminants to water bodies around the landfill. An understanding of the characteristics of the leachate is expected to be used to select appropriate management and economical.

References

[1] Eddy Santana Putra, 2012 *Solid Waste Management in Palembang City* , Presented at The 3rd High Level Seminar on Environmentaly Suistainable Cities, Siem Reap-Cambodia, March 2012.

[2] Damanhuri T P 1993 *Landfill lechate management in relation to prevention of environmental pollution*, a national seminar proceedings env.ironmental management Indonesia ITB, ISBN 979-8456-00-9

[3] Renou S, Givaudan J G, Poulain S, Dirassouyan F, Moulin P 2008 Landfill leachate treatment:review and oppurtunity *J.Hazard Mater.* **150** 468-493

[4] Aziz H A, Daud Z, Adlan M N, Hung Y T 2009 The use of polyaluminium chloride for removing colour, COD and ammonia from semi-aerobic leachate *Int.J.Environ.Eng.* **1** 20-35

[5] Foul A A , H A Isa, Hung Y T 2009 *J.Environ. Waste Manag.* **4** 282-298

[6] Aziz S Q, Yusoff M S, Bashir M J K 2010 Leachate Characterization in Semi Aerobic and Anaerobic Sanitary Landfills *J.of Env.Manaj.* **91** 2608-2614

[7]Tatsi A A, Zouboulis A I, Matis K A, Samaras P 2003 Coagulation flocculation pretreatment of sanitary lanfill leachates *Chemosphere* **53** 737-744

[8] Sanphoti N, Towprayoon S, Chaipraset P, Nopharatana A, 2006 The effects of leachate recirculation with supplemental water addition on methane production and waste decomposition in a simulated tropical landfill J. Environ.Manage. **81** 27-35

[9] Lin L, Lan C Y, Huang L N, Chan G Y S 2008 Antropogenic N_2O production from landfill leachate treatment. *J. Environ.Manage* **87** 341-349

[10] Oman C B, Junestedt C 2008 Chemical characterization of lanfill leachates 400 parameters and compounds *Waste Manag.* **28** 1876-1891

[11] Purwanta W 2007 Leachate Treatment Technology Review in Landfills (Landfill) MSW *JAI***3** No.1

[12] Kulikowska D, Kilimiuk E 2008 The effect of landfill age on municipal leachate composition *Bioresour Techl.* **99** 5981-5985

[13] Vazquez H A, Jefferson B, Judd S J, 2004 Membrane bioreactors vs conventional biological treatment of landfill leachate a brief review *J. Chem.Technol.Biotechnol* **79** 1043-1049

[14] Gotvajn A Z, Tisler T, Koncan T 2009 Comparison of different treatment stategies for industrial landfill leachate *J.Hazard.Mater.* **162** 1446-1456

[15] Foo K Y, Hameed B H 2009 An overview of landfill leachate treatment via activated carbon adsorption *J. Of Hazardous Materials* **171** 54-60

[16] Kostova I 2006 *Leachate from Sanitary Landfill Origin, Characteristic, Treatment* University of Architecture, Civil engineering and Geodesy, Iskar Summer School Borovetz 26-29 July

[17] Tchobanoglous G, Theisen H, Virgil S 1993 *Integral Solid Wate Management Engineering Principles and Management Issues*, first ed. Irwin McGraw-Hill, USA

[18] Komala P S 2008 Effect of Open System Development Waste Dumping at the Site Late to the Content of Heavy Metals in Ground Water *J.Teknika* **29** vol. 1

[19] Evanko C R and Dzomback 1997 *Remediation of Metals-Contaminated Soil and Groundwater, Ground-water Remediation Technologies Analysis Center,* Departmen of Civil and Environmental Engineering Carnegie Mellon University, Pittsburg.

Journal of Physics: Conference Series **423** (2013) 012049 doi:10.1088/1742-6596/423/1/012049

Sandwiched polymer fibre in fibrin matrices for the dictation of endothelial cells undergoing angiogenesis

I Sukmana[1*] and J R P Djuansjah[2]

[1] Medical Device and Implant Technology - MediTeg Research Group, Faculty of Biosciences and Medical Engineering, Universiti Teknologi Malaysia, Malaysia
[2] Department of Materials Engineering, Faculty of Mechanical Engineering, Universiti Teknologi Malaysia, UTM, Malaysia

E-mail: irza.sukmana@gmail.com

Abstract. We present here a three-dimensional (3D) sandwich system made by poly(ethylene terephthalate) (PET) fibre and fibrin extracellular matrix (ECM) for endothelial cell dictation and angiogenesis guidance. In this three-dimensional system, Human Umbilical Vein Endothelial cells (HUVECs) were firstly cultured for 2 (two) days to cover the PET fibre before sandwiched in two layer fibrin gel containing HUVECs. After 4 (four) days of culture, cel-to-cel connection, tube-like structure and multi-cellular lumen formation were then assessed and validated. Phase contrast and fluorescence imaging using an inverted microscope were used to determine cell-to-cell and cell-ECM interactions. Laser scanning confocal microscopy and histological techniques were used to confirm the development of tube-like structure and multi-cellular lumen formation. This study shows that polymer fibres sandwiched in fibrin gel can be used to dictate endothelial cells undergoing angiogenesis with potential application in cancer and cardiovascular study and tissue engineering vascularisation.

1. Introduction

The ability to dictate endothelial cells undergoing angiogenesis and growing microvasculature containing multi-cellular lumen has been a long-standing challenge in tissue engineering research [1]. Angiogenesis is the formation of new blood microvessel structure from the pre-existing vasculature. It is a key mechanism to promote microvascular network inside engineered tissue construct that mimic the development of blood vessel functions. To date, there are several three-dimensional assays have been developed for angiogenesis study. They are mainly aimed to assess the multiple steps during angiogenesis progression, including: cell-to-cell connection, tube-like structure development, extracellular matrix (ECM) degradation, and multi-cellular lumens formation [1-2].

The formation of microvascular network inside the engineered scaffold will maintain an adequate tissue oxygenation, nutrient transfer and waste removal, thus secure tissue regeneration [2]. One of the main problems faced by tissue engineering researchers now is how find suitable in vitro assays and methods to study angiogenesis regulations. Example, some angiogenesis studies failed to report the maturation of tube-like structure and the present of multi-cellular lumen structures [3-4].

This study presents an *in vitro* culture system using polymer monofilament, poly-(ethylene terephthalate) (PET), for the guidance of human umbilical vein endothelial cells (HUVECs) undergoing angiogenesis. Step-by-step developmental process of angiogenesis up to the stage of multi-cellular lumen formation will also be presented.

Journal of Physics: Conference Series **423** (2013) 012049 doi:10.1088/1742-6596/423/1/012049

2. Materials and Methods

2.1. Materials

Poly(ethylene terephthalate) 100-µm diameter monofilaments were used and fixed onto polycarbonate frame that fitted into traditional 6-well plates used in cell culture. Phosphate buffered saline solution (PBS 1X, pH 7.4) and Hank's balances salt solution (HBSS, H1387) were purchased from Sigma-Aldrich Inc. and used in these experiments.

Human umbilical vein endothelial cells (HUVECs) were purchased from Cambrex (cc-2519, Walkersville, MD USA). HUVECs were and cultured at 37°C and 5% CO_2 in modified M199 culture medium (M5017, Sigma-Aldrich). HUVECs between passages 2 and 6 were used in all experiments.

2.2. Three-dimensional Angiogenesis assays

Fibrin gels were firstly prepared to be used as the attachment bench, using 1 mL/well of fibrinogen solution (4.0 mg mL^{-1}) in HBSS that was mixed directly with 1 mL of a thrombin solution (2UmL^{-1}) in HBSS for the polymerization process of fibrinogen into fibrin (5 min at room temperature followed by 20 min at 37°C and 5 % CO_2). To evaluate cell adhesion on PET fibres, fibres-covered HUVEC was fixed after 4hrs and preceded for further nucleus staining process. Two different fibre-to-fibre distances (i.e., 100 and 200 micro meter) were tested in this study.

After 2 days of attachment process, fibres-covered cells were then transferred to the new well plate for the sandwich system for further angiogenesis assay. The numbers of cells as well as their connections were evaluated at day 2 and 4 of culture.

2.3. Angiogenesis images

Frame bearing fibres embedded in fibrin gel was firstly fixed with formaldehyde solution (3.75%, wt/v) for 20 min, followed by permeabilization process with Triton X-100 solution (0.5% v/v in PBS) for 5 min, and incubated 1 hr in a PBS solution containing 2% (wt/v) BSA.

Finally, the sample was stained with a mixture of TRITC-phalloidin (1:300 dilution, cat.#P1951, Sigma-Aldrich) and Hoechst (H33258, Sigma-Aldrich) (1:10,000 dilution). Pictures were taken either with epi-fluorescence or laser scanning confocal microscopy (LSCM).

3. Results and Discussions

Although endothelial cell adhesion on PET surface was known as negligible [5], the use of fibrin gel has been successfully increased cell adhesion. Data of cell adhesion after 4 hrs and following 2 days of culture allowing cells to proliferate and spread along the fibre axis, as presented on Table 1.

Table 1. Number of cell attachment on PET fibre surface

Adhesion Time	**Fibre-to-fibre distances**	
	0.1mm	**0.2mm**
4 hours	147.57	67.93
2 days	454.57	132.03
SD 4hrs	14.29	13.26
SD 2dys	60.97	23.91
SEM	34.07	
CD p<5%	74.98	

When HUVECs-covered fibres were sandwiched in the fibrin gel containing 100K of HUVECs, the cell-covered polymer fibres was able to dictate endothelial cell to develop cell-to-cell connection and tube-like structure (Figure 1). Further 4 days of culture, shows a better maturation of microvessel (Fig. 2a). Histological test on the microvessel prove the evidence of multi-cellular lumen (Fig. 2b).

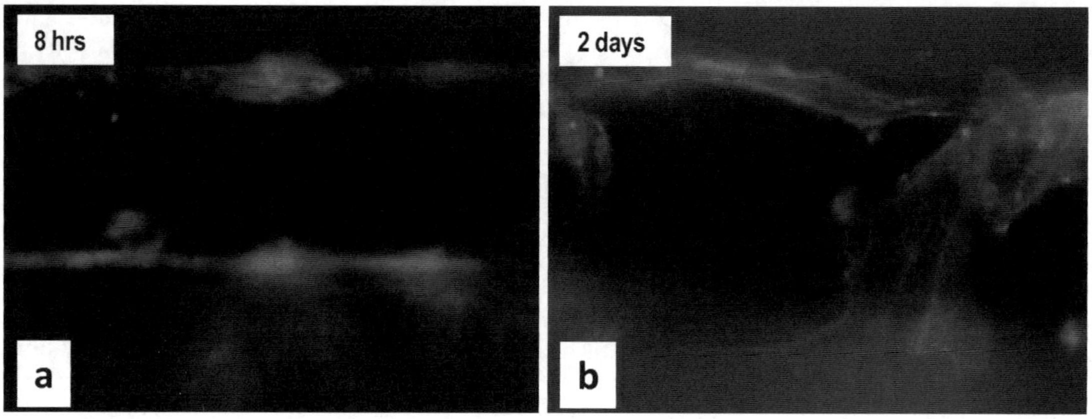

Figure 1. Cells-covered fibre dictates HUVECs undergoing angiogenesis. (A) Cell-to-cell connection. (B) Tube-like structure.

Based on those finding, it can be hypothesized that cell-covered fibres play a significant role as a bioactive sources and cellular guidance for the new cell inside the fibrin gel, which was also in the same agreement with other [6]. During the first 2 days of culture period, cell-covered fibre will migrate along the fibre axes and some are toward to the gel in order to build cell-to-cell connections.

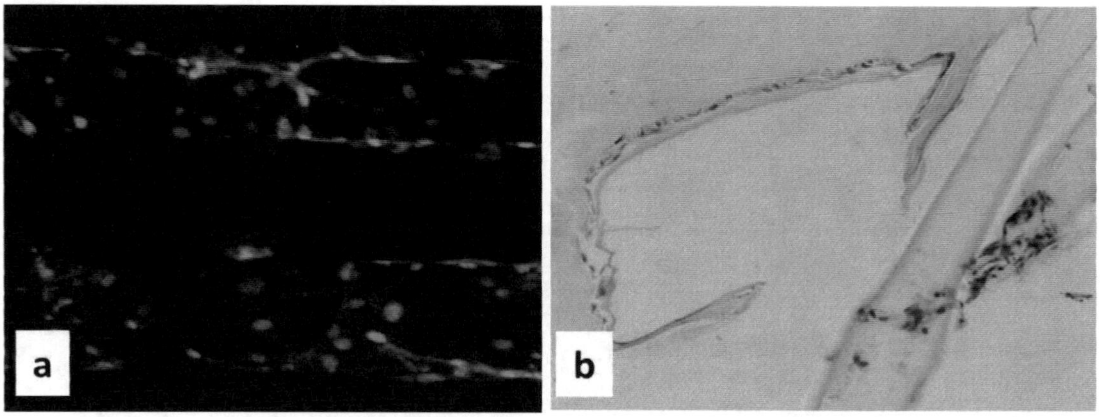

Figure 2. Assessment of microvessel formation. (A) Tube-like structure. (B) Multi-cellular lumen formation of HUVECs guided by polymer fibres.

4. Conclusions

This finding suggests that the direction of angiogenesis can be guided by the physical interactions involved between cells and polymer fibres to form a microvessel wall, allowing lumen formation. The system opens further potential applications in cardiovascular, cancer, and tissue engineering research. We hypothesize that the guidance of endothelial cell undergoing angiogenesis would subsequently facilitate the flow induction in order to vascularised engineering tissue construct.

Acknowledgments
The authors wishing to acknowledge assistance from Prof. Patrick Vermette (Department of Chemical and Biotechnological Engineering, University of Sherbrooke, Canada) for his comment and supervisions. This research work was supported by FRGS (vote# 4F128) and UTM Tier-1 (vote# 03H12) grants.

References

[1] Sukmana I 2012 *J. Artif. Organs* **15** 215
[2] Mooney D J, Sano K, Kaufmann P M, Schloo B, Vacanti J P and Langer R 1997 *J. Biomed. Mater. Res.* **37** 413
[3] Atala A 2004 *Rejuvenation Res.* **7** 15
[4] Nakatsu M N, Saison R C, Aoto J N, Taylor K L, Aitkenhead M, Perez-del-Pulgar S, Carpenter P M and Hughes C C 2003 *Microvasc. Res.* **66** 102
[5] Sukmana I and Vermette P 2010 *J. Biomed. Mater. Res. A* **92** 1587
[6] Hadjizadeh A, Doillon C J and Vermette P 2007 *Biomacromol.* **8** 864.

Journal of Physics: Conference Series **423** (2013) 012050 doi:10.1088/1742-6596/423/1/012050

Influence of stack plate thickness and voltage input on the performance of loudspeaker-driven thermoacoustic refrigerator

Nandy Putra[1] **and Dinni Agustina**[2,3]

[1] Heat Transfer Laboratory, Department of Mechanical Engineering, University of Indonesia, Kampus UI Depok 16426, Indonesia
[2] Graduate student, Mechanical Engineering, University of Indonesia
[3] Department of Mechanical Engineering, Syiah Kuala University, Darussalam, Banda Aceh, Indonesia

E-mail : nandyputra@eng.ui.ac.id

Abstrack. A loudspeaker-driven thermoacoustic refrigerator has been built and tested to gain understanding of its thermal performance and the cooling rate. The influence of plate thickness made of acrylic sheet was experimentally investigated by varying plate thickness of the stack, 0.15 mm, 0.5 mm and 1 mm, respectively. The experiments were conducted with various voltage input to the driver starting from setting 4 to 9 voltage peak-to-peak. The temperatures at both ends of the stack were acquired. For all variations, thermoacoustic cooling effect occurred in seconds and escalated rapidly in two minutes and became stable in ten-minute time. The experimental results showed that higher voltage input yielded higher thermal performance and faster cooling rate. For each set of experiment, the operating frequency and other parameters of the stack were kept unchanged. The experimental results show that the thermal performance and cooling rate increase with the decrease of plate thickness. The largest temperature difference, 14.8°C, was achieved with 0.1 mm plate thickness at voltage setting 9. However, the thermal performance gained for 0.5 mm plate thickness voltage setting of 9, was arguably the optimum thickness in terms of advantages in the ease of fabricating the stack and more consistent cooling.

1. Introduction

Refrigeration has been playing important role in almost every aspect of life to the level that living without it is impossible. It is even rated by the National Academy of Engineering as the10th position in "*Greatest engineering achievement in the 20th century*" [1]. Despite the advantages, refrigeration's high demand of energy and the use of harmful refrigerants have been some of major concerns in the obligatory need of greener technologies in the future. One of alternatives introduced is thermoacoustic technology devices which convert heat energy to sound energy and vice versa. The former is known as thermoacoustic engines or prime movers and the latter is called thermoacoustic heat pumps or refrigerators.

Over the last two decades various designs of thermoacoustic refrigerators [2] and engines [3] have been built and tested in order to increase their performance and reliability. A new spacecraft thermoacoustic cryocooler was developed and used in the space shuttle [4] and a temperature of – 65°C at cold side was achieved [5] by applying Rott's theories [6] and the modification of the theory by Hoffler and Swift [2,7]. These devices are attractive for harmless operation using air or inert gas and simplified manufacturing with much less moving parts in the system [8]. Since then, there are many studies in the existing literature that worked in an attempt to build and increase the performance of the refrigerator.

The operating condition, working fluid and the geometry of the refrigerator parts are main aspects to work with [9]. Experiments have been conducted to gain higher performance and efficiency by experimentally lowering the Prandtl number of the working fluid [10], increasing the ratio of acoustic pressure to mean pressure (drive ratio) numerically [11] and building an empirical equation for resonance frequency of the device [12]. The length of stack, the plate spacing and stack position in a resonator were analyzed [13] and the effects of blockage ratio, the ratio of plate thickness and plate spacing was simulated [14].

Stack is the heart of thermoacoustic refrigerator due to heat transfer occurs around it have great impact on the thermal performance of the device. Amplitude of sound wave input (in terms of acoustic energy) determines the efficiency of the refrigerator. Therefore, these two will be one of the first things to approach in constructing a thermoacoustic refrigerator. This paper will present the results of the experiments on this matters which emphasized on getting optimum plate thickness and input energy for a given resonator. The largest temperature difference across the stack, 14.7°C, which was realized in less than 10 minutes showed the feasibility of providing enough cooling for an environmentally friendly refrigeration system with a significant cooling rate. The potential application areas are commercial refrigerators and industrial air conditioning.

2. Design parameters of the thermoacoustic refrigerator

The schematic presentation of $\lambda/4$ wavelength thermoacoustic refrigerator is shown in figure 1. It operates by using high amplitude sound waves and inert gases to produce cooling. The frequency of the driver, such as loudspeaker, and the geometry of the resonator are calculated so as to get a standing sound wave in the resonator. A porous component called a stack is placed in the tube in such a way that a temperature difference due to thermoacoustic effect is created along the stack. One end of the stack starts to heat up while other end starts to cool down. By controlling temperature of hot side of stack (by means of a heat exchanger), the cold end of stack can provide lower temperature. There are five main components in a thermoacoustic refrigerator which are a resonator, a stack, a driver, working fluid and heat exchangers. In this experiment, there is no heat exchanger attached to device.

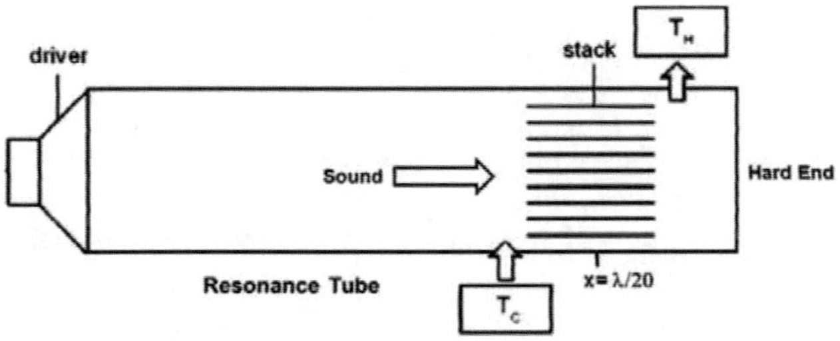

Figure 1. The schematic of $\lambda/4$ wavelength thermoacoustic refrigerator.

2.1. Resonator

A smooth, linear cylindrical resonator without steps, misalignments and abrupt transition should be used to avoid unwanted eddying or non- linear pressure variations that would greatly complicated the analysis [15]. The length and diameter of resonator are 40 cm and 5.3 cm, respectively. The resonance frequency of sound wave is mainly determined by the length and the diameter of resonator [16].

Journal of Physics: Conference Series **423** (2013) 012050 doi:10.1088/1742-6596/423/1/012050

$$f_n = \frac{nv}{4\left(L + \frac{14D}{23\pi}\right)} \tag{1}$$

However, these theories must meet the result of the frequency test of the thermoacoustic device which was performed by inserting a condenser microphone into the resonator and connected it to the oscilloscope to get a relation between the frequency and its connecting voltage. The one with the highest voltage result was the resonance frequency which is best operating frequency for the device.

The easiest way to decrease the acoustic power losses is to decrease the surface area of resonator wall; therefore the $\lambda/4$ wavelength resonator was chosen [2]. The material of resonator should be low in thermal conductivity and must endure vibration and pressure to the certain level. Acrylic tube was chosen because of these characteristics and its transparency help in order to arrange the stack and thermocouples and visualize them as well inside the resonator.

2.2. Stack

The dimension of the stack should be around 10% of the resonator volume and the spacing between the plates is determined by the thermal penetration depth of the stack which is 2 – 4 times the value of it [5]. In this experiment, acrylic sheet was cut into 6 cm length with the wide adjusted to the inside diameter of the resonator. The space between two plates ($2y_o$) is 0.55 mm ($3\delta_K$).

$$\delta_K = \sqrt{\frac{K}{\pi \rho f c_p}} \tag{2}$$

Low thermal conductivity and relatively large heat capacity is among the characteristics of the stack material to minimize heat transfer along the x-axis and maximize heat storing while phasing the sound wave in order to produce cooling effect [10]. Widely available acrylic sheet was chosen to meet these criterions. The variations of stack plate thickness ($2l_o$) are 0.15 mm, 0.5 mm and 1 mm. The stack was located at the distance of $\lambda/20$ from the closed end of the resonator [5] as this position will bring the optimum performance to the device because a well-balanced of pressure and velocity of the sound wave is achieved [3].

Figure 2(a) shows a cross sectional view of parallel plate stack, the blockage ratio is defined as the ratio of area available to gas in the stack to the total area of the stack. It is expressed as

$$BR = \frac{2y_o}{2y_o + 2l_o} \tag{3}$$

Figure 2(b) shows the gas parcels around the stack which experience displacement and temperature oscillation in association with the pressure variations. When the gas parcel compressed, it transfers heat to the plate so that temperature of the plate higher than that of the gas. Then the gas parcel expands and absorbs the heat of the plate. This phenomena which is induced by sound energy continues and generates temperature difference along the stack. Occurrence of the temperature difference will gradually end if the critical temperature gradient (∇T_{crit}) across the stack is equal to the local temperature gradient [17], where

$$\nabla T_{crit} = \frac{p}{\xi \rho c_p} \tag{4}$$

p is the acoustic pressure and ξ is the acoustic displacement amplitude. This temperature is important in determining the properties of a thermoacoustic device, since efficiency depends on a temperature differential caused by the sound waves that is larger than the critical temperature so that a large cooling effect is created [5].

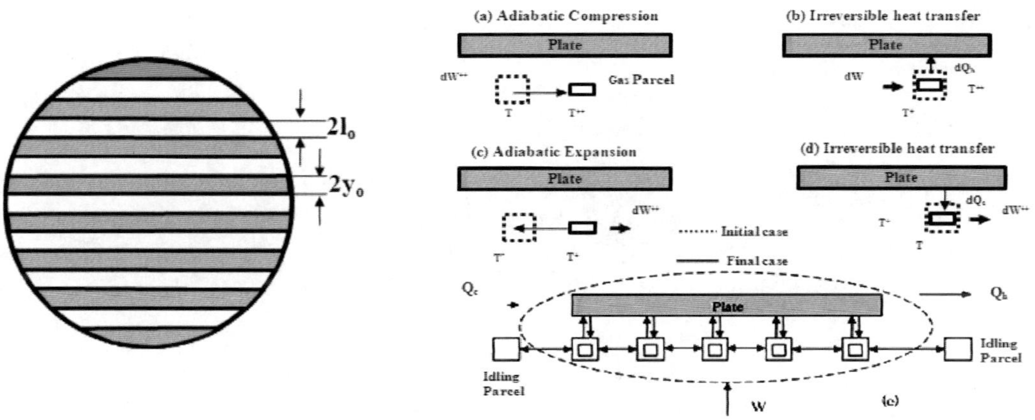

(a) (b)

Figure 2. (a). A cross sectional view of the stack (b). Heat transfer between plate and gas parcels [18]

2.3. Working gas
The thermal penetration depth in designing stack also put into consideration the properties of working gas, which in this experiment was air. Air was chosen mainly because of its availability and its Prandtl number falls into the acceptable category of working gas for thermoacoustic device [10].

3. Fabrication and procedure of the experiments
Components of the refrigerator was fabricated or chosen based on the literatures. The resonator was built from a 40 cm straight acrylic tube. The internal diameter of the tube is 5.3 and the wall thickness is 3 mm. The length of the resonator was set equal to quarter of wavelength of the acoustic wave based on equation 2. Air at atmospheric pressure was used as a working fluid, as mentioned in section 2.

An 8Ω loudspeaker with the maximum power of 30 W was used as acoustic driver and installed at one end of the resonator. Another end was closed tightly using the same material as the resonator. After determining the operating frequency experimentally, the square wave at frequency 180 Hz was set in built-in function generator of a National Instrument Elvis II and the sound wave was amplified.

The length and the thickness of each stack plate were 6 cm and 0.1 mm, respectively. The stack was made of acrylic sheet and cut manually to fit into a cylindrical stack holder. The spacing between the plates was realized by gluing 0.55 mm fishing line on to the surface of each plate. Then the plates were arranged and glued in a stack holder made of the same material as the plates. The stack holder was inserted into the resonator. The fabrication of the stack was done in the same procedures for other sets of experiments with 0.5 mm and 0.15 mm plate thickness, respectively.

Five type K thermocouples with accuracy of 0.05 were used; one for ambient temperature and two of those were inserted at different position at each end area of the stack. The temperature data from all thermocouples were acquired simultaneously every second by using National Instrument data acquisition module 9211. Both thermocouples at hot side read the same temperature while those at the cold side showed difference in reading of 1 – 2°C. The lowest reading was featured on the graphs.

3.1. Experimental Set-Up
Figure 3 shows the experimental setup of the thermoacoustic refrigerator constructed at the Heat Transfer Laboratory, University of Indonesia. The stack was placed at 6 cm from the closed end. The

Journal of Physics: Conference Series **423** (2013) 012050 doi:10.1088/1742-6596/423/1/012050

frequency and wave amplitudes were measured by a condenser microphone which was located at pressure antinode of the resonator. The microphone was connected to a built-in oscilloscope of National Instrument Elvis II via Labview software. The oscilloscope confirmed the form and frequency of the wave input.

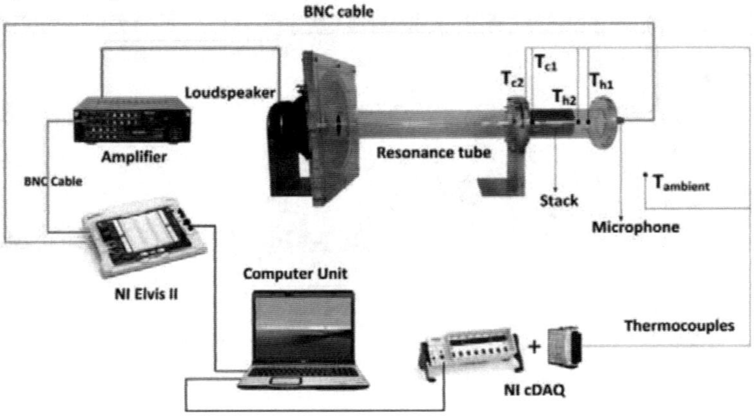

Figure 3. The experimental set-up of the experiment

The experiment was conducted with variations in voltage peak to peak setting from four to nine as the input voltage to the driver using a built in function generator in NI Elvis. The thickness of the plate stack was varied; 0.15 mm, 0.5 mm and 1 mm. Figure 4 is the capture of one of the experiment that shows (a) the function generator used to input the sound wave and input voltage to the driver and (b) the oscilloscope monitor.

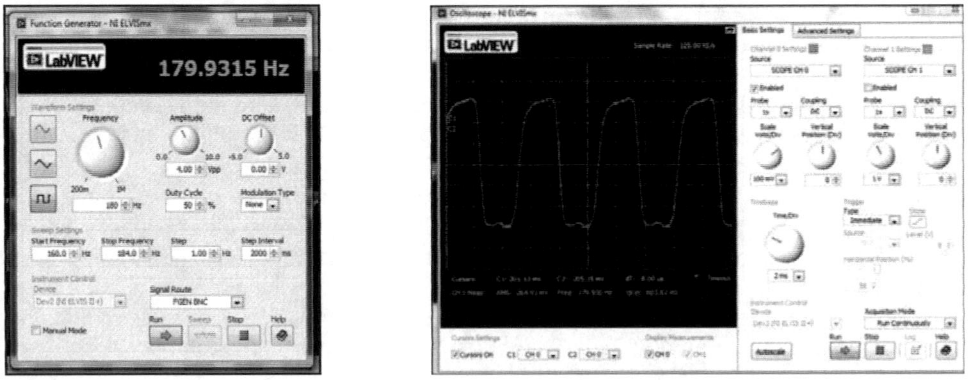

(a) (b)

Figure 4. Wave and voltage setting using National instrument Elvis II Function Generator (a) and (b) waveform and wave amplitude reading using NI Elvis II Osciloscope

4. Result and Discussion

The theoretical resonance frequency for the resonator based on Equation 2 is 209.6 Hz. The resonance frequency became around 180 Hz after installing the loudspeaker and placing the stack in the resonator. The existence of other structures and extra volume after the driver alter the resonance frequency [12]. 180 Hz is an approximation since the local temperature gradients established by the device alter the resonance frequency of the internal fluid [14].

Figure 5 shows the tested frequencies and their connecting voltage result at different positions; positions 1 to 6 which are near the edge of the tube, λ/20 from the driver, λ/8, at the middle of tube length, λ/20 from the tube end and the last one is the left end .The frequency that resulted higher voltage was considered as the resonance frequency which was consistently around 180 Hz.

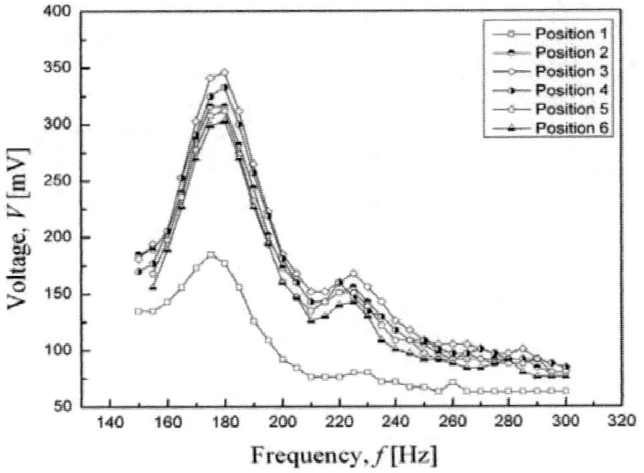

Figure 5. Frequencies and their connecting voltage output at different position in resonator

4.1. Effect of input voltage setting on thermal performance

To understand how sound wave amplitude affects the thermal performance of the resonator, sound wave amplitude was varied by setting the voltage peak to peak input from 4 to 9 with one increment for each experiment. The experiments were conducted for three different plate thicknesses; 1mm, 0.5 mm and 0.15 mm, respectively. Figure 6 shows the typical results for temperatures at the hot side of the stack (T_h) and the cold side of the stack (T_c) in a room with 26.5°C ambient temperature (T_a).

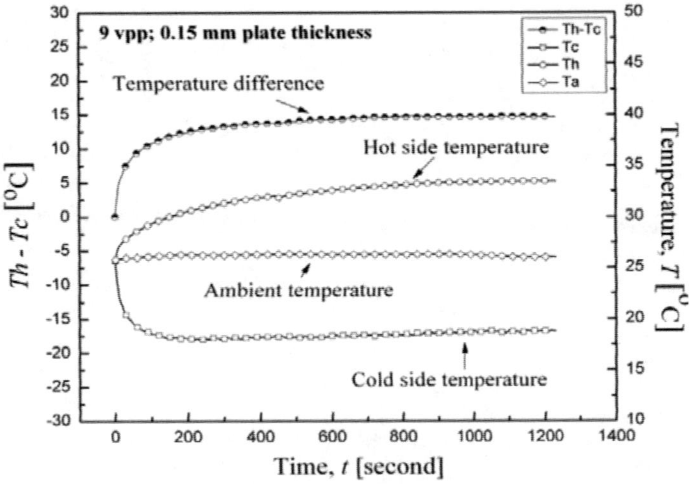

Figure 6. The largest temperature difference of hot side and cold side of stack

It is seen that the thermoacoustic phenomena inside the resonance tube yielded temperature difference between two extremities of the parallel plate stack. The result shown is the largest temperature difference, 14.8°C achieved in 10 minute time that occurred by using 0.15 mm plate thickness at input setting 9 vpp. Temperature at the cold side of the stack decreased rapidly to the lowest point, 18.4°C, in two minutes. Temperature at the hot side increased to the highest point, 34.2°C, in 10 minutes while at the same time the temperature at the cold side started to increase slightly. The trend in Figure 6 is similar to that found in the literature [19].

Figure 7 presents the temperature difference across the stack yielded with various input voltage using 0.15 mm plate thickness. As expected the plot shows that higher input voltage will produce linearly higher temperature difference. The difference grows rapidly more than 50% in 30 seconds and continues to around 80% in 4 minutes and becomes generally constant in 10 minutes. These trends are in agreement to those shown in literature [15]. It is observed that starting from 10 minutes duration of experiment the local temperature gradient is relatively equal to the critical temperature gradient so that the acoustic energy have been fully used to overcome the heat dissipation[17].

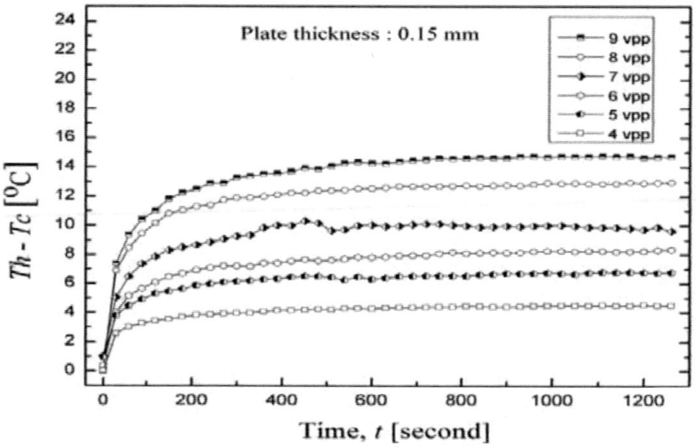

Figure 7. The thermal performance of the resonator from 4 to 9 vpp

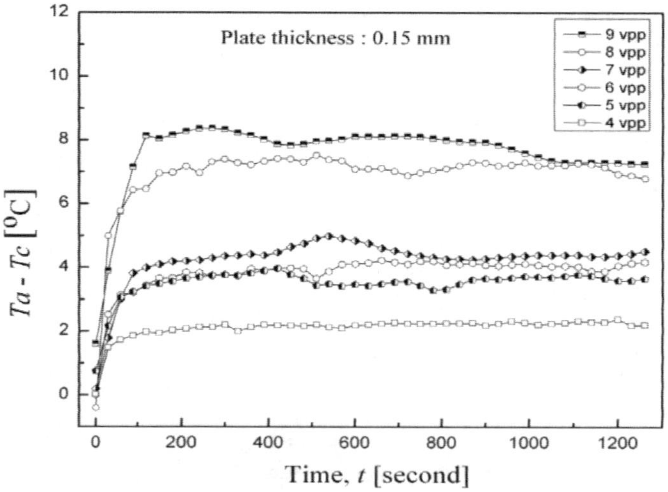

Figure 8. Cooling rate of cold side of the stack , plate thickness of 0.15 mm

The difference in temperature of the stack cold side (plate thickness of 0.15 mm) and the ambient temperature was plotted against time in figure 8. The data was recorded for 20 minutes with various input voltage to the loudspeaker. The figure shows that at 9 vpp input to the driver the temperature of cold side decreased the most, 8.4°C, in 90 seconds. The temperature then began fluctuating on the cold side of the stack due to drifting of thermocouples but it stayed lower than others. The figure also indicates that the cooling rate with input voltage 8 and 9 were faster than other input voltages although the trends are similar for all variations. The data was also recorded for different plate thickness, 0.5 mm and 1 mm with various input voltage.

4.2. The effect of plate thickness on the thermal performance and cooling rate

Figure 9 depicts that using stack plate of 0.15 mm thickness produced the largest temperature difference across the stack and the fastest cooling rate compared to plate thickness of 0.15 mm and 1 mm. The input voltage to the driver remained constant during the experiment. At 9 vpp, the largest temperature difference (T_h-T_c) for plate thicknesses of 0.15 mm, 0.5 mm and 1 mm were 14.8°C, 13.1°C and 9.2°C respectively. It indicates that for the given resonator geometry, thinner plates means lower blockage ratio and produces higher thermal performance as found in literature [20].

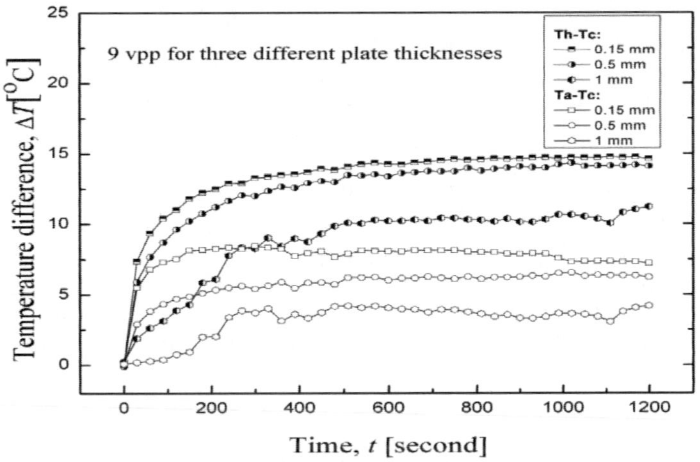

Figure 9. Thermal performance and cooling rate with variation in stack plate thickness at 9 Vpp

Figure 9 also shows that at the same condition of input voltage, decrease in temperature at cold side of the stack (T_c-T_a) were up to 8.1°C, 7°C and 4°C for plate thicknesses of 0.15 mm, 0.5 mm and 1 mm, respectively. It was observed that thinner stack plate shows faster cooling rate at the cost more effort in fabricating the stack plate and deterioration of the stack plate after several set of experiment due to less rigidity in material used for the stack.

5. Conclusions

In this study, experiments are conducted to observe the performance of the refrigerator in terms of temperature difference produced at the ends of the stack and the cooling rate. The results showed that higher amplitude of sound wave (ranging from 4 to 9 vpp setting) produce larger temperature difference, as expected. The experimental result showed that using plate thickness of 0.15 mm at 9 vpp yielded the largest temperature difference when compared to 0.5 mm and 1 mm plate for the given voltage input to the driver. Decrease in plate thickness leads to significant increase in cooling rate; which are for 0.15 mm plate thickness temperature dropped 8.5°C below the ambient temperature in two minutes. For 0.5 mm and 1 mm plate thicknesses, the temperature drops 5°C and 1°C, respectively, below ambient in the same amount of time. Thinner plates which provide more area for heat transfer relative to the given resonator ensure faster cooling rate. However, whether the thinnest is the most optimum one still needs further study since by using plate of 0.5 mm thickness the cooling effect was more consistent and the stack structure was much more rigid compared to others.

References :

[1] Poesse M E 2012 Handbook of Climate Change Mitigation *Springer US.* 1821 – 48

[2] Hofler T J 1986 Thermoacoustic refrigerator design and performance PhD thesis *Physics Department University of California San Diego*

[3] Swift G W 1992 Analysis and performance of a Large thermoacoustic engine *The Journal of Acoustical Society of America* **92** 1551 -63

[4] Garret S I, Adeff J A and Hofler T J 1993 Thermoacoustic refrigerator for space application *Journal of Thermophysics and Heat Transfer* **7** 595 – 9

[5] Tijani M E H, Zeegers J C H and De Waele A T A M 2002 Construction and performance of a thermoacoustic refrigerator *Cryogenics* **42** (4) 59 – 65

[6] Rott N, 1980 Thermoacoustics *Jou.Adv. Appl. Mech.*, **20** 135—175

[7] Swift G W 1988 Thermoacoustic Engines *J. Acoust.Soc.Am* **84***(4)* 1145 - 80

[8] Zink F, Vipperman J and Schaefer L 2010 CFD simulation of thermoacoustic cooling International Journal of Heat and Mass Transfer 53 3940 – 46

[9] Ghorbanian K, Hosseini H and Jafargholi M 2008 Design road-map for thermoacoustic refrigerators *The Journal of the Acoustical Society of America* **123** issue 5 3546

[10] Tijani M E H, Zeegers J C H and De Waele A T A M 2001 Prandtl number and thermoacousticrefrigerator *Journal Acoust. Soc. Am* **112**(1), 134-143

[11] Tasnim S H, Mahmud S, Fraser R A 2012 Effect of variation in working fluids and operating conditions on the performance of a thermoacoustic refrigerator *International Communications in Heat and Mass Transfer* **39**, 762 – 8

[12] Ghazali N M, Ghazali A D, Ali I S, Rahman M A A 2012 Geometry effects on cooling in a standing wave cylindrical thermoacoustic resonator *AIP Conf. Proc.* **1440** 1320 downloaded on December 1, 2012

[13] Nouh M A, Arafa N M and Rahman E A 2009 Stack Parameters effect on the performance of an anharmonic resonator thermoacoustic heat engine *3rd International Conference on Integrity, Realibility and Failure*, **S 2106**, 478, 2009

[14] Zoontjens L 2008 Numerical investigations of the performance and effectiveness of thermoacoustic couples Ph.D Dissertation, *School of Mechanical Engineering, University of Adelaide*downloaded from *digital.library.adelaide.edu.au*

[15] Hariharan N M, Sivashanmugam P, Kasthurirengan S 2012 Influence of stack geometry and resonator length on the performance of thermoacoustic engine *Applied Acoustics* **73** 1052 -58

[16] M J Moloney and D L Hatten 2001 Acoustic Quality Factor and Energy Losses in Cylindrical Pipes *Am. J. Phys* **69**(3) 311–314

[17] Xiao J H 1995 Thermoacoustic heat transportation and energy transformation. Part 3: Adiabatic wall thermoacoustic effect *Cryogenics* **35** 27

[18] Ghazali N M, Aziz A A and Rajoo S 2006 Environmentally friendly refrigeration with Thermoacustic Research Vote: 74166 University of Technology Malaysia

[19] Russell D A and Weibull P 2002 Tabletop thermoacoustic refrigerator for demonstrations *Am. J. Phys.* **70** (12)

[20] Akhavanbazaz M, Siddiqui M H K and Bhat R B 2007 The impact of gas blockage on the performance of a thermoacoustic refrigerator *Experimental Thermal and Fluid Science* **32** 231– 9

Journal of Physics: Conference Series **423** (2013) 012051 doi:10.1088/1742-6596/423/1/012051

Correlations in hydrothermal properties of building insulation

A Antonyová[1]

University of Prešov in Prešov, Faculty of Management, Department of Quantitative Methods and Managerial Informatics, Konštantínova 16, Prešov 08001, SR

E-mail: antonyova@gmail.com

Abstract. The contribution comprises analysis that is based on scientific work as a part of participation on the international research project carried out at the University of Prešov in Prešov and Vienna University of Technology entitled "Detection and Management of Risk Processes in Building Insulation" and numbered SRDA SK-AT-0008-10. Statistical approach with correlations among humidity, time and temperature values in the space between the wall and building insulation uses the set of data obtained during the measurement series as testing using a new technology with equipment that does not influence the environment properties in the space. Therefore such real mapping can bring a real picture of possible condensation as a risk process in the building envelope.

1. Introduction

Problem of indoor building environment is closely connected with the life quality of its inhabitants. Health of human being is directly influenced by the air components in his surrounding and is a subject matter in some of the research works with their specific points of view. The authors present for instance mass transfer model and its improvement analyzing environment interaction with both outdoor emissions and emissions from building materials, such as volatile organic compound emissions [8]. The main problem stresses indoor air quality and its influence on the health trying to use the mutual correlations of certain chemical elements for prediction. Specification of pollutants concentration, for instance inorganic gasses and bio aerosol [5], maps especially indoor quality under the outdoor influence.

All those issues are connected with the research on material quality [2], its thermal properties [6], stability under the external conditions for instance by setting the coefficient of moisture expansion at room temperature with comparison the materials according to the different thickness [7], based on laboratory experiments, using mostly numerical methods. Moisture problem also appears in connection with the building materials durability under the outer conditions [4]. Special attention is paid to the rising damp in historical buildings [3].

Another very serious problem is connected with microorganisms whose appearance is influenced with the high humidity level and the following condensation that is counted as a risk process for both building materials and health of the house inhabitants. Cavity in the space between the wall and building insulation is the place where air humidity can be concentrated in the high level. As building envelope is covered with different insulating materials also probability of condensation process appearance can reach different levels. Therefore detection of the humidity level in real situation without insulation destruction is of high importance. One of the signs that can influence the condensation process is the temperature leading together with the humidity level to dew point calculation. The paper deals with the correlation of humidity with the time and temperature in degrees of Celsius. The relationship is based on statistical analysis using the data obtained during the real measurement series using a new strategy with the equipment [1] constructed during the work on the international research project carried out at the University of Prešov in Prešov and Vienna University of Technology entitled "Detection and Management of Risk Processes in Building Insulation" and numbered SRDA SK-AT-0008-10.

2. Statistical analysis with setting the correlation dependency

The measurement series that was realized on building insulation made of polystyrene with the thickness of 8centimeters comprises more than 1680 points tested from January 17, 2012 to February 21, 2012 in the period of every 30 minutes. The obtained results are expressed using the graphical representation in figure 1 with the values of both temperature and humidity in the space between the wall and polystyrene and temperature in ambient surrounding of the building.

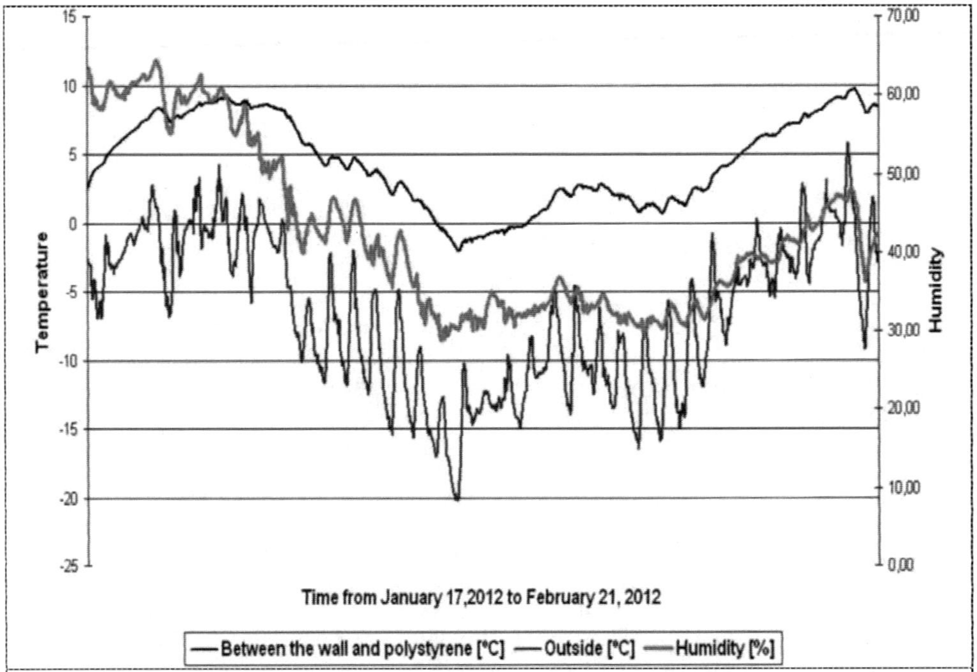

Figure 1. Graphical representation of both temperature and humidity values in the cavity between the wall and polystyrene and the temperature in ambient surrounding of the building.

2.1. Setting the correlation dependency

The values obtained during the measurement series were the basement for setting the correlation in statistical analysis. The correlation using the least square method expressed the dependency of humidity values on the time and temperature in the space between the wall and polystyrene. The decision was set for the linear dependency with the following testing of its statistical significance. The obtained least square regression line

$$z = a_0 + a_1 x + a_2 y, \tag{1}$$

where
x_i … values in the time of testing points,
y_i … the values of temperature in the space between the wall and insulation [°C],
z_i … the values of humidity level in the space between the wall and insulation [%].

Using the calculation, coefficient values were obtained as follows:

$a_0 = 30.98793132,$
$a_1 = 0.042121379,$
$a_2 = 2.479693774.$

2.2. Testing the statistical significance of obtained linear correlation dependency

To prove the reliability of the linear dependency among the values

$$s_{rez}^2 = \frac{SSE}{n-(p+1)}, \qquad (2)$$

$$SSE = \sum_{i=1}^{n} z_i^2 - a_0 \sum_{i=1}^{n} z_i - a_1 \sum_{i=1}^{n} x_i z_i - a_2 \sum_{i=1}^{n} y_i z_i \qquad (3)$$

$$s_{rez} = 6.7441803$$

The obtained values of residual sum of squares lead to the setting the characteristic t (4) for testing the statistical significance of the coefficients.

$$t = \frac{a_j}{s_{a_j}}, \qquad (4)$$

Hypotheses:
H_0: Variable Z is linearly dependent on variable X,
H_1: Variable Z is linearly dependent on variable Y.
The dependency is tested for the level $\alpha = 0.05$

$t_{0.95}(1680) = 1.645$

$$s_{a_i} = \frac{s_{rez}}{\sqrt{\sum_{i=1}^{n} x_i^2 - \frac{\left(\sum_{i=1}^{n} x_i\right)^2}{n}}} \qquad (5)$$

Statistical significance of the obtained value for coefficient a_1:

$t = 1.773655086.$

We can say that hypothesis H_0 is proved and variable Z is linearly dependent on variable X as for a_1:

$|t| > t_{0.95}(1680).$

$$s_{a_2} = \frac{s_{rez}}{\sqrt{\sum_{i=1}^{n} y_i^2 - \frac{\left(\sum_{i=1}^{n} y_i\right)^2}{n}}} \tag{6}$$

Statistical significance of the obtained value for coefficient a_2:

$$t = 49.28885992.$$

We can say that hypothesis H_0 is proved and variable Z is linearly dependent on variable Y as for a_2:

$$|t| > t_{0.95}(1680).$$

Confidence level in the form of intervals:

$$a_j \pm t_{1-\alpha(n-p-1)} \cdot s_{a_j}, \tag{7}$$

for coefficient a_1:

$$(0.0030553;0.0811874),$$

for coefficient a_2:

$$(2.3969348;2.5624528).$$

Using the statistical analysis, humidity values are linearly dependent on the time and temperature in the space between the wall and polystyrene with satisfactory significance.

3. Setting the correlation among the slope values with the following statistical significance testing

In this section the slope values of Δ humidity and Δ temperature are expressed in figure 2 and figure 3.

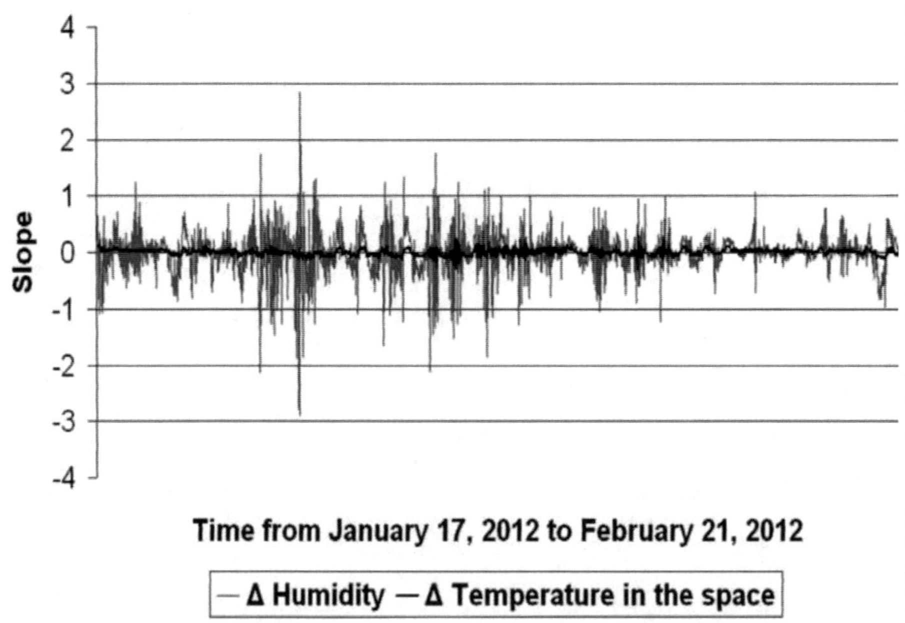

Figure 2. Graphical representation of both Δ humidity and Δ temperature values in the cavity between the wall and insulating material.

Journal of Physics: Conference Series **423** (2013) 012051 doi:10.1088/1742-6596/423/1/012051

The linear dependency of Δ humidity on time and Δ temperature in the space between wall and polystyrene

$$z = c_0 + c_1 x + c_2 y \tag{6}$$

where

x_i ... values in the time of testing points,
y_i ... the values of Δ temperature in the space between the wall and insulation,
z_i ... the values of Δ humidity level in the space between the wall and insulation.

Values obtained by calculation:

$c_0 = 0.027752551$,
$c_1 = -0.004041503$,
$c_2 = 2.855895657$.

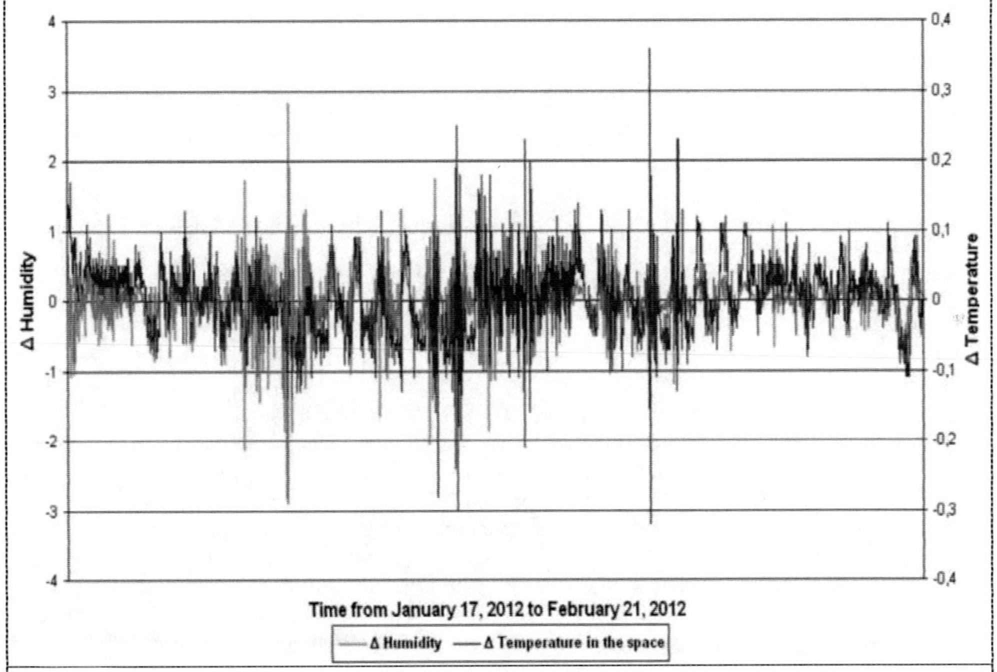

Figure 3. Graphical representation of both Δ humidity and Δ temperature values in the cavity between the wall and insulating material with two axes.

Hypotheses:
H_0: Variable Z is linearly dependent on variable X,
H_1: Variable Z is linearly dependent on variable Y.

The dependency is tested for the level $\alpha = 0.05$

$$t_{0.95}(1680) = 1.645$$

$$s_{rez} = 0.369101861$$

Statistical significance of the obtained value for coefficient c_1:

$$t = -3.106614541$$

We can say that hypothesis H_0 is proved and variable Z is linearly dependent on variable X as for c_1:

$$|t| > t_{0.95}(1680).$$

Statistical significance of the obtained value for coefficient c_2:

$$t = 17.66426465$$

We can say that hypothesis H_0 is proved and variable Z is linearly dependent on variable Y as for c_2:

$$|t| > t_{0.95}(1680)$$

Confidence level in the form of interval for coefficient c_1:

$$(-0.006182;-0.001901)$$

Confidence level in the form of interval for coefficient c_2:

$$(2.5899379;3.1218534)$$

Using the statistical analysis, the values of Δ humidity are linearly dependent on both time and Δ temperature in the space between the wall and polystyrene with satisfactory significance.

4. Summary

The paper is based on statistical analysis the real data obtained during the measurement series mapping the values of humidity and temperature in the cavity between the wall and insulating material on building envelope. The insulation was used in the form of polystyrene with the thickness of 8 centimetres. More than 1680 measuring results for every variable during the testing period from January 17, 2012 to February 21, 2012 were the basement for setting correlation using least square as statistical method for linear dependency of humidity with both time and temperature values and also for Δ humidity and both time and Δ temperature.

Hypotheses testing proved statistical significance of obtained values. It means that those correlations are important for decision regarding the quality of insulating material on the building envelope.

5. References

[1] Antonyová A and Antony P 2012 Heat Transfer Modelling through Building Insulation That Is Based on New Measurement Strategy *IEEE Symposium on Industrial Electronics and Applications(ISIEA 2012): Proc. of the scientific international conference (Bandung, Indonesia, 23-26 September 2012)* pp 208 – 213

[2] Acton Q A 2012 Issues in Materials and Manufacturing Research *(Atlanta: Scholarly Editions)*

[3] Delgado J M P Q (Ed.) 2012 *Heat and Mass Transfer in Porous Media* (Springer: Berlin)

[4] Lacasse M A and Vanier D J (Ed.) 1999 Durability of Building Materials and Components 8 *Service Life and Asset Management:Proc. Of the Eight International Conference on Durability of Building Materials and Components (Vancouver, Canada, 30 May – 3 June 1999) (National Research Council of Canada, Ottawa)*

[5] Li H 2008 A Modeling and Experimental Investigation of Coupled Heat, Air, Moisture and Pollutants Transport in Building Envelope Systems *(Ann Arbor: ProQuest Information and Learning Company)*

[6] Wang, J. T. and Lin, Ch. 1993 Thermal Stresses in Layered Cylinders *Composite Modelling and Processing Science: Proc. of the Ninth International Conference on Composite Materials (ICCM /9)* Miravete A (Ed.) *(Madrid 12 – 16 July 1993)*

[7] Wolff, E. G. (Ed.) 2004 *Introduction to the Dimensional Stability of Composite Materials* (Lancaster: DEStech Publications)

[8] Zhang, Y and Xu Y 2003 Characteristics and correlations of VOC emissions from building materials (*International Journal of Heat and Mass Transfer* vol 46) pp 4877–4883

Journal of Physics: Conference Series **423** (2013) 012052 doi:10.1088/1742-6596/423/1/012052

Statistical analysis as approach to conductive heat transfer modelling

A Antonyová[1] and P Antony[2]

[1]University of Prešov in Prešov, Department of Quantitative Methods and Managerial Informatics, Konštantínova 16, Prešov 08001, Slovak Republic
[2]Apmikro, Pavlovičovo námestie 19, Prešov 08001, Slovak Republic

E-mail: antonyova@gmail.com

Abstract. The main inspiration for article was the problem of high investment into installation of the building insulation. The question of its effectiveness and reliability also after the period of 10 or 15 years was the topic of the international research project carried out at the University of Prešov in Prešov and Vienna University of Technology entitled "Detection and Management of Risk Processes in Building Insulation" and numbered SRDA SK-AT-0008-10. To detect especially the moisture problem as risk process in the space between the wall and insulation led to construction new measuring equipment to test the moisture and temperature without the insulation destruction and this way to describe real situation in old buildings too. The further investigation allowed us to analyse the range of data in the amount of 1680 measurements and express conductive heat transfer using the methods of statistical analysis. Modelling comprises relationships of the environment properties inside the building, in the space between the wall and insulation and in ambient surrounding of the building. Radial distribution function also characterizes the connection of the temperature differences.

1. Introduction

To save energy especially during winter time of energy consuming heating systems is the main role of building insulation. Decision making for the proper material with reliable quality also after the period of 20 years can satisfy the high investment during the insulation construction. Suitable insulation property and prevention the case of high moisture level that can cost not only material destruction but also together with the condensation process the suitable environment for microorganisms especially in the space between the building wall and insulation means proper technology of both material and construction.

Regarding the heat transfer reliability important role play conduction, radiation and convention. Internal convective heat transfer modelling is usually expressed through convention coefficient of correlation using building simulation program [6], using the ESP-r [3] with the stress on internal building surface. Improving the modeling of surface convection during natural night ventilation in Building Energy Simulation Models [4] and model with numerous channels both parallel and perpendicular using fibrous porous material [9] try to predict effective thermal conductivity. The prediction results from the suggested models are compared with theory and experimental data.

Journal of Physics: Conference Series **423** (2013) 012052 doi:10.1088/1742-6596/423/1/012052

Sensitivity analysis of heat transfer formulations for insulated structural steel components [8] are based mostly on laboratory testing. Experimental is also optimal design of a photothermal radiometry method [2] using a crenel heating excitation for determination of the effective thermal conductivity. Numerical method, such as Gauss-Newton method and least squares objective function are used for optimization. Surfaces behavior of radiative exchange with the reflection from ideal dielectric surfaces to optical behavior of real surfaces is investigated [5] while thermal radiation is expressed both graphical view and analytical form with the stress on emission from gases, solids and liquids with absorption mechanisms.

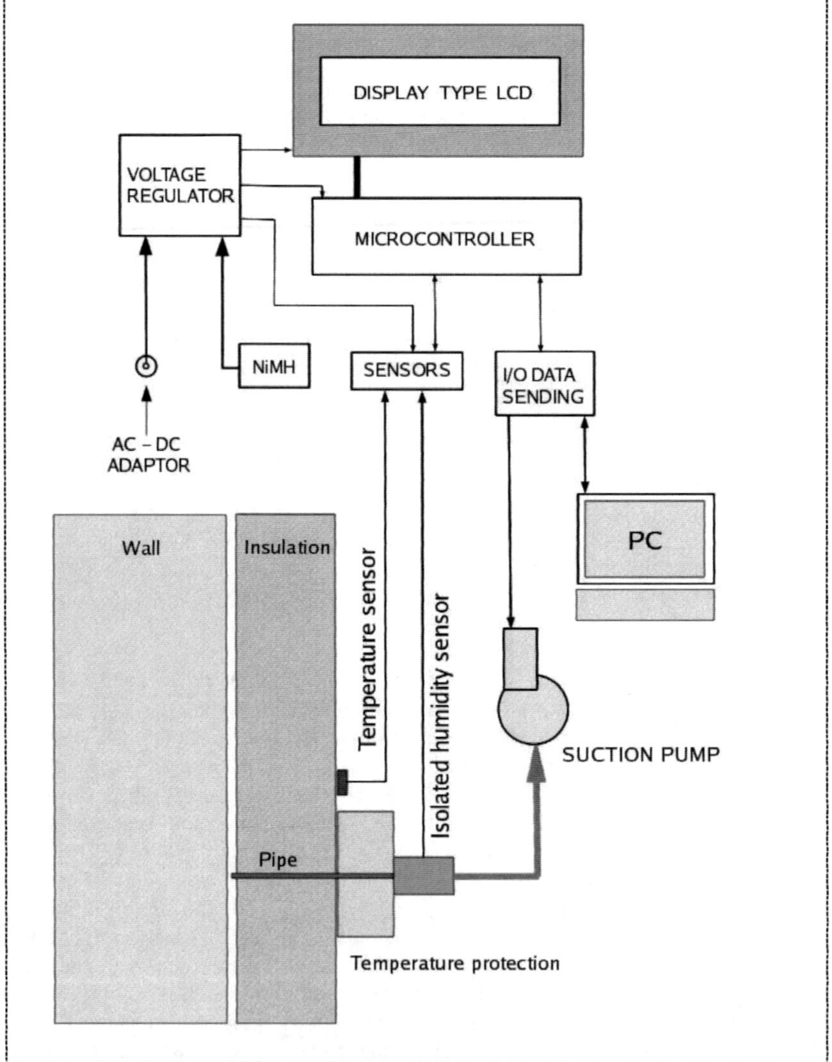

Figure 1. Scheme of the measuring equipment for testing both humidity and temperature values in the space between the wall and insulation.

Journal of Physics: Conference Series **423** (2013) 012052 doi:10.1088/1742-6596/423/1/012052

Effect of moisture and porosity on the thermal conductivity and thermal diffusivity of a conventional refractory concrete is tested [7] using the hot wire parallel technique. Experimental results showed a drastic influence of the moisture content on material. The author focuses primarily on thermal conductivity, thermal diffusivity and specific heat up to 1000 °C.

2. Measurement series description

Building insulation was tested using new equipment from January 17, 2012 to February 28, 2012 every half an hour. The used measuring equipment for testing, which is expressed in Figure 1, allows also detection the humidity level and temperature between wall and insulation without destruction using the humidity and temperature sensor SHT 15. Another temperature sensor was placed close to the insulation outside the building not to touch surface and not to be far from the surface, so that it is not influenced by the conditions of outside environment. More detailed description of measuring equipment and the testing process is given in Antonyová, Antony [1].

The set of more than 1680 data allowed presenting the graphical representation of the situation regarding the possibility of condensation as risk process in the space between wall and insulation using the calculation for dew point value.

The tested building with the walls made of bricks is heated regularly on the temperature 20 °C with polystyrene insulation of the thickness of 8centimetres. The experimental and calculated results proved that there is no danger of condensation process that is one of the conditions for the attack by microorganisms. Insulation made of polystyrene does not absorb water in comparison with for instance natural materials that are intensive water absorbent what highly influences their insulation properties. On the other hand if there is once the water between the wall and polystyrene it remains and can cost serious problems. Water can get there for instance through the holes for electric wires or during the flooding.

3. Conductive heat transfer performance

The temperature values which were obtained experimentally allow describing the process of conductive heat transfer for polystyrene insulation of 8centimetres. Conductive heat transfer is expressed by Fourier's Law (1).

$$\frac{q}{A} = k\frac{dT}{s} \tag{1}$$

where:

$\frac{q}{A}$ ... heat transfer per unit area (W/m^2),

k ... thermal conductivity (W/m.K),

dT ... temperature difference between: the temperature in the space between the wall and insulation and temperature in ambient surrounding of the building (°C),

s ... polystyrene panel thickness (m).

Taking into account the conditions of experimental testing when the polystyrene panels which were used as the insulating material on building envelope are with the thickness of 8centimeters and the value of thermal conductivity for polystyrene is approximately 0.03 W/m.K we get the following expression:

$$\frac{q}{A} = 0.03.\frac{dT}{0.08} \tag{2}$$

The equation (2) is the basement for graphical representation in Figure 2 with modelling the conductive heat transfer.

4. Statistical analysis

Conductive heat transfer between the surrounding inside the building and ambient surrounding of the building expresses equation (3) with its graphical representation in Figure 3.

$$\frac{q}{A} = \frac{T_l - T_n}{\frac{s_1}{k_1} + \frac{s_2}{k_2}} \quad (3)$$

where:

$\frac{q}{A}$ ... heat transfer per unit area (W/m^2),
T_l ... temperature inside the building that is 20℃,
T_n ... temperature in ambient surrounding of the building (℃),
S_1 ... thickness of building wall 0.4m,
S_2 ... thickness of polystyrene panel ... 0.08m,
k_1 ... thermal conductivity for bricks that is approximately 0.18 W/m.K,
k_2 ... thermal conductivity for polystyrene that is approximately 0.03 W/m.K.

We did not take into account air in the space between the wall and polystyrene in formula (3) as the space is very narrow and in some places the materials of wall envelope and polystyrene even touch one another. The relationship is expressed also using the least square by the polynomial of the 6th degree with the coefficient of determination in Figure 3.

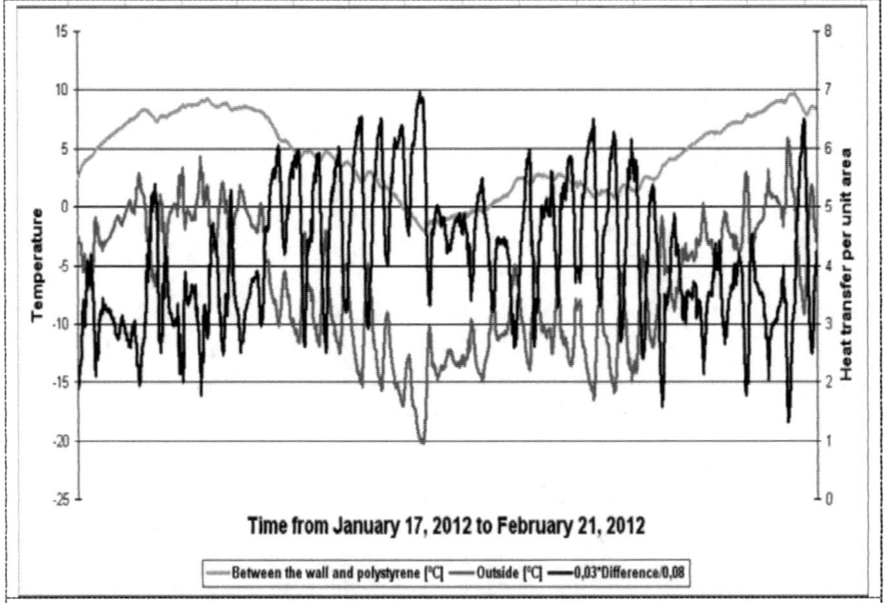

Figure 2. Graphical representation of modelling the conductive heat transfer through polystyrene with thickness of 8centimetres.

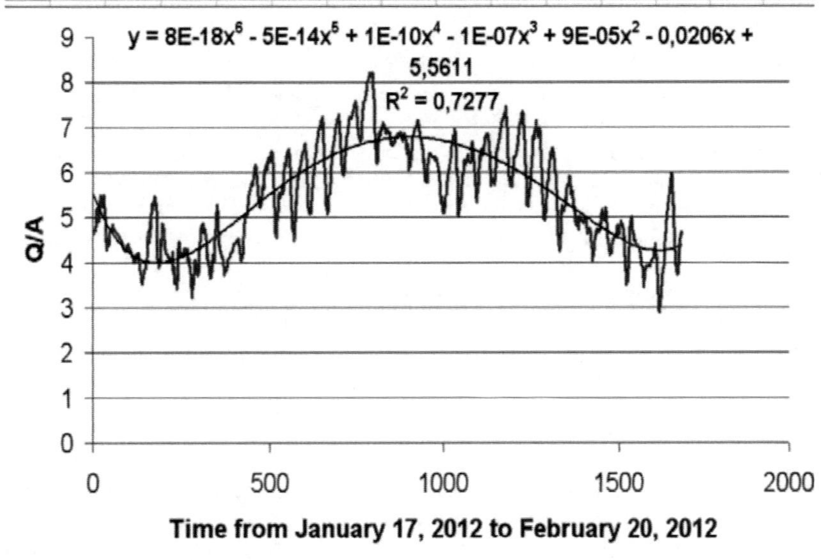

Figure 3. Graphical representation of conductive heat transfer modelling through the building envelope.

The next Figure 4 expresses the graphical representation of radial distribution function for the relationship between the temperature values in the space between the wall and insulation and in ambient surrounding of the building.

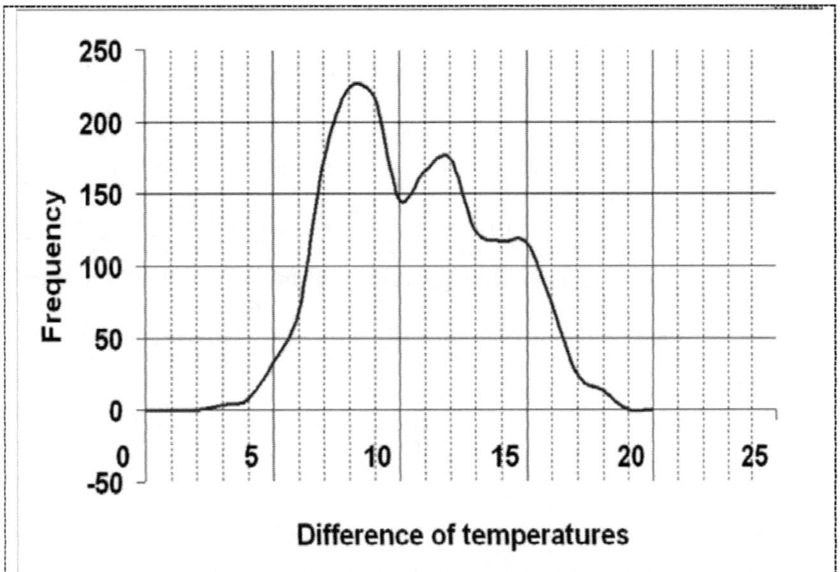

Figure 4. Graphical representation of radial distribution function for the relationship between the temperature values in the space between the wall and insulation and in ambient surrounding of the building.

5. Appendices

Conductive heat transfer was studied both experimentally and through statistical analysis. Experiment was carried out using the new technology with the measuring equipment which enables testing the humidity and temperature not only outside the building but also in the space between the wall and insulation. Polystyrene in the form of panels with thickness of 8centimeters was investigated as a sample of porous material.

Model comprises graphical representations of conductive heat transfer between inside the building and ambient surrounding of the building. The tested building is heated regularly on 20℃. The relationship was expressed also analytic way through the statistical method using the least square by the polynomial of the 6th degree with the coefficient of determination.

Statistical analysis comprises also radial distribution function for the relationship between the temperature values in the space between the wall and insulation and in ambient surrounding of the building.

6. References

[1] Antonyová A and Antony P 2012 Heat Transfer Modelling through Building Insulation That Is Based on New Measurement Strategy *IEEE Symposium on Industrial Electronics and Applications(ISIEA 2012): Proc. of the scientific international conference (Bandung, Indonesia, 23-26 September 2012)* pp 208 – 213

[2] Albouchi F, Fetoui M, Rigollet F, Sassi M and Nasrallah S B 2005 Optimal design and measurement of the effective thermal conductivity of a powder using a crenel heating excitation *(International Journal of Thermal Sciences* vol 44*)* pp 1090–1097

[3] Beausoleil-Morrison I 2002 The adaptive simulation of convective heat transfer at internal building surfaces *(Building and Environment* vol 37*)* pp 791–806

[4] Leenknegt S, Wagemakers R, Bosschaerts W and Sealens D 2011 Improving the modelling of surface convection during natural night ventilation in Building Energy Simulation models *Proc. Of the 12th Int. Conf of the International Building Performance Simulation Association* pp 2233-2240

[5] Mahan J R 2002 Radiation Heat Transfer: *A Statistical Approach* (New York: John Wiley & Sons, Inc.)

[6] Peeters L, Beausoleil – Morrison I and Novoselac A 2011 Internal convective heat transfer modeling: Critical review and discussion of experimentally derived correlations *(Energy and Buildings* vol 43*)* pp 2227-2239

[7] Santos W N 2003 Effect of moisture and porosity on the thermal properties of a conventional refractory concrete (*Journal of the European Ceramic Society* vol 23*)* pp 745–755

[8] Wong M B and Ghojel J I 2003 Sensitivity analysis of heat transfer formulations for insulated structural steel components (*Fire Safety Journal* vol 38*)* pp 187–201

[9] Zhu F, Cui S and Gu B 2010 Fractal analysis for effective thermal conductivity of random fibrous porous materials (Physics Letters A vol 374*)* pp 4411–4414

Journal of Physics: Conference Series **423** (2013) 012053 doi:10.1088/1742-6596/423/1/012053

Evaluation of environmental effect of coal stockpile in Muara Telang, Banyuasin, Indonesia

Rusdianasari[1,2], **Susila Arita**[3], **Eddy Ibrahim**[4] **and Ngudiantoro**[5]

[1] Candidate Doctor of Environmental Science, Sriwijaya University, Indonesia
[2] Department of Chemical Engineering, State Polytechnic of Sriwijaya
Jl. Srijaya Negara, Bukit Besar, Palembang 30139, Indonesia
email: diana_vsi@yahoo.com
[3] Department of Chemical Engineering, Faculty of Engineering, Sriwijaya University, Indonesia
[4] Department of Mining Engineering, Faculty of Engineering, Sriwijaya University, Indonesia
[5] Department of Mathematic, Faculty of Mathematics and Natural Sciences, Sriwijaya University, Indonesia

Abstract. Stockpile commonly serves as a temporary dump before the coal is transported through the waterways. This study investigated the effects of coal stockpiles on the surrounding environment: air, water, and soil. The location of the study is in the estuary of Telang, South-Sumatra, Indonesia, which is located at the edge of the river of Telang and close to the residential community. The monitoring of the environmental impact from the stockpile is intended to conduct an environmental assessment owing the existence and operations of coal accumulation. Enviromental impact analysis was conducted based on the value of the effluent, air pollution (dust), soil and water by determining the parameters of the coal wastewater pH, total suspended solid, ferrous dan ferrous metals contents. The results indicate that the total suspended particulate, total suspended solids, noise level, ferrous metal and manganese metal were 10-14 $\mu g/Nm^3$, 249-355 mg/L, 41.3 to 50.3 dBA, 6.074 to7.579 mg/L, and 1.987 to 2.678 mg/L, respectively. Meanwhile the pH of water and soil were 3 to 4 and 2.83 to 4.02 respectively. It is concluded that the pH value are beyond the threshold standard.

1. Introduction

World coal consumption in recent years has increased very rapidly. In 2007, world consumption reached 5,522 million tons, showing an increase of about 3.5% per year [1]. In the world coal trade, Indonesia plays an increasingly important over the years as a producer or exporter. In 2007, Indonesia was in seventh position as a coal producers which contributes up to 4.2% of the total world productions. On the other hand, as an exporter of coal Indonesia was in the second position with a total export capacity of 202 million tonnes [2]. South Sumatra has the potential coal resources. The total coal resources are about 22,240 billion tonnes or 41.5% of the total national reserve. The need for coal in South Sumatra is increasing with an average of about 5.06% annually [3].

Location of mining away from the coastline causing more expensive transportation costs. In order to reduce the cost, the transportation system was adjusted to a combination of land and sea transport. Coal from the mines is transported by road to the place of coal accumulation (stockpile) to further transported by barges. Stockpile serves as a temporary dump before the coal is transported through the waterways to be marketed at home and abroad. As the potential impacts by the opening of international port of Tanjung Api-Api, it is strongly effected to the potential development of coal stockpile area.

The coal stockpile in Muara Telang is one of the few existing stockpile in Tanjung Api-Api, which is commonly located in wetland areas affected by tides. The stockpile is located at (02°31'00.22" S - 02°31'05.5"S 104°48'00.7" E) and (02° 30'37.6"S - 02°31'24.3"S 104°48'18.11"E - 104°48'20.6"), with an area of about 61.09 hectares. The location is on the estuary of the Musi River where tidal variation ranging from 1.0 to 3.0 meters. The tidal effect is higher in this region compared to the influence of rainfall. In addition, altitude varies from 0.5 m to 2.0 m above sea level [4]. The stockpile location is very strategic because it can be passed by a barge as a transporter of coal to consumers. Although the location of the stockpile is far from the main road of the Tanjung Api Api and settlements, activities in the stockpile generate significant dust disturbing the surrounding community [5].

Suspended solids from the run-off in the coal stockpile are at levels of above 2,000 ppm even at 10,000 ppm. The existence of liquid waste from coal stockpile can reduce the degree of acidity (*pH*) and increase the content of total suspended solids (*TSS*), ferrous (*Fe*) and manganese (*Mn*) [6]. Relatively high suspended solids will reduce the penetration of light or sunlight into the water thus affecting the regeneration of oxygen in photosynthesis. On the other hand, high concentration of the *Fe* may affect the lives of aquatic organisms and may cause rust on the equipments made of metal. Meanwhile, the *Mn* metals are toxic at high concentrations. *pH* value of the water that is not in the normal range may affect organisms in the water, such as fish and other animals. In addition, an abnormal pH is corrosive to metals resulting in rust [7].

Monitoring the environmental impact on the stockpile is intended to conduct an environmental assessment of the impact arising due to the existence and operations of coal accumulation. The analysis was corresponded to the value of the effluent, air pollution (dust), soil and water by refering at the parameters of the coal wastewater *pH*, *TSS*, ferrous and manganese metals [8].

2. Method of Study

Data were collected from primary data and secondary data. Primary data obtained from the measurements of samples in the laboratory while secondary data were several reports from related study such as the analysis of the environmental impacts of coal stockpile in Muara Telang.

3. Results and Discussion

3.1 Air quality

Air quality parameters are including the temperature, carbon monoxide (CO), sulfur dioxide (SO_2), nitrogen oxides (NOx), total suspended particulate (TSP), and noise [9].

3.1.1 *Temperature*

Air temperature measurements are required, in which the gas content in the air is inversely proportional. At low temperatures, the concentration of gas pollutant in the air is considerably high (floating near the surface of the earth), while the air temperature increases, the gas in the lower air pollutants (gas rises into the atmosphere). The temperature of the air in coal stockpile locations ranged from 29.8 to 31.8°C.

3.1.2 *Carbon Monoxide (CO)*

Carbon monoxide is a toxic gas that is colorless, odorless, and has no taste. This component has a weight of 96.5% of the weight of the water and does not dissolve in water. Carbon monoxide is present in nature formed from one of the incomplete combustion of the carbon or carbon-containing components, the reaction between carbon dioxide and carbon-containing components at high temperatures, the carbon dioxide will decompose into carbon monoxide and oxygen.

Several studies have shown that the effect of CO on plants is usually not seen in practice. Whereas, the effect of CO on humans at high concentrations can cause death, whereas contact with CO at relatively low concentrations (30,000 ppm or less) can be detrimental to health. In addition, the effect of CO on the human body is mainly caused by the reaction of CO with hemoglobin (Hb) in the blood.

The results of measurements taken at four sampling points indicate that the CO level is below the established standards of 77-124 $\mu g/Nm^3$. We note that the CO concentration in the residential location is generally derived from the motor-vehicles activities and household activities.

3.1.3 *Sulfur dioxide (SO_2)*
Pollution by sulfur dioxide is mainly caused by two components of sulfur forms a colorless gas, namely sulfur dioxide (SO_2) and sulfur trioxide (SO_3), in which both are called as sulfur oxides (SOx). Sulfur dioxide has a characteristic pungent odor and non-flammable in air, while the sulfur trioxide is a component that is not reactive. SO_2 in the air always produced in large quantities, while the amount of SO_3 formed varies from 1-10% of the total SOx.

SO_3 in the air in the form of gas is possible only if the concentration of water vapor is very low. If the water vapor present in sufficient quantities, SO_3 and water vapor will soon join to form droplets of sulfuric acid (H_2SO_4) with the following reaction:

$$SO_3 + H_2O \longrightarrow H_2SO_4$$

The normal components contained in the air is H_2SO_4 not SO_3. But the amount of H_2SO_4 in the atmosphere is more than the result from SO_3 emissions. This suggests that there is other mechanism producing H_2SO_4 in the environment. After being in the atmosphere, SO_2 is converted into SO_3 (later to become H_2SO_4) by photolytic processes and catalytic. The total amount of SO_2 oxidized into SO_3 is influenced by several factors including the available water, intensity, time, and distribution of sunlight spectrum.

SOx influence on plants can be affected by two factors, namely the effect of SO_2 concentration and contact time. While, the effects of high concentration SOx on humans and animals is irritation of the respiratory system. We conducted measurement at four sample locations (*see* Table 1). The results show that the concentration of SO_2 about 5.26 to 30.26 $\mu g/Nm^3$, which is below the standards (900 $\mu g/Nm^3$).

3.1.4 *Nitrogen oxides (NOx)*
Nitrogen oxides (NOx) are a group of gases in the atmosphere of gas nitric oxide (NO) gases in the atmosphere. Although other forms of nitrogen oxides exist, but these two gases are most commonly found as an air pollutant. Nitrogen oxide is a gas that is colorless and odorless, whereas nitrogen dioxide has a reddish-brown color and pungent. Out measurement at four sample locations indicate that the concentration of NOx is about 2.08 to 21.73 $\mu g/Nm^3$, which is below the threshold standard.

3.1.5 *Total suspended particulate*
Dust particles generally contain a variety of different chemical compounds with different sizes and different shapes, depending on where the source emission. Naturally particulate dust can be generated from dry dust carried by the wind or derived from a volcanic activities. Particulate dust drift is also produced from the combustion of coal. The coal combustion is not perfect in such that the forming aerosol compound of tar beads. Compared with the burning of coal, oil and gas combustion generates fewer particulates floating dust. Motor-vehicle density can increase the total emission of black smoke particulate dust [9].

Size of dust particulate solid or liquid form in the air depends on the size. Size of dust particulates that endanger health generally range between 0.1 to 10 microns. In general, the size of about 5 microns of particulate dust is airborne particulate matter that can go directly into the lungs and settles in the alveoli. This situation does not mean that the size of particulates greater than 5 microns are not dangerous, because the larger particulates may interfere with the upper respiratory tract and cause irritation. This situation will worsen if the synergistic reaction with SO_2 gas contained in the air which comes from the activity of stone crusher and unloading coal. Besides particulate dust that floated and fluttered in the wind will cause irritation to the eyes and can impede penetration eye (visibility). The existence of spills of toxic metals contained in particulate dust in the air is the greatest danger to health. In general, the polluted air containing hazardous metals only about 0.01% to 3% of all particulate dust in the air. However, these metals can be cumulative and possibly synergistic reaction occurs in the body's tissues [10].

The results of measurements taken at four sample locations show that the content of particulate dust is10-44 $\mu g/Nm^3$ and it remains below the established standards.

Journal of Physics: Conference Series **423** (2013) 012053 doi:10.1088/1742-6596/423/1/012053

3.1.6 *Noise*

Noise in occupational health is defined as hearing voices. It can reduce both quantitatively (increased hearing threshold) and qualitatively (narrowing of the spectrum of hearing), factors related to the intensity, frequency, duration and timing pattern. Noise is measured with a sound level meter. With the mechanism of action, if any object vibrates, it will lead to changes in air pressure that can be captured by these tools, will move the meter pointer.

Noise causes various disorders on society, labor and animals around him. Noise can cause disruption, such as impaired physiological, psychological, communication disorders and deafness, or there is classification of disorders such as auditory disorders, such as disruption of hearing and non-auditory disturbances such as impaired communication, health hazards, decrease work performance, fatigue and stress.The results of conducted noise level measurements at four samples locations, the noise level is between 41.3 dBA to 50.3, still below the threshold standard.

The results of measurements taken at several locations in the stockpile area shown in Table 1. It is shown that the temperature has no significant difference, where the measurement takes place when the air temperature is quite bright.

Table 1. The results of measurements of air quality and noise levels in the coal stockpile.

No.	Parameter	Unit	Measurement Results				Threshold Standard[a]
			U-1[d]	U-2[e]	U-3[f]	U-4[g]	
1	Air temperature	^oC	29.8	30.6	31.8	30.5	-
2	CO	$\mu g/Nm^3$	77	124	92.4	90.2	30.000
3	SO_2	$\mu g/Nm^3$	5.26	30.26	10.16	10.10	900
4	NO_2	$\mu g/Nm^3$	2.08	21.73	6.55	6.50	400
5	TSP	$\mu g/Nm^3$	12	14	11	10	230
6	Noise	dBA	45.9	50.3	44.8	41.3	55[b] 70[c]

[a] South Sumatra Governor Regulation No.6 of 2012 [11]
[b] For residence area
[c] For industrial area
[d] Coal stockpile location
[e] Sri Tiga village
[f] Telang riverside
[g] Karang Anyar village

3.2 Water quality

Water quality can be seen from the physical and chemical parameters. Consequent changes in the physical parameters is seen from the water surface, water temperature and dissolved solids. The solids consist of organic and inorganic solids and suspended sediment. This material will settle at the bottom of water over time causing siltation of water bodies. Another consequence of this solid is a growth of toxic aquatic-plants that can influence other organisms. The number indicates sludge solids contained in the water. Chemical parameters include the level of acidity (*pH*) and heavy metals [12].

3.2.1 *Water temperature*

The standard surface water is set at normal temperatures. High surface water temperatures (>45^{0}C) will affect the speed of chemical reactions as well as lifes in the water. Temperature changes show activity of biological chemistry in solids and gases in water. The potential decay that occurs at high temperatures can cause the solubility of oxygen in the surface water. It is set to reduced, so that the process of aeration is needed in order to degrade organic matter that is inhibited. Furthermore, we will give effect to turn off the water biota in water bodies and vegetation lethal hit. Thus, the water temperature averaged 26.5°C is a good indicator, and in accordance to the conditions of the tropics. The average value of the temperature measured was still met the quality standard.

3.2.2 *Total suspended solids*

The solids consist of organic and inorganic solids dissolved, and suspended sediment. This material will settle at the bottom of water over time causing siltation in particular on surface water bodies. Another consequence is that the solid substances could pose a growth of particular aquatic plants that can be toxic to other creatures. The number indicates the amount of sludge solids contained in the water [12]. From the analysis of dissolved solids, dissolved solid substances classified as very high compared to environmental standards, the concentration of dissolved solids ranging from 249-355 mg/L, indicating that the location of water samples containing high dissolved solids than the Environmental Quality Standard of 200 mg/L.

3.2.3 *Acidity value (pH)*

Normal water *pH* value is between 6-8, while the polluted water such as waste water, has various pH depending on the type of exhaust. The changes in the acidity of the waste water either to the alkaline (higher *pH*) or to the acid (lower *pH*) would greatly disrupt the lives of fish and aquatic animals in the vicinity. In addition, waste water having very low *pH* is highly corrosive to steel and ferrous pipes. From Table 2, we found that almost all of the locations show low *pH* values (tends to be acidic) and beyond the environmental quality standards.

Table 2. Analysis data of river water.

No.	Parameter	Unit	Analysis Results				Threshold Standard[a]
			U-1[b]	U-2[c]	U-3[d]	U-4[e]	
1	Temperature	°C	26.5	26.5	26.5	26.5	-
2	pH	-	3.08	4.59	4.48	4.94	6-9
3	Total Suspended Solid	mg/L	249	310	355	289	200
4	Dissolved Ferrous	mg/L	6.074	7,579	6.551	6.505	7
5	Dissolved Manganese	mg/L	2.059	2.556	2.678	1.987	4

[a] South Sumatra Governor Regulation No.8 of 2012 [14]
[b] Main trench location
[c] Trench 3 location
[d] Telang river location
[e] Trench 4 location

3.2.4 *Disolved oxygen*

Dissolved oxygen is a basic requirement for plant and animal life in the water. Living beings in the water depends on the ability of water to maintain the minimum oxygen concentration required for life. Fish are aquatic organism that require the highest oxygen, followed by invertebrates. Bacteria are the aquatic organism that need the smallest amount of oxygen. Minimum dissolved oxygen concentration for microbial life can not be less than 5 ppm. Very low concentration of dissolved oxygen can cause a death on fish and results in a rapid corrosion process [13]. Dissolved oxygen levels of the aquatic ecosystem in the vicinity of the study showed a relatively low value (less than 6 mg/L). Thus we may conclude that the aquatic ecosystem in the study area is not good or the aquatic organism because its concentration is below the threshold standard.

3.2.5 *Heavy metal and toxic*

Water is often contaminated with inorganic components, including heavy metals. Manganese (*Mn*) and ferrous (*Fe*) is oxidized in water and insoluble brownish color. The contaminated water can not be used for domestic purposes and rocks containing compounds such as manganese and ferrous pyrite and hematite. In water bodies, ferrous (*Fe*) from the corrosion of heavy equipment and water pipes, metal materials as electrochemical reactions occurring on the surface. Water containing dissolved solids having electrically conductive properties to accelerate corrosion [8].

The results of laboratory tests on water samples show that the level of ferrous (*Fe*) in all locations are above the specified standards. This could be possible due to the oxidation of the pipes or objects that contain elements of ferrous along the river.

3.3 Soil quality

The process of soil formation in Muara Telang strongly influenced by the water (alluvial processes), with additional influences could be the intrution of saltwater or flooding tide in some areas. The condition of the soil in the area of activity is a thin peat marshland. Generally alluvial soil types with shrub land cover with gelam (*Melaleuca cajupati P*), palm, grass marsh tidal areas. Typology of ecosystem study area is peatland ecosystems, wetlands or marshy ground. Typology can be classified into a tidal marsh ecosystem. Ecosystems are usually located on the banks of the great rivers that form the ecosystem of freshwater peat swamp forest with water conditions affected by brackish.

3.3.1 *Physical properties of the soil*

Based on the interpretation of Land and Land Unit Map, published by the Agency for Survey and Mapping (Bakosurtanal) and Research Center of Land and Agro-climate Bogor [15], the coal stockpile area in the village of Muara Telang has alluvial soils. Based on the description of the physiographic, generally fall into three land units, namely:

- Land back swamp (marsh behind) beaches with fine sediments, flat (generally, 3%), dominated by swamp vegetation, forests and shrub gelam (*melaleuca cajupati P*), formed from parent material refined marine sediments from the quarter and some regions have been opened / created fields and some are still in the form of shrubs are still flooded. Alluvial soil found in both study sites in the rice paddies and natural land where there is a difference in the nature of morphologic.
- Tidal plains, mangrove vegetation, sediment smooth, flat (slope <3%).
- Peat land, with the thickness of 10-80 cm, formed from deposits of organic materials, drainage obstructed. Soil fertility is generally low to moderate, easy to shrink the physical properties not turn when in drainage, highly flammable when dry, and has a low mineral soil.

3.3.2 *Chemical properties of soil*

Observations were made at three points representing the existing land units, at ditch 4, ditch 3, and the center of the land. Composite sampling is done at a depth of 0-50 cm. Laboratory analysis of the soil chemical and physical properties is shown in the Table 3.

Table 3. The analysis of physical and chemical properties of soil.

No.	Parameter	Unit	Criteria		
			P-1[a]	P-2[b]	P-3[c]
A.	Chemical properties				
1	pH	Unit	4.02	2.83	2.72
2	C-organic	%	12.67	6.00	8.62
3	N-total	%	0.62	0.24	0.32
4	P-total	mg/L	0.049	0.022	0.023
5	K	me/100g	0.20	0.22	0.18
B.	Physical properties				
6	Porosity	%	45.55	54.23	41.55
7	Permeability	cm/hour	0.22	0.55	0.35
8	Texture: sand	%	15.94	17.77	19.24
	dust	%	37.25	35.21	31.12
	clay	%	46.31	47.02	49.64

[a] Trench 4 location
[b] Trench 3 location
[c] Coal stockpile location

Based on the results of the analysis as shown in Table 3, it is known that the chemical condition of the soil fertility is low to moderate, where the soil has a high acidity (low *pH*) is 2.38 to 4.02. Soil acidity (*pH*) is one of properties that affects absorption of nutrient elements by plants and it is an indicator of toxic elements that affect the growth of microorganisms. The content value of N, P, K total is low to very low, but it has swapped a relatively moderate base and C-organic content classified as moderate. The physically soil at the stockpile locations as low to moderate, with relatively moderate to high porosity, bulk density, and generally low to moderate clay fraction which is dominated by the component. Based on the chemical soil fertility, the fertility rate in the coal stockpile are relatively low.

4. Conclusion and recommendation

Coal stockpile activities have an impact on the surrounding environment. Location of stockpile in the river will have an impact on surface water and soil quality. Unloading-coal activities at the port also cause noise and influence the air quality at the site of stockpile. Air quality parameters including temperature, carbon monoxide (CO), sulfur dioxide (SO_2), nitrogen oxide (NOx), total suspended particulate and noise. Water quality is affected by temperature, *pH* value and metal content of ferrous and manganese. The low *pH* value of the soil will interfere with fertility soil in the stockpile.

The presence of high total suspended solids and ferrous metal content in Telang River causes effluent from the coal stockpile should be processed before it is discharged. Coal stockpile management system must consider the characteristics of coal stockpile in Muara Telang, Banyuasin.

References

[1] Pusdatin 2011 Coal Statistic Indonesia
[2] Miranti E 2008 Prospect of Coal Industry in Indonesia Economic Review **214**
[3] Arif T 2009 *J. of Human Development* **5th** edition
[4] Purbawaseso Bambang 1995 Threshold Approach in Urban Planning and the Environment
[5] District Planning Board Banyuasin 2008 Banyuasin in Figures 2008
[6] Ahrens Michael J 2010 Biological Effect of Unburnt Coal in the Marine Environment An Annual Review Volume **43** pp 69-122
[7] Axcil AKoldas 2006 Acid Mine Drainage (AMD): causes, treatment and case studies. *J. of Cleaner Production* **14** pp 1139-1145
[8] Fardiaz S 1992 Air and water pollution. Publisher Canicius
[9] Badr T Harion JL 2006 Effect of Aggregate Storage Piles on Dust Emission. *Atmospheric Environment* **41** pp 360-368
[10] Koninklijke 1980 *Dust Concentration around Coal Stockpiles* Amsterdam
[11] South Sumatra Governor Regulation No. 6 of 2012 on Quality standard emission stationary sources and limits of motor-vehicle exhaust emissions
[12] Diz H R 1997 *Chemical and Biological Treatment of Acid Mine Drainage for the Removal of Heavy Metals and Acidity*. Dissertation. Virginia Polytechnic Institute and State University.
[13] Sundstrom DW Klei H E 1979 Wastewater Treatment
[14] South Sumatra Governor Regulation No. 8 of 2012 on the Liquid Waste Quality Standard for Industrial Activities, Hotels, Hospitals, Domestic and Coal Mining
[15] Research Center of Land and Agro-climate Bogor 2008 Land and Land Unit Map

Bianchi Type IX Magnetized Stiff Fluid Cosmological Model

Rakeshwar Purohit [1] , Anita Bagora (Menaria)[2]

[1] *Department of Mathematics, M.L. Sukhadia University, Udaipur-313001, India*

[2] *Department of Mathematics, Jaipur National University, Jaipur-302027, India*

Email:anita_bagora@yahoo.com

Abstract. We have investigated homogeneous anisotropic tilted Bianchi type IX cosmological model for perfect fluid distribution with electro-magnetic field. The magnetic field is due to an electric current produced along x axis thus F_{23} is the only non-vanishing component of electromagnetic field tensor. To get a determinate model, we have assumed the condition $a=b^2$ between metric potentials. The various physical and geometrical aspects of the model viz pressure, density, shear, expansion factor, electromagnetic field tensor and rotation are also discussed.

1. Introduction

Bianchi type IX cosmological models are interesting because these models allow not only expansion but also rotation and shear and in general these are anisotropic. Many relativists have taken keen interest in studying Bianchi type IX universe because familiar solutions like Robertson-Walker universe, the de-sitter universe, the Taub-Nut solutions etc. are of Bianchi type IX space-times. Bianchi type IX universe include closed FRW models. The homogeneous and isotropic FRW cosmological models, which are used to describe standard cosmological models are particular cases of Bianchi type I, V and IX space-times according to the constant curvature of the physical three-space, t=constant, is zero, negative or positive. The cosmological models with rotation and expansion have been discussed by various researchers viz Novello and Reboucas[1],Reboucas and Lima[2-3], Rosquist[4],Bradley and Sviestins[5], Sviestins[6] and Patel et al.[7].

Also many authors have considered the behavior of individual Bianchi models that contain either a pure magnetic field or magnetic field plus fluid. The magnetic field has significant role in cosmological scale and is present in galactic and intergalactic spaces. Monaghan [8] has discussed the behaviour of magnetic field in stellar bodies. Lorenz [9] found Bianchi type cosmological models with electromagnetic field. Soares and Assad [10] have obtained anisotropic and spatially homogeneous Bianchi type VII and IX cosmologies in the presence of electromagnetic field. Vaidya and Patel [11] have studied spatially homogeneous space time of Bianchi type IX and they have outlined a general scheme for the derivation of exact solutions of Einstein's field equations presence of perfect fluid and pure radiation fields. Krori et al. [12] and Chakraborty and Nandy [13] have derived cosmological models of Bianchi type II, VIII and IX. There are many other researchers viz. Uggla, and Zur-Muhlen

Journal of Physics: Conference Series **423** (2013) 012054 doi:10.1088/1742-6596/423/1/012054

[14], Burd, Buric and Ellis [15], Kind [16] and Paternoga et al. [17] have studied Bianchi type IX space-time in different context.
Besides this, there has been a considerable interest in spatially homogeneous and anisotropic cosmological models in which the fluid flow is not normal to the hypersurface of homogeneity. These are called tilted universes. The tilted cosmological models in which the fluid vector is not normal to the hypersurface of homogeneity are more complicated than those of non-tilted one. Therefore, in recent years there has been a considerable interest in investigating such type of models. The general dynamics of these cosmological models have been studied in detail by King and Ellis [18], Ellis and King [19], Collins and Ellis [20], Ellis and Baldwin [21]. Dunn and Tupper [22] investigated that tilting universe is possible when electromagnetic field is present. Tilted Bianchi type I cosmological model for perfect fluid distribution in presence of magnetic field has been obtained by Bali and Sharma [23]. They have shown that tilted nature of the model is preserved due to magnetic field.
As far as matter concern the solution with stiff fluid has great importance. Electromagnetic Bianchi type I model with stiff matter has been studied by Lorenz [24]. The relevance of the stiff equation of state $\in = p$ to the matter content of the universe in its early stages has been discussed by Barrow [25]. Hajj Boutros [26] has obtained a stiff matter solution to Einstein equation which are spherically symmetric and state with a perfect fluid distribution of matter satisfying equation of state $\in = p$, where p is the pressure of fluid and $\in$ is the density using the condition $\in = np$, $n \geq 1$. Bagora [27-28] have also investigated Bianchi type I, III tilted cosmological model in different context. Also, Bagora [29] obtained tilted dust Bianchi type IX bulk viscous cosmological model.
Motivated by these studies, in this paper, we propose to find magnetized stiff tilted Bianchi type IX cosmological models. To get a determinate model, we have assumed the condition $a=b^2$ between metric potentials. The various physical and geometrical aspects of the model viz pressure, density, shear, expansion factor, electromagnetic field tensor and rotation are also discussed. The physical and geometrical aspects of the model together with singularities are discussed.

2. The Metric and Field Equations

We consider the homogeneous anisotropic Bianchi type IX metric in the form

$$ds^2 = -dt^2 + a^2(t)\, dx^2 + b^2(t)dy^2 + (b^2\sin^2 y + a^2\cos^2 y)\, dz^2 - 2a^2\cos y\, dx dz, \tag{1}$$

where a and b are functions of t alone.

The energy-momentum tensor for perfect fluid distribution with heat conduction given by Ellis [30] is given by

$$T_i^j = (\in + p)v_i v^j + p g_i^j + q_i v^j + v_i q^j + E_i^j, \tag{2}$$

together with

$$g_{ij}\, v^i v^j = -1, \tag{3}$$

$$q_i q^j > 0, \tag{4}$$

$$q_i v^i = 0. \tag{5}$$

Here E_i^j is the electromagnetic field given by Lichnerowicz [31] as

$$E_i^j = \bar{\mu}\left[|h|^2 (v_i v^j + \frac{1}{2} g_i^j) - h_i h^j\right], \tag{6}$$

where $\bar{\mu}$ is magnetic permeability and h_i is the magnetic flow vector defined by

$$h_i = \frac{\sqrt{-g}}{2\bar{\mu}} \in_{ijk\ell} F^{k\ell} v^j. \tag{7}$$

Here $F_{k\ell}$ is the electromagnetic field tensor and $\in_{ijk\ell}$ the Levi-Civita tensor density.

From (7) we find that $h_1 \neq 0$, $h_2 = 0$, $h_3 = 0$, $h_4 \neq 0$. This leads to $F_{12} = 0 = F_{13}$ by virtue of (7). We also find that $F_{14} = 0 = F_{24}$ due to the assumption of infinite conductivity of the fluid. We take the incident magnetic field to be in the direction of x-axis so that the only non-vanishing component of F_{ij} is F_{23}.
The first set of Maxwell's equation

$$F_{ij;k} + F_{jk;i} + F_{ki;j} = 0 .$$

leads to

$$F_{23} = H\sin y \text{ (say)} .$$

Now $h_1 = \dfrac{H}{\bar{\mu} b}\cosh\lambda$,

$$h_4 = \frac{H}{\bar{\mu} b}\sinh\lambda .$$

Since

$$\begin{aligned} |h|^2 &= h_\ell h^\ell \\ &= h_1 h^1 + h_4 h^4 \\ &= g^{11}(h_1)^2 + g^{44}(h_4)^2 \\ &= \frac{H^2\cosh^2\lambda}{\bar{\mu}^2 b^2} - \frac{H^2\sinh^2\lambda}{\bar{\mu}^2 b^2} \\ &= \frac{H^2}{\bar{\mu}^2 b^2} . \end{aligned}$$

Equation (6) leads to

$$E_1^1 = \frac{-H^2}{2b^2} = -E_2^2 = -E_3^3 = E_4^4. \tag{8}$$

Here p is the isotropic pressure, $\in$ the matter density and q_i the heat conduction vector orthogonal to v^i. The fluid flow vector v^i has the components $\left(\dfrac{\sinh\lambda}{a}, 0, 0, \cosh\lambda\right)$ satisfying the condition (3) and λ is the tilt angle.

The Einstein's field equation

$$R_i^j - \frac{1}{2} R g_i^j = -8\pi T_i^j . \qquad (\Lambda=0 \qquad c = G = 1)$$

The field equation for the line element (1) leads to

$$\frac{2\ddot{b}}{b} + \frac{\dot{b}^2}{b^2} - \frac{3}{4}\frac{a^2}{b^4} + \frac{1}{b^2} = -8\pi\left[(\in + p)\sinh^2\lambda + p + 2q_1\frac{\sinh\lambda}{a} - \frac{H^2}{2\bar{\mu} b^4}\right], \tag{9}$$

$$\frac{\ddot{a}}{a} + \frac{\ddot{b}}{b} + \frac{\dot{a}\dot{b}}{ab} + \frac{a^2}{4b^4} = -8\pi\left[p + \frac{H^2}{2\bar{\mu} b^4}\right], \tag{10}$$

$$\frac{2\dot{a}\dot{b}}{ab} + \frac{\dot{b}^2}{b^2} + \frac{1}{b^2} - \frac{a^2}{4b^4} = 8\pi\left[(\in + p)\cosh^2\lambda - p + 2q_1\frac{\sinh\lambda}{a} + \frac{H^2}{2\bar{\mu} b^4}\right], \tag{11}$$

$$(\in + p)a\sinh\lambda\cosh\lambda + q_1\cosh\lambda + q_1\frac{\sinh^2\lambda}{\cosh\lambda} = 0 . \tag{12}$$

Here (.) denotes the ordinary differentiation with respect to cosmic time 't'.

3. Solution of the Field Equations

Equations from (9) to (12) are four equations in six unknown a, b, $\in$, p, λ and q_1. For the complete determination of these quantities we need two extra conditions.

Firstly, we assume that the model is filled with stiff fluid distribution which leads to

$$\in = p. \tag{13}$$

Secondly,we assumed relation between metric potentials as

$$a = b^2. \tag{14}$$

Equations (9) and (11) lead to

$$\frac{2\ddot{b}}{b} + \frac{2\dot{b}^2}{b^2} + \frac{2\dot{a}\dot{b}}{ab} + \frac{2}{b^2} - \frac{a^2}{b^4} = 8\pi(\in - p) + \frac{K}{b^4}. \tag{15}$$

Using stiff fluid condition from (13) in equation (15), we have

$$\frac{2\ddot{b}}{b} + \frac{2\dot{b}^2}{b^2} + \frac{2\dot{a}\dot{b}}{ab} + \frac{2}{b^2} - \frac{a^2}{b^4} = \frac{K}{b^4}, \tag{16}$$

where $K = \frac{8\pi H^2}{\bar{\mu}}$.

Again using (14) in equation (16), we have

$$\frac{2\ddot{b}}{b} + \frac{6\dot{b}^2}{b^2} + \frac{2}{b^2} - 1 = \frac{K}{b^4}. \tag{17}$$

Let $\dot{b} = f(b)$ and $\ddot{b} = ff'$. (18)

Using (18) in equation (17), we have

$$ff' + \frac{6f^2}{b} = \frac{K + b^4 - 2b^2}{b^3}, \tag{19}$$

Equation (19) leads to

$$\dot{b} = \frac{1}{2b^3}\sqrt{\frac{6Kb^4 + 3b^8 - 8b^6 + 24N}{6}}, \tag{20}$$

$$\ddot{b} = \frac{b^8 - 2Kb^4 - 24N}{4b^7}, \tag{21}$$

where 'N' is constant of integration.
The metric (1) reduces to the form

$$ds^2 = -\left(\frac{dt}{db}\right)^2 db^2 + b^2(b^2dx^2 + dy^2) + b^2(\sin^2 y + b^2\cos^2 y)dz^2 - 2b^4\cos y\,dxdz. \tag{22}$$

Using (20) in the metric (22), we have

$$dS^2 = \frac{-24T^6dT^2}{[6KT^4 + 3T^8 - 8T^6 + 24N]} + T^2(T^2dX^2 + dY^2) + T^2(\sin^2 Y + T^2\cos^2 Y)dZ^2 - 2T^4\cos Y\,dXdZ, \tag{23}$$

where b =T, x=X, y=Y and z=Z.

4. Some Physical and Geometrical Features

The matter density and isotropic pressure for the model (23) are given by

$$8\pi \in = 8\pi p = \frac{120N + 32T^6 - 27T^8 - 18KT^4}{24T^8}. \tag{24}$$

The tilt angle λ is given by

$$\cosh\lambda = \sqrt{\frac{120N + 8T^6 - 9T^8}{120N - 16T^6 + 9T^8 + 18KT^4}}, \tag{25}$$

$$\sinh\lambda = T^2\sqrt{\frac{6(4T^2 - 3T^4 - 3K)}{120N - 16T^6 + 9T^8 + 18KT^4}}. \tag{26}$$

The scalar expansion θ calculated for the flow vector v_i is given by

$$\theta = \frac{4\psi_4}{T^4}\sqrt{\frac{6KT^4 + 3T^8 - 8T^6 + +24N}{6\psi_1\psi_3{}^3}}. \tag{27}$$

The flow vectors v^i and heat conduction vectors q_i for the model (20) are given by

$$v^1 = \sqrt{\frac{6\psi_2}{\psi_3}}, \tag{28}$$

$$v^4 = \sqrt{\frac{\psi_1}{\psi_3}}, \tag{29}$$

$$q_1 = \frac{-\psi_1}{24\pi T^4}\sqrt{\frac{6\psi_2}{\psi_3}}, \tag{30}$$

$$q_4 = \frac{6\psi_2}{24\pi T^2}\sqrt{\frac{\psi_1}{\psi_3}}. \tag{31}$$

The non-vanishing components of shear tensor (σ_{ij}) and rotation tensor (ω_{ij}) are given by

$$\sigma_{11} = 8\psi_5\sqrt{\frac{\psi_1(6KT^4 + 3T^8 - 8T^6 + 24N)}{6\psi_3{}^5}}, \tag{32}$$

$$\sigma_{22} = \frac{-2\psi_5}{3T^2}\sqrt{\frac{6KT^4 + 3T^8 - 8T^6 + 24N}{6\psi_1\psi_3{}^3}}, \tag{33}$$

$$\sigma_{33} = \frac{-2(\psi_6 T^2\cos^2 y - \psi_5\sin^2 y)}{3T^2}\sqrt{\frac{6KT^4 + 3T^8 - 8T^6 + 24N}{6\psi_1\psi_3{}^3}}, \tag{34}$$

$$\sigma_{44} = 8\psi_2\psi_5\sqrt{\frac{6KT^4 + 3T^8 - 8T^6 + 24N}{6\psi_1\psi_3{}^5}}, \tag{35}$$

$$\sigma_{14} = \frac{-4\psi_5}{3T^2}\sqrt{\frac{\psi_2(6KT^4 + 3T^8 - 8T^6 + 24N)}{\psi_3{}^5}} \tag{36}$$

$$\omega_{14} = \sqrt{\frac{\psi_2(6KT^4 + 3T^8 - 8T^6 + 24N)}{\psi_3}}. \tag{37}$$

In the absence of magnetic field i.e. K=0, the above quantities lead to

$$8\pi\in = 8\pi p = \frac{120N + 32T^6 - 27T^8}{24T^8}, \tag{38}$$

$$\theta = \frac{4\psi_7}{T^4}\sqrt{\frac{3T^8 - 8T^6 + +24N}{6(120N + 8T^6 - 9T^8)(120N - 16T^6 + 9T^8)^3}}, \tag{39}$$

$$\cosh\lambda = \sqrt{\frac{120N + 8T^6 - 9T^8}{120N - 16T^6 + 9T^8}}, \tag{40}$$

$$\sinh\lambda = T^3\sqrt{\frac{6(4 - 3T^2)}{120N - 16T^6 + 9T^8}}, \tag{41}$$

$$\sigma_{11} = 8\psi_8\sqrt{\frac{(3T^8 - 8T^6 + +24N)(120N + 8T^6 - 9T^8)}{6(120N - 16T^6 + 9T^8)^5}}, \tag{42}$$

$$\sigma_{22} = \frac{-2\psi_8}{3T^2}\sqrt{\frac{3T^8 - 8T^6 + +24N}{6(120N + 8T^6 - 9T^8)(120N - 16T^6 + 9T^8)^3}}, \tag{43}$$

$$\sigma_{33} = \frac{-2(\psi_9 T^2 \cos^2 y - \psi_8 \sin^2 y)}{3T^2} \sqrt{\frac{(3T^8 - 8T^6 + +24N)}{6(120N + 8T^6 - 9T^8)(120N - 16T^6 + 9T^8)^3}}, \tag{44}$$

$$\sigma_{44} = 8T^2(4 - 3T^2)\psi_8 \sqrt{\frac{3T^8 - 8T^6 + +24N}{6(120N + 8T^6 - 9T^8)(120N - 16T^6 + 9T^8)^3}}, \tag{45}$$

$$\sigma_{14} = \frac{-4\psi_8}{3T} \sqrt{\frac{(4 - 3T^2)(3T^8 - 8T^6 + 24N)}{(120N - 16T^6 + 9T^8)^5}}, \tag{46}$$

$$\omega_{14} = \sqrt{\frac{(4 - 3T^2)(3T^8 - 8T^6 + 24N)}{(120N - 16T^6 + 9T^8)}}, \tag{47}$$

$$v^1 = T\sqrt{\frac{6(4 - 3T^2)}{120N - 16T^6 + 9T^8}}, \tag{48}$$

$$v^4 = \sqrt{\frac{120N + 8T^6 - 9T^8}{120N - 16T^6 + 9T^8}}, \tag{49}$$

$$q_1 = \frac{-(120N + 8T^6 - 9T^8)}{24\pi T^3} \sqrt{\frac{6(4 - 3T^2)}{120N - 16T^6 + 9T^8}}, \tag{50}$$

$$q_4 = \frac{(4 - 3T^2)}{4\pi} \sqrt{\frac{120N + 8T^6 - 9T^8}{120N - 16T^6 + 9T^8}}. \tag{51}$$

Here

$\psi_1 = 120N + 8T^6 - 9T^8$,

$\psi_2 = 4T^2 - 3T^4 - 3K$,

$\psi_3 = 120N - 16T^6 + 9T^8 + 18KT^4$,

$\psi_4 = 7200N^2 - 600NT^6 - 1080NT^8 + 540KNT^4 - 104.5KT^{12} + 90KT^{10} + 117T^{14} - 64T^{12} - 40.5T^{16}$

$\psi_5 = 3600N^2 + 1920NT^6 - 2160NT^8 - 540KNT^4 - 121.5KT^{12} + 72KT^{10} + 72T^{14} - 32T^{12} - 20.25T^{16}$

$\psi_6 = 3600N^2 - 2400NT^6 + 2160NT^8 + 1620KNT^4 + 40.5KT^{12} + 36T^{14} - 32T^{12} - 20.25T^{16}$,

$\psi_7 = 7200N^2 - 600NT^6 - 1080NT^8 + 117T^{14} - 64T^{12} - 40.5T^{16}$,

$\psi_8 = 3600N^2 + 1920NT^6 - 2160NT^8 + 72T^{14} - 32T^{12} - 20.25T^{16}$,

$\psi_9 = 3600N^2 - 2400NT^6 + 2160NT^8 + 36T^{14} - 32T^{12} - 20.25T^{16}$.

The rates of expansion H_i in the direction of x, y and z axes are given by

$$H_1 = \frac{1}{T^4}\sqrt{\frac{(6KT^4 + 3T^8 - 8T^6 + 24N)}{6}}, \tag{52}$$

$$H_2 = H_3 = \frac{1}{2T^4}\sqrt{\frac{(6KT^4 + 3T^8 - 8T^6 + 24N)}{6}}. \tag{53}$$

5. Conclusion

The model (21) represents a tilted model. The model starts with a big-bang at T=0 and the expansion in the model decreases as time increases. The model has cigar type singularity at T=0

MacCallum, [32]). At the initial stage energy density $\in \to \infty$ but it vanishes asymptotically. Tiltness of the model increases when magnetic field increases. The velocity components vanish at initial stage and tend to infinity on later stage. The heat conduction vectors tend to infinity at initial stage but vanish asymptotically. The model represents expanding, shearing, rotating and tilted universe in general. Since $\underset{T \to \infty}{\mathrm{Lt}} \left(\frac{\sigma}{\theta}\right) \neq 0$, the model does not approach isotropy for large value of T.The Hubble components become infinitely large at initial stage whereas vanish asymptotically. The model completely follows the restrictions imposed by Ellis [30] and Dunn and Tupper [22] for tilted models. Therefore, this model is a realistic model.

In the absence of magnetic field, density is the decreasing function of time. The tilt angle is zero at T=0 so that model is non-tilted. Also, the rotation is throughout uniform In general, the model represents an anisotropic universe.

References

[1] Novello M and Reboucas M J 1978 *Astrophys.J.**225*** 719
[2] Reboucas M J and Lima J A S 1982 *J.Math.Phys.**23*** 1547
[3] Reboucas M J and Lima J A S 1982 *J.Math.Phys.**29*** 2699
[4] Rosquist K 1983 *Phys.Lett. A* **97** 145
[5] Bradley M and Svienstins E 1983 *Gen.Rel.Gravit.**16** 1119*
[6] Svienstins E 1985 *Gen .Rel. Gravit.* **17** 54
[7] Patel L K and Yadav S P 1987 *Ind. J. Pure & App. Math.**18(9)*** 840
[8] Monaghan J J 1966 *Mon. Not. Roy. Astron. Soc. (G.B.)* **134(3)** 275
[9] Lorenz D 1981 *Astrophys Space Sc.* ***85(1-2)*** 69
[10] Damio Soares I and Assad M J D 1983 *Phys.Rev.D* ***28(8)*** 1858
[11] Vaidya P C and Patel L K 1986 *Pramana-J. Phys.* **27(1-2)** 63
[12] Krori K D, Chaudhury T, Mahanta C R and Mazumdar A 1990 *Gen. Rel. Gravit.* **22(2)** 123
[13] Chakraborty S and Nandy G C 1992 *Astrophys. Space Sci.* **198** 299
[14] Uggla C and Zur-Muhlen H 1990 *Class. Quant. Grav.,* **7** 1365
[15] Burd A B, Buric N and Ellis G F R 1990 *Gen. Rel. Gravit.* **22** 349
[16] King D H 1991 *Phys. Rev.* D **44** 2356
[17] Paternoga R and Graham R 1996 *Phys. Rev.* D **54** 4805
[18] King A R and Ellis G F R 1973 *Comm. Math. Phys.* **31** 209
[19] Ellis, G F R and King, A R 1974 *Comm. Math. Phys.* **38(2)** 119
[20] Collins C B and Ellis G F R 1979 *Phys. Rep.* **56** 65
[21] Ellis G F R and Baldwin J E 1984 *Mon. Not. Roy. Astron. Soc.* **206** 377
[22] Dunn K A and Tupper B O J 1980 *Astrophys. J.* **235** 307
[23] Bali R and Sharma K 2000 *Astrophys. Space Sci.* **271** 227
[24] Lorenz D 1981 *Phys. Lett. A* **83** 155
[25] Barrow J D 1978 *Nature* **272** 211
[26] Joseph Hajj-Boutros 1989 *Int. J. Theor. Phys.* **28** 1
[27] Bagora A 2010 *EJTP* **4(14)** 373
[28] Bagora A 2009 *Astrophys. Space Sci.* **319** 155
[29] Bagora A 2012 *Isrn Astronomy and Astrophys.* **Article ID 954043**, 5 pages
[30] Ellis G F R 1971 General Relativity and Cosmology; ed. R.K. Sachs; Academic Press; New York, 116.
[31] Lichnerowicz A 1967 Relativistic Hydrodynamics and Magnetohydrodynamics; Benjamin, New York , 13.
[32] MacCallum M A H 1971 *Comm.Maths.Phys.* **20** 57

Journal of Physics: Conference Series **423** (2013) 012055 doi:10.1088/1742-6596/423/1/012055

TiO_2 Dye Sensitized Solar Cells Cathode Using Recycle Battery

I Daut[1], M Fitra[1], M Irwanto[1], N Gomesh[1], Y M Irwan[1]
[1]Centre of Excellent for Renewable Energy School of Electrical System Engineering Universiti Malaysia Perlis, Kangar 01000, Perlis, Malaysia

E-mail:mhdfitra@gmail.com

Abstract. This paper proposed an alternative source of carbon material from recycled battery for the fabrication of cathode of a dye sensitized solar cell. Carbon from recycle battery is extracted and used to fabricate the cathode part of a dye sensitized solar cell by using TiO_2 as the anode. The both anode and cathode is fabricated on a 6 x 2cm^2 indium tin oxide (ITO) coated glass and then tested under the solar irradiance of 693.69 w/m2 and average temperature 44.4 oC. Result shows that by using carbon from a recycle battery a Voc of 0.333V and short circuit current of 166.04 µA is produce.

1. Introduction

Dye-sensitized solar cells (DSSCs) have been studied since the 1970s. The first DSSC design was based on single crystal zinc oxide. Due to the low film surface area for dye molecule adsorption, the performance of the first generation DSSC is 3 very poor with solar-to-electric conversion efficiencies less than 1%. In 1991, Grätzel and O'Regan modified the DSSC design. They used mesoporous TiO2 nanoparticle film to replace the single crystal ZnO film to increase the dye loading. With this modification, the DSSC performance has been improved significantly to ~ 11%, which is nearly comparable to the performance of traditional solid state *p-n* junction solar cells. Different from the conventional solar cell systems, where the semiconductor performs both the task of light absorption and charge carrier transport, in DSSCs these two functions are separated into two different materials [1]. This fundamental difference has the practical consequence that the materials used in DSSCs do not need to be as defect-free as the crystalline silicon and DSSCs are more defect-tolerant than *p-n* junction solar cells. Therefore, DSSCs can be made with less expensive materials that can be produced using cheap methods such as low temperature solution crystallization and growth technique, which could eventually lower the solar cell fabrication cost2 [2].

With the increasing demand for utilization of energy resources, there has been a deflation in the availability of resources and there have been many methods of recycling of energy resources. Resources recycling has been an increasingly essential issue for the environmental protection and sustainable social development especially in the East Asia, one of the most rapid economic growth areas in the world.

1 Corresponding author. Tel.: +60164962652 ; fax : +6049798903

2. Methodology

2.1. Tools

Equipment used in this study are a steel sheet, a transparent glass, binder clip, tape, cutter, glass beads, mortars (grinding), tissue paper, filter paper, alligator clip wire, glass beaker, pipette drops, mumps pipette, measuring cup, oven, furnace, blueberries, stirrer rod, aluminium foil, multimeter, as shown in Figure 1.

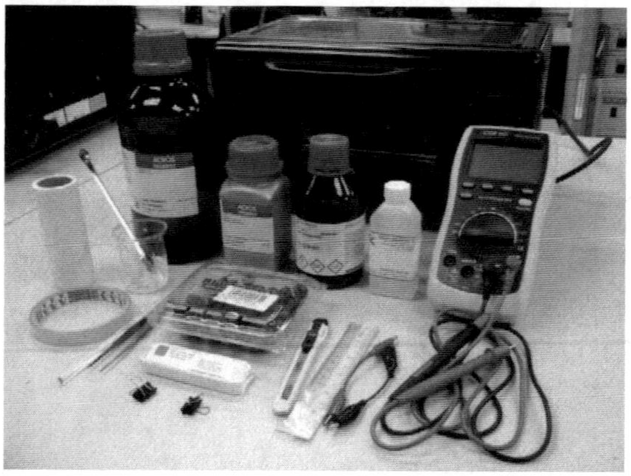

Figure 1. Equipments used in this study

2.2. Materials

The materials used in this study were nanotitanium dioxide TiO_2 powder, triton X-100, Recycle battery, Methoxypropionitrile, Indium tin oxide coated glass slide, 30 ~ 60 Ω/sq surface sensitivity, as shown in Figure 2.

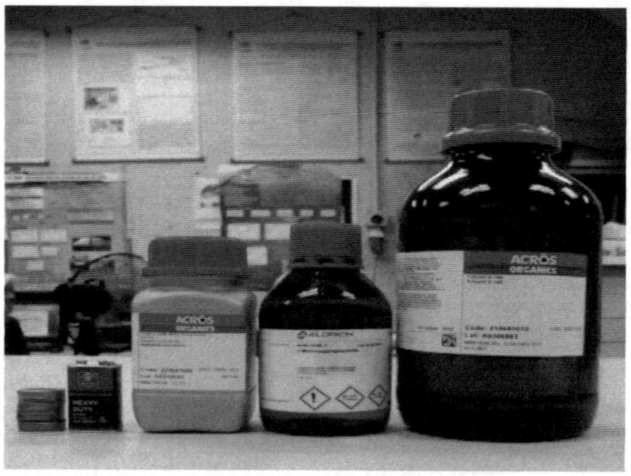

Figure 2. Materials used in this study

2.3. Proposed Technique

2.3.1. Glass Preparation

This section focuses on the preparation of the glass substrates for those who would like to prepare their own electrodes from scratch. It includes cutting the glass to the desired size and gives advice for cleaning prior to making any deposition. The typical substrate for making Dye Solar Cells is a glass plate coated on one side with a Transparent and Conductive Oxide (TCO) layer. The most frequently used conductive material for that purpose is fluorine-doped tin oxide (FTO). The terms TCO or FTO glass are often used to refer to this kind of substrate [3].

2.4. Identifying the Conductive Side

The TCO glass plates are only conductive on one side. Make sure to clearly identify the appropriate side while making your cells. The deposition of materials discussed in this tutorial will always be performed on the conductive side. A multimeter set to measure resistance should typically read about 58.4 Ohms when the two probe tips are placed one millimeter apart in the middle of the plate. This value may vary depending on the TCO glass type. There's also an easy trick that requires no instrument. This is particularly handy for identifying which is the conductive side at any time during the assembly. Scratch each side of the glass with a plastic spatula or even your fingernail. The conductive side is the one that feels a little sticky. The TCO coating has a rougher surface than bare glass. Visually, the coated side is also the one that looks hazy. Show in Figure 3.

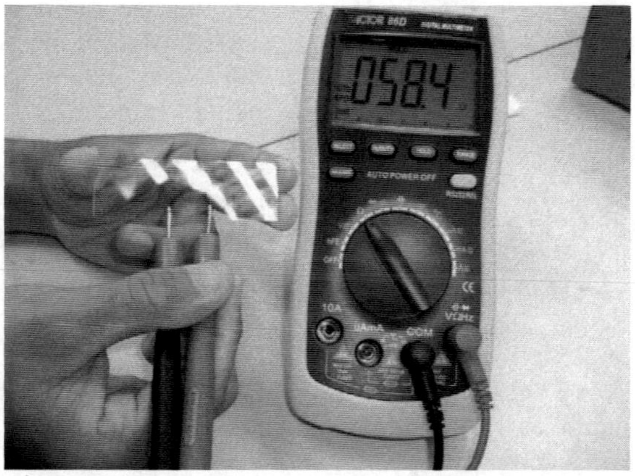

Figure 3. Indium tin oxide coated glass slide, 30 ~ 60 Ω/sq surface sensitivity

2.5. Preparing the Substrate for Deposition

The simplest and most widely used method for depositing titania paste on a substrate is the so-called doctor-blade method. The technique is also known as slot-coating in its mechanized version. It uses a hard squeegee, or doctorblade, to spread a portion of titania paste onto the glass. With this technique, the thickness of the titania layer is determined by the thickness of a spacer placed on both sides [4]. Using the Scotch tape which has a thickness of about 0.04 mm, as shown in Figure 4. (a) with the conductive side facing up, apply two parallel strips of tape on the edges of the glass plate, covering about 5 to 7 mm of glass. The area of uncovered glass in the middle of the glass is where the titania will be deposited. Edges masked by the tape will give room for future sealing and electrical contacts. Take advantage of the adhesive tape to hold the glass plate in position on the workbench. This will prevent the plate from moving while making the deposition stroke, as shown in Figure 4 (b).

Journal of Physics: Conference Series **423** (2013) 012055 doi:10.1088/1742-6596/423/1/012055

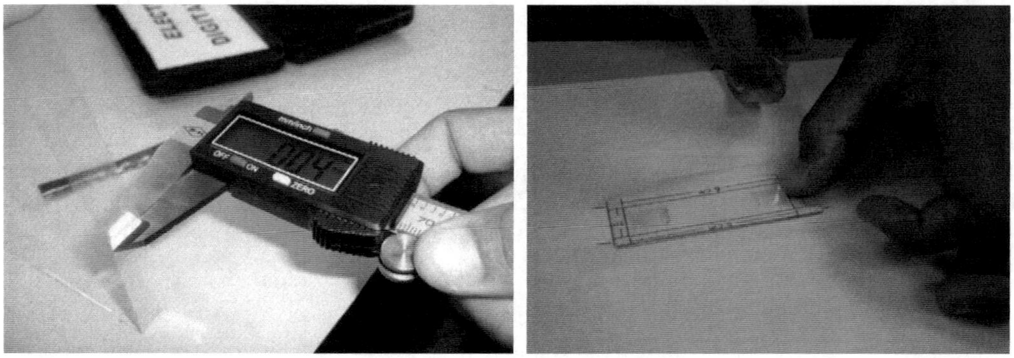

(a) (b)

Figure 4. (a) Thickness of 1 layer scoth tape; **(b)** Tape the transparent glass on four sides using Scotch tape

2.6. Preparation of Tio_2 and carbon suspension

TiO_2 and carbon from recycle battery powder as much as 3 grams of crushed using a mortar and then added into it as much as 0.2 mL of triton X-100. This mixture was stirred until evenly distributed. Then stirring to evenly. If it has been uneven. Settling of the titania paste can occur over time [5]. Make sure to start with a homogenous paste, stir well before use with a glass rod or similar tool, as shown in Figure 5.

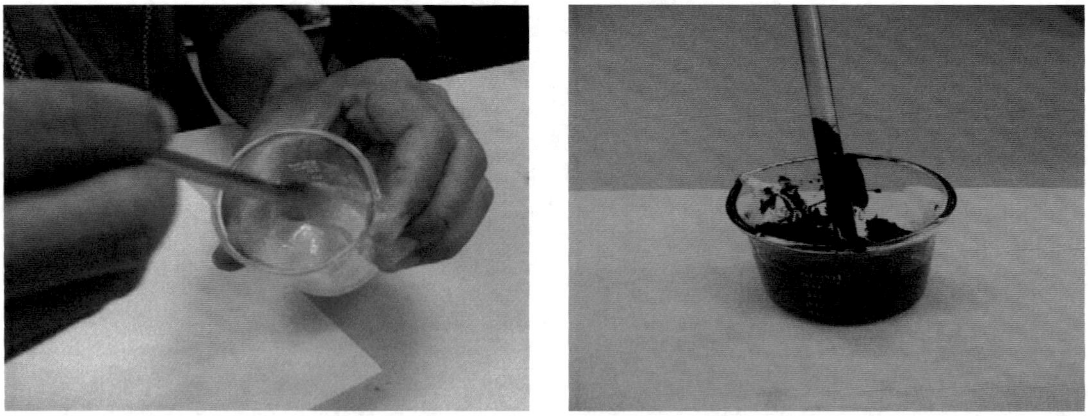

(a) (b)

Figure 5. (a) $Ti0_2$ mixture was stirred; **(b)** Carbon of recycle battery mixture until evenly distributed

2.7. Doctor-blading the titania paste

Apply a portion of paste near the top edge of the TCO glass between the two pieces of tape. With a rigid squeegee, such as a microscope slide or a glass rod, spread the paste across the plate with the support of the adhesive tapes on both sides. The gap between the strips of tape should be filled with a layer of titania paste. Repeat the operation until you have a reasonably homogenous layer. Coating was done by using doctor-blade until it reaches one thickness of Scotch tape [6], The cathode of a Dye

Solar Cell is made with carbon also. is easier and cheaper to realize. This makes carbon electrodes of great interest for research purposes. From a recycle battery of making a carbon counter-electrode are described here, as shown in Figure 6 (a) and (b).

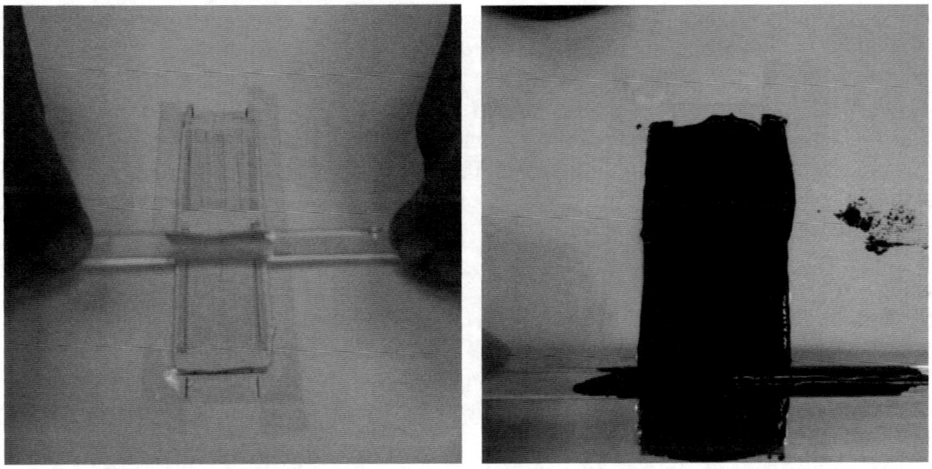

(a) (b)

Figure 6 (a) and (b) A reasonably homogenous layer titanium dioxide and carbon of recycle battery

The titania electrode is completed by firing the deposited layer. The vehicle of the paste burns away, leaving the titania nanoparticles sintered together. This process ensures electrical contact between particles and good adhesion to the TCO glass substrate. Sintering can occur in an oven, or on a programmable hotplate. First remove any adhesive tapes used for doctorblade deposition, and place the freshly coated glass plate in or on the heating device at room temperature. Set the desired temperature to 230°C. While heating up, you may observe that the titania layer turns brown/yellow and releases fumes. This corresponds to the evaporation and combustion of the non-toxic chemicals used in the paste formulation. After 40 to 45 minutes at 230°C, the baked titania layer looks white again or transparent, indicating that the sintering process is over. Allow the heating device to cool down before removing the fired electrode. A sudden change in temperature can cause the glass to break. The resulting nanoporous layer made from the sintered particles can absorb moisture from ambient air, make sure to store the sintered electrodes in a sealed environment soon after firing [7].

2.8. Staining the titania

Titanium dioxide is a white semiconductor that doesn't absorb visible light. Therefore, it is necessary to colour, or sensitize, the titania electrode with a dye that can absorb as much light as possible in the visible light spectrum. This section discusses the different methods for staining a titania electrode with a sensitizing dye.

2.9. Staining with a natural dye

Green plants and their fruits are a fantastic source of natural dyes that absorb visible light. The red pigments found in blueberries can work especially well in Dye Solar Cells. Red fruits are very handy for experimenting. The sensitization of titanium dioxide by natural dyes consists of soaking the titania electrode in mashed fruits [8]. Complete staining can take from several minutes to several hours, while the dye molecules from the fruit juice naturally adsorb onto the titania particles. The longer the electrode soaks in the dye, the better dyed the titania will be. Start crushing fresh fruits in a petri dish or similar container. The fruits must be juicy enough to get the titania electrode completely soaked. Place the electrode onto the mashed fruits with the titania surface facing down. If necessary, apply slight pressure on the glass plate so that the whole titania area is soaked by the mixture. The sensitization process takes an hour or more. Wait as long as possible, so that more dye molecules can attach to the titania. Remove the stained electrode and clean it carefully with tissue. Wait a few minutes for the ethanol to evaporate or use a hair-dryer to gently dry the electrode faster. The resulting stained titania should now look red almost all over its surface. If not, put it back in the fruit juice for further dyeing. Note that the mashed fruits don't offer a homogenous medium, and may cause a pattern on the stained titania electrode. This is not a problem with the operation of the solar cell. Place the Scotch tape around the titania area on the anode. The dimensions of the gasket and the stained titania shall match but not overlap. There should be one edge of the electrode that's not completely covered by the gasket to leave room for electrical contacts.

2.10. Putting electrodes together

Have seen how to prepare a titania anode and a counter-electrode made from carbon. Now let's see how to assemble the two electrodes into a solar cell. When the electrodes are put together, the active sides of the anode and the cathode will be facing each other. In other words, the stained titania will face the platinum or carbon of the counter-electrode. The gap left between the two glass plates will be filled with electrolyte during the next step.

This step can be accomplished using two different approaches [9]. First, electrodes can be pressed together, and the electrolyte soaked in the resulting stack by capillary effect. Second, the two electrodes can be sealed together and the electrolyte injected via gap two glass, as shown in Figure 7. (a) and (b).

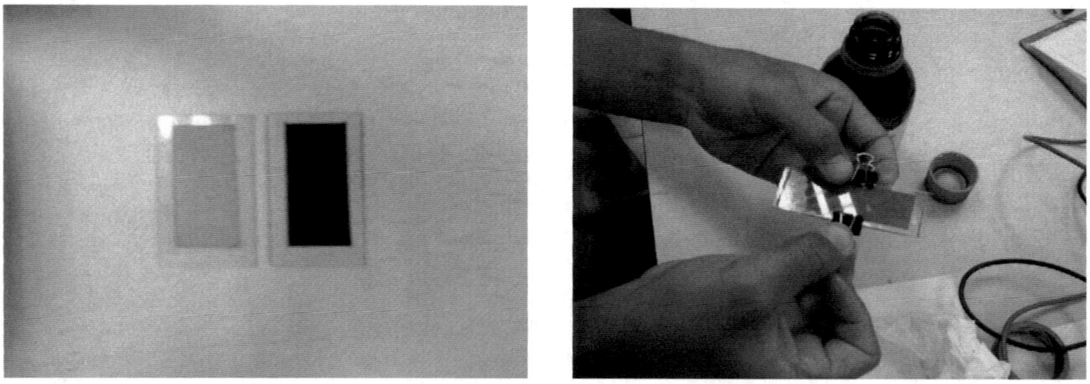

(a) (b)

Figure 7. (a) Ti02 as anode and Carbon as catode; (b) Assembled with two clip paper

3. Results

3.1. Performance of dye sensitized solar cells using recycle battery

The performance of dye sensitized solar cells using recycle battery has been measured from 09.00AM ~ 04.30 PM (Malaysia Time) with Graphtec and Solar power meter and get the result, given in Figure 8, Figure 9 and table 1.

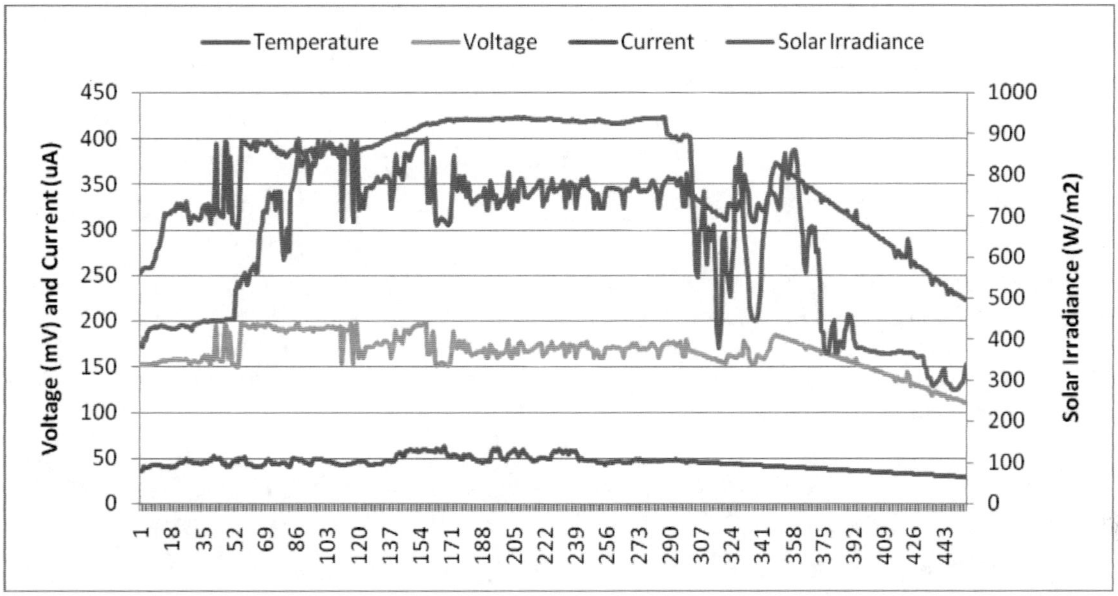

Figure 8. Graphic TiO_2 dye sensitized solar cells performance using recycle battery as function of solar irradiance and temperature

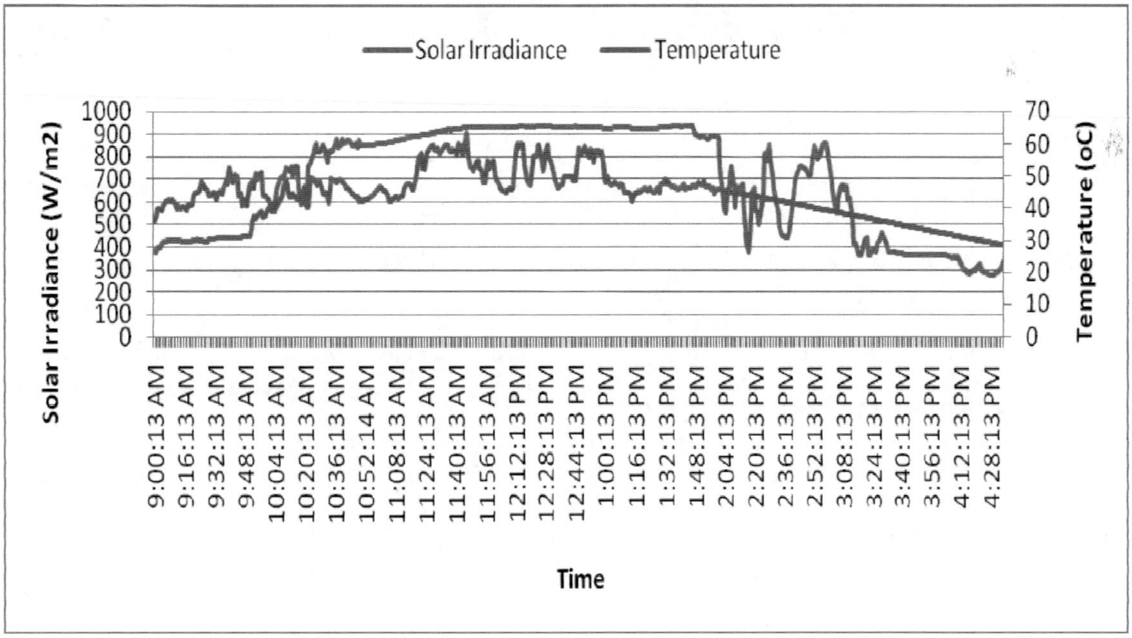

Figure 9. Grafic solar irradiance and tempereture as function of time

Table 1. Measurement result of dye sensitized titanium dioxide solar cells using recycle battery

Size of solar cells	**Average**		**FF (Fill Factor)**	**Temperature (^0C)**	**Solar Irradiance (w/m^2)**
	Voc (Volt)	Isc (µA)			
6 x 2 cm^2	0.33	166.04	1.19	44.49	693.69

3.2. Scanning Electron Microscope (SEM)

Titanium dioxide coated and Carbon coated with thickness 0.04 mm was image by scanning Electron Micriscope, show in Figure 10 (a) and (b).

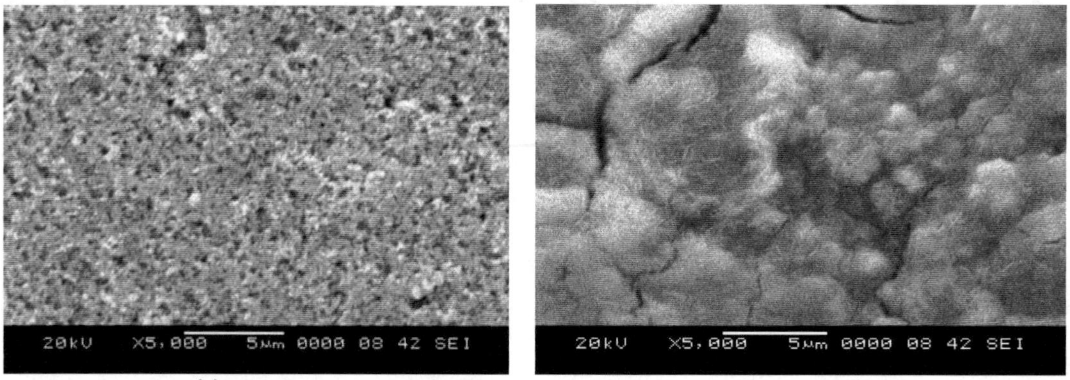

(a) (b)

Figure 10 (a). SEM image of Titanium dioxide surface at zoom x 5,000 and 5 µm; **(b).** SEM image of carbon surface at zoom x 5,000 and 5 µm

4. Conclusions.

1. In this paper has been successfully created a dye sensitized solar cells using recycle battery which is used to convert solar energy into electrical energy.
2. Battery recycling can reused to make it as enhancing performance of dye-sensitized solar cells, and this is a green energy that utilizes recycled. of the results obtained for dye sensitized titanium dioxide solar cells using recyle battery can get 0.33 Volt with an average solar irradiance 693.69 w/m^2 and average temperature 44.4 °C.
3. Carbon from a battery can produce negative electrode called cathode, which serves as the provider of electron, electrons from the carbon will increase the amount of energy produced by the solar cell.
4. Scanning Electron Microscope (SEM) showed that Titanium dioxide surface and carbon surface on tin oxide coated glass by naked eyes can not see it, so it can help us in research.

Acknowledgement

This work was supported by a grant from the University Malaysia Perlis. Aid in manuscript preparation by Prof. DR. Ismail Bin daut and experimental assistance by Muhammad Irwanto Bin Misrun, Gomesh Nair A/L Shasidharan, Mohd Irwan Yosuff is gratefully acknowledged.

References

[1] Huizhi Zhou, Liqiong Wu, Yurong Gao, Tingli Ma* *"Dye-sensitized solar cells using 20 natural dyes as Sensitizers"* Journal of Photochemistry and Photobiology A: Chemistry 219 (2011) 188–194.
[2] Bhattacharaya, Pallab, (1997), *"Semiconductor Optoelectronic Devices"*, Prentice Hall International, Inc., New Delhi.
[3] Information on http://www.solaronix.com.
[4] Zion Tachan, SvenR¨ uhle _, ArieZaban *"Dye-sensitized solar tubes: A new solar cell design for efficient current collection and improved cell sealing"* Zion Tachan,SvenR¨ uhle _, ArieZaban.
[5] Mayo, Elizabeth I, (2004), *" Kinetics and Thermodynamics of Dye (Group VIII Metal)- Sensitized Nanocrystalline Titanium Dioxide Photoelectrodes"*, Thesis PhD, California Institute of Technology, Pasadena, California.
[6] Jin-A Jeong, Han-KiKim* *"Thickness effect of RF sputtered TiO2 passivating layerontheperformanceof dye-sensitizedsolarcells"* Solar Energy Materials & Solar Cells 95 (2011) 344–348.
[7] Smestad, G., et al., (1994), *"Testing of Dye-Sensitized TiO2 Solar Cells I : Experimental Photocurrent Output and Conversion Efficiencies"*, Solar Energy Materials & Solar Cells.
[8] Qifeng Zhang, Guozhong Cao* *"Nanostructured photoelectrodes for dye-sensitized solar cells"* journalhomepage: www.elsevier.com/locate/nanotoday.
[9] Seigo Ito, Takurou N. Murakami 1, Pascal Comte 1, Paul Liska 1, Carole Grätzel, Mohammad K. Nazeeruddin 1, Michael Grätzel, *"Fabrication of thin film dye sensitized solar cells with solar to electric power conversion efficiency over 10%"*, Available online at www.sciencedirect.com

Journal of Physics: Conference Series **423** (2013) 012056 doi:10.1088/1742-6596/423/1/012056

A Critical Protection Level Derived from Dengue Infection Mathematical Model Considering Asymptomatic and Symptomatic Classes

N Anggriani, A K Supriatna, E Soewono*
Department of Mathematics
Faculty of Mathematics and Natural Sciences, Universitas Padjadjaran
Jl. Raya Bandung-Sumedang km 21 Jatinangor 45363, Indonesia
*Industrial and Financial Mathematics Group
Faculty of Mathematics and Natural Sciences, Institut Teknologi Bandung
Jl. Ganesha 10, Bandung 40132, Indonesia

E-mail: n_anggriani@unpad.ac.id and nsanti_a@yahoo.co.id

Abstract. In this paper we formulate a model of dengue fever transmission by considering the presence of asymptomatic and symptomatic compartments. The model takes the form as a system of differential equations representing a host-vector SIR (Susceptible – Infective - Recovered) disease transmission. It is assumed that both host and vector populations are constant. It is also assumed that reinfection of recovered hosts by the disease is possible due to a wanning immunity in human body. We analyze the model to determine the qualitative behavior of the model solution and use the concept of effective basic reproduction number ($\Re p$) as a control criteria of the disease transmission. The effect of mosquito biting protection (e.g. by using insect repellent) is also considered. We compute the long-term ratio of the asymptomatic and symptomatic classes and show a condition for which the iceberg phenomenon could appear.

1. Introduction

Dengue fever has long been endemic in tropical and sub-tropical areas. It is caused by four dengue virus serotypes and transmitted via the bites of female *Aedes aegypti* mosquitoes. The clinical manifestations of the disease vary from asymptomatic to fatal. Based on the symptoms, the clinical manifestations of the dengue virus infection can be broadly grouped into two categories: asymptomatic (without symptoms) and symptomatic (with symptoms). The latter may lead to the common cold (viral syndrome), Dengue Fever (DF), or Dengue Hemorrhagic Fever (DHF) including dengue Shock Syndrome (DSS) [1]. Figure 1 shows the diagram for the clinical manifestations of the dengue virus infection.

Fever is common in infants, children and some adults who are infected with dengue virus for the first time (a primary dengue infection). It is usually indistinguishable from the fever caused by other viral infections. DF can affect both children and adults, but it commonly affects children below the age of 15. Basically it is an acute biphasic fever characterized by headache and skin rash. Although it is basically harmless, the disease can reduce the function of the body, for instance causing muscle characterized by fever followed by signs of blood circulatory failure causing serious bleeding in many parts of body [1].

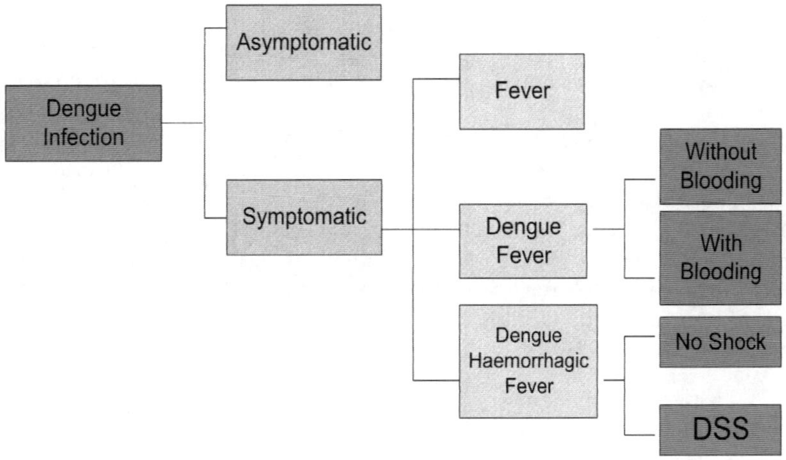

Figure 1. Diagram of Dengue Infection manifestation [16].

Dengue virulence occurred in 1998 in Indonesia, when thousands of clinically diagnosed DHF cases were found [2]. The maximum number of DHF/DSS cases reported during the month when the dengue fever peaked supports the DHF/DSS iceberg theory [3]. In relation to many dengue outbreaks, Balasubramanian, a senior consultant pediatrician at the Child Trust Hospital India, pointed out that hospitals only see the tip of the iceberg. He argued that for every one case in the hospital, there should be at least nine cases among the population, indicated by mild fever and almost no symptoms [4].
A person infected by the dengue virus for the first time generally does not enter the DHF or DSS stage. A primary or first infection usually results in no symptoms or only shows symptoms similar to the common cold, or at worst DF. However, a second infection and the subsequent infections may lead to DHF or DSS. Evidence shows that a person can be exposed to the DHF stage more than twice if their immunity is reduced, i.e. when the body immunity is waning [5].

To date, there have been a lot of mathematical models used in the study of the transmission of the dengue fever. Many of the models only use one strain of the dengue virus. These models include dengue models using constant population [7], varying population [8], and models incorporating the vaccination factor [6, 10, 11]. Although many mathematical models for the dengue fever transmission have made great contribution [6, 7, 8, 9, 10, 12], almost all of them neglect the details, such as the iceberg phenomenon, the waning immunity that causes re-infection by the same/different strain, and the presence/absence of symptoms.

An extensive review has been done in [13]. In this study, we develop a mathematical model for the dengue transmission by simultaneously incorporating these details. It is assumed that when a person has contracted dengue fever, after the recovery they can be re-infected due to the waning immunity in their body, regardless of the virus strain. We generalize an SIR (Susceptible - Infective - Recovered) dengue transmission model by dividing the infected class in the SIR model into the asymptomatic class and symptomatic class. The purpose is to facilitate the study of the condition in which the iceberg phenomenon can appear. The details of the model are given in the following section.

2. Model formulation

In constructing the mathematical model, we assume that the host and vector populations have constant sizes and constant rates for birth and death, which are μ_h and μ_v, respectively. In [7], the majority of individuals in the host population eventually become permanently immune with long-term immunity. In our model, individuals in the host population either become immune or lose their immunity.

In this model, we assume that susceptible hosts are divided into two categories: S_0 (susceptible virgin) from newborns and Z_0 (susceptible with previous infection) from recovered individuals who have lost their immunity. This division is based on the fact that, using the IgG and IgM test, dengue infection in a person can be detected when they were previously infected with the dengue virus. The infected class is divided into two more classes: I, the asymptomatic infection in humans, and Y, the symptomatic infection in humans (DF, DHF, DSS). There is no dengue transmission from Y since we assume that infection in this class is detectable, hence it can be easily isolated easily. Z refers to recovered patients. The vector population, due to its short lifespan, is divided into the susceptible V_0 and the infective V_1. The model is represented by the diagrams shown in Figures 2 and 3, with the description of the parameters given in Table 1.

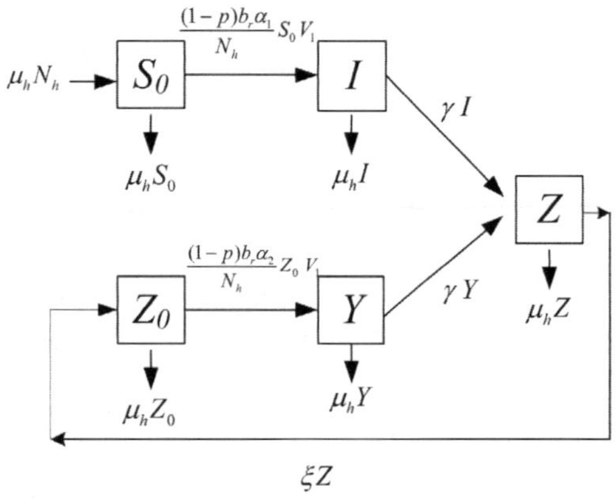

Figure 2. Host Transmission Diagram.

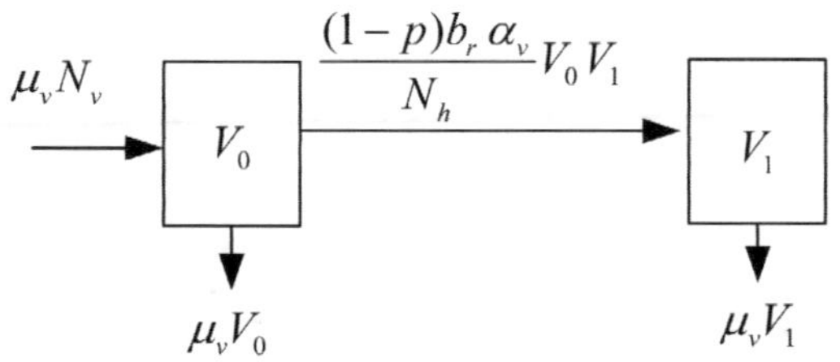

Figure 3. Vector Transmission Diagram.

We also assume that the susceptible host may use a level of protection against being bitten by mosquitoes, for example by applying an insect repellent. In this case $p = 0$ means no protection at all, while $p = 1$ means full protection (everybody is fully protected).

Table 1. Definition of parameters used in the model.

Parameter	Description
μ_h	Natural death rate for host
μ_v	Natural death rate for vector
α_i	Probability of successful transmission from in host
α_v	Probability of successful transmission from in vector
ξ	Wanning immunity rate
b_r	Biting rate
γ	Recovery rate

The dynamical equation for host and vector are :

$$
\begin{aligned}
\frac{dS_0}{dt} &= \mu_h N_h - \frac{b_r\,\alpha_1}{N_h}\,S_0\,V_1 - \mu_h\,S_0 \\
\frac{dZ_0}{dt} &= \xi Z - \frac{b_r\,\alpha_2}{N_h}\,Z_0\,V_1 - \mu_h\,Z_0 \\
\frac{dI}{dt} &= \frac{b_r\,\alpha_1}{N_h}\,S_0\,V_1 - (\gamma + \mu_h)\,I \\
\frac{dY}{dt} &= \frac{b_r\,\alpha_2}{N_h}\,Z_0\,V_1 - (\gamma + \mu_h)Y \\
\frac{dZ}{dt} &= \gamma\,(I+Y) - (\mu_h + \xi)Z \\
\frac{dV_0}{dt} &= \mu_v N_v - \frac{b_r\,\alpha_v}{N_h}\,V_0\,I - \mu_v V_0 \\
\frac{dV_1}{dt} &= \frac{b_r\,\alpha_v}{N_h}\,V_0\,I - \mu_v V_1
\end{aligned}
\qquad (1)
$$

Note that $N_h = S_0 + Z_0 + I + Y + Z$ is the total population of host, and $N_v = V_0 + V_1$ is the total population of vectors in the region of Ω, with $\Omega = \{(S_0; Z_0; I; Y; V_0; V_1) \in [0; 1]^7 \mid 0 \le S_0 + Z_0 + I + Y + Z \le N_h,\ 0 \le V_0 + V_1 \le N_v\}$. The description of the parameters used in the model is given in Table 1.

The model in this paper differs from known similar models that include the asymptomatic and symptomatic. In [14] and [15] the asymptomatic and symptomatic come from the same compartment, i.e. from susceptible people who get infected, through the introduction of a different transmission probability β_{ha} (the dengue virus's transmission probability from vector population to human population, changing it into an asymptomatic infectious human population) and β_{hs} (the dengue virus's transmission probability from vector population to human population, changing it into a symptomatic infectious human population). Other models such as in [16] cover the possibility of sequential infections with all four different serotype, by considering apparent (symptomatic) and non-apparent cases (asymptomatic), but assume that apparent cases may also appear as a first infection. The model also assumes that a re-infection cannot occur more than twice. The model in [17] presents a three-state infection. The three states of infection are respectively asymptomatic, partially asymptomatic, and fully asymptomatic. However, according the analysis's conclusion, these are not conditions in which the iceberg phenomenon may occur.

3. Analysis of the model

3.1. Basic Reproduction Number

The basic reproduction number (BRN) is an important threshold number in epidemiology. It is defined as the number of secondary infections caused by one primary infection in an entirely susceptible population [18] and usually written as $\mathcal{R}_0$. The value can be obtained through the construction of the next generation matrix [19, 20] of the infectious disease processes. of the infectious disease processes. Note that the next generation matrix is applied in cases when initially there is no infected individual in the host population. The BRN is also defined as the spectral radius or the largest eigen value of the next generation matrix, evaluated at the steady state population sizes in the absence of infection. It is widely known that, in many epidemiological models, $\mathcal{R}_0 > 1$ ndicates an epidemic case in which the infection progresses and grows away from zero infective. On the other hand, if $\mathcal{R}_0 < 1$ then the disease will die out. It will be shown that such is the case for the present model.

The next generation matrix G for system (1) is given by

$$MD^{-1} = \begin{bmatrix} 0 & 0 & \frac{(1-p)b_r\,\alpha_1}{\mu_v} \\ 0 & 0 & 0 \\ \frac{(1-p)b_r\,\alpha_v\,N_v}{N_h\,(\gamma+\mu_h)} & 0 & 0 \end{bmatrix} \quad (2)$$

If $X \in [I;\ V_1]$ where $i;\ j \in \{1;\ 2\}$, then G_{ij} is the number of new cases in X_j produced by one X_i during the transmission period. For example the first row and the second column of G show the number of infected hosts, produced by one infected vector during the period of transmission of V_1. Direct computation shows that the effective reproduction number (the BRN with protection p) is given by

$$\mathcal{R}_p = \sqrt{\frac{((1-p)\,b_r)^2\,N_v\,\alpha_1\,\alpha_v}{N_h\,(\gamma+\mu_h)\mu_v}} \quad (3)$$

and without protection we have the natural BRN.

$$\mathcal{R}_0 = \sqrt{\frac{b_r^2\,N_v\,\alpha_1\,\alpha_v}{N_h\,(\gamma+\mu_h)\mu_v}} \quad (4)$$

3.2. Equilibrium points

In this section we will discuss the equilibrium points of system (1) in the Ω region. It is shown that the positive equilibrium depends on the magnitude of $\mathcal{R}_p$. We present some results regarding the existence of equilibrium points of system (1), which are as follows:

Proposition 3.1 *The system (1) admits two equilibrium points: the disease free* $E_0 = (N_h,0,0,0,0,N_v,0)$ *and the endemic equilibrium* $E_e = (S_0^*,\ Z_0^*, I^*, Y^*, Z^*, V_0^*, V_1^*)$, *where*

$$S_0^* = \frac{N_h(A_1+\mathcal{R}_p^2\,K)}{\mathcal{R}_p^2\,(A_1+K)}, \quad (5)$$

$$Z_0^* = \frac{A_1\,C\,(\mathcal{R}_p^2-1)\,(A_1+K\,\mathcal{R}_p^2)\,\xi}{M(A_1+K)\mathcal{R}_p^2}, \quad (6)$$

$$I^* = \frac{A_1\,(\mathcal{R}_p^2-1)K}{\mathcal{R}_p^2\,(A_1+K)C}, \quad (7)$$

$$Y^* = \frac{A_1\,A_2\,K\,(\mathcal{R}_p^2-1)^2\,\xi\,\gamma}{\mathcal{R}_p^2\,(A_1+K)\,C\,M}, \quad (8)$$

$$Z^* = \frac{A_1\,\gamma\,(\mathcal{R}_p^2-1)\,(A_1+A_1\,(\mathcal{R}_p^2-1)+K\mathcal{R}_p^2)}{M(A_1+K)\mathcal{R}_p^2}, \quad (9)$$

$$V_0^* = \frac{N_v(A_1+K)}{A_1+K\,\mathcal{R}_p^2}, \quad (10)$$

$$V_1^* = \frac{K(\mathcal{R}_p^2-1)\,N_v}{A_1+K\,\mathcal{R}_p^2}, \quad (11)$$

with

$$A_i = (1-p)\,b_r\alpha_i\,N_v\,,\ i=1{,}2;\ K = N_h\,\mu_h\ ;\ C = \gamma+\mu_h\ ;\ D = \xi+\mu_h\,;\alpha_2 = \delta\,\alpha_1 \quad (12)$$

$$M = K\,\mathcal{R}_p^2\,D\,C + \mu_h\big(A_1 + A_2(\mathcal{R}_p^2-1)\big)D + A_2(\mathcal{R}_p^2-1)\gamma\,\mu_h + A_1\,D \quad (13)$$

Proposition 3.1 is easily proven as a consequence of direct computation. The next proposition provides a condition for the local stability of the disease-free equilibrium state E_0. The proof is deduced from the stability theorem for the disease-free equilibrium of general compartmental model [21].

3.3. Stability of the disease-free equilibrium

The following proposition provides the stability criteria for the disease-free equilibrium *E*o.

Proposition 3.2 *E_0 is a locally asymptotically stable disease-free equilibrium whenever $\mathcal{R}_p < 1$. E_0 is an unstable disease-free equilibrium whenever $\mathcal{R}_p > 1$.*

Proof. The Jacobian matrix *J*(*E*o) of system (1) is given by

$$J_{E_0} = \begin{bmatrix} -\mu_h & 0 & 0 & 0 & 0 & 0 & -(1-p)b_r\alpha_1 \\ 0 & -\mu_h & 0 & 0 & \xi & 0 & 0 \\ 0 & 0 & -\gamma-\mu_h & 0 & 0 & 0 & (1-p)b_r\alpha_1 \\ 0 & 0 & 0 & -\gamma-\mu_h & 0 & 0 & 0 \\ 0 & 0 & \gamma & \gamma & -\mu_h-\xi & 0 & 0 \\ 0 & 0 & -\dfrac{(1-p)b_r N_v\alpha_v}{N_h} & 0 & 0 & -\mu_v & 0 \\ 0 & 0 & \dfrac{(1-p)b_r N_v\alpha_v}{N_h} & 0 & 0 & 0 & -\mu_v \end{bmatrix}$$

The eigenvalues of *J*(*E*o) are $-\mu_h$ with multiplicity 2, $-(\mu_h+\gamma)$ with multiplicity 1, and the roots of polynomial $p_1(x) = a_1x^2 + b_1x + c_1$, where :

$$a_1 = N_h > 0$$
$$b_1 = \gamma + \mu_h + \mu_v > 0$$
$$c_1 = \mu_v(\gamma+\mu_h)N_h\left(1 - \mathcal{R}_p^2\right)$$

The coefficient $c_1 > 0$ if $\mathcal{R}_p < 1$, hence all the roots of polynomial *pi* have negative real part when $\mathcal{R}_p < 1$. This means that *E*o is locally asymptotically stable when $\mathcal{R}_p < 1$, with a saddle point where $\mathcal{R}_p > 1$. This proves the proposition. ■

Note that from Propositions 3.1 and 3.2, if $\mathcal{R}_p > 1$ then there exists an endemic equilibrium point. On the other hand, if $\mathcal{R}_p < 1$ then the endemic equilibrium disappears and the disease-free equilibrium becomes stable. Hence, to eliminate the disease it is enough to make the value of $\mathcal{R}_p < 1$. This can be done by controlling the value of protection level *p*. The condition $\mathcal{R}_p < 1$ is equivalent to $p > 1 - \frac{1}{\mathcal{R}_0}$. Here $p_c = 1 - \frac{1}{\mathcal{R}_0}$ is called the Critical Protection Level. In an endemic case, Proposition 3.1. also gives the ratio of asymptomatic and symptomatic classes that is,

$$\frac{I^*}{Y^*} = \frac{M}{A_2\,\gamma\,\xi\,(\mathcal{R}_p - 1)}$$

where *M* and *A*2 are given in equations (12) and (13). The above ratio indicates the presence or absence of the iceberg phenomenon. If *L* is a sufficiently big number (e.g. nine, as in the introduction), then the condition in which the iceberg phenomenon may exist is

$$M > L\,(\mathcal{R}_p - 1)\,A_2\,\gamma\,\xi \qquad (15)$$

The numerical example in Figure 4 shows that the condition is met when the probability of a successful dengue transmission for the previously infected and susceptible is sufficiently less than the probability of a successful dengue transmission for the naive and susceptible, i.e. $\alpha_2 << \alpha_1$.

4. Numerical Simulation

Numerical examples are shown in this section. Figure 4 shows the plot of I^*/Y^* as a function of the mosquito bite rate b_r. It reveals that the ratio for asymptomatic and symptomatic decreases when the mosquito bite rate increases, as suggested by equation (14).

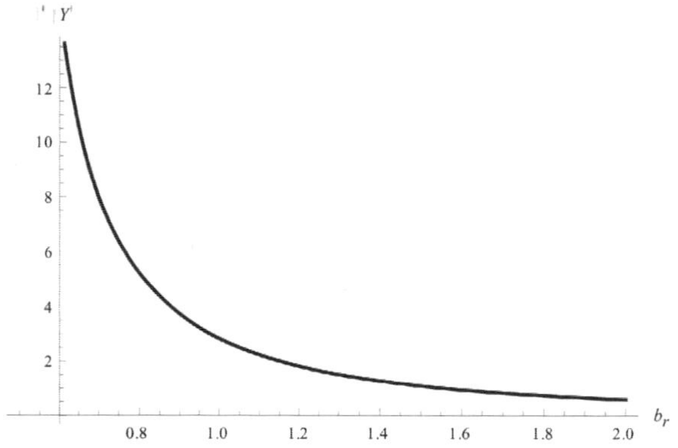

(b)

Figure 4. The ratio of asymptomatic and symptomatic as a function of *br*. The parameter in the simulation are $N_v = 10000$, $N_h = 1000$, $\gamma = 1/7$, $\alpha_1 = 0.75$, $\alpha_2 = 0.5$, $\alpha_v = 0.3$, $\xi = 0.4$, $\mu_v = 1/30$, $\mu_h = 1/(60 \times 365)$.

Figure 5 shows that for a fixed ξ the ratio for asymptomatic and symptomatic decreases when the mosquito bite rate increases. It also shows that for a fixed b_r the ratio for asymptomatic and symptomatic decreases when the body immunity increases. This suggests that the iceberg phenomenon may occur when the rates of both the mosquito bite and waning immunity are sufficiently small. The parameter in the simulation for Figure 5 are $N_v = 10000$, $N_h = 1000$, $\gamma = 1/7$, $\alpha_1 = 0.75$, $\alpha_2 = 0.5$, $\alpha_v = 0.3$, $\xi = 0.4$, $\mu_v = 1/30$, $\mu_h = 1/(60 \times 365)$. Figures 6 shows the effect of protection on the dynamics of the infected host population. Figure 7 show that sufficient protection (higher than the critical protection level) may eliminate both the disease and the iceberg phenomenon.

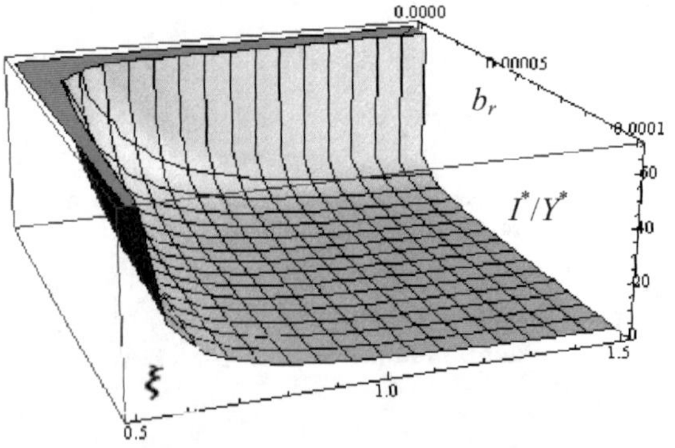

(a)

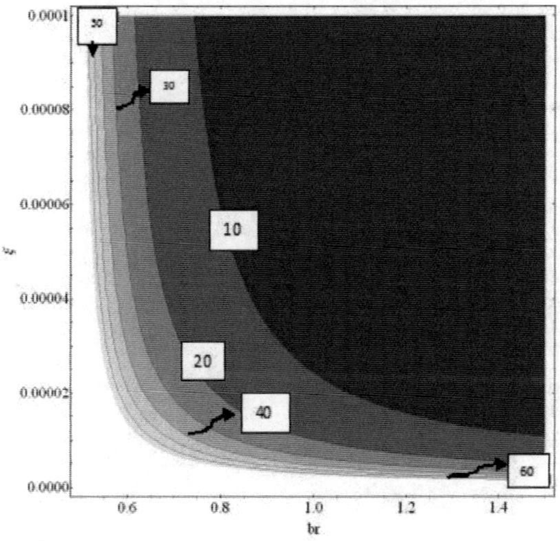

(b)

Figure 5. (a) The ratio of asymptomatic and symptomatic as a function of b_r and ξ with the level curve given in (b).

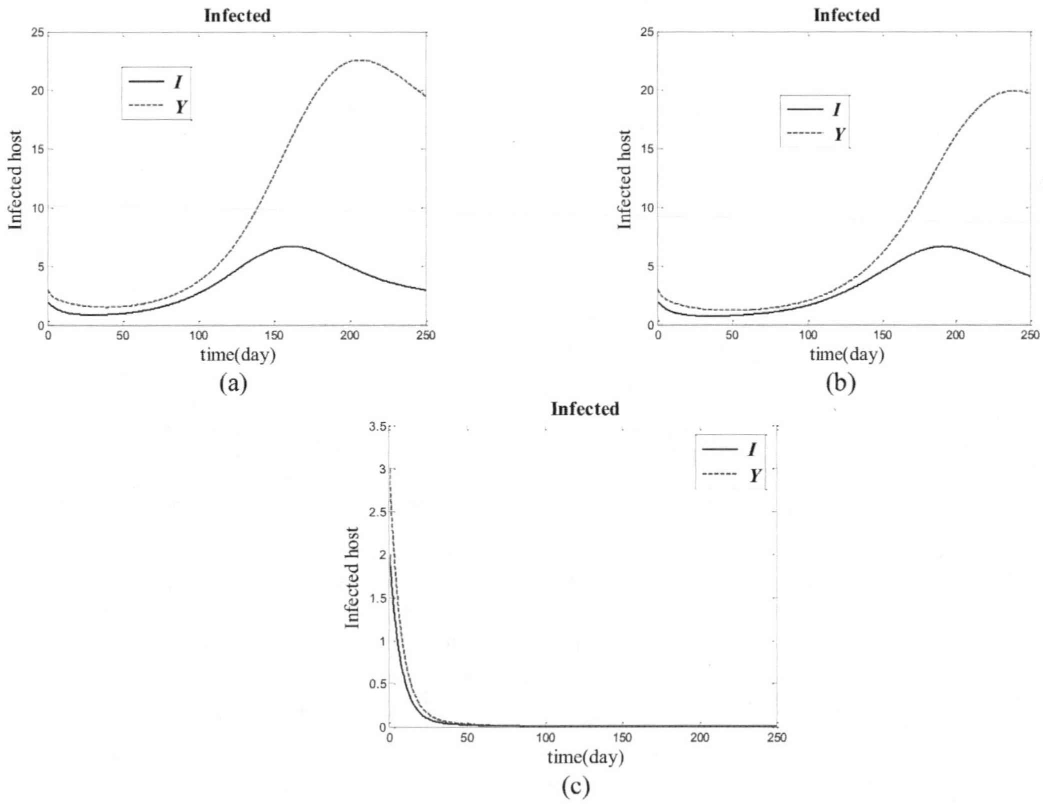

Figure 6. The Dynamics of compartments I and Y. (a) without protection (p = 0), (b) with not enough protection (p = 0.1), (c) with a sufficient protection (p = 0.948).

The parameter for Figure 6 are $N_v = 10000$, $N_h = 50000$, $\gamma = 1/7$, $\alpha_1 = 0.75$, $\alpha_2 = 0.5$, $\alpha_v = 0.3$, $\xi = 0.4$, $\mu_v = 1/30$, $\mu_h = 1/(60 \times 365)$, $S_0(0) = 30$, $Z_0(0) = 50$, $I(0) = 2$, $Y(0) = 3$, $Z(0) = 50$, $V_0(0) = 50$, $V_1(0) = 50$, $\mathcal{R}_0 = 19.21$, $\mathcal{R}_p = 0.998$.

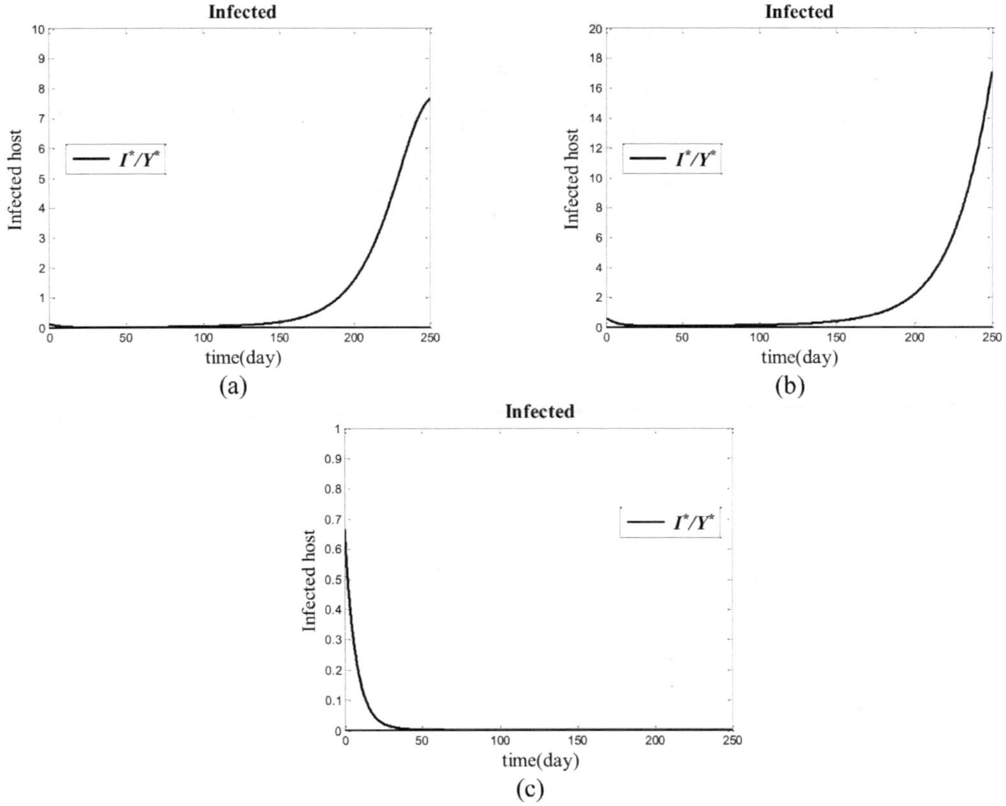

Figure 7. The graphs show the ratio of asymptomatic and symptomatic compartments. (a) without protection ($p = 0$), (b) with not enough protection ($p = 0.1$), (c) with a sufficient protection ($p = 0.948$). The parameter in simulation are $N_v = 100000$, $N_h = 500000$ another parameter are the same as in Figure 6.

5. Conclusion

In this paper we devoleped a mathematical model of dengue transmission by separating the infected class in the SIR model into the asymptomatic and symptomatic classes, where the division is done to facilitate the analysis on which condition for the iceberg phenomenon could appear. The model and the numerical simulation show that an iceberg phenomenon may appear when both the biting rate and wanning immunity is sufficiently small and they also reveal that enough protection (higher than the critical protection level) may eliminate the disease and the iceberg phenomenon simultaneously.

Acknowledgments

The authors thank to the Indonesian Government who funded this work through a competitive research grant (Penelitian Sentralisasi Dikti - Hibah Kompetensi 2012).

References

[1] WHO 2009 *Dengue Haemorrhagic Fever, Diagnosis, Treatment and Control* France(2009) http://whqlibdoc.who.int/publications/2009/9789241547871*eng*.pdf.

[2] K. Porter et al 2005 *Epidemiology Of Dengue and Dengue Hemorrhagic Fever in a Cohort of Adults Living in Bandung West Java Indonesia* J. Ameri Soc. of Tropic. Med. and Hyg 72(1):60-66.

[3] D Gubler 1998 *Dengue and Dengue Hemorrhagic Fever* J. Clinic Microbiol 11: 480-496.

[4] The Hindu 2005 Online edition of India's, *National Newspaper* Thursday Oct 20, http://www.hindu.com/2005/10/20/stories/2005102012730100.htm.

[5] S B Halstead and J Deen 2002 *The future of dengue vaccines* The Lancet 360.

[6] E Soewono and A Supriatna 2001 *A two-dimensional Model for The Transmission of Dengue Fever Disease* Bull. Malay. Math Sci. Soc 24 : 49-57

[7] L Esteva and C Vargas 1998 *Analysis of a dengue disease transmission model* J. Math. Biosci 150:131-151.

[8] L Esteva and C Vargas 1999 *A Model for Dengue Disease with variable human population* Journal of Mathematical Biology 38:220-240.

[9] L Esteva and C Vargas 2003 *Coexistence of different serotypes of dengue virus* J. Math. Biol 46:31-47.

[10] M Derouich and A Boutayeb 2003 *A Model of Dengue Fever* BioMed. Engineer online 2: 4 (doi:10.1186/1475-925x-2-4).

[11] M Derouich and A Boutayeb 2006 *Dengue Fever : Mathematical Modelling and Computer Simulation* Applied Mathematics And Computation 177:528-544.

[12] N Nuraini, E Soewono and K A Sidarto 2007 *Mathematical Model of Dengue Disease Transmission with Severe DHF Compartment* Bull. Malay (2007) 30 (2).

[13] M. Andraud, N Hens, C Marais and P Beutels 2012 *Dynamics Epidemiological Models for Dengue Transmisson:A Systematic Review of Structural Approach* Plos One 7(11).

[14] P Pongsumpun 2007 *Age Structured Model for Symptomatic and Asymptomatic Dengue Infections* Proceeding of ASM'07 The 16th IASTED International Conference on Applied Simulation and Modelling 348-352.

[15] C S Vargas 2009 *Mathematical Model for Dengue Epidemics with Differential Susceptibility and Asymptomatic Patients Using Computer Algebra* Springer-Verlag, Berlin, Lecture Notes in Computer Science 5743:284-298.

[16] E Chikaki and H Ishikawa 2009 *A dengue transmission model in Thailand considering sequential infections with all four serotypes* J Infect Dev Ctries 3(9):711-722.

[17] D Hincapie and J Ospina 2012 *Mathematical model for Dengue with three states of infection* Proc. SPIE 8371 Sensing Technologies for Global Health Military Medicine Disaster Response and Environmental Monitoring II and Biometric Technology for Human Identification IX 83710H (May 1, 2012) doi:10.1117/12.919063; http://dx.doi.org/10.1117/12.919063.

[18] R M Anderson and R M May 1992 *Infectious Disease of Humans: Dynamics and Control*, Oxford University Press.

[19] C Chavez and Z Feng 2000 *On The Computation of* $\mathcal{R}_0$ *and its Role On Global Stability*, In Mathematical Approaches for Emerging and Reemerging Infectious Disease: An Introduction., IMA 125 : 229- 250, Spriger-Verlag

[20] O Diekmann and J A P Heesterbeek 2000 *Mathematical Epidemiology of Infectious Diseases*, John Wiley & Son., New York.

[21] P Van den Driessche and J Watmough 2002 *Reproduction numbers and sub-threshold endemic equilibria for compartemental models of disease transmission Math. Biosci.* 180: 29-48.

ScieTech 2013 IOP Publishing
Journal of Physics: Conference Series **423** (2013) 012057 doi:10.1088/1742-6596/423/1/012057

Selective SHI irradiation for mesa type edge termination in semiconductor planar junction

Vibhor Kumar[1,2], A.S. Maan[2] and J. Akhtar[*1]

[1]Sensors and Nano-Technology Group, CSIR-CEERI, Pilani-333 031, India

[2]Physics Department, M.D.University, Rohtak, Haryana, India

[*]jimmyakhtar@gmail.com

Abstract. A mesa type edge termination in planar junction structures has been devised by involving vibrational energy localisation in crystalline medium leading to atomic scale reordering of host atoms on the boundary of enclosed irradiated region. A fabrication process based on selective MeV ion beam irradiation on a square shaped planar abrupt p-n junction is presented to realize semiconductor junction structures with near ideal electrical characteristics. The experimental validation of the process provides compatibility with microelectronics and suits for Schottky diode structures employing silicon carbide (SiC).

1. Introduction

Planar semiconductor abrupt junction structure is the backbone of larger section of solid state devices and circuits without which electronics as a whole ceases in its present form. Planarity is an advantage to support continuous miniaturization of the electronic structure on one hand while associated edge effects tend to reduce operating capability and long term stability due to unavoidable design and fabrication related deficiencies [1]. Suppression of curvature effects associated with active region of junction region is a challenge in planar structures [2]. This problem becomes grim with continuous decreasing size of the structures. Mesa type edge termination of semiconductor junction has been considered an ideal way to suppress curvature and edge effects but it is limited to larger size of discrete structures [3,4]. A number of planar junction termination techniques have been developed but for partial improvement [5,6,7,8,9]. There is no mesa type junction termination techniques so far suggested for miniaturized planar junction structures. With the advancement in semiconductor technology and improved understanding in ion-solid interactions, localized modifications in the crystalline structures at nano level are suggested and experimentally shown to introduce for their further applications as junction edge termination [10,11]. This type of termination resembles with mesa and has been found compatible with microelectronics involving miniaturized planar structures.

Basic process of selective SHI irradiation and mechanism responsible for localized reordering of discrete system at nano level with modified electronic properties is explained in next section. Curvature effects in planar semiconductor junction structures and conventional approaches adopted for their suppression is presented in the followed section. Experimental results are presented thereafter on a model planar square shaped abrupt p-n junction to validate selective SHI irradiation in mesa type termination. The results are discussed with emphasis on selective SHI irradiation as viable planar junction termination technique suitable for SiC schottky diode.

2. Selective SHI irradiation and localized reordering

Interaction of swift heavy ions with crystalline semiconductor can be classified into two broad categories : blank exposure to the entire medium and selective exposure using thick metal mask. Blank exposure of MeV ion beams is usually carried out and reported in published literature for imparting common ion-fluence initiated modifications in different materials [9]. Employment of selective irradiation of few hundreds of MeV ion beam so that the exposed region is enclosed by protective mask of thick metal, results however, in hot and cold regions [12]. The very nature of electronic energy loss mechanism in the exposed region leaves the crystalline medium unchanged while the ions are implanted deep into the bulk. The energy transfer from the incident ions to the local atoms result in energy localization, which also traverse along the atomic high ways. The localized energy is finally dissipated at the interface of irradiated and masked regions of the medium and results in the modification of atomic arrangement within few tens of nano-meters. The reordering of atoms at the interface of closed exposed region has been experimentally shown in crystalline silicon using 200 MeV Ag+14 ion beam with a metal mask of nickel in the shape of a square mesh of 40 μm grid size. The reordered nanostructure having modified electronic properties isolate the exposed region with bulk material by encircling it in the lateral direction and therefore offers a mesa type edge termination [10, 12].

The swift heavy ions equipped with energy of hundreds of MeV interact with crystalline silicon atoms close to the surface and lose their energy through inelastic collision. The sudden impact of large energy results in localized excitation, which moves from atoms to atoms leaving no sign of structural modification to the exposed region of the silicon target. However any dislocation or defect present in the crystalline silicon is also swept away along with the moving localized energy packet. This way the silicon crystal becomes more clean from unwanted defects and dislocations which are invariably generated during crystal growth. The continuous exposure of the swift heavy ions to the enclosed silicon region using thick metal mask, initiates a region full of moving energy packets towards the boundary of the exposed region. At the interface of the exposed and un-exposed region, these energy packets release their localized energy, which is dissipated in terms of rearrangements of local atoms under the thick metal mask. The rearrangement of silicon atoms under the impact of large energy dissipation induces reordering of silicon atoms.

Balance in nonlinearity and discreteness in the system causes energy localization. Presence of defects and dislocations within a lattice is a known source of energy localization termed as Anderson localization. Nonlinear science provides theoretical proofs of energy localization even in perfect crystals; short range dynamics, Intrisic localized mode (ILM), Discrete Breathers, Solitary waves. Nonlinear periodic lattices occure in a large variety of systems, such as biological molecules, nonlinear optical waveguides, solid-state systems and Bose-Einstein Condensates. The underlying dynamics in these systems is dominated by the interplay tunneling between adjacent potential wells and non-linearity. A balance between these two effects can result in a self localised state: a lattice or discrete soliton. The discreteness of space is important in; solid state physics, photonic crystals, coupled optical wave guides, coupled Josephson junctions, Bose-Einstein condensates in optically induced lattices and micromechanical cantilever systems [11,13,14].

3. Planar junction

A planar p-n junction is a key electronic entity in the fabrication of integrated circuits. Using planar technology, each individual electronic component can be realized on the semiconductor wafer by processing individually selected regions. A p-n junction fabricated using planar process is always terminated at the surface of the semiconductor wafer where it is exposed to the outer environment.

Termination of a planar junction on the surface is therefore associated with a curvature in the diffused p-n junction geometry [1,6]. Secondly, the window through which a planar p-n junction is formed, introduces another curvature to the junction geometry [2]. The shape of the junction geometry on the semiconductor surface and the curvature associated with the metallurgical p-n junction inside the semiconductor, cause multidimensional effects which lead to higher leakages and reduced avalanche breakdown voltages in planar abrupt p-n junctions. However in case of linearly graded junction the electric field enhancement due to metallurgical junction curvature is not found to be crucial but its termination at the surface can not be avoided due to the planar process. The impurity profile is responsible for the realisation of abrupt and linearly graded junctions.

On the basis of symmetry, planar junctions are classified into three categories; plane junction, cylindrical junction and the spherical junction. A plane junction is the ideal form of a p-n junction having one dimensional doping profile on either side of the junction. This kind of a plane junction model is helpful to identify the limits of a practical p-n junction. A planar p-n junction formed by using a long rectangular window results in a cylindrical junction, while a junction formed through a sharp corner is modeled by a spherical junction. Electric field enhancement in a planar p-n junction with spherical symmetry is maximum, in comparison to cylindrical junction symmetry. A planar junction formed using a square or a rectangular window is associated with spherical symmetry at the corners, which is the deciding region for the entire junction characteristics. Square or rectangular geometries of the junction window are preferred in order to avoid semiconductor material wastage. However degradation in the reverse characteristics restricts such sharp corner geometries for planar junction formation. This is the reason that junction geometries with rounded corners are preferred for improved characteristics with less semiconductor area wastage [2]. The rounded corner structures provide junction breakdown in between spherical and cylindrical with less wastage of device material thus making it more economical.

In planar junctions, the edge region governs controlling of electrical characteristics of the device made therefrom, mainly due to localisation of maximum electric field in the region. Surface vicinity of the edge region further adds to the sensitivity of the device performance. At the early stages of device fabrication, when size of the devices used to be in cm or mm and no integrated circuit was introduced, mechanical or chemical bevelling was used to terminate the junction boundary at a suitable angle so that resulting electric field is not enhanced. However with the advent of IC technology and demand of miniaturized devices, planar technology restricts the use of bevelling techniques for junction edge termination. Here again, bevelling of the edge remained the limit for all possible edge terminations in planar technologies. Mesa etching has been a widely used edge termination for most of transit time device fabrications based on p-n junction and Schottky barrier [3,4] as shown in figure-1 and figure-2 respectively.

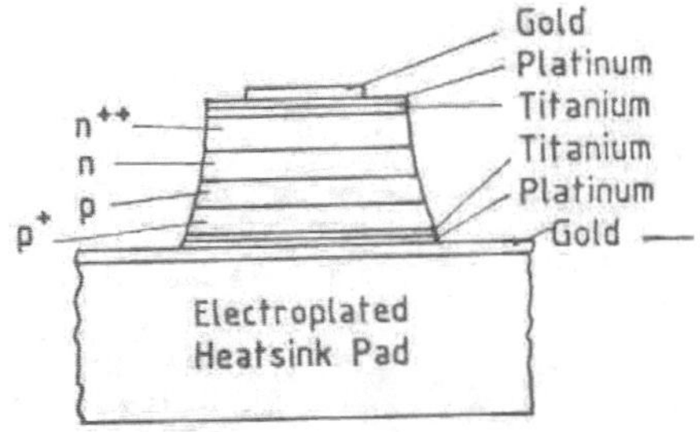

Fig-1 Mesa etching in IMPATT Diode fabrication [4]

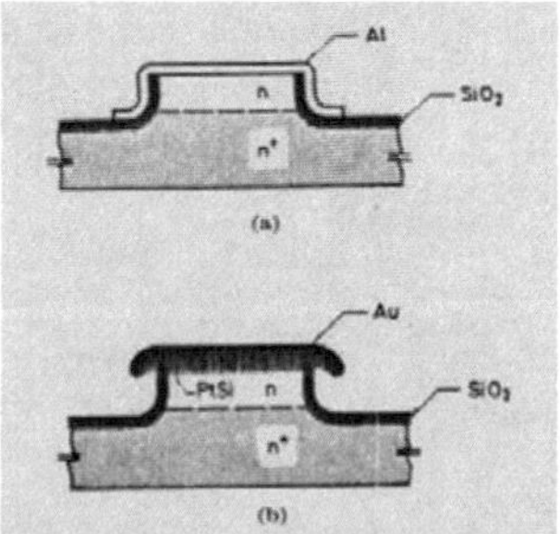

Fig-2 Mesa Schottky diode structures (a) Al and (b)Au metallisations [3]

4. Experimental

The experimental evidence of MeV ion irradiation induced reordering phenomena along the edges of exposed and virgin region of the targeted single crystalline silicon with modified electronic properties in terms of electronic band gap has immediate potentials for the edge termination in planar p-n junction structures. The reordered region can be used for isolating the edge region on the surface of a planar p-n junction in order to enhance its breakdown voltage as shown in figure-3. Also leakage current can be improved due to edge region isolation. Introduction of impurity for the junction formation therefore meets a sort of discontinuity around the edges. This eventually diminishes the multidimensional nature of the edge region, giving rise to a mesa type termination. Selective irradiation of MeV ion to the closed active device window provides a technique for edge effect termination without utilizing a significant chip area. Besides, selective MeV ion irradiation has potentials for dragging away defects and dislocations from the active region of the device. The implanted ions deep in to the bulk provide a shield to the surface active region from substrate.

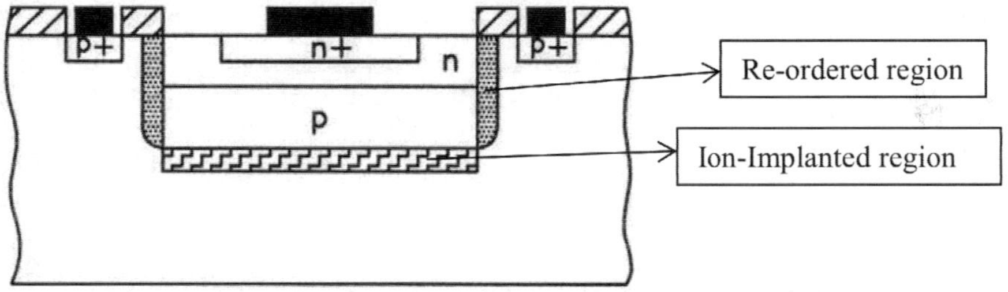

Fig.3 Schematics of selective irradiation of n-region of a semiconductor n-p junction

4.1 Planar p-n junction fabrication with MeV ion irradiation

A sharp corner square window is the ideal layout for a planar p-n junction with negligible wastage of silicon material. However with these square geometries multidimensional edge effects become very prominent to degrade junction characteristics. A square geometry of a planar junction is therefore a worst shape from device behaviour point of view. Spherical symmetry associated with the sharp corner in a square geometry is advantageous to accurately estimating the breakdown voltage of the planar junction. Taking up the worst geometry of a planar junction in the shape of a square window provides two fold advantages; easy to monitor any improvement due to edge termination and accurate theoretical estimation of the breakdown characteristics. Besides, a square geometry planar junction provides a cost effective layout design in a production environment. A square of 2mm × 2mm size has been considered for the planar junction formation. A 9×9 matrix of these squares with a pitch of 4.1 mm in x and y directions has been set for a 2" diameter silicon wafer.

4.2 Process sequence

Realisation of a p-n junction is the first process step for which a 2mm × 2mm square window in oxidized silicon surface is opened. Thermally grown oxide acts as a mask for defining the junction geometry. For a p-type of silicon substrate, phosphorous implantation is made through the window for n-type conversion leading to a step junction formation. Oxide thickness on silicon wafer is set about 1.0 μm sufficient to masking the underneath p-type silicon region. MeV ion irradiation of Ag^{+14} is done on n-region of 2mm×2mm through an aligned metal mask using a special jig for this purpose. After MeV ion irradiation metal mask is removed. As the bulk region of the silicon substrate is going to be blocked by the MeV ions, ohmic contact region for the p-type silicon and n-region of the junction are to be taken from the surface. A square annular window around the main junction with sufficient distance from the junction is required to be opened for p^+ region definition. Similarly a window for n^+ region on the main junction region is required for ohmic contact. Both the regions on p and n– type silicon are realized by ion implantation using photoresist as a mask for selectivity. Metallisation on the p^+ and n^+ regions is done with smaller size contact regions. The fabrication process sequence is shown in figure-4.

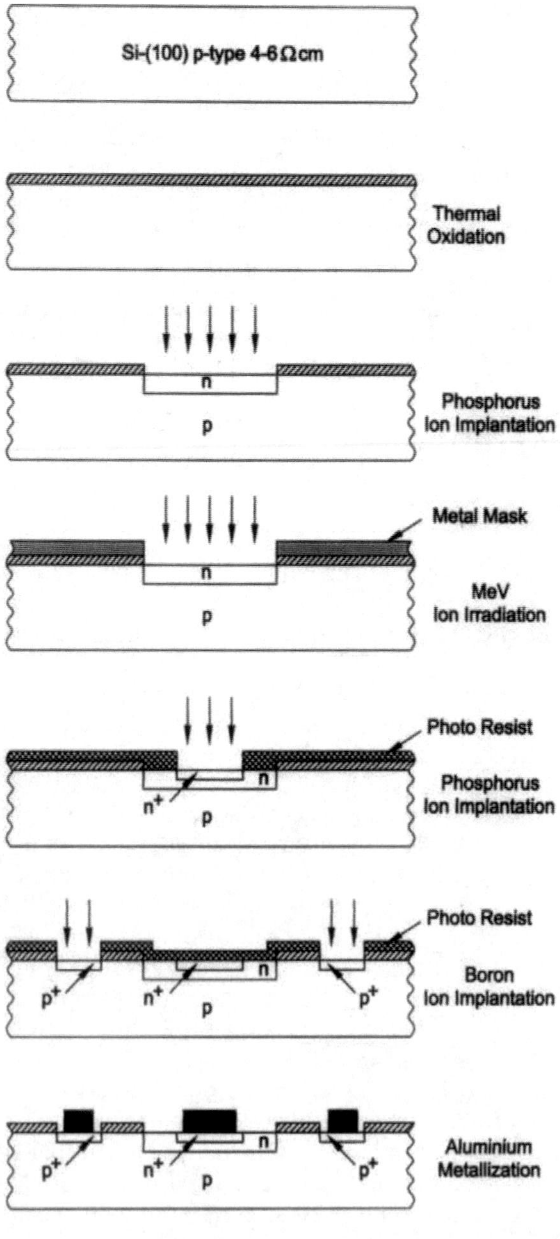

Fig-4 Fabrication process flow of planar p-n junction with MeV ion irradiation

5. Results and discussion

A planar p-n junction is characterized by its I-V and C-V characteristics, which can be further used for estimating parameters such as substrate doping, leakage currents, on resistance, avalanche breakdown voltage. For the electrical characterization of the junction structure, wire was bonded on the wafer itself using conducting silver paste. A computer controlled test facility comprising of LCR meter and current source was employed for on line data acquisition for I-V and C-V measurements on virgin and irradiated devices. A frequency variation from 0.1 MHz to 1.0 MHz was used for C-V measurements. A typical silicon wafer carrying planar junction structure with and without irradiation is shown in figure-5. The reference of the devices on the wafer has been used for further analysis of the measured characteristics. The matrix of the device is 9×9 and if counted from the left upper corner the rows and columns will identify the location of the device on the wafer. The 5×5 matrix of the devices in the centre have been irradiated on this wafer. Device No. 55 is the centrally located and irradiated with 200 MeV Ag^{+14}beam. Device No. 22 and 28 are located outside the irradiated region.

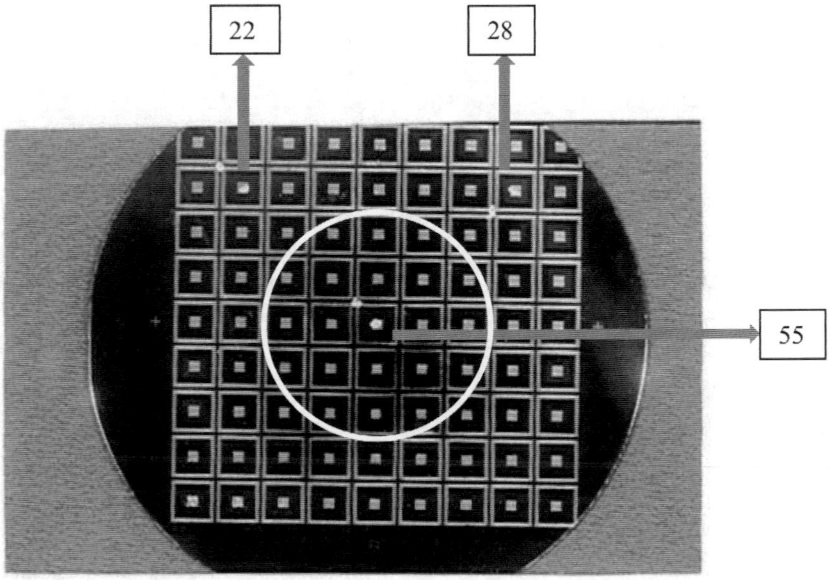

Fig-5 A processed silicon wafer with planar junction structures. Diode No. 22 and 28 are virgin while diode No. 55 is 200 MeV Ag^{+14} irradiated

The circled region on the wafer shows the devices which have been irradiated using metal mask having 5×5 matrix of windows. The square grid around the centre region shows the p^+ contact for the diode while centre square region shows the n^+ contact pad. Dot of silver paste on the contact pads can be seen in the above figure. The device structures have also been diced using diamond saw for individual diode testing.

5.1 Forward I-V Characteristics

Forward I-V characteristics of the three devices out of which device No.-22 and device No.-28 are virgin while device No.-55 is 200 MeV Ag^{+14}irradiated have been shown in figure-6 (a). The measured data has been taken up with an upper limit of 2.0 mA for the current source. The on voltage for the irradiated device is less than 1.0 volts while in case of virgin devices it is about 1.0 volt. The series resistance for the irradiated device is low than virgin devices. A 0.1 volt of incremental step was used for taking up the I-V data at room temperature in an open environment. The ln(J) versus forward voltage curves for virgin and irradiated p-n junction structures have been shown in figures-6(b) and 6(c). The curves clearly shows the G-R and diffusion dominated regimes with varying voltage. The

G-R region is narrowed down in case of irradiated device indicating removal of defects from the active region due to energy packet movement.

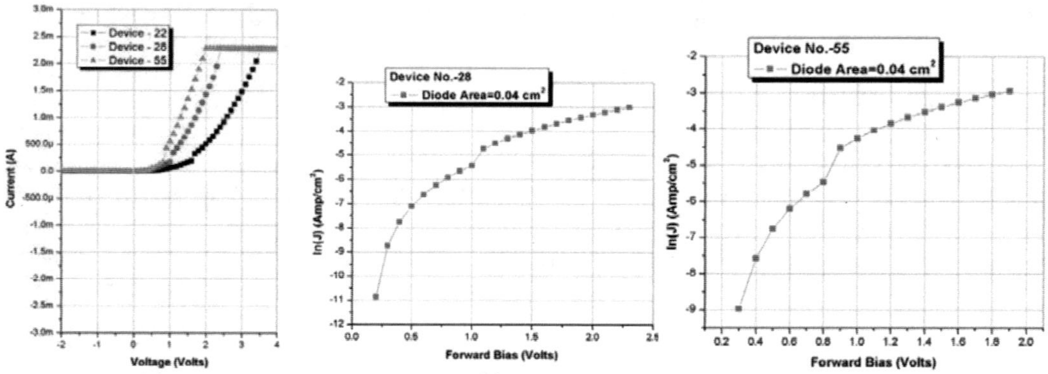

(a) (b) (c)

Fig-6 (a) Measured forward characteristics of fabricated planar p-n junction (b) Virgin diode (c) irradiated diode

5.2 Reverse I-V characteristics

The reverse characteristics of the devices have been measured for leakage currents and avalanche breakdown voltages. The virgin diodes have shown breakdown at 6.0 volts with leakage currents around 2.0 μA at 1.0 volt. The 200 MeV Ag^{+14} beam irradiated device has shown an improved leakage current with no breakdown at 6.0 volts. At 6.0 Volts leakage current can be found below 1.0 μA in case of irradiated diode. Characteristics of the three devices have been shown in figure-7 for the details of leakage currents. The device treated with selective 200 MeV Ag^{+14} beam shows marked improvements in the leakage and the breakdown voltage.

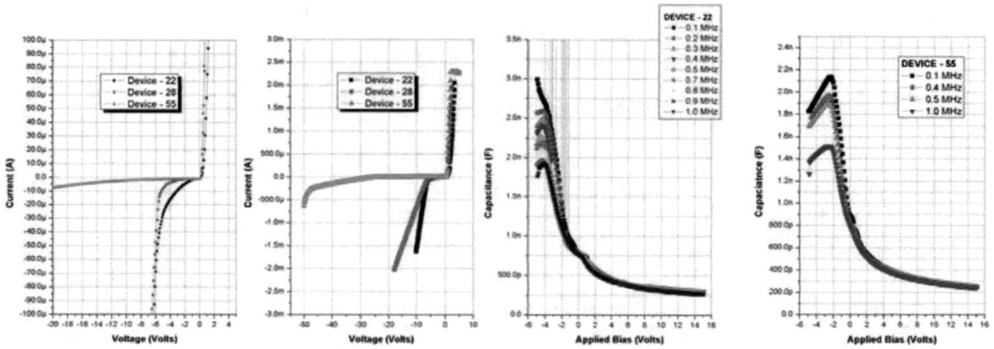

(a) (b) (c) (d)

Fig-7 Measured I-V and C-V characteristics of irradiated and virgin diodes (a) reverse leakage currents (b) complete I-V Characteristics (c) C-V of virgin diode and (d) C-V of irradiated diode

The one sided abrupt planar junction of spherical symmetry with 0.1 μm radius of curvature shows breakdown voltage of 6.0 volts for a silicon substrate of 2-4 Ω cm resistivity. The virgin devices are in agreement with the theoretical limits of the breakdown voltage.

5.3 C-V Measurement

Measurement for the C-V of the fabricated planar junction devices have been performed using Agilent LCR meter 4284A connected to a computer through GPIB. The provision of frequency variation from 0.1 MHz to 1.0 MHz has been used in order to determine frequency influence on the device capacitance. Depletion and diffusion capacitances of the devices have been measured by biasing the structures in reverse and forward mode respectively. The capacitance of forward biased p-n junction is due to excess minority carriers and therefore has relevance with the presence of defects. The C-V characteristics shown in figure-7(c) show a shooting behaviour in diffusion capacitance of virgin diode at typical frequency and forward voltage. However, no shooting behaviour is observed in case of irradiated diode C-V, as shown in figure-7(d).

5.4 Theoretical estimation of breakdown voltage

Theoretically avalanche breakdown voltage of an abrupt planar junction with either symmetry among spherical, cylindrical and plane, can be determined for a given uniform substrate doping and junction depth. In the case of under consideration, silicon substrate resistivity is of 2-4 Ω cm. The corresponding doping level can be determined using Irvin's curve as shown in figure-8 (a). For a p-type silicon at room temperature a doping level of 4 -6 $\times$ 10^{15} cm^{-3} can be extracted from the Irvin's curve for a p-type resistivity of 2-4 Ω cm. The coloured arrow has been put on the curves for clarity.

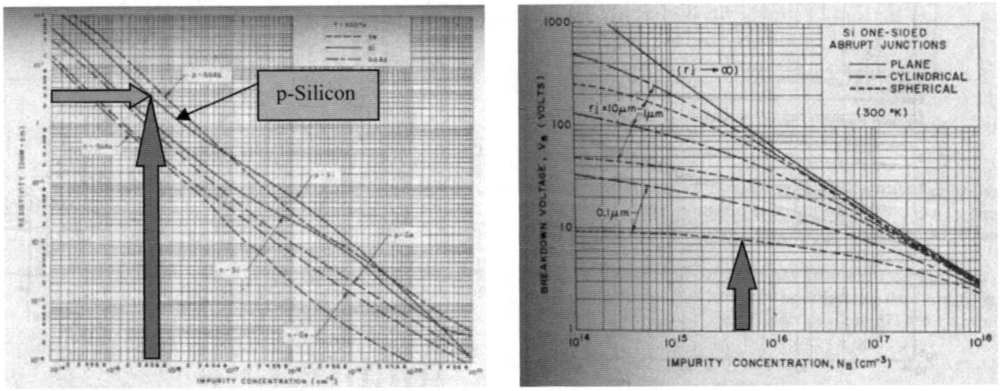

(a) (b)

Fig-8 **(a)** Resistivity vs. impurity concentration for p-type silicon (b) Avalanche breakdown voltage versus impurity concentration for one-sided abrupt doping profile with spherical junction geometries [6]

The junction geometry considered in the present work is a square with sharp corners. Along the sharp corner doping diffusion follows the spherical symmetry in the formation of a p-n junction. Any p-n junction geometry associated with spherical symmetry is dominated by the spherical junction limits. Accordingly the breakdown voltage determined for the spherical junction of 0.1 junction depth and a substrate doping level of 4-6 $\times$ 10^{15} cm^{-3} will be the breakdown for the p-n junction structure as a whole. From the computed data for one – sided abrupt silicon junction breakdown voltage of 6-7 volts is estimated for the given background doping. Coloured arrow on the curve given in figure-8 (b) shows the avalanche breakdown voltage of a spherical junction. The measurement of I-V characteristics of virgin devices has shown a breakdown voltage of around 6-7 volts, which coincides with the computed values. The improvement in the breakdown voltage and the leakage currents in case of irradiated devices is due to junction termination at its edges by reordered structures formed by the release of energy moving from the location of incident ion. The defects present in the silicon have been further flowed away by the energy wave towards the edges, resulting into diffusion capacitance without shooting up phenomenon.

Schottky diode is the special case of one sided abrupt p-n junction to require edge termination. Selective SHI irradiation has potential applications for introducing mesa type edge termination for Schottky barrier based structures employing materials like silicon carbide (SiC) and GaAs.

6. Conclusion

The measurement results of the fabricated planar abrupt p-n junction in p-type silicon with sharp corner geometry has demonstrated experimentally the usage of selective MeV ion irradiation for depositing large amount of energy on phonons through localization of vibrational energy and its further movement in the lateral direction to its release at the interfaces of virgin and irradiated silicon causing reordering in a narrow dimensions. The termination of the planar junction edges has been carriedout by the presence of reordered structures at and near the surface of the junction. A mesa type edge termination has been realized by the reordered structures. The technique is unique in its implementation as the incident ions are not in direct use for the modification of the material properties rather the energy released by them. A 8-10 times improvement in the breakdown voltage is due to elimination of the multidimensional effects. A sharp corner planar junction geometry is the worst case for the multidimensional effects based degradations in the breakdown characteristics. The improvement in the breakdown voltage from 6 volts to 50 volts has been a very effective usage of the reordered structures for junction termination. The movement of defects and disorder from the active region of silicon by the energy waves has been confirmed by the improved C-V. The technique can be utilized for producing high quality of single crystalline materials with a controlled treatment of few hundreds of MeV ions irradiation particularly in case of thin layers of materials.

7. Acknowledgement

Director CEERI, Dr Chandra Shekhar and Prof Prasenjit Sen of JNU New Delhi are thanked for their constant support and encouragement. Management of MDU Rohtak is thanked for academic support to one of the authors (VK).

References

[1] S.M.Sze and G.Gibbons, " Effect of junction curvature on breakdown voltages of abrupt cylindrical and spherical junctions", Solid State Electronics, Vol.9, p-831, 1966
[2] J. Akhtar and S. Ahmad, " Breakdown voltage of rectangular planar diffused junction with rounded corners", IEEE Trans. on Electron Devices, Vol.ED-31(12),p-1781, 1984
[3] P. Gutknecht and M.J.O. Strutt, " Thermally oxidized mesa Schottky barrier diodes", IEEE Trans. on Electron Devices, Vol.ED-21, No.2,p-172, February 1974
[4] R. Gopal, I. Ahmad and M. Kumar, Batch fabrication of mm-wave silicon Impatts, IETE Journal of Research, Vol.-41, No.1, Jan.-Feb. 1995
[5] B. Jayant Baliga, " High-Voltage device termination techniques-A comparative review
[6] S.M. Sze, Physics of Semiconductor Devices, Wiley Eastern Limited, 1979, pp 84-85
[7] A. Itoh, T. Kimoto and H. Matsunami, " Excellent reverse blocking characteristics of high-voltage 4H-SiC Schottky rectifiers with boron implanted edge termination", IEEE Electron Device Letters, Vol.17. No.3,pp 139-141, March 1996
[8] S.K. Gupta, N. Pradhan, C. shekhar and J. Akhtar," Design, Fabrication and Characterization of Ni/4H-SiC (0001) Schottky Diodes Array Equipped with Field Plate and Floating Guard Ring Edge Termination Structures", IEEE Transactions on Semiconductor Manufacturing, Vol.25, No.4, November 2012
[9] B.L. Crowder, J.F. Ziegler, F.F. Morehead, G.W. Cole, "Ion implantation in semiconductors and other materials", B.L. Crowder, Editor, p.267, Plenum Press, New Yourk (1973)
[10] P. Sen, J. Akhtar and F.M. Russell, " MeV ion induced movement of lattice disorder in single crystalline silicon", Europhysics Letters, Vol.51, No.4, pp401-406, 2000
[11] J. Cuevas, C. Katerji, J.F.R. Archilla, J.C. Eilbeck and F.M. Russell, " Influence of moving breathers on vacancies migration", Physics Letters A 315, pp 364-371, 2003
[12] P. Sen and J. Akhtar, "Nonequilibrium processes for generating silicon nanostructures in single-crystalline silicon, Current Science, Vol. 85, No.12, 25 December 2003
[13] A. Scott, Nonlinear science; Emergence and dynamics of coherent structures (Oxford University Press, 1999)
[14] A.J. Sievers and S. Takeno, Phys. Rev. Lett. 61, 970 (1988) S. Ahmad, A.S.V. Sarma, J. Akhtar,

Journal of Physics: Conference Series **423** (2013) 012058 doi:10.1088/1742-6596/423/1/012058

Modelling the life insurance needs using the human life value revision method

Haslifah Hashim[1] **and David Service**[2]
[1] Faculty of Computer and Mathematical Sciences, Universti Teknologi MARA, Malaysia
[2] School of Finance, Actuarial Studies and Applied Statistics, Australian National University, Australia

E-mail: haslifah@tmsk.uitm.edu.my

Abstract. There are numerous methods to determine the appropriate amount of life insurance a person needs – it can be scientific or simplistic. Many life insurance agents and financial advisors simply rely on traditional rules of thumb using the multiple of income method. The more scientific methods are the needs analysis and the human life value. The needs analysis is regarded as the most commonly used sales tool and the human life value is the most agreed academic expression for the purpose of life insurance. However, there are several weaknesses of using both methods. By using needs analysis as a sales tool, the recommendation amount of life insurance would leave a person underinsured. Similar goes to the human life value method. Nevertheless, both methods can be improved with a few revisions. The post-death needs under the needs analysis must be revised to incorporate the reality that the family's standard of living changes over time. The projection of a changing standard of living is a part of human life value analysis. Therefore, this research looked into both methods and combines both concept of needs analysis and human life value to create a powerful methodology that provide adequate life insurance protection – a method we name it as '*the Human Life Value Revision Method*'.

1. Introduction

"*How much life insurance does a person need?*" This is a common question, but the answer can be quite complicated. This has been at the core of the dilemma of life insurance agents and financial advisers. They can merely guess using a simple multiple of current income or some of them turn to tools like life insurance or financial calculators that used scientific methods. The fact is, not all life insurance agents and financial advisers scientifically evaluate the life insurance cover needed for their clients and hence a professional basis has been lacking in life insurance selling. This is the reason why a person is frequently under or overinsured.

There are many different methods of calculating how much life insurance a person should carry, ranging from the simple to the complex. Many life insurance agents and financial advisors simply rely on traditional rules of thumb. A common rule is that a person life should be insured for a minimum of six to ten times of annual income [1]. An older person would use the lower end of the range and a younger person should consider insurance coverage at the higher end since they typically have less savings, fewer investments and modest amounts in their retirement plans. Clearly, this method is very simple, but it fails to take into account a person individual needs and obligations such as household demographics, past savings, social security offsets and housing expenses. It also ignores expected life changes and individual preferences about sustaining the living standard of survivors. In addition, different life insurance agents may suggest different multiples.

The more scientific methods to determine the appropriate amount of life insurance a person should carry are the human life value and the needs analysis. The human life value concept was invented in 1927 by Solomon Huebner, to establish what a family would lose in income upon the death of the breadwinner of a family. This concept is not limited to life insurance in its application and has been used in many other area including judicial awards in accidental death and disability liability cases and key value of a man. This method uses four factors to determine how much life insurance will be needed to protect a family and assets. These factors are annual income and expenses, years left until retirement, and the expected value of the current ringgit at the end of this period due to inflation. Using these factors, a reasonably accurate assessment of the value of an income can be made. In mathematical terms, it is the net present value of a person potential future earnings for the rest of their working life.

The other scientific method commonly used by life insurance agents and financial advisors in calculating the amount of life insurance should be purchased is the needs analysis. It also has been used by most financial planning websites. A needs analysis looks at the sources of income and assets such as investments, social security and pension as a portfolio and relates those assets to needs that would have to be covered and determine any additional life insurance requirements.

Another way to calculate life insurance needs is using the economic approach which is based on the life-cycle model of saving and the canonical model of life insurance. The goal of the economic approach is to smooth household's living standards over their life-cycle and to ensure comparable living standards for potential survivors. In the economic approach, spending targets are endogenous. The targets are derived by calculating the most that each household can afford to consume in the present given that it wants to preserve that living standard into the future [2]. Economic Security Planner, known as *ESPlanner* software was developed by Douglas Bernheim, Jagadeesh Gokhale and Laurence Kotlikoff based on this approach [3]. The software calculates a person and his family's highest sustainable living standard and the amounts they need to save and insure to maintain their standard of living through time. However, the software requires a tremendous amount of data entry about the user's past, present and future financial life to begin the financial planning process.

2. Human Life Value and Needs Analysis as Scientific Methods to Estimate Life Insurance Needs

This research focuses on two scientific methods in determining the life insurance needs – the human life value and needs analysis. The human life value approach is more comprehensive insofar as it covers both specific and unforeseen needs. Whereas, the needs analysis covers only specific needs and its accuracy relies completely on the assumptions it uses to calculate the needs. Both methods are accepted practices used within the life insurance industry. According to John Scarborough in his article *the Balance of Life* that was published in *Advisor Today*, the most commonly used sales tool is the capital needs analysis and the human life value is the most agreed academic expression for the purpose of life insurance. This section presents an extensive review of both methods.

2.1. Human life value

The human life value concept is generally associated with life insurance. The concept however is not limited to insurance in its application and has been used in many other areas – an analysis of population to estimate the value of nation, life insurance programs, judicial awards in accidental death and disability liability cases, assessment of public policies and key value of a man.

One of the first attempts to estimate the monetary value of a human being was made in year 1691 by an English statistician, William Petty. He used the concept of human capital to determine the capital value of nation. The earliest reference to the human life value concept in life insurance was made by Jacob Greene in his book *An Agent's Work: How to Represent the Connecticut Mutual Life Insurance Company*. The book that was published in 1885 discussed the money value of human life as a foundation of life insurance. According to Greene, the money value of a person's life is the present value of what he may fairly be able to earn in the future and during his probable life. If he dies, his family loses that sum just as really, as if they lost that much in any other form of property. His life is their property and that sum is the value. An insurance practitioner, John Holcombe had been one of the first to recognise the possibilities of the human life value concept as a basis for marketing insurance when he pointed it in a speech given at the National Association of Life Underwriters (NALU) convention of 1900. Three years later, when he had become president of Phoenix Mutual, Holcombe was invited to join in Yale University's first course in Insurance. He discussed the application of the human life value concept to life insurance in his first lecture and again in an address before the Economic Club of Providence, Rhode Island in January 1906.

The name most often associated with the human life value is Solomon Huebner. It was mainly through his efforts that the concept gained widespread recognition and acceptance. He began his discussion of the application of human life value concept to life insurance in his *Life Insurance* book published in 1915. In this book, he devoted a section of the *Capitalization of the Value of a Human Life and Indemnification of that Value*. The philosophy that governs his theory is that the family must be seen as a business, run according to business principles and protected against financial impairment or bankruptcy due to the loss of the spouse's life, which has a certain economic value. Huebner developed this concept further in 1927 with the publication of his *Economics of Life Insurance* book. In this book, he discussed the human life value concept as the economic basis of life insurance, the monetary importance of human life values, the need for scientific treatment and methods for appraisal. Huebner capitalised a person's contribution to his family over the period to retirement age, or to the end of his life expectancy.

In 1930, Louis Dublin and Alfred Lotka published a book in this area, *the Money Value of a Man*. They developed for life insurance business uses a series of tables, which should give the money value of persons at various ages, according to the amount of their earnings. These tables would be useful as a guide to life insurance agents in advising their clients, as a measure of what is in his effort to protect his family and as a measure of the cost of disease and premature death. The works of Dublin and Lotka represents a real contribution in that it was the first to be devoted entirely to an estimation of the human life values of the population and one of the best expositions available. Unfortunately, the statistical base of the work is essentially a reflection of the limited availability of usable data at the time it was written.

An American researcher, Alfred Hofflander has continued the research in human life value. He defined the concept of human life value both verbally and mathematically and showing how the concept might be applied to life in different fields. He developed a mathematical model of the human life value for life insurance by finding a single sum, which represents the amount necessary to replace the amount a person would normally have provided for his family had he lived. He also used the same techniques to establish a mathematical model of the money value of a person in cases of recovery for wrongful death [4].

In life insurance industry, the human life value concept is a subject which has been of interest for many years, since the time of Huebner. The human life value was agreeable as the accurate approach to writing life insurance on people's lives because of full economic replacement. Also in 1928, the National Association of Life Underwriters (NALU) met in Los Angeles, California to determine the best methods for evaluating life insurance needs for individuals. They agreed that the human life value approach was an appropriate way to solve for the death benefit amount for consumers. However, later on the life insurance industry began to slowly drift away from this approach and adopt the needs analysis method that generates smaller amount of life insurance recommendations than the human life value. This was viewed as easy to sell by many marketing and sales department.

2.2. Needs analysis

Because human life value concept has essentially been ignored in life insurance and financial services, the industry has created alternatives. Needs analysis has been around since the early days but it was refined in the late 1960's by Thomas Wolff. At the 1966 Million Dollar Round Table meeting, Wolff introduced the concept of capital preservation in estate planning and called it needs analysis. The purpose was to address the real as well as the immediate needs of the family after the breadwinner's death. In 1981, he designed a new financial needs analysis to suit the needs of single households and a two-income family. The plan is flexible so the life insurance agent can sell the type and amount of insurance best suited to the customer's lifestyle and income.

This method attempts to determine the amount of life insurance a person needs by looking at two categories – post death needs and asset available to meet post death needs. Post death needs consist of the amount of capital needed after death of the breadwinner, for instance children education fund, household debts and funeral expenses. Existing assets including life insurance and post-death sources of income that could satisfy some of those requirements are identified. Any shortfall implies a need for additional life insurance.

The needs analysis method is the life insurance agent's most commonly used sales tool and has been used by most life insurance agents, financial advisors and financial-planning websites. There are hundreds of websites provide various forms of calculators and make calculations based on different methodologies. There are at least two different types of calculators – needs analysis and human life value, whereby the needs analysis method is by far more popular [5]. However, the life insurance recommendations generated by internet websites generally were unreliable and wide variations in advice [6]. Two possibilities exist for bad output from these calculators is because of there is an inadequate input or the underlying mathematical model is inaccurate or illogical [7].

2.3. Weaknesses of both human life value and needs analysis

There are several weaknesses of using human life value method alone, or needs analysis alone as a life insurance selling tool. The weakness of using human life value to estimate the amount of life insurance is that it does not take into consideration there may be capital needed immediately upon the death of the breadwinner to satisfy certain financial obligations, for instance funeral expenses, children education cost, mortgages and debts.

The same applies to needs analysis. There are some weaknesses of using needs analysis in estimating the amount of life insurance needed. A needs analysis ignoring future changes and only identifies the needs of a family at the current situation. Needs inevitably change in the future as family characteristics and finances change. By ignoring these factors, needs analysis would leave a person underinsured. The needs analysis also raises several concerns. If a person sets a spending target too high for survivors, the method will generate a larger amount of life insurance than is appropriate. This will cause the person too much in life insurance coverage. If the spending target is set too low, the recommended amount will leave the person underinsured. Moreover, a needs analysis does not explicitly take account of other financial planning goals such as having an adequate income in retirement. As an analytical engine, needs analysis is really no matter than asking three basic questions: "*What money do you think your family will want after you die?*", "*what do you think they will have after you die?*" and "*how much of the difference do you want to pay for while you live?*" [8]. It is suggested that needs must incorporate the reality that the family's standard of living changes over time and not simply because of inflation. The projection of a changing in standard of living is an integral element of human life value method.

Both human life value and needs analysis do not explicitly mean having an adequate income in retirement. It would also be affected by surviving spouse's age and ability or inability to work. By using needs analysis alone as a sales tool, the recommendation amount of life insurance would leave a person over or underinsured. Similar goes to the human life value method. However, both methods can be improved with a few revisions. The post-death needs under the needs analysis must be revised to incorporate the reality that the family's standard of living changes over time. The projection of a changing standard of living is a part of human life value analysis.

This research looked into both methods and combine both concept of needs analysis and human life value to create a powerful methodology that provide adequate life insurance protection – a method that we name the '*Human Life Value Revision Method*'.

3. Formulation of the Human Life Value Revision Method

The primary function of life insurance is to cover the economic loss to a family caused by the death of the breadwinner. A breadwinner's demand for life insurance depends on the demographic structure of his household. In Malaysia, generally the breadwinner is the husband and the beneficiaries are his wife and children. In this research, three major areas are emphasised in developing the human life value revision method model:

- Modelling of income: The breadwinner's total annual income over the remaining years of his working lifetime as well as his retirement benefits.
- Modelling of income tax: The income tax is calculated using the existing tax tables and rules produced by the Inland Revenue Department in Malaysia.
- Modelling of expenditures: The household expenditures of the breadwinner and his family members are calculated using the data published by the Department of Statistics, Malaysia. Any capital needed immediately upon the death of the breadwinner is identified, for example children's education cost.

The central feature of the model is that individuals are born, work for a period of time, retire and die. During their working lifetime, a person has to save for retirement. This type of model has a long tradition since the original work of Franco Modigliani and Albert Ando in their life-cycle hypothesis. Therefore, all major stages of a life-cycle were analysed in our model – unattached, marriage, parenting from babies through adolescents, launching adult children, retirement and senior years.

3.1. Modelling of income

To begin the human life value revision method formulation, we estimated the breadwinner's total income for each year from projection of his current salary over the remaining years of his working lifetime, and income after retirement such as retirement benefits. For married couples, we estimated the income of both the husband and wife. In Malaysia, the head of the household is the husband. He is the source of an income or earnings stream to his family. We followed Goldsmith's method that includes wife's income in determining the amount of life insurance needs on the husband [9]. In his research, Goldsmith considered the wife's human capital accumulation in determining purchases of life insurance on the husband, taking into consideration the wife's labour force participation and the household size.

The factors involved in calculating the present value of income are,

x = Current age
$I_{h(x+t)}$ = Total income for the husband at age $x+t$
$(e_x^{ow})_h$ = Husband's working life expectancy
$(EPF\text{ Savings})_{x+(e_x^{ow})_h}$ = Husband's Employee Provident Fund (EPF) savings on retirement
S_x = Current monthly salary at age $x+t$
EPF = The statutory Employee Provident Fund (EPF) contribution rate (employee)
$SOCSO$ = Social Security Organisation's contribution rate
s = Constant future salary increment rate per annum
p = Constant promotional increment rate
${}_tP_x$ = Probability that a person age x will survive for t more years
i = Interest rate
V^t = Present value at time t

The formula to estimate the total income for the husband at age $x+t$ is as follows,

$$I_{h(x+t)} = 12S_x(1+s)^t(p)\,{}_tP_x[1-EPF-SOCSO]$$

And, (1) $SOCSO = 0$, when S > RM3,000

(2) $s = 0.06$, when $x + t > 1$
(3) For the promotional scale, we define $1 \leq n \leq 9$
$p = (1 + p_1)(1 + p_2)(1 + p_3)\text{.........}(1 + p_9)$, where $x + t \geq a_n$
And where, $p_1 = 0.099$, when $a_1 = 25$
$p_2 = 0.094$, when $a_2 = 28$
$p_3 = 0.049$, when $a_3 = 31$
$p_4 = 0.032$, when $a_4 = 34$
$p_5 = 0.021$, when $a_5 = 40$
$p_6 = 0.021$, when $a_6 = 46$
$p_7 = 0.021$, when $a_7 = 52$
$p_8 = 0.021$, when $a_8 = 56$
$p_9 = 0.021$, when $a_9 = 58$

The income for the breadwinner, which is the husband is estimated for each year until retirement age. On the retirement age, the husband receives the full amount of *EPF* savings. Therefore, the present value of income for the husband is as follows,

$$PV(I_h) = I_{h(x+0)}V^0 + I_{h(x+1)}V^1 + I_{h(x+2)}V^2 + \dots + I_{h[x+(e_x^{ow})_h - 1]}V^{(e_x^{ow})_h - 1} + (EPF\text{ Savings})_{x+(e_x^{ow})_h} V^{(e_x^{ow})_h}$$

$$PV(I_h) = \sum_{x}^{x+(e_x^{ow})_h} I_{h(x+t)}V^t + (EPF\text{ Savings})_{x+(e_x^{ow})_h} V^{(e_x^{ow})_h} \quad t = 0, 1, 2, 3 \dots t < (e_x^{ow})_h \qquad (1)$$

Where, $V = \dfrac{1}{1+i}$

And (*EPF* Savings) is the projected EPF savings at the end of *n* years,

$$= 12Xk(1+d)^{n-1} * \left[\frac{1 - \left(\frac{1+s}{1+d} \right)^n}{1 - \left(\frac{1+s}{1+d} \right)} \right], \text{ if } s < d$$

$$= 12Xk(1+d)^{n-1} * \left[\frac{\left(\frac{1+s}{1+d} \right)^n - 1}{\left(\frac{1+s}{1+d} \right) - 1} \right], \text{ if } s > d$$

$$= 12Xk(1+d)^{n-1}, \text{ if } s = d$$

Next, again on a year-to-year basis, the income for the wife is projected. Let,

x = Husband's current age
y = Wife's current age
$I_{w(y+t),\, h(x+t)}$ = Total income for the wife at age $y+t$ (while the husband's age is $x+t$)
$(e_x^{ow})_h$ = Husband's working life expectancy
$(EPF\text{ Savings})_{x+(e_x^{ow})_h}$ = Husband's Employee Provident Fund (EPF) savings on retirement
S_x = Husband's current monthly salary at age $x+t$
EPF = The statutory Employee Provident Fund (EPF) contribution rate (employee)
SOCSO = Social Security Organisation's contribution rate

s = Constant future salary increment rate per annum
p = Constant promotional increment rate
${}_tP_x$ = Probability that a husband age x will survive for t more years
${}_tP_y$ = Probability that a wife age y will survive for t more years
f_{y+t} = Women's labour force participation rate at age $y+t$
i = Interest rate
V^t = Present value at time t

The formula to estimate the total income for the wife at age $y+t$ is as follows,

$$I_{w(y+t),h(x+t)} = 12S_x(1+s)^t(p)\,{}_tP_x(f_{y+t})\,{}_tP_y[1-EPF-SOCSO]$$

And, (1) $SOCSO = 0$, when S > RM3,000

(2) $s = 0.06$, when $x+t>1$

(3) For the promotional scale, we define $1 \le n \le 9$

$$p = (1+p_1)(1+p_2)(1+p_3).........(1+p_9), \text{ where } x+t \ge a_n$$

And where, $p_1 = 0.099$, when $a_1 = 25$

$p_2 = 0.094$, when $a_2 = 28$

$p_3 = 0.049$, when $a_3 = 31$

$p_4 = 0.032$, when $a_4 = 34$

$p_5 = 0.021$, when $a_5 = 40$

$p_6 = 0.021$, when $a_6 = 46$

$p_7 = 0.021$, when $a_7 = 52$

$p_8 = 0.021$, when $a_8 = 56$

The income for the wife is estimated for each year until retirement age. Therefore, the present value of income for the wife is as follows,

$$PV(I_w) = I_{w(y+0)}V^0 + I_{w(y+1)}V^1 + I_{w(y+2)}V^2 + ... + I_{w[y+(e_y^{ow})_w]}V^{(e_y^{ow})_w - 1} + (EPF\text{ Savings})_{y+(e_y^{ow})_w}V^{(e_y^{ow})_w}$$

$$PV(I_w) = \sum_{y}^{y+(e_y^{ow})_w} I_{w(y+t)}V^t + (EPF\text{ Savings})V^{(e_y^{ow})_w} \quad t = 0, 1, 2, 3... \; t < (e_y^{ow})_w \tag{2}$$

Lastly, the general formula for the total present value of income is derived as follows,

$$PV(I) = (1) + (2), \text{ where } PV(I_w) = 0 \text{ for single person} \tag{3}$$

3.2. Modelling of income tax

A portion of individual incomes is paid as taxes. These tax calculations are made using the existing tax tables and rules published by The Inland Revenue Department Malaysia as stipulated in the Income Tax Act 1967. An individual is taxed on his chargeable income at graduated rates from 1% to 28% after the deduction of tax reliefs. Projecting income taxes requires performing tax calculations for each year in which the total family income is projected.

The factors involved in calculating the present value of tax are,

x = Current age
y = Wife's current age
$T_{(xy+t)}$ = Total tax payable for married couple at age $x+t$ (husband) and $y+t$ (wife)
$T_{(x+t)}$ = Total tax payable for a single man before getting married at age $x+t$
$T_{(y+t)}$ = Total tax payable for a single woman before getting married at age $y+t$

$(e_{xy}^{\ ow})$ = Working life expectancy of husband or wife whichever occur last
$(e_x^{\ om})$ = Age at first marriage for men
$(e_y^{\ om})$ = Age at first marriage for women
V^t = Present value at time t
ded = Personal deductions
reb = Rebates

The formula to calculate the tax is as follows,

Taxable income in year $x+t = I_{x+t} - ded$

Tax charged in year $x+t$ = Tax on the first taxable income + Tax on the next taxable income

Therefore, income tax payable in year $x+t$, T_{x+t} = Tax charges in year $x+t - reb$

To conduct the analysis of income tax for the purpose of this human life value revision method model, payable income tax is estimated for each year until retirement age. Thus, the present value of the total income tax payable for the husband is,

$$PV(T) = T_{xy+0}V^0 + T_{xy+1}V^1 + T_{xy+2}V^2 + \dots\dots\dots + T_{xy+(e_{xy}^{\ ow})}V^{(e_{xy}^{\ ow})}$$

$$= \sum_{xy}^{xy+(e_{xy}^{\ ow})} T_{(xy+t)}V^t \qquad t = 0, 1, 2, 3\ldots t < (e_{xy}^{\ ow}) \tag{4}$$

For a single person, the income tax payable is estimated for each year until the age at first marriage. Then the analysis continues on the combined account assessment as shown above.

For a currently single man,

$$PV(T) = T_{x+0}V^0 + T_{x+1}V^1 + T_{x+2}V^2 + \dots\dots\dots + T_{x+(e_x^{\ om})_h}V^{(e_x^{\ om})_h} + \dots\dots\dots + T_{xy+(e_{xy}^{\ ow})}V^{(e_{xy}^{\ ow})}$$

For a currently single woman,

$$PV(T) = T_{y+0}V^0 + T_{y+1}V^1 + T_{y+2}V^2 + \dots\dots\dots + T_{yx+(e_y^{\ om})_w}V^{(e_y^{\ om})_w} + \dots\dots\dots + T_{xy+(e_{xy}^{\ ow})}V^{(e_{xy}^{\ ow})}$$

3.3. *Modelling of expenditures*

Expenditures in this model consist of the projection of the total amount of household expenditures and cost of university education for children. Firstly, we estimated the total amount of household expenditures for each year until the expected age of death of the husband or wife whichever occur last.

The factors involved in calculating the expenditures are,

x = Husband's current age
y = Wife's current age
E_{x+t} = Total expenditure costs at age $x+t$
HEI = Household expenditure Index
$HE_{(xy+t)}$ = Average monthly household expenditure for married couple at age $x+t$ for the husband and at age $y+t$ for the wife
$HE_{(x+t)}$ = Average monthly household expenditure for a single man before getting married at age $x+t$
$HE_{(y+t)}$ = Average monthly household expenditure for a single woman before getting married at age $y+t$
$(e_{xy}^{\ o})$ = Life expectancy of husband or wife whichever occur last
$(e_x^{\ om})$ = Age at first marriage for men
$(e_y^{\ om})$ = Age at first marriage for women
V^t = Present value at time t

The formula to calculate the household expenditure is as follows,

$$E_{x+t} = 12\,HE_{x+t}\;HEI\,(1+i)^t \qquad t = 0, 1, 2, 3\ldots t < (e_{xy}{}^{o})$$

The present value of household expenditure for the husband is determined with this formula,

$$PV(E) = E_{xy+0}V^0 + E_{xy+1}V^1 + E_{xy+2}V^2 + \ldots\ldots\ldots + E_{xy+(e_{xy}{}^{o})}V^{(e_{xy}{}^{o})}$$

$$= \sum_{xy}^{xy+(e_{xy}{}^{o})} E_{(xy+t)}V^t \qquad t = 0, 1, 2, 3\ldots t < (e_{xy}{}^{o}) \tag{5}$$

For a single man and woman, we used the similar method above. For a currently single man,

$$PV(E) = E_{x+0}V^0 + E_{x+1}V^1 + E_{x+2}V^2 + \ldots\ldots\ldots + E_{x+(e_x{}^{om})_h}V^{(e_x{}^{om})_h} + \ldots\ldots\ldots + E_{xy+(e_{xy}{}^{o})}V^{(e_{xy}{}^{o})}$$

For a currently single woman,

$$PV(E) = E_{y+0}V^0 + E_{y+1}V^1 + E_{y+2}V^2 + \ldots\ldots\ldots + E_{yx+(e_y{}^{om})_w}V^{(e_y{}^{om})_w} + \ldots\ldots\ldots + E_{xy+(e_{xy}{}^{o})}V^{(e_{xy}{}^{o})}$$

The estimation of the amount of income, personal tax and household expenditures is a part of human life value analysis. This method may not fully anticipate other needs that arise with the death of a person, for instance funeral expenses, children education cost, mortgages and debts. An estimate of these additional concerns is obtained through a needs analysis approach.

This human life value revision method model also considered household debts as post-death needs the family have to meet following the death of the breadwinner. *Bank Negara Malaysia's Annual Report 2010* reported that 45% of total household debts in Malaysia were made of mortgages, while another 20% were hire-purchase financing. Credit card debts made up only 5% of household debts. As home loans are a big commitment, repaying it can be a major burden if calamity strikes. Although it is not compulsory to have Mortgage Reducing Term Assurance (MRTA) to secure home loans, most financial institutions in Malaysia make it compulsory to insure against such an event. Alternatively, borrowers can opt for various types of life policies to secure the home loan. For the purpose of this research, we compiled information from 15 financial institutions in Malaysia offering home loans and it shows that 13 financial institutions impose as a condition of the home loan to take up an MRTA and 2 financial institutions offer lower interest rates to encourage borrowers to have MRTA. Therefore, we will not include mortgage repayment in our human life value revision method model to avoid redundant insurance coverage.

The second largest category that made up household debts in Malaysia was hire purchase financing for transport vehicles. However, we will not include this in the model since the estimation has been included in the average monthly expenditure per household based on Household Expenditure Survey. The data under the category of 'Transport and Communication' covers the expenditure for transport including vehicle purchase, road tax, motor insurance, maintenance and parking fee.

Therefore, we incorporated funeral expenses and children's university education cost into our model, as the cost is not included in the household expenditures. Let,

x = Current age
y = Wife's current age
OE = Total of other expenses
b_i = Median age at birth for the i child according to age of mother, where $i = 0, 1, 2\ldots\ldots\ldots n$
$(e_x{}^{o})_h$ = Husband's life expectancy
$(e_y{}^{o})_w$ = Wife's life expectancy
V^t = Present value at time t

Therefore, the present value of other expenses is as follows,

$$\begin{aligned} PV(OE) = {} & (\text{Education cost for the first child})_{y+b_1} V^{b_1-y} \\ & + (\text{Education cost for the second child})_{y+b_2} V^{b_2-y} + \dots\dots\dots \\ & + (\text{Education cost for the } n^{\text{th}} \text{ child})_{y+b_n} V^{b_n-y} \\ & + (\text{Funeral expenses for the husband})_{x+e_x^o} V^{e_x^o} \\ & + (\text{Funeral expenses for the wife})_{y+e_y^o} V^{e_y^o} \end{aligned} \quad (6)$$

Then, the present value of net future income flow for the breadwinner can be calculated using the equation below,

$$\text{Present Value of Net Future Income} = (3) - (4) - (5) - (6) \quad (7)$$

Finally, we can estimate the amount of life insurance required for a person in times of current income by using the equation below. Any existing sources of funds available to cover those needs such as the social security benefit from the Employee's Provident Fund (EPF) and current amount of life insurance will be deducted.

$$(\text{Life Insurance Required in Times of Income})_x = \frac{(7) - (\text{EPF saving}) - (\text{Current Life Insurance})}{\text{Annual Income}}$$

4. Conclusion

The primary function of life insurance is to cover the economic loss to a family caused by the death of the breadwinner. A breadwinner's demand for life insurance depends on the demographic structure of his household. In Malaysia, generally the breadwinner is the husband and the beneficiaries are his wife and children. The main purpose of this research is to identify the strength and weakness of the life insurance sales tool currently used in practice and develop a scientific method as a correct solution and hence provide adequate protection to relate the amount of life insurance a person needs to carry to the real value of a person's earnings. The research combined two methods currently used in practice – the needs analysis and human life value, and with a few revisions, we developed a method of determining the amount of life insurance needs. A method that we name the '*Human Life Value Revision Method*'.

In this research, a human life value revision method model has been developed to determine the amount of life insurance required to protect the family against economic loss in the event of the breadwinner's death. To the end, we suggest a model to compute the amount of life insurance coverage in a multiple of current annual income that can be used in the life insurance selling industry or as an underwriting guideline.

References

[1] Cordell D M 2004 *J. Financial Planning* **17** 32–4
[2] Gokhale J, Kotlikoff L J and Warshawsky M 2001 *Essays On Saving, Bequests, Altruism and Life-cycle Planning* (Cambridge: The MIT Press)
[3] Bernheim D B, Gokhale J, Kotlikoff L J and Williams L 2000 *Esplanner* (Cambridge: The MIT Press)
[4] Hofflander A E 1966 *J. Risk and Insurance* **33** 381–91
[5] Elger J F 2003 *J. Financial Service Professionals* **57** 38–40
[6] Brayman S 2009 *J. Financial Planning* **22** 48
[7] Dorfman M S and Adelman S W 2002 *Risk Management and Insurance Rev.* **5** 135–54
[8] Daily G S 2000 *Capital Strategies Conference* (New York)
[9] Goldsmith A 1983 *J. Risk and Insurance* **50** 33–43

Journal of Physics: Conference Series **423** (2013) 012059 doi:10.1088/1742-6596/423/1/012059

Mathematical Model Of Tuberculosis Transmission With Reccurent Infection And Vaccination

J. Nainggolan[1,2], Sudradjat Supian[2], A. K. Supriatna[2], and N. Anggriani[2]
[1]Department of Mathematics, Universitas Cenderawasih, Indonesia
[2]Department of Mathematics, Universitas Padjadjaran, Indonesia

E-mail: jonn_cesil@yahoo.co.id

Abstract: This paper presents a model of tuberculosis transmission with vaccination by explicitely considering the total number of recovered individuals, either from natural recovery or due to vaccination. In this paper the endemic and nonendemic fixed points, basic reproduction number, and vaccination reproduction number are given. Some results regarding the stability of the fixed points and the relation to the basic reproduction numbers are analysed. At the end of this study, the numerical computation presented and it shows that vaccination is capable to reduce the number of laten and infectious populations.

1. Introduction

Tuberculosis (TB) is an infection disease caused by *Mycobacterium tuberculosa*. The disease is able to spread quickly through the air medium and becoming the second most killer among all transmission disease. It is also regarded to be the third grade in the list of the ten highest killer disease in Indonesia [13]. The World Health Organization (WHO) data in 2009 shows that the total number of tuberculosis victims in Indonesia reach up to 429,000 persons, and lies in the fifth position in the world after India, China, South Africa, and Nigeria [1]. Hence, TB is one of the disease that has been raising a serious problem for a long time in Indonesia [13].

Some researchers have been studying the tuberculosis transmission using modeling approach [2, 4, 9]. Some of them have include intervention to over come the spread of TB in their models [3, 4, 6]. An example of WHO recommended intervention is by giving BCG (*Baccilus Calmete Guerin*) vaccination to newborn baby an in general population [1, 7, 13, 15, 16]. The person that is vaccinated with BCG could be protected 70% - 80% from tuberculosis infection [13].

In this paper, to evaluate the effectiveness of vaccination program we develop a mathematical model and use the concept of the *basic reproduction number*. It is well known in the literature that, the free infection condition is globally stable if the *basic reproduction number* value is less than one, but if the *basic reproduction number* value is greater than one the tuberculosis disease will be endemic [5, 8]. In this paper we will show that in the present of vaccination, a modified *threshold* parameter which is called the *vaccine reproduction number* will have the value less than one for a certain level of vaccination. The vaccination model had been studied by some researchers in various forms, among others: vaccination in *SVI* model [11], where *S* is a *susceptible* compartment, *V* is a vaccination compartment, and *I* is an infected compartment; vaccination in *SVIR* model [12], where *R* is a *recovered* compartment; vaccination in *SVIER* model by considering the existence of an *exposed* phase of infected individuals *(E)*

*Corresponding author. E-mail address: jonn_cesil@yahoo.co.id

[10]. In this paper we modify the model in [3] by considering that the *recovered* individuals can only be reinfected by making contact with the infectious individuals.

2. Dynamical System model of tuberculosis spreading with vaccination

State variable on the dynamical model is given as in [3] that is: The number of *susceptible* individual compartment on the time t denoted by $S(t)$. The number of *exposed* individual compartment on the time t denoted by $E(t)$. The number of *infected* individual compartment on the time t denoted by $I(t)$, and the number of *recovery* individual compartment on the time t denoted by $R(t)$. The model of tuberculosis transmission with recurrent infection is as follows:

$$\left.\begin{aligned}
\frac{dS}{dt} &= \Lambda - \lambda S - \mu S \\
\frac{dE}{dt} &= f\lambda S - \delta_1\lambda E - (\mu + k)E + f_1\delta_2 R \\
\frac{dI}{dt} &= (1-f)\lambda S + \delta_1\lambda E + kE - (\mu + p + d)I + (1-f_1)\delta_2\lambda R \\
\frac{dR}{dt} &= pI - \mu R - \delta_2\lambda R
\end{aligned}\right\} \tag{1}$$

where

$$\lambda = \frac{\beta c I}{N} \tag{2}$$

is the force of infection. The parameters used in the model are as follow. Λ is recruitment rate into susceptible compartment, c is the per capita contact rate, and β is the probability that one susceptible individual becomes infected by an infectious individual. Susceptible individuals infected with Mycobacterium tuberculosis bacteria (Mtb) moved to latently infected compartment at a rate $f\lambda$, where f is the probability that the infected enters the latent compartment. Susceptible individuals infected with Mtb move into the infective compartment at a rate $(1-f)\lambda$. The latently infected progress to active TB at rates k for endogenous reactivation and $\delta_1\lambda$ for exogenous re-infection respectively. Once in active compartment of the disease, an individual may recovered naturally at rate p and move into the recovered compartment R. Individuals in R are not totally immune to Mtb infection and are infected at rate $f_1\delta_2\lambda$ and move into E, where f_1 is the probability that relapse back into the latent compartment, $(1-f_1)\delta_2\lambda$ relapse back into the active infected compartment. Infectives have an additional TB induced death rate $d > 0$, and the natural death rate in each class is assumed to be $\mu > 0$.

In this paper we extend the model above to include vaccination compartment (V) as expressed on the schematic transmission diagram in Figure 1, with the description of the parameters given in Table 1.

*Corresponding author. E-mail address: jonn_cesil@yahoo.co.id

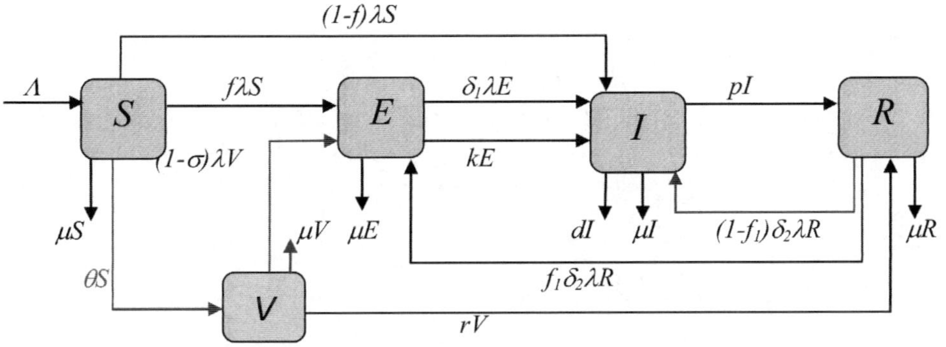

Figure 1. Schematic diagram of tuberculosis transmission

Based on the schematic diagram on the Figure 1, we have the following model

$$\left.\begin{aligned}
\frac{dS}{dt} &= \Lambda - \lambda S - (\mu+\theta)S \\
\frac{dV}{dt} &= \theta S - (1-\sigma)\lambda V - (\mu+r)V \\
\frac{dE}{dt} &= f\lambda S + (1-\sigma)\lambda V - \delta_1\lambda E - (\mu+k)E + f_1\delta_2 R \\
\frac{dI}{dt} &= (1-f)\lambda S + \delta_1\lambda E + kE - (\mu+p+d)I + (1-f_1)\delta_2\lambda R \\
\frac{dR}{dt} &= rV + pI - \mu R - \delta_2\lambda R
\end{aligned}\right\} \tag{3}$$

Here we assume that BCG vaccine moves individuals from the susceptible compartment to the vaccinated compartment at a rate θ, and an individual into recovered compartment because immune due to vaccination at a rate r. In this case, the effective contact rate, βc is multiplied by a scaling factor $(1-\tau)$, where $0 \le \tau \le 1$ is the efficacy of the vaccine.

The *nonendemic* equilibrium point of the equation (3) is given by

$$Q_0 = \left(\frac{\Lambda}{\mu+\theta}, \frac{\Lambda\theta}{(\mu+\theta)(\mu+r)}, 0, 0, \frac{\Lambda\theta r}{\mu(\mu+\theta)(\mu+r)} \right) \tag{4}$$

The *endemic* equilibrium point of the equation (3) is given by

$$\begin{aligned}
Q_1 &= (S_1, V_1, E_1, I_1, R_1) \\
&= \left(\frac{\Lambda}{\mu+\theta+\lambda_1}, \frac{\Lambda\theta}{(\mu+\theta+\lambda_1)(\mu+r+[1-\sigma]\lambda_1)}, E_1, I_1, R_1 \right)
\end{aligned} \tag{5}$$

where: $\lambda_1 = \dfrac{\beta c I_1}{N}$, $R_1 = \dfrac{r\theta\Lambda + p(\mu+\theta+\lambda_1)(\mu+r+[1-\sigma]\lambda_1)I_1}{(\mu+\delta_2\lambda_1)(\mu+\theta+\lambda_1)(\mu+r+[1-\sigma]\lambda_1)}$,

$$E_1 = \frac{f\lambda\Lambda(\mu+r+[1-\sigma]\lambda_1)(\mu+\delta_2\lambda_1) + [1-\sigma]\lambda_1\theta\Lambda(\mu+\delta_2\lambda_1) + f_1\delta_2\lambda_1(r\theta\Lambda + p(\mu+\theta+\lambda_1)(\mu+r+[1-\sigma]\lambda_1)I_1)}{(\mu+k+\delta_1\lambda_1)(\mu+\delta_2\lambda_1)(\mu+\theta+\lambda_1)(\mu+r+[1-\sigma]\lambda_1)},$$

$$I_1 = \frac{(1-f)\lambda_1 S_1 + (\delta_1\lambda_1 + k)E_1 + (1-f_1)\delta_2\lambda_1 R_1}{\mu+p+d}.$$

*Corresponding author. E-mail address: jonn_cesil@yahoo.co.id

The reproduction number of vaccinaion by using *Next Generation Matrix* [5, 8] is given by:

$$R_V = \frac{\beta c \Lambda \left(\mu[\mu + r][(1-f)\mu + k] + [(1-f_1)\mu + k]\delta_2 \theta r + (1-\sigma)k\theta\mu \right)}{\mu N(\mu + r)(\mu + k)(\mu + \theta)(\mu + d + p)} \tag{6}$$

To be biologically meaningful, the initial condition of the model are given by $S(0) = S_0 \geq 0$, $V(0) = V_0 \geq 0$, $E(0) = E_0 \geq 0$, $I(0) = I_0 \geq 0$, and $R(0) = R_0 \geq 0$. It can be shown that the solution of the model defined for all the time $t \geq 0$. Since the equation (3) have the solution in the nonnegative area

$$\left\{ (S,V,E,I,R) \in R_+^5 : S + V + E + I + R \leq \frac{\Lambda}{\mu + \theta} \right\}, \tag{7}$$

where $\frac{\Lambda}{\mu+\theta}$ is the number of total population at nonendemic time.

Table 1. Definition of parameter used in the model.

Parameters	Description	Estimation
Λ	Recruitment rate	3500 per year
θ	Vaccination rate from S	0.2 per year
μ	Natural mortality rate	0.01 per year
c	Contact rate of S with I	21 per day
σ	Vaccination effectiveness	0.8 per year
d	Death rate of I	0.05 per year
β	Transmission probabilities	0.17 per year
k	Natural rate of progression to I	0.00013 per year
p	Natural recovery rate from I	0.15 per year
r	Natural recovery rate from V	0.35 per year
δ_1	Probability of the individul entering I from E that recontact with I	0.2 per year
δ_2	Probability of the individual to be passive infected from R	0.15 per year
f	Probability of S entering into E	0.99 per year
f_1	Probability of R entering into E	0.7 per year

The solution of equation (3) with the positive initial condition is bounded for every time $t \geq 0$. Further we have the following proposition for the equilibrium and its stability analysis

PROPOSITION 1

The disease-free equilibrium of the system (3) is locally asymptotically stable when the vaccination reproduction number $R_V < 1$ *and unstable when* $R_V > 1$.

*Corresponding author. E-mail address: jonn_cesil@yahoo.co.id

 doi:10.1088/1742-6596/423/1/012059

Proof:

Let $\frac{dS}{dt}=F,\frac{dV}{dt}=G,\frac{dE}{dt}=H,\frac{dI}{dt}=K,\frac{dR}{dt}=L,$ be the Jacobian matrix on the point P is

$$J(P)=\begin{bmatrix} \frac{\partial F}{\partial S} & \frac{\partial F}{\partial V} & \frac{\partial F}{\partial E} & \frac{\partial F}{\partial I} & \frac{\partial F}{\partial R} \\ \frac{\partial G}{\partial S} & \frac{\partial G}{\partial V} & \frac{\partial G}{\partial E} & \frac{\partial G}{\partial I} & \frac{\partial G}{\partial R} \\ \frac{\partial H}{\partial S} & \frac{\partial H}{\partial V} & \frac{\partial H}{\partial E} & \frac{\partial H}{\partial I} & \frac{\partial H}{\partial R} \\ \frac{\partial K}{\partial S} & \frac{\partial K}{\partial V} & \frac{\partial K}{\partial E} & \frac{\partial K}{\partial I} & \frac{\partial K}{\partial R} \\ \frac{\partial L}{\partial S} & \frac{\partial L}{\partial V} & \frac{\partial L}{\partial E} & \frac{\partial L}{\partial I} & \frac{\partial L}{\partial R} \end{bmatrix}$$

$$=\begin{bmatrix} -\theta-\lambda-\mu & 0 & 0 & -\frac{c\beta S}{N} & 0 \\ \theta & -(1-\sigma)\lambda-r-\mu & 0 & -\frac{(1-\sigma)c\beta V}{N} & 0 \\ f\lambda & (1-\sigma)\lambda & -\delta_1\lambda-\mu-k & \frac{fc\beta S+(1-\sigma)c\beta V-\delta_1 c\beta E+f_1\delta_2 c\beta R}{N} & f_1\delta_2\lambda \\ (1-f)\lambda & 0 & \delta_1\lambda+k & \frac{(1-f)c\beta S+\delta_1 c\beta E}{N}-\mu-d-p+\frac{(1-f_1)\delta_2 c\beta R}{N} & (1-f_1)\delta_2\lambda \\ 0 & r & 0 & p-\frac{\delta_2 c\beta R}{N} & -\mu-\delta_2\lambda \end{bmatrix}$$

The Jacobian matrix on a point P_0 is

$$J(P_0)=\begin{bmatrix} -\theta-\mu & 0 & 0 & -\frac{c\beta\Lambda}{(\mu+\theta)N} & 0 \\ \theta & -r-\mu & 0 & -\frac{(1-\sigma)c\beta\theta\Lambda}{(\mu+r)(\mu+\theta)N} & 0 \\ 0 & 0 & -\mu-k & \frac{fc\beta\Lambda}{(\mu+\theta)N}+\frac{(1-\sigma)c\beta\theta\Lambda}{(\mu+r)(\mu+\theta)N}+\frac{f_1\delta_2 c\beta\theta r\Lambda}{\mu(\mu+r)(\mu+\theta)N} & 0 \\ 0 & 0 & k & \frac{(1-f)c\beta\Lambda}{(\mu+\theta)N}-\mu-d-p+\frac{(1-f_1)\delta_2 c\beta\theta r\Lambda}{\mu(\mu+r)(\mu+\theta)N} & 0 \\ 0 & r & 0 & p-\frac{\delta_2 c\beta\theta r\Lambda}{\mu(\mu+r)(\mu+\theta)N} & -\mu \end{bmatrix}$$

Det[$J(P_0)$]= $\beta c\Lambda\left(\mu[\mu+r][(1-f)\mu+k]+[(1-f_1)\mu+k]\delta_2\theta r+(1-\sigma)k\theta\mu\right)-\mu N(\mu+r)(\mu+k)(\mu+\theta)(\mu+d+p)$

Det[$J(P_0)$] > 0 when

$$\frac{\beta c\Lambda\left(\mu[\mu+r][(1-f)\mu+k]+[(1-f_1)\mu+k]\delta_2\theta r+(1-\sigma)k\theta\mu\right)}{\mu N(\mu+r)(\mu+k)(\mu+\theta)(\mu+d+p)}=R_V<1$$

The real part of eigen value of Det[$J(P_0)$ - λ] = 0 are all negative when $R_V < 1$ [3, 10]. As a result, the disease-free equilibrium point P_0 is locally asymptotically stable whenever $R_V < 1$. On the contrary, the real part of eigen value of Det[$J(P_0)$ - λ] = 0 is positive when $R_V > 1$ [3, 10]. As a result, the disease-free equilibrium point P_0 is locally asymptotically unstable whenever $R_V > 1$. □

*Corresponding author. E-mail address: jonn_cesil@yahoo.co.id

Journal of Physics: Conference Series **423** (2013) 012059 doi:10.1088/1742-6596/423/1/012059

Relation of the vaccination reproduction number to the basic reproduction number

The reproduction number without vaccination or *the basic reproduction number* is given by

$$R_0 = \frac{\beta c\Lambda((1-f)\mu+k)}{\mu N(\mu+k)(\mu+d+p)}. \tag{8}$$

As a result,

$$R_V = R_0\left(\frac{(\mu+r)([1-f]\mu^2+k\mu)+[(1-f_1)\mu+k]\delta_2\theta r+(1-\sigma)k\theta\mu}{(\mu+r)(\mu+\theta)[(1-f)\mu+k]}\right) < R_0,$$

since $(\mu+r)([1-f]\mu^2+k\mu)+[(1-f_1)\mu+k]\delta_2\theta r+(1-\sigma)k\theta\mu <$ $(\mu+r)(\mu+\theta)[(1-f)\mu+k]$. Hence, for ($\theta > 0$ and $0 < \sigma \leq 1$) , the effect of vaccine will always reduce the *basic reproduction number* from the disease.

Furthermore, let $R_V = 1$ and solve for θ so that the solution is θ^c. This solution is a *critical vaccination level* that given by:

$$\frac{\beta c\Lambda\left(\mu[\mu+r][(1-f)\mu+k]+[(1-f_1)\mu+k]\delta_2\theta r+(1-\sigma)k\theta\mu\right)}{\mu N(\mu+r)(\mu+k)(\mu+\theta)(\mu+d+p)} = 1$$

$$\beta c\Lambda\left(\mu[\mu+r][(1-f)\mu+k]+[(1-f_1)\mu+k]\delta_2\theta r+(1-\sigma)k\theta\mu\right) =$$
$$\mu N(\mu+r)(\mu+k)(\mu+\theta)(\mu+d+p)$$

$$\theta^c = \frac{\mu^2 N(\mu+r)(\mu+k)(\mu+d+p)-\beta c\Lambda\left(\mu(\mu+r)[(1-f)\mu+k]\right)}{\beta c\Lambda\left([(1-f_1)\mu+k]\delta_2\theta r+(1-\sigma)k\mu\right)-\mu N(\mu+r)(\mu+k)(\mu+d+p)}$$
$$= \frac{\mu(1-R_0)}{\frac{([1-f_1]\mu+k)\delta_2\theta r+(1-\sigma)k\mu}{(\mu+r)[(1-f)\mu+k]}R_0-1}. \tag{9}$$

We can see that if $\theta < \theta^c$ vaccination will not effective in eliminating the disease, and if $\theta > \theta^c$ vaccination will effectively eliminate the endemic condition. The following numerical example gives an illustration for this result.

3. Numerical Simulation

Numerical simulation for the model of the equation system (3) is done using Matlab with the four order Runge-Kutta numerical scheme . The parameter values are given in Table 1 (most of them taken from [3]) and the initial conditions *S(0)* = 95703, *V(0)* = 35703, *E(0)* = 13670, *I(0)* = 2000, and *R(0)* = 0. The trend of exposed, infected, and recovered populations are given in Figure 2, Figure 3 and Figure 4 below.

*Corresponding author. E-mail address: jonn_cesil@yahoo.co.id

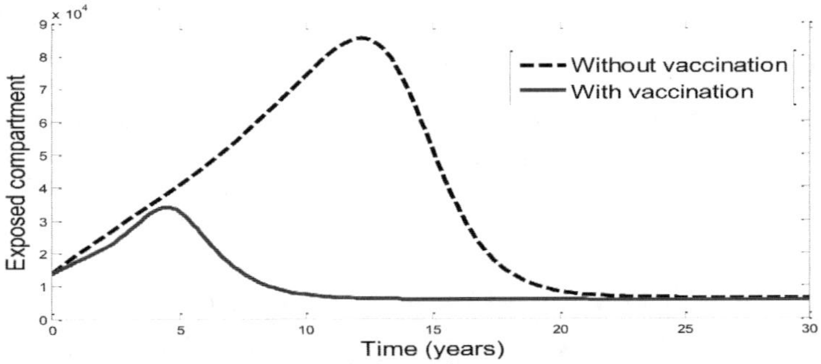

Figure 2. Exposed compartment

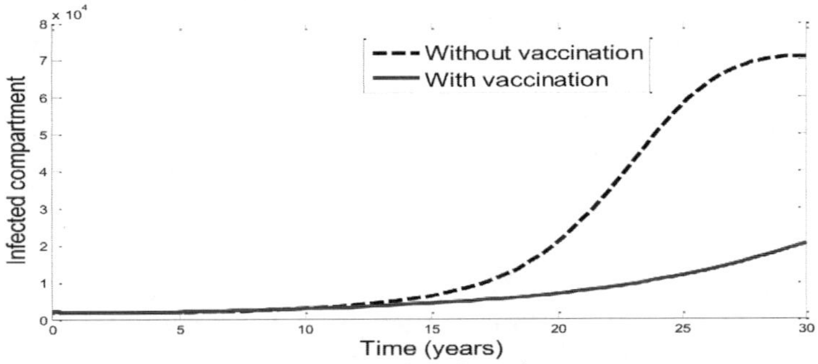

Figure 3. Tuberculosis infected compartment

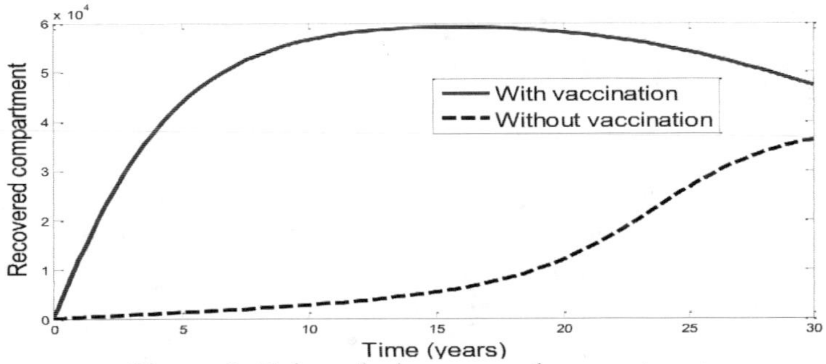

Figure 4. Tuberculosis recovered compartment

Assuming vaccination is effective in protecting susceptibles individuals, Figures 2 and 3 show that vaccination can reduce the peak of exposed and infected compartment drastically. Figure 4 shows different dynamics of recovered compartment with vaccination and without vaccination.

4. Conclusion

In this paper we have developed a mathematical model for the transmission of tuberculosis disease by considering recurrent infection and vaccination. The disease free equilibrium and its stability is presented and its relation to the basic reproduction number and vaccination reproduction number is discussed. Numerical examples show that vaccination is able to prevent the disease from spreading. Further study can be done for the optimal vaccination level (currently under investigation) as the function of recurrent rate of reinfection.

*Corresponding author. E-mail address: jonn_cesil@yahoo.co.id

5. Acknowledgments

The authors thanks to Irfan Wahyudi for reading and revising the grammar. Part of the work is funded by the Indonesian Government through Penelitian Hibah Kompetensi 2013 to AKS.

References

[1] Aditama T Y dan Subuh M 2010 *Strategi Nasional Pengendalian TB Di Indonesia 2010-2014* (Jakarta: Kemenkes RI Dirjen P2 & PL) p 12

[2] Aparicio J P and Castillo-Chavez C 2009 *Mathematical Modelling of Tuberculosis Epidemics,* Math. Biosci. **2** 209

[3] Bhunu C P, Garira W, Mukandavire Z, and Zimba M 2008 *Tuberculosis Transmission Model with Moprophylaxis and Treatment*, Bulletin of Math. Bio. **70** 1163

[4] Castillo-Chavez C and Song B 2004 *Dynamical Models of Tuberculosis and Their Applications*, Math. Biosci. and Eng. **2** 361

[5] Castillo-Chavez C, Fengy Z, and Huangzon W 2002 *The Computation of R_0 and its Role on Global Stability.* Mathematical Approaches for Emerging and Reemerging Infectious Disease: An Introduction, IMA, Springer-Verlag **125** 229

[6] Castillo-Chavez C and Feng Z 1997 *To Treat or not to Treat: the case of tuberculosis*, Math. Biol. **35** 625

[7] Crofton S J, Horne N and Miller F 2002 *Clinical Tuberculosis*, (London : MacMilan Education Ltd) p 16

[8] Driessche P v D and Watmough J 2002 *Reproduction Numbers and Sub-Threshold Endemic Equilibria for Compartmental Models of Disease Transmission*, Mat. Biosci. **180** 29

[9] Feng Z, Castillo-Chavez C and Capurro,A F 2000 *A Model for Tuberculosis with Exogenous Reinfection*, Theoretical Population Biology **57** 235

[10] Gumel A B, McCluskey C C and Watmough J 2006 *An SVEIR Model for Assessing Potential Impact of an Perfect Anti-Sars Vaccinne*, Math. Biosci. **3** 485

[11] Kribs-Zaleta C M and Velasco-Hernandez J X 2000. *A Simple Vaccination Model with Multiple Endemic States*, Math. Biosci. **164** 183

[12] Liu X, Takeuchi Y and Iwami S 2008 *SVIR epidemic models with Vaccination Strategies*, J. Theoretical Biology **253** 1

[13] Manaf A 2007 *Pedoman Nasional Penanggulangan Tuberkulosis* (Jakarta: Depkes RI) p 3-5

[14] Perko L 1991 *Differential Equation and Dynamical Systems* (New York: Springer Verlag) p 128

[15] Satcher D, Gayle H D, and Castro K G 1996 *The Role of BCG Vaccine in the Prevention and Control of Tuberculosis in the United States*. Centre for Disease Control and prevention, v.45, n. RR.4

[16] WHO *Global tuberculosis control 2011* (WHO Library Cataloguing-in-Publication Data) p 9-10

*Corresponding author. E-mail address: jonn_cesil@yahoo.co.id

Synthesis and characterization of calcium iodate, monohydrate crystals grown in silica gel

S. J. Shitole[1]

[1]Department of Physics, Z. B. Patil College, Dhule, Maharashtra, India

E-mail: sjshitole@hotmail.com

Abstract. Simple gel technique was used to grow single crystals of calcium iodate, monohydrate, by single diffusion method. For the growth of crystals, the optimum conditions were established. Morphologies and habit faces like prismatic, prismatic pyramidal crystals were obtained. Few crystals were opaque, some were translucent and some good quality transparent crystals were obtained. Doping of Cu+2 and Fe+3 was done, to study the effect on structure, thermal and nonlinear optical properties. Cell parameters were obtained from the X-ray diffractograms. Structural analysis was done by using FT-IR spectroscopy. TGA and DTA techniques were used to carry out thermal analysis. Nonlinear optical properties were studied using SHG measurements.

1. Introduction

The study of growth and characterization of single crystals is receiving increasing importance due to their various applications in solid-state technology and laser technology. With the absence of crystals, there would be no electronic industry and fiber optic communication Growth of crystals by gel method is a promising technique for growing single crystals of substances which are sparingly soluble in water and decompose before their melting point [1]. A variety of crystals of high quality can be grown in a large range of solubility and temperature. In gel growth, crystals are mostly formed at ambient temperature and hence are free from strain often present in crystals prepared from melt or vapour. Method is inexpensive and within the scope of small laboratories. Large scale movements like convection currents are almost completely suppressed, which otherwise could be harmful to the quality of crystal [2-7].

Single crystals of calcium iodate, which is non-centro-symmetric, exhibits prominent nonlinear optical property, and piezoelectric property [8]. In the present investigation, Single crystals of calcium iodate, monohydrate was grown by gel method. Since the compound decomposes before its melting point, conventional high temperature methods for its growth are not suitable. It is sparingly soluble in water; hence, gel method is the only alternative technique for growing the crystals of the size and quality as reported here at ambient temperature.

2. Experimental

To carry single and double diffusion experiments, test tubes and U-tubes were used as crystallizing vessels. Silica gel is used as growth medium. Acetic acid is used to acidify the gel. Proper chemical reaction is selected to get the desired product. The two soluble reactants are diffused into a gel, where they react to form an insoluble product. Dopants are incorporated either in gel or above the gel along with the supernatant. Optimum conditions were obtained by varying parameters such as density of gel, pH of gel, gel aging, concentration of reactants, concentration programming, concentration and amounts of dopants. Various concentrations of acetic acid and those of sodium metasilicate were used to prepare gel. For this purpose, 5cc, 2N acetic acid was taken in a beaker, to which sodium metasilicate solution having different densities was added drop by drop with constant stirring by using magnetic stirrer. It avoids premature local gelling. To this mixture, 5cc of potassium or sodium iodate solution was added with constant stirring. The pH of the mixture was maintained at 4.2. Experiments were performed to optimize suitable pH value for growth of good quality crystals. This mixture was then transferred to the test tube and it was closed with cotton plug. The gel was allowed to set. It took nearly 12 days for setting. This set gel was aged for 5 days. Aging helps in nucleation control due to reduction in the diameter of the capillaries in gel. Calcium chloride or calcium nitrate was used as supernatant. Supernatants having different molarities were carefully poured over the set gels. Experiments were also performed by interchanging positions the reactants.

The chemical reactions inside the gel can be expressed as,

$XCl_2 + 2YIO_3 \rightarrow X(IO_3)_2 + 2YCl$, or

$X(NO_3)_2 + 2YIO_3 \rightarrow X(IO_3)_2 + 2Y(NO_3)_2$,

Where X = Ca and Y = K or Na.

3. Result and discussion

Parameters such as gel density, gel setting time, gel aging time, concentration of reactants, pH of gel, concentration of impurities have considerable effect on growth of crystals. Crystals having different morphologies were obtained. Prismatic crystals of size 3 x 2 x 2 mm^3 far away from gel interface and prismatic pyramidal crystals of size 5 x 2 x 2 mm^3 away from gel interface were obtained. It was observed that the number of crystals growing diminished with the increase in the distance from gel interface. It may be due to reduced rate of diffusion of supernatant. Second reason may be attributed to the aging of gel, since crystals in this region nucleate in a comparatively older gel.

High concentration of reactants results into either needle shaped crystals of size 8 x 2 x 2 mm^3 or hopper crystals of size 10 x 3 x 2 mm^3 near the gel interface. The reason may be high diffusion gradient near the gel interface. Less concentration of reactants leads to the formation of platy crystals due to slow diffusion rate and insufficient supply of reactants. Figure1 shows few transparent needle shaped crystals, observed away from gel interface, at somewhat more concentration of reactants. Figure 2 shows hopper shaped crystals, grown inside the test tube near gel interface, due to higher concentration of reactants. Figure 3 shows different habits of calcium iodate crystals. Figures 4 and 5 show needle shaped crystal and its schematic. Figures 6 and 7 show prismatic pyramidal crystal and its schematics, resp.

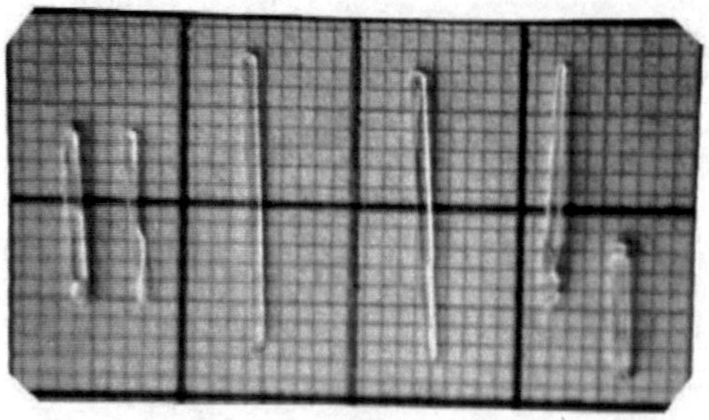

Figure 1. Needle shaped crystals

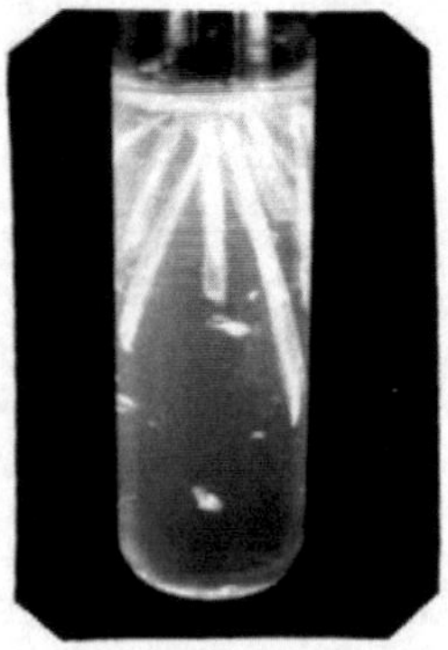

Figure 2. Hopper shaped crystals inside test tube

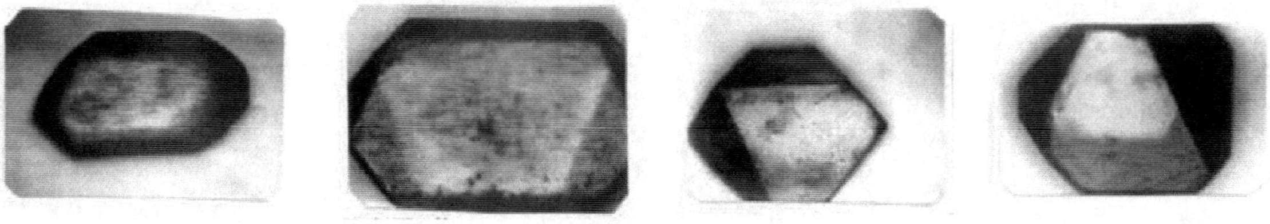

Figure 3. Different habits of Calcium iodate crystals

Figure 4. Prismatic needle shaped crystal

Figure 5. Schematic of needle shaped crystal

Figure 6. Prismatic pyramidal crystal

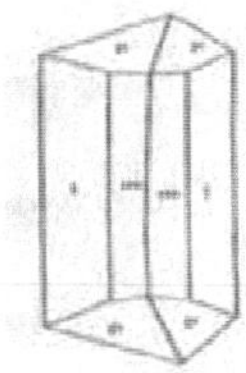

Figure 7. Schematic of prismatic pyramidal crystal

4. Characterization

Calcium iodate crystals were characterized by XRD (X-Ray diffractometry), FTIR (Infrared), and Thermal Analysis [TGA (Thermo Gravimetric Analysis), DTA (Differential Thermal Analysis) to study structure, and SHG Measurements to study NLO property.

4.1. XRD (X-Ray Diffractometry)

X-ray diffraction (XRD) is an excellent analysis technique for identifying unknown solid materials. The XRD pattern can act as a fingerprint to identify unknown materials in crystal and semi crystalline samples. The atoms are arranged in a regular pattern, and there is as smallest volume element, that by repetition in three dimensions describes the crystal. This smallest volume element is called a unit cell. The dimensions of the unit cell are described by three axes: a, b, c and the angles between them alpha, beta and gamma.

X- Ray diffractograms were recorded using powder rotation photograph method. The specifications of diffractometer are, Cukα radiation $\lambda = 1.5418$ A^0, 2θ range = 10^0 to 90^0 and scanning speed = 10^0 / min. Figures

8, 9, and 10 show diffractograms of undoped, Cu-doped and Fe-doped calcium iodate crystals respectively

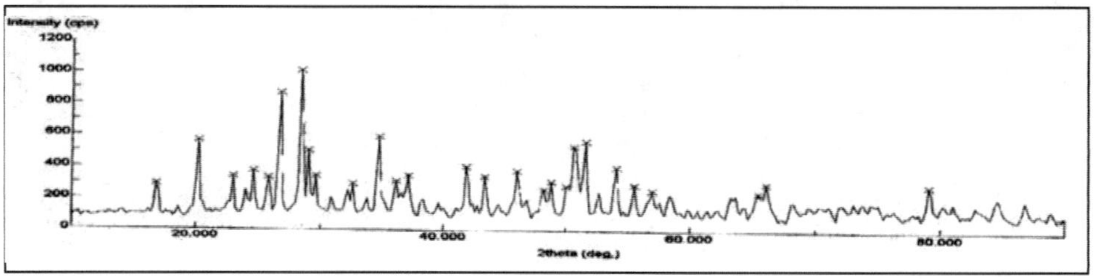

. **Figure 8.** X-ray diffractogram of undoped calcium iodate

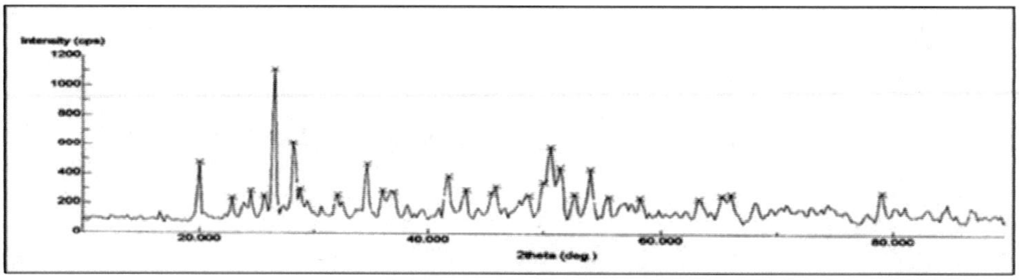

Figure 9. X-ray diffractogram of Cu-doped calcium iodate

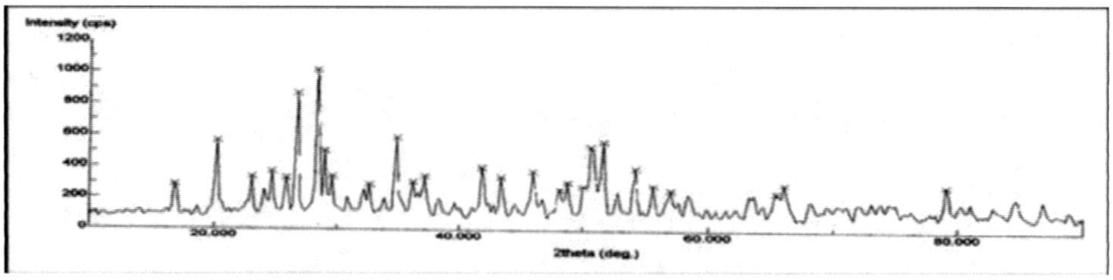

Figure 10. X-ray diffractogram of Fe-doped calcium iodate

From diffractograms, it is clear that impurities are induced only in certain planes, but it does not change the structure of crystal. Impurities only cause a slight change in lattice parameters. Change in volume of unit cell of doped crystals may be attributed to ionic radii of doped impurities (Cu^{+2}, Fe^{+3}). These crystals belong to monoclinic system with $a \neq b \neq c$ and $\alpha = 90^0 \neq \beta$. The calculated values of a, b, c, and β are in good agreement with the reported ones [13]. The calculated values of lattice parameters, β, and volume of undoped and doped crystals with respect to reported values are represented in table 1.

Table 1. Lattice parameters of calcium iodate, monohydrate

parameters	Reported	Observed	(Observed)	(Observed)
a A^0	8.509	8.494 (1)	8.543 (7)	8.554 (4)
b A^0	10.027	10.033 (2)	10.136 (7)	10.049 (2)
c A^0	7.512	7.512 (2)	7.549 (6)	7.501 (3)
β^0	95.27	95.00	94.00	95.00
V $(A^{0})^{3}$	638.213	637.457	651.673	642.349

4.2 FT-IR analysis

FTIR is an important technique for identification and characterization of a substance. Infrared Spectroscopy gives information on the vibrational and rotational modes of motion of a molecule.

In order for a vibrational mode in a molecule to be "IR active," it must be associated with changes in the dipole. A permanent dipole is not necessary, as the rule requires only a change in dipole moment. A molecule can vibrate in many ways, and each way is called a vibration mode. For molecules with N atoms in them, linear molecules have 3N 5 degrees of vibrational modes, whereas nonlinear molecules have 3N 6 degrees of vibrational modes, also called vibrational degrees of freedom. Simple diatomic molecules have only one bond and only one vibrational band. If the molecule is symmetrical, the band is not observed in the IR spectrum, but only in the Raman spectrum. Asymmetrical diatomic molecules, absorb in the IR spectrum. More complex molecules have many bonds, and their vibrational spectra are correspondingly more complex

FT-IR spectra of undoped and doped iodate crystals were scanned by using FT-IR spectrophotometer, Spectrum 2000, Perkin Elmer model by placing sample KBr pellet in the sample beam in two ranges, 300 710 cm^{-1} and 400 4000 cm^{-1}. FT-IR spectra of undoped, Cu-doped, and Fe-doped calcium iodate crystals represented in figures 11, 12 and 13 respectively

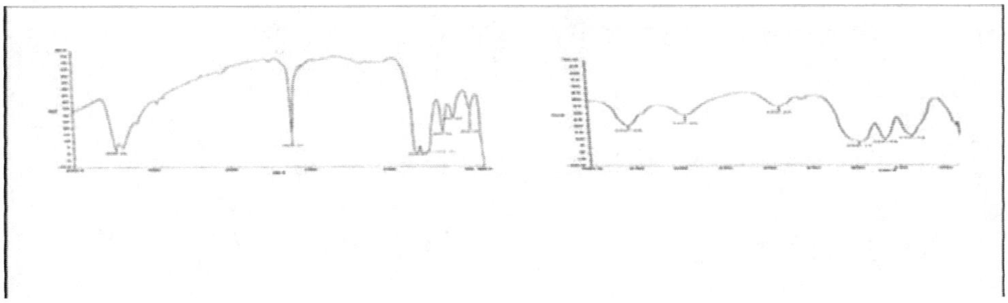

Figure 11. FT-IR spectra of undoped calcium iodate in the range 4000-400 cm^{-1} & 710-300 cm^{-1}

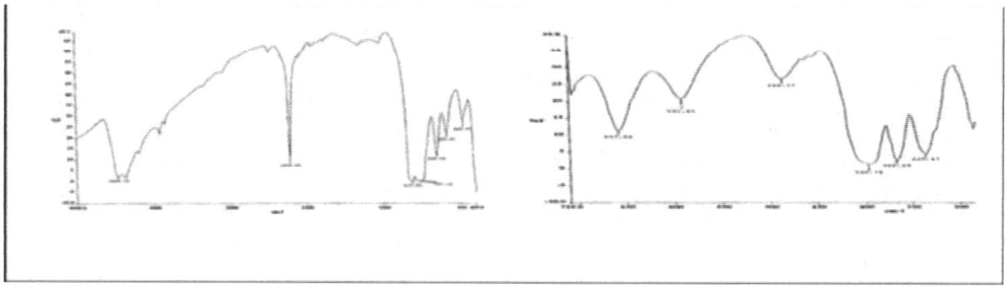

Figure 12. FT-IR spectra of Cu-doped calcium iodate in the range 4000-400 cm^{-1} & 710-300 cm^{-1}

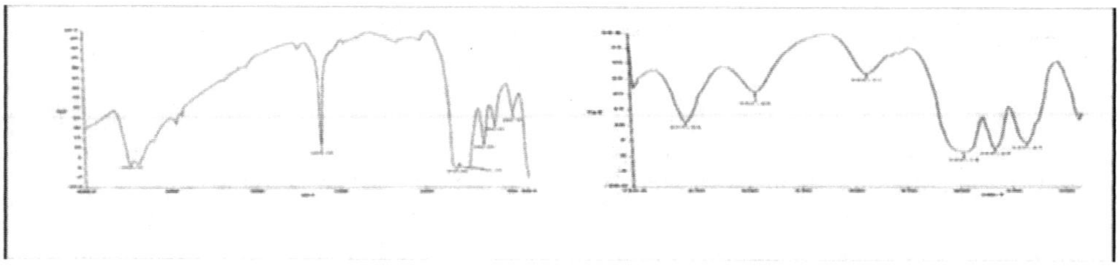

Figure 13. FT-IR spectra of Fe-doped calcium iodate in the range 4000-400 cm^{-1} & 710-300 cm^{-1}

Fundamental frequencies of the pyramidal iodate ions as have been reported earlier by Nassau et al[14], Dasent and Waddington [15], Balicheva and Petrova [16], Niquist and Kagal [17] Rocchiccioli [18], Dratovsky and Pacesova [19], and Nakamoto [20] are in good agreement with the observed values in the present investigation. Vibrational frequencies of undoped, Cu-doped, and Fe-doped calcium iodate crystals are given in table 2.

Table 2. Vibrational frequencies of undoped, Cu-doped, and Fe-doped calcium iodate crystals

Fundamental Frequencies	**Calcium iodate**		
	Undoped (cm^{-1})	**Cu-doped** (cm^{-1})	**Fe-doped** (cm^{-1})
Symmetric stretching frequency, γ_1	759.15	759.15	759.15
Symmetric bending frequency, γ_2	396.77	396.74	396.72
Asymmetric stretching frequency, γ_3	818.83	815.86	818.85
Asymmetric bending frequency, γ_4	335.04	335.47	335.15

4.2. Thermal Analysis

The principle of thermogravimetry is based on the simple fact that, the sample is weighed continuously as it is being heated to elevated temperature. In DTA, the study depends on the measurement of difference in heat content of a sample with reference to a standard substance as a function of temperature or time.

Thermal studies were carried doped out using Mettler Toledo Star system. Definite amount of sample was taken and heating was carried out from ambient to 900°C for TGA and DTA at the rate of 5°C /min in an air medium. Figure 14 includes TGA curve [Figure14 (a)] and DTA curve [Figure14 (b)] of undoped calcium iodate. Figures15 includes TGA [Figure15 (a)] and DTA curve [Figure15 (b)] of Cu-doped calcium iodate and Figures 16 includes TGA [Figure16 (a)] and DTA curve [Figure16 (b)] of Fe-doped calcium iodate crystals, respectively.

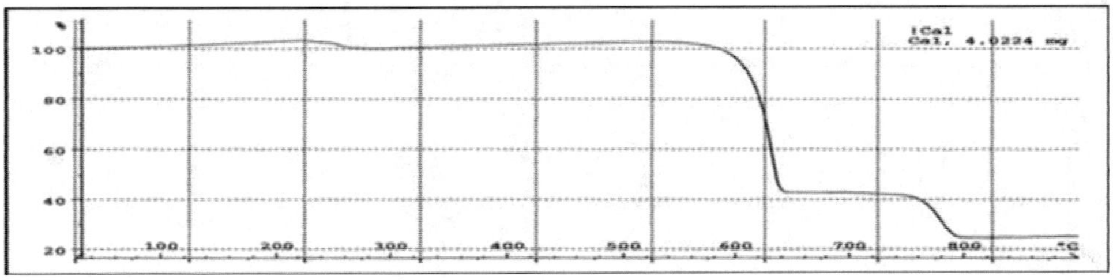

Figure 14(a). TGA curve of undoped calcium iodate

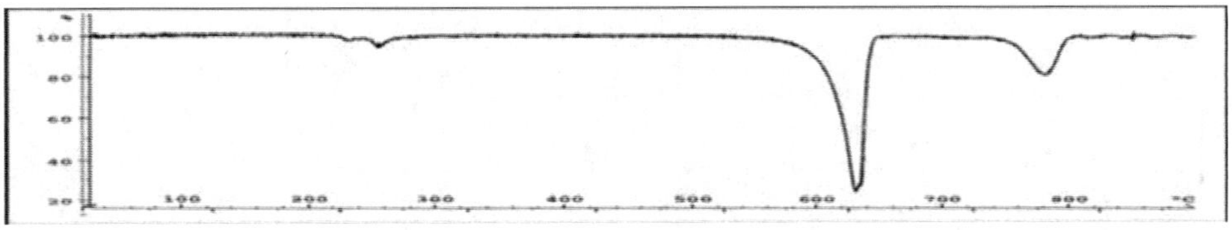

Figure 14(b). DTA curve of undoped calcium iodate

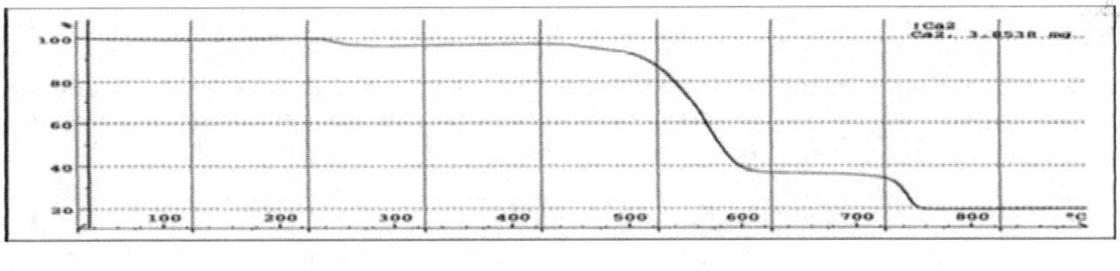

Figure 15(a). TGA curve of Cu-doped calcium iodate

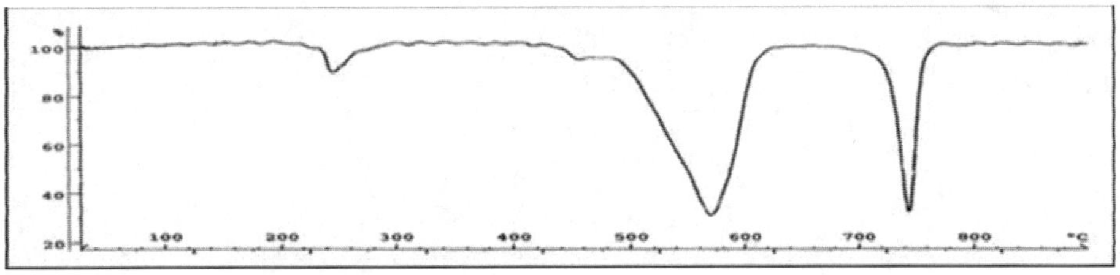

Figure 15(b). DTA curve of Cu-doped calcium iodate

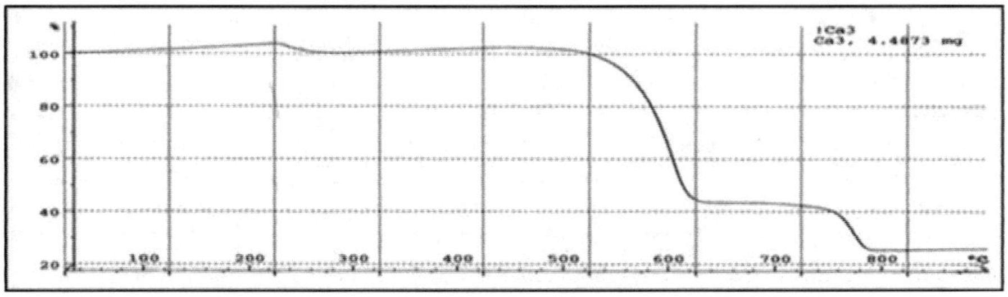

Figure 16(a). TGA curve of Fe-doped calcium iodate

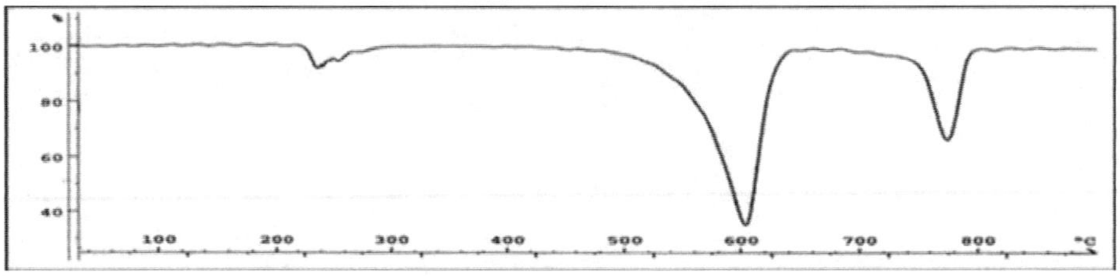

Figure 16(b). DTA curve of Fe-doped calcium iodate

These crystals exhibit three steps explicitly. Table 3 shows kinetic data from TGA of these crystals. Results obtained are in good agreement with the reported one [21]. The compound is stable up to 220°C. The weight loss begins at 220°C. There is 3.2% weight loss in the temperature range 220 - 280°C. This is the first step and is suggestive of loss of one water molecule from the crystal. This water loss step is not very sharp but extends over a wide temperature range of about 60°C and indicates that lattice water is strongly hydrogen bonded with iodate group, showing that calcium iodate crystals are monohydrated.

For Cu-doped and Fe-doped calcium iodate crystals, almost similar results are obtained with slight changes in the temperature range of decomposition reactions and values of % weight loss. Thus doping has no effect on the structure of crystal. Kinetic data from dynamic TGA of undoped, Cu-doped, and Fe-doped calcium iodate is shown in table 3.

Table 3. Kinetic data from dynamic TGA of undoped, Cu-doped, and Fe-doped calcium iodate

Compound	Step	Temp. Range (^{0}C)	% Weight loss	Probable product formed (solid)
Undoped calcium iodate, monohydrate	I	220 – 280	3.2	$Ca(IO_3)_2$
	II	580 – 640	58	$Ca_5(IO_6)_2$
	III	750 – 800	18	$Ca_5(IO_4)_2$
Cu-doped calcium iodate, monohydrate	I	240 – 280	2	$Ca(IO_3)_2$
	II	450 – 600	60	$Ca_5(IO_6)_2$
	III	720 – 750	18	$Ca_5(IO_4)_2$
Fe-doped calcium iodate, monohydrate	I	220 – 270	3.2	$Ca(IO_3)_2$
	II	500 – 620	58	$Ca_5(IO_6)_2$
	III	750 – 780	18	$Ca_5(IO_4)_2$

Journal of Physics: Conference Series **423** (2013) 012060 doi:10.1088/1742-6596/423/1/012060

4.3. Powder SHG measurements

Powder SHG technique [22, 23] makes it possible to predict the magnitude of the nonlinear coefficients of the powder samples in a very simple manner, should provide a substantial increase in the number of new materials for use in nonlinear optic applications.

In SHG, an optical beam of frequency 2ω can be generated from the interaction of a high power laser beam of frequency ω with a suitable material either in the form of crystal or powder sample. It cannot occur in isotropic medium or in centrosymmetric crystals. Hence, a noncentrosymmetric crystal is a must to get SHG [24, 25].

For performing experiments, sample plates were prepared from powder samples of KDP, undoped, Cu-doped, and Fe-doped calcium iodate. All the samples were sieved by a 75 μm sieve, in order to maintain uniform particle size. Each sample plate was placed in front of Nd: YAG laser beam. Before sample plate, beam splitter splitted the input signal into reference signal, collected by photo detector and second signal on sample plate, which in turn generated second harmonic signal. Reference signal and second harmonic signal were both applied to digital storage oscilloscope. V (ω) of reference signal and V (2ω) of SHG signal were recorded as a function of arc lamp voltage. Conversion efficiency was calculated by taking ratio V (ω)/V (2ω). Graphs of arc lamp voltage versus conversion efficiency were plotted and tanθ has been found. Nonlinear coefficients were calculated by using the relation,

$$d^2 \text{ known} / d^2 \text{ unknown} = \tan\theta \text{ known} / \tan\theta \text{ unknown}$$

Where, d known = nonlinear coefficient of KDP

d unknown = nonlinear coefficient of the sample under test, to be found out

tanθ known = slope of graph of arc lamp voltage versus conversion efficiency of reference sample (KDP)

tanθ unknown = slope of graph of arc lamp voltage versus conversion efficiency of sample under test

Table 4 gives the results of calculations of nonlinear coefficients (d s). Cu and Fe doped samples of calcium iodate do not show any nonlinear property. Output from these sample plates was completely stopped.

Table 4. Nonlinear coefficients of KDP and undoped calcium iodate crystals

Substance	**Nonlinear coefficient (d 's) m/v**
KDP	6.3×10^{-13}
Undoped calcium iodate	1.2535×10^{-13}

5. Conclusions

Gel technique can be successfully employed for the growth of calcium iodate crystals In case of gel method, various habits can be obtained by changing various parameters. Well known Liesegang phenomenon has been observed in case of calcium iodate crystals. Calcium iodate crystals exhibit the phenomenon of efflorescence. Suitable value of pH for calcium iodate crystals is 4.2. Doping of impurities has no significant effect on the morphology of crystals. XRD results match very well with the standard JCPDS data. Dopants have been incorporated only in certain planes. No major changes in structure have been observed. IR analysis confirms the presence of fundamental IR frequencies observed in all iodate compounds. Thermal analysis exhibits three steps explicitly on heating the samples. The first step involves dehydration at 250°C, second step shows decomposition at 580°C, and the third step involves again decomposition at 640°C. Powder second harmonic generation experiments exhibit the nonlinear nature of the substance. But the intensity of SHG is very small. Cu^{+2} and Fe^{+3} doped samples have negative effect on SHG property. These samples do not show nonlinear effect.

References

[1] Armington AF and O Connor JJ 1968 Gel growth of cuprous halide crystals *J. Cryst. Growth* **3-4** 367-71
[2] Shitole SJ and Saraf KB 2001 Growth and study of some gel grown group II single crystals of iodate *Bull. Mater. Sci.* **24(5)** 461-68
[3] Shitole SJ and Saraf KB 2002 Growth, Structural and microtopographical studies of calcium Iodate, monohydrate crystals grown by silica gel. *Cryst. Res. Technol.* **37(5)** 440-45
[4] Shitole SJ and Saraf KB 2007 Spherulitic growth of cadmium iodate in silica gel and its XRD and FTIR analysis. *J. Adv. Sci. And Tech.* **10(I & II)** 42-55
[5] Joshi MS and Trivedi SG 1981 Growth of single crystals of barium iodate and strontium iodate in silica gels. *Cryst. Res. Technol.* **16(1)** 19-26
[6] Garud SL, Mahajan NK and Saraf KB 2009 Study of gel grown mixed crystals of BaxCa (1 x)(IO3)4. *Bull. Mater. Sci.* **32(2)** 187-92
[7] Garud SL and Saraf KB 2008 Growth and study of mixed crystals of Ca Cd iodate. *Bull. Mater. Sci.* **31(4)** 639-43
[8] Morosin B, Bergman JG and Crane GR 1973 Crystal structure, linear and nonlinear optical properties of Ca (IO3)2.6H2O. *Acta. Crysta.* **B29** 1067-72
[9] De Kock and AJR 1980 *Handbook on Semiconductors* vol3, ed Kellar SP and Moss TS (Amsterdam: North Holland) p 247
[10] Nyrit J 1982 *Industrial Crystallization* (Amsterdam: Elsevier) p 203
[11] Dennis J and Henisch HK 1967 *J. Electrochem. Soc.* 114-263
[12] Dake HC 1938 *Quartz family minerals: a handbook for the mineral collector* (New York: Whittlesey House)
[13] *JCPDS Card for X-ray Diffraction Data* No. **26** p 1405
[14] Nassau K, Shiever JW and Prescot BE 1973 Transition metal iodates. I. Preparation and characterization of the 3d iodates. *J. Solid State Chem.* **7** 186-204
[15] Dasent WE and Waddington TC 1960 Iodine Oxygen Compounds. Part I. Infrared Spectra and Structure of Iodates *J. Chem. Soc.* 2429-32
[16] Balicheva TG and Petrova VA 1973 Vibrational spectra of the IO3-ion in alkali iodates. *J. Structural Chem.* **14(3)** 424-30
[17] Nyquist RA and Kagal RD 1971 *Infrared Spectra of Inorganic Compounds* (New York: Academic Press)
[18] Rocchiccioli, C. Compt. Rend. 250 (1960) 1232.
[19] Dratovsky, M. and Pacesova, L. Russian Chemical Reviews, Ed. Uspekhi Khimji (1968).
[20] Nakamoto, K. Infrared Spectra of Inorganic and Coordination Compounds, 2nd Ed., Wiley Interscience (1970).
[21] Sanyal GS and Nag K 1977 Thermal studies on the dimesoperiodates and iodates of Ca (II), Sr (II) and Ba (II) preparation and characterization of hexavalent iodates. *J. Inorg. Chem.* **39** 1127-30
[22] Chemla DS and Zyss J 1987 *Nonlinear Optical Properties of Organic Molecules and Crystals* (New York: AcademicPress)
[23] Kurtz SK and Perry TT 1968A powder technique for the evaluation of nonlinear optical materials. *J. Appl. Phys.* **39(8)** 3798-813
[24] Bloembergen N, Armstrong JA, Ducuing J and Pershan PS 1962 Interactions between light waves in a nonlinear dielectric. *Phys. Rev. Lett.* **127** 1918-39
[25] Franken PA, Hill AE, Peters CW and Wernreich G 1961 Generation of Optical Harmonics *Phys. Rev. Lett.* **7** 118-9

Journal of Physics: Conference Series **423** (2013) 012061 doi:10.1088/1742-6596/423/1/012061

Optimization of Power Consumption for Centrifugation Process Based on Attenuation Measurements

M. S. Salim[1], M. F. Abd Malek[2], Naseer Sabri [3], M. Iqbal bin Omar[1], Latifah Mohamed[2], and K. M. Juni*

[1]School of Mechatronics Eng., University Malaysia,Perlis,01000, Malaysia.
[2]School of Electrical System Eng., University Malaysia,Perlis,01000, Malaysia.
[3]School of Computer and Communication Eng.,University Malaysia Perlis, Perlis 01000, Malaysia.
*Electrical Engineering Department, Politeknik Tuanku Syed Sirajudain, Malaysia.

E-mail muhsabri1967@yahoo.com

Abstract. The main objective of this research is to produce a mathematical model that allows decreasing the electrical power consumption of centrifugation process based on attenuation measurements. The centrifugation time for desired separation efficiency may be measured to determine the power consumed of laboratory centrifuge device. The power consumption is one of several parameters that affect the system reliability and productivity. Attenuation measurements of wave propagated through blood sample during centrifugation process were used indirectly to measure the power consumption of device. A mathematical model for power consumption was derived and used to modify the speed profile of centrifuge controller. The power consumption model derived based on attenuation measurements has successfully save the power consumption of centrifugation process keeping high separation efficiency. 18kW.h monthly for 100 daily time device operation had been saved using the proposed model.

Index Terms- power consumption, attenuation measurements, velocity profile

1. Introduction

The centrifugal devices can use in a variety of medical and industrial applications. Typically, the power consumption, capacity, speed of rotation, separation precision and the centrifugation time are essential technical parameters of the centrifuge device, and hence their reliability may affect the system reliability and productivity. For this reason, centrifugation process is often obeying to developments and improvements continuously.

Continuing from our previous research [1], we derived a modified period of velocity profile model (spinning period) which leads to decreasing the power consumption of laboratory centrifuge devices. The new model has ability to estimate the power consumed for pre-defined separation efficiency. The

laboratory diagnosis of blood depends on separation efficiency which is done by means of centrifugal force and the spinning time for the blood sample.
The current method of separating contents of a blood sample is not accurate in terms of centrifugation force and time. These variables are evaluated based on sedimentation theory, which calculates the sedimentation time based on higher particle density (red blood cells) and depth of sedimentation [2]. Therefore, about 0.5ml volume of a 1ml blood sample, the current centrifugation time is five minutes (also as recommended by the centrifuge device recommended manufacturer). To minimize or eliminate such uncertainties (due to damage of some red blood cells) from the process, we have developed a new technique that accurately predicts the required time for the separation then control the velocity profile period of centrifuge device. This technique is based on measuring the attenuation of the wave propagated through liquid its density varies with time of spinning[3], [4]. Separation efficiency of blood and plasma is evaluated empirically by percentage count for the red blood cells, white blood cells, HCT, and platelets in a sample using blood analyzer device (cell-dyne 1800). Experimental data was used to calculate the power consumption during the time taken for accurate separation.
The schematic diagram of experimental attenuation measurements is shown in figure 1. Human blood was drawn from 84 healthy volunteers, 23 women and 61 men; their HCT values ranged from 21% to 53.7%; age range, 21 to 50 years. Samples of plasma were prepared using a serum separator tube (EDTA (K2); demophorius Ltd, cypus, EU, UK).
Figure 2 shows the mathematical model deriving structure. This method can also be applied to other purpose of centrifugal device, such as measurements of solid contents in liquid and sedimentation measurements [5].
Since the installation of additional measurement sensors may be costly and the sensors can also reduce the device reliability, they are not always the most cost-effective solution for the separation percentage monitoring of a centrifuge device. For this reason, the centrifuge velocity profile controller must be modified according to new mathematical model. The modified period of linear velocity profile and mathematical model derivation are seen in next sections. Another section of your paper
The first paragraph after a heading is not indented.

2. Linear Velocity Profile

The choice of method for any particular system is a complex decision, in which the conflicting demands of high performance and low cost are evident. For example, an exponential ramp may be required for optimum acceleration, but its implementation is expensive and so the designer may compromise with a linear ramp, which is available at low cost. The linear profile is the common practicing to approximate the velocity profile by linear functions figure 3a. However, this approximation

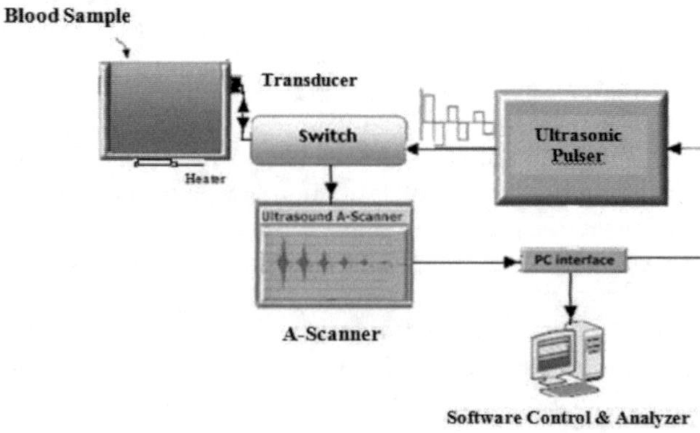

Fig. 1. Schematic diagram of experimental apparatus.

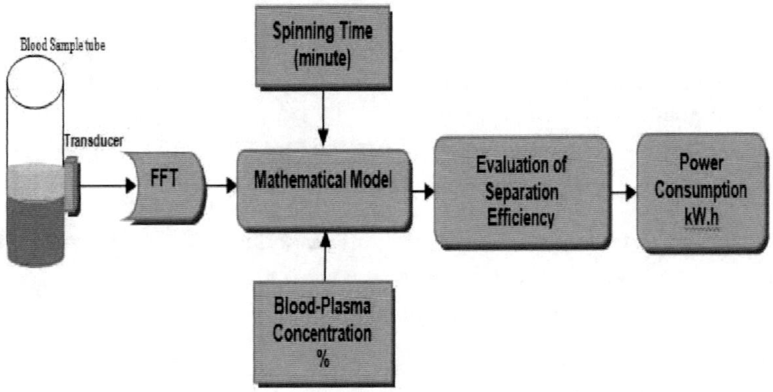

Figure 2. The deriving structure of mathematical model

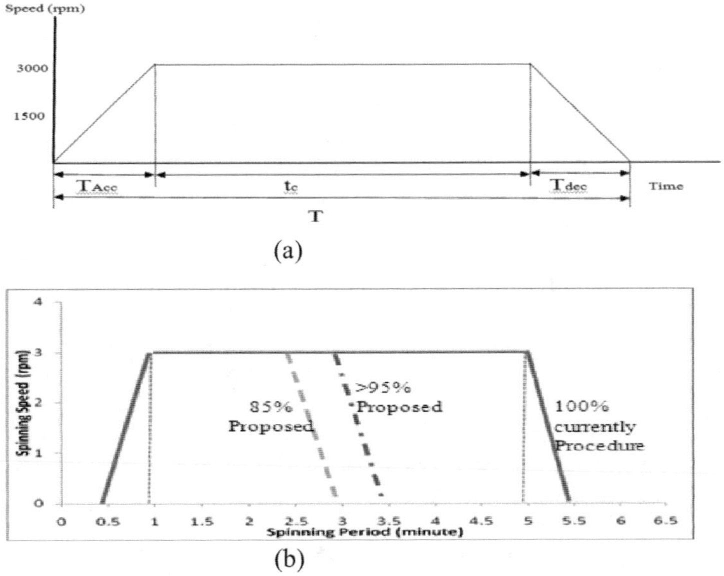

Figure 3. Linear velocity profile (a), and spinning period for currently and proposed procedure (b).

imposes a constraint on the system's maximum stepping rate, which becomes a function of the ramp gradient. For any speed, the maximum permissible acceleration the following equation [6], :

$$A = \frac{Q}{J} \qquad (1)$$

Where A is the acceleration, Q is the motor torque, and J is moment of Inertia. The time (T) taken to accelerate to the maximum speed (v) is:

$$T = v / A \qquad (2)$$

For linear profile, the acceleration velocity is constant and limit by the maximum permissible acceleration. In general, the motor is able to decelerate faster than it can accelerate. Therefore, If a single ramp used to control both acceleration and deceleration, the deceleration rate must be matched to the acceleration rate to prevent the damage may be happen to blood cells during acceleration

/deceleration process time, therefore the speed of acceleration and deceleration must be identical to produce smooth sedimentation for liquid contents.. The centrifuge controller will program according to ramp speed profile in which the distance and time of acceleration and deceleration are same. [7],[8], [9]

In practice, acceleration and deceleration period for centrifuge velocity is 16 second. In this research the laboratory centrifuge device type Kubota corporation model 2420, was used which the normal operation mode was selected to enable the device spin at 3000 rpm, with ramp speed profile that it's acceleration/deceleration have each 16 second time period.

Therefore, summation of constant velocity period (t_c), acceleration period (t_{Acc}), and deceleration period (t_{Dec}) are total period of linear velocity profile (T) as shown in figure 3a.

$$T = t_c + t_{Acc} + t_{Dec} \qquad (3)$$

3. Experimental Setup

The Mathematical Model of the shorter separation process time and the separation efficiency greater than 95% of 0.35ml plasma, evaluated based on the mathematical model of ultrasound attenuation measurements in previous research [1]. Utilize of attenuation measurements of wave propagate through a sample of blood during centrifugation process, is to calculate the separation efficiency of blood-plasma. In this research we divided the mathematical model into two models, first is the time duration of centrifugation as a function of separation efficiency model, while the second model is device power consumption as a function of time duration centrifugation technique.

The tools used in the experimental setup, as shown schematically in figure 1, consisted of a blood tube, the ultrasound pulser, Gampt–Echo Scan as a receiver, switch, and software analysis, while figure 4 shows the experimental procedure steps for evaluate the mathematical model. The transducer with center frequencies of 2 MHz was attached to the outside of the vessel. In the attenuation measurements reported here, data was obtained for the 2 MHz transducer, which gave the required detection depth and sufficient level of clarity of the separation efficiency measurements of the blood plasma. Table 1 was filled follow same procedure mentioned in our previous research for evaluating the optimization time for centrifugation process. Therefore the two part of mathematical model can be derived.

Table 1. Evaluation of the attenuation, separation efficiency, and power consumption for discrete five minute using 180 watt centrifuge device.

Number of Operation	Constant speed period (minute)	Total spinning period (minute)	Attenuation dB/cm	Plasma separation efficiency %	Power Consumption kW.h
1	0.5	1.033	-0.0257	20	0.003099
1	1	1.533	-0.0662	35	0.004599
1	1.5	2.033	-0.1202	55	0.006099
1	2	2.533	-0.1607	70	0.007599
1	2.5	3.033	-0.2012	85	0.009099
1	3	3.533	-0.2282	95	0.010599

Journal of Physics: Conference Series **423** (2013) 012061 doi:10.1088/1742-6596/423/1/012061

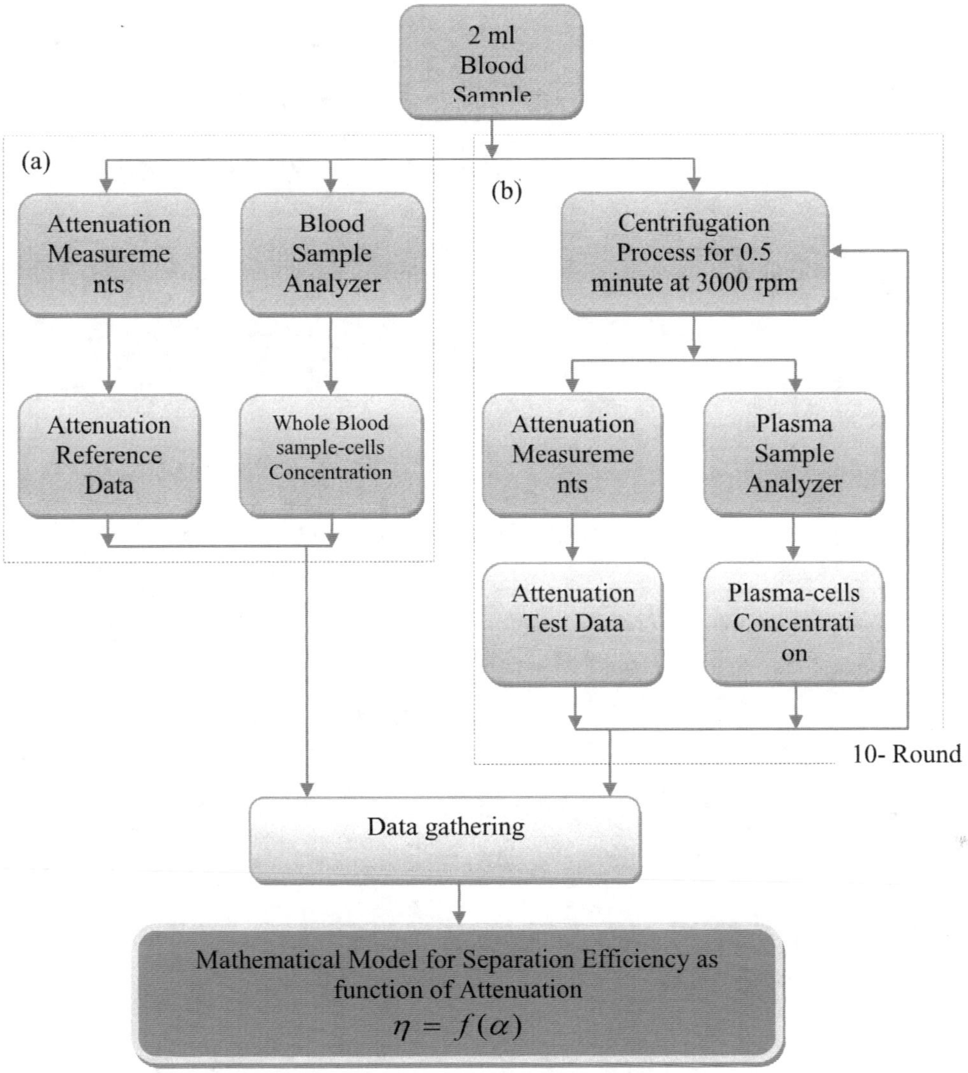

Fig.4. Experimental procedure sequence for one blood sample test.

Table 2 Monthly Electrical power Consumption for 180 watt centrifuge device.

Method Type	Plasma Separation Efficiency	Total spinning time $t_c + 2T_{Acc}$	Pc* for 100 time operating device daily kW.h	Monthly Pc* kW.h.
Currently	100%	5+2*16	1.659	49.77
Optimization time	95-100%	3+2*16	1.06	31.797

Pc* is power consumption.

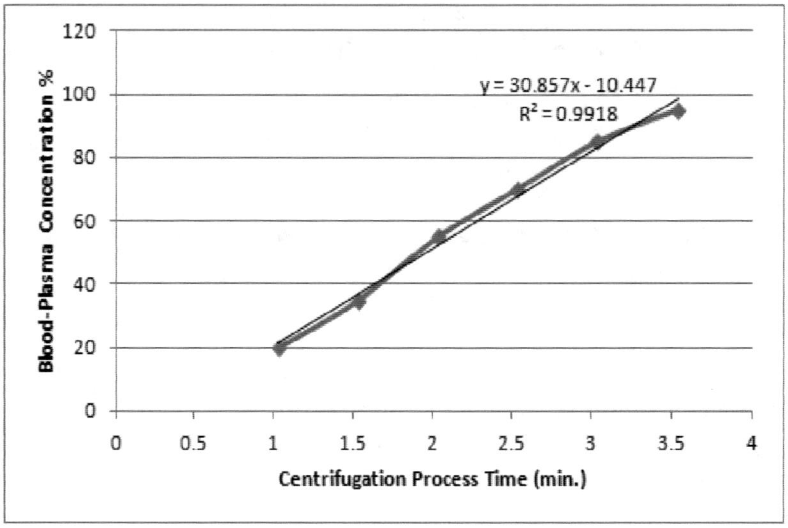

Figure 5. Blood-plasma concentration versus centrifugation time.

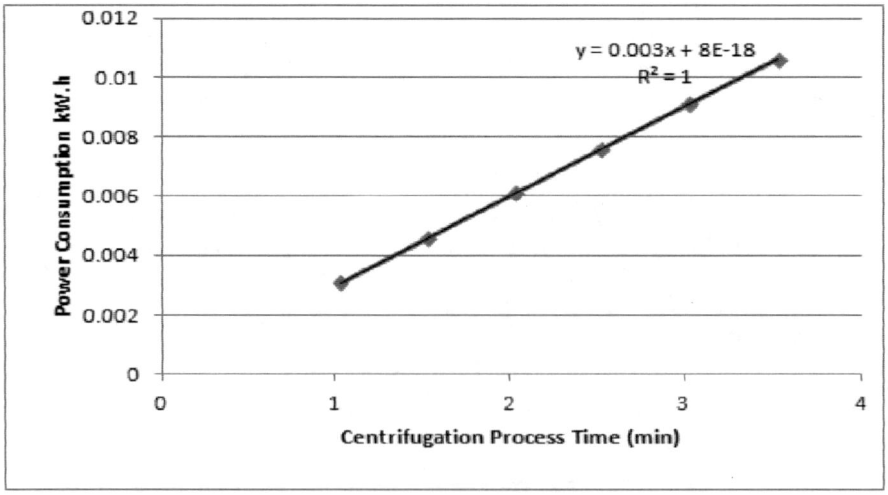

Figure 6. Power consumption versus 1ml blood sample centrifugation time.

3.1 *Centrifugation period model*

A mathematical relationship for the constant speed profile period (*tc*) as a function of separation efficiency (η) was derive, table 1.

$$t_c = 0.0531\eta + 0.0155 \tag{4}$$

Linear speed profile was use to control the spinning of centrifuge device The rapid mode (linear speed profile) in which acceleration and deceleration period are match and equal to 16 second.

Therefore, the total period for centrifugation process be

$$T = (0.0531\eta + 0.0155) + t_{Acc} + t_{Dec} \quad (5)$$

According to prescribed separation efficiency only the time of constant speed change figure 3b. Based on experimental results, 0.35ml with greater than 90% plasma concentration, the proper time is 3 minute, therefore, the total time of centrifugation be about 3.5 minutes.

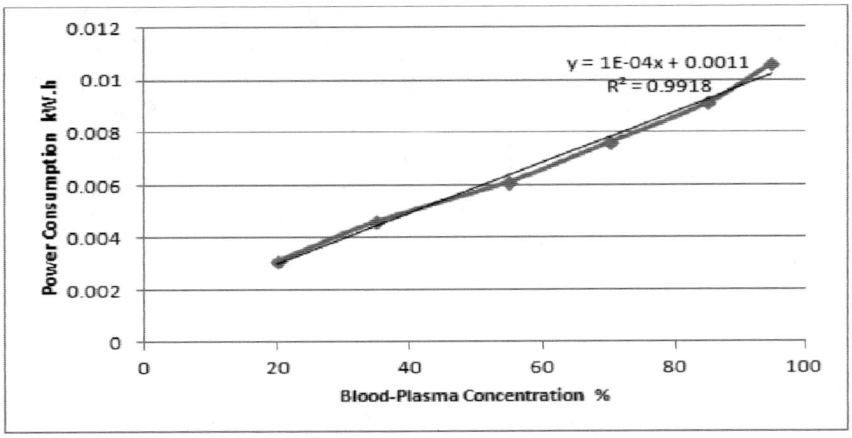

Figure 7. Centrifuge device power consumption measurements for (20-100)% Blood–plasma concentration.

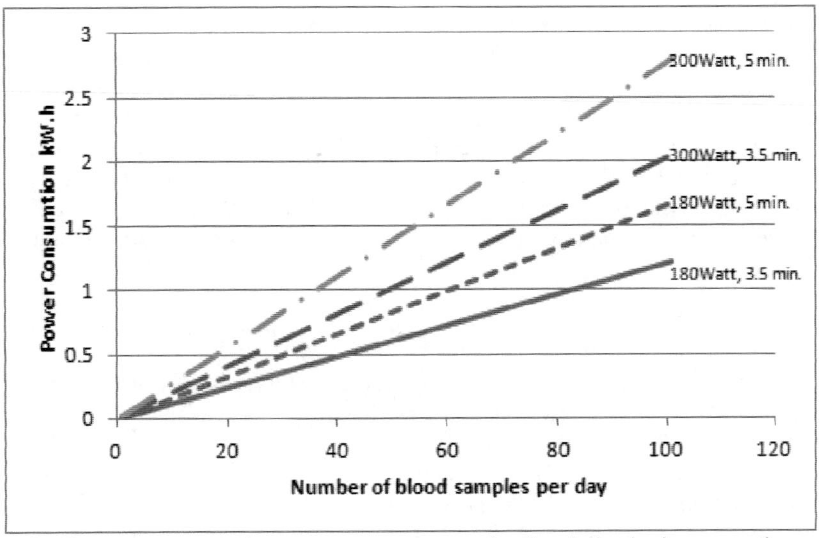

Figure 8. Power consumption measurements results for daily device operating using currently and optimize procedure of 180watt and 300watt laboratory centrifuge devices.

3.2 Power Consumption model

Electrical power consumption can be easily calculated using the following formula:

$$T_{PC}(kW.h) = D_{PC}(kW) \times O_t(h) \quad (6)$$

Where T_{PC} is the total power consumption, D_{PC} is device power consumption, and O_t is operation time period. The power consumption of Laboratory centrifuge device (Kubota corporation model 2420) used is 180watt [10]. The currently spinning time required to 0.35 ml plasma from 1 ml blood with separation efficiency greater than 90% , is five minute according to manufacture recommended, while based on proposed method (Attenuation measurements) is 3.5 minute, figure 5.

Therefore, the power consumption of currently centrifuge controller in which greater than 95% separation effieciny is 0.01659 kW.h (5 minutes sample pinning), while 0.0106 kW.h when using derived model (3.533 minute) ,figure 6, 7. Table 2 shows that about 18kW.h per month can be save using optimization time model. Figure 8 illustrates the comparison of the two different wattage centrifuges devices, using currently and proposed velocity profile.

Conclusion

The mathematical model of optimization time of centrifugation process is included the acceleration and deceleration period, because of , the blood sample subjected to accelerated centrifugal force during acceleration period, same thing for deceleration period. These forces are contributes the main centrifugal force (at 3000 rpm) to precipitation of blood cells from plasma. Based on the derived model, the currently centrifuge controller can be modified by reprogram it with equation (5). This low cost modification leads to decrease time duration of blood test, and device power consumption, keeping separation efficiency same as currently procedure. In addition, Equation (5) increases the reliability for centrifuge device to estimate the centrifugation period for predefined separation efficiency figure 4. The relationship between device power consumption and centrifugation time and plasma separation efficiency is linear as shown in figure 4 and 5 respectively, therefore, the power consumption for any centrifuge device wattage can be estimate. The mathematical model derived based on attenuation measurements has successfully save the power consumption of centrifugation process. 18kW.h monthly for 100 daily time device operation had been saved using the proposed model. Figure 7 shows that a comparison for power consumption using currently and modified controller of centrifuge device.

Reference

[1] M. S. Salim, M. F. Abd Malek, R. B. W. Heng, Naseer Sabri Salim and K. M. Juni, "A new measurement method of separation percentage for human blood plasma based on ultrasound attenuation", IJPS, Vol. 6, 30, 6891 – 6898,2011.
[2] Theory of sedimentation and centrifugation, 2009, www.bbka.org.uk/local/iceni/bm~doc/pollensuspension-2.pdf
[3] M.S. Greenwood, J.D. Adamson, J.A. Bamberger, "Long-path measurements of ultrasonic attenuation and velocity for very dilute slurries and liquids and detection of contaminates", Ultrasonics, 44, 461-466,2006.
[4] Bain BJ. Blood cells: A practical guide. Malden: Blackwell Publishing. 476p,2006.
[5] Bamber JC. Attenuation and absorption. In: Hill CR, Bamber JC, ter Haar GR, (eds). Physical principles of medical ultrasonics.Chichester: John Wiley and Sons Ltd, pp. 93-166, 2004.
[6] P.P. Acarnnley, "Stepping Motors Guide To Modern Theory and Practice", peter pereginus,

LTD ,1982.
[7] K. Ramu, Electronic Control of Machines. ECpE 4324, A-1 Copies, Blacksburg, VA1998. From the forthcoming book Electronic Control of Machines, Prentice-Hall, USA.
[8] W. Leonard, Control of Electric Drives, Verlag, Berlin, 1985.
[9] S. Chung et. al. "A Robust Speed Control of Brushless Direct Drive Motor Using Integral Variable Structure Control with Sliding Mode Observer," Conf. Record IAS Annual Meeting. v1 1994, IEEE, Piscataway, NJ, pp. 393-400, 1994.
[10] Kubota corporation –Laboratory centrifuge catalogue, 2010, http://www.centrifuge.jp/product/2420/index.html

A New Ultrasound Pulser Technique for Wide Range Measurements

M. S. Salim[1], M. F. Abd Malek[2], N.M.Noaman[3], Naseer Sabri*, Latifah Mohamed[2], and K. M. Juni**

[1]School of Mechatronics Eng., University Malaysia,Perlis,01000, Malaysia.
[2]School of Electrical System Eng., University Malaysia,Perlis,01000, Malaysia.
*Electrical and Electronics Engineering Department, Gulf University, Bahrain Kingdom.
*School of Computer and Communication Eng.,University Malaysia Perlis, Perlis 01000, Malaysia.
**Electrical Engineering Department, Politeknik Tuanku Syed Sirajudain, Malaysia.

E-mail muhsabri1967@yahoo.com

Abstract. The objective of this research was to design and implement a new ultrasonic pulse-power–decay technique that transmits multiple ultrasound pulses through slurry to determine the lowest concentration that can provide an accurate attenuation measurement. A wide measurement range is obtained using the pulsed-power-decay transmission technique, and regardless of the material used to construct the container. A signal in the receiver transducer provides the attenuation measurements, for each echo, a fast Fourier transform (FFT) of the appropriate signal was obtained and compared with the water signals to yield the attenuation as a function of frequency. The data show the feasibility of measuring a kaolin concentration of 5% wt. When using a commercial pulser with the same device setting, no detectable echo was observed. Therefore, new technique measurements may prove useful in detecting solid content in liquid. This study demonstrated that the proposed pulsed-power transmission technique is promising for evaluating low concentrations of solids in fluids and for measuring sedimentation in solid-liquid systems.

Index Terms — Pulse-power decay system, Attenuation, Contaminates detection, process control.

1. Introduction

Measuring the concentration of solids in liquid is desirable, as many industries use this information for quality control and to monitor their product contents. Different pulsed ultrasound power levels should be used to increase the ability to measure the solid weight percentage of low-concentration (less than 5%) slurry and also to provide an accurate measurement of the velocity of sound. The analysis of multi-echoes can be used to measure slurry density. The classic methods used depends on the power, medium, frequency, quantity of the transmitted pulses and the reflection coefficient of the material from which the container is made of. Therefore, if the power, medium, frequency and number of the transmitted pulses are constant, then the container material is an essential parameter that is directly proportional to the measurement range. M.S. Greenwood et al. [1] used a stainless steel container, which had a high reflection coefficient and increased the weight percentage measurement of the slurry concentration. Also many researchers used a pitch-catch method to measure the liquid density and the attenuation of waves propagated through a suspension of solid particles in liquids (slurry). In such as these method, a constant train of power pulses was injected into the liquid-filled container, based on the attenuation measurement of the solid concentration. The concentration measurement was limited by the power and number of the transmitted pulses [2]. Clearly, the two methods have disadvantages related to power consumption and they depend on the type of container material is made of. Therefore, the number of echoes received depends upon the power level transmitted and the container material is made of. This means that when the container material has a low reflection coefficient or if the liquid has a low impedance value, then the number of echoes and the measurement range also decrease. However, we designed and implemented a new ultrasonic pulser technique by combining the useful features of the two methods currently used. The proposed technique is included in new ultrasound pulses transmission models and hardware designs, Firstly, the new pulser technique lack dependence on container material is made of; it changes the path measurement technique used for different concentrations and consumes less power than a commercial pulsed ultrasound uses to measure the same concentrations, And secondly, the new pulser technique is a programmable pulser , which it generates a pulsed-power decay (PPD) set that is separated by an adjustable time interval. In the remaining sections, further details are provided about the PPD device, and the performance of the new technique is evaluated and compared with the performance of a commercial pulser.

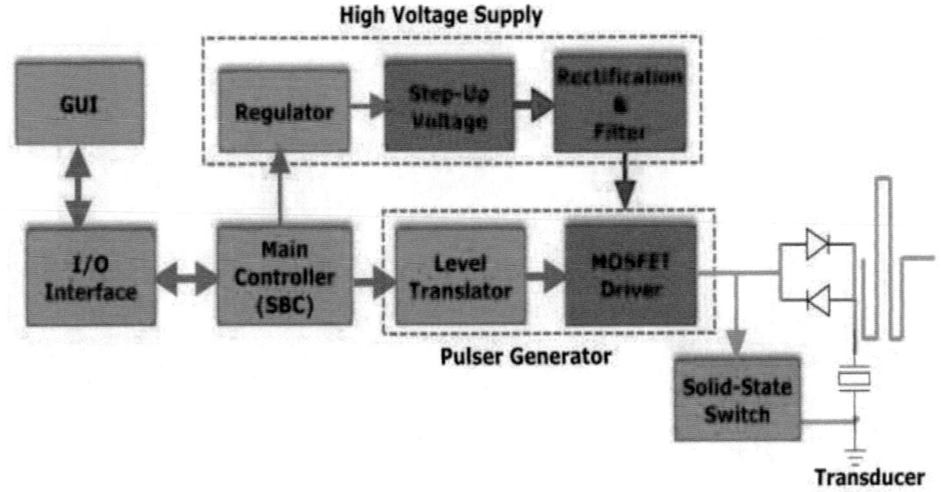

Figure 1. A block diagram of the pulse-power decay ultrasonic pulser system.

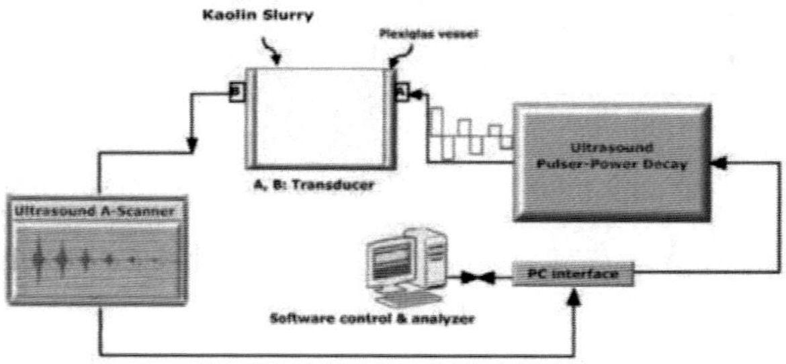

Figure 2. Sketch experimental setup.

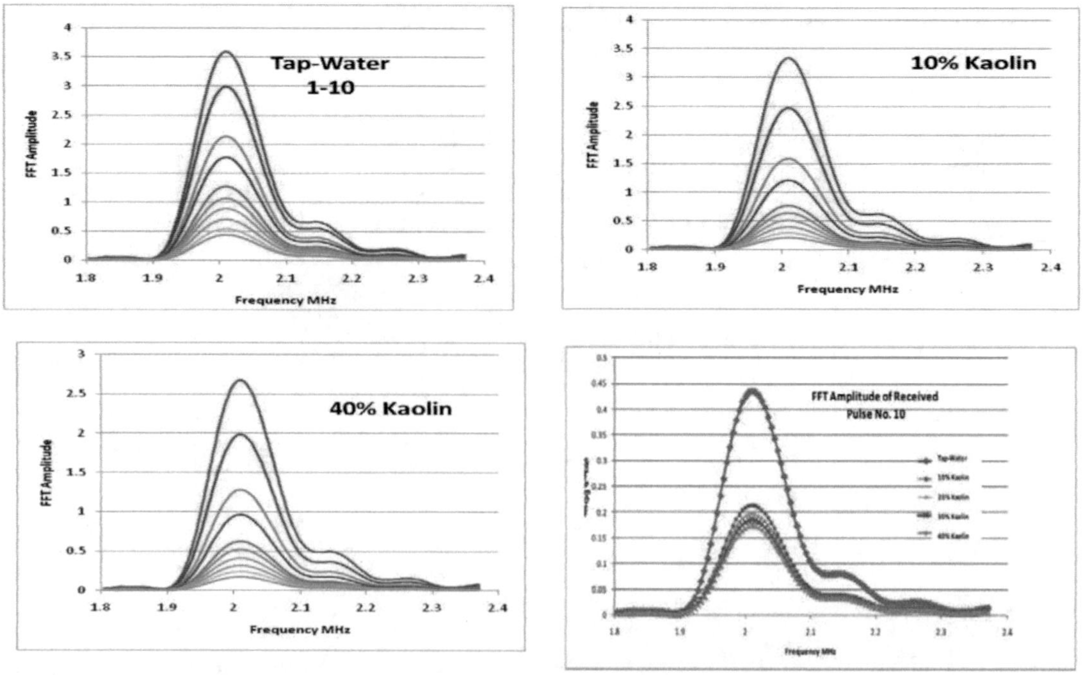

Figure 3. FFT amplitude for one transmitted set size 10, propagated through water, 10% kaolin, and 40% kaolin (a). FFT amplitude comparison of 10th interest signal received for water, 10%, 20%, 30%, and 40% (b).

Table I. One set size 10, FFT amplitude of IS for tap-water and 10% kaolin at 2MHz

Pulse No.	Tape-Water,FFT amplitude at 2MHz	10% Kaolin, FFT amplitude at 2MHz
1	3.51	3.262
3	2.089	1.558
5	1.243	0.758
7	0.863	0.505
10	0.425	0.208

Table II. The detection comprise between commercial and proposed technique for 1% kaolin.

Set Pulse No.	Amplitude of Commercial Pulser (Volt)	Amplitude of PPD Pulser (Volt)	Tap water FFT Amplitude of		1% Kaolin FFT amplitude of	
			Commercial Pulses	PPD Pulses	Commercial Pulses	PPD Pulses
1	75	75	3.51	3.51	3.496	3.494
5	75	55	3.51	1.243	3.496	1.05
10	75	30	3.51	0.425	3.496	0.281

2. PPD Ultrasound System Design

The pulse-power decay device (PPD) ultrasound has a new technique that generates multi-pulse transmissions. The PPD starts with constant amplitude train pulses (commercial pulsed ultrasound) until the set size and time interval of the decay-power pulses are set. A block diagram of the proposed ultrasonic pulser system is shown in figure 1. The PPD device consists of six hardware parts, panel control software, an I/O serial interface, the main controller (single board computer [SBC]), a high voltage power supply, a bipolar pulses generator, and a solid-state switch. The transducer is driven by pulses from a decay power generator, which delivers sets of high-voltage bipolar pulses. The advantage of using bipolar pulses is that the peak-to-peak pulse voltage can measure twice the voltage rating of the coaxial cable connecting the pulser and the transducer. The pulse frequency, decay power factor, set size, time interval, set mode and start amplitude are programmed

$$P_n = P_p - P_{df} \quad (1)$$

Where P_n *is* the next pulse amplitude, P_p is the previous pulse amplitude, and P_{df} is the decay power factor. The pulse transmission style can be easily modified via the control panel software. The PPD technique can follow the commercial pulsed ultrasound by setting both the time interval and decay power factor values to zero; therefore, the PPD device generates a sequential set of constant amplitude pulses.

3. Experimental Setup

The new PPD technique can be adaptive with wide application fields. In this research, the objective of the experiments was to evaluate the performance of PPD technique to gather ultrasound measurements for different percentages of solid concentrations in liquid. Invasive measurement is the most direct method of achieving an accurate measurement. Hence, this method was used for all the experiments carried out for this paper. Using the assumption that 1 ml of tap water equals 1gram in weight, kaolin was mixed with tap water in concentrations of 10%, 20%, 30% and 40% to create the different sets of slurries. These concentrations were measured by the weight of kaolin and tap water.

The vessel, shown schematically in figure 2, consists of a Plexiglas container and two transducers with

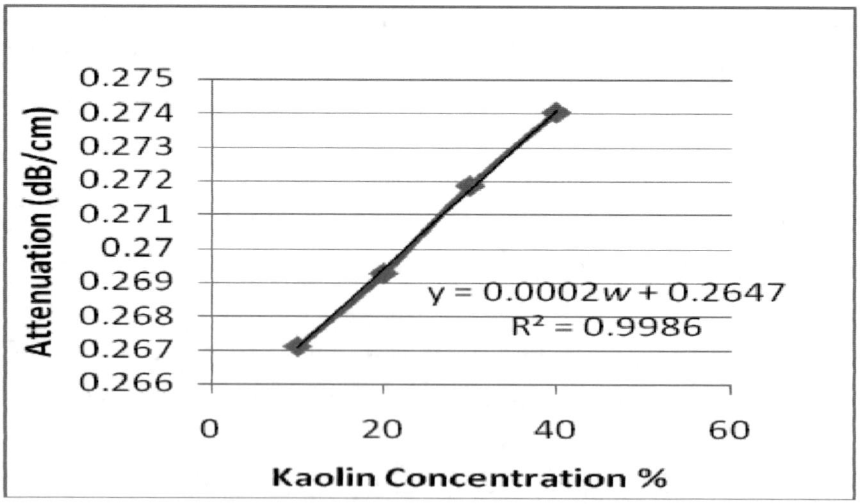

Figure 4. Relationship between attenuation measurement (dB/cm) and solid concentration wt%.

a centre frequency of 2 MHz affixed to the outside of the container on opposite sides, using a pitch-catch method. The vessel walls have a thickness w of 3.2 mm and the inside walls are separated by a distance D of 4 cm. The temperature, measured by temperature sensor (LM35), is recorded. For simplicity, the temperatures for all the experiments were kept at room temperature (20°C), between ±1°C [3]. When the vessel is filled with water, less than 10% of the ultrasound is reflected at the Plexiglas-water interface and the rest is transmitted into the water. At the opposite wall, 91% is reflected at the water-Plexiglas interface. With the pitch-catch mode operating, the receive transducer records only the first individual signal, as it corresponds to travelling through the slurry. The ''signal of interest'' (IS) occurs at the shortest time and the path length for the Nth IS (for example the interest received signal of the fifth pulse of set) in the pitch-catch mode is ($2w+D+v\times T$), where v is the speed of sound through material and T is the total time delay of Nth pulse; and calculated using the equation (2):

$$T=(N-1)(t_S+t) \qquad (2),$$

Where the t_S is the pulse spacing time and t is the bipolar pulse width.
To obtain attenuation measurements as a function of frequency, the fast Fourier transform (FFT) of the peak of data were corrected for the receiver gain. The results of the FFT analysis of the ''peak of interest'' are shown for water, 10%, and 40% kaolin (figure 3 (a)). The effects of attenuation for water are clearly observed as the peak amplitude shifts to a smaller frequency as the IS number increases.

4. Data analysis

Data similar to were obtained for the kaolin slurries of 10%, 20%, 30%, and 40% kaolin by weight. The attenuation due to the 10% kaolin slurry can be observed in Table 1 by the increasing separation between the FFT amplitudes for water and the slurry. The data for kaolin slurry wt10% concentration were compared with that for water by evaluating the natural logarithm of the ratio (FFT amplitude slurry/ FFT amplitude water) for a specified frequency (2 MHz). The data exhibit straight lines and the interest signal (IS) for each set pulse was obtained for water and for the slurry. The data were obtained in ten steps, using ap appropriate decay factor for the transmitted pulses and the receiver gains. Overlapping interest signals (IS) were used to determine the effect of changing the pulser voltage travelling through slurry, and the data were normalized to a pulser voltage of ±75 V. Additionally, the slopes (S) were obtained. The lower pulses-power such as 40, 35, 30 volts greatly affected by low concentrations of slurry than the higher pulse power for one transmitted set as shown in figure 3(b).for example, the received pulse number 10 has 0.425 volt while the amplitude of same pulse travelled through 10% kaolin was 0.208 volt (table 1); therefore, to detect low concentrations with a long range of measurement, the set size should be increased (i.e., decreasing the decay factor leads to incremental increases in the amount of pulsed-power decay). Decreasing the decay factor leads to an increase in the number of pulses in one transmitted set and causes more generated lower-power pulses; this increase is limited by the high voltage supply. Due to the attenuation of the ultrasound waves propagated through slurry, the relationship between kaolin concentration and the attenuation was determined. The attenuation (α) was calculated using equation 3, [4].

$$\alpha=\frac{-20}{D}\log_{10} e^S \qquad (3)$$

For a given frequency, a plot of the attenuation (α) versus the kaolin weight percentage (w) displays a straight line (figure 4), as expected. For example, when the values of the attenuation at 2 MHz are extracted from figure18, such a plot shows the following linear relationship: α = 0.0002w + 0.2647, where the slope has units of dB/cm wt%. Table 2 illustrates the differences between a commercial pulser and the PPD technique.

5. Discussion

In this study, we developed a self-contained prototype of ultrasonic decay pulser system with computer control for solid concentration content in fluid detection. Incorporating the new pulser technique and pitch-catch method, we demonstrated that the system could be used to distinguish the presence of lower than 5% concentration in fluid, while the current method cannot detect with the same device settings. The PPD ultrasound system, built in-house, generates a multi-pulsed powered decay train for transducer excitation instead of a single negative pulse, which is used in traditional pulsers. It also delivers variable transmitted ultrasound energy that would increase the received interest signal level for SNR improvement, and the concentration accuracy measurement was increased. The number of pulses per transmitted set is determined by the user based on the concentration detection level required. Our comparison results reveal that PPD ultrasound transmission through the fluid conversion for solid concentration detection is more efficient, especially for very low concentration (5%) of solid content in fluid than the currently pulsed ultrasound. Although the attenuation coefficient of the last numbered pulses of the transmitted set is higher than that of the first numbered pulses due to the decay power technique used (from the A-Scanner device) the concentration is very clearly effects by high coefficient container wall. This technique can be used to detect different concentration in fluid and can use in control process.

Conclusion

A PPD ultrasound system incorporating a pitch-catch technique was proposed and successfully demonstrated solid detection of 5% concentration in water. The presence of 5% concentration in fluid kaolin slurry was clearly distinguished with using decay waves propagation through water, but the currently pulsed ultrasound method failed to detect the same concentration with same device setting. The PPD technique is adaptive with all types of container material is made of, wide range of transducer frequency operation, and concentration measuring.

References

[1] M.S. Greenwood, J.D. Adamson, J.A. Bamberger. Long-path measurements of ultrasonic attenuation and velocity for very dilute slurries and liquids and detection of contaminates. Ultrasonic 44 (2006) e461–e466.

[2] Chen Q., S. Freear, D. M. J. Cowell. Measurement of Solid in Liquid Content Using Ultrasound Attenuation. 5th World Congress on Industrial Process Tomography, Bergen, Norway.

[3] Parker D., LEC R.M., PENDSE H.P., VETELINO J.F. (1990), Ultrasonic Sensor for the Characterisation of Colloidal Slurries, IEEE Ultrasonics Symposium, pp.295-2981.

[4] M. S. Salim, M. F. Abd Malek, R. B. W. Heng, Naseer Sabri Salim and K. M. Juni4. A new measurement me-thod of separation percentage for human blood plasma based on ultrasound Attenuation. International Journal of the Physical Sciences Vol. 6(30), pp. 6891 - 6898, 23 November, 2011.

Journal of Physics: Conference Series **423** (2013) 012063 doi:10.1088/1742-6596/423/1/012063

Path Loss Analysis of WSN Wave Propagation in Vegetation

Naseer Sabri[1], S A Aljunid[1], M S Salim[2], R Kamaruddin[3], R B Ahmad[1] , M F Malek[1]
[1] Computer and Communication Engineering School, UniMAP, Malaysia
[2] Mechatronics Engineering School, UniMAP, Malaysia
[3] Bioprocess Engineering School, UniMAP, Malaysia

E-mail: nasseersabri@yahoo.com

Abstract. Deployment of a successful wireless sensor network requires precise prediction models that provide a reliable communication links of wireless nodes. Prediction models fused with foliage models provide sensible parameters of wireless nodes separation distance, antenna height, and power transmission which affect the reliability and communication coverage of a network. This paper review the line of sight and the two ray propagation models combined with the most known foliage models that cover the propagation of wireless communications in vegetative environments, using IEEE 802.15.4 standard. Simulation of models is presented and the impacts of the communication parameters, environment and vegetation have been reported.

1. Introduction

Pervasive networks, represented by wireless sensor actor network, emerge the promises of automation the environment with sensors-actors wireless nodes which yielding a new era of environment automation applications [1] [2]. Currently, Wireless sensor network platform and applications are in an exponential growing. Initially, they were oriented for indoor applications such as, home automation and industrial control or for medical applications such as elderly monitoring [3]. Currently WSANs are adopted in a wide range of applications including telemedicine, home and building automation, surveillance systems and agricultural, avoiding the expensive retrofit necessary in wired systems. WSN nodes are highly constrained in terms of physical size, power consumption of CPU and transceiver, memory size and bandwidth. For remote nodes, the power consumption is considered a crucial parameter and have to care precisely where energy efficiency is paramount and batteries may have to last for years. Indeed, challenging physical environment like heat, dust, moisture, rain and interference [4].

WSN applications are growing exponentially and one of the most promising applications is the precision agriculture. Precision agriculture is one of the promising domains where wireless sensor networks could be exploited, for example, by observing the micro-climate within a field so that, ultimately, plant-specific farming can be realized. Agricultural application relies on the use of new era of WSAN to endorse monitoring and controlling management of a field based on site conditions [4]. WSNs offer a great flexibility when deploying new systems and when updating existing systems [5].

Since vegetation areas are cover a large portion of the Earth's surface, the propagation of radio waves in a forest has long been of researcher's interest. Radio waves propagating in vegetation usually experience much higher path loss than in environments without vegetation. Therefore, well known of the propagation mechanisms through a forest is critical for communication and sensing in such

Journal of Physics: Conference Series **423** (2013) 012063 doi:10.1088/1742-6596/423/1/012063

environments. However, WSN nodes are spatial distribution in the field and must consider all the parameters that may have effects the wireless channel communication. Currently, the most promising of WSN technology in agricultural field is ZigBee [6-8]. This protocol has been widely adopted by WSN developer's community; it relies on the IEEE 802.15.4 standard [9]. ZigBee technology is a low rate, low cost, low power consumption wireless node protocol aiming to remote and automation application systems.
ZigBee is expected to provide low power and cost connectivity for nodes that need long operation of battery of several years with low data rate. The IEEE-ZigBee are expected to transmit over 10-75 meters based on the RF power output consumption for specific application and the environment, and operates in the unlicensed RF worldwide 2.4 GHz with 250 Kbps data rate. To determine the behaviour of electromagnetic waves , a precise model of propagation must be adopted, however models normally used in wireless communication might not be precise describe the wireless sensor network. WSN node are spatially located, usually, near the earth surface thus may induce absence of main ray between sender-receiver nodes which is known of no line of sight (NLOS) status occur, although WSN nodes have spatially short distance distribution. Therefore, WSN propagation waves may face obstacle like trees, fence, building and dense foliage.

2. Radio Wave Propagation

The wireless radio channel faces a severe challenge as a medium for reliable high speed communication. It is susceptible to interference, noise, and other channel impediments which may change over time in unpredictable ways due to nodes-user movement. Path loss is caused by dissipation of the power radiated by the transmitter as well as effects of the propagation channel. Generally, path loss models propose that path loss is same for a given distance between sources of transmit and receive. Shadowing is caused by obstacles between the transmitter and receiver which will attenuate the signal power through reflection, diffraction, absorption, and scattering. For high level of attenuation, the signal will be blocked. Variation due to path loss occurs over very large distances (100-1000 meters), whereas variation due to shadowing occurs over distances proportional to the length of the obstructing object (10-100 meters in outdoor environments and less in indoor environments) [10, 11]. Such an environmental parameters not foreseen by developers and neither considered by theoretical models, and not accounted by simulators [11, 12] are affecting the wireless communication link. This is especially true for an arable farming environment in which foliage, growing crops, and ever changing weather conditions have an unknown effect on the exact propagation of the radio waves. The wireless channel is modelled based on several key parameters. These parameters vary significantly with the environment, rural versus urban, or flat versus mountainous. Different kinds of fading can be categorized in three types [13] [14]:

Distance Dependence of path loss (measured in *dB*) is approximated by $L(d) = L_0 + 10n \times \log(d/d0)$, where n is the path loss exponent, which varies with terrain and environment, and L_0 is the path loss at an arbitrary reference distance d_0.
Large-scale Shadowing *causes* variations over larger areas, and is caused by terrain, building, and foliage obstructions; the large-scale fading due to various obstacles is commonly accepted to follow a log-normal distribution [15]). This means that its attenuation x measured in *dB* is normally distributed $N(m, \sigma)$, with mean m and standard deviation σ. The probability density function of x is given by the usual Gaussian formula.
Small-scale fading causes great variation within a half wavelength. It is caused by multipath and moving scatters. Resulting fades are usually approximated by Rayleigh, Rican, or similar fading statistics – measurements also show good fit to *Nakagami*-m and *Weibull* distributions.

Radio systems rely on diversity, equalizing, channel coding, and interleaving schemes to mitigate its impact. There are different propagation characteristic of different spectrum bands which require different prediction models. Some propagation models are well suited for computer simulation in

presence of detailed terrain and building data; others aim at providing simpler general path loss estimates [14, 15]. Wireless sensor network of mobile nodes are burdened with particular propagation complications, making reliable wireless communication more difficult than fixed communication between and carefully positioned wireless nodes. The antenna height at a mobile terminal is usually very small, typically less than a few meters. Hence, there is very little clearness, so obstacles and reflecting surfaces in the vicinity of the antenna have a substantial influence on the characteristics of the propagation path. Moreover, the propagation characteristics change from place to place and, if the terminal moves, from time to time.
Path loss models define the attenuation of a signal as a function of the propagation distance between a transmit and a receive antennas and other parameters Some models include many details of the terrain profile to estimate the path loss, while others just consider the signal frequency and the separation distance. Antenna heights are other critical parameters [13].

2.1. *Free Space Propagation*
The free space propagation model assumes the influence of the earth surface to be entirely absent. This model assumes that the transmitter and receiver antennas to be located in an otherwise empty environment. Neither absorbing obstacles nor reflecting surfaces are considered [13][14]. The radiated energy is spread over the surface of a sphere of radius d, so as the propagation distance increases, the power received decreases proportional to d^{-2}. The received power is

$$P_{dB} = P_o - 20\log_{10}(d/d_o) \qquad (1)$$

Hence, the free space propagation model assumes ideal propagation condition that there is only one clear line-of-sight path between the transmitter and receiver. A physical argument of conservation of energy leads to the Friis' power transmission formula in free space. A transmitted power source P_t radiates spherically, with an antenna gain G_t; the power density W at distance d from a transmitter with power P_t and antenna gain G_t is

$$W = P_T G_t / 4\pi d^2 \qquad (2)$$

The portion of that power impinging an effective area A_e, hence at a distance d, the power is $P_r=P_tG_tA_e / (4\pi d^2)$. The effective area of an antenna is related to antenna gain by $A_e/\lambda^2=G_t/4\pi$, which is used for the receiving antenna, and thus yields [13, 14]:

$$P_r = P_t G_t G_r (\lambda / 4\pi d)^2 \qquad (3)$$

P_t, G_t and P_r, G_r are the transmitted and received power and gain respectively, λ is the wavelength of the signal, and d is the separation distance. The product G_tP_t is called the effectively radiated power (ERP) of the transmitter. The path loss reflects how much power is dissipated between transceiver and receiver antennas (without counting any antenna gain). Path loss is often expressed as a function of frequency (*f*), distance (*d*), and a scaling constant that contains all other factors of the formula. Equation 3 shows a free-space dependence in $1/d^2$, and is sometimes loss is expressed in decibels (dB): $L(dB) = 10 \times log(P_T/P_r)$. For instance, when $f_o = 1MHz$, and $d_o = 1km$, the loss expressed as;

$$L(dB) = 32.44 + 20\log_{10}(f/f_o) + 20\log_{10}(d/d_o) \qquad (4)$$

The free space model basically represents the communication range as a circle around the transmitter. If a receiver is within the circle, it receives all packets. Otherwise, it loses all packets.

2.2. *Two-ray ground reflection model*
Ray tracing is a method that uses a geometric approach, and examines what paths the wireless radio signal takes from transmitter to receiver as if each path was a ray of light (possibly reflecting off surfaces). Ray-tracing predictions are good when detailed information of the area is available. But the

predicted results may not be applicable to other locations, thus making these models site-specific.
The well-known two-ray model uses the basic FSP model with a function that combine the reflected signal which shows a fading effects on received signal. The fact that for most wireless propagation cases, two paths exist from transmitter to receiver: a direct path and a reflected off the ground. That model alone shows some important variations of the received signal with distance [13]. The two-ray ground reflection model considers both the direct path and a ground reflection path. This model gives more accurate prediction at a long distance than the FSM as shown [13, 14]. The received power at distance d is predicted by

$$P_r = \frac{G_T G_r P_t \lambda^2}{(4\pi D)^2}[2\sin\frac{2\pi}{\lambda}\frac{h_t h_r}{d}]^2 \qquad (5)$$

$$d_{th} = 4\pi h_t h_r / \lambda \qquad (6)$$

Where ht and hr are the heights of the transmitter and receiver antennas respectively, d_{th} is defined as a turnover point where the argument of the sine tends to be equal to $\pi/4$. For example, carrier frequencies of *2400 MHz,* a base station height of 4 meter and a fixed wireless node of antenna height of 1 meter, the turnover distance is about 128 meters as shown in figure 1. Hence for wireless sensor network nodes, distances less than the turnover distance are more relevant. The direct path (using the first term only of equation (5)) leads to the simple on-slope free-space model; the complete expression leads to the two-ray model, which shows interesting characteristics:

- The presence of a second ray causes great variations: signals can add up or nearly cancel each other, causing deep fades over small distances.
- In close proximity, the overall envelope of power decay varies in $1/d^2$.
- After a certain cutoff distance (approximately $4h_t h_r/\lambda$) the model approaches power decay in $1/d^4$.

The full path loss expression shows an interference pattern of the line-of-sight and the ground-reflected wave for relatively short ranges, and a rapid decay of the signal power beyond the turnover distance. It can be shown that for distances substantially greater than turnover point d, $d >> (h_t h_r)^{0.5}$, the reflection coefficient tends to -1, path loss tends to the fourth power distance law (7) and in dB, we have (8):

$$P_r = \frac{P_t G_T G_r h_t^2 h_r^2}{d^4} \qquad (7)$$

$$P_r(dBm) = P_t(dBm) + 10\log_{10}(GT) + 20\log_{10}(h_t h_r) - 40\log_{10}(d) \qquad (8)$$

The above equation shows a faster power loss than equation (3) as distance increases. However, the two-ray model does not give a good result for a short distance due to the oscillation caused by the constructive and destructive combination of the two rays. Instead, the free space model is still used when d is small. Therefore, a turnover distance d_{th} is calculated in this model. When $d<d_{th}$, equation (3) is used. When $d>d_{th}$, equation (7) is used. At the cross-over distance, equation (3) and equation (7) give the same result. The power fall off with distance in the two-ray model can be approximated by averaging out its local maxima and minima. This results in a piecewise linear model with three segments, which is also shown in figure 2, is slightly offset from the actual power falloff curve for illustration purposes. In the first segment, the power fall off is constant and proportional to $1/(d^2+h_t^2)$, for distances between ht and d_{th} power falls off at *-20 dB/decade*, and at distances greater than d_{th} power falls off at *-40 dB/decade*.

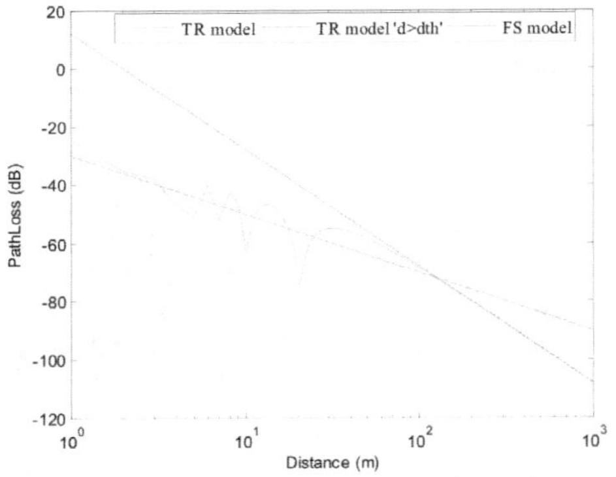

Figure 1. Two ray propagation where transmit antenna is at 4m and receive antenna is at 1m

2.3. Shadowing

In real environment, the received power at certain distance is a random variable due to multipath propagation effects, which is also known as fading effects while the FSP model and the TRP model predict the received power as a deterministic function of distance. They both represent the communication range as an ideal circle. In fact, the above two models predicts the mean received power at distance *d*. A more general and widely-used model is called the shadowing model [13]. The shadowing model is based on two parts. The first part is known as path loss model, which is predicts the mean received power at distance *d*, denoted by $\overline{P_{r(d)}}$.

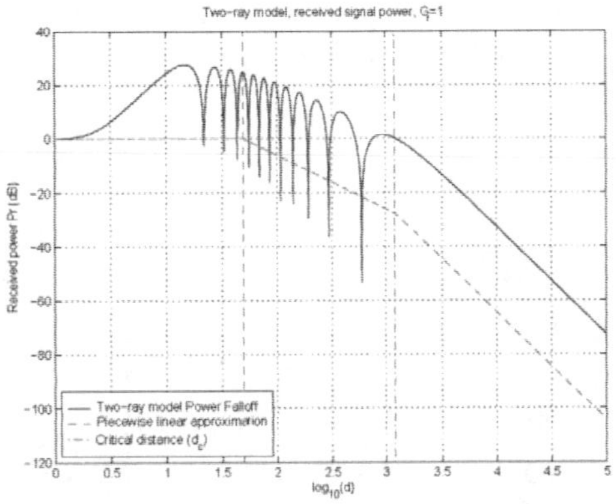

Figure 2. Received Power versus Distance for Two-Ray Model

It uses a close-in distance do as a reference. $\overline{p_r(d)}$ is computed relative to $P_r(d_o)$ as follows;

$$\frac{P_r(d)}{P_r(d_o)} = \left(\frac{d}{d_o}\right)^n \qquad (9)$$

α is called the path loss exponent, and is usually empirically determined by field measurement. From equation (3) we know that $\alpha = 2$ for free space propagation. $P_r(d_o)$ can be computed from (3).The path loss in *dB* is;

$$\left[\frac{P_r(d)}{P_r(d_o)}\right]_{dB} = -10\alpha\log_{10}\left(\frac{d}{d_o}\right) + \sigma_{dB} \qquad (10)$$

The second part of the shadowing model reflects the variation of the received power at certain distance. It is a log-normal random variable, that is, it is of Gaussian distribution if measured in dB. The overall shadowing model is represented by;

$$P_r(d)_{dB} = P_r(d_o) - 10\alpha \log_{10}\left(\frac{d}{d_o}\right) + \sigma_{dB} \qquad (11)$$

Where σ_{dB} is a Gaussian random variable with zero mean and standard deviation. It is called the shadowing deviation, and is also obtained by measurement. Equation 11 is also known as a log-normal shadowing model. The shadowing model extends the ideal circle model to a richer statistic model: nodes can only probabilistically communicate when near the edge of the communication range.

3. Foliage Loss

Research on propagation loss due to foliage has focused mainly on forests [16, 17]. There are several models to evaluate the path loss of a propagation path due to the presence of foliage. Most of these models are based on working frequency (*f*) and foliage depth (d_f) which is represented by the expression of equation (12) [18]. The excess path loss L_{Veg} due to foliage is based on the frequency of the travelled signal f and the depth of the foliage along this distance, while *A, b, c* are constant empirically determined.

$$L_{Veg} = A \times f^b \times d^c \qquad (12)$$

The Weissberger MED (Modified Exponential Decay) model [18]; Equation 13 expresses the excess path loss due to foliage, in dB,

$$L(dB) = \begin{cases} 1.33 \times d_f^{0.588} \times f^{0.284} \cdots\cdots 14 \le d_f \le 400 \\ 0.45 \times d_f \times f^{0.284} \cdots\cdots 0 < d_f < 14 \end{cases} \qquad (13)$$

where f is in GHz and d_f is in meter ; There are other propagation models that are in the form of (4), Based on same parameters and units as the MED model are: Early ITU model, equation (14), Fitted ITU-R model, equation (15);[19-22].

$$L(dB) = 0.2 f^{0.3} d_f^{0.6} \qquad (14)$$

$$L(dB) = 0.39 f^{0.39} d^{0.25} \qquad (15)$$

For single vegetation obstruction model [22], the foliage influence in dB is expressed as a function of distance d in meter and the specific attenuation for short vegetative paths in dB/m in equation (16),

$$SV = d\gamma \qquad (16)$$

4. WSN Wave Propagation Based on Agricultural Applications

WSN applications require RF or microwave propagation from point to point very near the earth's surface and in the presence of various impairments propagation losses over terrain, foliage, and/or buildings may be attributed to various phenomena, including diffraction, reflection, absorption, or scattering. As a result the received signal power is attenuated and hence the communication range will

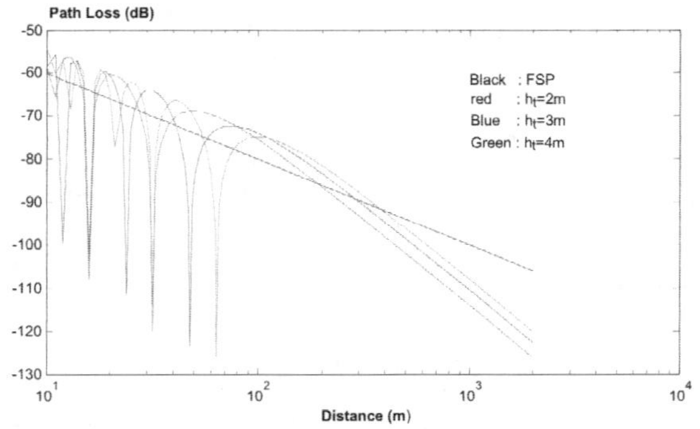

Figure 3. Path loss versus distance for various antenna heights for 2.4GHz, Hr=1m.

be limited. When WSN deployed, almost the distance between the wireless nodes is small, so the earth curvature could be neglected [21]. The wireless nodes antenna height within agricultural is less compared to the transmission range of WSN hardware. In this case the TR model [13] can be applied. Equation 7, which reads in logarithmic form, for isotropic antennas where the propagation loss follows the inverse fourth law with distance. The received power falls by *12dB* when the distance is doubled. Therefore, for short range near ground vegetation environment, the total loss of propagation is modelled by fusion of the foliage imposed effect and the reflections effect of the radio wave bounce off the ground. Hence The total accumulated path loss (TAP_{Loss})is expressed as the summation of the path loss (FSP_{Loss}/TRP_{Loss}) due to wave spreading and the foliage excess loss (Veg_{Loss}) in the propagation path, equation (17);

$$TAP_{LOSS} = FSP_{LOSS} + VEG_{LOSS} \qquad (17)$$

Figure 3 shows that the loss follow $1/d^2$ for first 100m for various antenna height 4, 3, 2 with receiver antenna of 1m height. Thus, WSN applications which uses like these heights for nodes antennas will suffer the FSP model attenuation which is approximately equal to the average of maxim-minim fluctuation of the TR model for fist hundred meters. Figure 4 shows comparisons of the Weissberger and ITU models, for foliage depths of *2* and *4m*. Note that the frequency scale is in GHz for each plot, but the frequency used in the model is MHz for the ITU model as specified. The plots indicate a moderate variation between the models, particularly as frequency increases. The amount of foliage loss is monotonically increasing with foliage depth and frequency as expected based on equations (13) and (14), where the MED model yields of less losses. Therefore, modelling the path losses based on foliage models yield different losses, hence those models can be used only when their results are validate with an experimental work to find the right model that reflect the losses of the application.

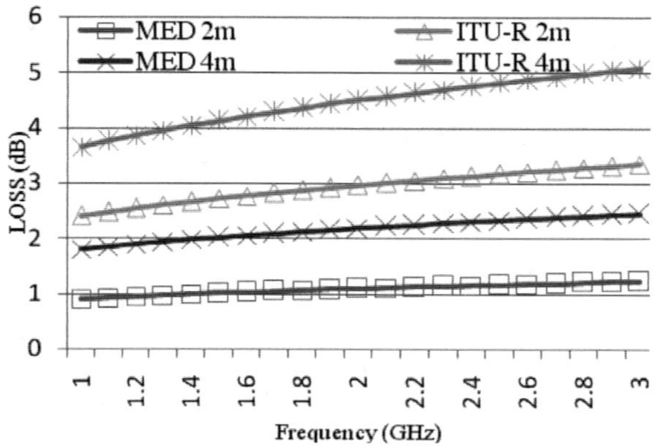

Figure 4. Vegetation loss versus frequency for 2 and 4-m foliage depth

5. CONCLUSIONS

Path loss prediction model is very important step in designing a WSN, the precise prediction models are needed to determine the parameters that affects the radio waves and could lead to efficient and reliable coverage of a specified service area. The effect of the foliage in the propagation of WSN waves can be predicted using propagation models. Thus, an appropriate propagation model based on the environment condition and the WSN application should be used.

This paper review FSP model and the TRP model, and introduce the fusion of propagation models with foliage models to predict the path loss within vegetation area. Using of TRP model depend on dc, so either reception power decay proportional to inverse of d^2 for distances less than d_{th}, or proportional to inverse of d^4. The separation distance of WSN nodes are almost around tens of meters and antenna height range from few centimetres above ground to few meters based on the application requirement. Thus, TRP model is used for path loss prediction for application need like different height of antennas and near ground propagation environment, therefore, it is highly occurs that a line of sight combined with a reflected components received at the receiver antenna. Different propagation models have different results and this yield to prior careful study of path modelling for WSN applications.

WSN requires a proper spatial node distribution strategy to achieve high communication reliability between sensor nodes. Therefore, a precise path loss propagation model should be adopted to find antenna heights and path loss within the limits of network connectivity and coverage range.

References

[1] R. Morais, M. A. Fernandes, S. G. Matos, C. Serˆodio, P. Ferreira, and M. Reis, "*A zigbee multi-powered wireless acquisition device for remote sensing applications in precision viticulture,*" Computers and Electronics in Agriculture, vol. **62**, no. 2, pp. 94 – 106, 2008.

[2] D. Estrin, D. Culler, K. Pister, and G. Sukhatme, "*Connecting the physical world with parvasive networks,*" Pervasive Computing, January- March, 2002.

[3] N.K. Suryadevara, A. Gaddam, R.K. Rayudu, S.C. Mukhopadhyay 2012. "*Wireless SensorsNetwork Based Safe Home to Care Elderly People: Behaviour Detection* ". Procedia Engineering Volume **25**, 2011, Pages 96–99.

[4] Naseer Sabri, S.A. Aljunid M.F. Malek, A.M. Badlishah, M.S. Salim, R. kamaruddin. "*Wireless Sensor Actor Networks Application for Agricultural Environment,*" IEEE Symposium on Wireless Technology and Applications(ISWTA), 2011.

[5] C. Serodio, J. B. Cunha, R. Morais, C. Couto, and J. Monteiro, "*A networked platform for agricultural management systems,*" Computers and Electronics in Agriculture, vol. **31**, no. 1, pp. 75–90, March 2001.

[6] ZigBee Alliance, ZigBee Specification, v1.0, 2006A.

[7] Camilli, C. E. Cugnasca, A. M. Saraiva, A. R. Hirakawa, and P. L. Correa, "*From wireless sensors to field mapping: Anatomy of an application for precision agriculture,*" Computers and Electronics in Agriculture, vol. **58,** no. 1, pp. 25–36, 2007.
[8] N.Wang, N. Zhang, and M.Wang, "*Wireless sensors in agriculture and food industry–recent development and future perspective,*" Computers and Electronics in Agriculture, vol. **50**, no. 1, pp. 1–14, January 2006.
[9] IEEE standard 802.15.4 – Wireless Medium Access Control (MAC) and Physical Layer (PHY) Specifications for Low-Rate. Wireless Personal Area Networks (LR-WPANs), IEEE, 2003.
[10] T. Sarkar, J. Zhong, K. Kyungjung, A. Medouri, and M. Salazar-Palma, "*A survey of various propagation models for mobile communication,*" IEEE Antennas and Propagation Magazine, vol. **45**, no. 2, pp. 51–82, June 2003.
[11] T. J. Andersen, T. Rappaport, and S. Yoshida, "*Propagation Measurements and Models for Wireless Communications Channels,*" IEEE Communications Magazine, vol. **33**, no. 1, pp. 42–49, January 1995.
[12] B. Sklar, Digital Communications Fundamentals andApplications, 2nd ed., Prentice- Hall, 2000
[13] River, NJ, 2001, pp. 286–290.P. Mestre, C. Serodio, R. Morais, J. Azevedo, and P. Melo-Pinto, "*Vegetation growth detection using wireless sensor networks,*" Proceedings of The World Congress on Engineering 2010, WCE 2010 Lecture Notes in Engineering and Computer Science, vol. I, 2010, pp. 802–807.
[14] T. Rappaport, Wireless Communications: Principles and Practice, 2nd ed., Upper Saddle River, N.J.: Prentice Hall, 2002.
[15] Goldsmith, Wireless Communications, New York: Cambridge University Press, 2005.
[16] H.L. Bertoni, Radio Propagation for Modern Wireless Systems, (Upper Saddle River, NJ: Prentice-Hall Inc., 2000).
[17] Y.S.Meng, Y.H.Lee, and B.C.Ng, "*Path loss modeling for near-ground vhf radio-wave propagation through forests with tree-canopy reflection effect,*" Progress In Electromagnetics Research M, vol. **12**, pp. 131–141, 2010.
[18] K. Sarabandi and I. Koh, "*Effect of canopy-air interface roughness on hf-vhf wave propagation in forest,*" IEEE Transactions on Antennas and Propagation, Vol. **50**, No. 2, February, 2002.
[19] M. A. Weissberger, "*An Initial Critical Summary of Models for Predicting the Attenuation of Radio Waves by Trees,*" Electromagnetic Compatibility Analysis Center, Annapolis MD, Final report, 1982.
[20] CCIR, "Influences *of terrain irregularities and vegetation on tropospheric propagation,*" tech. rep., CCIR Report, July 1986.
[21] J. D. Parsons, The Mobile Radio Propagation Channel, 2nd ed.,Wiley,West Sussex, 2000
[22] ITU-R Recommendations, Attenuation in vegetation, ITU-R P.833-3

Toward Optical Sensors: Review and Applications

Naseer Sabri[1], S A Aljunid[1], M S Salim[2], R B Ahmad[1], R Kamaruddin[3]
[1] Computer and Communication Engineering School, UniMAP, Malaysia
[2] Mechatronics Engineering School, UniMAP, Malaysia
[3] Bioprocess Engineering School, UniMAP, Malaysia

E-mail: nasseersabri@yahoo.com

Abstract. Recent advances in fiber optics (FOs) and the numerous advantages of light over electronic systems have boosted the utility and demand for optical sensors in various military, industry and social fields. Environmental and atmospheric monitoring, earth and space sciences, industrial chemical processing and biotechnology, law enforcement, digital imaging, scanning, and printing are exemplars of them. The ubiquity of photonic technologies could drive down prices which reduced the cost of optical fibers and lasers. Fiber optic sensors (FOSs) offer a wide spectrum of advantages over traditional sensing systems, such as small size and longer lifetime. Immunity to electromagnetic interference, amenability to multiplexing, and high sensitivity make FOs the sensor technology of choice in several fields, including the healthcare and aerospace sectors. FOSs show reliable and rigid sensing tasks over conventional electrical and electronic sensors. This paper presents an executive review of optical fiber sensors and the most beneficial applications.

1. Introduction

The field of fiber optics has undergone tremendous growth and advancement over the last 25 years. The fiber optic communication industry has literally revolutionized the telecommunication industry by providing higher performance, more reliable telecommunication links with ever decreasing bandwidth cost. This revolution is bringing about the benefits of high volume production to component users and a true information superhighway built of glass. Developments fiber optic sensor [1] technology has been a major user of technology associated with the optoelectronic and fiber optic communication industry. Since the past decades, various devices, including fiber optic gyroscopes; sensors of temperature, pressure, and vibration; and chemical probes has been under development of Fiber optic sensors (FOS) technology. FOS offer a number of advantages, such as increased sensitivity compared to existing techniques and geometric versatility, which permits into arbitrary shapes [2].

The ability of fiber optic sensors has been enhanced to substitute traditional sensors for acoustics, vibration, electric and magnetic field measurement, acceleration, rotation, temperature, pressure, linear and angular position, strain, humidity, viscosity, chemical measurements and a host of other sensor applications. They can be used in high voltage, high temperature, or corrosive environments due to its dielectric property; In addition, these sensors are compatible with communications systems and have the capacity to carry out remote sensing. Recently, investigation in the field has focused on the development of new materials with non-linear optical properties for important potential applications in photonics. Examples of these materials are the conjugated semiconducting polymers that combine optical properties with the electronic properties of semiconductors. In addition, these conducting polymers have photo luminescent and electroluminescent properties, making them attractive for applications in optoelectronics [3].

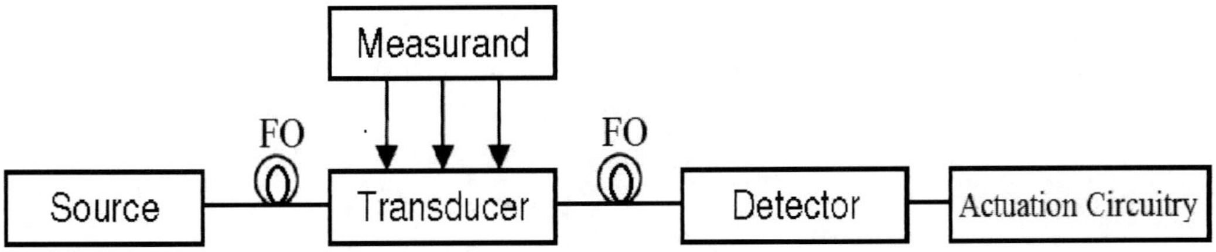

Figure 1. Optical fiber sensor system basic components.

Fiber optic sensor technology in turn has often been driven by the development and subsequent mass production of components to support these industries. The inherent advantages of fiber optic sensors which include their ability to be lightweight, of very small size, passive, low attenuation, and low power, immunity to electromagnetic interference (EMI), high sensitivity, wide bandwidth and environmental ruggedness were heavily used to offset their major disadvantages of high cost and unfamiliarity to the end user[4][5].

2. Optical Fiber Sensors Advantages Over Traditional Sensors

The advances in research and development of FOS devices has extended their applications to various fields of technology, such as medical, chemical, and telecommunications industries. They have been developed to work on a wide variety of physical properties, like temperature, chemical changes, electric and magnetic fields, vibrations, strain, displacement (position), flow, pressure, rotation, radiation, liquid level, light intensity, and colour. For performance in harsh environments, FOS proves reliable and rigid sensing devices over conventional electrical and electronic sensors where they have difficulties [1][6][7]. Contrary, optical fiber sensors exhibit a number of advantages over other types of sensors; they are;

- Enable small sensor sizes
- Are non-electrical devices
- Offer sensitivity to multiple environmental parameters
- Require small cable sizes and weights
- Permit remote sensing
- Allow access into normally inaccessible areas
- Often do not require contact
- Immune to radio frequency interference (RFI) and electromagnetic interference (EMI)
- Do not contaminate their surroundings and are not subject to corrosion
- Provide high sensitivity, resolution and dynamic range
- Can be interfaced with data communication systems

3. Fiber Optic Sensor Principles

The general structure of an optical fiber sensor system is shown in figure 1 which it composed of mainly of four components that are source, fiber optic, transducer and detector. Figure 2 depicts these component in more details, an optical source (Laser, LED, Laser diode etc), optical fiber (single or multimode), sensing or modulator element (which transducers the measure and to an optical signal), an optical detector and actuating circuitry (processing electronics, oscilloscope, optical spectrum analyzer etc). In the FOSs devices, the optical parameters that can be modulated are the amplitude, phase, colour (spectral signal), and state of polarization [6][8]. The optical modulation methods of the sensors involve the following:

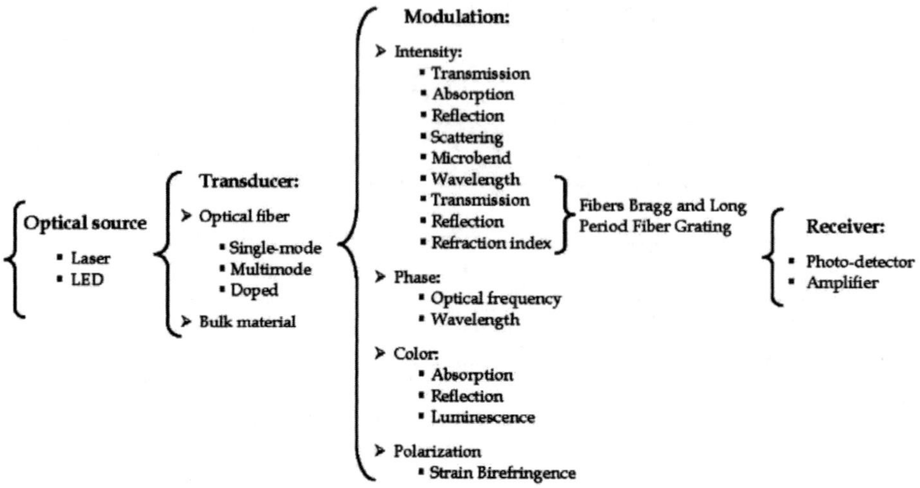

Figure 2. Basic components of an optical fiber sensor

Intensity-modulated sensors: The variation of the light intensity that is proportional to the perturbing environment can be detected by sensors. The concepts associated with intensity modulation include transmission, reflection, and micro bending. For this, a reflective or transmissive target can be incorporated in the fiber. Other mechanisms that can be used independently or in conjunction with the three primary concepts include absorption, scattering, fluorescence, and polarization. Intensity modulated sensors normally require more light to function than phase modulated sensors; as a result, they employ large core multimode fibers or bundles of fibers.

Phase-modulated sensors: The phase of the light in a sensing fiber is compares to a reference fiber in a device known as an interferometer. These sensors employ a coherent laser light source and two single-mode fibers. The light is split and injected into the reference and sensing fibers. If the light in the sensing fiber is exposed to the perturbing environment, a phase shift occurs between them. The phase shift is detected by the interferometer. Phase modulated sensors are much more accurate than intensity-modulated sensors. The change in colour is proportional to the changes in the absorption, transmission, reflection, or luminescence of the optical signal, whereas the polarization is related to the strain birefringence.

4. Intensity Based Fiber Optic Sensors

Fiber optic sensors are often loosely grouped into two basic classes according to the manner in which the optical fiber is used, sensor can be either an intrinsic one, if the modulation takes place directly in the fiber, or extrinsic, if the modulation is performed by some external transducer [4][6][7].

Intrinsic fiber optic sensor (IFOS): the intensity, phase, polarization, wavelength or transit time of light can be modulated. Sensors which modulate light intensity tend to use mainly multimode fibers, while only mono mode cables are used to modulate other parameters. An interesting feature of IFOSs is that they can provide distributed sensing over distances of up to 1 meter. In Intrinsic sensors, the variable of interest (physical perturbation) must modify the characteristics of the optical fiber to modify the properties of the light carried by the fiber, figure 3 (Intrinsic).

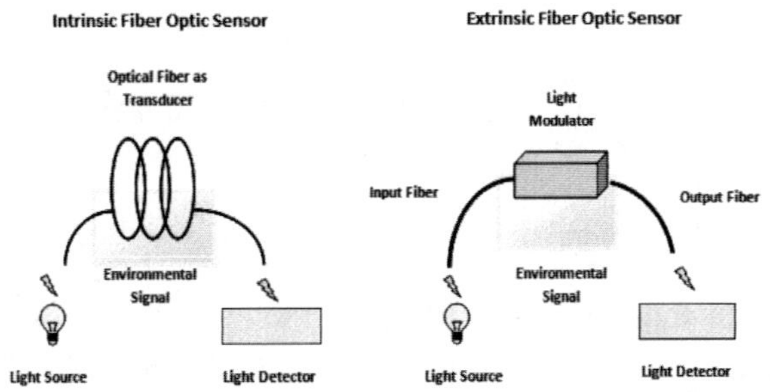

Figure 3. Types of FOSs. Intrinsic devices and Extrinsic devices.

Intrinsic FOSs directly employs an optical fiber as the sensitive material, sensor head, and also as the medium to transport the optical signal with information of the perturbation environment to be measured. They operate through the direct modulation of the light guided into the optical fiber. The light does not leave the fiber, except at the detection end, the output, of the sensor. These sensors can use interferometric configurations, Fiber Bragg Grating (FBG), Long Period Fiber Grating (LPFG), or special fibers (doped fibers) designed to be sensitive to specific perturbations. Light intensity is the simplest parameter to manipulate in intrinsic sensors because only a simple source and detector are required.

Extrinsic fiber optic sensor (EFOS): the optical fiber is simply used to guide the light to and from a location at which an optical sensor head is located. The sensor head is external to the optical fiber and is usually based on miniature optical components, which are designed to modulate the properties of light in response to changes in the environment with respect to physical perturbations of interest figure 3 (Extrinsic). Thus, in this configuration, one fiber transmits optical energy to the sensor head. Then this light is appropriately modulated and is coupled back via a second fiber, which guides it to the optical detector. Extrinsic fiber optic sensors use a fiber optic cable, normally a multimode one, to transmit modulated light from a conventional sensor. A major feature of extrinsic sensors, which makes them so useful in such a large number of applications, is their ability to reach places which are otherwise inaccessible. For example, the insertion of fiber optic cables into the jet engines of aircraft to measure temperature by transmitting radiation into a radiation pyrometer located remotely from the engine. Fiber optic cable can be used in the same way to measure the internal temperature of electrical transformers, where the extreme electromagnetic fields present make other measurement techniques impossible. Extrinsic fiber optic sensors provide excellent protection of measurement signals against noise corruption. An executive comparison of fiber optic sensors classes are shown in table 1.

Table 1. Comparison of the two types of optical sensors

	Extrinsic	**Intrinsic**
Applications	Temperature, Pressure, Liquid level and Flow	Rotation, Acceleration, Strain, Acoustic pressure and Vibration
Sensitivity	Less sensitive	More sensitive
Multiplexed	Easily	Tougher to multiplex
Connection problems	Ingress/ egress	Reduces
Cost	Less expensive	More expensive
Flexibility	Easier to use	More elaborate signal demodulation

5. Optical Sensor Types

Fiber optical sensors are sensors that can work in harsh environments where conventional electronic and electrical sensors have difficulties. They can sense various physical properties, such as pressure, position, strain, chemical changes, magnetic and electric fields, flow, vibration, light level, radiation and colour [3][6][9].

- Chemical sensors; remote spectroscopy, groundwater and soil contamination.
- Temperature sensors; largest commercially available sensors, range -40 °C to 1000 °C.
- Strain sensors; fiber Bragg gratings (FBG) technology, senses as little as 9 micro strain.
- Biomedical sensors; spectroscopic biomedical sensors, CO_2, O_2 and PH can be measured simultaneously, flow monitoring by laser dopplerimetry
- Electrical and magnetic sensors; appealing inherent dielectric nature, less sensitive to electromagnetic interference, small size and safer, they are almost always hybrid.
- Rotation sensor; based on the sagnac effect, two types ring laser gyroscope (RLG) and fiber optic gyroscope (FOG).
- Pressure sensors; based on piezoresistive technique or movable diaphragm, high performance (polarization based sensors); operating pressure ranges from 0-70,000 torr.
- Displacement and position sensors; one of the first optoelectronic sensors to be developed, simple sensors rely on the change in retro reflectance due to a proximal mirror surface, also referred as liquid level sensors.

6. Optical Fiber Sensing Applications

For harsh environmental condition and sensing over long distances, Optical sensing is consider as the perfect solution for applications where conventional electrical sensors such as foil strain gages, thermocouples, and vibrating wires have proven ineffective or difficult to use due to those conditions. The features and benefits of fiber Bragg grating (FBG) optical sensing include the nonconductive, electrically passive, immune to electromagnetic interference (EMI), induced noise, sensor measurements over very long distances (10+ km), daisy chain multiple sensors on a single fiber. Large and lightweight structures can benefit from the distributed single fiber architecture to simplify installation and reduce weight [10]. The nonconductive and noncorrosive nature of the fiber benefits outdoor and industrial applications where hazardous gases and voltages might be present. Also, the immunity to EMI removes the need for expensive and often difficult signal conditioning required for measurements near noisy sources such as power transformers. FBG optical sensing can benefit many applications in areas such as energy, civil infrastructure, and transportation monitoring [6] [10].

Medical field; Optical-chemical and biochemical sensing is being researched extensively all over the world, and these sensors are finding an increasing number of applications in industry, environmental monitoring, medicine, biomedicine, and chemical analysis [2]. The main physical phenomena exploited for optical chemical sensing are absorption and fluorescence, although chemical luminescence, Raman scattering, and Plasmon resonance have also been used. Health-care is unquestionably the application field which seems to have the most potential for future development [11][12]. Optical biosensors are finding ever-increasing applications in different branches of medicine for several reasons.

Energy field; FBG optical sensing has unique attributes that make it suitable for traditionally difficult applications. Monitoring structures that generate, produce, distribute, and convert electrical power introduces many challenges that can be addressed with FBG optical sensing. Whether it is a windmill requiring a lightweight solution or a hydroelectric turbine needing an EMI-resistant system, optical sensing has unique feature that match perfectly with these traditionally difficult applications. Partial discharge detection uses optical fiber sensors optical fiber sensors are being tested for use in detecting partial discharges in electrical transformers. Pinpointing such discharges is essential to preventing insulation breakdown and catastrophic failures. For example, monitoring the structural integrity of a wind turbine blade with electrical sensors would often result in noisy measurements because of long copper lead wires. With optical sensing, accurate and noise-free strain measurements on wind turbine blades are possible with little added weight to the structure. Furthermore, the nonconductive and distributed nature of optical fibers lends well for many uses in oil and gas applications, including pipeline monitoring and downhole monitoring. Applications of energy FOS devices are wind turbine blade monitoring, pipeline monitoring, power line monitoring, offshore platform monitoring and downhole monitoring.

Civil field; Structural health monitoring systems based on electrical sensors often face significant environmental challenges. An electrical monitoring system would require the installation of countless wires, a lightning grounding system, periodic external calibration, and the potential maintenance of corroded and/or degrading sensors. With an optical sensing solution, these downfalls are all eliminated. The ability to daisy chain multiple sensors on a single fiber greatly reduces the weight and complexity of the system. Furthermore, optical fiber does not corrode or conduct like copper wire, which increases longevity and reduces the risk of damage due to lightning. These attributes, coupled with the fact that optical sensors and NI interrogators do not require calibration, drastically reduce the amount of maintenance required. Applications of civil FOS devices are large building monitoring, bridge and road monitoring, airport landing strip load monitoring and dam monitoring.

Transportation field; To ensure passenger safety, monitoring systems are increasingly deployed to ensure the proper operation of airplanes, railways, ships, and more. However, weight, size, and harsh environmental requirements can pose significant challenges to implementing an electrical monitoring system. FBG optical sensors alleviate these challenges by providing lightweight distributed sensor measurements that are immune to corrosion, high voltage, and EMI-induced noise. Also, because of the longevity and ease of installation of FBG optical sensors and lack of need for external calibration, these sensing systems can be deployed reliably for decades without needing any maintenance; this is especially beneficial for long-term railway and ship hull monitoring. The ability to have multiple sensors on a single, very thin fiber dramatically reduces the weight of the monitoring system, which is especially important in aerospace applications. Applications of transportation FOS devices are railway monitoring, fuel tank monitoring, airplane wing monitoring, and Ship hull monitoring.

7. Conclusions

Recently, the advances of technology and applications based on fiber optics are progressed very rapidly. As a physical medium, Optical fiber is subjected to various perturbation of at all times. Thus, it experiences size and shape changes and optical (refractive index, mode conversion) changes to a larger or lesser extent depending upon the nature and the magnitude of the perturbation. It is crucial for communication application to be reliable all times and to minimize various effects on transmitted signals. On the other hand in fiber optic sensing, the response to external influence is deliberately enhanced so that the resulting change in optical radiation can be used as a measure of the external perturbation. Since light is characterized by amplitude (intensity), phase, frequency and polarization, any one or more of these parameters may undergo a change. The fiber acts as a modulator in sensing phase, while in communication, the signal passing through a fiber is already modulated. It also works as a transducer that converts measurands into a corresponding change in the optical radiation.

FOSs offer many advantages such as freedom from EMI, wide bandwidth, compactness, geometric versatility and economy. Indeed of dielectric construction, passive, and high sensitivity when compared to other types of sensors. Specially prepared fibers can withstand high temperature and other harsh environments. In telemetry and remote sensing applications it is possible to use a segment of the fiber as a sensor gauge while a long length of the same or another fiber can convey the sensed information to a remote station. Deployment of distributed and array sensors covering extensive structures and geographical locations is also feasible.

References

[1] Kashyap, R., "*Photosensitive Optical Fibers: Devices and Applications*", Op. Fiber Tech., Vol.**1**, 1994.

[2] R. Narayanaswamy, O. S. Wolfbeis, Optical Sensors: Industrial, Environmental and Diagnostic Applications, 2004.

[3] P. Ficocelli, "Fiber optic Thermal Monitoring Systems for SAGD", PTAC Technology Information Session, March, 2008

[4] Keiser, G. (1991). Optical Fiber Communications. (2ed), McGraw-Hill, ISBN 0070336172, NewYork USA.

[5] Ghatak, A. & Thyagarajan, K. (2000). "An Introduction to Fiber Optics", Cambridge University Press, ISBN 0521571200, New York USA.

[6] Yu, F. T. S. Yin, S. Ruffin, P. B. (2008). "Fiber Optic Sensors," CRC Press Taylor&Francis Group, ISBN 9781420053654, United State of America.

[7] "Distinctive advantages give optical sensors the edge over conventional systems" www.eurekalert.org/pubreleases/ 2003-08/ti-dag081303.php

[8] D. Krohn, "Overview of Fiber Optic Sensors", KMI's 26TH Annual Newport Conference, October, 2003.

[9] O. Sezerman, "Fiber Optics Distributed Strain and Temperature Sensor (DSTS)", www.ozoptics.com, September, 2008.

[10] NI, "Fiber-Optic Sensing",http://www.ni.com/opticalsensing/, 2012.

[11] F. Baldini, A. Giannetti, "Optical chemical and biochemical sensors: new trends", Proc. SPIE,Vol: **5826**, pp. 485-499, 2005.

[12] F. Baldini, A. Falai, A. R. De Gaudio, D. Landi, A. Lueger, A. Mencaglia, D. Scherr, W. Trettnak, "Continuous monitoring of gastric carbon dioxide with optical fibres", Sensors and Actuators, Vol: **90**, pp. 132-138, 2003.

Power Input to non-deterministic Subsystems via Piezoelectric Patch Actuators: Effect of the patch size and location

Azni N Wahid [1], Asan G A Muthalif [2] and Khairul A M Nor
Department of Mechatronics Engineering, International Islamic University Malaysia, Jalan Gombak 53100, Kuala Lumpur, Malaysia

E-mail: [1] azni.nabela.wahid@gmail.com , [2] asan@iium.edu.my

Abstract. Many engineering systems such as aircraft and automotive are considered built-up structures, fabricated from components that are classified as deterministic subsystems (DS) and non-deterministic subsystems (Non-DS). The response of Non-DS is sensitive to uncertainties therefore presents problems in vibration control. Therefore, analytical solution to estimate average power delivered by a piezoelectric (PZT) patch actuator when attached to a Non-DS needs to be established. The response of Non-DS is estimated using statistical modelling technique such as statistical energy analysis (SEA), in which any external input to the subsystem must be represented in terms of power input. In this paper, ensemble average of power given by a PZT patch actuator to a simply-supported plate when subjected to structural uncertainties is studied using Lagrangian method. The effects of size and location of the PZT actuators on the power delivered to the plate are investigated. It is found that changing the patch location on the structure will not affect the average power supplied while changing the patch size will change the power magnitude proportionally but with some variations at higher frequency.

1. Introduction

Most of engineering systems are subjected to forms of excitation with frequencies that range from low to high which later determine the type of response of the system. If vibrational response characteristics such as displacement, acceleration, and stress are known precisely as functions of time, the vibration is known as *deterministic vibration* (DS) i.e. can be mathematically described using deterministic method. In addition, systems such as aircraft are usually subjected to excitation at high frequency. The modelling of such systems presents the problem of *high-frequency* vibration [1]. The physical characteristic of a high-frequency vibration is that the precise natural frequencies and shapes of high-order modes are impossible to be calculated hence causing difficulties in vibration analysis and therefore any control effort. The system response subjected to this type of vibration is also very sensitive to uncertainties due to the short wavelength deformation, in which identical structures can produce very different dynamic response properties [2]. Subsystem subjected to this type of vibration falls into *non-deterministic* (Non-DS) subsystem category and therefore statistical modelling method such as Statistical Energy Analysis (SEA) are commonly employed [3] [4].

PZT patch actuator has received significant attention for vibration analysis and control due to its unique direct-and-inverse PZT effect. Adding a PZT actuator on a structure is equivalent to adding an external moment to the dynamics of the structure. Conveniently, the influence of adding moment on a DS is known; mathematical model describing power supplied by PZT patch actuator is available.

Journal of Physics: Conference Series **423** (2013) 012065 doi:10.1088/1742-6596/423/1/012065

However, analytical solution to estimate average power delivered by a PZT actuator when attached to a Non-DS needs to be established [5]. Such analytical solution is also required when attempting to control vibration of Non-DS as built-up structure using PZT actuator.

In this paper, ensemble average of power given by the PZT patch actuator to a plate when subjected to structural uncertainties is simulated. The effects of size and location of PZT actuator on the power delivered to the plate are investigated and presented. These two parameters are on interest considering the practicality in real application. A simply-supported plate with attached PZT patch and distributed point masses is taken as a benchmark model. The distributed point masses are used to introduce uncertainties in the response of the plate. Equation of motion for the benchmark model is derived using Lagrangian method [6].

2. Derivation of Equation of motion

2.1 A rectangular plate with distributed point masses attached with a PZT patch actuator

The constitutive equations of a general PZT material are [7]:

$$\sigma_{kl} = c_{ijkl}^{E}\varepsilon_{ij} - e_{kij}E_k \tag{1}$$

$$D_i = e_{ikl}\sigma_{kl} + \xi_{ik}^{T}E_k \tag{2}$$

whereσ_{kl} and $\varepsilon_{ij}$are components of stress and strain tensors, respectively, c_{ijkl}^{E} is the elastic stiffness under constant electric field (Hooke's tensor), e_{kij} is the PZT constant, E_k is the electric field, D_i is the electric displacement andξ_{ik}^{T} is the dielectric constant under constant strain. Consider a rectangular plate with length *a*, width *b* and thickness *h*. The plate has a PZT actuator patch located at distance x_1 along the length and distance y_1 along the width. The patch has x_2-x_1 length, y_2-y_1 width and thickness t_p. A number of point masses are distributed randomly across the plate.

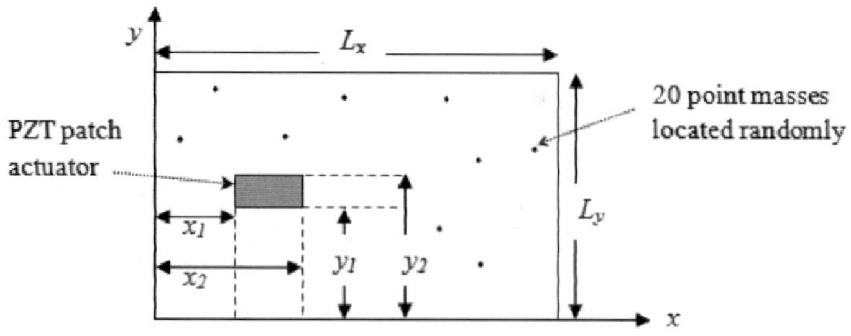

Figure 1. Schematic diagram of a rectangular plate attached with PZT patch and 20 randomly-located point masses.

The stresses corresponding to normal strains and the shear strain are represented as:

$$\sigma_x = \frac{E}{1-v^2}(\epsilon_x + v\epsilon_y) = \frac{Ez}{1-v^2}\left(\frac{\partial^2 w}{\partial x^2} + v\frac{\partial^2 w}{\partial y^2}\right) \tag{3}$$

$$\sigma_y = \frac{E}{1-v^2}(\epsilon_y + v\epsilon_x) = \frac{Ez}{1-v^2}\left(\frac{\partial^2 w}{\partial y^2} + v\frac{\partial^2 w}{\partial x^2}\right); \quad \tau_{xy} = G\gamma_{xy} = \frac{Ez}{(1+v)}\left(\frac{\partial^2 w}{\partial x \partial y}\right) \tag{4}$$

However, with the existence of PZT patch on the plate, the cross section of the plate will not be uniform therefore stress distribution along the thickness of the plate needs to be modified [8][9]. The neutral axis of the system is shifted towards the PZT patch and this shift can be calculated by balancing the force and moment across the patch thickness:

$$\int_{-\frac{h}{2}}^{\frac{h}{2}} E(z - z_n)dz + \int_{\frac{h}{2}}^{\frac{h}{2}+t_p} E_{PZT}(z - z_n)dz = 0 \tag{5}$$

which resulted to:

$$z_n = \frac{E_{PZT}t_p(h + t_p)}{2(Eh + E_{PZT}t_p)} \tag{6}$$

wherez_n is the shifted neutral axis measured from the plate centre.

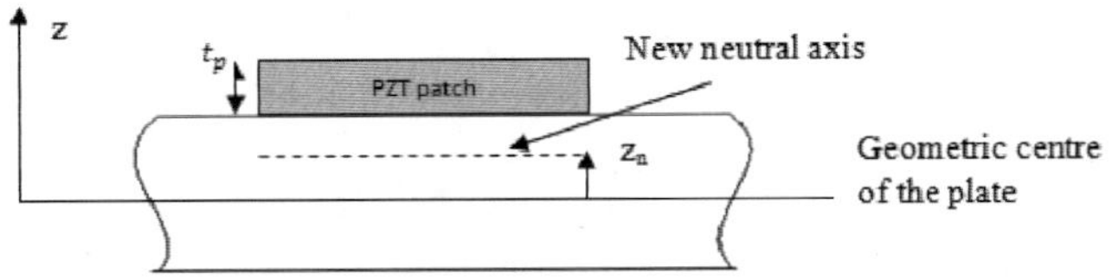

Figure 2. Cross section along xy-axes to show the shift of neutral axis due to PZT patch attachment [8]

Hence, the linear strain displacement relationships are modified as [6]:

$$\epsilon_x = -(z - z_n)\frac{\partial^2 w}{\partial x^2} \qquad \epsilon_y = -(z - z_n)\frac{\partial^2 w}{\partial y^2} \qquad \gamma_{xy} = -2(z - z_n)\frac{\partial^2 w}{\partial x \partial y} \tag{7}$$

To obtain the equation of motion of the system, Lagrangian method will be employed:

$$\frac{\partial}{\partial t}\left(\frac{\partial L}{\partial \dot{W}_{mn}}\right) - \frac{\partial L}{\partial W_{mn}} = F_{mn} \quad , \text{ where } L = T - U \tag{8}$$

where T and U are the total kinetic energy and total potential energy of the system, respectively. The total kinetic energy of the system has contributions from the plate, PZT actuator and distributed point masses:

$$T_{total} = T_{plate} + T_{PZT} + T_{dist.point\ mass} \tag{9}$$

$$T_{total} = \frac{1}{2}\rho h \iint \left(\frac{\partial w}{\partial t}\right)^2 dA + \frac{1}{2}\rho_{pe} t_p \iint \left(\frac{\partial w}{\partial t}\right)^2 S(x,y)dA_p + \frac{1}{2}\iint \sum_r m_r \delta(x - x_r, y - y_r)\left(\frac{\partial w}{\partial t}\right)^2 dA \tag{10}$$

Where ρ_{pe} is the density of the PZT actuator, A_p is the area of the PZT actuator and $S(x,y) = [H(x-x_1) - H(x-x_2)][H(y-y_1) - H(y-y_2)]$ is a product of Heaviside function which takes care of the presence of PZT patch on the plate. The total potential energy of bending for the system has contributions from the plate and PZT actuator:

$$U_{total} = U_{strain\ plate} + U_{strain\ PZT} \tag{11}$$

$$\begin{aligned} U_{total} &= \frac{D}{2}\iint\left\{\left(\frac{\partial^2 w}{\partial x^2}+\frac{\partial^2 w}{\partial y^2}\right)^2 - 2(1-v)\left[\left(\frac{\partial^2 w}{\partial x^2}\frac{\partial^2 w}{\partial y^2}\right)-\left(\frac{\partial^2 w}{\partial x \partial y}\right)^2\right]\right\}dA \\ &+ \frac{D_1(x,y)}{2}\iint\left\{\left(\frac{\partial^2 w}{\partial x^2}+\frac{\partial^2 w}{\partial y^2}\right)^2 - 2(1-v)\left[\left(\frac{\partial^2 w}{\partial x^2}\frac{\partial^2 w}{\partial y^2}\right)-\left(\frac{\partial^2 w}{\partial x \partial y}\right)^2\right]\right\}dA \\ &+ \frac{D_2(x,y)}{2}\iint\left\{\left(\frac{\partial^2 w}{\partial x^2}+\frac{\partial^2 w}{\partial y^2}\right)^2 - 2(1-v_{pe})\left[\left(\frac{\partial^2 w}{\partial x^2}\frac{\partial^2 w}{\partial y^2}\right)-\left(\frac{\partial^2 w}{\partial x \partial y}\right)^2\right]\right\}dA \\ &+ B(t)\iint S(x,y)\left[(d_{31}+v_{pe}d_{32})\frac{\partial^2 w}{\partial x^2} + (d_{32}+v_{pe}d_{31})\frac{\partial^2 w}{\partial y^2}\right]dA \end{aligned} \tag{12}$$

Where,

$$D_1(x,y) = d_1 S(x,y) = \frac{E}{(1-v^2)}\left(\frac{h^3}{12} + z_n^2 h\right)S(x,y) \tag{13}$$

$$D_2(x,y) = d_2 S(x,y) = \frac{E_{pe}}{(1-v_{pe}{}^2)}\left[\frac{t_p{}^3}{3} + \frac{h^2 t_p}{4} + \frac{h t_p{}^2}{2} - z_n\left(h t_p + t_p{}^2\right) + z_n^2 t_p\right]S(x,y) \tag{14}$$

$$B(t) = \frac{E_{pe}V(t)}{2\left(1-v_{pe}{}^2\right)}\left(h + t_p - 2z_n\right) \tag{15}$$

Solving equation (8) using equations (10) and (12) and evaluating the integrations lead to the general form of equation of motion:

$$\left(-\omega^2[M] + [K]\right)W_{mn} = [F_{mn}] \tag{16}$$

Equation (16) will be used to simulate vibration response of a Non-DS rectangular plate attached with a PZT patch actuator.

3. Power delivered to a non-deterministic rectangular plate

To obtain the power delivered to the plate, the following equations for energy of the plate due to vibration are first attained [10]:

$$E = \frac{1}{2} W_{mn}{}^{T} K_C W_{mn} \quad \text{or} \quad E = \frac{1}{2} \omega^2 W_{mn}{}^{T} M W_{mn} \tag{17}$$

which uses complex modal stiffness matrix, $K_C = K(1 + j\eta)$ and mass matrix, M respectively. ω is frequency, η is modal loss factor and K is stiffness matrix. The extra term $(1 + j\eta)$ is incorporated by assuming a hysteretic damping model, introduced for proportional structural damping effect to the system [7]. Since the system is not coupled to any other subsystem (single system), power input, P_{in} is equal to power dissipated from the system.

$$P_{in} = \eta\omega E = \frac{1}{2} \eta\omega W_{mn}{}^{T} K_C W_{mn} \quad or \quad P_{in} = \frac{1}{2} \eta\omega^3 W_{mn}{}^{T} M W_{mn} \tag{18}$$

4. Simulation studies

A rectangular plate made of steel is used for simulation studies. The properties of the plate and PZT patch actuator used are tabulated in Table 1. The centre of the PZT actuator is located at 0.25 of the length and width of the plate. Twenty point masses with total mass 20% of the plate mass are located randomly on the plate surface. To simulate high-order mode response, high modal overlap (M) factor are captured by using high numbers of half-waves in the x direction and y direction; m=15 and n=15 respectively. Modal loss factor of 0.05 is used. Responses for fifty ensembles (location of point masses is randomized for each simulation) are taken, equation (18) is then used to obtain power delivered by the patch and then the responses are averaged to get the ensemble average power. Table 2 shows frequency range and its corresponding modal overlap factor for the plate.

Table 1. Properties of plate and PZT patch actuator used.

Properties	Plate (Steel)	PZT actuator (PZT-5A)
Young's Modulus	$200\text{x}10^9$ (Pa)	$70\text{x}10^9$ (Pa)
Density	7850 (kg/m^3)	7750 (kg/m^3)
Length x Width x thickness	0.7x0.6x0.001 (m)	6x5x0.02 (cm)
Poisson's ratio	0.33	0.31
Piezoelectric constant, d_{31}	-	$-1.71\text{x}10^{-10}$ (V/m)

Table 2. Modal overlap factor and its corresponding frequency value.

MOF	1	2	3	4	5	6	7	8
Frequency (Hz)	118	237	355	473	592	710	828	947

For each ensemble in Figure 3, the response exhibit small magnitude difference at M less than unity, but significant as frequency increases. When modal overlap factor, M is less than unity, the ensemble average exhibits distinct narrow peaks which indicate resonant modal response of the system. As M approaches unity, the individual modal responses begin to overlap and as it increases beyond unity, the peaks gradually become broader and indistinct i.e. resonant modal response is not clearly observable. The frequency response of multi-mode system in the high frequency range is very sensitive to small disturbance due to the randomly located point masses in the system. Hence the detail of high frequency response is unpredictable i.e. non-deterministic.

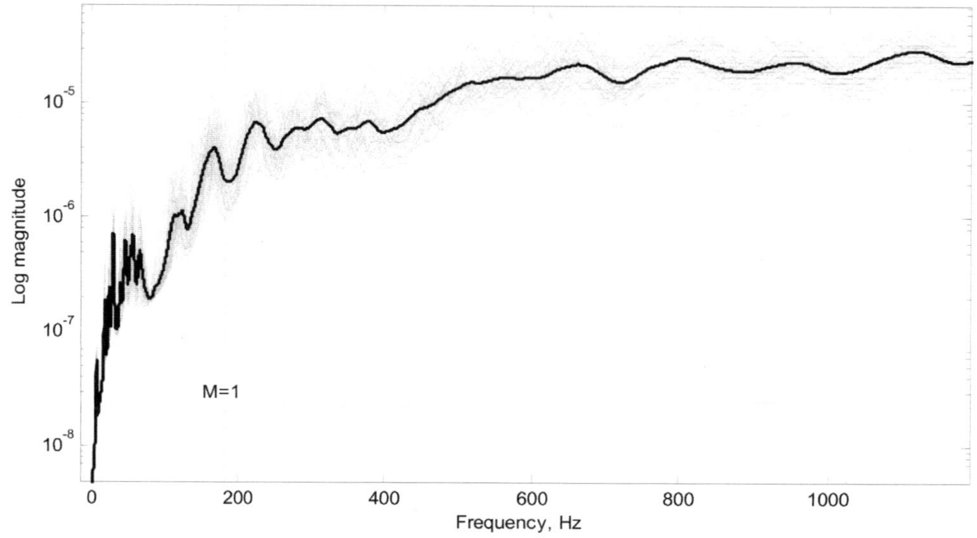

Figure 3. Ensemble average of power delivered by a PZT actuator to plate (solid black line) for 50 ensembles.

4.1 Influence of PZT patch actuator location and size on the estimation of average power

Two case studies are considered in order to further investigate how the average power delivered by PZT patch actuator to the plate is affected when physical parameter of the system is changed.

4.1.1 Case 1: Changing PZT actuator location on plate.

Nine different grid locations on the plate, as shown in Figure 4, are taken to locate the patch and for each location, ensemble average power of 50 simulations is plotted in Figure 5. From Figure 5, it can be seen that narrow peaks are exhibited at $M<1$ to indicate resonant modal response of the system. At higher frequency range, it is observed that the peaks become broader and indistinct. In addition, the difference in magnitude between responses at higher frequency range is not significant. An important observation can be made that different PZT actuator location on structure does not affect the ensemble average power obtained.

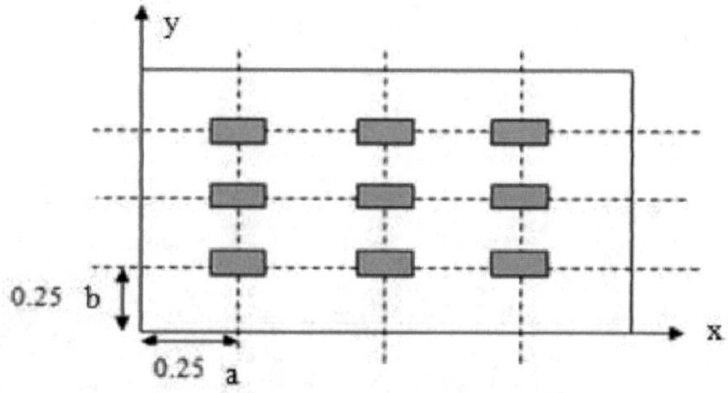

Figure 4. Schematic figure of nine different location of PZT actuator on plate.

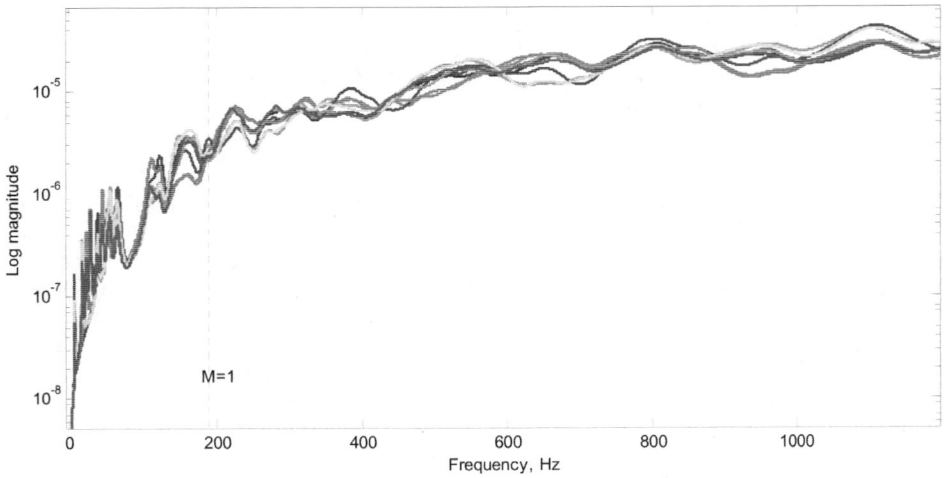

Figure 5. Ensemble average power delivered (50 ensembles each) by a PZT actuator located at 9 different locations on the plate

4.1.2 Case 2: Changing PZT actuator size.

Five different sizes of PZT actuator are taken; the original size, 0.25, 0.5, 1.5 and twice of the original size are considered separately. Ensemble average power of 50 simulations is plotted in Figure 6.

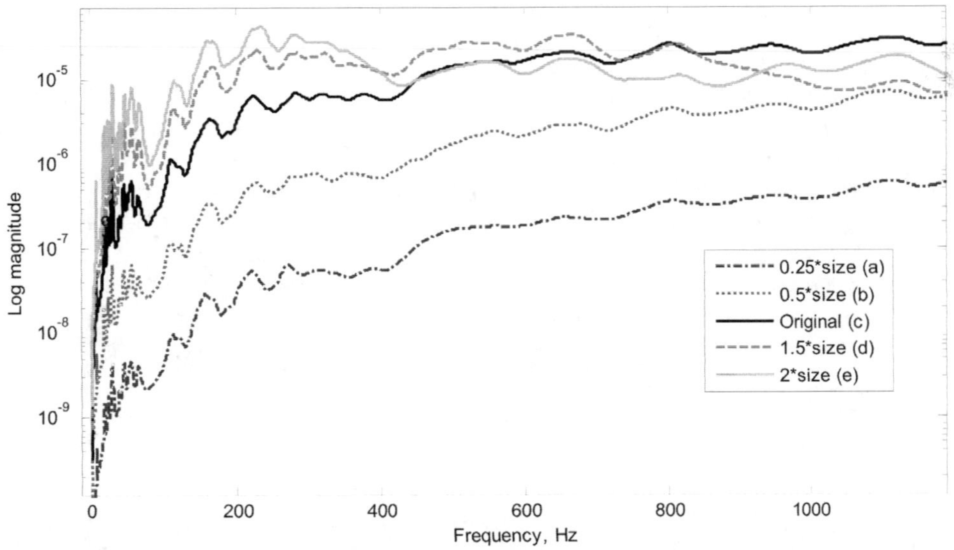

Figure 6. Average power delivered (50 ensembles each) by different sizes of PZT patch actuator to the plate.

From figure 6, a general observation can be made that the magnitude of response decreases as the PZT actuator size decreases however the response shape between different PZT sizes has no significant difference. There are some variations when the patch size is made too big or too small from the original size which indicates that mass effect becomes important at high frequency. The inertia effect tends to pull down the amplitude of the response in proportion to frequency which explains the increasing variance as patch size is made bigger.

5. Conclusions

PZT material is used widely as actuators for excitation or structural vibration control because it is compact and has wide frequency range. Understanding the way vibrational response behaves at high frequency that is by knowing the average power delivered by PZT actuator to a non-deterministic structure is a huge advantage for analysing structural response using SEA method. This paper has presented simulation studies on the ensemble average power delivered by a PZT actuator to a non-deterministic structure. Parametric studies showed that changing the patch location on the structure will not affect the average power supplied. On the contrary, changing the patch size will change the power magnitude proportionally as larger patch means larger force is applied. However, there are some variations for ensemble average which becomes more significant as the patch size is made too big or too small. This is due to the inertial effect which pulls down the amplitude of the response in proportion to frequency.

Acknowledgements

This work was supported by Ministry of Higher education Malaysia under, Fundamental Research Grant Scheme (FRGS-MOHE) and IIUM endowment.

References

[1] Fahy F and Gardonio P 2007 *Sound and Structural Vibration – Radiation, Transmission and Response 2nd Ed*', (Amsterdam: Academic Press)

[2] Muthalif A G A and Langley R S 2012 Active Control of High-Frequency Vibration: Optimization Using The Hybrid Modelling Method *J. of Sound and Vibration* **331** 969 –2983

[3] Lyon R H 1975 *Statistical energy analysis of dynamical systems: Theory and applications*, (Cambridge, MA: MIT Press)

[4] Woodhouse J 1981 An approach to the theoretical background of statistical energy analysis applied to structural vibration *J. Acoustical Society of America* **69**(6) 1695-1709

[5] Muthalif A G A 2010 *Investigating the Average Power Supplied by a Piezoelectric Patch Actuator When Attached to Random Structures (Application to Vibration Control)* CLA Presentation of High Impact Research Proposal Kuala Lumpur, Malaysia

[6] Weaver W, Timoshenko S P and Young D H 1990 *Vibration Problems in Engineering 5th Ed* (Wiley-Interscience)

[7] Preumont A 2011 *Vibration Control of Active Structures: An Introduction 3rd Ed* (Berlin: Springer)

[8] Jalili N 2010 *Piezoelectric Based Vibration Control* (New York: Springer)

[9] Malili P 2011 Vibration Analysis of Thin Plates Subject to Piezoelectric Actuation: A New Perspective in Modeling and Numerical Analysis *Proc. of 2011 ASME International Design Engineering Technical Conferences & Computers and Information in Engineering Conference (Washington, DC, August 28-31, 2011)*

[10] Gardonio P and Brennan M J 2004 Mobility and Impedance Methods in Structural Dynamics *Advanced Applications in Acoustics, Noise and Vibration: Chapter 9* (UK: Spon Press)

Journal of Physics: Conference Series **423** (2013) 012066 doi:10.1088/1742-6596/423/1/012066

Prediction of solubilities for ginger bioactive compounds in hot water by the COSMO-RS method

Syaripah Zaimah Syed Jaapar[1], Noor Azian Morad[2], Yoshio Iwai[3]
[1,2] Center of Lipids and Engineering Research (CLEAR), Malaysia-Japan International Institute of Technology (MJIIT), Universiti Teknologi Malaysia (UTM), Jalan Semarak, 54100 Kuala Lumpur, Malaysia.
[3] Department of Chemical Engineering, Faculty of Engineering, Kyushu University, 744 Motooka, Nishi-ku, Fukuoka 819-0395, Japan.

E-mail: szsj_im@yahoo.com[1]; azian@ic.utm.my[2] ; iwai@chem-eng.kyushu-u.ac.jp[3]

Abstract. The solubilities in water of four main ginger bioactives, 6-gingerol, 6-shogaol, 8-gingerol and 10-gingerol, were predicted using a conductor-like screening model for real solvent (COSMO-RS) calculations. This study was conducted since no experimental data are available for ginger bioactive solubilities in hot water. The σ-profiles of these selected molecules were calculated using Gaussian software and the solubilities were calculated using the COSMO-RS method. The solubilities of these ginger bioactives were calculated at 50 to 200 °C. In order to validate the accuracy of the COSMO-RS method, the solubilities of five hydrocarbon molecules were calculated using the COSMO-RS method and compared with the experimental data in the literature. The selected hydrocarbon molecules were 3-pentanone, 1-hexanol, benzene, 3-methylphenol and 2-hydroxy-5-methylbenzaldehyde. The calculated results of the hydrocarbon molecules are in good agreement with the data in the literature. These results confirm that the solubilities of ginger bioactives can be predicted using the COSMO-RS method. The solubilities of the ginger bioactives are lower than 0.0001 at temperatures lower than 130 °C. At 130 to 200 °C, the solubilities increase dramatically with the highest being 6-shogaol, which is 0.00037 mole fraction, and the lowest is 10-gingerol, which is 0.000039 mole fraction at 200 °C.

1. Introduction

Recently, various alternative and green solvents, such as sub- and supercritical water or fluids have been developed in order to substitute common solvents that have a high level of toxicity. Hot water or subcritical water is an alternative green solvent that has been developed and applied in the extraction process of natural and medicinal plants. This shows that the properties of water have high potential to be discovered, for example, the dielectric constant and viscosity of water. Water is the most abundant,

[1] Center of Lipids and Engineering Research (CLEAR), Malaysia-Japan International Institute of Technology (MJIIT), Universiti Teknologi Malaysia, 54100 Kuala Lumpur, Malaysia (e-mail: szsj_im@yahoo.com).
[2] Center of Lipids and Engineering Research (CLEAR), Malaysia-Japan International Institute of Technology (MJIIT), Universiti Teknologi Malaysia, 54100 Kuala Lumpur, Malaysia (e-mail: azian@ic.utm.my).
[3] Department of Chemical Engineering, Faculty of Engineering, Kyushu University, 744 Motooka, Nishi-ku, Fukuoka 819-0395, Japan (e-mail: iwai@chem-eng.kyushu-u.ac.jp (Y. Iwai).

safe, cheap and environmentally friendly pure solvent. Because of this, water seems to be one of crucial importance for further development of the chemistry of subcritical and supercritical fluids [1].

The current study, which was done by Sarip [2] for ginger using subcritical water extraction, indicates that the extracted bioactives by water depend on the temperature of the process. For example, at 135°C, the bioactive extracted was just 6-gingerol. Whereas, when the extraction was carried out at 165 and 200°C, more bioactives, such as 6-shagaol and 10-gingerol, were extracted. These temperature conditions are a function of the dielectric constant, which reduces as the temperature elevates [3].

The equilibrium and transport properties, such as solubility and diffusion coefficient, are very useful for process design [4]. In order to establish the feasibility of an extraction, the solubility of a substance is a fundamental property [5]. There are four main components of ginger bioactives, which are 6-gingerol, 6-shogaol, 8-gingerol and 10-gingerol. The identification technique of ginger bioactives are well established, so these can be the reference to validate the extraction process using water as the solvent at subcritical conditions or at lower dielectric constant properties. However, physical properties of the bioactives for engineering application such as solubility and diffusivity in solvents are not published.

The activity coefficient is a basic physical-property value needed for the design of separation equipment [6]. The COSMO-RS method [7-8] has the capability of predicting the thermo-physical properties of liquids, such as activity coefficients from a statistical approach by quantum calculations without experimental data, based on a unimolecular quantum chemical calculation, which is an alternative to the structure-interpolating group contribution methods (GCMs). Basically there are two steps for the COSMO-RS calculations standard method: first, to calculate the quantum chemical COSMO for the molecular species involved by extracting the information about the solutes and solvents, and, second, to perform the COSMO-RS statistical calculations with the COSMOtherm program [7-9].

In this work, the COSMO-RS method was applied for the calculation of the activity coefficients for the binary mixtures. NRTL (Non-Random Two-Liquid) was used to fit the activity coefficients calculated by the COSMO-RS method. The interaction parameters in NRTL equation were determined from the calculated activity coefficients by using the COSMO-RS method. The liquid-liquid equilibria (LLE) were calculated by NRTL. Then the solubilities were obtained from LLE. The solubilities of binary mixtures for hydrocarbon + water systems were calculated and compared with the data in the literature. The calculated results are in good agreement with the experimental data. The solubilities of ginger bioactive compounds in water for the binary systems were then predicted.

2. Calculation Method

2.1. Computational Programs:

The predictions of solubilities were carried out using the software package COSMOtherm version 3.2 (COSMOlogic GmbH & Co KG, Leverkusen, Germany) and Gaussian software package version 3.09 (GausView, Gaussian, Inc. USA).

2.2. Calculation Procedures

2.2.1. The calculations were performed on the Density Functional Theory (DFT) level using Gaussian software for each molecule to obtain the COSMO input files for water, hydrocarbon and ginger bioactive molecules. In these calculations, a charge density σ of a segment on the molecular surface was calculated in a virtual conductor. The ideally screened molecule energy with screening charge distribution on the surface was used as the database for COSMO files.

2.2.2. The distribution function $P_i(\sigma)$ of the obtained σ (figure 1), which is called σ-profile, for each molecule can give the σ-profile of the mixture $P(\sigma)$ by using $P_i(\sigma)$ and the mole fraction x_i of component i in the mixture [8, 12]. Detailed equations and the method of calculation have been described by Shimoyama et al., [12]. The σ-profiles and σ-potential for each molecule were calculated with the COSMOtherm program. The activity coefficients for the binary systems were predicted using the COSMO-RS calculation.

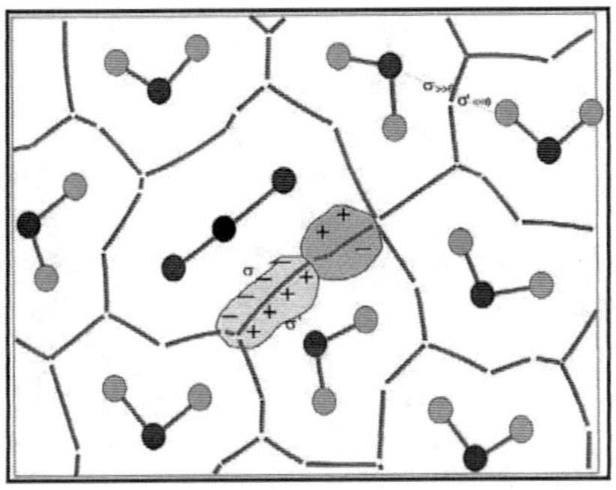

Figure 1. Schematic illustration of contacting molecular cavities and contact interaction. Source: Klamt & Eckert [8].

2.2.3. The activity coefficients were fitted with the NRTL model and the binary parameters in NRTL were obtained. The equality of activities between the two liquid phases in LLE is written as follows:-

$$x_i^I \gamma_i^I = x_i^{II} \gamma_i^{II} \qquad (1)$$

where γ is the activity coefficient, and superscripts I and II mean phases I and II, respectively [10].

The LLE was calculated using the NRTL model. The solubilities were obtained from the LLE. The COSMO-RS calculations standard method [8] was applied in this research and modified to obtain the solubilities of the molecules from the LLE diagram. Figure 2 shows the flowchart for the calculation of solubilities.

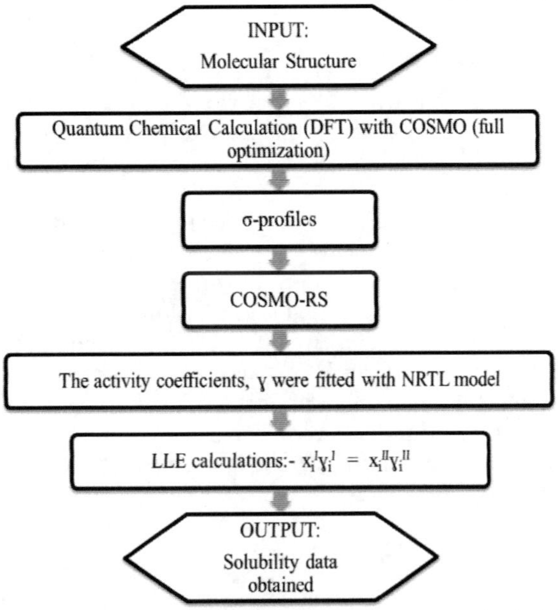

Figure 2. Flowchart of solubility prediction from COSMO-RS

3. Results and Discussion

3.1. Validation of Calculated Results.

Since no experimental data for ginger bioactives are available, the calculations were done part-by-part for the smaller parts of the ginger bioactives. 3-pentanone was selected as it has double O bonding, 1-hexanol has –OH in the formula structure, benzene has ring structure, while 3-methylphenol and 2-hydroxy-5-methylbenzaldehyde have –OH and $-CH_3$ connected to the ring structures. Table 1 shows the properties of the selected hydrocarbon molecules. These five hydrocarbon molecules were selected because they have a similar formula structure of smaller parts of ginger bioactives and the experimental data are available to prove the reliability of the COSMO-RS method. The solubilities of the selected hydrocarbon molecules in five binary systems were then calculated with COSMO-RS and compared with the experimental data, which were 3-pentanone in water, 1-hexanol in water, benzene in water, 3-methylphenol in water, and 2-hydroxy-5- methylbenzaldehyde in water.

Table 1. Hydrocarbon molecules selected properties.

Product Name	3-pentanone	1-hexanol	benzene	3-methylphenol	2-hydroxy-5-methylbenzal dehyde
Formula structure	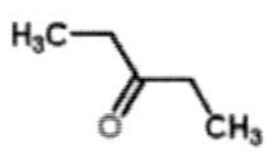		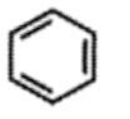	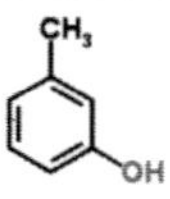	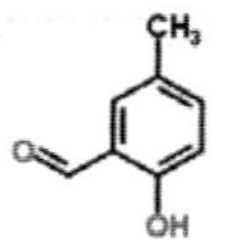
Formula	$C_5H_{10}O$	$C_6H_{14}O$	C_6H_6	C_7H_8O	$C_8H_8O_2$
Molecular Weight	86.13	102.17	78.11	108.14	136.15

3.1.1. σ-profiles

Figure 3 shows the σ-profiles and COSMO-cavities of six selected compounds. The σ-profile of water is very broad, almost symmetric, with two peaks from both the negative and positive side, which are around -0.016 e/A^2 and +0.018 e/A^2, resulting from the two polar hydrogen and lone pair of oxygen atoms, respectively. This σ-profile of water was the same as reported by Zhou et al. [11] and almost the same as the results reported by Klamt [8].

The interaction energies were determined by the screening charge density σ. The notation of "σ-profile" is used to describe the distribution ps(σ), which describes the amount of surface in the ensemble having a screening charge density between σ and $\sigma+d\sigma$. Regarding the definition of the σ-profile, there are two important σ-moments related to the hydrogen bonds that represent the capability of hydrogen bond (HB) donor (-) and acceptor (+) [8, 11]. A molecule is polar enough if the charge density goes beyond ± 0.008 e/A^2. The σ-profile of 1-hexanol is narrowed in the middle region due to no significant electrostatic moments and non-polar character with the peak being -0.003 e/A^2, which is different to 2-hydroxyl-5-methylbenzal-dehyde, benzene and 3-methyphenyl, which have quite a symmetric σ-profile and are less broad than water. They also show no intensity in both the hydrogen bonding regions, which is beyond ± 0.01 e/A^2.

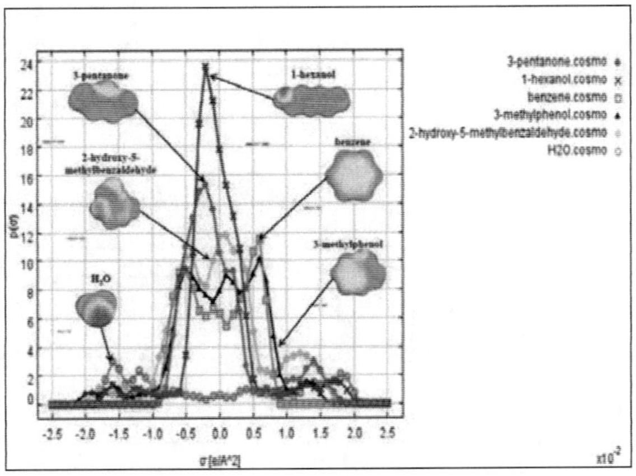

Figure 3. σ-profiles of selected hydrocarbons and COSMO-cavities [red:- positive surface screening charges resulting from negative partial charges within molecule, blue:-negative surface charges and green:-neutral charges]

3.1.2. LLE phase diagram for solubilities of hydrocarbons in water

The calculated results from the COSMO-RS method are compared to experimental data for the selected binary systems as shown in figures 4 to 8. Figure 4 shows the comparison of experimental data and calculated results at temperature 20-120 °C, for the solubilities of 3-pentanone in water where the trend of the data almost same and really close to each other with average percentage of difference is 9%. In figure 5, there about 4% difference between experimental data and calculated results for the solubilities of 1-hexanol in water at temperature 20-110 °C. Figure 6 shows the solubilities of benzene in water the percentage of difference is about 19.27% for the temperature 0-80 °C. While the solubilities of 3-methylphenol in water show about 22.85 % difference between experimental data and calculated results. However, in figure 8, the percentage of difference for the solubilities of 2-hydroxy-5-methylbenzaldehyde in water is about 59.11% which is high maybe because less of experimental data are available to compare.

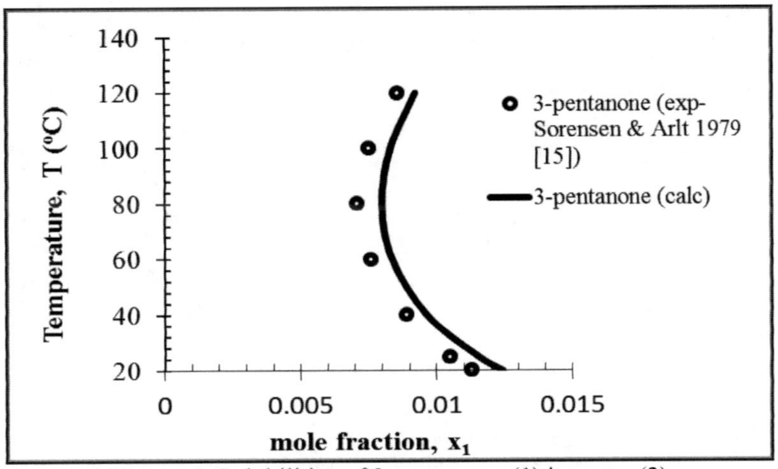

Figure 4. Solubilities of 3-pentanone (1) in water (2).

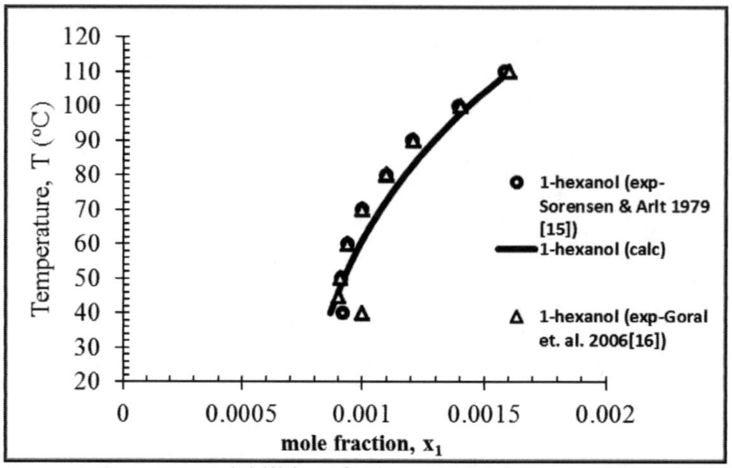

Figure 5. Solubilities of 1-hexanol (1) in water (2).

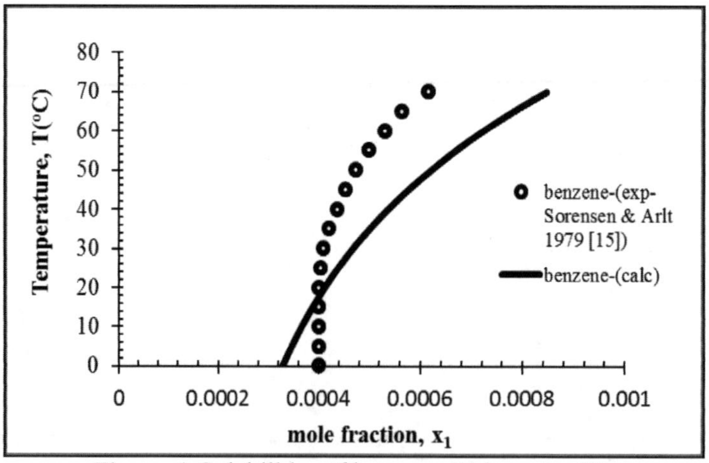

Figure 6. Solubilities of benzene (1) in water (2).

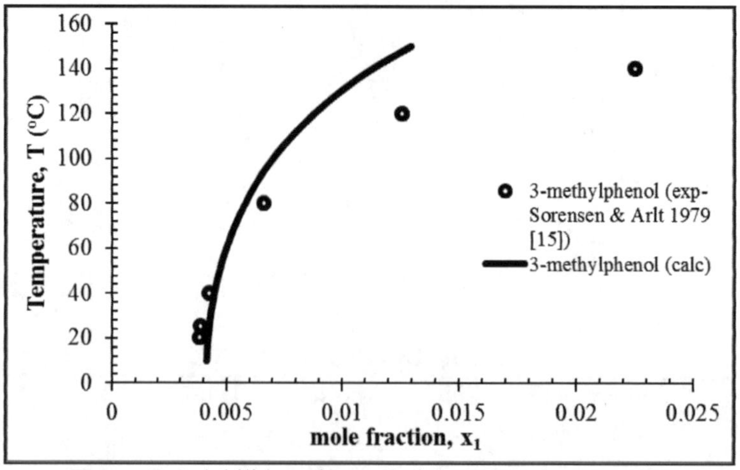

Figure 7. Solubilities of 3-methylphenol (1) in water (2).

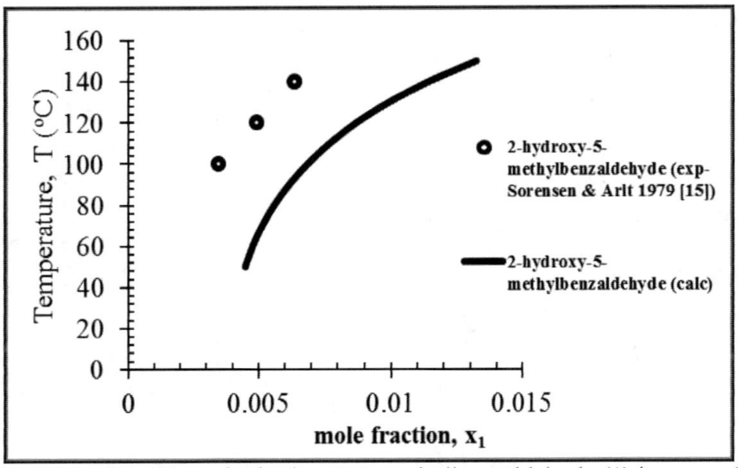

Figure 8. Solubilities of 2-hydroxy-5-methylbenzaldehyde (1) in water (2).

The results show that the trends of experimental data and calculated results are almost the same for the binary system. The agreement between the experimental and calculated results in this study is not perfect, but it is satisfactory. In addition, a few papers reported that the calculated and experimental results are in good agreement when using COSMO-RS to predict the thermophysical properties of certain compounds [9, 12-14].

3.2. Calculated Results for Solubilities of Ginger Bioactives in Water.

The solubilities of four main components of ginger bioactives were calculated at 50 to 200 °C where the ginger bioactives are in liquid form at this range of the temperature. Additionally, this range of temperature was selected in order to compare with the Subcritical Water Extraction (SWE) of ginger bioactives, which was done by Sarip [2], and in consideration of the operational cost if higher temperatures were applied as well as the stability of the ginger bioactive itself from the degradation. Table 2 shows the formula structures and molecular weights of four main ginger bioactive compounds. The solubilities of these ginger bioactives in water were calculated separately in the binary systems which were 6-gingerol in water, 6-shogaol in water, 8-gingerol in water, and 10-gingerol in water using COSMO-RS method.

Table 2. Ginger Bioactive properties

Product Name	6-Gingerol	6-Shogaol	8-Gingerol	10-Gingerol
Formula structure	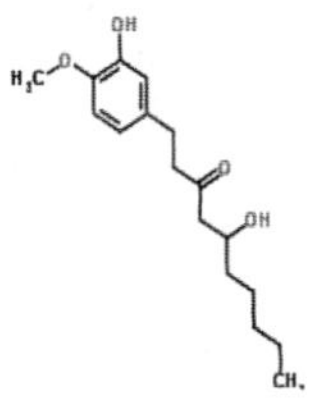	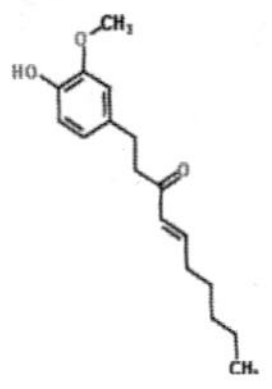	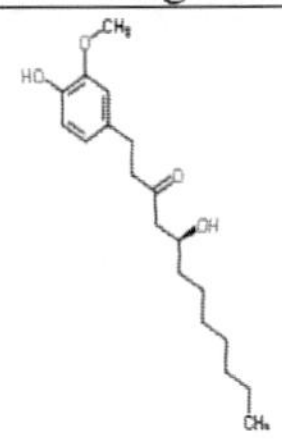	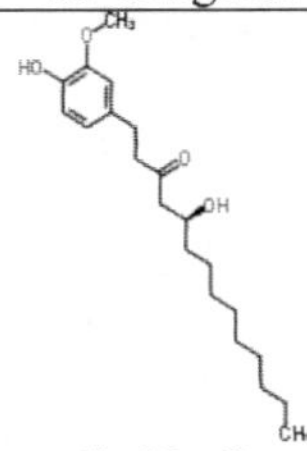
Formula	$C_{17}H_{26}O_4$	$C_{17}H_{24}O_3$	$C_{19}H_{30}O_4$	$C_{21}H_{34}O_4$
Molecular Weight	294.39	276.37	322.44	350.49

3.2.1. σ-profiles

The σ-profiles and COSMO-cavities of four ginger bioactive compounds and water are shown in figure 9. The σ-profile of water is used as the reference for comparison with the previous results. The σ-profile for each compound of ginger bioactives is almost the same, which is narrow in the middle region. 10-Gingerol has the highest peak among the compounds. The σ-profile of 10-gingerol is almost the same as 1-hexanol, as shown in figure 3. This σ-profile shows the molecular polarization or interactions [11]. They show that there is no electrostatic moment or non-polar character in the main parts of the ginger bioactives. However, all the ginger bioactives have a peak beyond +0.01 e/A^2, which represents the presence of the hydrogen bond. All the ginger bioactives show a higher screening charge density, which is in the region of HB acceptor region rather than the HB donor region.

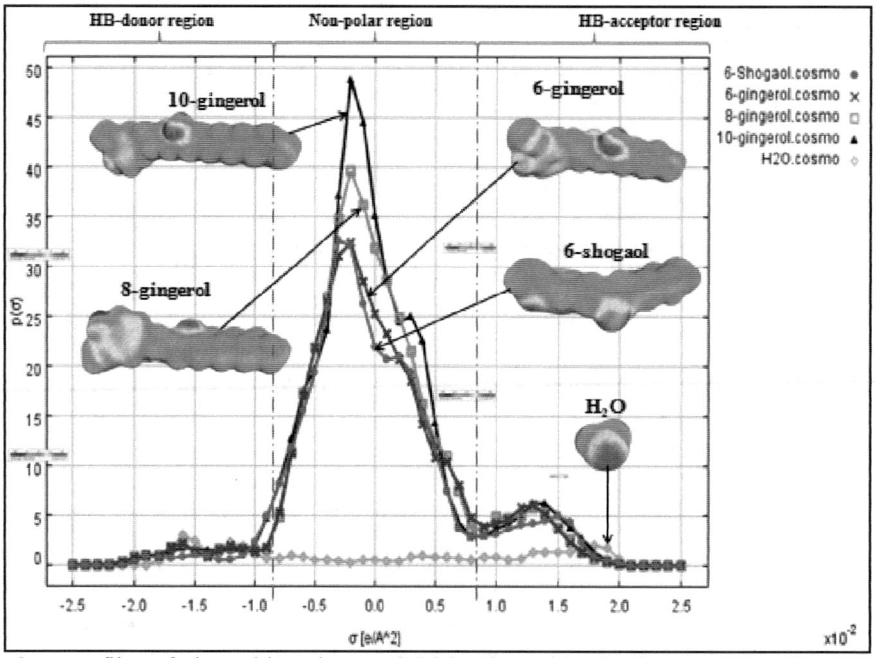

Figure 9. σ-profiles of ginger bioactives and COSMO-cavities [red:- positive surface screening charges resulting from negative partial charges within molecule, blue:-negative surface charges and green:-neutral charges]

3.2.2. LLE phase diagram for solubilities prediction of ginger bioactives in water.

The solubility for each ginger bioactive compound was calculated separately in the binary systems (6-gingerol in water, 6-shogaol in water, 8-gingerol in water and 10-gingerol in water). The calculated results of the solubilities for ginger bioactives are plotted in the LLE phase diagram for comparison, as shown in figure 10. These solubilities of ginger bioactives were calculated from 50 to 200 °C. The solubilities of 6-gingerol and 6-shogaol are higher than those of 10-gingerol and 8-gingerol. This is because the molecular structures or molecular weights of 10-gingerol and 8-gingerol are bigger than those of 6-gingerol and 6-shogaol. The trends of solubilities for all ginger bioactives are almost the same. The solubilities of these ginger bioactives are very low being less than 0.0001 mole fractions when the temperature is below 130 °C. As the temperature increases from 130 to 200 °C, the solubilities of these ginger bioactives increase dramatically. The highest solubility obtained is from 6-shogaol, followed by 6-gingerol, 8-gingerol and 10-gingerol, which are 0.00037, 0.00029, 0.00007 and 0.00004 mole fraction, respectively, at 200 °C. Ginger bioactives are also sensitive to heat and will degrade as the temperature increases [2].

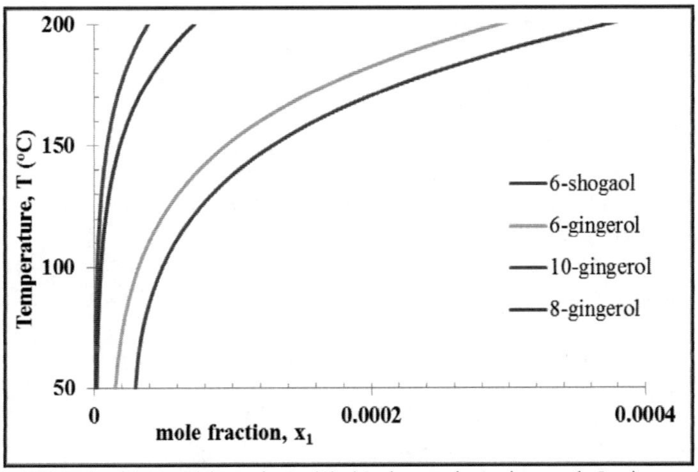

Figure 10. Solubilities for ginger bioactives (1) (6-shogaol, 6-gingerol, 8-gingerol and 10-gingerol) in water (2).

4. Conclusion

The solubilities of 6-gingerol, 6-shogaol, 8-gingerol and 10-gingerol were calculated with COSMO-RS in order to establish the feasibility of water as the solvent at high temperature with lower dielectric constant. The σ-profiles of 6-gingerol, 6-shogaol, 8-gingerol and 10-gingerol show that the compounds have the slight possibility of polar characteristic with the existence of hydroxide group in the structure. The molecules interaction and polarization are also shown in σ-profile. The COSMO-RS method can be used to calculate the LLE phase diagram. The solubilities of these ginger bioactives increase dramatically after the temperature increase from 130 to 200 °C. This shows that at high temperature, the dielectric constant of water decreased and mimic the properties of common non-polar solvent in engineering application especially extraction. The solubilities of ginger bioactives in hot water can be predicted with COSMO-RS method without extensive experimental work.

5. Acknowledgments

Our sincere gratitude goes to Mr Aokawa in Department of Chemical Engineering, Kyushu University, Japan. He helped a lot in the COSMO-RS and Gaussian simulation. The authors gratefully acknowledge the financial support from Malaysia-Japan International Institute of Technology (MJIIT), UTM Kuala Lumpur, Malaysia for the Fellowship Program with Kyushu University, Fukuoka, Japan.

6. References

[1] Galkin A A, and Lunin V V 2005 *Russian Chemical Reviews* **74** 21-35

[2] Sarip M S M 2012 *Subcritical water extraction of 6-gingerol and 6-shogaol from Zingiber officinale* (M. S. thesis, Dept. of Chemical Eng., Universiti Teknologi Malaysia (UTM), Skudai, Johor, Malaysia)

[3] Carr A G, Mammucari R and Foster N R 2010 Solubility and Micronization of Griseofulvin in Subcritical Water *Ind. Eng. Chem. Res.* **49** 3403–3410

[4] Higashi H, Iwai Y and Arai Y 2001 Solubilities and diffusion coefficients of high boiling compounds in supercritical carbon dioxide," *Chemical Engineering Science* **56** 3027-3044

[5] Carr A G, Mammucari R, and Foster N R 2011 A review of subcritical water as a solvent and its utilisation for the processing of hydrophobic organic compounds. *Chemical Engineering Journal* **172** 1-17

[6] Iwai Y and Yamamoto Y 2013 Concentration dependent surface area parameter model for

calculation of activity coefficients *Fluid Phase Equilibria* **337** 165-173
[7] Klamt A 2003 Prediction of the mutual solubilities of hydrocarbons and water with COSMO-RS *Fluid Phase Equilibria* **206** 223-235
[8] Klamt A and Eckert F 2000 COSMO-RS: a novel and efficient method for the a priori prediction of thermophysical data of liquids *Fluid Phase Equilibria* **172** 43-72
[9] Freire M G, Santos L M N B F, Marrucho I M and Cautinho J A P 2007 Evaluation of COSMO-RS for the prediction of LLE and VLE of alcohols+ionic liquids," *Fluid Phase Equilibria* **255** 167-178.
[10] Assael M J, Martin Trusler J P and Tsolakis T F 1996*An Introduction To Their Prediction: Thermophysical Properties Of Fluids* (London: Imperial College Press)
[11] Zhou T, Chen L, Ye Y, Chen L, Qi Z, Freund H and Sundmacher K 2012 An Overview of Mutual Solubility of Ionic Liquids and Water Predicted by COSMO-RS *Industrial & Engineering Chemistry Research* **51** 6256-6264
[12] Shimoyama Y, Iwai Y, Takada S, Arai Y, Tsuji T and Hiaki T 2006 Prediction of phase equilibria for mixtures containing water, hydrocarbons and alcohols at high temperatures and pressures by cubic equation of state with GE type mixing rule based on COSMO-RS *Fluid Phase Equilibria* **243** 183-192
[13] Ferreira A R, Freire M G, Ribeiro J C, Lopes F M, Crespo J G and Cautinho J A P 2011 An Overview of the Liquid−Liquid Equilibria of (Ionic Liquid + Hydrocarbon) Binary Systems and Their Modelling by the Conductor-like Screening Model for Real Solvents *Industrial & Engineering Chemistry Research* **50** 5279-5294
[14] Freire M G, Ventura S P M, Santos L M N B F, Marrucho I M and Coutinho J A P 2008 Evaluation of COSMO-RS for the prediction of LLE and VLE of water and ionic liquids binary systems *Fluid Phase Equilibria* **268** 74-84
[15] Sorensen J M and Arlt W 1979 *Liquid-liquid Equilibrium Data Collection; Binary Systems. Chemistry Data Series,* vol. V (Frankfurt, Germany; Dechema :Deutsche Gesellschaft fur Chemisches Apparatewesen)
[16] Goral M, Wisniewska-Goclowska B and Maczynski A 2006 Recommended Liquid-Liquid Equilibrium Data. Part 4. 1-Alkanol-Water Systems *Journal of Physical and Chemical Reference Data* **35** 1391-1414

Journal of Physics: Conference Series **423** (2013) 012067 doi:10.1088/1742-6596/423/1/012067

Earthquake hazard zonation using peak ground acceleration (PGA) approach

E Irwansyah[1], E Winarko[2], Z E Rasjid[3], R D Bekti[4]

[1,3,4] School of Computer Science, BINUS University, Jl. KH. Syahdan No 9 Palmerah Jakarta 11480 Indonesia

[2] Department of Computer Science and Electronic, Gadjah Mada University, Sekip Utara, Bulak Sumur, Yogyakarta 55281, Indonesia

edirwan@binus.ac.id

Abstract. The objective of this research is to develop seismic hazard area zones in the building infrastructure of the Banda Aceh City Indonesia using peak ground acceleration (PGA) measured using global and local attenuation function. PGA is calculated using attenuation function that describes the correlation between the local ground movement intensity the earthquake magnitude and the distance from the earthquake's epicentre. The data used comes from the earthquake damage catalogue available from the Indonesia meteorology, climatology and geophysics agency (BMKG) with range from year 1973 – 2011. The research methodology consists of six steps, which is developing the grid, calculation of the distance from the epicentre to the centroid of the grid, calculation of PGA values, developing the computer application, plotting the PGA values to the centroid grid, and developing the earthquake hazard zones using kriging algorithm. The conclusion of this research is that the global attenuation function that was developed by [20] can be applied to calculate the PGA values in the city of Banda Aceh. Banda Aceh city in micro scale can be divided into three hazard zones which is low hazard zone with PGA value of 0.8767 gals up to 0.8780 gals, medium hazard zone with PGA values of 0.8781 up to 0.8793 gals and high hazard zone with PGA values of 0.8794 up to 0.8806 gals.

1. Introduction

Indonesia is one of the countries in the world with a high intensity of earthquakes. The United States Geological Survey catalog (USGS) recorded four large earthquake incidence in Indonesia which is Banda earthquake (8.5 Mw) in 1983, Sumatera-Andaman Islands earthquake (9.1 Mw) in 2004, North Sumatera earthquake (8.6 Mw) in 2005 [17] and Sumatera West Coast earthquake (8.6 Mw) in 2012 [18]. The high intensity of earthquake is the main characteristics of tectonic of the Islands of Indonesia which is located between three main plates, which is the Eurasia plate at the North, Indo-Australia plate at the South and Pacific plate at the Northeast. Earthquake with a specific intensity and magnitude as a response of plate movements can cause physical infrastructure damage and casualties.

The most dominant physical Infrastructure damage due to earthquake is the damage of the buildings, whether is due to the poor quality of constructions (internal) or due to the building conditions (external) where the building is located. The number of building damages that is significant is recorded in Banda Aceh city Indonesia as the effect of the Sumatera-Andaman Islands earthquake in 2004 with a total building damage up to 35% out of all buildings [8]. The condition

Journal of Physics: Conference Series **423** (2013) 012067 doi:10.1088/1742-6596/423/1/012067

also occurred in other towns as an effect of a different earthquake with a total damage on 140,000 building units caused by the Bantul, Indonesia earthquake in 2006 [11]. The most casualties as a result of earthquake followed by tsunami are recorded at the Province of Aceh and North Sumatera occurred in December 2004 with total casualties of 110,229 people died, 12,123 people missing and 703,518 people evacuated [5]. The study to assess building damages caused by earthquakes conducted by [8] in Banda Aceh city concluded that the building damages shows a certain directed spatial pattern on the northwest-northeast relatively parallel to the coast line.

On an earthquake case, the damage pattern on buildings is determined by the earthquake factor itself and the environment conditions where the building is built. The relationship between earthquake factor and the effect it causes can be studied by the attenuation function (attenuation relationship) where this function serves as a key component to assess earthquake hazard [13]. The earthquake hazard assessment and its relationship with the damage of building infrastructure can be performed using the acceleration value in the ground or peak ground acceleration (PGA).

In order to assess the earthquake hazard especially for Banda Aceh city, this research is aimed to develop an earthquake hazard zonation for the building infrastructure using the PGA value approach which is adopted from the global attenuation function by [20], and the local attenuation function that is adopted from [12]. The information resulted from the earthquake hazard assessment in the city is necessary nowadays for development in order to develop a city with earthquake resistance physical infrastructure and to minimize the casualties in an earthquake event.

2. Attenuation Function

The information reagarding the PGA characteristic caused by earthquake can be obtained from the records of earthquake events in the past. The records of ground acceleration allow to extract the main characteristics from the ground motion recordings such as the peak ground velocity, can be calculated as the acceleration of the ground (ground acceleration) as well as the acceleration of earthquake on the ground surface. The PGA values can be calculated using the attenuation function. The attenuation function is a function that represents the colleration between the intensity of local ground movement (a), the erathquake's magnitude (M), and the distance between one point in the source of the earthquake (r).

Experts has formulated the attenuation function where the attenuation function which is applied in one location does not necessarily mean that it can be applied elsewhere, because the attenuation function is highly dependent on the natural condition in one location. The selection of attenuation function is based on the equality geological condition and tectonic in the area where attenuation function is developed [7].

Currently variants of attenuation function exists, such as a function for shallow earthquake source, deep earthquake background and attenustion function for the source of the earthquake caused by subduction, as published by[1][4][6][15][10][13][15][20].

A regression attenuation function using the earthquake data catalogue betwen plates with variety of magnitude between 5–8.2 Mw recorded in subduction area in Alaska, Chile, Cascadia, Japan, Mexico, Peru and the Solomon Islands suggested by [20]. The attenuation function is modified by [15] by comparing the observed value and the predicted value using earthquake data catalogue year 1991 – 2001 in the area covering New Ireland, New Britain, Kamchatca, Santa Cruz Islands, Peru, Kurile Islands, Japan and Sumatra. The modification of the attenuation function is concentrated on the case with the distance of the epicenter greater than 200 Km with a magnitude 6.8 – 8.3 Mw.

An attenuation function for the subduction in Cascadia from the function recommended by [20] using stochastic finite-fault ground motion model with a variant on the high magnitude ranging from 8.0 – 9.0 Mw developed by [6]. The advantage using this model is that there are no empirical attenuation relation, which require a sample and geometri based on a series of available strong motion, the effect of fintie-fault such as rupture propagation, directio and source to the geometical site, can be systematically calculated with the stohastic finite-fault model.

Maximum similarity regression method with moment magnitude 5.0 - 8.3 Mw coming from subduction area in the world such as Alaska, Japan, Mexico and Central America is used by [1] to develop the attenuation function. The result of the analysis of the regional variables from ground motion amplitude using an available global database to support facts that there exists a significat difference between reagionals as shown by the different amplitudeson two or more factos between the area of Cascadia and Japan. This model is only using the closest distance from the epicenter on the range 10 – 500 Km as used by [20] and [6].

In the same year [4] used a combination of empirical model that uses the estimated ground motion value, whether it is stochastic or theoritical model to develop a typical regression model to be implemented in the east of North America zone (ENA) using a regression model which has been developed earlier using earthquake data in the west of North America (WNA). Attenuation function by [4] is developed mostly using earthquake witha avriety of magnitude 5.0 – 7.5 with the closest straight line distance from the epicenter (1 – 1000 Km).

[10] Using an earlier reference of attenuation function that was developed by [20] a regression attenuation function model using the recordings of earthquake at the surface in a subduction between plates in Northeast Taiwan and other area with a low magnitude around 4.1 up to 8.7 on the Richter scale. The usage of low magnitude (< 5.0 Mw) is somewhat different from attnuation function that was developed by earlier researcher.

The development of the latest local attenation function was developed by [13] from the attenuation function also developed by [12] in the form of synthetic movement regression of surface rocks in the earthquake zone cause by subduction between plates on the West coast of Sumatra using finite-fault kinematic model as adopted from [6]. The validation of this model is using megathrust Sumatra data covering earthquakes with a magnitude up to 9.0 Mw.

3. Methodology

Research data is obtained from the earthquake catalogue from the Indonesia meteorology, climatology and geophysics agency (BMKG) with damages (< 5.0 Mw) spanning from year 1973 – 2011 (39 years) in a radius of 500 Km from Banda Aceh city. Data attribute consist of the date and time of the earthquake, the coordinates of earthquake center, depths and the earthquake magnitude.

The data processing in this research consists of six steps which is (1) grid development, (2) the calculation of distance from the epicenter to the grid centroid, (3) computer application development, (4) PGA value calculation, (5) plotting the PGA values in each of the grid centroid, and (6) development of hazard zone using kriging algorithm. Grids with a size of 1 x 1 square Km with a total of 255 grids covering all of the area and surroundings, each grid is determined by the centroid. The distance from the epicenter to the grid centroid is calculated by comparing the epicenter coordinate grid centroid and the following equation:

$$D = \sqrt{(A_{long} - B_{long})^2 + (A_{lat} - B_{lat})^2} \qquad (1)$$

Where D is the distance from the epicenter to the centrid (Km); A_{long} longitudinal coordinate of the epicenter; B_{long} is the longitudinal coordinate of the grid centroid; A_{lat} serves as the latitude of the epicenter and B_{lat} serves as a latitude of the grid centroid.

The PGA value is calculated using two global attenuation functions which is adopted from [20] and the local attenuation function from [12]. The equation of Youngs' attenuation function can be elaborated as follows:

$$\ln(PGA)=0.2418+1.414M+C1+C2(10-M)^3+C_3 \ln(r_{rup}+1.^{7818e0.554M})+0.0060H+ 0.3846Z_r \qquad (2)$$

where PGA is the *peak ground acceleration* (gals) value; M is the earthquake magnitude value (Richter Scale); r_{rup} is the horizontal distance from the epicenter to the grid centroid; H is the depth of

the earthquake center (Km); Zr is the type of epicenter (0 for interface and 1 for intraslab) and C_1, C_2, C_3 represent the regression coefficient from the function (C_1 and C_2 value 0 and C_3 = -2.552)
The equation for the local attenuation function from Megawati et al, 2005 can be elaborated as follows:

$$\ln(\text{PGA}) = \alpha 0 + \alpha 1 Mw + \alpha 2 Mw2 + \alpha 3 \ln(R) + \alpha 4 R + \alpha 5 H + \varepsilon H \quad (3)$$

where PGA is the *peak ground acceleration* (gals) value; M is the earthquake magnitude (Richter Scale); R is the horizontal distance from th epicenter to the centroid; H represents the depth of the earthquake center (Km). The variation of PGA is highly dependent on the εH value, which is a random number from the source's parameter (εH=0.4413); $\alpha 3$ represents the average value of geometrical regression ($\alpha 3$ = -1); $\alpha 4$ represenst the inelastic attenuation coefficient ($\alpha 4$ = -0.001548) and other coefficient values which is $\alpha 0$ = -7.198, $\alpha 1$ = 2.3691, $\alpha 2$ = -0.013856 and coefficient $\alpha 5$ = 0.08909.

The calculation of the distance as well as the PGA values for the whole grid is performed by the computer application which is developed especially for this research. The application is developed using Visual C with a database is connected to an MS. Excell spreadsheet.

Next step is to plot the average PGA on each grid, using ArcMap Software, and the development of zonation using kriging algorithm. Kriging is a method of optimal prediction or estimation in geographical space, often known as a best linear unbiased predictor (BLUP). It is the geo-statistical method of interpolation for random spatial processes. Kriging provides a solution to a fundamental problem faced by environmental scientists of predicting values from sparse sample data based on a stochastic model of spatial variation. One methods of kriging is ordinary kriging. Consider that a random variable, *Z*, has been measured at sampling points or locations, *xi*, *i* = 1, … *n*, and we want to use this information to estimate its value at a point *x*0 (punctual kriging) with the same support as the data.

$$\hat{Z}(x_0) = \sum_{i=1}^{n} \lambda_i z(x_i) \quad (4)$$

Kriging is calculate the best estimation of $\dot{Z}(x_0)$ which based on spatial dependencies stochastic model which good perform by semi-variogram $\gamma(h)$. The value of λ_i estimated by:

$$\lambda = \mathbf{C}^{-1}\mathbf{D} \quad (5)$$

$$\lambda = \begin{bmatrix} \lambda_1 \\ \lambda_2 \\ .... \\ \lambda_n \\ \mu \end{bmatrix} \quad \mathbf{C} = \begin{bmatrix} \overline{C}_{11} & \overline{C}_{12} & \cdots & \overline{C}_{1n} & 1 \\ \overline{C}_{21} & \overline{C}_{22} & \cdots & \overline{C}_{2n} & 1 \\ \vdots & \vdots & \vdots & \cdots & \vdots \\ \overline{C}_{n1} & \overline{C}_{n2} & \cdots & \overline{C}_{nn} & 1 \end{bmatrix} \quad \mathbf{D} = \begin{bmatrix} \overline{C}_{10} \\ \overline{C}_{20} \\ .... \\ \overline{C}_{n0} \\ 1 \end{bmatrix} \quad (6)$$

where **C** is the matrix of semi-variances between data points and **D** is the vector of semi variances between data points and the target. The empirical semi-variances can be estimated from data, $z(x_1)$, $z(x_2)$, …, by

$$\hat{\gamma}(h) = \frac{1}{2n(h)} \sum_{\alpha=1}^{n(h)} [z(x_i + h) - z(x_i)]^2 \quad (7)$$

where $z(x_i)$ and $z(x_i+h)$ are the actual values of Z at places (x_i) and $(_{xi}+h)$, and $m(h)$ is the number of paired comparisons at lag h

4. Result and Discussion

4.1. Ground Acceleration and Attenuation Function

PGA in the research area is calculated using attenuation function adopted from [20] to represent the global attenuation function and the attenuation functio by [12] representing local attenuation. Local attenuation function has been validated using the Sumatera earthquake incident in 2004, which also represents one of the data series used in this research. The usage of two different attenuation function in this research is to realize the difference and similarities of data and results in order to achieve the goal of this research.

To realize the data consistency and the resulting PGA value, plotting is performed to recognize the regression relationship between the resulting PGA values against the distance of the epicenter and depth of earthquake for both attenuation functions. The use of the local attenuation function where the relation of the PGA values and the epicenter distance showed the data distribution with a considerable amount of maximum and minimum differences, although it still showed the reduction of linear PGA values against the increase of distance from the measured location against the epicenter (Fig. 1). The graphs measuring the PGA and the depth of the earthquake showed a data distribution that is inconsistent where the data distribution showed that the high value of the depth of the earthquake showed a high value of PGA (Fig. 2).

This fact still requires further reassessments especially for the calculation of the PGA with a series of large data in an area of destructive earthquake.

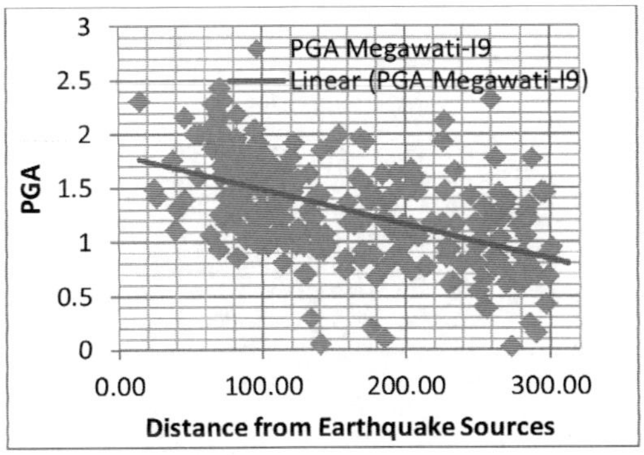

Figure 1. Relation between PGA Values and Distance from the epicenter using [12] attenuation function

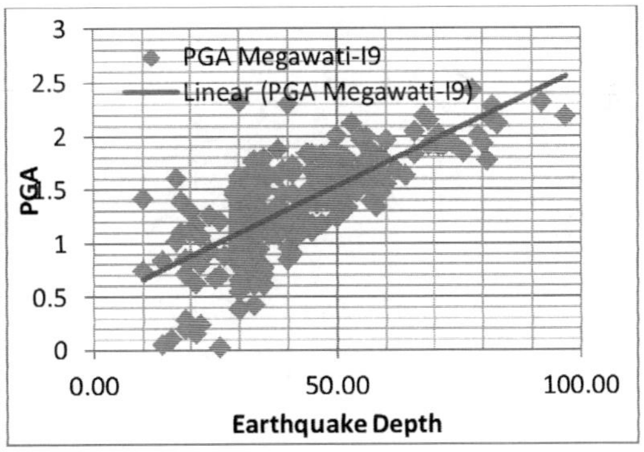

Figure 2. Relation between PGA Values and depth of earthquake using [12] attenuation function

Plotting the calculated data that uses attenuation function [20] shows the relation between the PGA values and the epicenter's distance with data distributed with a small difference between the maximum and minimum values and also shows that there is a strong relation between the decrease of linear PGA value against the increase of the distance to the measured location against the epicenter. (Fig. 3)

Plotting the PGA relation and the depth of the earthquake also shows that the data distributed is consistent, where the graphs shows the increase of the linear earthquake's depth against the PGA values which is decreasing (Fig. 4). The consistency of these relations represents facts that the use of global attenuation function as explained by [20] can be applied accordingly to the local condition

(Banda Aceh City) where its tectonic conditions may differ from those where the model was developed.

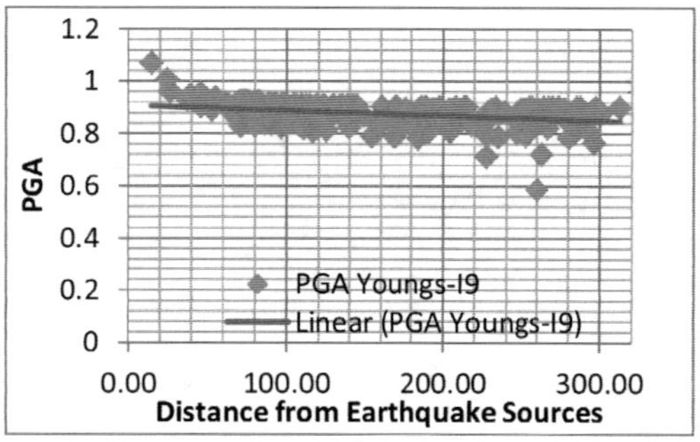

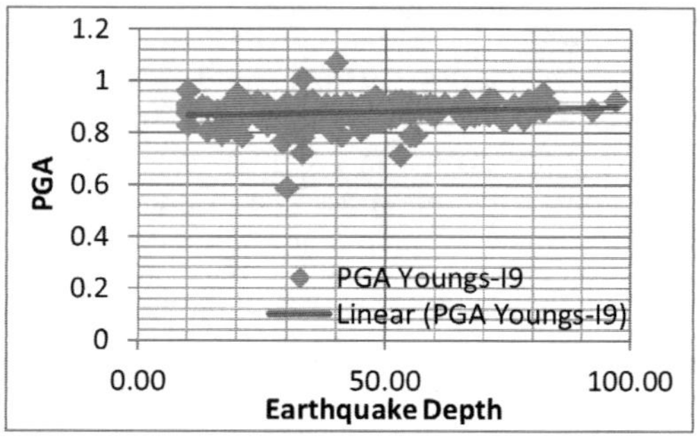

Figure 3. Relation between PGA Values and Distance from the earthquake's epicenter using [20] attenuation function

Figure 4. Relation between PGA Values and Distance from the earthquake's depth using [20] attenuation Function.

The validity of the PGA values with the global attenuation function is also supported by the model developed by [15] which is using the Sumatra earthquake catalogue with various global as well as regional attenuation functions. The linear relation resulted from the model (Fig.5) is similar with the relation of PGA values and the distance from the earthquake's epicenter which is the result of this research (Fig.3).

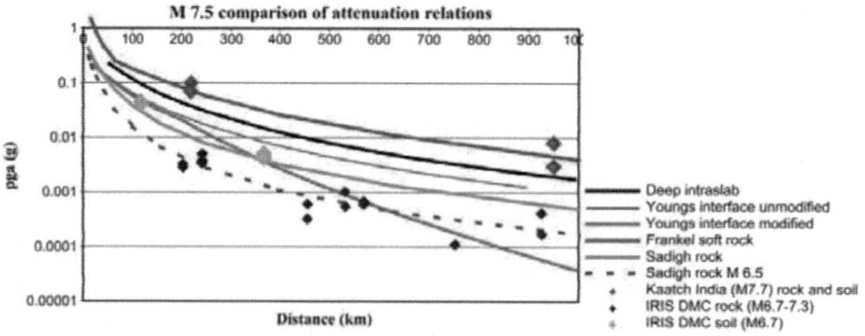

Figure 5. Model of the relation of PGA and the distance to the earthquake's epicenter from [15] attenuation function

4.2. Peak Ground Acceleration in Banda Aceh City Indonesia
The PGA value calculated by the global attenuation function varies with a minimum of 0.8767 gals and maximum of 0.8806 gals with the average of 0.87858 gals. The PGA which is calculated using the local attenuation function has a variation with minimum value 1.23758 gals and maximum value of 1.27293 gals with average value of 1.25568 gals. The PGA values resulted from the two different attenuation function shows a significant difference of average 0.37710 gals. The global PGA value has the tendency to be lower than the PGA value resulting from the local attenuation function (Table 1)

[3] In their research, experimented using two different methods, that is calculating PGA from the attenuation function adopted from [12] using the data from the Sumatera Earthquake Incident 2004 and performed a direct measurement. The PGA resulting from the calculation is 1.11 gals and from direct measurement is 0.3 gals. The PGA values from this measurement is slightly higher compared to

the result that was published by [2] where Banda Aceh city has PGA value in the *range* of 0.2 gals up to 0.25 gals and lower than the result of direct measurement from the area by [16] in the district of Meuraxa, city of Banda Aceh, using *cone penetration test* with a result between 0.325 gals up to 0.4 gal.

[7] In their proposal to improve the Sumatra and Java earthquake hazard map using attenuation function [20] and 3-Dimensional (3D) fault source model resulting in PGA value from 0.6 to 0.7 gals for Banda Aceh city and its surroundings. The proposal for the improvement of this value is based on the Sumatera earthquake 2004, North Sumatra earthquake 2005 and Bantul earthquake 2006.

The difference of PGA value whether it is calculated using local or global attenuation function and the value resulted by the earlier researchers and the differences by using a variety of methods is shown in Table 1.

Table 1. Average PGA value calculated using global attenuation function by [20] and local by[12] and its corresponding differences from the calculated values and actually meaasured by [3]

Adopted Attenuation Function	PGA value [12]	PGA value [20]	Difference [12] and [20]	Difference [12] and [3] - Calculated	Difference [12] and [3] - Measured	Differecne [20] and [3] - Calculated	Difference [20] and [3] - Measured
PGA Minimum	1.23758	0.87670	0.36087	0.12758	0.93758	0.23330	0.57670
PGA Maksimum	1.27293	0.88064	0.39229	0.16293	0.97293	0.22936	0.58064
PGA Average	1.25568	0.87858	0.37710	0.14568	0.95568	0.23142	0.57858

The difference in the calculated result of PGA value from two attenuation function local and global, also represented by [3] in the research of earthquake hazards conducted in [10] uses attenuation function from [10] as a reference in the global attenuation function. In that research, a difference of result is obtained, where the results from the researcher states that using attenuation function to calculate the PGA value, the predicted value is higher compared to the result obtained using the earlier attenuation function to calculate the PGA values and lower compared to using the attenuation equation used globally in particular for earthquake zones of subduction.

4.3. Earthquake Hazard Zonation in Banda Aceh City Indonesia

Banda Aceh and its surroundings, based on the result of PGA calculations from a series of earthquake data within the last 39 years, can be divided into three different hazard zones for each of the attenuation functions used. Zonation which resulted from the attenuation function [20] Banda Aceh can be divided into low hazard zone, with PGA value 0.8767 gals up to 0.8780 gals, medium hazard zone with PGA value 0.8781 gals up to 0.8793 and high hazard zone with PGA value 0.8794 gals up to 0.8806 gals. The spatial distribution of each hazard zone class can be seen on Fig.6.

As the zoning of attenuation function [20], the zoning obtained from the attenuation function Megawati et al, 2008, Banda Aceh can also be divided into three hazard zones which is low hazard zone with PGA value ranging from 1.2375 gals up to 1.2493 gals, medium hazard zone with PGA values ranging from 1.2494 gals up to 1.2611 and high hazard zones with PGA value ranging from 1.2612 gals till 1.2729 gals. The spatial distribution on each hazard zone class can be seen in Fig.7.

In more detail and using actual measuremenet from the incident on the field, [16] performed micro -zonation of earthquake hazard, particularly in district of Meuraxa in Banda Aceh city and divided the region into three zones, whic is hard soil, medium soil and soft soil, each having PGA value 0.325 gals, 0.325 gals up to 0.375 gals and 0.375 gals up to 0.40 gals respectively. Spacially, the zonation propoposed is within the medium eartquake hazard zone based on the zonation developed by using the calculated value using [20] attenuation function as well as the attenuation function from [12].

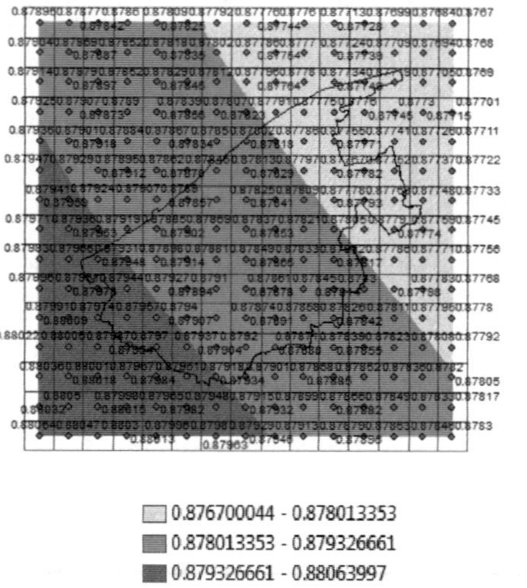

Figure 6. Average PGA at each centroid calculated using [20] attenuation function and earthquake hazard zone obtained

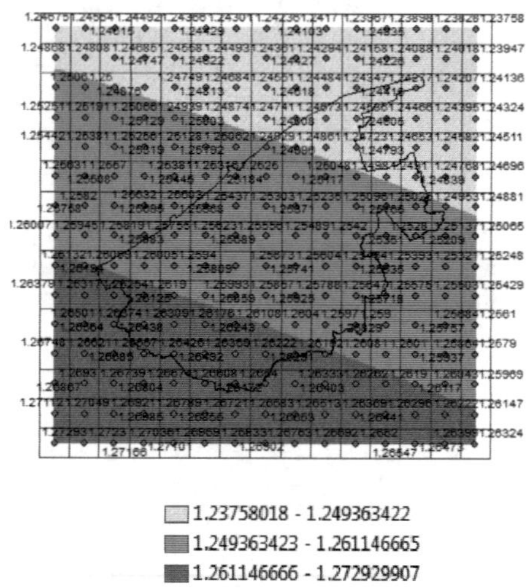

Figure 7. Average PGA at each centroid calculated using [12] attenuation function and earthquake hazard zone obtained

5. Conclusion

The global attenuation function that was developed by [2] can be applied accordingly to calculae the peak ground acceleration (PGA) in Banda Aceh City Indonesia using the earthquake data from Sumatera. The attenuation function from [12] requires forther assessment to be able to apply it especially for the Sumatera earthquake for destructive earthquake.

The PGA value in Banda Aceh City which is calculated using global attenuation function varies from a value of minimum 0.8767 gals and maximum 0.8806 gals with average value 0.87858 gals. There is a discrepancy in the result obtained between the PGA value calculated by global and local attenuation function with a difference of 0.37710 gals.

Banda Aceh can be divided into three earthquake harzard zones which is low hazard zone with PGA value 0.8767 gals up to 0.8780 gals, medium hazard zone with PGA value 0.8781 gals up to 0.8793 and high hazard zone with PGA value 0.8794 gals up to 0.8806 gals

References

[1] Atkinson, G. M., & Boore, D. M 2003 Empirical ground-motion relations for subduction-zone earthquakes and their application to Cascadia and other regions *Society*, *93*(4), 1703-1729

[2] Badan Standarisasi Nasional 2002 (In Bahasa) Standar Nasional Indonesia-Tata Cara Perencanaan Ketahanan Gempa Untuk Bangunan Gedung (SNI 03-1726-2002).

[3] Balendra, T., & Li, Z 2008 Seismic Hazard of Singapore and Malaysia *Earthquake* 57-63.

[4] Campbell, K. W 2003 Prediction of Strong Ground Motion Using the Hybrid Empirical Method and Its Use in the Development of Ground-Motion (Attenuation) Relations in Eastern North America. *Society*, *93*(3), 1012-1033.

[5] Indonesia. State Ministry for National Planning Development Agency/BAPPENAS 2005 Preliminary damage and loss assessment-the December 26, 2004 natural disaster. Jakarta: Government Printer.

[6] Gregor, N. J., Silva, W. J., Wong, I. G., & Youngs, R. R 2002 Ground-Motion Attenuation Relationships for Cascadia Subduction Zone Megathrust Earthquakes Based on a Stochastic

Finite-Fault Model. *Society*, *92*(5), 1923-1932.

[7] Irsyam, M., Dangkua, D. T., Hendriyawan., Hoedajanto. D., Hutapea.B.M., Kertapati, E.K., Boen. T., Petersen. M.D 2008 Proposed seismic hazard maps of Sumatra and Java islands and microzonation study of Jakarta city Indonesia. *Earth System Science*, (November), 865-878.

[8] Irwansyah. E 2010 Building Damage Assessment Using Remote Sensing, Aerial Photograph and GIS Data: Case Study in Banda Aceh After Sumatera Earthquake 2004. *Proc. Int. Conf on Intelligent Technology and Its Application-SITIA 2010* vol 11, pp.57.

[9] Kramer. S.L 1996 *Geotechnical Earthquake Engineering.* Prentice Hall, Upper Saddle River, New Jersey, USA.

[10] Lin, P.-S., & Lee, C.-T 2008 Ground-Motion Attenuation Relationships for Subduction-Zone Earthquakes in Northeastern Taiwan. *Bulletin of the Seismological Society of America*, *98*(1), 220-240. doi:10.1785/0120060002

[11] Miura, H., Wijeyewickrema, A. & Inoue, S 2006 Evaluation of Tsunami Damage in the Eastern Part of Sri Lanka due to the 2004 Sumatra Earthquake using High-Resolution Satellite Images. *Proc.Int. Work. on Remote Sensing for Post-Disaster Response* [Online]. Available: http://ares.tu.chiba-u.jp/workshop/ChibaRS2005/Paper_Miura.pdf [Sep.13, 2005]

[12] Megawati K, Pan TC, Koketsu K 2005 Response spectral attenuation relationships for Sumatran-subduction earthquakes and the seismic hazard implications to Singapore and Kuala Lumpur. *Soil Dynamics and Earthquake Engineering,* vol.25, no.1, pp11-25.

[13] Megawati, K., & Pan, T.-chien 2010 Ground-motion attenuation relationship for the Sumatran megathrust earthquakes. *Earthquake.* 827-845. doi:10.1002/eqe

[14] MM. Fischer and A. Getis 2010 *Handbook of applied spatial analysis: software tools, method, and applications*". New York:Springer.

[15] Petersen, M. D., Dewey, J., Hartzell, S., Mueller, C., Harmsen, S., Frankel, A. D., & Rukstales, K 2004 Probabilistic seismic hazard analysis for Sumatra , Indonesia and across the Southern Malaysian Peninsula. *Tectonophysics*, *390*, 141-158. doi:10.1016/j.tecto.2004.03.026

[16] Sengara I.W 2008 Seismic Hazard And Microzonation For A District In Banda Aceh City Post 2004 Great Sumatra Earthquake. *Proc.Int.Conf on Earthquake Engineering* October 12-17, 2008, Beijing, China

[17] USGS 2009 Historic World Earthquakes [Online]. Available: http://earthquake.usgs.gov/earthquakes/world/historical_country.php#indonesia. [November 23th, 2009]

[18] USGS 2012 Magnitude 8.6 - Off The West Coast Of Northern Sumatra. [Online]. Available: http://earthquake.usgs.gov/earthquakes/recenteqsww/Quakes/usc000905e.php. [May 13th, 2012]

[19] Villaverde. R 2009 *Fundamental Concepts of Earthquake Engineering*, CRC Press-Taylor and Francis Group, Boca Raton, FL, USA

[20] Youngs et al 1997 Strong Ground Motion Attenuation Relationship For Subduction Zone Earthquake.pdf. (n.d.).